THE BIG PICTURE

MEDICAL BIOCHEMISTRY

THE BIG PICTURE

MEDICAL BIOCHEMISTRY

Written and Edited by

Lee W. Janson
Edinburgh, Scotland

Marc E. Tischler
University of Arizona
College of Medicine
Department of Biochemistry
& Molecular Biophysics
Tucson, Arizona

New York Chicago San Francisco Lisbon London Madrid Mexico City
New Delhi San Juan Seoul Singapore Sydney Toronto

The McGraw-Hill Companies

The Big Picture: Medical Biochemistry

1 2 3 4 5 6 7 8 9 0 CTP/CTP 16 15 14 13 12 11

ISBN 978-0-07-163791-6
MHID 0-07-163791-5

This book was set in Minion by Aptara, Inc.
The editors were Michael Weitz and Brian Kearns.
The production supervisor was Catherine H. Saggese.
Project management was provided by Samir Roy at Aptara, Inc..

China Translation and Printing Services, Ltd. was printer and binder.

This book is printed on acid-free paper.

Library of Congress Cataloging-in-Publication Data

Janson, Lee W., 1964-
The big picture : medical biochemistry / Lee Janson, Marc Tischler.
p. ; cm.
Includes index.
ISBN-13: 978-0-07-163791-6 (soft cover : alk. paper)
ISBN-10: 0-07-163791-5
1. Biochemistry. 2. Clinical biochemistry. I. Tischler, Marc. II. Title.
[DNLM: 1. Biochemical Phenomena. QU 34]
QP514.2.J36 2012
612′.015—dc23

2011026718

CONTENTS

DEDICATIONS

To my family: my dear wife, Meryl, who has unfailingly supported my professional endeavors and endured my countless evening hours at the computer on this project; my daughters, Rebecca, Laura, and Miriam, who bring me incredible "naches"; my mother and father (may they rest in peace) for their support in my formative years; my brothers Howard and Matthew; my mother-in-law Martha for her ever-present accolades; my father-in-law Ed (may he rest in peace); and my father-in-law Lou for the professional respect always accorded me.

Finally to the more than 3000 medical students I have taught who inspired my success as a teacher and educator and who preceded those students who I trust will benefit from this textbook

—Marc E. Tischler

To my parents, family, and friends who persevered throughout the writing, proofing, and publication of this book as well as the many instructors, from high school to university to graduate school to medical school and beyond, who instilled in me not only a love for learning but also for teaching.

This book is personally dedicated to one such person, Cassie Murphy-Cullen, PhD (may she rest in peace), who served as a teacher, counselor and, most-of-all, constant and dedicated friend to all of her students, including myself.

—Lee W. Janson

ACKNOWLEDGEMENTS

Sincere thanks to the current authors of Harper's Illustrated Biochemistry for their reviews and comments with special thanks to Dr. Robert Murray, editor of Harper's, for his patience, kindness, and exceptional efforts in reviewing this book. The authors also wish to thank Andrea Aguirre, Chineyne Anako, Martin Benjamin, Natasha Bhuyan, Joseph Carroll, Katharine Flannery, Silvija Gottesman, Michael Ori, Charles Rappaport, Christopher Riley, Alan Schumacher, and Karen Stern, medical students at the University of Arizona, who served on a focus group to provide valuable insight for this text from a medical student perspective.

ABOUT THE AUTHORS

Marc E. Tischler received an undergraduate degree in biology from Boston University, a master degree in chemistry from the University of South Carolina, and his doctorate degree in biochemistry from the University of Pennsylvania. After serving in a postdoctoral position in physiology at Harvard Medical School, he joined the faculty at the University of Arizona in 1979 where he is currently a professor of biochemistry and molecular biophysics holding joint appointments in physiology and in internal medicine. Having served as coordinator of the medical biochemistry course for a decade, he was recruited to play a major role in the development of the revised medical curriculum at the University of Arizona, which debuted in 2006 and offers an integrated, organ-based approach akin to the second half of this textbook. In that new curriculum, he designed and serves as director of the medical block entitled Digestion, Metabolism, and Hormones. He has taught more than 3000 medical students during his tenure in Arizona.

Lee W. Janson received a BS in Biochemistry/minor in Latin from the University of Rochester followed by a PhD in biological sciences/biochemistry from Carnegie Mellon University. After a postdoctoral fellowship at NASA-Johnson Space Center doing research on immunological activation in microgravity, he entered medical school at the University of Texas Southwestern Medical Center at Dallas, continuing with a family practice residency in both Dallas and San Antonio. He entered the active duty Air Force and served as a family physician and a flight surgeon, including tours of Korea, Iraq, and Afghanistan. After military service, he permanently moved to Edinburgh, Scotland in 2007, where he practices in the National Health Service as an emergency room doctor and a general physician in the United Kingdom with occasional work in Australia and other parts of the world. In his free time, he writes and does research. Past book publications include *Brew Chem 101: The Basics of Homebrewing Chemistry* (Storey Publishing).

SECTION I

THE BASIC MOLECULES OF LIFE

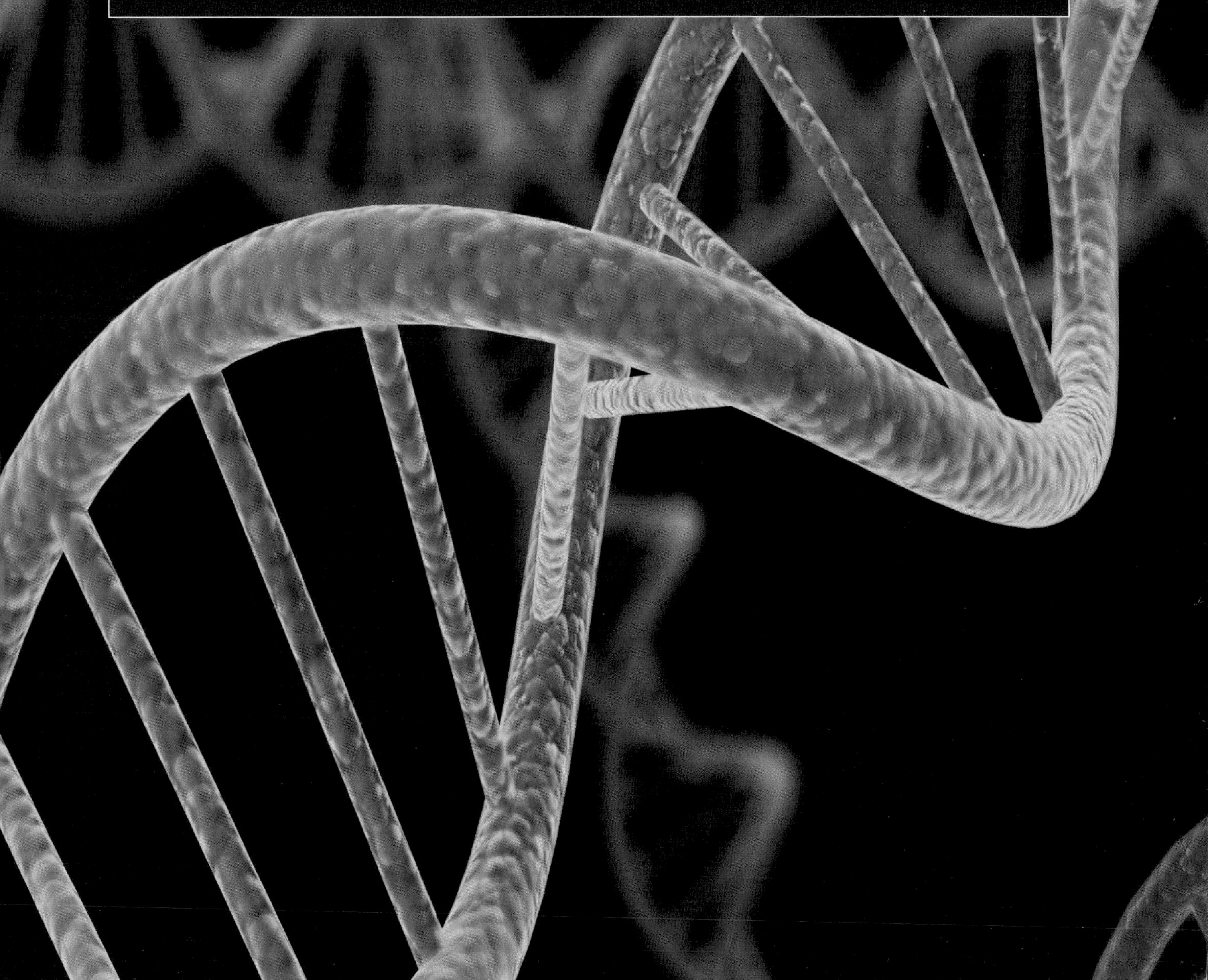

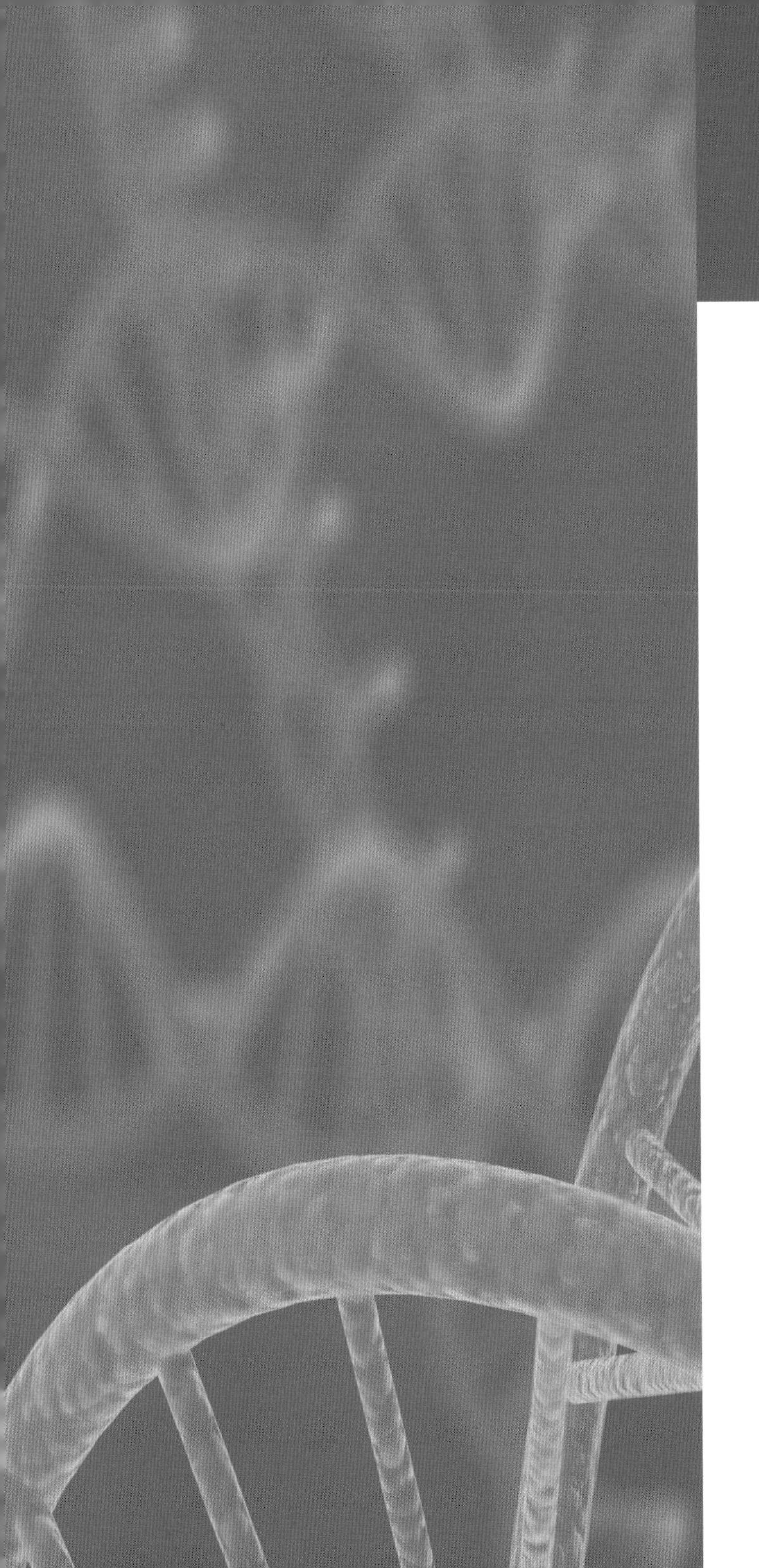

CHAPTER 1

AMINO ACIDS AND PROTEINS

OVERVIEW

Amino acids are the basic building blocks of proteins and serve as biological molecules in their own right with a variety of functions. Amino acids are often categorized as essential or nonessential, depending on the ability of the body to manufacture each amino acid versus requirement for ingestion in the diet. Although several hundred amino acids exist, 20 play the predominant role in the human body. Each amino acid has a characteristic R-group that determines its chemical nature and, therefore, how it will interact with other amino acids, other molecules, and with its environment.

Amino acids link together via peptide bonds to form peptides and proteins. These peptides and proteins fold into their final three-dimensional shape as the result of hydrophobic, hydrophilic, hydrogen bonding, and ionic bonding forces (among others) that result from the amino acids in the peptide chain, including the characteristics of their R-groups. Proteins may be categorized as enzymes, structural proteins, motor proteins, and transport/channel proteins. The specific roles of amino acids and proteins in the synthesis of other molecules and in the functions of organ systems and the human bodies will be explored in detail in subsequent chapters.

AMINO ACIDS—STRUCTURE AND FUNCTIONAL GROUPS

ESSENTIAL AND NON-ESSENTIAL

Amino acids are the basic building blocks of proteins. Twenty amino acids make up proteins in living organisms; several hundred more amino acids perform specialized functions in human and non-human biology. Amino acids are often described as

- **Essential** (must be obtained directly from food)
- **Non-essential** (the human body is able to produce them on its own).

There is some debate about the exact definitions of these terms, but 8–10 amino acids are usually deemed essential and 10–12 are non-essential (see Table 1-1).

Succotash, a combination of corn and lima beans, was used by Native American hunters and warriors. Light in weight, easy to carry, and simple to prepare, succotash contains all the essential required amino acids needed by humans and kept these travelers well nourished and healthy during long trips away from their settlements.

BASIC STRUCTURE

Every amino acid has four components linked together with a central carbon atom (Figure 1-1):

1. Amino group
2. Carboxylic acid group
3. Hydrogen atom
4. **R-group**, which varies with each amino acid

CHARACTERISTICS OF R-GROUPS

As amino acids are identical except for their R-group, four R-group characteristics classify the amino acids (see also Table 1-1).

1. **Hydrophobic (Water Hating) R-groups:** These amino acids prefer to be inside a folded protein or covered by another part of a protein or a lipid membrane (Chapter 8) where they are hidden from the external water environment.
2. **Hydrophilic (Water Loving) R-groups:** With partial charges from the hydroxyl (OH^-) or amino (NH_3^+) parts of these R-groups, these amino acids prefer to be at or near the surface of a protein where they interact with surrounding water molecules. An exception would be the surface portion of a membrane protein that interacts with the hydrophobic region of the phospholipid bilayer (Chapter 8). Hydrophilic R-groups are often important at the active site of an enzyme (Chapter 5).
3. **Charged R-groups:** Either positively or negatively charged, these amino acids prefer to be at the surface of a folded protein or in contact with other charged atoms/molecules.
4. **Special R-groups:** There are four amino acids with special quality R-groups.
   - Proline has a glutamate R-group that has bonded onto itself (see Figure 1-2A) forming an "imino" acid. Proline is often found at sharp turns of folded proteins (Figure 1-2B).
   - Cysteine has a sulfhydryl group that can bond with another cysteine sulfhydryl group to form a cystine "disulfide" bond (Figure 1-3). This bond can be either within one protein or between two different proteins and is important in making strong structures such as the protein keratin found in fingernails.
   - Methionine has a sulfur atom contained within its R-group. Although it does not make disulfide bonds, this sulfur is seen at the site of some enzyme reactions or at special areas of protein structure.
   - Histidine has a unique imidazole ring containing two nitrogen atoms, which can be uncharged or positively charged. This unique characteristic makes the histidine R-group important in enzyme reactions that make or break bonds.

Hair and Cystine Double Bonding: Hair, which is composed of linear protein sequences, is curly or straight depending on the number of cystine disulfide bonds. The number and location of cysteine residues and accessory proteins specific for each person affect the number of disulfide bonds resulting in this very individualized characteristic. Chemicals that promote these disulfide bonds are used for "perms" and, alternatively, chemicals that break these bonds are used in hair straightening treatments.

TABLE 1-1. Amino Acids—R-Group Classifications

Class	Name	R-Group	Notes
Hydrophobic	Glycine	$H-$	Smallest R-group
Hydrophobic	Alanine	CH_3-	Methyl R-group that normally folds easily within a protein
Hydrophobic	**Valine**	$(H_3C)_2CH-$	Bulky structure can impact folding of protein
Hydrophobic	**Leucine**	$(H_3C)_2CH-CH_2-$	Bulky structure can impact folding of protein
Hydrophobic	**Isoleucine**	$CH_3-CH_2-CH(CH_3)-$	Bulky structure can impact folding of protein
Hydrophobic	**Phenylalanine**	(benzene ring)$-CH_2-$	Aromatic ring
Hydrophobic	**Tryptophan**	(indole ring, N–H)$-CH_2-$	Indole ring
Hydrophilic	Serine	$CH_2(OH)-$	Hydroxyl (OH) group with partial negative (−) charge (not shown); may be phosphorylated
Hydrophilic	**Threonine**	$CH_3-CH(OH)-$	Hydroxyl (OH) group with partial negative (−) charge; may be phosphorylated
Hydrophilic	Asparagine	$H_2N-C(=O)-CH_2-$	Amino (NH_2) group with partial positive (+) charge (not shown)
Hydrophilic	Glutamine	$H_2N-C(=O)-CH_2-CH-$	Amino (NH_2) group with partial positive (+) charge (not shown)
Charged	Tyrosine	$HO-$(benzene ring)$-CH_2-$	Aromatic ring with hydroxyl group, giving partial or full negative (−) charge (not shown); may be phosphorylated
Charged	Aspartic acid	$^{-}OOC-CH_2-$	Negative (−) charge from COO^-
Charged	Glutamic acid	$^{-}OOC-CH_2-CH_2-$	Negative (−) charge from COO^-
Charged	**Lysine**	$CH_2(NH_3^+)-CH_2-CH_2-CH_2-$	Positive (+) charge from NH_3^+
Charged	**Arginine**	$H-N(C(=NH_2^+)NH_2)-CH_2-CH_2-CH_2-$	Positive (+) charge from NH_2^+

TABLE 1-1. Amino Acids—R-Group Classifications (Continued)

	Name	R-Group	Notes
Special	Proline	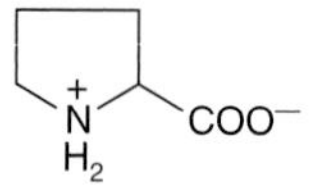 (pyrrolidine ring, $\overset{+}{N}H_2$, COO^-)	β-turns
	Cysteine	CH_2—, SH	Disulfide bonds
	Methionine	CH_2-CH_2-, $S-CH_3$	Sulfur atom
	Histidine	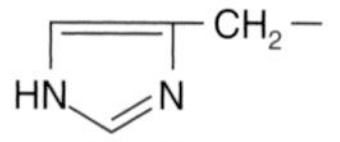 (imidazole ring, HN, N) CH_2-	Partial or full positive (+) charge from NH^+ (not shown)

Note: Ten essential amino acids that must be obtained from food sources are noted in bold. Arginine generally is only considered essential in children. Reproduced with permission from Murray RA, et al.: Harper's Illustrated Biochemistry, 28th edition, McGraw-Hill, 2009.

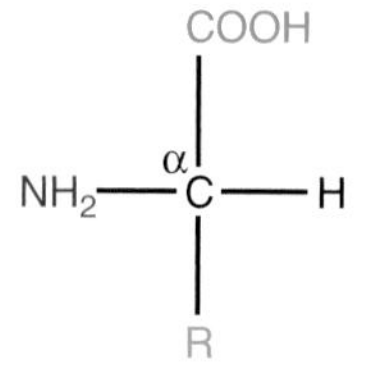

Figure 1-1. Basic Structure of an Amino Acid. A central alpha (α) carbon is bonded to an amino group (NH_2), a carboxyl group (COOH), and an R group. [Reproduced with permission from Naik P: Biochemistry, 3rd edition, Jaypee Brothers Medical Publishers (P) Ltd., 2009.]

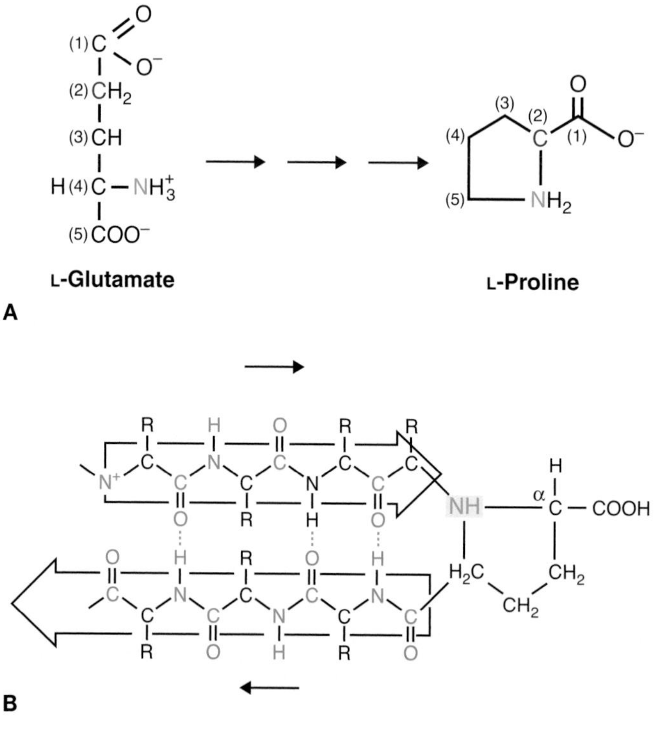

Figure 1-2. A–B. Proline. A. Proline forms from the amino acid glutamate when the amino group nitrogen (blue) creates a bond with carbon 2 on the R-group (red), making an amino acid that is "folded" onto itself. The carbon atoms are individually numbered to allow the reader to follow the process that takes three steps (indicated by arrows). **B.** Proline's special structure allows the formation of a "hairpin" β-turn via formation of an imino bond. The amino group from proline is shaded in blue. [Adapted with permission from Naik P: Biochemistry, 3rd edition, Jaypee Brothers Medical Publishers (P) Ltd., 2009.]

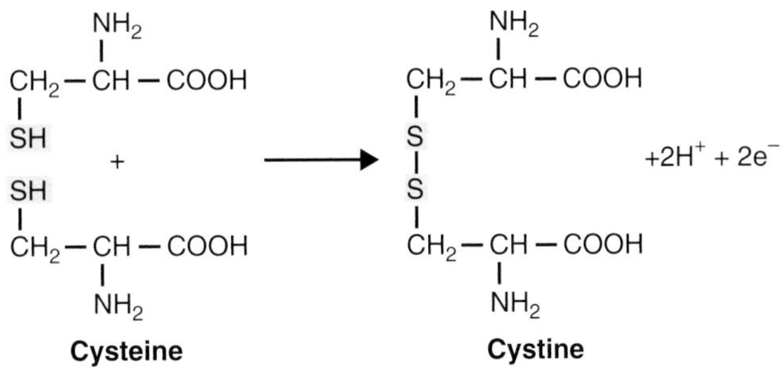

$$\begin{array}{c}NH_2\\ |\\ CH_2-CH-COOH\\ |\\ SH\\ +\\ SH\\ |\\ CH_2-CH-COOH\\ |\\ NH_2\end{array} \longrightarrow \begin{array}{c}NH_2\\ |\\ CH_2-CH-COOH\\ |\\ S\\ |\\ S\\ |\\ CH_2-CH-COOH\\ |\\ NH_2\end{array} + 2H^+ + 2e^-$$

Cysteine **Cystine**

Figure 1-3. Formation of Cystine Disulfide Bond. The sulfhydryl groups (SH) from two cysteine amino acid residues from different parts of a single or two separate amino acid sequences may lose one hydrogen atom each to become a cystine residue by the formation of a disulfide bond (S—S). [Adapted with permission from Naik P: Biochemistry, 3rd edition, Jaypee Brothers Medical Publishers (P) Ltd., 2009.]

BASIC PROTEIN STRUCTURE

Several factors affect basic protein structure, including the following:

- **Amino Acid Composition:** Whether the R-group of each amino acid wants to be away from, near, or in contact with water at the surface of the folded protein.
- **Special Amino Acids:** Proline, cysteine, and/or methionine exert effects on protein folding due to resulting bends and disulfide bonds.
- **Functional Sites:** Structural proteins or proteins that catalyze a reaction (i.e., enzymes; Chapter 5) usually contain specific amino acids that are important for that protein's particular function. The R-groups at these sites will affect the protein's final folded shape.
- **Final Modifications:** Most proteins are initially made with extra amino acids at their beginning and/or end. These extra amino acids may be modified or removed during the maturation of the protein, resulting in changes to the final structure.
- **Final Destination:** Whether the protein will end up in an aqueous solution or in a membrane will also change the folding, as proteins that go to membranes will ultimately want their hydrophobic, "water hating" R-groups on the outside in contact with the hydrophobic membrane (see below, Chapter 8).

LEVELS OF PROTEIN STRUCTURE

When describing protein structure, we use the following terms: primary (1°), secondary (2°), tertiary (3°), and/or quaternary (4°) structure.

- **Primary (1°) Structure:** The particular sequence of amino acids in a protein (also called a **polypeptide**) is termed as **primary structure**. The amino acids within a polypeptide are termed **residues** and are linked via **peptide bonds** (Figure 1-4A), formed when the carboxylic acid group of one amino acid interacts with the amino group of a second amino acid. The combination produces the peptide bond and one molecule of water. Repeated peptide bonds form a chain of peptide bond linkages with the nitrogen from the amino group, the central carbon from the amino acid, and the carbon from the carboxylic acid forming the protein "backbone." The primary structure is often shown as "beads on a string" with each bead representing an amino acid residue (Figure 1-4B).

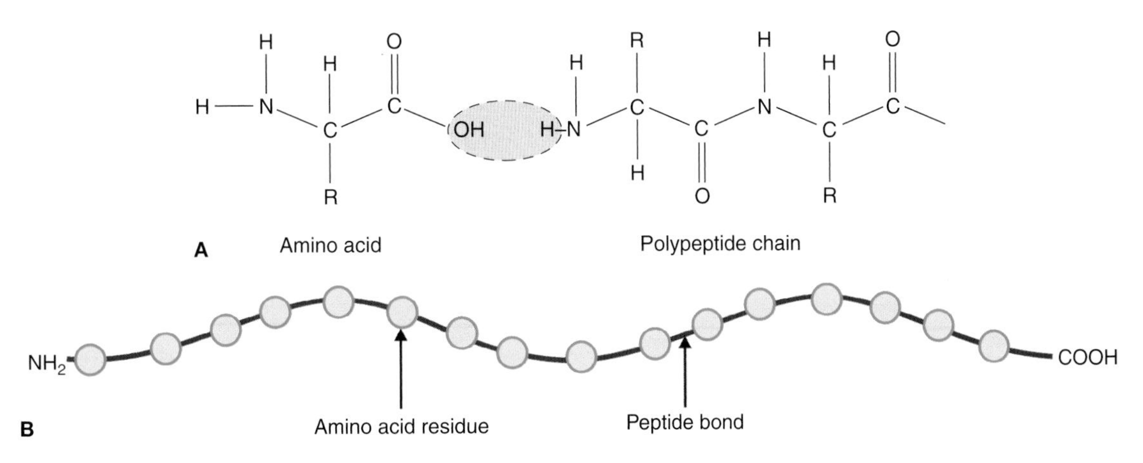

Figure 1-4. A. Formation of a Peptide Bond. A peptide bond is formed between the carboxylic acid (COOH) group and the amino nitrogen (HHN) group to form a new C—N bond. The process releases one molecule of water (OH and H as shown in pink circle). By convention a protein is always drawn starting with the N-terminus on the left. [Reproduced with permission from Barrett KE, et al.: Ganong's Review of Medical Physiology, 23rd edition, McGraw-Hill, 2010.] **B. Primary (1°) Structure.** The primary structure of a protein is the chain of amino acids from the N-terminal (NH_2 group of amino acid 1) to its C-terminal end (COOH group of the final amino acid). Individual amino acids are denoted as green "beads" and bonds are red "strings" in this stylized representation. Proteins vary greatly in length from a few amino acids to several hundred. [Reproduced with permission from Naik P: Biochemistry, 3rd edition, Jaypee Brothers Medical Publishers (P) Ltd., 2009.]

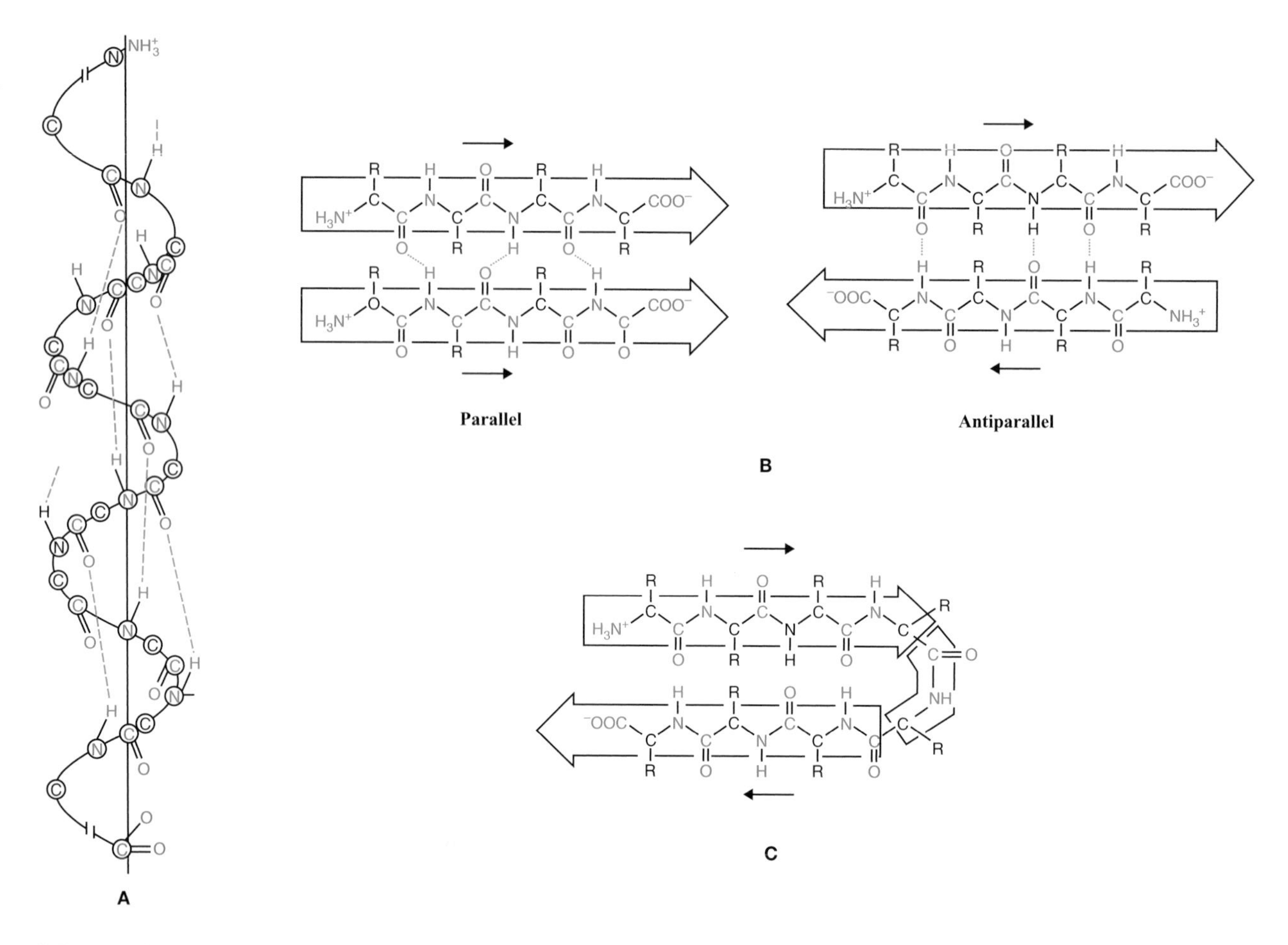

Figure 1-5. A–C. Secondary (2°) Structure. A. The α-helix structure is stabilized by hydrogen bonds (blue dotted line) between the N—H of one amino acid and the C═O of another amino acid. The hydrophobic and hydrophilic nature of the R-groups from each amino acid in the helix also plays a part in formation of the α-helix. **B.** The β-strands form when the one sequence of successive amino acids form hydrogen bonds between the N—H and the C═O of another group of successive amino acids. R-groups of these amino acids also influence the formation of β-strands by both charge and steric forces. β-Strands can be either parallel (left) or antiparallel (right) as noted by the arrows. **C.** A β-turn forms at the juncture between two antiparallel β-strands and usually involves amino acids with small R-groups, including glycine, alanine, and valine. β-turns can also form via proline's unique structure (Figure 1-2B). [Adapted with permission from Naik P: Biochemistry, 3rd edition, Jaypee Brothers Medical Publishers (P) Ltd., 2009.]

- **Secondary (2°) Structure:** From the linear chain of amino acids, the C—N and C—C bonds rotate around the central carbon atom (see Figure 1-1). This rotation forms **secondary structures** (Figure 1-5A–C) called an **α-helix**, **β-strand**, or **β-turn** (see also Figure 1-2B above) depending on the hydrophilic and hydrophobic charge and size influences of the R-groups as well as hydrogen and ionic bonding. Parts of the peptide that do not form conventional secondary structures are referred to as "random coil."
- **Tertiary (3°) Structure:** Secondary structure and the remaining amino acid sequences continue to fold and interact with other parts of the amino acid sequence to form tertiary (3°) structure. These processes again are driven by hydrophobic and hydrophilic forces of the individual parts of the sequence, as well as hydrogen bonding and ionic bonding between charged amino acid R-groups. In particular, R-groups and the partial positive charge of the nitrogen and negative charge of oxygen (OH^- and $C{=}O^-$) molecules often form hydrophilic and hydrophobic "sides" of the α-helix or β-strand. As a result, these secondary structures will position themselves with each other and with other similar areas of the folded protein to keep their hydrophobic areas away from and their hydrophilic areas exposed to the external water environment. Examples are shown in Figure 1-6. Additionally, tertiary structure starts to develop at active sites of proteins where critical actions and interactions will take place. These active sites will be discussed in subsequent chapters.

Globular vs. Fibrous Proteins: Proteins are often grouped into the broad categories of "globular"—having an approximately spherical three-dimensional shape—or "fibrous"—being long and fairly straight overall. A vast majority of proteins are globular, with fibrous proteins often playing structural or very specialized functional roles. Examples of fibrous proteins include actin, collagen, and keratin, which play structural roles in muscle, connective tissue, and skin/nails.

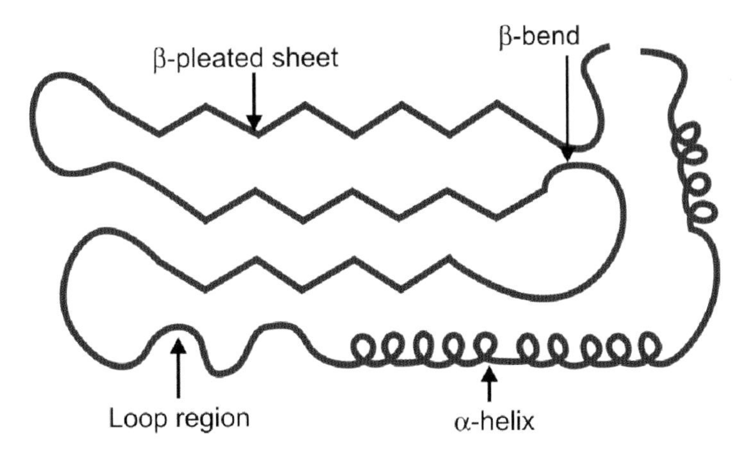

Figure 1-6. Examples of Tertiary (3°) Structure. Secondary structures, including α-helices, β-sheets, β-turns/bends, and loop regions combine to form tertiary structure domains. Several common forms of tertiary structure have been characterized and are illustrated in the figure. [Reproduced with permission from Naik P: Biochemistry, 3rd edition, Jaypee Brothers Medical Publishers (P) Ltd., 2009.]

- **Quaternary (4°) Structure:** Although many proteins are made of only one peptide chain (called "monomers" or "subunits"), two or more folded chains may combine together to make a final active "oligomer" protein. The subunits in a multimeric protein may have identical (**homo-oligomer**) or different (**hetero-oligomer**) amino acid sequences and interact to optimize hydrophobic and hydrophilic areas and hydrogen/ionic bonding. The combination of the multiple protein subunits makes quaternary (4°) structure (Figure 1-7).

The process of protein folding is not a truly linear one; a protein never exists as a long amino acid string. Instead, protein folding is a complex interaction of these processes and secondary and tertiary structures actually form somewhat in parallel. Many more complex factors not discussed here also occur to help in determining the final protein conformation. The process of making and trafficking a protein will be explored more fully in Chapters 5 and 9.

CATEGORIES OF PROTEINS

AMINO ACID AND PEPTIDE-DERIVED HORMONES AND NEUROTRANSMITTERS

Amino acids and the peptides/proteins that they form serve several critical roles in human biochemistry and life. The major role of amino acids is to provide the building blocks for proteins

Dystrophin and Muscular Dystrophy: Muscular dystrophies (including Duchenne and Becker muscular dystrophies) are diseases in which skeletal muscle rapidly breaks down, resulting in muscle weakness and wasting, decreased motor skills, and, eventually, the inability to walk (usually by the age of 12 years). Of approximately, 30 different types of muscular dystrophies, the Duchenne and Becker forms show X chromosome-linked inheritance and, therefore, almost always affect males at a rate of approximately one in 3500 boys worldwide. Both are caused by the absence or mutation of the protein **dystrophin**, which provides strength for skeletal muscle cells by anchoring the internal cell components to the extracellular matrix.

The interaction of dystrophin with several other proteins involved in this structural support network is a prime example of protein–protein interactions involving the forces of their hydrophobic and hydrophilic regions. More importantly, the mutations in the dystrophin protein are believed to significantly alter one or more of these regions, resulting in breakdown of the anchoring complex and thus disease.

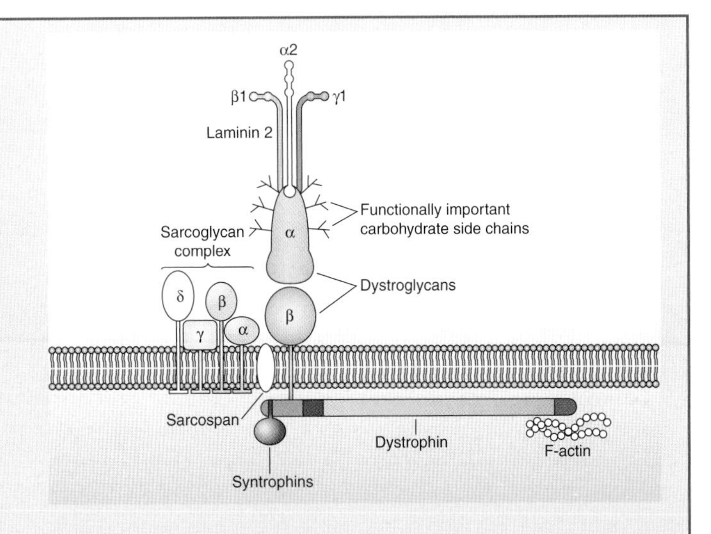

Reproduced with permission from Kandell E, et al.: Principles of Neuroscience, 4th edition, McGraw-Hill, 2000.

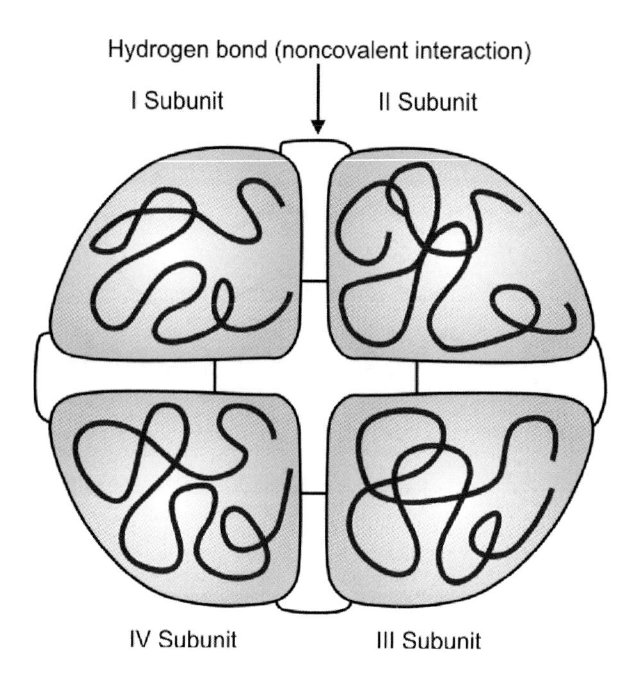

Figure 1-7. Quaternary (4°) Structure. Protein monomers may join, held together by hydrogen bonding (shown) and/or hydrophobic–hydrophilic interactions, to make oligomers. Identical protein subunit monomers may join together to make a homo-oligomer. Nonidentical protein subunit monomers may join together to make a hetero-oligomer. [Reproduced with permission from Naik P: Biochemistry, 3rd edition, Jaypee Brothers Medical Publishers (P) Ltd., 2009.]

Quaternary (4°) Structure: Quaternary (4°) structure is exemplified in many other ways than those given above, where two secondary structural elements are simply in close proximity to each other. Many fibrous proteins expand on this concept, with two or more α-helices wound around each other for most of their amino acid sequence. Examples include F-actin and the tail section of myosin, keratin in hair, and intermediate filaments, which play an important structural role inside all cells.

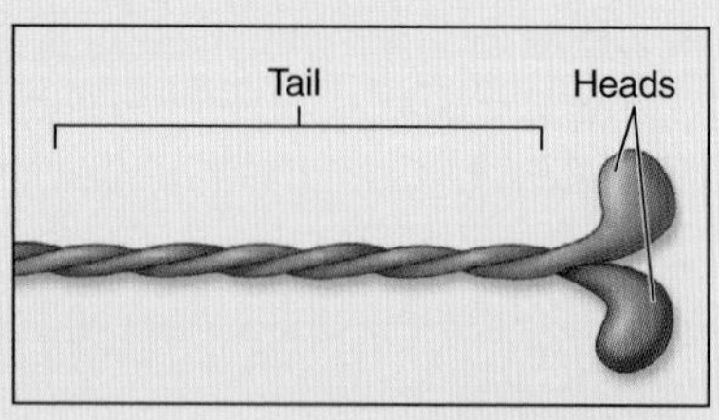

Reproduced with permission from Mescher AL: Junqueira's Basic Histology Text and Atlas, 12th edition, McGraw-Hill, 2010

Collagen, another fibrous structural protein of skin and connective tissue, is composed of three intertwined helical protein sequences (see the figure on right), which differ from the α-helical structure. Unlike an α-helix where hydrogen bonding within the same protein sequence predominates, collagen utilizes hydrogen bonding between the three helical protein chains. Collagen optimizes this very unique structure by having the small amino acid glycine as every third amino acid to allow the three helices to fit very close together (see the figure below). In addition, collagen has an abundance of proline residues to promote the helical protein structure and a special form of proline (hydroxyproline) with an extra hydroxyl (OH^-) to form the interprotein hydrogen bonds with the glycine NH groups. Hydroxylysine residues help to stabilize the structure as well.

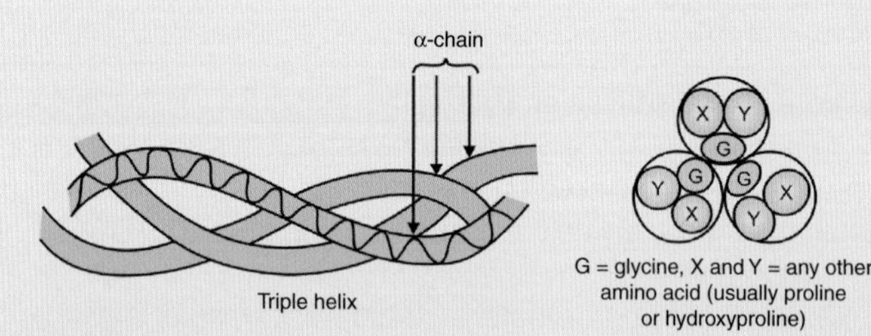

Adapted (left) and reproduced (right) with permission from Naik P: Biochemistry, 3rd edition, Jaypee Brothers Medical Publishers (P) Ltd., 2009.

Osteogenesis imperfecta (OI) is a genetic disease that affects collagen-containing tissues such as bone, skin, joints, eyes, ears, and teeth because of point mutations that destabilize or alter collagen's important triple helix structure. Patients with OI often display frequent fractures and easy bruising (sometimes mistaken for child abuse); weak joints; a bluish color to the normally white part of their eyes; hearing loss due partly to abnormalities of the inner ear bones; and poorly shaped, small, blue-yellow teeth.

TABLE 1-2. Amino Acids—Components of Various Biological Molecules

Name	Role	Amino Acid Source
DNA, RNA	Genetic material	Glycine, glutamine, aspartate
Dopamine, norepinephrine, epinephrine	Neurotransmitters	Tyrosine
Serotonin	Neurotransmitter	Tryptophan
Histamine	Blood vessel dilation, allergic response	Histidine
Heme/porphyrins	Hemoglobin/myoglobin, several metabolic molecules	Glycine
Thyroxin	Thyroid hormone	Tyrosine
Sphingosine	Precursor for important sphingolipids	Serine
Melanin	Skin pigment	Tyrosine
NAD^+	Coenzyme	Tryptophan, glycine
Glutathione	Anti-oxidant (sulfur)	Glutamate, cysteine, glycine
Nitrous oxide (NO)	Signaling molecule	Arginine

and a vast majority of the body's amino acids are involved in this function. However, amino acids are also the major precursors of several biologically important molecules, as noted in Table 1-2.

As a result, either the excess or deficiency of amino acids can lead directly to disease that may result in central nervous system defects, dietary and metabolism problems, liver and kidney failure, skin and eye lesions, and even death.

ENZYMES

Enzymes are specialized proteins that accelerate a chemical reaction by serving as a biological catalyst. By catalyzing these reactions, enzymes cause them to take place one million or more times faster than in their absence. Enzymes are usually identified by the ending of "ase" to their name (e.g., hexokinase, the first enzyme in the breakdown of glucose). There are exceptions for enzymes that were discovered before this naming scheme was adopted (e.g., trypsin, pepsin, and thrombin). Enzyme reactions will be discussed in greater detail in Chapter 5.

STRUCTURAL PROTEINS

Proteins also serve an important role as structural elements of cells and tissues. The best examples of these proteins are actin and tubulin, which form actin filaments and microtubules, respectively (Figure 1-8A-B). In skeletal muscle, actin filaments provide the "scaffolding" against which the motor protein myosin can generate force to produce muscle contraction. In smooth muscle and non-muscle (e.g., skin and immune system), actin filaments create the mechanical structure of the cell and are directly associated with linkages to surrounding cells allowing intercellular signaling. The actin filaments also provide tracks on which specialized myosin molecules move vesicles and organelles (see Chapter 12). Finally, actin filaments are intimately involved in cell motility, a wide array of cellular movements such as wound healing (the movement of skin cells into cuts), the immune response (the process of white blood cells contacting and recognizing each other in the highly selective process of immune reactions), and cytokinesis (the division of one cell into two during mitosis).

The structural protein tubulin creates microtubule tracks for the movement by two molecular motor proteins, dynein and kinesin, of vesicles, granules, organelles, and chromosomes. Microtubules are also the structural component in flagella and cilia involved in functions such as sperm motility, the movement of the egg down the Fallopian tubes, and the expulsion of dirt and mucus out of the lungs and trachea. Nonmotile cilia are also important in rod cells in the eye and neurons involved in olfaction (smell). Microtubules also serve a mechanical–structural role in the cell similar to actin microfilaments and are responsible for the movement and separation of chromosomes during mitosis (see Chapter 12).

MOTOR PROTEINS

"Motor" proteins transport molecules inside a cell, provide movement of certain parts of individual cells involved in specialized functions (e.g., immune responses and wound healing),

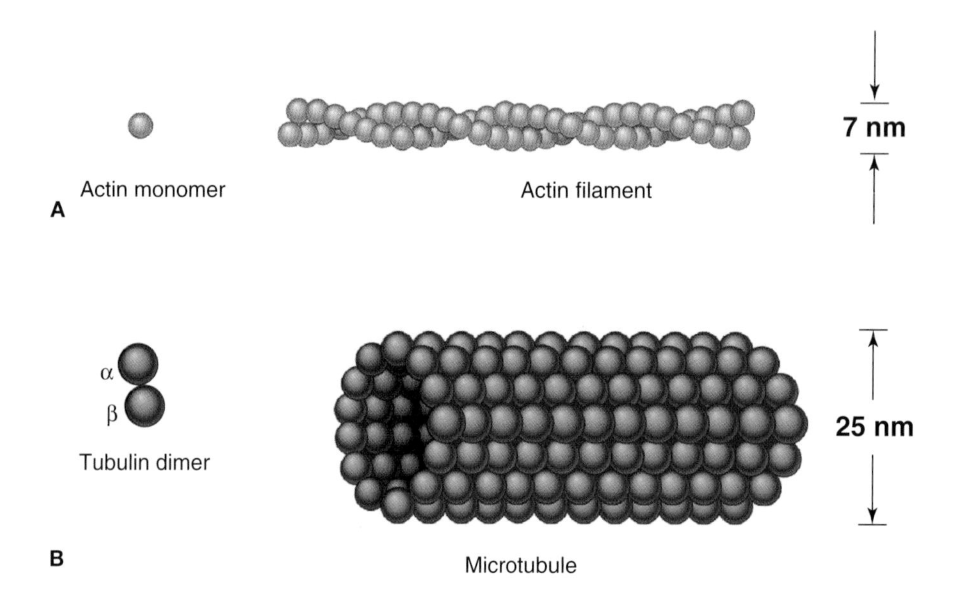

Figure 1-8. A–B. A. Actin and B. Tubulin in Monomer and Filamentous Forms. [Adapted with permission from Barrett KE, et al.: Ganong's Review of Medical Physiology, 23rd edition, McGraw-Hill, 2010.]

generate larger scale movements of fluids and semisolids such as the circulation of blood and movement of food through the digestive tract, and finally provide movement of the human body through their roles in skeletal muscles.

Myosin is a protein with a hydrophobic tail; a head group, which can attach and detach from actin filaments; and a "hinge" section, which moves the head group back and forth resulting in movement. Two major types of myosin exist. Myosin I is composed of one molecule with an additional area on its short tail that can bind to other proteins and membranes. Myosin II (several subtypes exist) is composed of a long tail that binds via hydrophobic interactions to other myosin II molecules, resulting in a composite molecule that can shorten skeletal muscles by its interaction with actin filaments in those muscles. Kinesin and dynein are very much like myosin I in form and function (see Chapter 12).

TRANSPORT/CHANNEL PROTEINS

Another group of proteins fold into a tertiary or quaternary structure that creates channels for the movement of molecules into and out of the nucleus, various cell organelles, and from cells to their outside environment such as the blood stream. The hydrophobic and hydrophilic nature of the amino acids that make up these channel proteins allows an exterior of the protein

Microtubules, Cilia, and Flagella—Roles in Disease Processes: Although usually rare, defects in cilia/flagella, known as ciliopathies, lead to several diseases/syndromes including the following:

Kartagener Syndrome/Primary Ciliary Dyskinesia—defective cilia in the respiratory tract, Eustachian tube, and Fallopian tubes leading to chronic lung infections, ear infections and hearing loss, and infertility. Possible association with "situs inversus," a condition in which major internal organs are "flipped" left to right.

Senior–Loken Syndrome/Nephronophthisis—eye disease and formation of cysts in the kidneys leading to renal failure.

Bardet–Biedl Syndrome—dysfunction of cilia throughout the body leading to obesity due to inability to sense satiation, loss of eye pigment/visual loss and/or blindness, extra digits and/or webbing of fingers and toes, mental and growth retardation and behavioral /social problems, small and/or misshaped genitalia (male and female), enlarged and damaged heart muscle, and kidney failure.

Alstrom Syndrome—childhood obesity, breakdown of the retina leading to blindness, hearing loss, and type 2 diabetes.

Meckel–Gruber Syndrome—formation of cysts in kidneys and brain leading to renal failure and neurological deficits, extra digits and bowing/shortening of the limbs.

Increased Ectopic (Tubal) Pregnancies/Male Infertility—deficient cilia in Fallopian tubes or deficient flagella/sperm tail motility.

Autosomal Recessive Polycystic Kidney Disease—much rarer than the autosomal dominant form, dysfunction of basal bodies and cilia in renal cells leads to alterations of the lungs and kidneys leading to a variety of secondary medical conditions and often death.

Parkinson's and Alzheimer's diseases—Although work is still ongoing, researchers now feel that some forms of Parkinson's and Alzheimer's diseases may result, in part, from damage to microtubules and associated proteins. Treatments aimed at stabilizing microtubules may help many sufferers of these maladies.

that can exist inside the extremely hydrophobic environment of a membrane bilayer and a hydrophilic interior that can allow charged molecules to move through the membrane (Chapter 8). Channels are essential for the transportation of nutrients into and out of cells as well as for nerve signals and the selective filtration of molecules in the kidneys. These specialized functions of channels will be discussed in detail in Chapter 8 and Section III.

REVIEW QUESTIONS

1. What is the meaning and significance of essential and nonessential amino acids?
2. What is the significance of each amino acid R-group (hydrophobic, hydrophilic, and charged)?
3. What are the four major types of structural elements of proteins and how are they defined?
4. What is an enzyme and how do the terms catalyst and active site relate to enzymes?
5. What is the basic structure of amino acids?
6. How do the elements peptide bond, peptides, α-helix, β-strand, β-turn, hairpin turn, and disulfide bond relate to the structure of proteins?
7. How does an amino acid sequence fold?
8. What are the roles of R-groups and primary to quaternary structure in the final conformation of proteins?
9. What are the different categories of proteins and how are they defined?

CHAPTER 2

CARBOHYDRATES

OVERVIEW

Carbohydrates are vastly important in human biology, including roles as a major energy source, structural molecules when combined with other carbohydrates, proteins, and other molecules, and binding and signaling between molecules and cells. As a result of all these important functions, carbohydrate biochemistry is involved in a large number of disease states. Although multiple carbohydrates exist, only a few sugar molecules and polysaccharides are important to human physiology (e.g. only eight different carbohydrates are found as constituents of glycoproteins and glycolipids). However, a number of additional molecules created by linkages of carbohydrates to proteins play various roles in cell–cell interactions and biological structures.

BASIC CARBOHYDRATE STRUCTURE AND FUNCTION

Carbohydrates, whose names end in "*-ose*," have a formula of $(CH_2O)_x$ where x is a number from three to seven (giving the names of **tri*ose***, tetrose, **pent*ose***, **hex*ose***, and hept*ose*). All carbohydrates contain a ketone or an aldehyde group, as well as one or more hydroxyl groups (Figure 2-1A–B; Appendix III). The oxygen atoms of the ketone and aldehyde groups have similar reactive qualities to that of the carboxylic acid group seen in amino acids and are the sites of chemical reactions within the carbohydrate molecule, as well as with other carbohydrate, protein, or lipid molecules. Often, the ketone

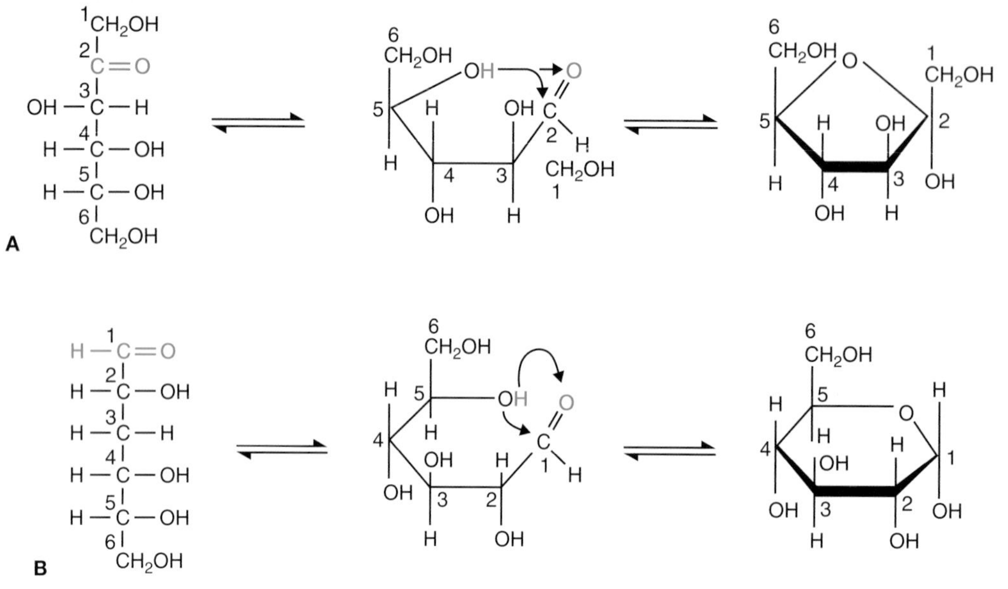

Figure 2-1. A–B. Basic Carbohydrate Structures. A. The reactive ketone group of carbon 2 (green carbon group) from the hexose fructose reacts with the hydroxyl group of carbon 5 to form a new bond and a five-sided (pentose) ring structure. All carbon atoms are numbered for clarity. This reaction is fully reversible as indicated by the bidirectional arrows. As a result, the linear and ring structures are constantly changing in solution. **B.** The reactive aldehyde group of carbon 1 (green carbon group) from the hexose glucose reacts with the hydroxyl group of carbon 5 to form a new bond and a six-sided (hexose) ring structure. All carbon atoms are numbered for clarity. This reaction is fully reversible as indicated by the bidirectional arrows. As a result, the linear and ring structures are constantly changing in solution. [Adapted with permission from Naik P: Biochemistry, 3rd edition, Jaypee Brothers Medical Publishers (P) Ltd., 2009.]

or aldehyde reacts with a hydroxyl group from the same sugar molecule to form a carbohydrate ring structure as shown.

Carbohydrates play a major role in humans as energy sources and storage, and their role in diet and nutrition, although sometimes controversial, is always one of supreme importance. However, carbohydrates play other roles as noted in Table 2-1.

MONOSACCHARIDES AND DISACCHARIDES

Although there are multiple trioses, pentoses, hexoses and heptoses depending on the various arrangements of hydroxyl groups and hydrogens around the central carbon backbone, only a few of these single residue sugars are commonly seen in human biology. Common single sugar groups, called **monosaccharides**, include the triose **glyceraldehyde**; the pentose **ribose**; and the hexoses **fructose**, **glucose**, and **galactose** (shown in Figure 2-2).

Hydroxyl groups (OH) are often locations of enzyme reactions, especially the formation of a new bond between two carbohydrate molecules with the resulting release of a water molecule. When monosaccharides form such bonds, the resulting molecules are a **disaccharide**, **trisaccharide**, and so on. The various combinations of all the different monosaccharides would produce a vast mixture of these new molecules but, in fact, only a few are common in humans (Figure 2-3),

TABLE 2-1. Biochemical Roles of Carbohydrates

Biochemical Role	Notes	Examples
Energy source	Carbohydrate metabolism	ATP is a phosphorylated derivative of a carbohydrate
Structural	Nucleic acids Cell wall (bacteria and plants)	DNA (deoxyribonucleic acid)
Binding	Points of binding between separate molecules	Binding of influenza virus to target cells
Signaling	Recognition of specific sugars during binding elicits particular biological responses in and between cells	Lymphocyte–monocyte recognition and immune response

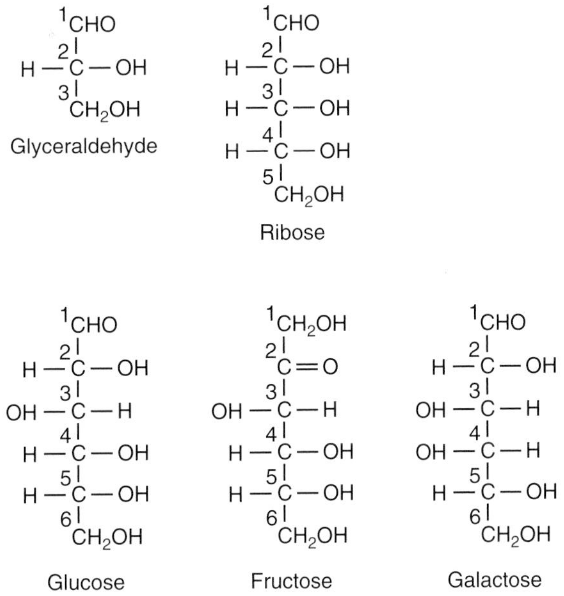

Glyceraldehyde:
^{1}CHO
$H-{}^{2}C-OH$
$^{3}CH_2OH$

Ribose:
^{1}CHO
$H-{}^{2}C-OH$
$H-{}^{3}C-OH$
$H-{}^{4}C-OH$
$^{5}CH_2OH$

Glucose:
^{1}CHO
$H-{}^{2}C-OH$
$OH-{}^{3}C-H$
$H-{}^{4}C-OH$
$H-{}^{5}C-OH$
$^{6}CH_2OH$

Fructose:
$^{1}CH_2OH$
$^{2}C=O$
$OH-{}^{3}C-H$
$H-{}^{4}C-OH$
$H-{}^{5}C-OH$
$^{6}CH_2OH$

Galactose:
^{1}CHO
$H-{}^{2}C-OH$
$OH-{}^{3}C-H$
$OH-{}^{4}C-H$
$H-{}^{5}C-OH$
$^{6}CH_2OH$

Figure 2-2. Common Monosaccharides in Human Biology. The common monosaccharide carbohydrates found in humans are shown above, including the three-carbon triose glyceraldehyde; the five-carbon pentose ribose; and the six-carbon hexoses fructose, glucose, and galactose. Note the only structural difference between glucose and galactose is the placement of the hydrogen atom and hydroxyl group at carbon 4. Carbon atoms are numbered for clarity. [Adapted with permission from Naik P: Biochemistry, 3rd edition, Jaypee Brothers Medical Publishers (P) Ltd., 2009.]

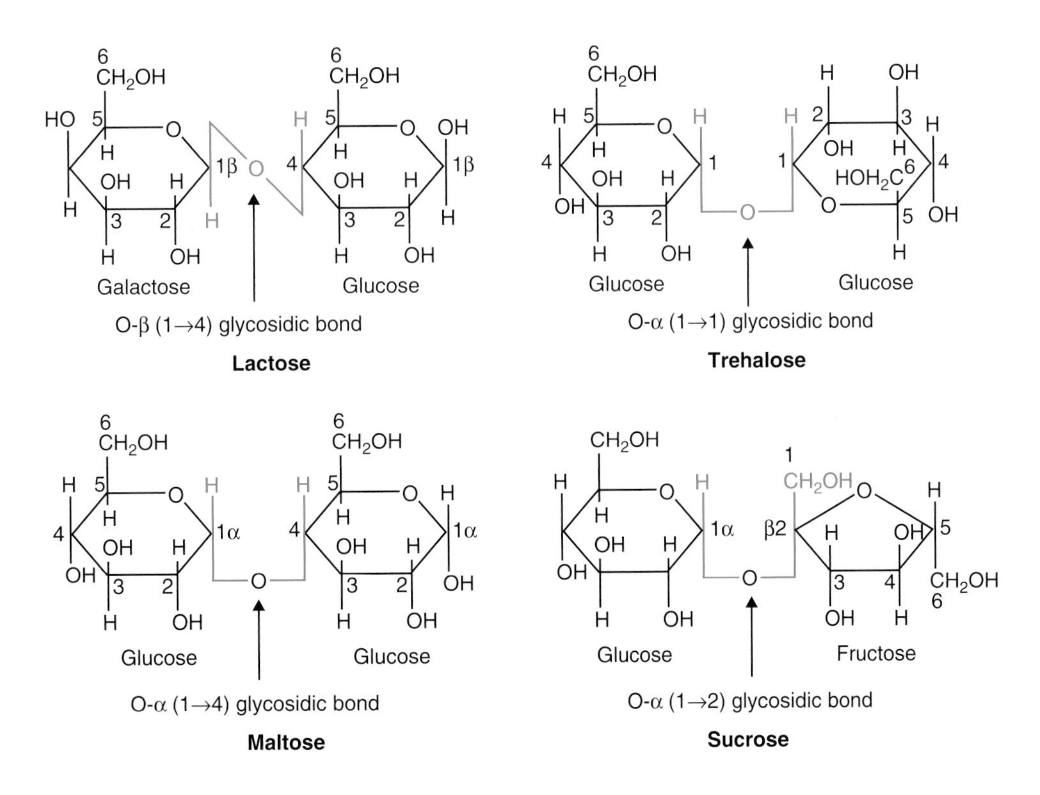

Figure 2-3. Common Disaccharides in Human Biology. The common disaccharide carbohydrates found in humans are shown above, including lactose, trehalose, maltose, and sucrose. Component monosaccharides and specific α- and β-bond configurations are indicated immediately under each disaccharide. Note that the second glucose molecule in trehalose and the fructose molecule in sucrose are flipped horizontally in the final disaccharide molecule (see carbon numbers). [Adapted with permission from Naik P: Biochemistry, 3rd edition, Jaypee Brothers Medical Publishers (P) Ltd., 2009.]

namely **lactose** (the primary sugar found in milk), **trehalose** [found in plants (e.g., sunflower seeds), animals (e.g., shrimp), Baker's yeast, and several types of mushrooms)], **maltose** [(found in many foods made from grains (e.g., barley)], and **sucrose** (found naturally in plants; usually artificially made for human consumption as common table sugar). The ring structure of monosaccharides also has reactive hydroxyl groups at each of their carbon atoms, especially at the first and sixth carbons, which can bond to other sugar molecules and amino groups and proteins (discussed below and in the following chapters).

Carbohydrate Intolerance: The digestion of carbohydrates from food involves several processes relying on proteins, both enzymes and transport/channel types. Defects in any of these proteins lead to disease states in which a particular carbohydrate cannot be tolerated in the patient's diet. One of the best known examples is **lactose intolerance**, which develops in adolescence or adulthood and is caused by the inability to digest the sugar lactose, the primary carbohydrate in cow's milk. Decreased digestion increases bacterial fermentation of the excess sugar molecules producing intestinal gas, bloating, nausea, and painful cramping. The osmotic effect of the excess monosaccharide and disaccharide molecules leads to increased water retention and absorption in the large intestine, causing watery diarrhea.

In fact, historical evidence indicates that humans only recently evolved the ability to digest lactose when cow's milk became a staple of their diet. In addition, some races of humans are less able to digest lactose, leading to increased incidence of lactose intolerance in that racial group.

Other types of carbohydrate intolerance include enzyme deficiencies in the digestion of sucrose, maltose, and trehalose—the latter seen predominately in Inuit and Greenland populations—and the absence or decreased action of transporter/channel proteins that can cause profound effects on digestion of glucose, fructose, and galactose. Some carbohydrate intolerance can lead to serious problems of failure to thrive and kidney and/or liver disease. Treatment of carbohydrate intolerances is normal by avoidance of the offending sugar or by supplementation of the affected enzyme.

GLYCOGEN AND STARCHES

Linkages of monosaccharides and disaccharides form long carbohydrate chains called polysaccharides. The common polysaccharides found in nature are **glycogen** and **starch**. In fact, over half of all the carbohydrates in the human diet are starch molecules. Although glycogen is always a branched polysaccharide molecule, starch can be either branched (called **amylopectin**) or unbranched (called **amylose**). The glycogen molecule and starch molecules are shown in Figure 2-4.

Glycogen Storage Diseases: Glycogen storage diseases include 11 classes of inborn errors in the production and breakdown of these long carbohydrate chains. The genetic errors, although rare, result in an inability of the body to respond to the increased need for glucose molecules. As a result, patients suffering from glycogen storage diseases are limited in their ability to exercise or develop low blood sugar after a relatively short period of food deprivation. These glycogen storage diseases will be explored in fuller detail in later chapters.

Cellulose, the another major plant polysaccharide besides starch, differs from **amylose** only in the bonds between carbons 1 and 4. Changes in the way the aldehyde and hydroxyl groups of the two glucose molecules form the bond results in either an amylose α-bond [indicated by a downward bond (Figure 2-4A)] or a cellulose β-bond [indicated by an upward bond (Figure 2-5)]. Although the difference may seem insignificant, this β-bond results in a markedly different overall structure of the polysaccharide. Although α-linkages create an overall helical structure of the chain, β-linkages create a straight chain. The α-linked helical structure is important for accessing the carbohydrate molecules for metabolism, whereas the β-linked linear chain is stronger and, therefore, well suited for cellulose-containing structures such as plant walls (e.g., wood). In addition, the β-linkages cannot be digested by humans but can be digested by animals such as cows and termites, and it is the reason humans cannot live on grass or wood, both of which are composed of β-bonded glucose molecules.

GLYCOPROTEINS

Carbohydrates may also form bonds with proteins via any of the carbohydrate hydroxyl groups combining with the amino acid hydroxyl groups of serine or threonine or the amine nitrogen of asparagine. The resulting carbohydrate–protein molecules are referred to as **glycoproteins**, and approximately half of the proteins in the human body are estimated to be glycoproteins. The diversity of glycoproteins seen in nature and humans is immense with complex mixtures of proteins, amino acids, and sugars (monosaccharides, disaccharides, and trisaccharides) linked together in a vast number of linear and branched formations generically called **oligosaccharides**. These oligosaccharides play a diverse number of functional roles in binding, signaling, and regulation that will be more fully explored in Section III.

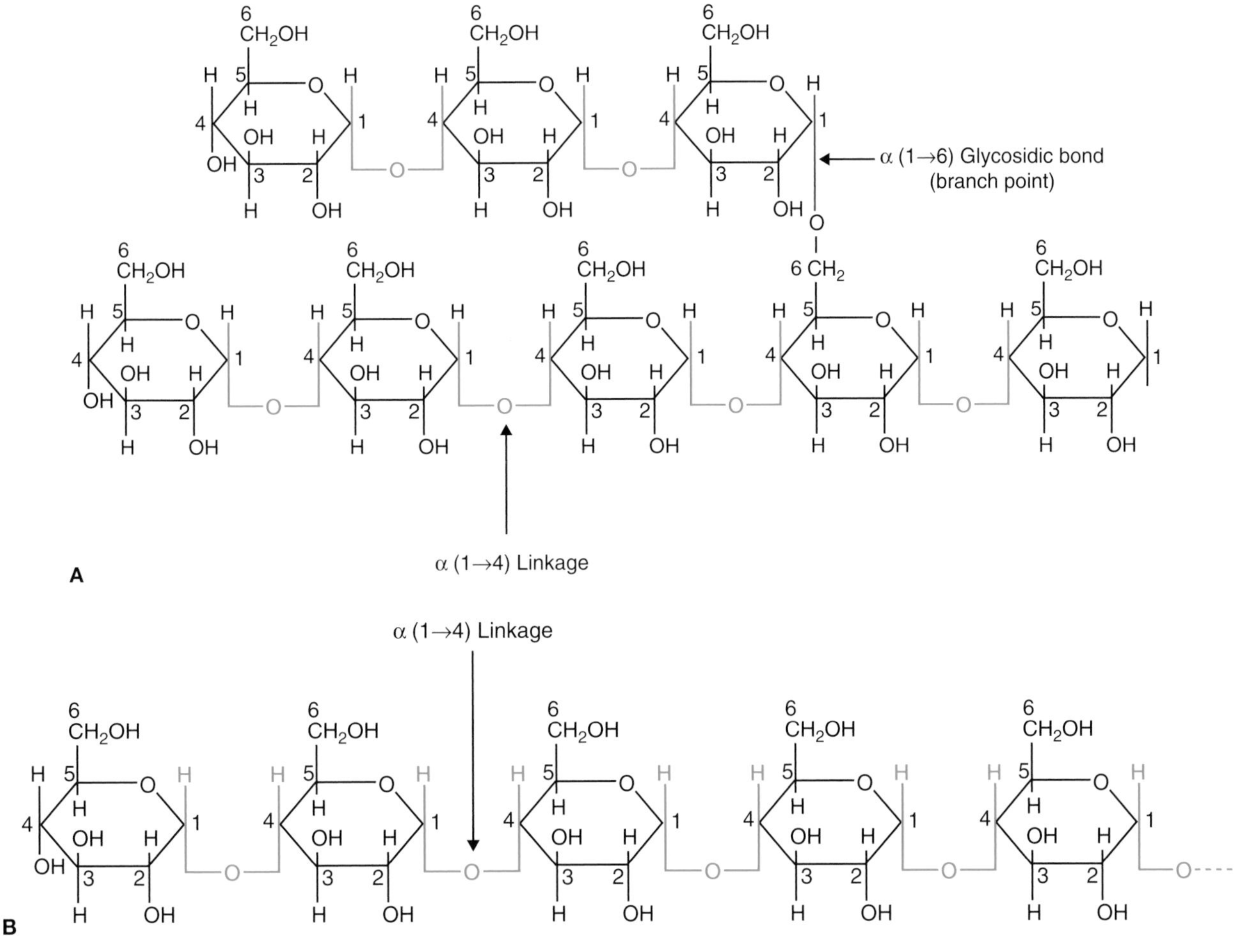

Figure 2-4. A–B. Glycogen and the Plant Starch Forms Amylopectin and Amylose. A. Glucose molecules, bonding both linearly between the carbons 1 and 4 (green bonds) and as branches between carbons 1 and 6 (red bond) of successive glucose molecules, create glycogen (branches approximately every 10 glucose residues) or the form of plant starch called amylopectin (branches approximately every 30 glucose molecules). Glycogen and starch are usually thousands of glucose molecules long and are extremely important forms of carbohydrate storage in the human body and plants, respectively. **B.** Glucose molecules, bonding only linearly between the carbons 1 and 4 (green bonds) of successive glucose molecules, create the form of plant starch called amylose, which has no branching. Amylose is an important form of carbohydrate storage in plants. [Adapted with permission from Naik P: Biochemistry, 3rd edition, Jaypee Brothers Medical Publishers (P) Ltd., 2009.]

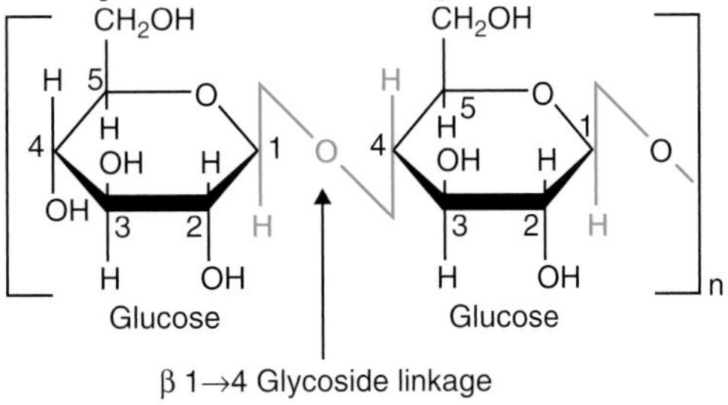

Figure 2-5. Cellulose. Note the β-linkages (upward-going green bond) between carbons 1 and 4 of the two glucose molecules that differ from the α-linkages seen in glycogen and starches. Cellulose is composed of multiple repeats of this basic unit as indicated by the subscript "*n*." [Adapted with permission from Naik P: Biochemistry, 3rd edition, Jaypee Brothers Medical Publishers (P) Ltd., 2009.]

GLYCOSAMINOGLYCANS

Even more complex than simple oligosaccharides are the **glycosaminoglycans** (often referred to as "**GAGs**"), which contain repeating disaccharide chains of a modified glucose and/or galactose. Prior to their linkage to the GAG molecule, these carbohydrate molecules have an amine group added with an additional acetyl (e.g., $NHCOCH_3$) or negatively charged sulfate (e.g., $NHSO_3^-$) group, producing **glucosamine** and **galactosamine** molecules. In addition, negatively charged sulfate (SO_3^-), via the reactive hydroxyl groups, and/or carboxylate (COO^-) groups are linked to at least one of the sugars of the repeating chain (Figure 2-6). A very well known GAG is **heparin**, a potent inhibitor of blood clot formation, used in patients with heart attacks, strokes, and clotting diseases. Other GAGs include **chondroitin** and **hyaluronic acid** (the longest of the GAGs), which bond with a central linear core protein to create ***proteo*glycans**. These molecules are important as strong structural elements in connective tissue and cartilage (Figure 2-6C and Table 2-2).

Glycosaminoglycans are responsible for a vast number of functions in human biology, mostly involving structure outside

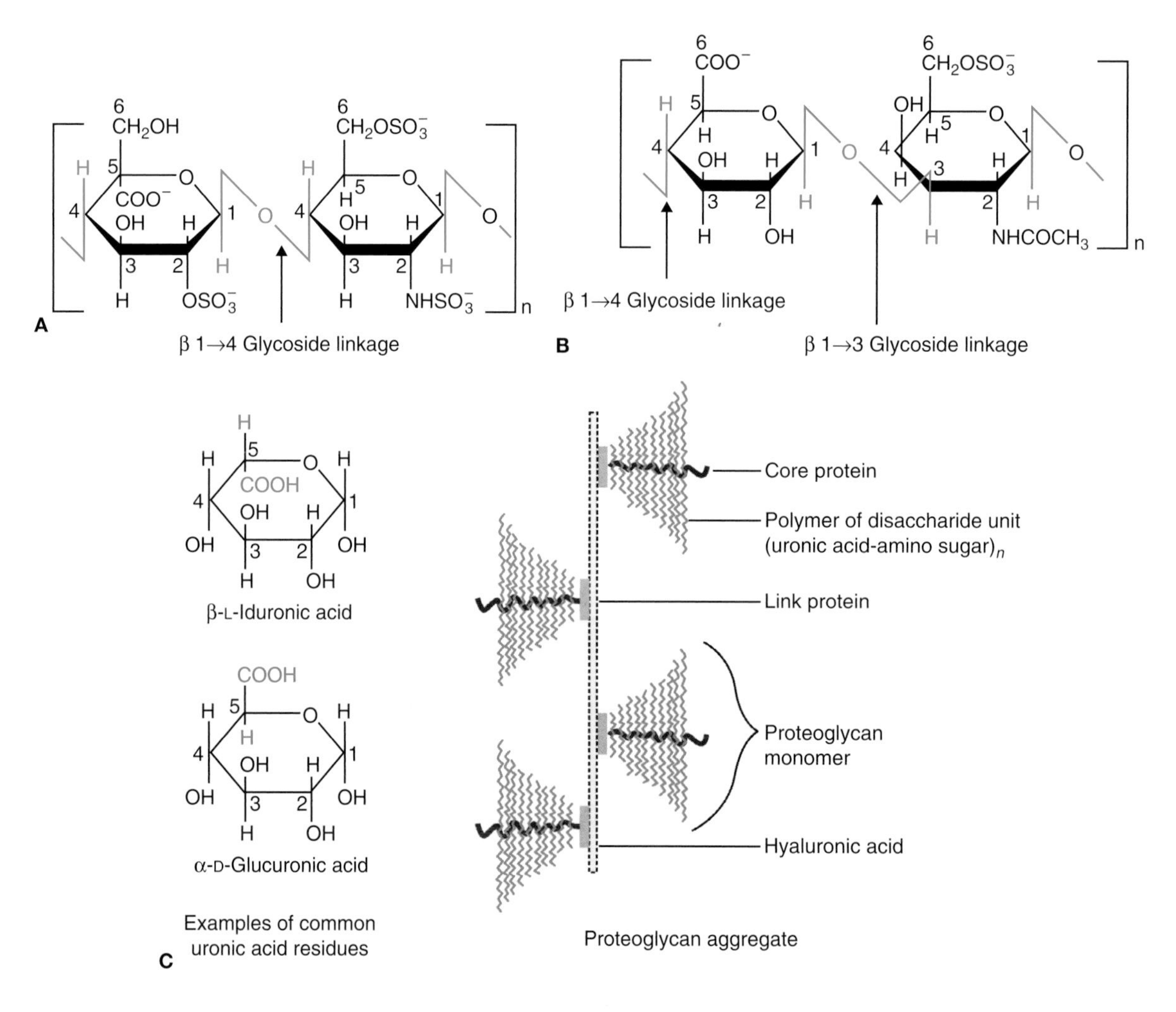

Figure 2-6. A–C. Examples of Glycosaminoglycans (GAGs) and of Proteoglycan Associated with Collagen. A. Heparin, a GAG and an important molecule in blood clot regulation, is composed of two carbohydrate molecules linked via a β-linkage between carbons 1 and 4 and the addition of a COO^- group, as well as sulfate and nitrogen and sulfate groups at various other carbon atoms. Although multiple forms of heparin exist, depending on the particular carbon that is modified, one of the most common is shown above. **B.** Chondroitin, a GAG important in cartilage, tendon, and bone structure, is composed of alternating carbohydrate molecules (glucuronic acid N-acetyl-galactosamine; note COO^- and nitrogen groups) linked via an α-linkage between carbons 1 and 3. Each carbohydrate molecule may be sulfated (once or twice) or left unsulfated. The more common sulfated forms are referred to as chondroitin sulfate and are felt to be mainly responsible for the molecule's biological activity. Chondroitin has recently become popular to ingest in pill form by some patients with knee and other joint pain in hopes of "replacing" cartilage that has been depleted by time and wear and tear. **C.** Collagen, the main protein in connective tissue, is strongly associated with proteoglycans composed of several GAGs. Common components include hyaluronic acid, and long, linear chains of repetitive units of uronic acids (left) and chondroitin sulfate or keratan sulfate bound together by link and core proteins. The resulting protein and carbohydrate molecules as well as interactions between the charged GAG groups and surrounding water molecules create an overall structure (right) that is both strong and rigid but also flexible and conformable. These qualities help to make collagen the perfect substance to provide structure but also allow movement in joints. [Adapted with permission from Naik P: Biochemistry, 3rd edition, Jaypee Brothers Medical Publishers (P) Ltd., 2009.]

Carbohydrates and Fertilization: The fertilization of a human egg relies on carbohydrate binding and signaling. Binding of a receptor on the sperm's surface to a *galactose* molecule within an oligosaccharide on the egg's surface signals the sperm to release molecules that allow sperm entry into the egg, resulting in fertilization.

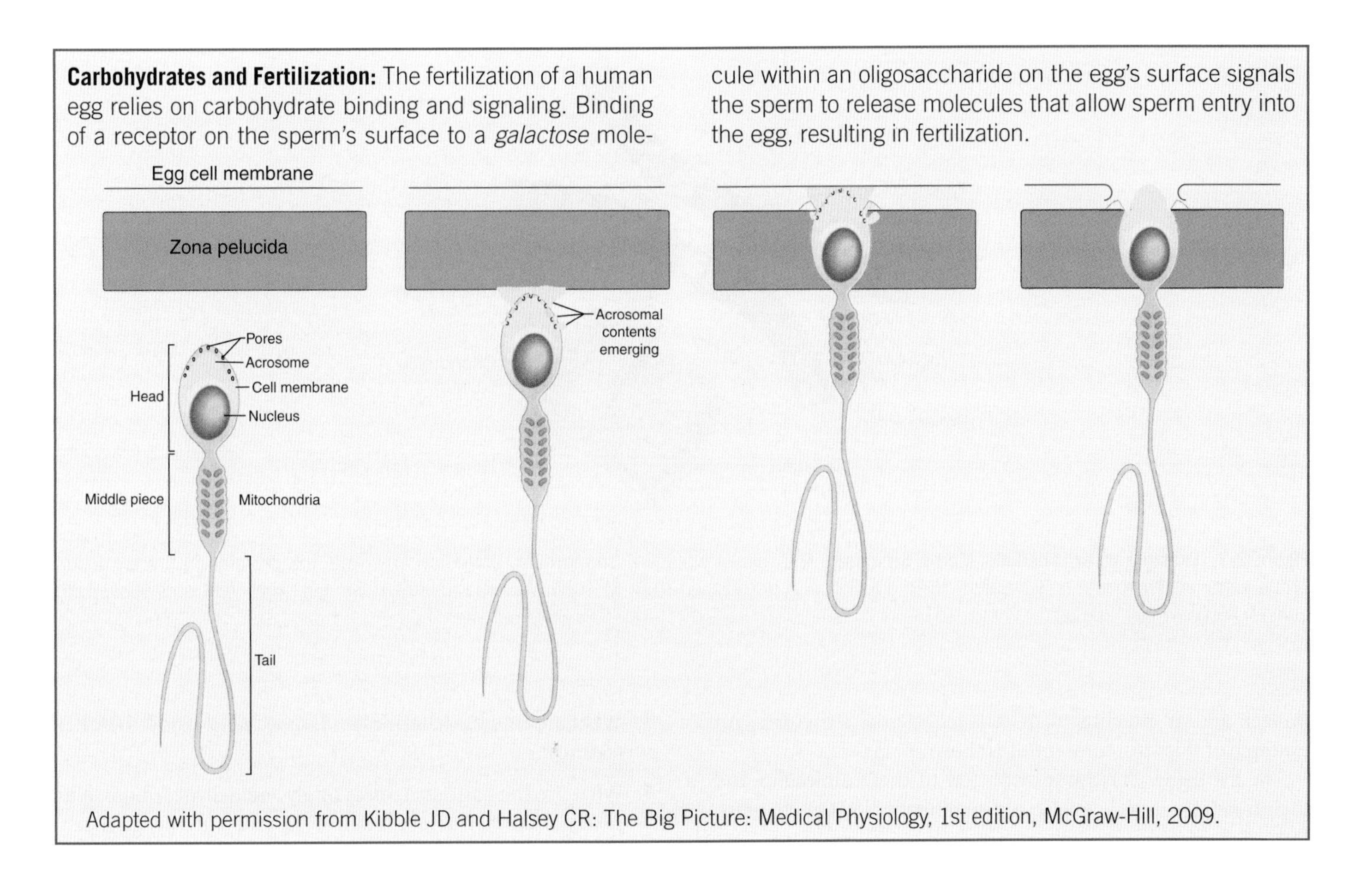

Adapted with permission from Kibble JD and Halsey CR: The Big Picture: Medical Physiology, 1st edition, McGraw-Hill, 2009.

TABLE 2-2. Biochemical Roles of Glycosaminoglycans

Glycosaminoglycan	Location in Human Body	Function(s)
Hyaluronic acid	Joints Eye	Component of proteoglycan binding to collagen Extracellular matrix in joints Binding of collagen in the cornea
Chondroitin	Joints and bone	Structural component and attachment point to collagen
Dermatan	Skin Eye Circulatory system	Structural component of skin Binding of collagen in the cornea Binding of proteins Regulation of clot formation (blood vessel walls and heart valve surfaces)
Heparan	Outer surface of cells	Attachment to collagen for various cells including liver cells
Heparin	Mast cells Liver cells	Regulates immune response Regulates blood clot formation
Keratin	Cartilage Eye	Component of cartilage Cornea of the eye

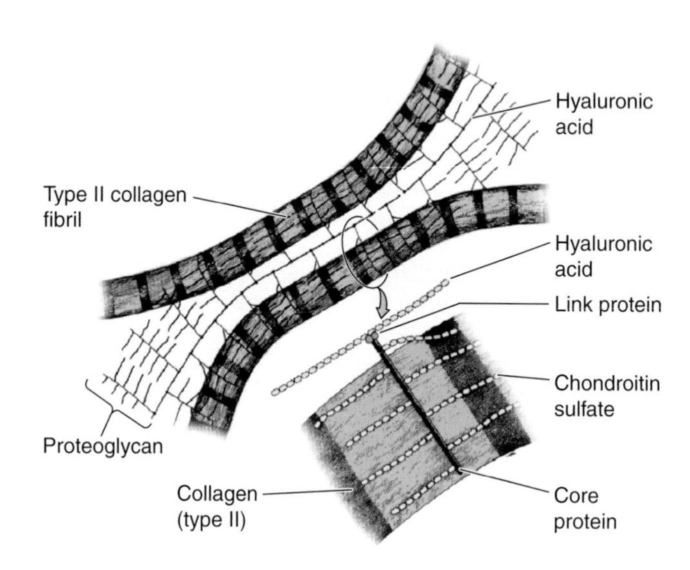

Figure 2-7. Example of Extracellular Matrix Structure. The role of collagen, chondroitin, proteoglycans, and link proteins in the formation of the extracellular matrix is illustrated. [Reproduced with permission from Mescher AL: Junqueira's Basic Histology Text and Atlas, 12th edition, McGraw-Hill, 2010.]

the cell and/or attachment of cells to external structures. An example of this structural motif is shown in Figure 2-7.

The list below illustrates just a few of these molecules and their functions, which will be explored further in later chapters.

REVIEW QUESTIONS

1. What are triose, pentose, and hexose carbohydrates and their key features?
2. What are ketone and aldehyde groups and their key features?
3. What are monosaccharides and disaccharides and their key features?
4. What are the basic structures of glyceraldehyde, ribose, glucose, galactose, fructose, maltose, lactose, sucrose, glycogen, starch, amylopectin, amylase, glycoproteins, and glycosaminoglycans?
5. What are the basic roles and functions of carbohydrate molecules, including the various other types of molecules to which they may bond and the resulting structural characteristics?

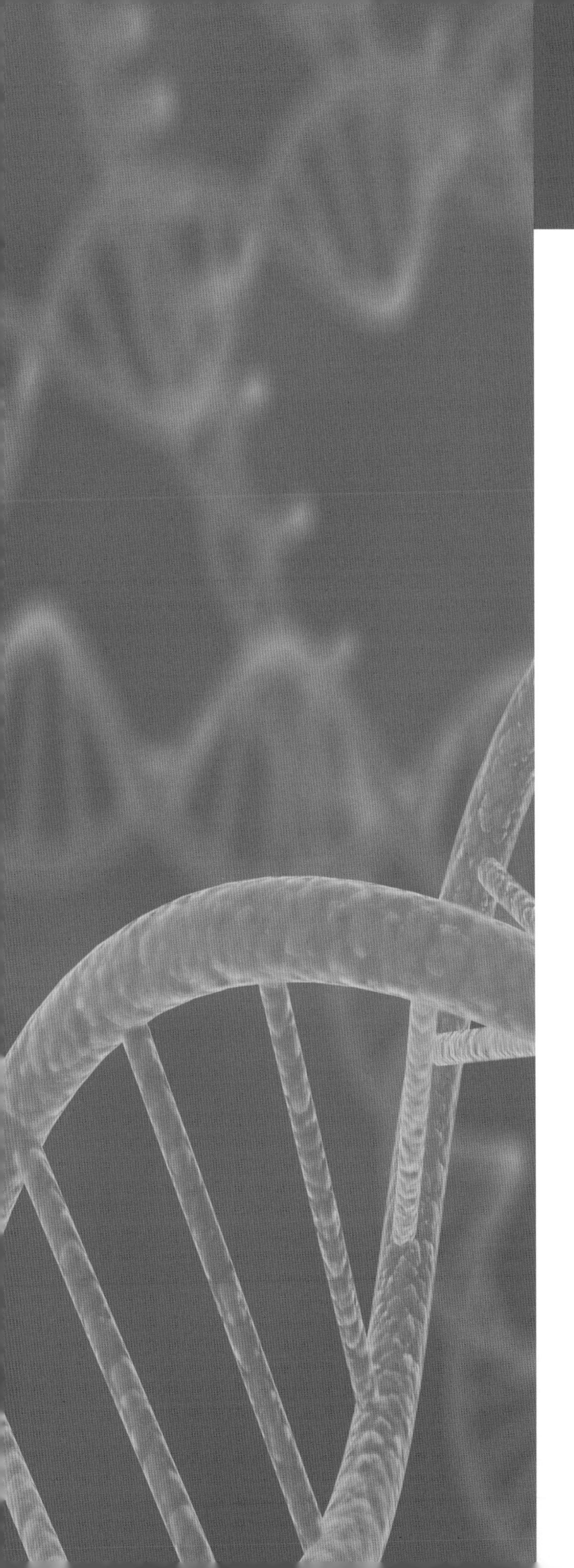

CHAPTER 3

LIPIDS

OVERVIEW

Lipids are the third major type of biochemical molecule found in humans. Although one of their major functions relates to the formation of biological membranes (phospholipids and cholesterol), lipid molecules are also essential for energy storage and transport (triacylglycerols), cellular binding and recognition and other biological processes (glycolipids), signaling (steroid hormones), digestion (bile salts), and metabolism (fatty acids, ketone bodies, and vitamin D). Lipid molecules are mainly hydrophobic and are, therefore, found in areas away from water molecules or are involved in mechanisms such as lipoprotein complexes that allow their movement in and through water environments. The smaller hydrophilic parts of lipids are, themselves, important in formation of biological membranes and in the several specific functions of lipids and lipid-derived molecules.

BASIC LIPID FUNCTIONS

A major role of lipid molecules is to provide the building blocks for biological membranes, including **phospholipids**, **glycolipids**, and **cholesterol**. However, other lipids known as **triacylglycerols** (also referred to as triglycerides or fats) function in the storage of biological energy and **bile salts**, derived in the liver from cholesterol and serve in the digestion of dietary fat. Finally, several lipid-derived molecules serve as important hormones and intracellular messengers.

It is important to note that the major part of every lipid molecule is hydrophobic in nature and, like the hydrophobic parts of proteins discussed earlier (Chapter 1), prefers to be away from and protected against interaction with water molecules. This hydrophobic character is fundamental in membrane formation, lipid transport, and in many of the functions that the various types of lipid molecules perform.

BASIC MEMBRANE LIPID STRUCTURE

A membrane lipid is composed of three basic components that are as follows:

1. **Fatty acids** are composed of long chains of carbon molecules with a carboxylic acid (COOH) at carbon 1 and a CH_3 (methyl) group at the end of the chain (Figure 3-1A). The carboxylic acid group is involved in bonding of the fatty acid to the other components of a lipid molecule. In humans, fatty acids are usually 12–24 carbons long and most often the number of carbons in the fatty acid backbone is even. Fatty acids can contain single (C—C), double (C=C), or triple (C≡C) carbon–carbon bonds.
   - **Saturated** fatty acids contain only single carbon–carbon bonds, and all of the carbon molecules are bonded to the maximum number of hydrogen molecules.
   - **Unsaturated** fatty acids have at least one double carbon–carbon bond with the potential for additional hydrogen atom bonding still existing for some of the carbon atoms in the backbone chain. If more than one double bond is present, the term polyunsaturated is used. These double bonds can exist in either a "kinked" *cis* double bond or a more linear *trans* double bond (Figure 3-1B-C).
2. **Glycerol** is a simple three-carbon molecule with hydroxyl groups at each carbon (Figure 3-2A). These hydroxyl groups are the reactive location where fatty acids and other components of a lipid molecule bond to form diacylglycerol (Figure 3-2A) and triacylglycerol (Figure 3-2B) molecules.

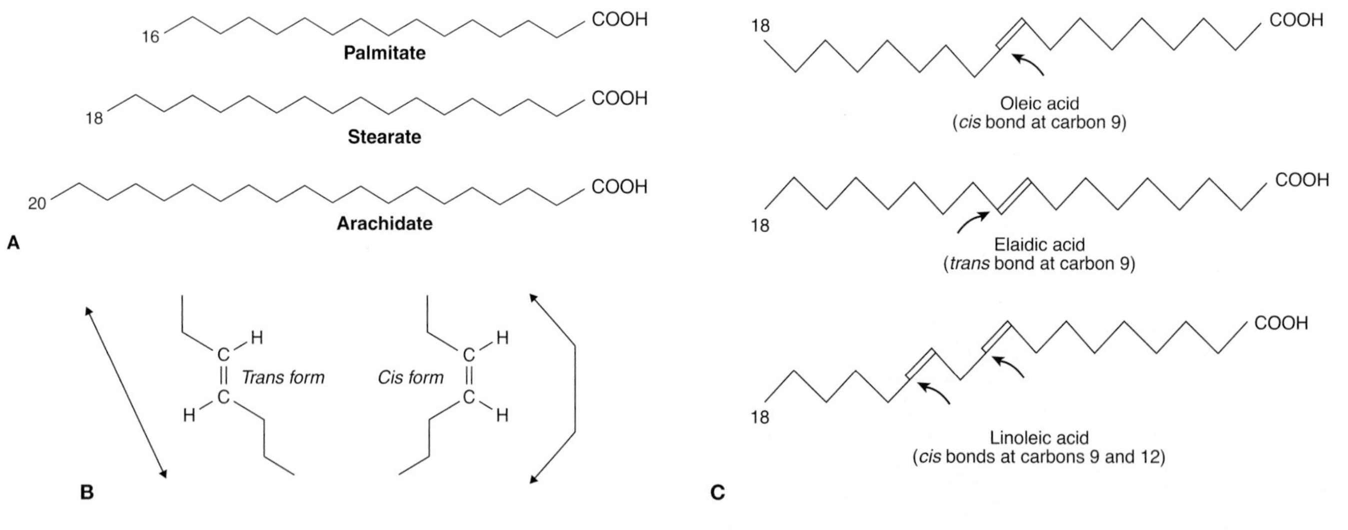

Figure 3-1. A. Common Saturated and Unsaturated Fatty Acids. Palmitate (16-carbon), stearate (18-carbon), and arachidate (20-carbon), fatty acids, shown by the carbon backbone and in "stick diagram" form for arachidate often used for simplicity. The carbon atoms of fatty acids are numbered from the carboxylic acid (COOH) to the terminal methyl (CH_3) group. Hydrogen atoms are not shown for clarity. **B. Fatty Acid Chain Double Bonding.** Detail of unsaturated fatty acid carbon chain, illustrating *trans* double bond (left: hydrogen atoms on opposite sides of the bond and resulting linear carbon chain) and unsaturated *cis* double bond (right: hydrogen atoms on the same side of the bond and resulting "kinked" carbon chain). **C. Common Unsaturated Fatty Acids.** (Top and middle) Two 18-carbon unsaturated acids showing a *cis* double bond (oleic acid) and a *trans* double bond (elaidic acid) both at carbon 9. Arrows illustrate different conformations of fatty acid chain that result from the two types of saturated bonds. (Bottom) An 18-carbon unsaturated fatty acids with two *cis* double bonds (linoleic acid) at carbons 9 and 12 (see arrows). Note how the addition of a second *cis* double bond creates an even more nonlinear fatty acid chain, which causes increased disorder of packing and, therefore, increased fluidity of biological membranes. [Adapted with permission from Murray RA, et al.: Harper's Illustrated Biochemistry, 28th edition, McGraw-Hill, 2009.]

Double Bonds and Melting Temperatures: The number and type of double bonds in a particular lipid molecule's fatty acid affects how that fatty acid "packs" with other fatty acids and, therefore, the temperature at which the particular lipid molecule melts. For example, the saturated fatty acids listed in Table 3-1 have melting points between 44°C and 77°C. The unsaturated fatty acids have far lower melting points, ranging from 13°C to –50°C, decreasing as the number of double bonds (polyunsaturation) increases. "Kinked" *cis* double bonds make the packing of fatty acids even more disorganized and lower the melting temperature even further. The effects of saturated/unsaturated/polyunsaturated and *cis*/*trans* fatty acids are readily seen in the different melting temperatures of butter, composed of a highly saturated lipids, and margarine, composed of unsaturated lipids. The linear nature of *trans* fatty acids makes them similar to the structure of saturated fatty acids. This structural feature, relative to the *cis* configuration, seems to make metabolism of *trans* fats difficult. Consequently, *trans* fats remain longer in the circulation, thereby contributing to arterial deposition and subsequent development of coronary heart disease. In general, unsaturated fats are healthier for the human body and, therefore, dietary fats composed mostly of *cis* double bonded, polyunsaturated fats are recommended by dieticians and clinicians to help avoid heart disease and other medical problems.

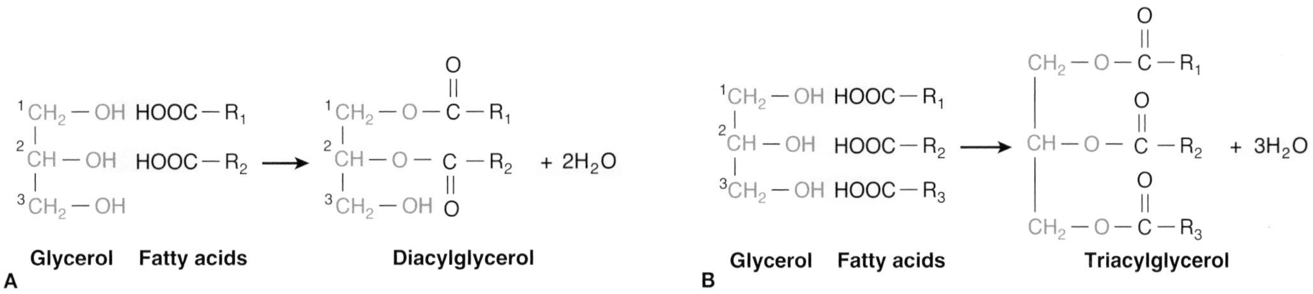

A

$$^{1}CH_2-OH \quad HOOC-R_1$$
$$^{2}CH-OH \quad HOOC-R_2 \longrightarrow \quad ^{1}CH_2-O-C(=O)-R_1, \; ^{2}CH-O-C(=O)-R_2, \; ^{3}CH_2-OH \; + 2H_2O$$
$$^{3}CH_2-OH$$

Glycerol Fatty acids Diacylglycerol

B

$$^{1}CH_2-OH \quad HOOC-R_1$$
$$^{2}CH-OH \quad HOOC-R_2 \longrightarrow \quad CH_2-O-C(=O)-R_1, \; CH-O-C(=O)-R_2, \; CH_2-O-C(=O)-R_3 \; + 3H_2O$$
$$^{3}CH_2-OH \quad HOOC-R_3$$

Glycerol Fatty acids Triacylglycerol

Figure 3-2. A–B. A. Glycerol, Diacylglycerol, and Triacylglycerol. Glycerol is a simple, three-carbon chain molecule (green) with a hydroxyl group (OH) bonded to each of the carbon atoms. The hydroxyl groups at carbons 1 and 2 of glycerol bond react with the carboxylic acid groups (COO^-) of the fatty acid chains, resulting in two new bonds and two water (H_2O) molecules. In general, unsaturated fatty acids bond to carbon 1, whereas saturated fatty acids bond to carbon 2. The resulting molecule is called "diacylglycerol" and is involved in important signaling pathways (Chapter 8). **B.** Triacylglycerol is formed when a third fatty acid bonds to the third glycerol hydroxyl group. This resultant molecule is an important storage form of energy (Chapter 8). [Adapted with permission from Naik P: Biochemistry, 3rd edition, Jaypee Brothers Medical Publishers (P) Ltd., 2009.]

Essential Fatty Acids: Much like there are *essential* amino acids that the body can only get from dietary sources, certain fatty acids are also deemed essential. Two particular essential fatty acids, linoleate and linolenate, have double bonds at the sixth and third carbon atoms counting from the methyl end of their chains, respectively, and are needed to produce certain 20-carbon long fatty acids containing double bonds. These two fatty acids are known as **omega-6 (ω-6)** and **omega-3 (ω-3)** fatty acids. Humans do not have the ability to produce double bonds at these locations and, therefore, must obtain these two required fatty acid building blocks from vegetable oils. Arachidonate with a 20-carbon chain and four *cis* double bonds is also an essential fatty acid involved in several important biological functions. Recently, longer carbon chain ω-3 fatty acids have been proposed to decrease heart attacks and strokes; supplements and some food products are now available, which contain these fatty acids. Interestingly, excessive dietary intake of ω-6 fatty acids has been implicated in an increased risk of heart attacks, strokes, some cancers, and even depression.

3. **Head Group** The final component of a lipid molecule varies with each type of lipid and, along with the two specific fatty acids, defines each particular lipid. This third part of the lipid molecule is often called the "head group," aptly named if one envisions the end methyl group of the fatty acid chains to be the tail of the lipid molecule (Figure 3-3A). Most lipids in a biological membrane have a phosphate group (PO_4^{-3}) attached to the third glycerol carbon and are, therefore, called **phospholipids.** Usually, an additional molecule (several common examples found in humans and the resulting phospholipid molecules are shown in Figure 3-3B) is attached to the phosphate molecule, resulting in the final head group of the lipid molecule. This head group is usually charged, creating a part of the lipid that is hydrophilic, and wants to be near water, a quality that is essential for the formation of biological membranes (Chapter 8) and many lipid functions.

The common fatty acids found in humans are listed in Table 3-1.

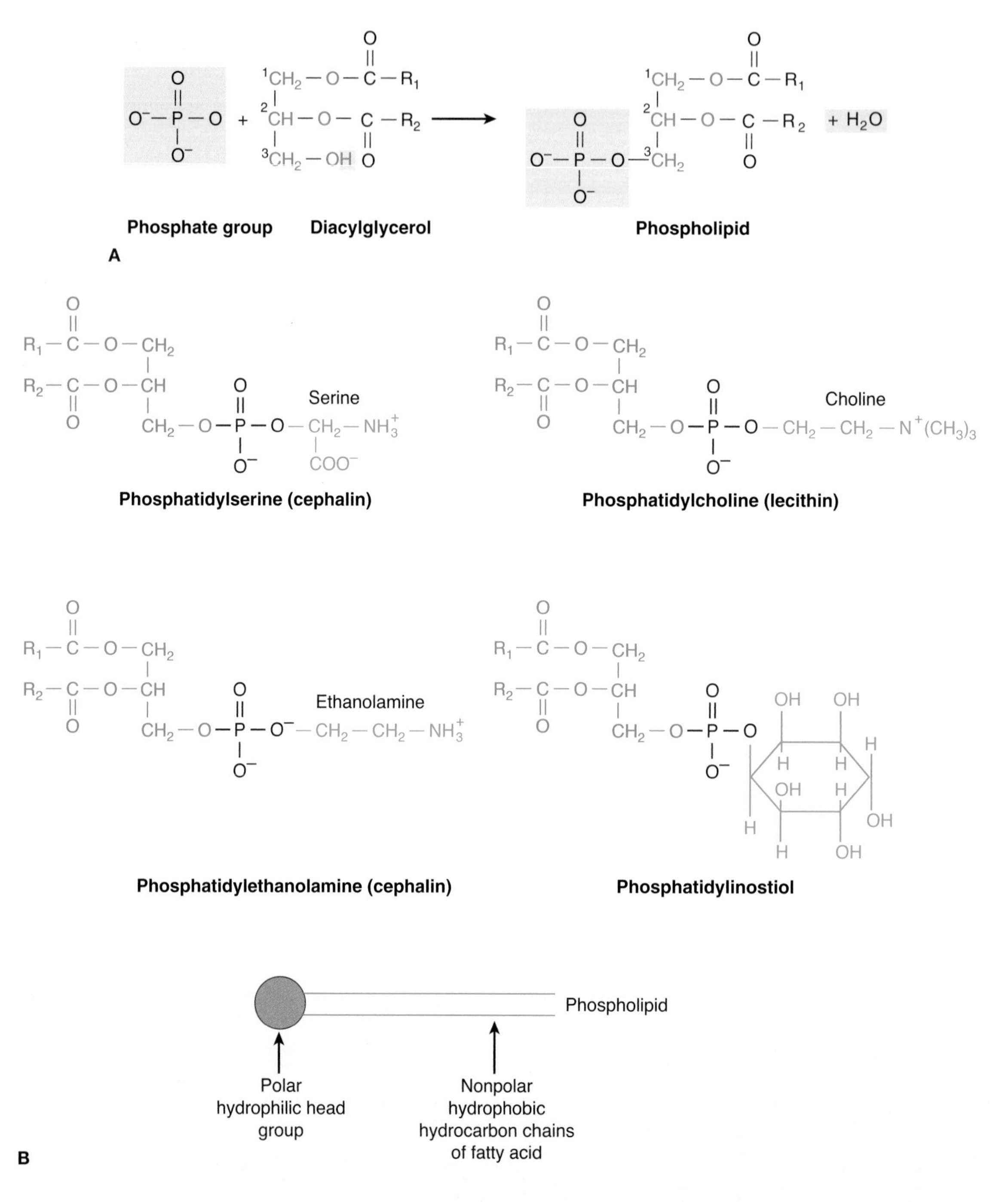

Figure 3-3. A. Phospholipid Components and Formation. Bonding between hydroxyl (OH) of a phosphate group with the hydroxyl (OH) of glycerol carbon 3 (green) results in a phospholipid molecule and one water molecule. The basic phospholipid illustrated above is termed phosphatidic acid and is the building block of common phospholipids found in the cell membrane (see below). Fatty acid chains are depicted in black. **B. Common Phospholipids Found in Humans.** Common head groups, which form phospholipids (top and middle) and are shown in blue, bonded to the hydroxyl (OH) of a phosphate group (purple), which itself is bonded to the hydroxyl (OH) of glycerol carbon 3 (green). This structure results in a phospholipid molecule that is generally found in membranes. Lipid molecules are often represented in "stick diagram" form (bottom) with the charged phosphate group and hydrophilic head group shown as a circle and oval and with the fatty acid "tails," depicted as either straight or jagged lines, forming the hydrophobic region. [Adapted with permission from Naik P: Biochemistry, 3rd edition, Jaypee Brothers Medical Publishers (P) Ltd., 2009.]

TABLE 3-1. Common Fatty Acids Found in Humans

	Name	Carbon Atoms	Shape
Saturated	Laurate	12	
	Myristate	14	
	Palmitate	16	
	Stearate	18	
	Arachidate	20	
	Behenate	22	
	Lignocerate	24	

	Name	Carbon Atoms	Double Bonds	Shape
Unsaturated	Palmitoleate	16	1 (*cis*) at carbon 9	
	Oleate	18	1 (*cis*) at carbon 9	
	Linoleate	18	2 (*cis*) at carbons 9 and 12	
	Linolenate	18	3 (*cis*) at carbons 9, 12, and 15	
	Arachidonate	20	4 (*cis*) at carbons 5, 8, 11, and 14	

[Adapted with permission from Wikipedia: Fatty acid, 2002.]

COMPLEX LIPIDS

GLYCOLIPIDS/SPHINGOLIPIDS

Just as carbohydrate and protein molecules can bind together (discussed in Chapter 2), carbohydrates can also bind to lipids to form a **glycolipid**. However, in a human glycolipid, the glycerol backbone is generally replaced by a backbone of **sphingosine** [made from the amino acid serine and the 16-carbon fatty acid palmitate (Figure 3-4A)] and is, therefore, referred to as a **sphingolipid**. Sphingosine can bind two other molecules with the remaining hydroxyl (OH) and amino (NH_3) groups from the serine amino acid. In human sphingolipids, the amino group is always bound to another fatty acid to make the molecule **ceramide**. From ceramide, the particular molecule(s) attached to the remaining hydroxyl group defines both the name and the characteristics of the resulting sphingolipid. For example, the molecule **sphingomyelin**, which can make up to 20% of the total phospholipid in many biological membranes, is made of ceramide and a phosphoryl choline head group (Figure 3-4B).

From the base molecule of a sphingolipid (e.g., ceramide), carbohydrate molecules may also be attached to form a **glycosphingolipid**. In general, human glycosphingolipids are grouped into four categories.

1. **Cerebrosides**—Ceramide (see above) attached to only a *single* glucose or galactose residue, producing glucosylceramide or galactosylceramide, respectively (Figure 3-4C). Cerebrosides are important in the membranes of muscle and nerve cells and are located in myelin, which covers nerve axons and enables fast and efficient conduction of nerve impulses. Cerebrosides may also be involved in the binding of morphine and other opiates.
2. **Sulfatides**—Galactose-based glycosphingolipid molecules, which contain a sulfur atom-containing sulfate (SO_4^{-2}) group in place of the phosphorous atom (Figure 3-4D). Sulfatides are found mainly in the brain and central and peripheral nervous systems but are seen in trace amounts in other tissues. Sulfatides are believed to be involved in the regulation of cell growth and signaling and may serve to both help form and, alternatively, break down blood clots possibly by affecting the transportation of sodium and potassium in and out of cells such as platelets. Sulfatides also may play roles as an adhesion molecule, including the recruitment of immune cells to inflamed tissue and the binding and replication of influenza viruses. Changes in the production of sulfatides have been noted as one of the earliest indicators of Alzheimer's disease.
3. **Globosides**—Glycosphingolipid molecules with the carbohydrate molecule ***N*-acetyl-galactosamine** (a.k.a. GalNAc) along with two or more other carbohydrate molecules (Figure 3-4E). Globosides are found in several organs including red blood cells, serum, liver, and spleen. Although the functions of globosides are not well understood, they are believed to play an important role in cell receptors. Interestingly, the binding of *Escherichia coli*, a bacterium common in

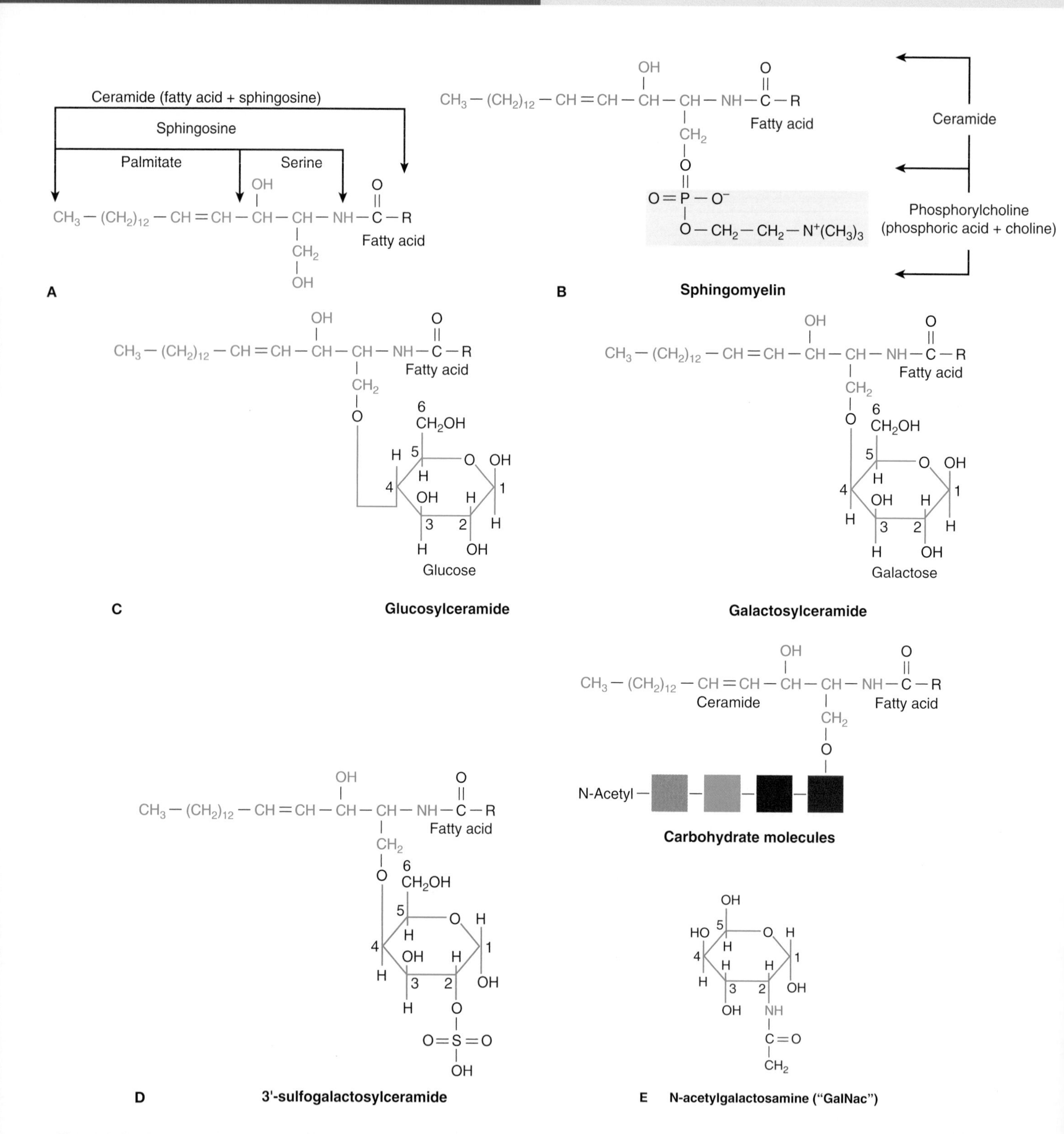

Figure 3-4. A-B. Ceramide and Sphingolipids. A. Ceramide is produced from the combination of the amino acid serine and the 16-carbon, fatty acid palmitate (green) with the subsequent addition of a second fatty acid (black) to the amino (NH_3) group (referred to as *N*-acetylation). Other molecules can then bind to the ceramide hydroxyl (OH) to produce a wide range of sphingolipids important to humans, an example of which is sphingomyelin (**B**) shown in the right panel and discussed in the text. **C. Common Cerebrosides.** Cerebrosides are composed of sphingosine and a single glucose or galactose molecule attached via the hydroxyl group at carbon 4. **D. A Sulfatide.** Sulfatides are simply cerebrosides with the addition of a carbohydrate-linked sulfate molecule. Compare with galactocerebroside in Figure 3-4B. **E. A Globoside and GalNAc Carbohydrate.** Globosides are composed of ceramide bonded to a fatty acid via the NH group (green) and several carbohydrate molecules, including GalNAc (light blue square), bonded via the serine OH (upper panel). The structure of *N*-acetyl-galactosamine or GalNAc is shown in the lower panel with the NH attachment highlighted in green. *(continued)*

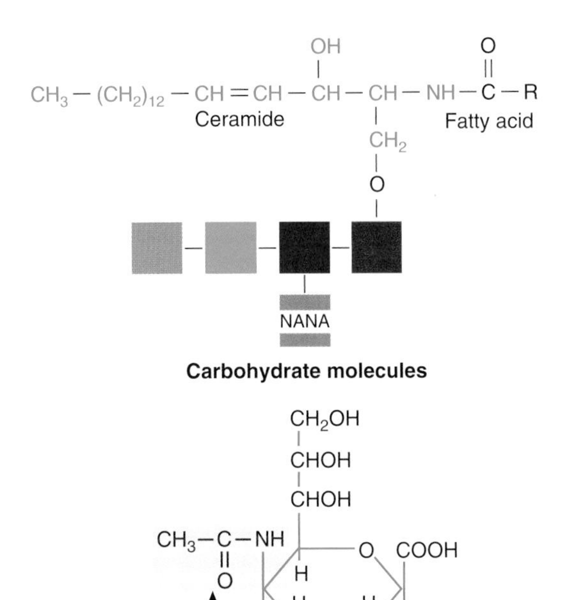

Figure 3-4. (Continued) F. A Ganglioside and NANA. The structure of N-acetylneuraminic (NANA), a type of sialic acid, is shown in the lower panel (*N*-acetyl group indicated in blue). [Adapted with permission from Naik P: Biochemistry, 3rd edition, Jaypee Brothers Medical Publishers (P) Ltd., 2009.]

urinary tract infections, to cells in the urinary tract is believed to occur through globosides.

4. **Gangliosides**—Glycosphingolipid molecules with one or more attached **sialic acid** molecules, most often ***N*-acetylneuraminic acid (a.k.a. "NANA")**, a complex, nine-carbon carbohydrate molecule (Figure 3-4F). There are a large number of different gangliosides, which differ in their structure depending on the number and location of the carbohydrate and NANA molecules. Gangliosides are involved in binding, recognition, and signaling between cells. Although gangliosides are abundant in the nervous system, they are also believed to play an important role in binding immune cells. Gangliosides are also found in less abundance in other cell types. Gangliosides are also believed to play a role in the binding and entrance into cells of the influenza virus and the toxin that causes cholera.

EICOSANOIDS

Eicosanoid is the general term for molecules that are all composed of 20-carbon fatty acids and involved in signaling. The four major groups of eicosanoids include the **prostaglandins (PGs)**, the **prostacyclins (PGIs)**, the **thromboxanes (TXs)**, and the **leukotrienes (LTs)**. Functions of eicosanoids include promotion of inflammation (usually ω-6 derived), immune response, neurological signaling, regulation of blood pressure, control of platelet aggregation/disaggregation, and modulation of levels of triacylglycerols. They also have direct/indirect effects on a variety of diseases, including cardiovascular and rheumatoid pathologies, among other roles. The signaling actions of eicosanoids are mainly transmitted via G-protein receptors (Chapter 8).

All eicosanoids are produced from either **ω-6 acids** (**dihomo-gamma-linolenic acid (DGLA)**, **arachidonic acid (AA)**, or an **ω-3 acid [eicosapentaenoic acid (EPA)]**. The predominant synthetic pathway for eicosanoids is via the ω-6, AA pathway whose products are denoted by the subscript "2" (Figure 3-5). LTs are produced from AA via a different pathway illustrated in the figure. Importantly, eicosanoids derived from ω-6/DGLA and the ω-3/EPA precursors are much less inflammatory/anti-inflammatory in nature. Increased intake of DGLA (dietary supplements), especially EPA ("ω-3" fish oils), results in lowered associated diseases via direct competition with the AA pathway. Medications such as aspirin, nonsteroidal anti-inflammatory drugs, and cyclooxygenase (COX)-2 inhibitors also decrease the inflammatory effects of PG, PGI, and TX (collectively known as **prostanoids**) by direct inhibition of **COX-1 or -2**, which are key enzymes in prostanoid synthesis (Figure 3-5). Corticosteroids (e.g., prednisolone) inhibit phospholipase A_2 (Figure 3-5).

More specifically, PGs are eicosanoids whose major functions include regulation of inflammation, smooth muscle contraction (blood vessels, gastrointestinal, bronchial, and uterus), neurological pain, platelet function, hormone activity, and cell growth. PGIs are derived directly from a prostaglandin precursor (PGH_2) and are important in decreasing platelet function and, therefore, blood clot formation as well as dilation of blood vessels. TXs are also derived from PGH_2 and, opposite of PGIs, act as vasoconstrictors to promote platelet aggregation and blood clot formation. LTs, the fourth class of eicosanoids, are produced via a pathway that deviates from the other eicosanoids' production at AA (Figure 3-5). LTs are mainly seen in inflammatory processes of the lung, including asthmatic reactions and the inflammation associated with bronchitis. Specific functions of the major eicosanoids and their role in disease and treatments are reviewed in Table 3-2 and will be covered further in Section III.

ABO Blood Groups: The human blood groups O, A, B, and AB are determined by specific glycosphingolipids found in the red cell membrane. The basic glycosphingolipid, called **"H-substance,"** is composed of a sphingolipid connected to three carbohydrate molecules and a membrane-bound protein. The addition of specific, extra carbohydrate molecule to a galactose residue on H-substance produces the four common blood types as follows:

Blood Type	Glycosphingolipid Carbohydrate
A	*N*-acetygalactosamine (GalNAc)
B	Galactose
AB	GalNAc and Galactose
O	No additional sugar residues

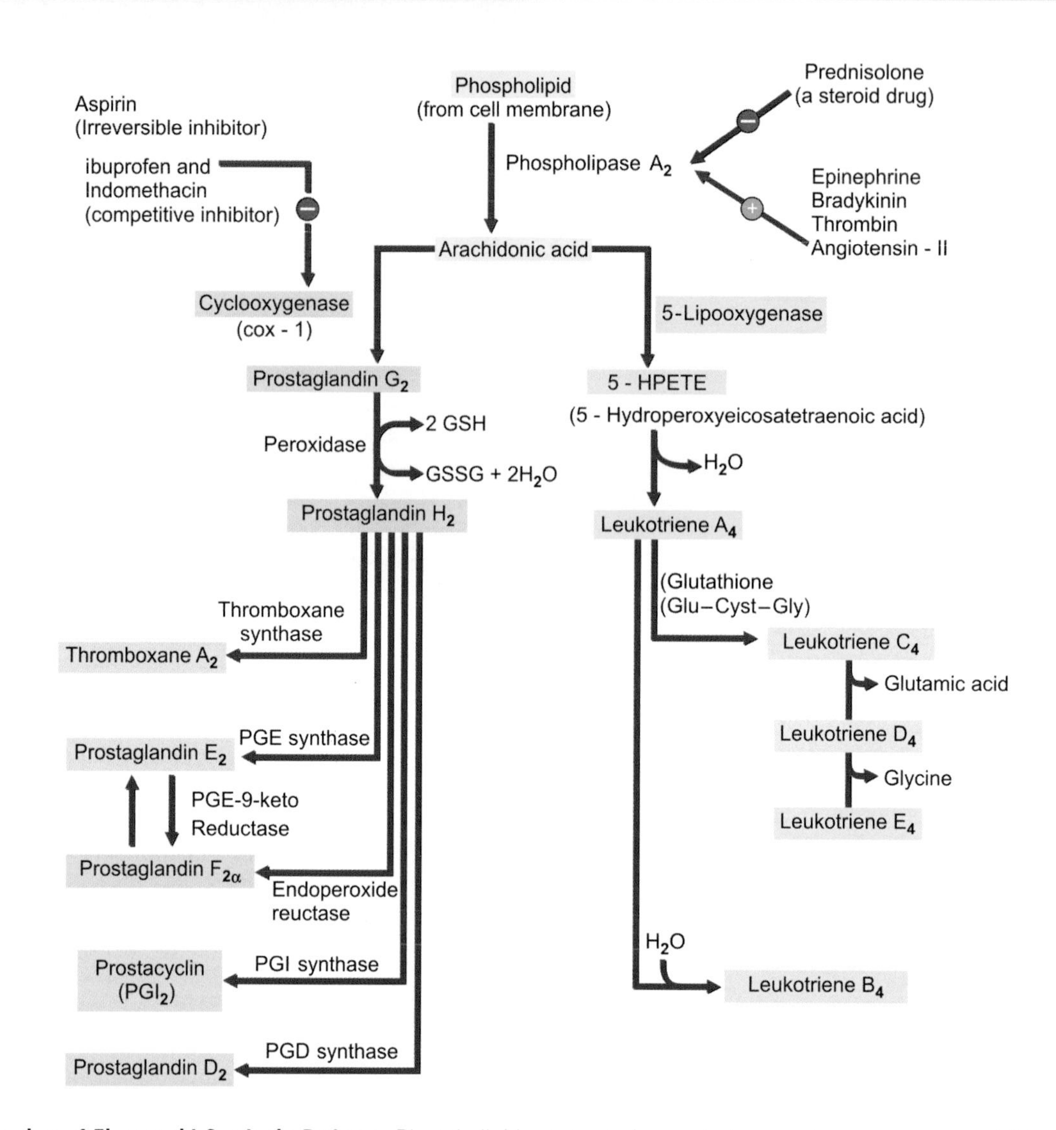

Figure 3-5. Overview of Eicosanoid Synthetic Pathway. Phospholipids generated as noted above (see also Figure 3-3A) provide the basic building blocks for the formation of arachidonic acid (AA), the source of all eicosanoids. The varying synthetic pathways coming from AA are shown. See also Table 3-2 for a review of the major known eicosanoids and their functions. [Adapted with permission from Naik P: Biochemistry, 3rd edition, Jaypee Brothers Medical Publishers (P) Ltd., 2009.]

Eicosanoids and Inflammation: Inflammation is often characterized by the four signs of calor, dolor, tumor, and rubor. Each of these features results, at least in part, from the action of one or more eicosanoids. **Calor** (warmth) is caused by the prostaglandin PGE_2. **Dolor** (pain) is heightened via the action of PGE_2, which also increases the sensitivity of neurons. **Tumor** (swelling) results from the leakage of plasma from blood vessels whose permeability is increased by the leukotriene LTB_4. Inflammatory **rubor** (redness) results from the action of thromboxane TXA_2, initially released after injury, which subsequently increases the concentration and, therefore, the blood vessel dilation activity of PGE_2 and LTB_4, resulting in engorged vessels and reddening.

CHOLESTEROL

Cholesterol is an extremely important molecule found only in eukaryotic organisms with a variety of functions in the human body. Although humans can make cholesterol, they have to rely on a complex mechanism of binding and modifications by other molecules so that cholesterol may be eliminated from the body (see below). As a result, properly controlling the day-to-day amount of cholesterol in the body—a balance between production, elimination, and the external influence of dietary intake—can play a key role in health and illness. The functions of cholesterol are listed below.

1. Cholesterol is one of the essential lipid components of biological membranes where it is a modulator of the fluidity of membranes. The ability of membranes to modify their

TABLE 3-2. Overview of Eicosanoids

Class	Subclass	Name	Functions	Synthesis	Medical Applications
Eicosanoid	Prostaglandins	PGG_2	Intermediate in production of PGH_2	*Cyclooxygenase 1 or 2*	—
Eicosanoid	Prostaglandins	PGH_2	Central intermediate, "**H**ead of Pathway"	*Cyclooxygenase 1 or 2*	—
Eicosanoid	Prostaglandins	PGD_2	Central nervous system neuromodulator, affects sleep ("**D**rowsy") and body temperature (opposite effect of PGE_2), contraction of bronchial tubes, inhibits platelet aggregation	*PGD_2 synthase* in brain, mast, and fat Cells	Believed to be critical in development of allergy-induced asthma (concentration 10-fold higher in patients vs. controls)
Eicosanoid	Prostaglandins	PGE_1	Dilates blood vessels ("**E**rection")	*Unknown enzyme*	***Alprostadil/Caverject:*** Treatment of erectile disorder (suppository/injection) and maintaining a patent ductus arteriosus in the newborn
Eicosanoid	Prostaglandins	PGE_2	Wakening ("**E**ye opener"), pain, fever, inflammation, renal arteriolar dilation, smooth muscle activity (intestines, lungs, and uterus), increases bone reabsorption by osteoclasts	*Prostaglandin E synthases* (membrane bound; many cell types)	***Dinoprostone/(Cervdil):*** Used topically to soften cervix in preparation for labor
Eicosanoid	Prostaglandins	$PGF_{2\alpha}$	Uterus contraction ("**F**etus"), bronchoconstriction	*Prostaglandin F synthase*	***Dinoprost:*** labor (abortion) induction
Eicosanoid	Prostacyclins	PGI_2	Platelet disaggregation ("**I**nhibits aggregation"), vasodilation, bronchodilation, and decreased stomach acid secretion	*Prostacyclin synthase* (endothelial cells)	***Ventavis (Iloprost)/Treprostinil (Remodulin)/ Epoprostenol (Flolan):*** inhaled or injected PGI_2 used to treat pulmonary hypertension or Raynaud's phenomenon
Eicosanoid	Thromboxanes	TXA_2	Platelet aggregation ("gather **T**ogether"), vasoconstriction that leads to increased blood pressure	*Thromboxane synthase* (platelets)	***Aspirin*** is used to inhibit TXA_2 platelet aggregation activity to reduce heart attacks, TXA_2 may be involved in vasoconstriction seen in Prinzmetal's angina
Eicosanoid	Leukotrienes	LTA_4, LTB_4, and LTF_4	Promote inflammation in **L**ung (asthma and bronchitis) as well as other tissues	*Lipoxygenase* (leukocytes, including mast cells, monocytes, neutrophils, eosinophils, and basophils	***Leukotriene receptor antagonists (Monteleukast, Zafirlukast):*** used to treat asthma
Eicosanoid	Leukotrienes	LTC_4, LTD_4, LTE_4	Anaphylactic reaction, including prolonged constriction of bronchi	Mast cells and basophils	Constituents of slow-reacting substance of anaphylaxis (SRS-A)

structure and/or the ability for other molecules to be able to move within the membrane is critical for cell signaling, binding, wound healing, immune response, and so on.

2. Cholesterol serves as the primary source for the production of steroid hormones (see next section below), bile salts, and even vitamin D.
3. Cholesterol metabolism is also important in the regulation of the transportation of lipids throughout the body, and disturbance of this system leads to the deposition of cholesterol in the artery wall causing atherosclerosis, which ultimately leads to heart attacks and strokes. An understanding of this regulation has led to important treatments for decreasing the risk of these diseases.

The cholesterol molecule has four rings made of carbon atoms—three rings have six sides and one has five sides—with a six-carbon ring tail (Figure 3-6A). Most of the carbons are single bonded and, therefore, have their full complement of hydrogen atoms. The three-dimensional structure of cholesterol is approximately a flat plane, exposing all of the hydrophobic parts of the cholesterol molecule to the environment. Only one hydroxyl group with its hydrophilic nature creates any charged quality to the cholesterol molecule. As a result, cholesterol does not like to be exposed to water environments, preferring to be shielded by other hydrophobic molecules such as lipids or hydrophobic parts of proteins. The impact of cholesterol's unique structure and its role in plasma membrane structure and fluidity will be examined further in Chapter 8.

However, some cholesterol is also found outside of biological membranes in adrenal glands, blood, and other tissues where it is often bonded via the hydroxyl group to a long fatty acid as a cholesterol ester (Figure 3-6B). These cholesterol esters are highly hydrophobic and insoluble and can form fatty lesions or "plaques" in the artery wall that can lead to heart attacks or strokes. Fortunately, the human body has developed a unique way of addressing this problem, namely by forming lipoproteins (see below and Figure 3-7).

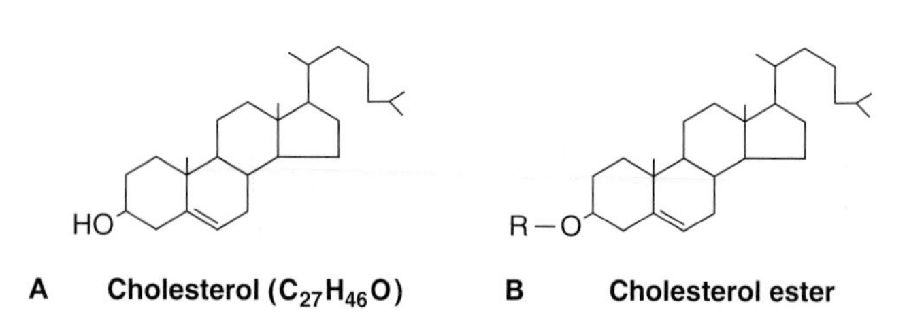

Figure 3-6. Cholesterol Molecule (A) Unesterified and (B) Esterified. A. Unesterified cholesterol molecule with free end hydroxyl group creating a hydrophilic charge important in biological membrane packing. **B.** Esterified cholesterol molecule with R group bonded by ester bond to oxygen from end hydroxyl group. The R group is usually a fatty acid bonded to cholesterol by the enzyme lecithin cholesterol acyl transferase (LCAT). [Adapted with permission from Barrett KE, et al.: Ganong's Review of Medical Physiology, 23rd edition, McGraw-Hill, 2010.]

Cholesterol, Phospholipids, Gangliosides, and Cancer: Emerging research has shown that some cancer cells (e.g., meningosarcoma and certain leukemias) have significant changes in their cell membrane cholesterol, phospholipid, and ganglioside content that alters the membrane fluidity as well as lipid signaling that produces unregulated growth or carcinogenesis. These changes can also cause resistance to agents used to treat cancer. Understanding how changing membrane composition leads to these findings may help to develop new and different cancer-fighting agents with a unique mechanism of lipid attack.

LIPOPROTEINS

Lipoproteins, as the name implies, are formed from the complexes of lipid and protein molecules. Unlike glycolipids, in which simple bonds connect the principal component molecules, lipoproteins are far more complex and form large particles containing several lipid classes and protein. Their complexity emerges from the primary function of lipoproteins, namely the transportation and delivery of fatty acids, triacylglycerol, and cholesterol to and from target cells in a variety of organs (see also Chapter 7 for further discussion). Thus, although glycolipids once produced and in their final location remain there for relatively long periods of time, lipoproteins are more transient. The bonding and structure of lipoproteins reflect this characteristic.

A simplified model of a lipoprotein includes a center core composed of cholesterol ester and triacylglycerol molecules surrounded by an outer shell of phospholipids and cholesterol molecules with their hydrophobic areas inward toward the lipid core and their charged, hydrophilic areas facing outward toward

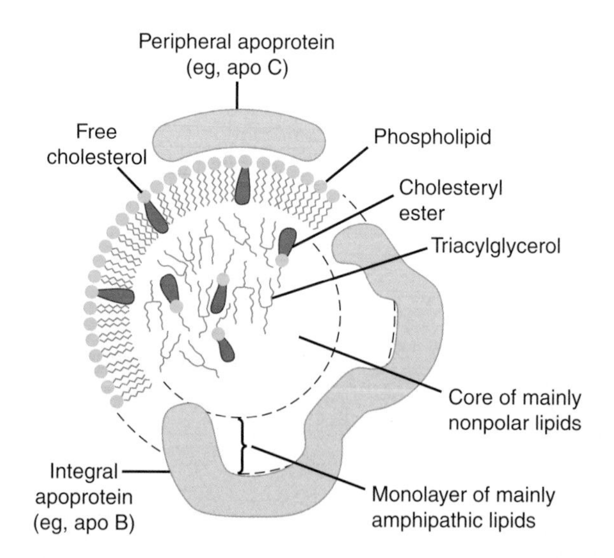

Figure 3-7. Basic Lipoprotein Structure. Lipoprotein complexes are composed of a central lipid core (cholesterol ester and triacylglycerol molecules), a charged lipid outer shell (phospholipid and cholesterol molecules), and surrounding apoproteins that help to transport hydrophobic lipid molecules throughout the body. [Adapted with permission from Murray RA, et al.: Harper's Illustrated Biochemistry, 28th edition, McGraw-Hill, 2009.]

TABLE 3-3. Basic Lipoprotein Characteristics

Lipoprotein Name	Density (g/mL)	Major Components	Diameter (μm)
Chylomicrons	<0.95	Dietary triacylglycerols (90%) and cholesterol	75–1,200
Very-low-density lipoprotein	0.95–1.006	Endogenous triacylglycerols and cholesterol	30–80
Intermediate-density lipoproteins	1.006–1.019	Triacylglycerols and cholesterol	25
Low-density lipoproteins	1.019–1.063	Cholesterol	18–25
High-density lipoproteins	1.063–1.210	Phospholipid and protein	5–12

the aqueous environment (Figure 3-7). Specialized proteins, known as **apoproteins**, wrap around the outer shell of the lipoprotein particle also involved in interactions with external water. The composition of lipoprotein particles varies but they can be broadly categorized depending on their density as shown in Table 3-3. Lipoproteins, their components, function, metabolism, and medical implications will be discussed in much more detail in Chapters 11 and 16.

BILE SALTS

Bile salts, composed of bile acids conjugated with glycine or taurine (Figure 3-8A), are produced in the liver directly from cholesterol and are important in solubilizing dietary fats in the mainly watery environment of the small intestine. After production in the liver and prior to secretion into the gall bladder and/or digestive system, they are often bonded to the amino acid glycine (Chapter 1) or taurine, a derivative of the common amino acid cystine (Chapter 1), to increase their water solubility (Figure 3-8B). Glyco- and tauro-bile acids are also called as conjugated bile acids.

Bile Salt Sequestrants: Medicines that bind with or "sequester" and help excrete bile salts are sometimes used in patients with high cholesterol levels in their blood. The removal of these bile salts causes further production of replacement bile acids from the body's cholesterol store to allow fat digestion, thereby lowering the total cholesterol in these patients. However, such treatments can interfere with normal lipid absorption affecting dietary requirements as well as resulting in the unpleasant side effect of fatty, foul-smelling bowel movements.

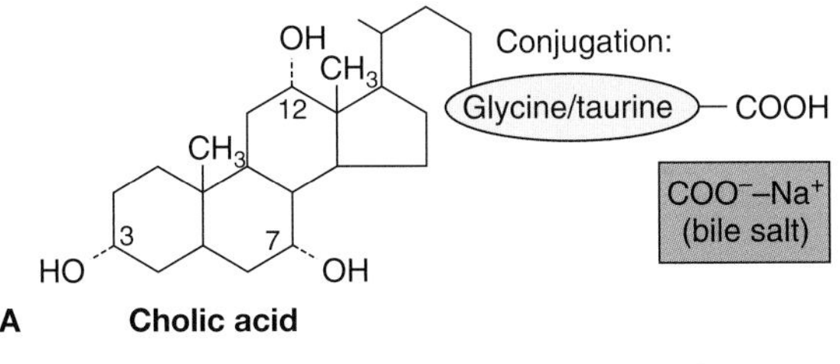

	Group at Position			% in Bile
	3	7	12	
Cholic acid	OH	OH	OH	50
Chenodeoxycholic acid	OH	OH	H	30
Deoxycholic acid	OH	H	OH	15
Lithocholic acid	OH	H	H	5

B — Glycocholic acid ($C(=O)-NH-CH_2COOH$, Glycine); Taurocholic acid ($C(=O)-NH-(CH_2)_2-SO_3H$, Taurine)

Figure 3-8. A-B. Major Bile Acids and Bile Salts Found in Humans. A. Bile salts, including the primary forms cholic acid and chenodeoxycholic acid, are produced by the liver and commonly found in humans. Removal of hydroxyl groups (OH) by intestinal bacteria produces the secondary bile acids deoxycholic acid and lithocholic acid. [Adapted with permission from Kibble JD and Halsey CR: The Big Picture: Medical Physiology, 1st edition, McGraw-Hill, 2009.] **B.** The addition of the amino acid glycine (left) or the cystine-related taurine (bottom right) increases the solubility of the bile salts. [Adapted with permission from Murray RA, et al.: Harper's Illustrated Biochemistry, 28th edition, McGraw-Hill, 2009.]

LIPID-DERIVED HORMONES/VITAMIN D

Steroid hormones, which are all produced from cholesterol (Figure 3-9), perform a variety of different functions in the human body as listed below.

CORTICOSTEROIDS (ADRENAL GLAND)

Mineralocorticoids. The primary mineralocorticoid in humans is **aldosterone**, which is produced in the outer layer of cells of the adrenal cortex. Synthesis and secretion of aldosterone is controlled by the renin–angiotensin system (see Chapter 18). In certain disease states, a minor mineralocorticoid, **11-deoxycorticosterone**, may instead be produced in excess in the adrenal cortex and substitute for aldosterone. The role of the mineralocorticoids is to maintain blood volume and blood pressure by increasing the reabsorption of Na^+ from the urine to the blood. The increase of Na^+ reabsorption is accompanied by the excretion of K^+ and protons (H^+) and leads to reabsorption of water by the action of **antidiuretic hormone** (see Chapter 18).

Glucocorticoids. The primary glucocorticoid in humans is **cortisol**, which is produced in the middle and innermost layer of cells of the adrenal cortex. Glucocorticoids have a variety of unique

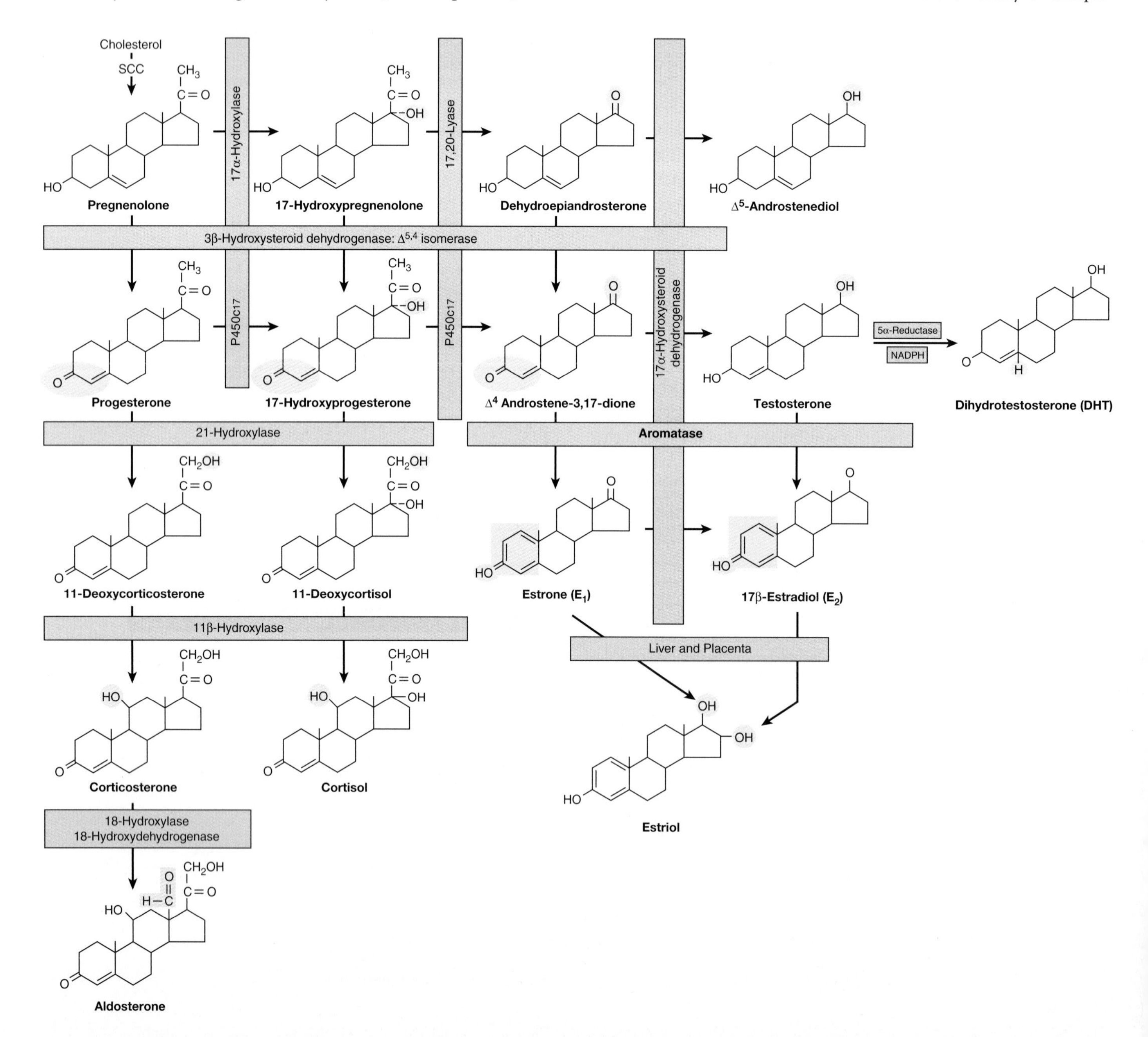

Figure 3-9. Major Steroid Hormones Found in Humans. Steroid hormones, derived from cholesterol (upper left corner), control several important bodily functions essential to life. Their synthesis involves multiple, interrelated steps, which occur in the adrenal cortex (mineralocorticoids and glucocorticoids), testes (androgens), and ovaries (estrogens). The synthesis of the various steroid-derived hormones can also be categorized by the specific enzyme (yellow boxes) involved and/or the number of carbon molecules in the substrate or product. Blue highlight indicates position of chemical change. Several diseases related to steroid production involve these enzymes. [Adapted with permission from Barrett KE, et al.: Ganong's Review of Medical Physiology, 23rd edition, McGraw-Hill, 2010.]

functions. They play an important role in fuel homeostasis during starvation and other stress-related conditions by promoting the mobilization of fatty acids from triacylglycerols (Figure 3-2B) stored in fat cells, promote the breakdown of muscle protein, and increase liver glucose synthesis (gluconeogenesis) using the amino acids derived from muscle and lactic acid (see Chapter 10). Additionally, they diminish inflammatory responses by decreasing production of PGs (see Figure 3-5).

Androgens. The cells in the innermost layer of the adrenal cortex produce androgens, primarily **androstenedione**, as a major product. The adrenal cortex is the sole source of androgens in females but a minor source in males who produce large amounts of androgens in the testes (see Chapter 20). Adrenal androgens, particularly in females, contribute to certain secondary sexual characteristics such as the growth of axillary and pubic hair, as well as contribute to libido.

Progestogens. The primary progestogen in humans is **progesterone**. Progesterone produced in the adrenal cortex is primarily an intermediate in the production of the other adrenal steroid hormones. Its physiological role is implantation of a fertilized egg in the uterine lining and maintenance of pregnancy. Its primary production occurs in the corpus luteum of the ovaries and the placenta following fertilization.

ANDROGENS (TESTES) AND ESTROGENS (OVARIES)

Responsible for sexual functions such as puberty (onset, development, and changes in secondary sexual characteristics), menopause, ovulation, and sperm formation.

VITAMIN D

Vitamin D, one of the fat-soluble vitamins, is required for calcium metabolism. A small amount of vitamin D is obtained from dietary sources but a majority is produced by the conversion of cholesterol through a unique process involving sunlight and three separate organs (Figure 3-10). Cholesterol, now dehydrocholesterol, having had a second carbon–carbon double bond formed in the liver, is transported to the skin where ultraviolet

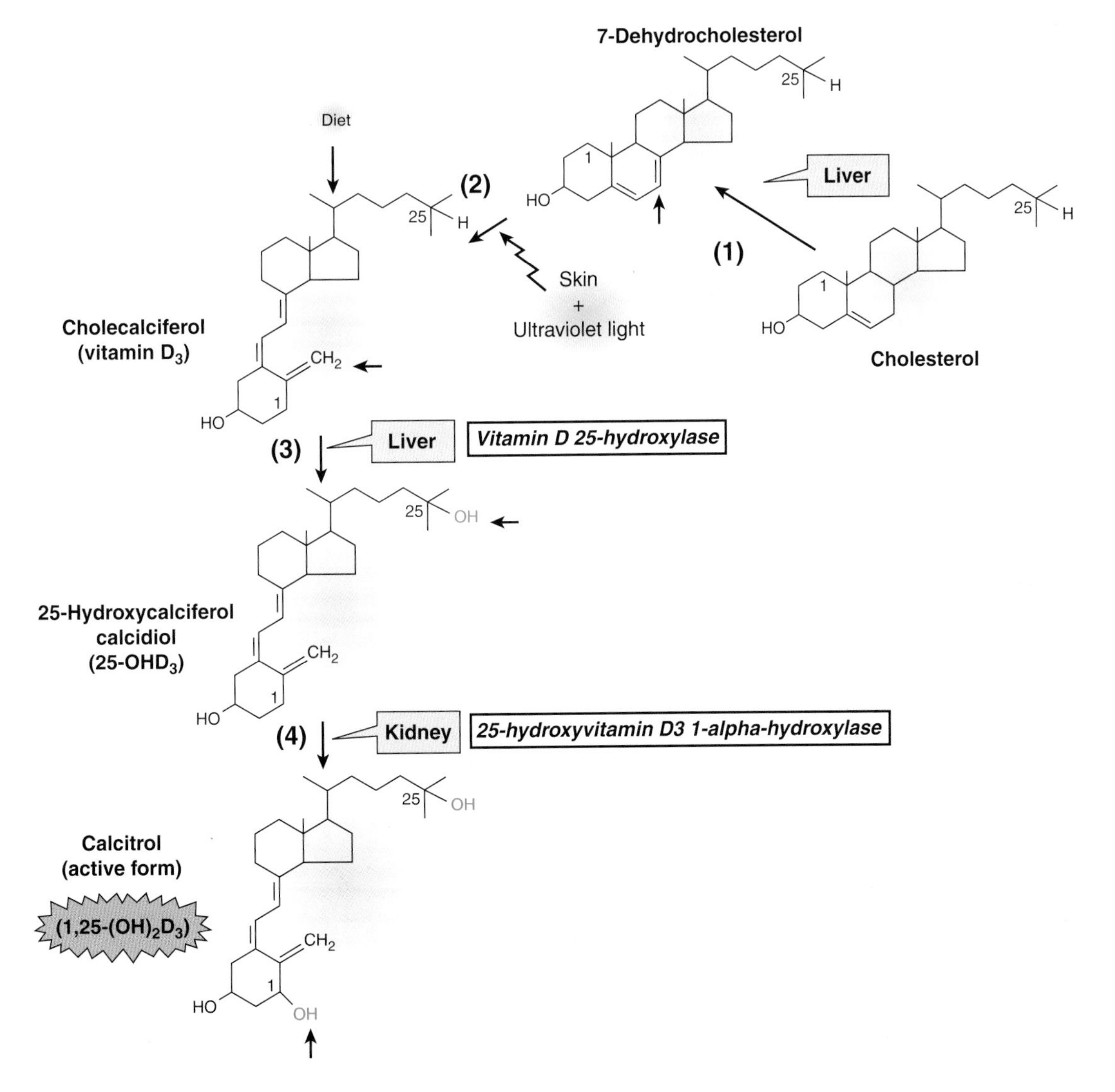

Figure 3-10. Production of Vitamin D from Cholesterol. Active vitamin D (calcitrol) is produced from cholesterol by means of four steps taking place in the liver, skin, liver, and kidney, respectively. Dietary sources can also provide cholecalciferol directly. Small, dark arrows indicate the location where molecular change takes place in each step. Added hydroxyl groups (OH) are indicated in green. [Adapted with permission from Kibble JD and Halsey CR: The Big Picture: Medical Physiology, 1st edition, McGraw-Hill, 2009.]

rays break open the second and third rings. Traveling back to the liver, a second hydroxyl group is added to the formerly hydrophobic "tail." Finally, the molecule travels to the kidney where another hydroxyl group is added near the site of the lone cholesterol hydroxyl group. The resulting molecule, "calcitrol," is the active form of vitamin D whose functions will be discussed more fully in Chapters 10 and 13.

REVIEW QUESTIONS

1. What are fatty acids, triacylglycerols, and phospholipids?
2. What are lipid head groups and the key characteristics of common ones found in man?
3. What are cerebrosides, sulfatides, globoside, and gangliosides and their key features?
4. What are lipoproteins, cholesterol, bile salt (acid), steroid hormones, and calcitrol and their key features?
5. What are the basic structures and characteristics of saturated, unsaturated, cis, trans fatty acids, glycerol, and phospholipid?
6. What are the basic structures and characteristics of the various glycolipids (i.e., cerebrosides, sulfatides, globosides, and gangliosides)?
7. What are the basic structures and characteristics of the lipoproteins (high-density lipoproteins, low-density lipoprotein, intermediate-density lipoproteins, and very-low-density lipoprotein)?
8. What are the basic structures and characteristics of bile salts, steroid hormones, and vitamin D?
9. What are the basic functions of the four types of glycolipids?
10. How do lipoproteins transport cholesterol and lipid molecules throughout the body?
11. What are the basic functions of bile salts, steroid hormones, and vitamin D that are produced from cholesterol?

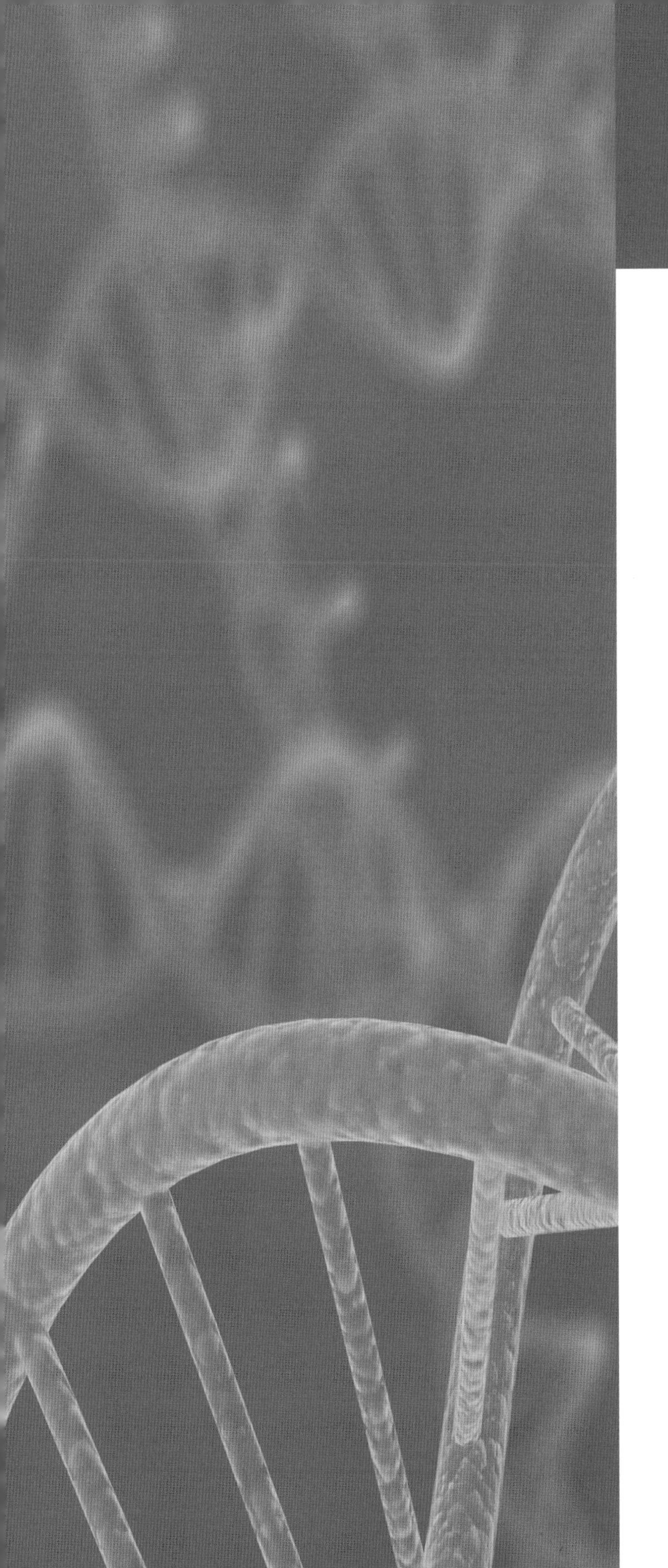

CHAPTER 4

NUCLEOSIDES, NUCLEOTIDES, DNA, AND RNA

OVERVIEW

Nucleosides and nucleotides are the fourth and final major group of biochemical molecules and are essential for numerous biological functions in humans, including maintaining and transferring genetic information, playing a major role in energy storage, and acting as signaling molecules. These molecules can be divided into two major families—purines, which include adenosine and guanine, and pyrimidines, which include cytosine, thymidine, and uracil. The unique structures and interactions of these molecules serve as the fundamental building block of RNA and DNA molecules and allow fundamental processes of gene replication and protein synthesis to occur. Many other functions of the various nucleosides and nucleotides will be explored in later chapters.

NUCLEOSIDES AND NUCLEOTIDES

Nucleosides and **nucleotides** are closely involved in the preservation and transmission of the genetic information of all living creatures. In addition, they play roles in biological energy storage and transmission, signaling, regulation of various aspects of metabolism, and even an important role as an antioxidant. Mistakes or deficiencies in their synthesis usually lead to death. Overproduction or decreased elimination of nucleic acid derivates also lead directly to medical conditions.

Nucleosides have a nitrogenous base and a five-carbon carbohydrate group, usually a ribose molecule (see Chapter 2). Nucleotides are simply a nucleoside with one or more phosphate groups attached (Figure 4-1). The resulting molecule is found in **ribonucleic acid** or **RNA**. If one hydroxyl (OH) group has been removed from the ribose, the *deoxy* versions of the nucleoside and nucleotide form the building blocks of **deoxyribonucleic acid** or **DNA** (Figure 4-1). Each component of nucleosides and nucleotides is discussed below.

COMPONENTS OF NUCLEOSIDES AND NUCLEOTIDES

1. Nitrogenous base—The nitrogenous base of a nucleoside or nucleotide (named because of the nitrogen atoms found in its structure) may be either a purine or a pyrimidine. **Purines**, including **inosine** (I), **adenine** (A), and **guanine** (G), are two-ring structures and **pyrimidines**, including **uracil** (U),

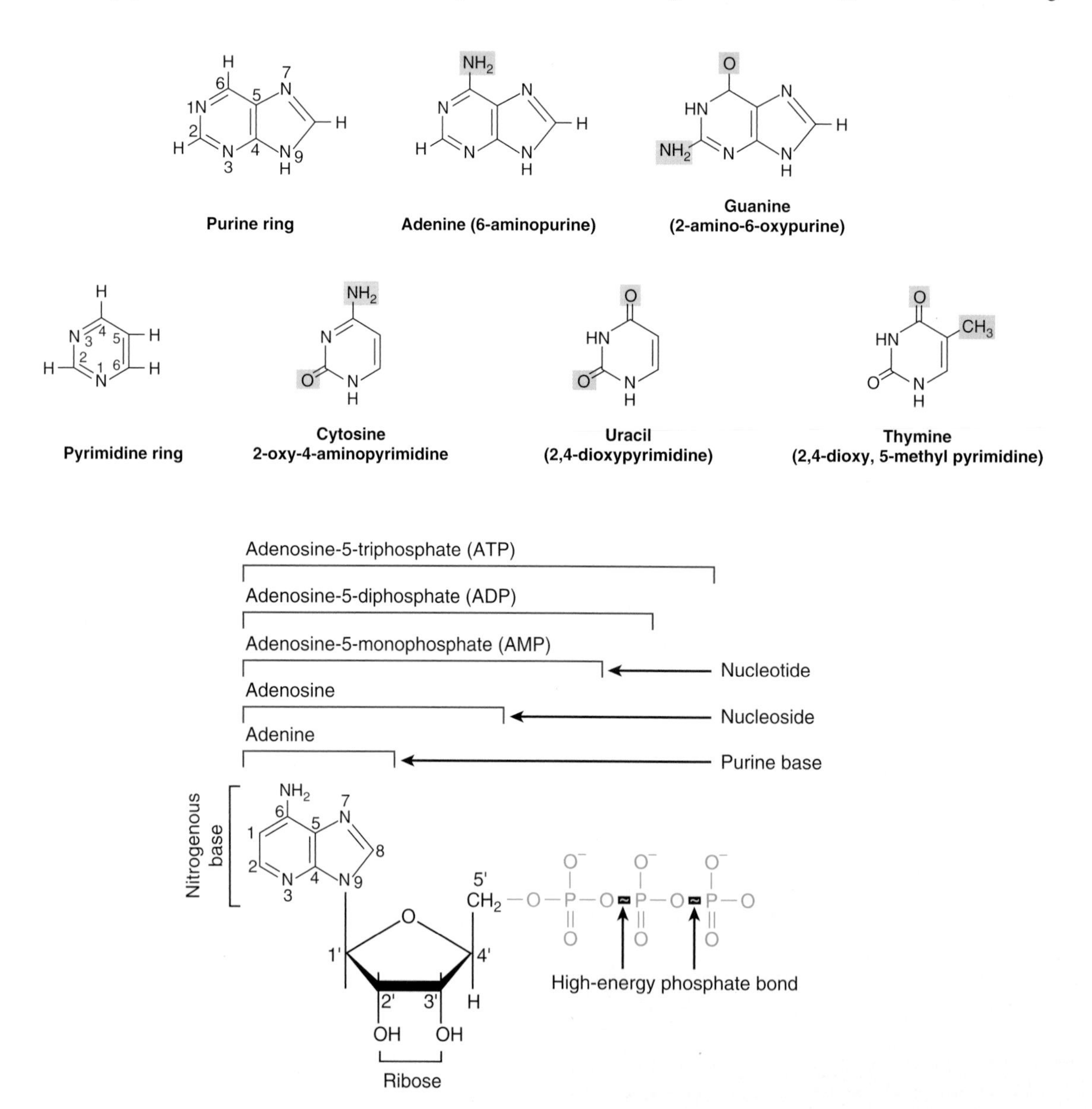

Figure 4-1. Basic Structure of Nucleosides and Nucleotides. Five major nucleoside bases are common in human biology, including the purines (two-ring structure) adenine and guanine (top) and the pyrimidines (one-ring structure) cytosine, uracil, and thymine (middle). Nucleosides (bottom) are made of a nitrogenous base, usually either a purine or pyrimidine, and a five-carbon carbohydrate, usually ribose. A nucleotide is simply a nucleoside with an additional phosphate group or groups (blue); polynucleotides containing the carbohydrate ribose are known as ribonucleotide or RNA. If 2′ hydroxyl group (OH) is removed, the polynucleotide *deoxy*ribonucleic acid (DNA) results. [Adapted with permission from Naik P: Biochemistry, 3rd edition, Jaypee Brothers Medical Publishers (P) Ltd., 2009.]

cytosine (C), and **thymine** (T), have only one ring (Figure 4-1). Both purine and pyrimidine nitrogenous bases are made, in part, from amino acids as shown in Figures 4-2 and 4-3, respectively.

2. Carbohydrate—The carbohydrate component of nucleosides and nucleotides is usually the sugar **ribose**. When the hydroxyl group is removed from carbon 2 (normally removed *after* the addition of the nitrogenous base to the carbohydrate), ***deoxyribose***, the sugar molecule found in DNA, results.
3. Phosphate Group—One or more phosphate groups (PO_4^{-3}) may be attached to the carbon 5 of the carbohydrate molecule. The phosphate molecules are important in energy storage and signaling functions of nucleosides and nucleotides.

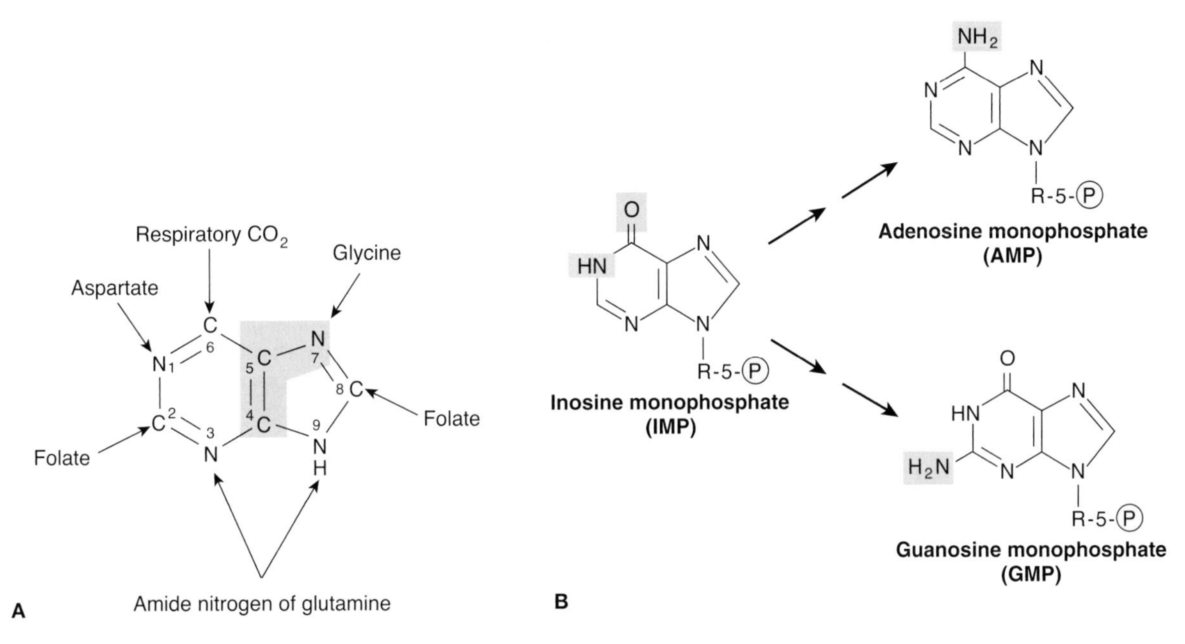

Figure 4-2. A. Constituents of the Purine Ring. The sources of carbon, nitrogen, and oxygen molecules that form the nitrogenous base of inosine are indicated. Atoms from glycine are indicated in blue shade. **B. Synthesis of Adenosine and Guanosine.** The nucleosides and nucleotides adenosine and guanosine are produced from the molecule IMP. Synthesis of adenosine requires one additional nitrogen molecule from the amino acid aspartate (blue shade). Guanosine synthesis utilizes one additional nitrogen molecule from the amino acid glutamine (blue shade). [Adapted with permission from Murray RA, et al.: Harper's Illustrated Biochemistry, 28th edition, McGraw-Hill, 2009.]

SYNTHESIS OF PURINE NUCLEOSIDES AND NUCLEOTIDES

The adenosine- and guanine-based nucleosides and nucleotides are formed by first producing another nucleotide called **inosine monophosphate (IMP)**. IMP is produced from a ribose, parts of one glycine, one aspartate, two glutamine amino acids, two tetrahydrofolate molecules (a modified form of the vitamin folate), and one carbon dioxide (CO_2) molecule (Figure 4-2A). After IMP is formed, **adenosine monophosphate** or **guanosine monophosphate** may be produced depending on the body's needs (Figure 4-2B).

The synthesis of purines starts by transference of a phosphate group from an existing adenosine triphosphate (ATP) molecule to a new ribose molecule. As a result of this requirement, the concentrations of ATP must be high for purine synthesis to proceed—a method of regulating when to produce or not produce more of these molecules. Thus, the body only produces more nucleosides and nucleotides when energy levels (from food) are high. Purines can be produced as above, or can be obtained from the diet or by breaking down and recycling existing nucleosides and nucleotides. Mammals, including humans, preferentially use the recycling pathway when appropriate starting materials are available.

SYNTHESIS OF PYRIMIDINE NUCLEOSIDES AND NUCLEOTIDES

Unlike purines, the pyrimidine nitrogenous bases, uracil and cytosine, are formed before bonding to the carbohydrate portion of the nucleotide. The process starts by producing the uracil ring (Figure 4-3A) in a three-step process utilizing one each of the amino acids glutamine and aspartate and one bicarbonate molecule (HCO_3^-). Next, **cytidine triphosphate (CTP)** is derived from the nucleotide uridine monophosphate (UMP) after the conversion of UMP to the triphosphate form (uridine triphosphate); an extra nitrogen is gained from one additional glutamine amino acid (Figure 4-3B, far right). Cytidine monophosphate (CMP) and diphosphate (CDP) can be produced from (CTP) by the subsequent loss of two or one phosphate groups, respectively. The final pyrimidine-derived nucleotide, thymidine, found only in DNA, is produced by a separate process involving the deoxy form of UMP (dUMP). This process will be discussed below.

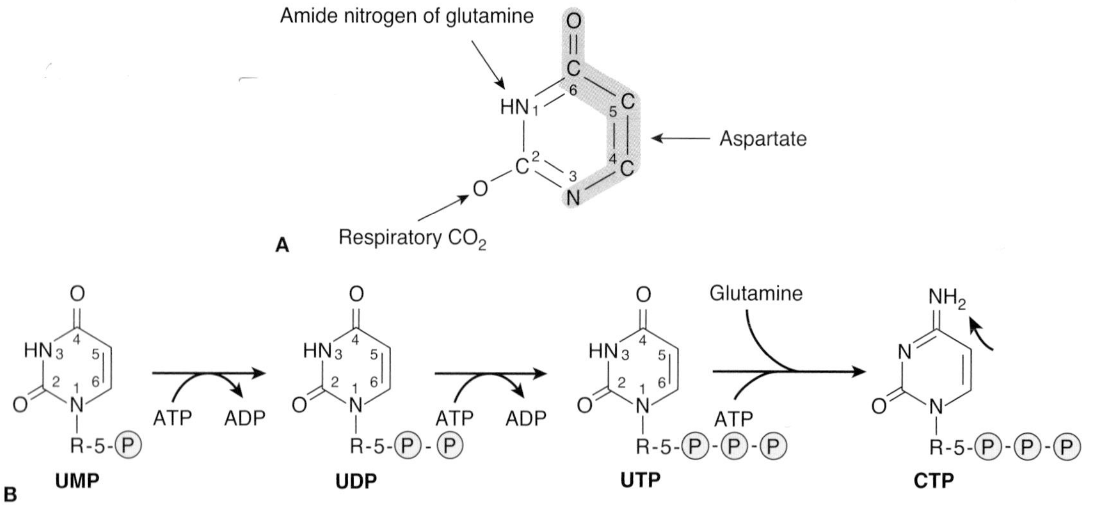

Figure 4-3. Synthesis of Pyrimidine-Derived Nucleosides Uridine and Cytidine. A. Constituents of the Pyrimidine Uracil Ring. The nucleoside uridine is produced from one aspartate amino acid, one glutamine amino acid, and a bicarbonate molecule from CO_2 respiration. Atoms from aspartate are indicated in blue shade. **B. Uridine and Cytidine Nucleoside Synthesis.** Cytidine synthesis utilizes one additional nitrogen molecule from the amino acid glutamine (indicated by arrow) after the addition of two additional phosphate groups to the UMP molecule. [Adapted with permission from Murray RA, et al.: Harper's Illustrated Biochemistry, 28th edition, McGraw-Hill, 2009.]

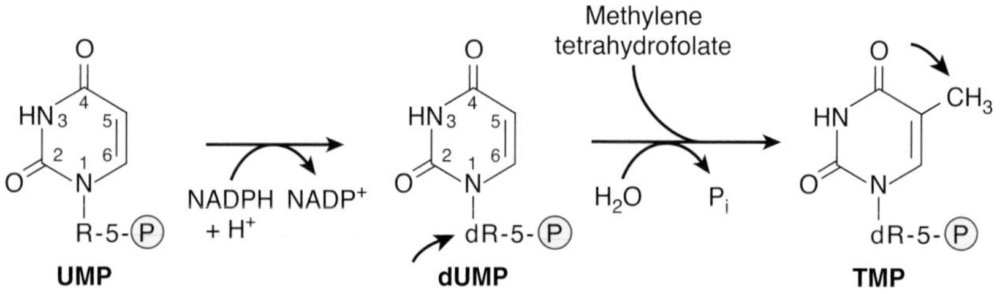

Figure 4-4. Synthesis of Thymidine Deoxyribonucleotide. UMP is deoxygenated at the 2′ hydroxyl of the ribose ring (see Figure 4-1 above) and subsequently converted to dTMP (TMP) by the addition of a methyl group (indicated by arrow) from methylene tetrahydrofolate. [Adapted with permission from Murray RA, et al.: Harper's Illustrated Biochemistry, 28th edition, McGraw-Hill, 2009.]

FORMATION OF DEOXY NUCLEOSIDES AND NUCLEOTIDES

Adenosine, guanosine, and cytidine deoxyribonucleotides found in DNA are formed directly from their corresponding, double-phosphorylated (nucleoside diphosphates) ribonucleotides by removal of the hydroxyl (OH^-) group from the carbon 2 (e.g., ADP → dADP, GDP → dGDP, CDP → dCDP). **Thymidine deoxyriboncleotide (dTMP)**, found only in DNA, is formed from dUMP (UMP → dUMP → dTMP) by the addition of a methyl (CH_3) group (Figure 4-4). Because there is no oxyribonucleotide form of dTMP, the "d" designation is often not used (TMP). See summary in Table 4-1.

BREAKDOWN OF PURINES AND PYRIMIDINES

Nucleotides can be broken down into their purine or pyrimidine nitrogenous bases by removal of the phosphate groups to form the respective nucleosides and removal and transfer of the ribose carbohydrate back into carbohydrate metabolism. Once the free purine bases remain, they are changed into a slightly different purine called **xanthine**, which is subsequently converted into uric acid and excreted mainly in urine by the

Nucleoside and Nucleotide Analogues as Chemotherapy Agents and Antibiotics: Nucleosides and/or nucleotides are essential components of genetic material and are, therefore, intimately involved in the proliferation of cells. One method of cancer treatment is to stop the reproduction of cancerous cells by inhibiting the production of nucleosides and/or nucleotides. Examples include the pyrimidine analogue of dUMP, **fluorodeoxyuridine (FdUMP)**, which blocks the production of dTDP. Because dTDP has its own special synthetic process, no other nucleotides and nucleosides are affected by FdUMP. Another analogue, **fluorocytosine**, can be used as an antibiotic because it is converted into an active "nucleoside or nucleotide" only in bacteria—human cells are not affected by its presence. **Aminopterin** and **methotrexate** are also chemotherapy agents, which act by blocking folate addition to purine rings. Unlike fluorocytosine, though, these agents also affect rapidly dividing human cells such as hair and intestines, resulting in the common chemotherapy side effects of hair loss and nausea/vomiting.

TABLE 4-1. Summary of Nitrogenous Bases, Nucleosides, and Nucleotides

Base	Ribonucleoside	Ribonucleotide (monophosphate form)	Base	Deoxyribonucleoside	Deoxyribonucleotide (monophosphate form)
Adenine (A)	Adenosine	Adenylate (AMP)	Adenine (A)	Deoxyadenosine	Deoxyadenylate (dAMP)
Guanine (G)	Guanosine	Guanylate (GMP)	Guanine (G)	Deoxyguanosine	Deoxyguanylate (dGMP)
Uracil (U)	Uridine	Uridylate (UMP)	Thymine (T)	Deoxythymidine	Deoxythymidylate (dTMP)
Cytosine (C)	Cytidine	Cytidylate (CMP)	Cytosine (C)	Deoxycytidine	Deoxycytidylate (dCMP)

kidneys. Breakdown of pyrimidine nitrogenous bases is simply a reverse of the steps of synthesis with the resulting molecules being directed into regular metabolism.

Gout and Lesch–Nyhan Syndrome: Gout is a medical condition typified by excessive amounts of uric acid in the body. These high levels result in the formation of needle-shaped uric acid crystals, which then become lodged in soft tissues, especially joints such as those in the first toe, resulting in severe "gouty arthritis" pain. Gout is treated by several different types of medications, including allopurinol, which directly decreases the conversion of purine nitrogenous bases to xanthine and uric acid. ***Uloric*** (***febuxostat***) has been approved for use in the treatment of chronic hyperuricemia. This drug appears to be more effective than allopurinol in preventing acute attacks and reducing the size of the crystal deposits. **Lesch–Nyhan syndrome** is a genetic disease, affecting almost solely males, of excessive synthesis of purines because of defective recycling and, therefore, uric acid production from their breakdown. Lesch–Nyhan syndrome is characterized by gouty arthritis but, in addition, affects the brain, resulting in mental retardation, loss of control of arm/leg/face movements, aggressive behavior, and self-mutilation by biting and scratching. Successful treatment of this disorder is still being sought.

Nucleotide molecules are able to form RNA and DNA strands by the bonding of the phosphate group of one nucleotide and the ribose sugar molecule of the next nucleotide (Figure 4-5). This linkage is called a "**phosphodiester bond**" and helps to form the fundamental structure essential for the storage, maintenance, and transmission of the genetic code of living creatures. For purposes of nomenclature, RNA and DNA molecules have a 5′ and 3′ end, depending on the number of the carbon bonded to the remaining phosphate group (Figure 4-5).

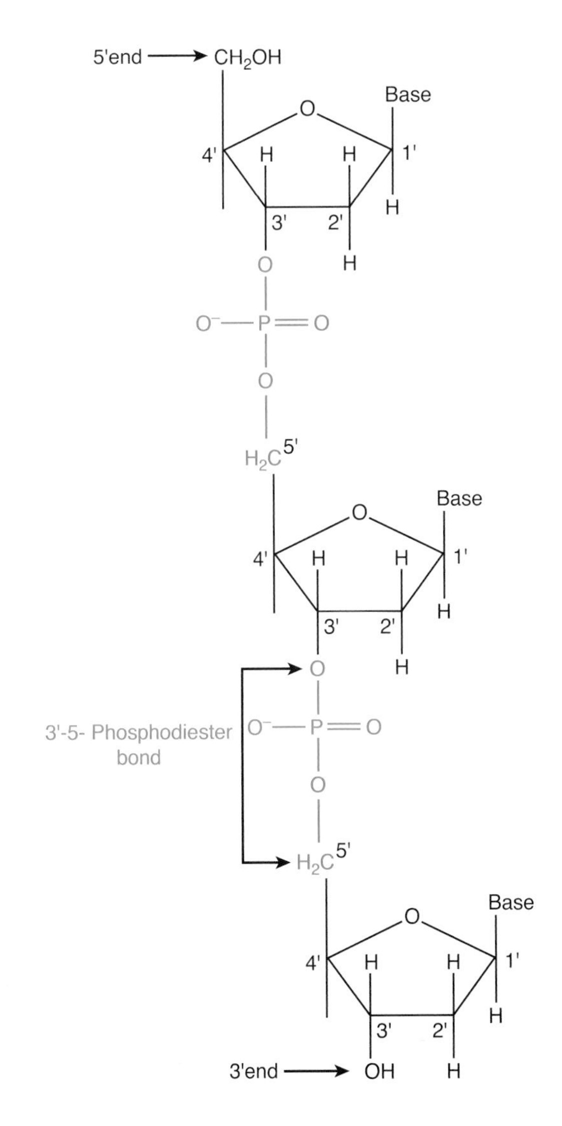

Figure 4-5. Structure of Phosphodiester Bond Found in RNA and DNA. Phosphodiester bond (bold arrows) formed in RNA and DNA between the hydroxyl on the third carbon of ribose and the hydroxyl from the fifth carbon. Successive bonds form the backbone of these molecules and link successive nucleotides to form the organism's genetic code. Formation of the bond results in the production of one molecule of water (not shown). Both RNA and DNA are written by convention 5′ and 3′ ends (indicated). [Reproduced with permission from Naik P: Biochemistry, 3rd edition, Jaypee Brothers Medical Publishers (P) Ltd., 2009.]

RNA AND DNA—BASIC STRUCTURE AND FUNCTION

RNA

RNA molecules are single strands **containing** the nucleotides: adenine (A), guanine (G), cytosine (C), and uracil (U). RNA molecules often form secondary (2°) structures much like proteins and may interact with DNA, other RNA molecules, and proteins. These interactions help to define the particular function of each type of RNA.

In general, there are four distinct types of RNA molecules, each with a particular function (Table 4-2). **Messenger RNA (mRNA)** molecules, which can be 100s–1000s of nucleotides long, function as the transmitter of genetic information from the DNA genetic code to the resulting protein. In this task, mRNA is often bound by proteins that help to protect and regulate these important genetic messages. **Transfer RNA (tRNA)** molecules, normally 65–110 nucleotides long, carry individual amino acids and match them with a specific mRNA sequence during protein synthesis. tRNA molecules have a very characteristic "T" shape that is optimized for this function. **Ribosomal RNA (rRNA)** is associated with proteins and makes up the actual working "machinery" responsible for the synthesis of protein molecules. The exact mechanism of protein synthesis and the roles of each individual type of RNA will be discussed in more detail in Section II. A fourth type of RNA, often referred to as **regulatory RNA**, is involved in regulation of DNA expression, posttranscriptional mRNA processing, and the activity of the transcribed mRNA message (see Chapter 9).

DNA

DNA molecules are composed of two single strands of deoxynucleotides, **adenine (A)**, **guanine (G)**, **thymidine (T)**, and **cytosine (C)**, in which the two strands are paired to form a ladder-type molecule referred to as a **double helix** (Figure 4-6A). This double helix forms when atoms in the nitrogenous bases of the nucleotides form hydrogen bonds (G bonding with C and A bonding with T) (Figure 4-6B) while the hydrophilic phosphate and hydroxyl groups of the sugar "backbone" are exposed to the water environment.

The **double-helix** structure of DNA is essential for its function because these two bonded strands can temporarily separate at specific parts of the DNA molecule to allow for DNA replication (shown in Figure 4-6C). mRNAs and associated proteins can also access the DNA strands to copy the contained genetic sequence as the first step, leading to protein synthesis (see Chapter 9). This process is discussed in more detail in Section II. The unique double helix can then rebond to again protect the vital DNA message. A separate type of DNA is found in mitochondria. This **mitochondrial DNA (mtDNA)** codes for only a few proteins (13 total proteins in human mitochondria) involved in the production of energy within the mitochondria. Unlike DNA found in the nucleus, mtDNA forms a small circular structure with the same bonding and strand separation seen in linear double-helix DNA molecules.

The specific sequence of A, G, C, and T nucleotides or "**genome**" defines all the functional molecules of a living creature and, as such, DNA is the blueprint of life. The genome of humans is estimated to contain approximately 20,000–25,000 different genes. Contained within the chromosomes (Figure 4-7) are the nucleotide strands, which contain the message or "code" for every single protein. The sequences of A's, G's, C's, and T's or "**genes**" that code for mRNA molecules and, subsequently, these proteins are referred to as "*ex*pressed sequences" or "***ex*ons**." Sequences that do not code for a protein are called "*in*tervening sequences" or "***in*trons**." In fact, introns comprise over 90% of the total DNA sequence found in humans and are believed to be leftover, nonfunctional remnants of evolutionary changes or, perhaps, important regulatory sequences whose functions are yet to be determined. Introns are removed or "spliced" from the sequence during the early stages of protein synthesis (see Section II).

Knowing the exact sequence of an organism's DNA genome would allow one to duplicate or "clone" that organism exactly. "**Cloning**" is simply the ability to copy the DNA sequence and replicate it to form the particular organism. Isolating specific sequences that are the code for a particular protein allows researchers to study and manipulate these proteins (see Appendix II), and is vital for scientific and medical purposes.

Ribozymes: Ribozymes (<u>ribo</u>nucleic acid e<u>nzymes</u>) represent a unique departure from the original thought that enzymes can only be proteins. Like their protein counterparts, particular RNA sequences possess secondary or tertiary structure that enables them to catalyze a reaction. Most ribozymes act on either themselves or another RNA molecule. However, some ribozymes, including those in ribosomes (see Chapter 9), catalyze the transfer of amino groups to a growing protein sequence and assist new proteins to fold into their appropriate conformation. The development of potential scientific and medical applications of ribozymes is ongoing, including possible treatments against initial HIV infection. Theories suggesting that RNA, not DNA, molecules were the original genetic code molecules also infer that ribozymes may have been some of the initial enzymatic molecules that allowed propagation of early life.

Adenosine Deaminase Deficiency and Gene Therapy: Deficiencies in the breakdown of nucleotides and nucleosides lead to often fatal diseases early in life in which the immune response is markedly decreased. **Severe combined immunodeficiency syndromes** encompass several examples of these diseases but almost half of patients have a deficiency in the breakdown of adenosine by **adenosine deaminase**. Patients have been treated by bone marrow transplant but also by the emerging technology of gene therapy with some success. Gene therapy offers the potential for producing a "good" copy of a gene and uses it to replace the defective copy. The implications of this treatment are vast.

TABLE 4-2. Characteristics of mRNA, tRNA, and rRNA

Name of RNA	Abbreviation	Size	Structure	Function
Messenger RNA	mRNA	500–6,000 nucleotides	Single strand of ribonucleotides with some secondary structures (e.g., loops as shown below) 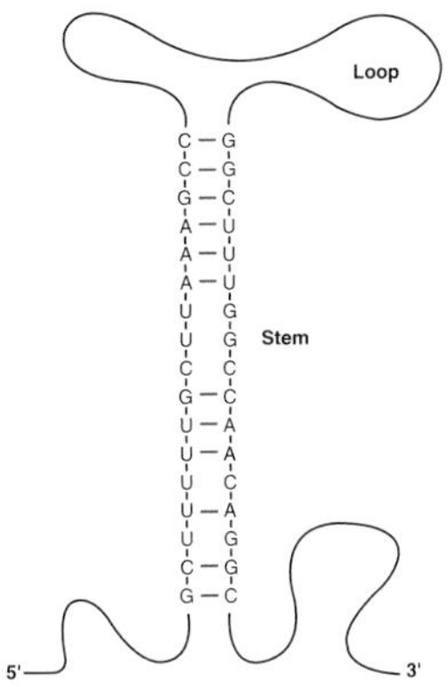	Carries genetic code to produce a protein from a specific DNA sequence
Transfer RNA	tRNA	65–110 nucleotides	Single strand of ribonucleotides form "cloverleaf" structure as shown below 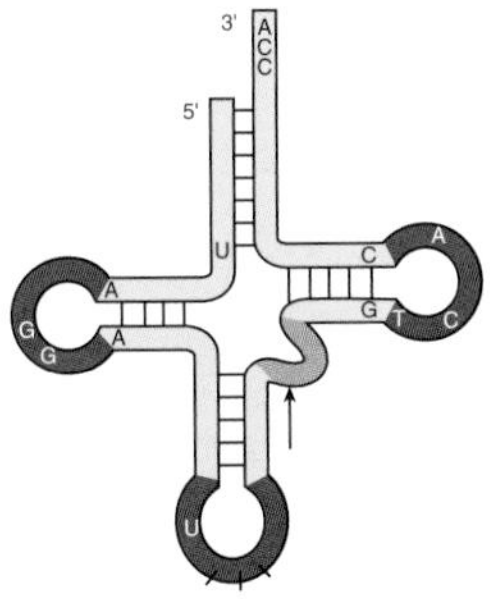	Transports specific amino acid to its place in a growing protein sequence
Ribosomal RNA	rRNA	Four different rRNA sizes: 120, 160, 1,500, and ~3,000 nucleotides	Four rRNAs combine with several ribosomal proteins to form the active ribosome 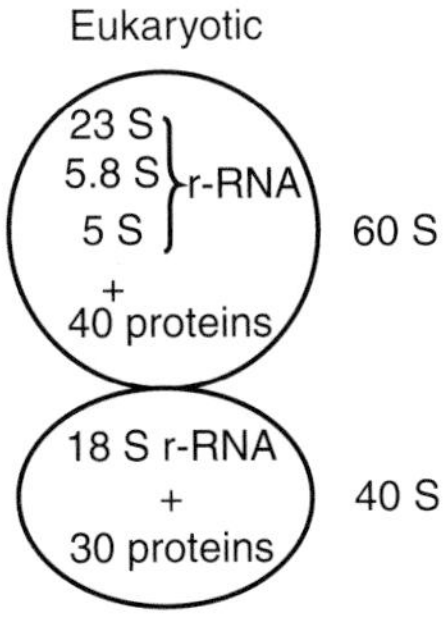	Along with ribosomal proteins, an essential part of the "machinery" of protein synthesis
Regulatory RNA	–	Examples include microRNA (miRNA), small interfering RNA (siRNA), Piwi-interacting RNAs (piRNA), and antisense RNA	Multiple structures	Regulation of DNA expression, mRNA posttranscriptional processing, and mRNA activity

[(Single strand of ribonucleotides with some secondary structures) Reproduced with permission from Murray RA, et al.: Harper's Illustrated Biochemistry, 28th edition, McGraw-Hill, 2009.]
[("Cloverleaf" structure and active ribosome structure) Adapted with permission from Naik P: Biochemistry, 3rd edition, Jaypee Brothers Medical Publishers (P) Ltd., 2009.]

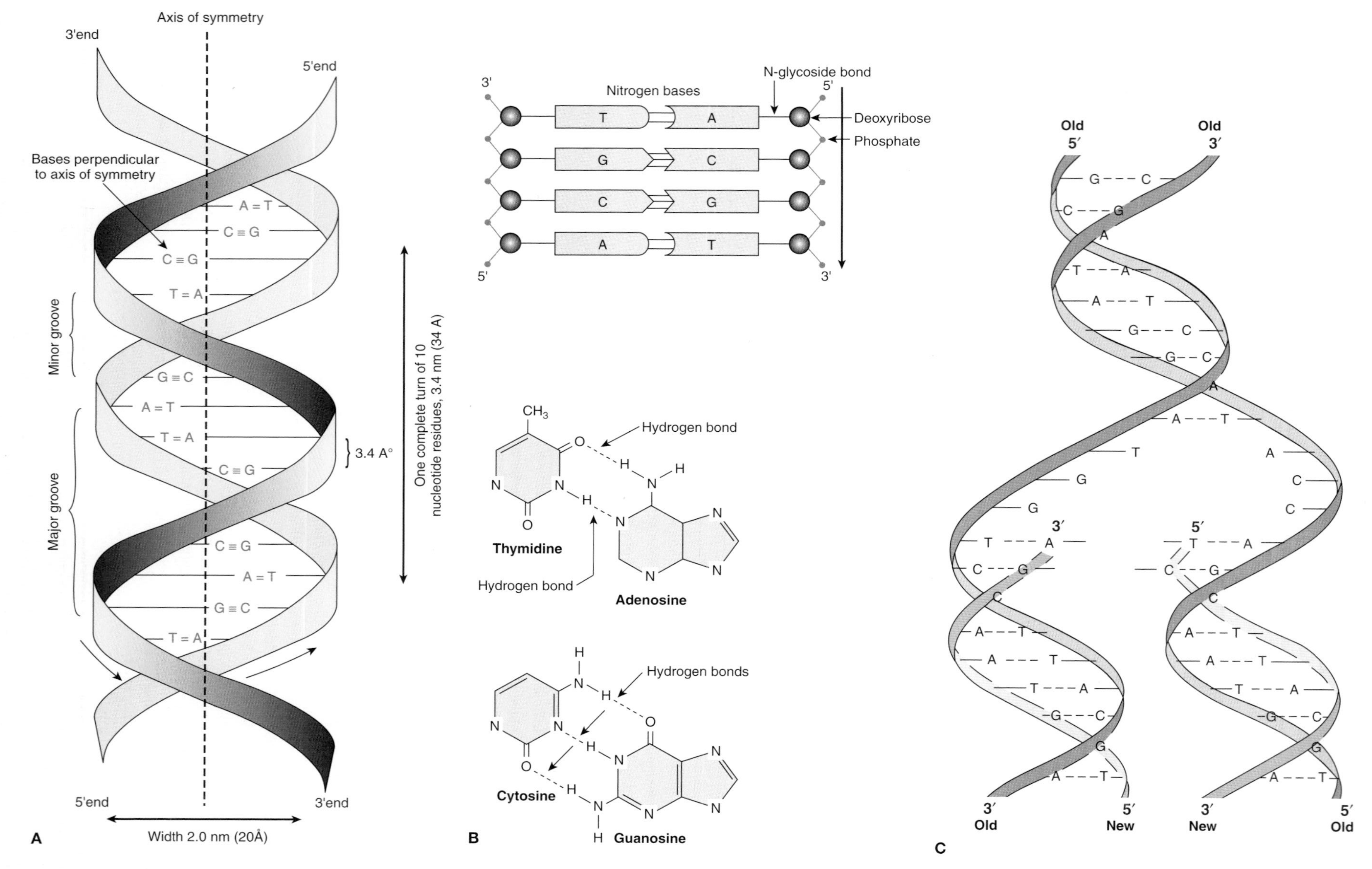

Figure 4-6. Basic Structure of DNA. A. Double helix model (left figure), illustrating the winding "ladder" structure with the inside "rungs" created by nucleotide pairing and the outside "runners" created by charged phosphate groups. **B.** Detail (top) of how the double helix structure protects the inner hydrophobic area and allows hydrogen bonding between paired G–C and A–T nucleotides while exposing the outer hydrophilic deoxyribose and phosphate groups to the water environment. Bonding (bottom) in DNA only between T and A and G and C, which helps to dictate the genetic code and fidelity in its replication and expression. [Reproduced with permission from Naik P: Biochemistry, 3rd edition, Jaypee Brothers Medical Publishers (P) Ltd., 2009.] Detail (bottom) of hydrogen bonding (arrows) between purine and pyrimdine pairs. **C.** Mechanism of DNA Replication. Unwinding of DNA to allow access to new nucleotides during DNA replication. [Reproduced with permission from Murray RA, et al.: Harper's Illustrated Biochemistry, 28th edition, McGraw-Hill, 2009.]

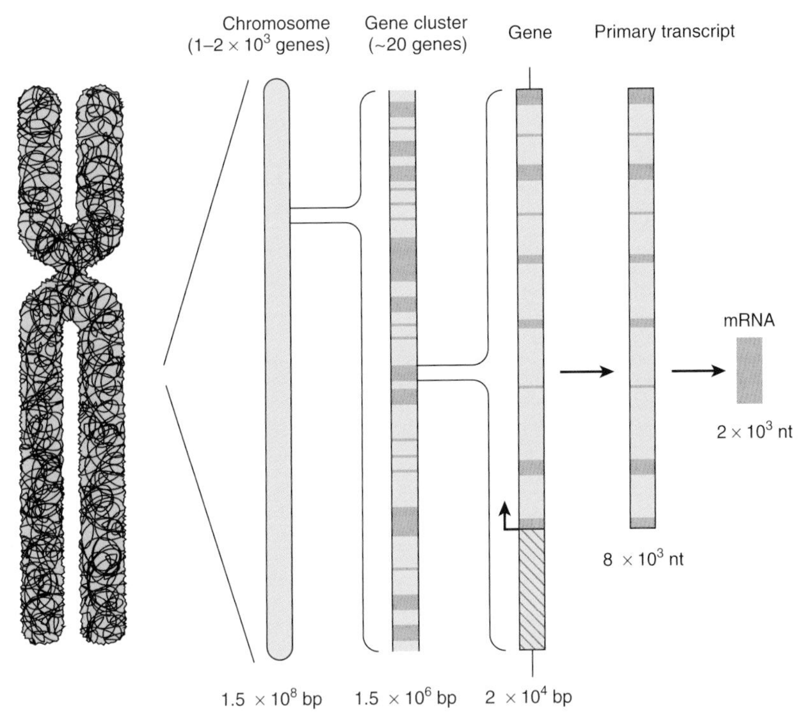

Figure 4-7. Relationship between Chromosomes (right) and a Gene. Each human chromosome (far left, whose DNA content is graphically illustrated as far left bar) contains approximately 1000–2000 genes arranged in clusters of approximately 20 genes. Within the gene are "expressed sequences" (exons, dark orange) and "intervening sequences" (introns, light orange). The intervening sequences are removed during expression of the primary mRNA transcript and processing to the final mRNA product. Introns do not appear to code for any protein and their role, if any, is still unknown. Length of molecules is indicated as bp (base pairs) or nt (nucleotides). [(Left part) Reproduced with permission from Mescher AL: Junqueira's Basic Histology Text and Atlas, 12th edition, McGraw-Hill, 2010. (Right part) Adapted with permission from Murray RA, et al.: Harper's Illustrated Biochemistry, 28th edition, McGraw-Hill, 2009.]

REVIEW QUESTIONS

1. What are nucleosides, nucleotides, deoxynucleotides, purines, and pyrimidines, their key features and basic structure?
2. How do you distinguish between adenine, guanine, cytosine, thymidine, and uracil structure?
3. What are ribonucleic acid (RNA) and deoxyribonucleic acid (DNA), their basic structure, and their roles in human biology?
4. What are genes, exons, and introns and their key features?
5. What is the relationship between amino acids and the nitrogenous bases of purines and pyrimidines?
6. What is the role of folate in the synthesis of nitrogenous bases of purines?
7. What is the overall pathway of synthesis of the five major nucleosides and nucleotides including the deoxy forms?

SECTION I

INTEGRATED USMLE-STYLE QUESTIONS AND ANSWERS

QUESTIONS

I-1. Several families were studied whose affected individuals have nephrogenic diabetes insipidus. This disease causes childhood symptoms of polyuria (frequent urination), polydipsia (constant thirst and frequent drinking), poor growth, and hypernatremia (increased serum sodium concentration). Administration of antidiuretic hormone was not curative, focusing attention on a renal water loss due to a transport defect. A gene named aquaporin-2 was cloned from renal tubular epithelium, its amino acid sequence derived, and structural domains hypothesized to facilitate separation of mutations from benign variants. The hypothesized structure contained several transmembrane domains demarcated by β-turns, and these potential water channels were found to be mutated in affected individuals. Which of the following amino acids is most suggestive of β-turns?

A. Arginine and lysine
B. Aspartic acid and glutamic acid
C. Glycine and proline
D. Leucine and valine
E. Tryptophan and tyrosine

I-2. A 7-year-old female presents with dehydration after 3 days of explosive diarrhea after consuming a large amount of ice cream. Analysis of her fecal matter reveals a large amount of lactose. Which of the following sugar pairs would be found in this disaccharide?

A. Fructose and glucose joined by a α-1, 2 bond
B. Galactose and fructose joined by a β-1, 2 bond
C. Galactose and glucose joined by a β-1, 4 bond
D. Two molecules of glucose joined by a α-1, 1 bond
E. Two molecules of glucose joined by a α-1, 4 bond

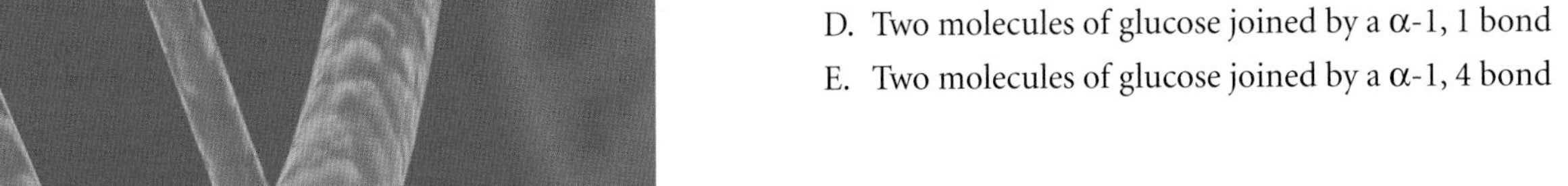

I-3. A deficiency of which of the following vitamins would alter metabolism of calcium?

A. A

B. C

C. D

D. E

E. K

I-4. Which of the following characteristics distinguish most RNA molecules from DNA?

A. 3′-phosphate group linked to a pentose sugar

B. 5′-phosphate group linked to a pentose sugar

C. Adenine base on RNA, whereas a uracil base on DNA

D. Purine or pyrimidine base linked to a pentose sugar

E. Uridine base on RNA, whereas a thymidine base on DNA

I-5. A severe form of osteogenesis imperfecta (OI) was noted, in which newborn infants have extremely deformed limbs and chest, causing them to die shortly after birth because their chest wall was not adequate for respiration. This "type II" severe form was initially thought to be autosomal recessive, but was later shown to involve mutations in type I collagen like other forms of OI. Collagen is a fibrous protein consisting of three peptide chains entwined in a triple helix, formed by a repeating amino acid motif (where X or Y can be any amino acid). Which of the following shows that the repeating 3-amino acid motif is most compatible with collagen triple helix formation and the mutation most likely to cause severe OI?

A. Ala–X–Y mutated to Gly–X–Y in one repeat

B. Ala–X–Y mutated to Leu–X–Y in one repeat

C. Gly–X–Y mutated to Ala–X–Y in one repeat

D. Gly–X–Y mutated to Pro–X–Y in one repeat

E. Pro-X-Y mutated to Gly-X-Y in one repeat

I-6. A 60-year-old man is brought to his physician from an institution for severe mental deficiency. The physician reviews his family history and finds he has an older sister in the same institution. Their parents are deceased but reportedly had normal intelligence and no chronic diseases. The man sits in an odd position as though he was sewing, prompting the physician to obtain a ferric chloride test on the man's urine. This test turns color with aromatic (ring) compounds, including certain amino acids, and a green color confirms the physician's diagnosis. Which of the following amino acids was most likely detected in the man's urine?

A. Glutamine

B. Glycine

C. Methionine

D. Phenylalanine

E. Serine

I-7. Children with severe vitamin A deficiency develop accumulation of a glycosaminoglycan on the corneal epithelium and is accompanied by "dry eye." Which of the following types of glycosaminoglycan would you expect to find accumulated on the surface of the cornea?

A. Chondroitin

B. Heparan

C. Heparin

D. Keratin

E. Perlecan

I-8. Which of the following is an essential fatty acid?

A. Arachidonic acid (C-20:4-$\Delta^{5,8,11,14}$)

B. Eicosatetraenoic acid (C-20:3-$\Delta^{8,11,14}$)

C. Linoleic acid (C-18:2-$\Delta^{9,12}$)

D. Oleic acid (C-18:1-Δ^{9})

E. Palmitic acid (C-16:0)

I-9. A 47-year-old man complains of pain in the joints of his left big toe, which are obviously swollen and tender. The pain has been chronic but became intolerable the day after Thanksgiving when he had a large meal and several glasses of red wine. He is obese, and his past medical history is significant for removal of kidney stones. Which of the following is involved in the pathophysiology of this patient's condition?

A. Deficiency of folic acid

B. Elevated orotic acid

C. Elevated uric acid

D. Loss of red blood cells

E. Low blood glucose

I-10. A child presents with severe vomiting, dehydration, and fever. Initial blood studies show acidosis with low bicarbonate. Preliminary results from the blood amino acid screen show two elevated amino acids, both with nonpolar side chains. A titration curve performed on one of the elevated species shows only two ionizable groups: one that is acidic and the other that is basic (i.e., no charged side chain). Which of the following pairs of elevated amino acids is most likely elevated?

A. Arginine and isoleucine

B. Aspartic acid and glutamine

C. Glutamic acid and threonine

D. Histidine and valine

E. Leucine and isoleucine

I-11. Certain amino acids are not part of the primary structure of proteins but are modified after translation. In scurvy, which of the following amino acids that is normally part of collagen cannot be hydroxylated after translation?

A. Alanine

B. Histidine

C. Proline

D. Tryptophan

E. Tyrosine

I-12. A woman returns from a year long trip abroad with her 2-week-old infant, whom she is breastfeeding. The child soon starts to exhibit lethargy, diarrhea, vomiting, jaundice, and an enlarged liver. The pediatrician prescribed a switch from breast milk to infant formula containing sucrose as the sole carbohydrate. The baby's symptoms resolve within a few days. Which of the following was the most likely diagnosis?

A. Deficiency of an enzyme in the metabolism of pentose sugars

B. Deficiency of an enzyme in the pathway that metabolizes glucose

C. Galactosemia

D. Intolerance to dietary fructose

E. Intolerance to lactase

I-13. Despite the fact that *trans* fatty acids are unsaturated, their contributions to atherosclerosis are similar to those of saturated fats. This similarity in physiological action can be attributed to which of the following?

A. Relatively linear structures

B. Similar rates of metabolism

C. Similar tissue distributions

D. Solubilities in water

E. Tendency to form triglycerides

I-14. Methotrexate is a potent anticancer agent that starves dividing cells of deoxyribonucleotides through direct inhibition of which of the following processes?

A. Degradation of purine bases to uric acid

B. Deoxygenation of the ribose ring to produce deoxyribose for DNA

C. Metabolism of folic acid

D. Synthesis of purine bases

E. Synthesis of thymidine

I-15. An infant is normal at birth but becomes lethargic after several feedings; the medical student describes an unusual smell to the urine but is ignored. Infection (sepsis) is suspected, and blood tests show normal white blood cell counts with a serum pH of 7.0. Evaluation for an inborn error of metabolism shows an abnormal amino acid screen. The report states that branch-chain amino acids are strikingly elevated. Which of the following amino acids does the report refer to?

A. Arginine

B. Aspartic acid

C. Isoleucine

D. Lysine

E. Threonine

I-16. A 10-month-old white boy is being evaluated for weakness, pallor, hemorrhages under the fingernails, and bleeding gums. Radiographs indicate that bone near the growth plates shows reduced osteoid formation and grossly defective collagen structure. What would be the most effective treatment for this patient's condition?

A. Exclusion of dairy products from the diet

B. Growth hormone treatment

C. Oral iron supplementation

D. Oral vitamin A

E. Oral vitamin C

I-17. A 9-month-old girl is suffering from vomiting, lethargy, and poor feeding behavior. Her mother reports that the symptoms began shortly after the baby was given a portion of a popsicle and mashed bananas by her grandparents. The baby's discomfort seemed to resolve after breastfeeding was resumed. Which of the following is the most likely diagnosis?

A. Deficiency of an enzyme in the metabolism of pentose sugars

B. Deficiency of an enzyme in the pathway that metabolizes glucose

C. Galactosemia

D. Intolerance to dietary fructose

I-18. During starvation glucagon, a peptide hormone, is important for activating glucose synthesis and mobilizing fats. Besides glucagon, which of the following is a lipid-derived hormone that regulates sugar and fat metabolism in starvation and other stress states?

A. Glucocorticoids

B. Mineralocorticoids

C. Prostagens

D. Vitamin A

E. Vitamin D

I-19. A 2-year-old boy's mother is concerned about his tendency to bite himself to the point of bleeding. The boy's fingers show scarring and several scabs, and his lips are swollen and bruised. He exhibits poor coordination, poor muscle tone, and frequent jerking movements of his arms and legs. He is significantly delayed in speech. His urine is orange in color and "gritty." Which of the following is the most likely diagnosis?

A. Adenosine deaminase deficiency

B. Cerebral palsy

C. Gout

D. Lesch–Nyhan syndrome

E. Tay–Sachs disease

ANSWERS

I-1. The answer is C. A β-turn structure consists of four amino acids in which the first residue is hydrogen bonded to the fourth residue of the turn (see Figure 1-5C). Glycine residues are small and flexible, whereas proline residues assume a *cis* or flattened conformation, making these residues amenable to tight turns. Transport proteins often have several membrane-spanning domains demarcated by β-turns that allow them to exit and return back into the membrane. These transmembrane domains form channels that regulate transport of ions and water in organs such as lung, gut, and kidney. Nephrogenic diabetes insipidus results when the kidney is less responsive to antidiuretic hormone excreted by the posterior pituitary, causing abnormal water excretion, dehydration, and electrolyte disturbances.

I-2. The answer is C. Disaccharides are carbohydrates composed of two sugar molecules. Lactose in this case is composed of galactose and glucose joined by a β-1,4 glycosidic bond. Sucrose is a disaccharide of a fructose and a glucose molecule joined by a α-1,2 glycosidic bond. Maltose is a disaccharide of two glucose molecules joined by a α-1,4 glycosidic bond. There is no disaccharide composed of fructose and galactose (see Figure 2-3).

I-3. The answer is C. Vitamin D is a fat-soluble vitamin essential for the absorption of dietary calcium and the reabsorption of calcium by the kidney. Vitamin A is also a fat-soluble vitamin essential for vision, reproduction, and development but does not specifically affect any minerals. Vitamin C is a water-soluble vitamin important in two enzymes in collagen formation and as an antioxidant. Vitamin E is a fat-soluble vitamin important as an antioxidant. Vitamin K is fat-soluble and important in blood clotting.

I-4. The answer is E. RNA and DNA are composed of nucleoside units where a purine or pyrimidine base is linked to a pentose sugar. The 1′ carbon of the pentose is linked to the nitrogen of the base. In DNA, 2′-deoxyribose sugars are used; in RNA, ribose sugars are used that contain 2′- and 3′-hydroxyls. The nitrogenous bases are adenine, thymine, guanine, and cytosine in DNA, with thymine replaced by uridine in RNA. Nucleotide polymers are chains of nucleotides with single phosphate groups, joined by bonds between the 3′-hydroxyl of the preceding pentose and the 5′-phosphate of the next pentose. Polymerization requires high-energy nucleotide triphosphate precursors that liberate pyrophosphate (broken down to phosphate) during joining. The polymerization reaction is given specificity by complementary RNA or DNA templates and rapidity by enzyme catalysts called polymerases.

I-5. The answer is D. Collagen is the most abundant fibrous protein found in connective tissue, bone, cartilage, skin, ligaments, and tendons as well as other tissues. Collagen consists of two strands of α_1 and one strand of α_2 collagen intertwined in a triple helix. The triple helix has 3.3 residues per turn. The triple helix is stabilized by hydrogen bonds between residues in different strands. The presence of glycine in every third residue allows for very tight packing of each strand in the triple helix because glycine residues are small and confer flexibility. Mutations substituting a larger or differently configured amino acid for the glycine of one repeat would disrupt the tight packing at that position of the collagen strand and alter triple helix conformation; proline substitutions for glycine (answer D) would alter chain direction and be more disruptive than those with alanine (answer C). Substitutions of amino acids at the X–Y positions of a repeat would have less influence on packing, better preserving the triple helix structure. Mutations at positions in the α_1 chain would also be more severe because two of these strands are incorporated with each triple helix. Milder mutations alter bone collagen to decrease bone thickness and cause susceptibility to fractures. More severe mutations interfere with bone formation during development, causing severe deformities of limbs and chest wall that are incompatible with life. These mutations also inhibit collagen synthesis in the sclerae (whites of the eyes), causing them to be thinner so the underlying blue choroid shows through.

I-6. The answer is D. Before phenylketonuria was recognized and before the advent of routine newborn metabolic screening, children with this autosomal recessive disorder developed severe mental retardation without other identifying symptoms. Although few children with developmental disability are being placed in institutions today, the movement for release into home or halfway house care in the 1960s and 1970s was complicated by long-term residents who had not developed life skills. Thus, some older individuals in institutions have disorders that could have been detected by modern screening and would never have prompted institutionalization with modern care. Phenylalanine, with its benzene ring, is an essential amino acid that is converted to tyrosine by phenylalanine hydroxylase, the enzyme that is deficient in phenylketonuria. This disease can also be caused by a deficiency of the enzyme's cofactor, biopterin.

I-7. The answer is D. Keratin is a component of cartilage and cornea of the eye. In vitamin A deficiency, keratinization of the skin and of the mucous membranes in the respiratory, GI, and urinary tracts can also occur. Drying, scaling, and follicular thickening of the skin and respiratory infections can result from this deficiency. Chondroitin is a structural component and attachment point to collagen. Heparan is found attached to collagen for various cells, including liver cells. Heparin regulates the immune response and blood clot formation. Perlecan is a structural component of the kidney membrane.

I-8. The answer is C. The essential fatty acid linoleic acid (C-18:2-$\Delta^{9,12}$), with 18 carbons and two double bonds at carbons 9 and 18, is desaturated to form β-linolenic acid (C-18:3-$\Delta^{6,9,12}$), which is sequentially elongated and desaturated to form eicosatrienoic acid (C-20:3-$\Delta^{8,11,14}$) and arachidonic acid (C-20:4-$\Delta^{5,8,11,14}$). Many of the eicosanoids (20-carbon compounds)—prostaglandins, thromboxanes, and leukotrienes—are derived from arachidonic acid. Arachidonic acid can only be synthesized from essential fatty acids obtained from the diet. Palmitic acid (C-16:0) and oleic acid (C-18:1-Δ^{9}) can be synthesized by the tissues (Table 3-1).

I-9. The answer is C. This patient shows many symptoms consistent with an episode of gout. His joint pain is due to gouty arthritis, an inflammatory condition arising from deposition of sodium urate crystals. The swelling in the joints of his big toe (tophaceous gout) is also a manifestation of this phenomenon. In his case, the episode seems to have been triggered by excessive eating at Thanksgiving dinner along with alcohol consumption, leading to degradation of large quantities of purine nucleotides and consequent increased flux through the pathway that produces uric acid. Whether his gout arises from impaired excretion of uric acid or is due to a mutation of phosphoribosylpyrophosphate synthetase cannot be determined from the data. Analysis of his blood may confirm the gout if high concentrations of uric acid (hyperuricemia) are present.

I-10. The answer is E. Leucine and isoleucine have nonpolar methyl groups as side chains. As for any amino acid, titration curves obtained by noting the change in pH would show an acidic ionizable group for the primary carboxyl group and a basic ionizable group 5 for the primary amino group; there would be no additional ionizable side chain. Aspartic and glutamic acids (second carboxyl group), histidine (basic imino group), glutamine (second amino group), and threonine (hydroxyl group) all have ionizable side chains that would show an additional group on the titration curve. The likely diagnosis here is maple syrup urine disease, which involves elevated isoleucine, leucine, and valine together with their ketoacid derivatives. The ketoacid derivatives cause the acidosis, and the fever suggests that the metabolic imbalance was worsened by an infection.

I-11. The answer is C. Proline and lysine are hydroxylated after the synthesis of new collagen molecules. The hydroxylation of proline and lysine residues occurs in reactions catalyzed by prolyl and lysyl hydroxylase enzymes that require the reducing agent ascorbic acid (vitamin C). In scurvy, which results from a deficiency of vitamin C, insufficient hydroxylation of collagen causes abnormal collagen fibrils. The weakened collagen in teeth, bone, and blood vessels causes tooth loss, brittle bones with fractures, and bleeding tendencies with bruising and bleeding gums.

I-12. The answer is C. The patient's symptoms and course in response to a lactose-containing formula are consistent with a diagnosis of galactosemia. Because the child can consume sucrose, that contains glucose and fructose, the problem cannot be in pathways that metabolize either glucose or pentose sugars. Additionally, the child can tolerate dietary fructose as evidenced by the ability to consume sucrose. Finally, while one can be intolerant to lactose, the sugar, the same is not true regarding lactase, the enzyme. Although genetic screening tests required in most states identify newborns with galactosemia, these tests may not have been performed on a child born outside the United States.

I-13. The answer is A. Saturated fatty acids and *trans* fatty acids are structurally similar; their hydrocarbon tails are relatively linear. This allows them to pack tightly together in semicrystalline arrays such as the membrane bilayer. Such arrays have similar biochemical properties in terms of melting temperature (fluidity). Although some of the other properties listed are also shared by saturated and *trans* fats, they are not thought to account for the tendency of these fats to contribute to atherosclerosis (Figure 3-1B).

I-14. The answer is C. Methotrexate is an analog of folic acid that binds with very high affinity to the substrate-binding site of dihydrofolate reductase, the enzyme that catalyzes conversion of dihydrofolate to tetrahydrofolate, which is used in various forms by enzymes of both the purine and pyrimidine de novo synthetic pathways. Thus, synthesis of deoxythymidine monophosphate from deoxyuridine monophosphate catalyzed by thymidylate synthetase and several steps in purine synthesis catalyzed by formyltransferase are indirectly blocked by the action of methotrexate because both those enzymes require tetrahydrofuran coenzymes. Xanthine oxidase that catalyzes degradation of purine bases to uric acid is unaffected.

I-15. The answer is C. Amino acids are classified as acidic, neutral hydrophobic, neutral hydrophilic, or basic, depending on the charge or partial charge on the R-group at pH 7. Hydrophobic (water-hating) groups are carbon–hydrogen chains like those of leucine, isoleucine, glycine, or valine. Basic R-groups, such as those of lysine and arginine, carry a positive charge at physiologic pH owing to protonated amide groups, whereas acidic R-groups, such as glutamic acid, carry a negative charge owing to ionized carboxyl groups. Threonine with its hydroxyl side chain is neutral at physiologic pH. Leucine, isoleucine, and valine are amino acids with branched side groups, and they share a pathway for degradation that is deficient in children with maple syrup urine disease. Their amino groups can be removed, but the resulting carboxylic acids accumulate with resulting acidosis, coma, and death unless a diet free of branch-chained amino acids is instituted.

I-16. The answer is E. The patient shows many signs of vitamin C deficiency or scurvy, which is seen most frequently in infants, the elderly, and in alcoholic patients. Particularly indicative of vitamin C deficiency are the multiple small hemorrhages that occur under the skin (petechiae) and nails and surrounding hair follicles. Bleeding gums are a classic indicator of scurvy.

I-17. The answer is D. The main sugar in mother's milk is lactose. When the baby was given the fruit and the artificially sweetened popsicle, she was exposed to fructose for the first time and apparently is intolerant to dietary fructose. Because the child can consume lactose, which contains glucose and galactose, the problem cannot be in pathways that metabolize either glucose or pentose sugars. The symptoms are also consistent with galactosemia, but would be expected as a reaction to lactose intake. This diagnosis of fructose intolerance should be confirmed by genetic testing. Essential fructosuria is a benign condition that would not have produced such severe symptoms.

I-18. The answer is A. Glucocorticoids are made in the adrenal cortex, the principal one being cortisol in humans. Cortisol promotes breakdown of proteins in starvation to provide precursors for the synthesis of glucose (blood sugar) and also promotes the breakdown of stored lipids (triglycerides) to provide the energy for this process. Cortisol performs a similar function in chronic stress. Mineralocorticoids, the principal one being aldosterone, controls sodium reuptake by the kidney in exchange for potassium and thereby regulates and changes blood volume and pressure. Prostagens are required for implantation of a fertilized egg in the uterine lining and maintenance of a pregnancy. Vitamin A is not a lipid-derived hormone. It is involved in vision, reproduction, and tissue development. Vitamin D controls the blood calcium–phosphate matrix and thereby promotes bone formation.

I-19. The answer is D. This patient's self-mutilation behavior, neurologic symptoms, and developmental delay are consistent with diagnosis of Lesch–Nyhan syndrome. This disorder is due to the deficiency of hypoxanthine–guanine phosphoribosyl transferase, which prevents salvage of hypoxanthine and guanine to their respective nucleotides, inosine monophosphate and guanosine monophosphate. This leads in turn to hyperactivity of the purine synthesis pathway, excessive purine degradation, and overproduction of uric acid. The gritty substance and orange color of the patient's urine are because of excretion of both dissolved uric acid and precipitated sodium urate. Gout might account for the excessive uric acid production but not the neurologic symptoms. Self-mutilation is not characteristic of Tay–Sachs disease, cerebral palsy, or adenosine deaminase deficiency.

SECTION II

FUNCTIONAL BIOCHEMISTRY

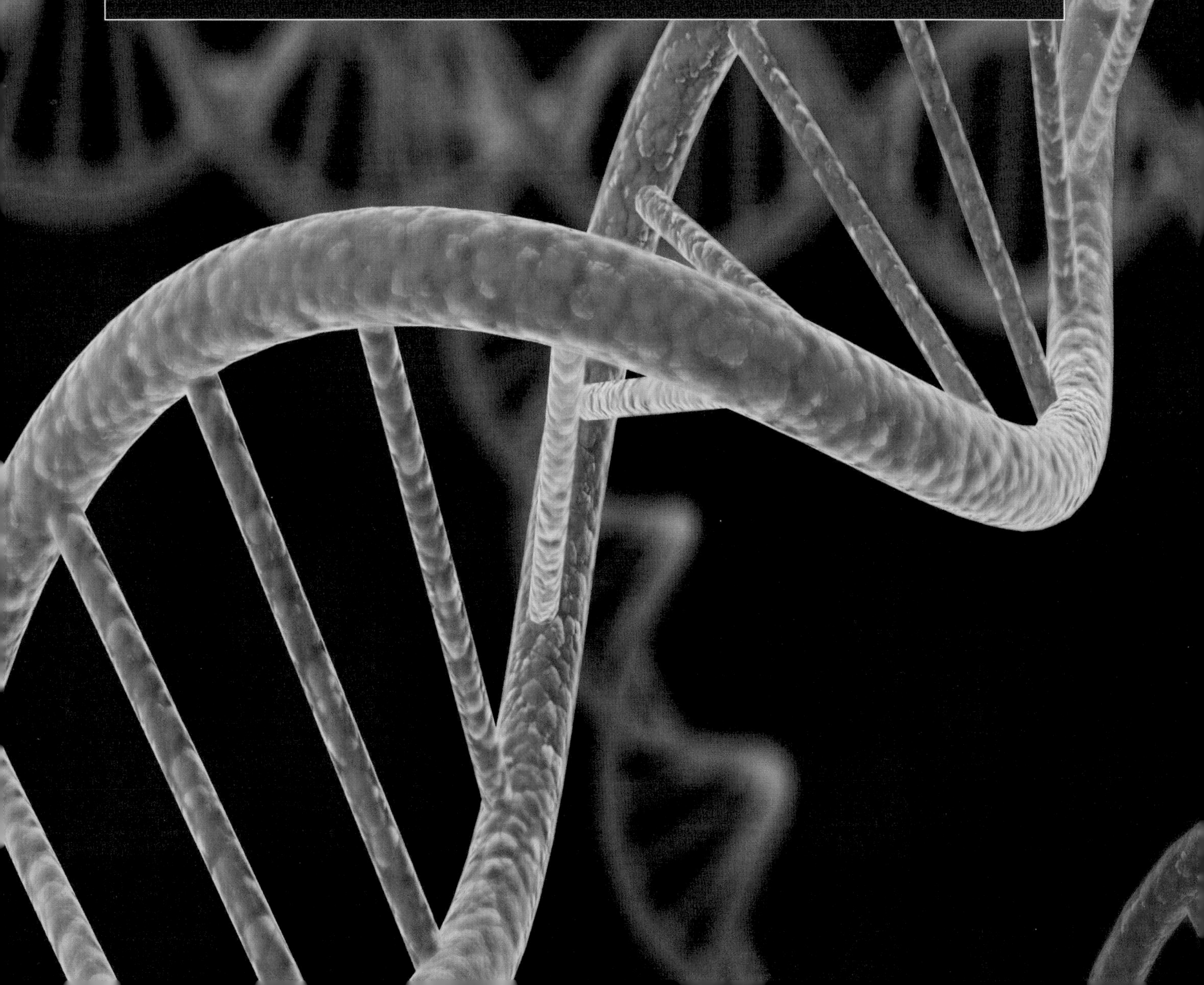

CHAPTER 5

ENZYMES AND AMINO ACID/PROTEIN METABOLISM

OVERVIEW

Enzymes are specialized proteins, which accelerate or **catalyze** a biochemical reaction. Each enzyme catalyzes a specific reaction and is regulated by competitive and noncompetitive inhibitors and/or by allosteric molecules. Multiple enzymes can catalyze a series of consecutive reactions, known as **pathways**, to produce and/or break down complex biological molecules. Examples include amino acid synthesis and degradation, the coordinated reactions involved in protein synthesis, and the urea cycle. Problems with enzyme pathways can not only lead to disease but also offer the opportunity for disease treatment via medications, which target specific points in these pathways.

ENZYMES

ENZYME REACTIONS

Enzymes bring together one or more molecules, called **substrates**, to form a resulting molecule called a **product** (Figure 5-1). Most enzymes catalyze one specific reaction. However, some multipart enzyme complexes catalyze a series of step-by-step reactions—the first enzyme passes its product, now a new substrate, to a second enzyme that is part of the complex, the second passes its product to a third and so on. Enzymes are responsible for many essential reactions in the human body; in fact, there are from 20,000 to 25,000 total human genes, with about 25% of them producing enzymes. It is not surprising that problems with enzymes are caused by or result in diseases.

The concept of enzyme **kinetics** allows an exact description of the enzymatic reaction, including the influence of substrate and product molecules and how fast the enzyme catalyzes the reaction and the impact. More advanced enzyme kinetics allows the mathematical expression of how other molecules such as **cofactors**, **inhibitors**, and **activators** (see below) affect the enzyme reaction.

For example, as discussed in Chapter 1, specific amino acids form an enzyme's primary structure and, therefore, secondary to quaternary structure. These structures, in turn, form **substrate-binding sites** (also known as the "**active site**") to accommodate the substrates, their chemical reaction, and the departure of the product. Substrates and products can both be a **hydrophobic**, **hydrophilic**, **charged**, **uncharged**, or **neutral** molecule, or a combination of the above. Mutations that result in the change of an amino acid or amino acids that make the substrate pocket can drastically change the enzyme's activity. For example, a hydrophobic substrate will easily enter an enzyme pocket that is lined with hydrophobic amino acids. However, if one of these hydrophobic amino acids is changed to a highly charged amino acid, the substrate may not be able to enter the pocket and the enzyme may no longer function. How well the substrate interacts with the substrate-binding site represents the "**affinity**" of the enzyme for the substrate. The stronger the affinity, the less substrate is needed to achieve a certain rate of reaction. This concept is important if a mutation changes the substrate-binding site and lowers the affinity. This concept is illustrated in Figure 5-1.

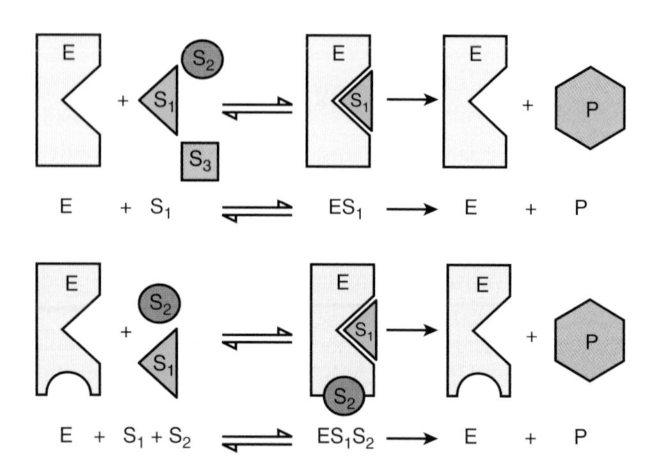

Figure 5-1. Illustration of Simple One and Two Substrate Enzyme Reaction. Components include enzyme (E), substances (S_1, S_2, S_3), and product (P). If the substance can bind to the enzyme's substrate-binding site (e.g., S_1, top figure, and S_1 and S_2, bottom figure), then it acts as a substrate. Molecular shape as determined by secondary, tertiary, and quaternary structure as well as the hydrophobic/hydrophilic and neutral or charged nature influences which substrate molecules can bind as substrates (represented graphically by differing shapes of S_1, S_2, and S_3 and the triangular and circular binding pockets). The enzyme binds the substrate molecule(s), catalyzes the enzymatic reaction, and releases the product after which the enzyme is ready to catalyze the same reaction again. The rate of the overall reaction is influenced by each step in the process, including substrate binding, rate of the reaction, and product release as discussed in the text below. [Adapted with permission from Naik P: Biochemistry, 3rd edition, Jaypee Brothers Medical Publishers (P) Ltd., 2009.]

Substrate Concentration and Enzyme-based Diseases: Phenylketonuria (PKU) is an important autosomal recessive disease occurring in an average of 1 in 15,000 births. This genetic disease is usually caused by the deficiency of the enzyme **phenylalanine hydroxylase**, required to convert phenylalanine to tyrosine (see reaction below; hydroxyl group indicated by small arrowhead), important in the production of neurotransmitters and skin pigment (melanin). However, the disease does not result only from the decrease of tyrosine levels; rather, the resulting high *concentration* of phenylalanine leads to the activity of an otherwise minor enzymatic pathway, which produces phenylketones, including phenylacetate, phenylpyruvate, and phenylethylamine. The presence of these phenylketones is normally screened for in the blood as part of standard neonatal testing and, again, at 2 weeks of age.

$C_6H_5-CH_2-CH(COOH)-NH_2 \xrightarrow{\text{Phenylalanine hydroxylase}} HO-C_6H_4-CH_2-CH(COOH)-NH_2$

[Adapted with permission from Kibble JD and Halsey CR: The Big Picture: Medical Physiology, 1st edition, McGraw-Hill, 2009.]

Patients with PKU usually express traits of albinism, i.e., very fair skin; white-blond hair, and notable, pale blue eyes; a "musty" odor from phenylacetate in their sweat, urine, skin, and hair; and sometimes develop eczema. Untreated PKU results in decreased brain development (microcephaly), hyperactivity, brain damage, seizures, and severe learning disabilities/mental retardation. PKU is usually treated with a diet low in phenylalanine content, although some feel that residual phenylalanine levels may still cause neurological damage. As a result, additional treatments are being developed, which further decrease phenylalanine levels.

The number of molecules of substrate increases or decreases the likelihood of the substrate interacting with the enzyme and, therefore, the rate of the reaction. When all of the enzyme molecules have substrate occupying the binding sites, the enzyme is said to be **saturated** by substrate and the rate of the reaction is maximized. If lower amounts of substrate are present, the reaction would also be slower than normal, potentially leading to clinical manifestations. If a disease state causes a molecule to abnormally increase in concentration, it may become a substrate for an enzyme with which it would normally not react.

The final factor that decides maximum rate of the reaction is the efficiency at which an enzyme can catalyze a single reaction. Enzyme mutations can diminish this efficiency, although, in rare instances, a mutation can increase activity. Likewise, the cell has regulatory molecules that can change the rate of the reaction by decreasing or increasing the efficiency of the enzyme.

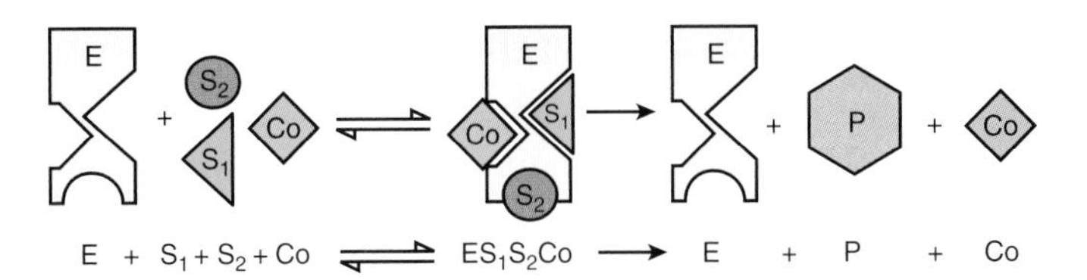

Figure 5-2. Enzymatic Reaction Involving a Cofactor. Components include enzyme (E), substrates (S_1, S_2), cofactor (Co), and product (P). The cofactor is necessary for the creation of the final product but is unchanged at the end of the reaction. [Adapted with permission from Naik P: Biochemistry, 3rd edition, Jaypee Brothers Medical Publishers (P) Ltd., 2009.]

COFACTORS

Enzyme reactions often require additional molecules called **cofactors** (Figure 5-2). Examples of these cofactors include vitamins, nicotinamide adenine dinucleotide (NADH), nicotinamide adenine dinucleotide phosphate (NADPH), flavin adenine dinucleotide ($FADH_2$), and coenzyme A (CoA). For example, during an enzyme reaction, NADH or NADPH donates a hydrogen atom as well as electrons to the product and becomes NAD^+ or $NADP^+$, respectively, and are then released. $FADH_2$ also donates electrons and two hydrogen atoms to produce FAD. Conversely, NAD^+, $NADP^+$, and FAD accept electrons to become the respective reduced forms. Some inorganic metal ions such as iron (Fe), copper (Cu), calcium (Ca), manganese (Mn), magnesium (Mg), or zinc (Zn) also serve as cofactors for reactions. Metal ions may contribute or accept an electron (e.g., $Fe^{2+} \rightarrow Fe^{3+}$) or their associated charge may simply provide essential stabilization for the chemical reaction without changing the metal.

Diseases of Copper Deficiency: The importance of proper levels of copper (Cu^{2+}) to the human body is often not appreciated. For example, the incorporation of copper into several enzymes in mitochondria is essential for oxidative phosphorylation (Chapter 6), and low levels can adversely affect the production of ATP. The opposite, an excess of copper, leads to the important clinical conditions of **Wilson's disease** and **Menke's "kinky hair" disease**.

Wilson's disease results from low activity of **ATPase, Cu^{2+} transporting, beta polypeptide (ATP7B)**, which is responsible for copper transport out of cells. The transported copper is carried by ceruloplasmin in the circulation. In Wilson's disease, copper ion concentrations are increased in liver, brain, eyes, and other tissues. Liver deposits result in fatigue, confusion from high ammonia levels (hepatic encephalopathy), heightened blood pressure (portal hypertension), an increased risk of bleeding from blood vessels in the esophagus (esophageal varices), and, eventually, liver failure and/or cancer. Accumulation in the brain can lead to deterioration of memory and thought processes, loss of muscle control and tone, seizures, migraines, depression, anxiety, psychosis, and, often, the tremors and slow movement of Parkinson's disease. Deposition in eyes results in the diagnostic "Kayser–Fleischer ring," a brownish-green ring evident in the cornea. Kidneys, heart, and parathyroid glands can also be adversely affected by copper deposits.

Menke's "kinky hair" disease is a very rare, X-linked recessive disorder that mutates **ATPase, Cu^{2+} transporting, alpha polypeptide (ATP7A)** and, therefore, affects males more often than females. The disease usually strikes infants (although a childhood version is also known), slowly causing growth failure, developmental delay, and mental retardation. The subtle onset often leads to no treatment and death in the first decade. Excess copper accumulation mainly in the kidneys and the digestive tract leads to symptoms similar to Wilson's disease. However, low levels of copper in bone, brain, skin, blood vessels, and hair affect copper-containing enzymes. The ATP7A enzyme is absent in the liver. Other symptoms particular to Menke's disease include abnormal body temperature; rupture or blockage of arteries in the brain; weakened bones and increased fractures; and the sparse, coarse, fragile, kinky, colorless, or steel-colored hair that gives the disease its alternative name. The only treatment is daily injection of copper as well as symptomatic and supportive care.

REGULATION

The speed or "**rate**" at which an enzyme catalyzes a reaction depends on numerous factors, which can increase (**activate**) or decrease (**inhibit**) this rate. This effect is called **regulation** and is essential not only for single enzyme reactions but also for the overall coordination of multiple, and often competing, reactions in the human body. Like any complicated process, resources need to be conserved and used in an efficient manner. It is this regulation that allows the body to use energy or store energy and/or produce certain proteins or carbohydrates or lipid molecules when particular needs arise.

Enzyme regulation takes many forms. The simplest is by varying the concentration of the substrate. As discussed above, the increase or decrease of the amount of available starting material increases or slows the reaction, respectively, and forms the basis of not only the single enzyme regulation but also global regulation of metabolism as well. Examples will be seen in other chapters in Sections II and III.

Enzyme regulation can also occur as the result of a molecule other than substrate or product interacting separately with the enzyme to modify its reaction rate. This molecule, whether an activator or inhibitor, works in one of two ways: (1) **competitive**—binding to the same site as the substrate molecule(s) or (2) **noncompetitive**—binding to a completely separate part of the enzyme, which changes the shape of the enzyme and, as a result, the efficiency of the reaction (Figure 5-3). A third type of enzyme competition, **uncompetitive**, is rare and will not be covered here. The binding of a regulatory molecule may either increase or slow the reaction by increasing/decreasing substrate binding and/or the actual chemical reaction and/or product release. (Note: The terms "competitive" and "noncompetitive" are normally not used to describe activation; however, for simplicity, these terms are used here.) Both activators and inhibitors of enzymatic reactions are important in disease processes. However, they are even more important in disease treatment because most medicines are designed to increase the production of a beneficial product or decrease/stop the formation of a harmful product.

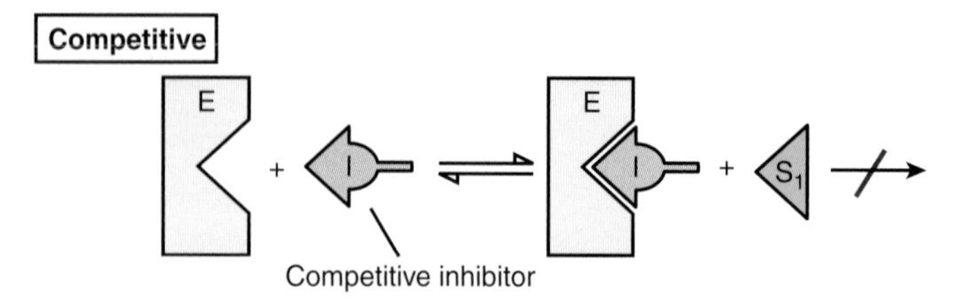

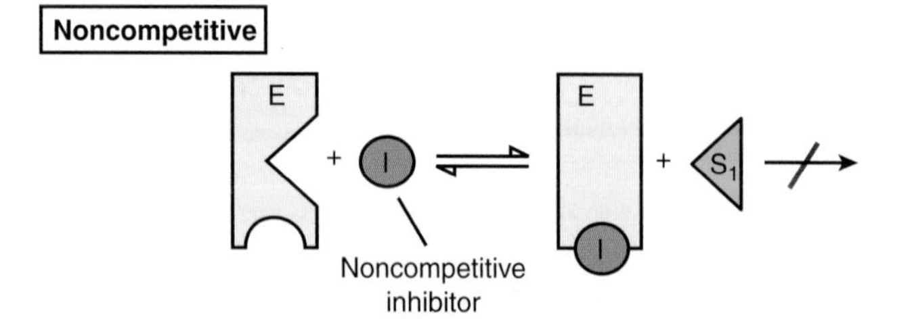

Figure 5-3. Competitive and Noncompetitive Inhibition. Competitive inhibitor (upper panel) binds to and blocks the enzyme's substrate-binding site and hence the binding of the substrate (S_1), thereby preventing production of the product. Noncompetitive inhibitor (bottom panel) binds to a site other than the substrate-binding site, leading to a conformational change of the enzyme, which alters the normal substrate-binding site. This change blocks substrate (S_1) binding and production of the product. [Adapted with permission from Naik P: Biochemistry, 3rd edition, Jaypee Brothers Medical Publishers (P) Ltd., 2009.]

β-Lactam Antibiotics and Competitive Inhibition: Gram-positive bacteria have cell walls composed of up to 40 layers of peptidoglycans, specialized carbohydrate, and amino acid structures, which offer support and protection. Synthesis of the peptidoglycan cell wall involves a final step, which links the individual peptidoglycans together and involves the cofactor penicillin-binding protein (PBP). **β-Lactam antibiotics** (see the figure below), of which penicillin is the archetype, mimic the end amino acid sequence of the unlinked peptidoglycans and irreversibly bind to the substrate site to competitively block the final cross-linking reaction of the PBPs. The disruption of this final step of cell wall synthesis leads to death of the bacteria. Many types of antibiotics are β-lactams, including cephalosporins, monobactams, carbapenems, and β-lactamase inhibitors (e.g., clavulanic acid, sulbactam, and tazobactam).

Penicillin | Cephalosporin

The structure of penicillins (left) and cephalosporins (right) includes the β-lactam component shown in red. These antibiotics mimic the terminal end of proteoglycans and block Gram-positive cell wall synthesis.
[Adapted with permission from Katzung BG, et al.: Basic and Clinical Pharmacology, 11th edition, McGraw-Hill, 2009.]

Feedback regulation is a particular type of simple regulation in which a product of a single or a series of enzyme reactions can bind to and activate or inhibit itself or an earlier enzyme in the series (Figure 5-4). Higher or lower concentrations of this product offer "feed back" regarding levels of the product and the need to increase or decrease the rate of the reaction(s) according to the needs of the body. Several examples of feedback regulation are seen in **metabolism**—the breakdown of biochemical molecules discussed in later chapters in this section and Chapter 10.

The activity of an enzyme can also be controlled by the concept of precursor or **proenzymes**. Proenzymes are often, but not always,

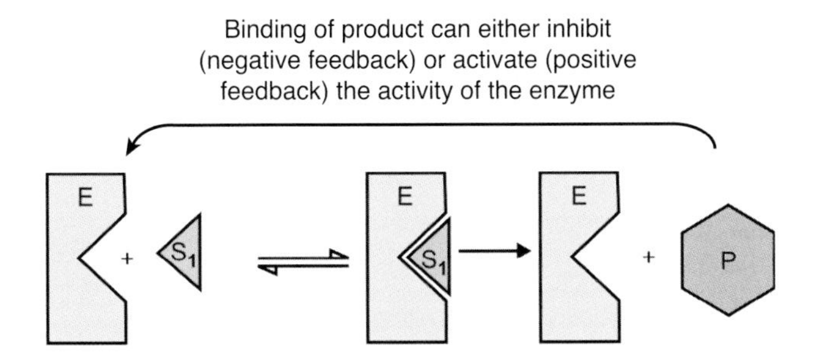

Figure 5-4. Feedback Inhibition and Activation. The product can bind to its own enzyme or another enzyme and either inhibit or activate the enzymatic reaction according to the body's needs. Components include enzyme, substrate (S_1), and products (P). [Adapted with permission from Naik P: Biochemistry, 3rd edition, Jaypee Brothers Medical Publishers (P) Ltd., 2009.]

Isozymes: Isozymes allow the body to differentially regulate the same enzyme reaction in different tissues of an individual. One of the best examples is hexokinase and glucokinase (see Chapter 6). **Hexokinase**, found in all human cells, is able to bind glucose even at very low concentrations. **Glucokinase**, found only in pancreas, liver, small intestine, and brain, requires glucose concentrations 100 times higher for binding and is not inhibited by glucose-6-phosphate. Even though hexokinase and glucokinase catalyze the same reaction, the different binding of glucose and regulation allows the body to properly regulate glucose use (e.g., release of insulin by beta cells of the pancreas) and storage (e.g., liver cells).

denoted by the suffix "*-ogen.*" Proenzymes contain the active enzyme and an additional amino acid sequence that keeps the activity "turned off." A separate enzyme, specific for the proenzyme to enzyme transformation, removes this additional sequence to allow the enzyme function to begin. Many enzymes involved in the digestion of food are controlled this way and are only turned on when specific signals are present (e.g., the presence of a recently ingested food source or a clot formation). Examples of proenzymes include pepsinogen → pepsin, fibrinogen → fibrin, and procarboxypeptidase → carboxypeptidase.

Isozymes (also known as isoenzymes) are two or more enzymes in the same individual that differ in amino acid sequence but which catalyze the exact same chemical reaction. Isoenzymes are usually found in different tissues as well as in different cell organelles, where they can be differently regulated. Examples of proenzymes and isozymes will be seen in future chapters.

Finally, **allosteric regulation** involves the binding of an **effector** molecule to a part of a single subunit ("monomeric") enzyme, or to one of several subunits, in the case of enzyme complexes with more than subunit ("multimeric"). For the more common allosteric regulation of multimeric enzymes, the effector is often the substrate for more than one active site contained within the complex. Binding of the effector causes a conformational change, which affects the active site of an enzyme (Figure 5-5). The binding either activates (positive cooperativity) or inhibits (negative cooperativity) additional substrate binding or enzyme activity—an effect known as allosteric regulation. Allosteric regulation has been classically illustrated in the sequentially increasing binding of oxygen by the four subunits of hemoglobin as well as the decreased binding of oxygen by the subunits by the molecule 2,3-bisphosphoglycerate (both allow increased but regulated release of oxygen near

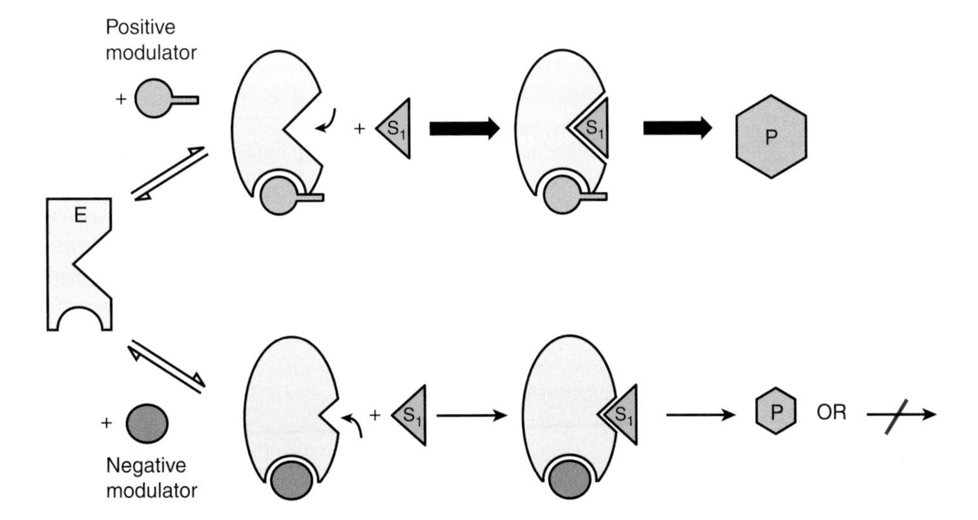

Figure 5-5. Allosteric Regulation of Enzyme Activity. Binding of a positive allosteric modulator (upper panel) leads to a conformational change of the enzyme (E), which enhances enzyme activity (bold arrows) by increased substrate binding (small curved arrow) and/or increased rate of product (P) formation. Binding of a negative allosteric modulator (lower panel) leads to a conformational change, which decreases enzyme activity (thin arrows) by decreasing substrate binding (small curved arrow) and/or leading to reduced or no product formation. [Adapted with permission from Naik P: Biochemistry, 3rd edition, Jaypee Brothers Medical Publishers (P) Ltd., 2009.]

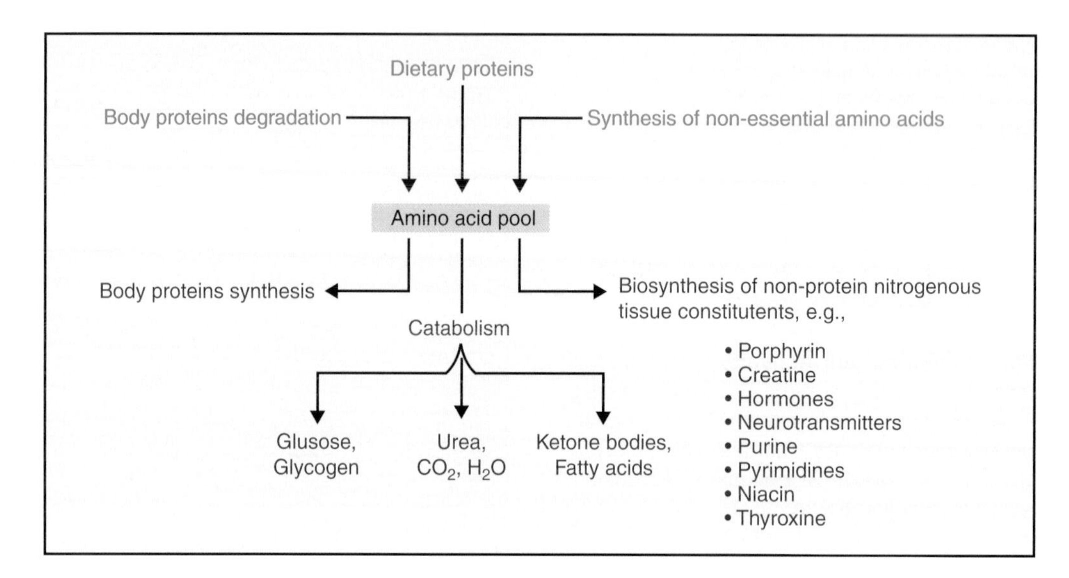

Figure 5-6. Sources and Fates of Amino Acids. Amino acids, from endogenous or dietary proteins as well as *de novo* synthesis, are used for synthesis of a variety of nitrogenous compounds as well as new proteins. Their carbon skeleton can make a variety of fuels, depending on energy needs, or be burned as a fuel, whereas the toxic nitrogen waste is largely excreted as urea. [Reproduced with permission from Naik P: Biochemistry, 3rd edition, Jaypee Brothers Medical Publishers (P) Ltd., 2009.]

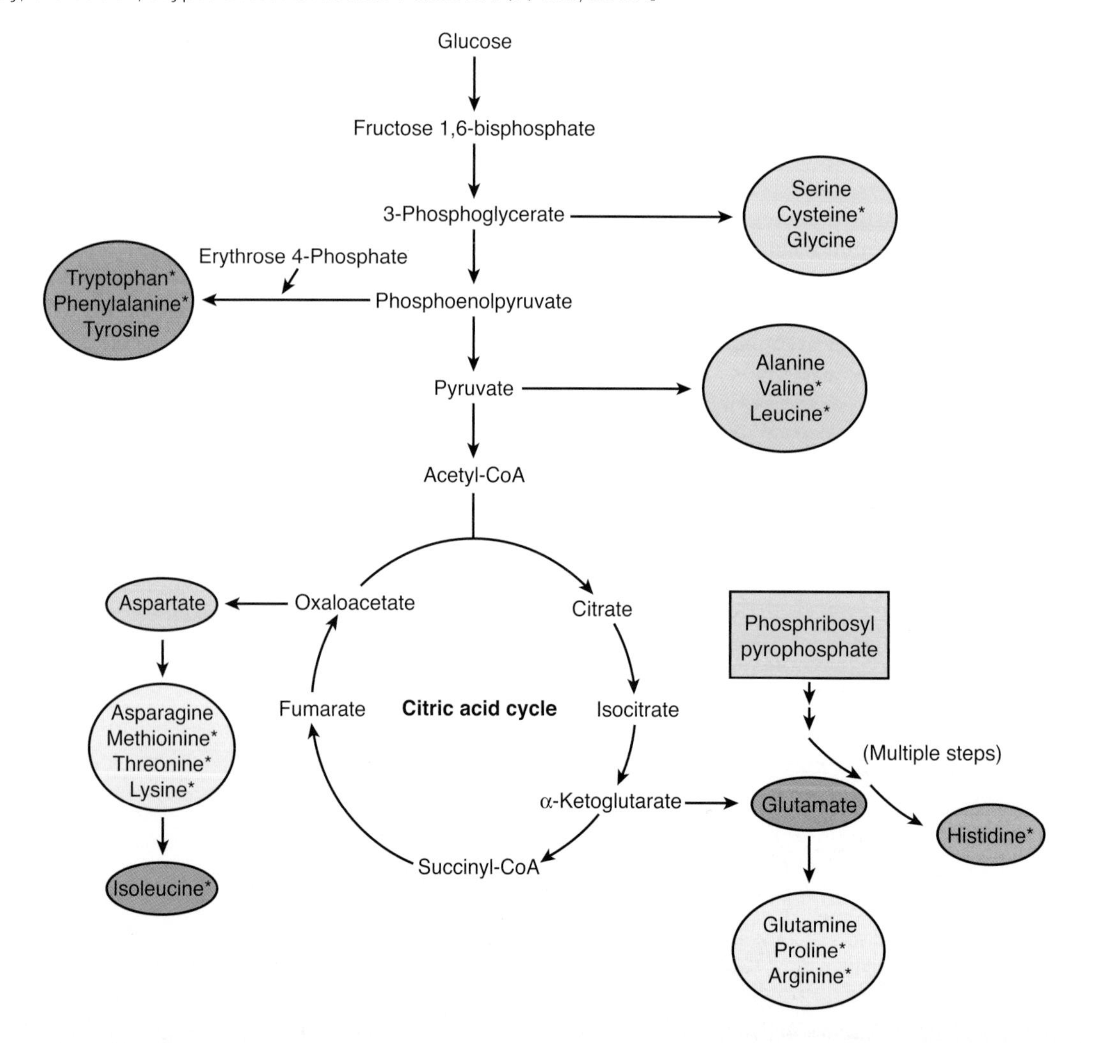

Figure 5-7. Synthesis of Amino Acids. Precursors of the 20 amino acids found in humans are predominately from carbohydrate metabolism (glycolysis and citric acid cycle, see Chapter 6). Essential amino acids required from the diet are indicated by asterisks (*). Histidine is derived partly from the nucleic acid precursor phosphoribosyl pyrophosphate and the amino acid glutamate. Specific medical conditions associated with the synthesis of amino acids are shown in Table 5-1.

tissues low in oxygen). An example of allosteric regulation of monomeric enzymes includes the effect of lactate produced during periods of prolonged physical effort to decrease oxygen binding by myoglobin to help maintain ATP concentration. Several drugs, including ibuprofen and acetaminophen, also allosterically affect the binding of the anti-anxiety class of drugs called benzodiazepines to albumin, thereby influencing the actual concentration of this medication in the blood.

AMINO ACID METABOLISM

AMINO ACID SYNTHESIS

The synthesis of amino acids is of paramount importance to the human body. Once synthesized or ingested, amino acids are used as building blocks not only for proteins but also for several other critical biological molecules such as nucleic acids (both purines and pyrimidines), hormones, neurotransmitters, antioxidants, and various signaling molecules (Figure 5-6; see also Chapter 1 and later chapters).

Although individual amino acids are usually obtained from the diet, especially the essential amino acids, nonessential amino acids can also be produced from other sources. The non-essential amino acids are all made from precursors obtained from the metabolism of carbohydrates to be studied in the next chapter. These amino acids and their precursor molecule(s) are shown in Figure 5-7.

AMINO ACID DEGRADATION

Protein and amino acid breakdown, known as "**degradation**," is essential as an energy source when carbohydrate and/or lipid sources are low, as well as to produce atoms and molecules required for the synthesis of other molecules. The breakdown of particular amino acids also helps the body to regulate how much nitrogen it contains and to excrete excess nitrogen in the form of urea in urine. Each amino acid undergoes a specific series of enzymatic reactions, leading to molecules involved with and, therefore, leading into carbohydrate or lipid metabolism and energy production (Figure 5-8). As with amino acid synthesis, deficiencies and decreased activity of the enzymes involved in these breakdown pathways lead to disease states. These diseases are usually present at birth and require adherence to specific dietary regimens to control the illness.

Several diseases involving the absence or inactivity of an enzyme involved in amino acid metabolism are listed in Table 5-1. These deficiencies in enzymes responsible for the synthesis, degradation, absorption, and transport of the amino acids generally cause failure or abnormal development, neurological/psychiatric problems, some quite unique (repetitive self-hugging and "lick and flip" activity—licking of fingers and flipping of book and magazine pages),

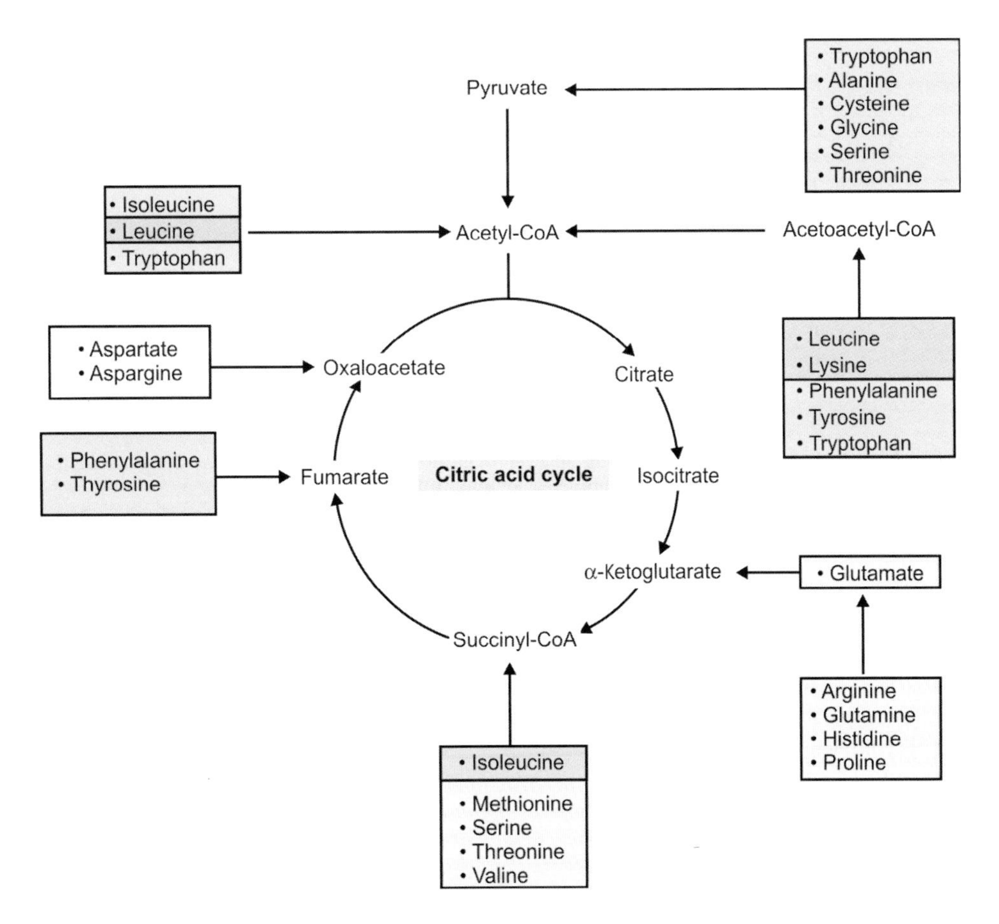

Figure 5-8. Overview of Amino Acid Degradation Pathways. Amino acids enter into and are metabolized via the carbohydrate metabolic pathways (citric acid cycle and glycolysis, see Chapter 6). [Reproduced with permission from Naik P: Biochemistry, 3rd edition, Jaypee Brothers Medical Publishers (P) Ltd., 2009.]

TABLE 5-1. Diseases of Amino Acid Metabolism

Amino Acid(s)	Disorder
Alanine	Alanine transaminase deficiency
Alanine, serine, threonine, valine, leucine, isoleucine, phenylalanine, tyrosine, tryptophan, and histidine	Neutral aminoaciduria (Hartnup's disease)
Arginine	Citrullinemia (type I), argininosuccinic aciduria/acidemia
Glutamine	Glutamine synthetase deficiency
Glycine	Smith–Magenis syndrome, ketotic hyperglycinemia/propionic acidemia, nonketotic hyperglycinemia/glycine encephalopathy
Glycine, proline, hydroxyproline	Iminoglycinuria I and II
Histidine	Histidinemia
Leucine	Isovaleric acidemia
Leucine, Isoleucine, and Valine (branched-chain amino acids)	Maple syrup urine disease (branched chain ketoaciduria)
Methionine	Homocystinuria I and II, methionine malabsorption syndrome (Smith–Strang/oasthouse urine disease)
Phenylalanine, tyrosine	Phenylketonuria, alkaptonuria (black urine disease)
Proline	Hyperprolinemia type I and II, Δ^1-pyrroline 5-carboxylate synthase deficiency
Serine	3-Phosphoglycerate dehydrogenase deficiency, phosphoserine aminotransferase deficiency, phosphoserine phosphatase deficiency, serine dehydratase deficiency
Threonine	Threonine dehydratase deficiency
Tryptophan	Drummond's (blue diaper) syndrome
Tyrosine, phenylalanine	Albinism, tyrosinemia, types I–III; hawkinsinuria
Valine	Hypervalinemia/valinemia

musculoskeletal disorders, abnormal function or failure of organ and organ systems, several unique odors ["maple syrup," "mousy," "cabbage-like," "oasthouse" (building for drying hops)], and colors of urine (black and blue). These conditions may also lead to death of the fetus or infant. A more complete description of each of these conditions is found in Appendix I.

THE UREA CYCLE

Urea provides a critical means of ridding the body of nitrogen waste. The **urea cycle** is the series of enzyme reactions in liver cell mitochondria and cytoplasm responsible for the removal of excess nitrogen and the potentially toxic by-product ammonia via the production of urea. More importantly, the urea cycle helps keep in balance nitrogen atoms in the human body. The urea cycle accepts nitrogen atoms in the form of amino groups (NH_4^+) exclusively from the amino acids glutamate and aspartate (Figure 5-9A). Other sources of excess nitrogen, including from other amino acids, feed through the glutamate and aspartate pathways.

Finally, the body also relies on an ammonia buffering system of α-ketoglutarate, glutamate, and glutamine for removal, production, and/or storage of ammonia groups (Figure 5-9B).

Regulation of the urea cycle is via the enzyme carbamoyl phosphate synthetase I (CPS-I), located within mitochondria. This enzyme catalyzes the reaction that converts ammonia from glutamate into carbamoyl phosphate, which then combines

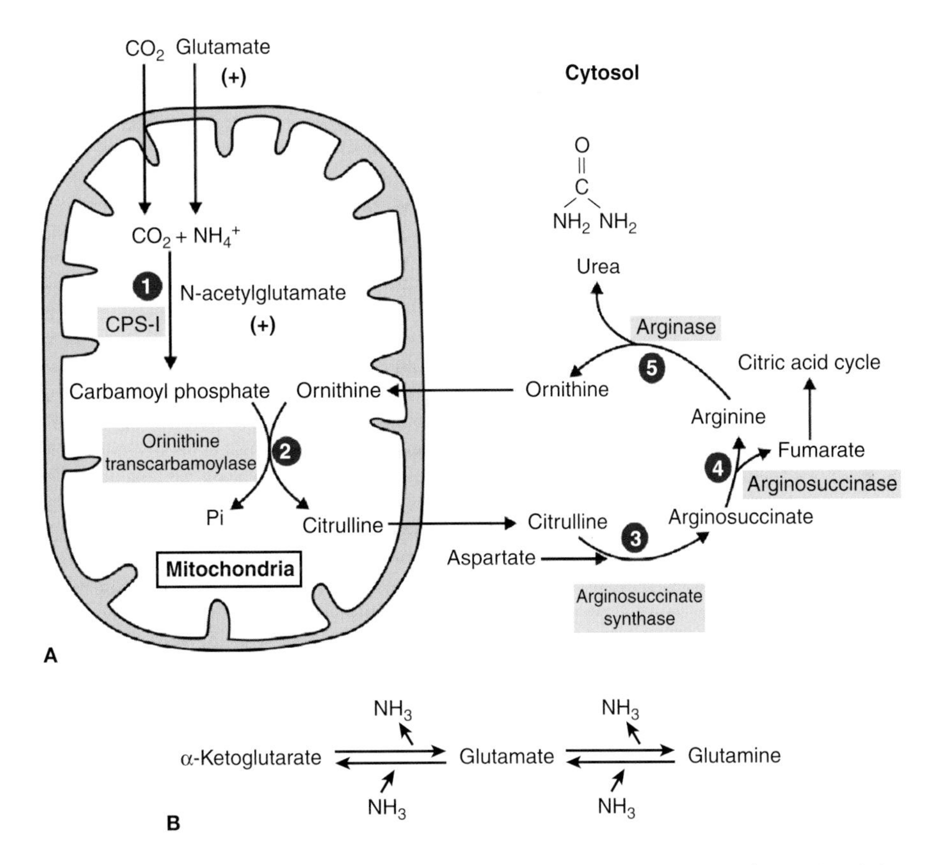

Figure 5-9A. Urea Cycle. A. The urea cycle converts two amino groups (NH_3) from glutamate and aspartate into urea as part of the body's nitrogen balance. Glutamate donates the first nitrogen atom (1) via the molecule carbamoyl phosphate that is synthesized via the regulated enzyme CPS. This enzyme is activated by increased glutamate, *N*-acetylglutamic acid, and/or arginine concentrations [indicated by (+)]. Carbamoyl phosphate then enters the urea cycle by combination with ornithine via ornithine transcarbamoylase. Aspartate donates the second nitrogen (3). Other amino acids contribute amino groups after conversion into glutamate or aspartate. [Adapted with permission from Naik P: Biochemistry, 3rd edition, Jaypee Brothers Medical Publishers (P) Ltd., 2009.] **B.** α-ketoglutarate/Glutamate/Glutamine Ammonia Buffering System. The interconversion of α-ketoglutarate to/from glutamate to/from glutamine offers the body an important system to utilize, generate, and/or store ammonia.

with ornithine to form citrulline. CPS is regulated by increased concentrations of *N*-acetylglutamic acid, produced from the combination of acetyl-CoA and glutamate, as well as increased concentration of arginine. Both of these amino acids are direct or indirect participants in the urea cycle, and increased concentrations of either reflect higher amino acid concentrations.

REVIEW QUESTIONS

1. What is an enzyme and how do the terms catalyze, substrate, and product relate to enzymes?
2. What is simple feedback and allosteric regulation of enzymes?
3. What is meant by amino acid synthesis and amino acid degradation?
4. What are the relationships between amino acid synthesis and degradation and other metabolic pathways?
5. For each basic enzyme reaction what is the role of cofactors and regulation?
6. What is the function of the urea cycle and how would you diagram this pathway?

CHAPTER 6

CARBOHYDRATE METABOLISM

OVERVIEW

The breakdown (**catabolism**) and synthesis (**anabolism**) of carbohydrate molecules represent the primary means for the human body to store and utilize energy and to provide building blocks for molecules such as nucleotides (Figure 6-1). The enzyme reactions that form the metabolic pathways for monosaccharide carbohydrates (Chapter 2) include **glycolysis**, the **citric acid cycle**, and **oxidative phosphorylation** as the main means to produce the energy molecule adenosine triphosphate (ATP). **Gluconeogenesis** and the **pentose phosphate pathway** represent the two main anabolic pathways to produce new carbohydrate molecules. Glycogen has its own metabolic pathway for lengthening, shortening, and/or adding branch points in the carbohydrate chain(s). Not surprisingly, all of these processes are highly regulated at multiple points to allow the human body to efficiently utilize these important biomolecules. Finally, many modified carbohydrates are part of a variety of surface and cytosolic signaling molecules, including glycoproteins and glycosaminoglycans (GAGs) (Chapter 2). These important carbohydrate molecules and the control points in carbohydrate and glycoprotein metabolism, therefore, present clinicians with opportunities to modify these many reactions to improve health or to fight disease.

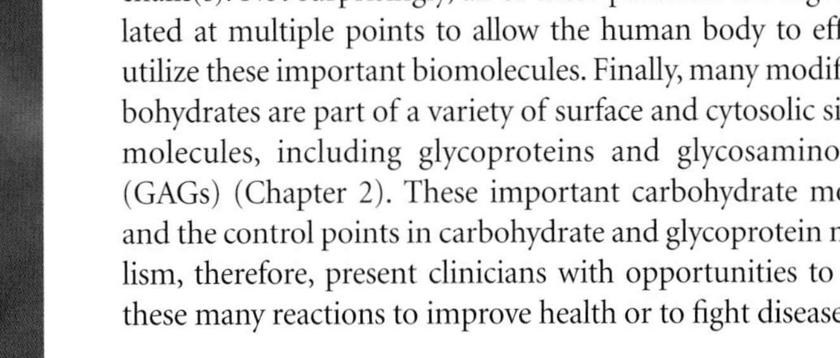

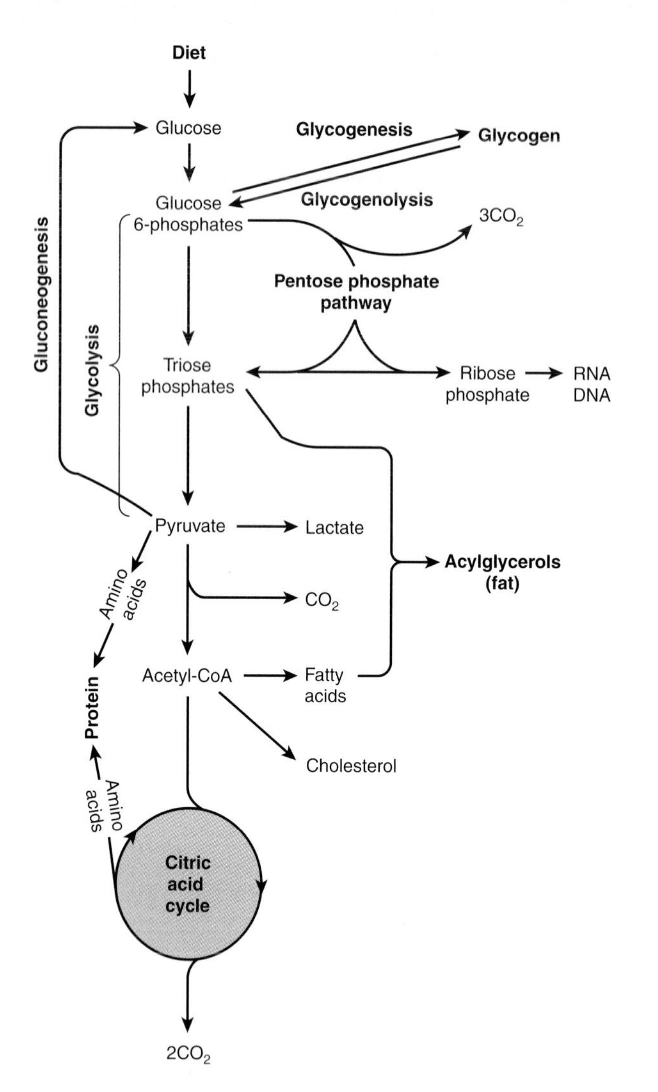

Figure 6-1. Overview of Carbohydrate Metabolism. Glucose from the diet can be metabolized via glycolysis or glycogenesis. Resulting metabolic products can return to glucose via gluconeogenesis or glycogenolysis, respectively, or proceed further along carbohydrate metabolism to the citric acid cycle. Alternatively, glucose products can be shunted off to fat or amino acid metabolism as indicated. Details are discussed in the text and other chapters. [Adapted with permission from Murray RA, et al.: Harper's Illustrated Biochemistry, 28th edition, McGraw-Hill, 2009.]

GLYCOLYSIS

Glycolysis is the metabolic pathway that breaks down (catabolism) hexose (six-carbon) monosaccharides such as glucose, fructose, and galactose into two molecules of pyruvate, two molecules of ATP, two molecules of NADH, two water (H_2O) molecules, and two hydrogen ions (H^+) (Figure 6-2). Glycolysis involves 10 enzyme-mediated steps and is best envisioned in two phases—**phosphorylation** and **energy production**—all of which occur in the cytoplasm. The phosphorylation phase (sometimes referred to as the preparatory phase) starts with the six-carbon carbohydrate glucose and involves two phosphorylations from ATP and the cleavage into two molecules of the trisaccharide (three-carbon sugar) glyceraldehyde-3-phosphate. The energy production phase involves the next five steps during which the two molecules of glyceraldehyde-3-phosphate are converted to two pyruvate molecules with the production of two NADH molecules and four ATP molecules. Glucose-6-phosphate, the first intermediate of glycolysis, cannot exit the cell-like glucose, so it also traps the glucose molecule in the cell for energy production via glycolysis or glycogen synthesis (see below). NADH represents an alternative energy storage form than ATP, which may be utilized by the **oxidative phosphorylation** pathway.

Not surprisingly, all three regulatory steps involve either an investment or production of ATP molecules, a committing process in the balance of energy production and storage. The first reaction of glycolysis (catalyzed by **hexokinase**) results in the addition of a phosphate group to form the molecule glucose-6-phosphate. In the liver and pancreas, hexokinase is replaced by the enzyme **glucokinase. Phosphofructokinase-1** adds a second phosphate group to produce fructose-1,6-bisphosphate. **Pyruvate kinase**, the final reaction in glycolysis, produces an ATP molecule. Although not regulated at the enzyme level such as the above enzymes, **phosphoglycerate kinase** is involved in the initial production of ATP from glycolysis and represents the "break even" point when ATP invested in the preparation phase is gained back. Although regulated in many different ways and by many different molecules, a theme emerges that, when energy stores are low, glycolysis is activated to produce more ATP. When energy levels are high, either the glycolytic pathway is inhibited or alternative pathways are promoted for storage of the glucose molecule for later use. Regulation of carbohydrate metabolism and these important regulatory enzymes will be discussed in further detail in later chapters (e.g., Chapter 10).

Phosphofructokinase Deficiency: Because of its importance in energy production, diseases involving enzymes from glycolysis are rare. One example, though, is the deficiency of phosphofructokinase, also known as glycogen storage disease type VII or **Tarui's disease**. It is inherited in an autosomal recessive manner. This disease leads to a buildup of glycogen in cells and impairs various red blood and muscle cells from using glucose and many other monosaccharides to produce ATP. There are three different variations of the disease, depending on the severity of impairment. The first, termed "infantile," presents in the first year of life usually starting with blindness/cataracts and retardation with death by the age of 4 years. A second "classic" variant presents in childhood with muscle cramping and weakness after exercise and death/breakdown of muscle cells as well as low red blood cell count (anemia). The third "late onset" variant presents in early adulthood with progressive weakness of arms and legs but without muscle cramping/breakdown as well as anemia. Diagnosis is by noting high levels of glycogen in muscle samples and measuring the activity of phosphofructokinase. The only treatment is avoidance of high carbohydrate meals and any vigorous exercise.

ATP is the primary energy molecule used by the human body (Figure 6-3). Three enzymes involved in glycolysis, hexokinase, phosphofructokinase-1, and pyruvate kinase, are regulated to

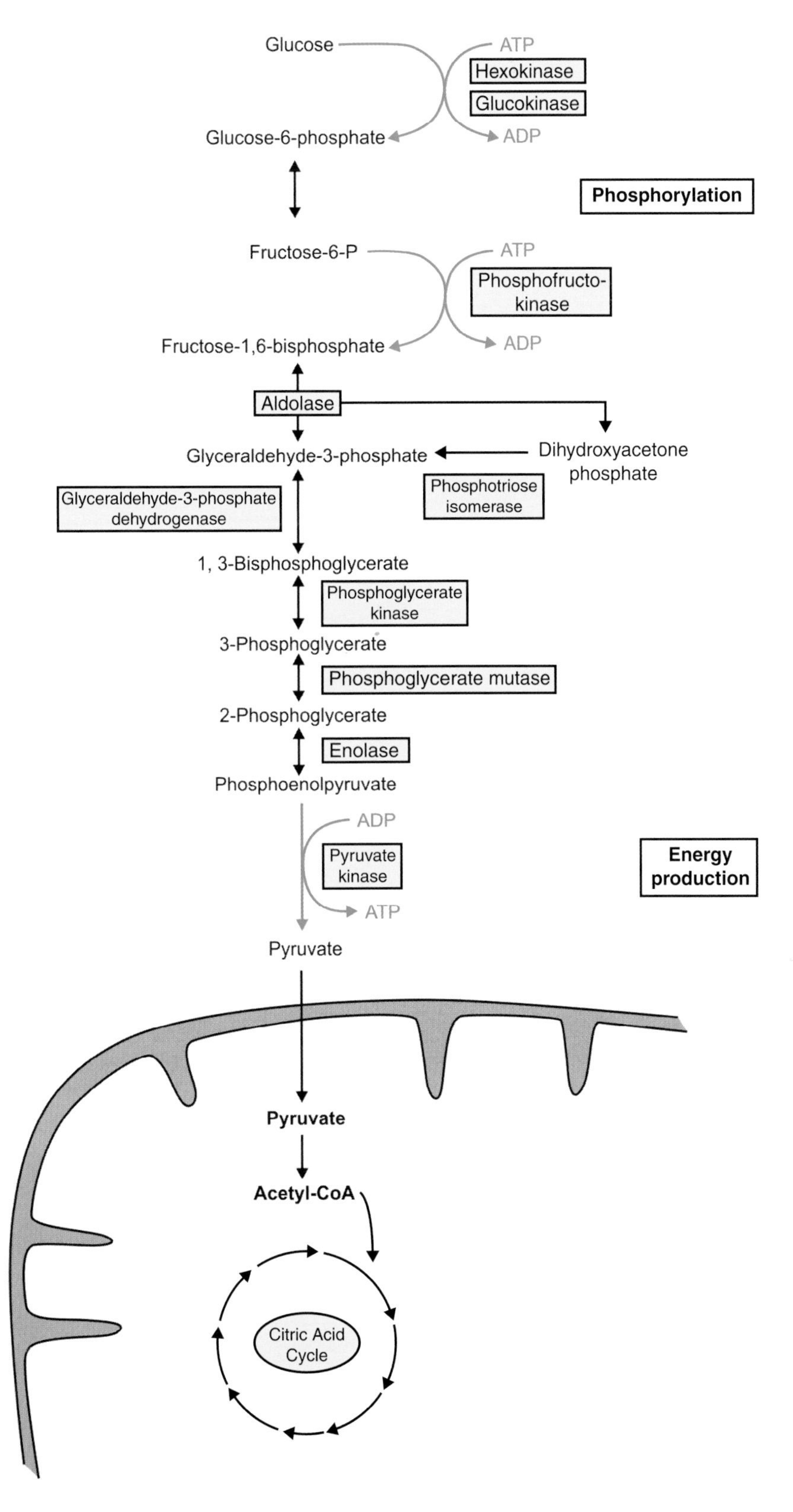

Figure 6-2. Glycolysis. The pathway of glycolysis includes 10 enzyme steps, which break down one molecule of glucose to two molecules of pyruvate, yielding two ATP and two NADH molecules. Pyruvate subsequently enters the mitochondria and the citric acid cycle (see below). Glycolysis can be divided into phosphorylation (Steps 1–5) and energy production (Steps 6–10) phases for better comprehension. Enzymes are indicated by yellow-shaded boxes. [Adapted with permission from Naik P: Biochemistry, 3rd edition, Jaypee Brothers Medical Publishers (P) Ltd., 2009.]

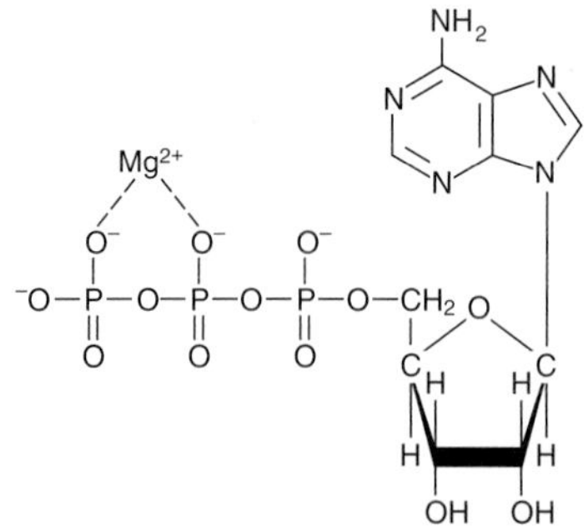

Figure 6-3. ATP Structure with Its Magnesium Cofactor. ATP serves as the primary energy molecule of the human body utilizing the energy contained in the phosphate bonds. Magnesium (Mg^{2+}) stabilizes these bonds and hence is necessary for ATP to be biologically functional. [Reproduced with permission from Murray RA, et al.: Harper's Illustrated Biochemistry, 28th edition, McGraw-Hill, 2009.]

allow the body to decide both whether to use its carbohydrate resources for ATP production and how much to produce. At each step, the cell decides to continue energy production or to channel the resulting intermediates into other molecules such as glycogen, lipids, and/or proteins. The ability of the human body to sense its biochemical surroundings and then respond to that environment enables intelligent and efficient use of its limited resources.

CITRIC ACID CYCLE

The **citric acid cycle** is the second metabolic pathway involved in the catabolism of carbohydrates into energy and involves eight enzyme steps, starting and ending with the molecule **oxaloacetate** (Figure 6-4). In humans, all of the steps of the citric acid

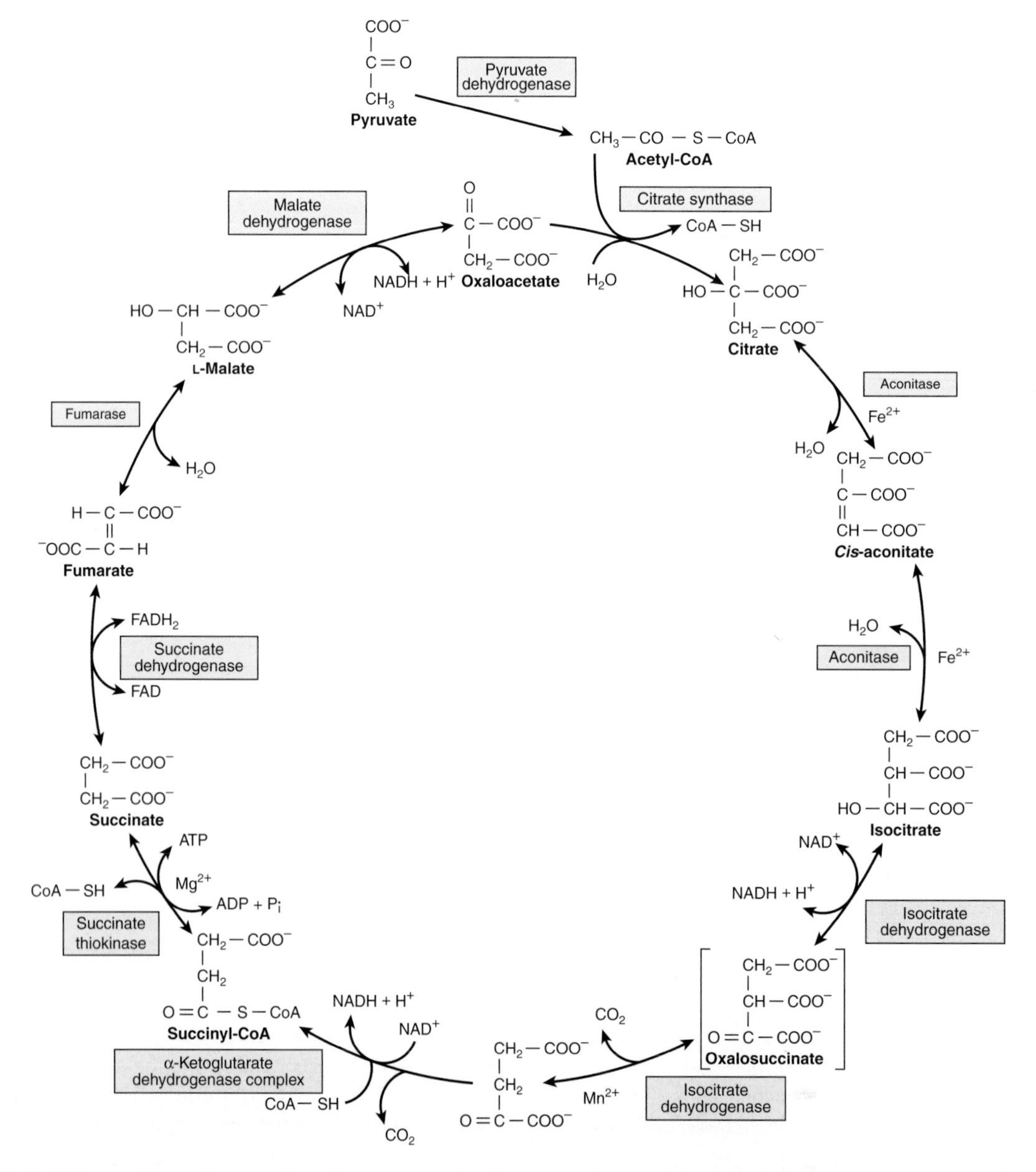

Figure 6-4. Citric Acid Cycle. The citric acid cycle includes eight enzymatic steps (tan-shaded boxes), which receive pyruvate from glycolysis via the acetyl-CoA molecule. The cycle produces one GTP, three NADH, and one $FADH_2$ as it progresses to oxaloacetate at which point the cycle repeats. [Adapted with permission from Murray RA, et al.: Harper's Illustrated Biochemistry, 28th edition, McGraw-Hill, 2009.]

cycle, including the linking reaction of pyruvate conversion to acetyl-CoA, take place within the mitochondrial matrix, the area within the inner mitochondrial membrane. **Pyruvate** enters the citric acid cycle following its conversion to acetyl-CoA; the reaction is catalyzed by pyruvate dehydrogenase that links glycolysis to the citric acid cycle and produces one molecule of NADH. By means of acetyl-CoA and intermediates in this cycle, the body can also divert fats and amino acids into energy production (discussed later in this chapter). In addition, the citric acid cycle provides precursor molecules for the synthesis of amino acids and/or receives these amino acids when proteins are used for energy production (see Chapter 5, Figure 5-7).

Acetyl-CoA, a two-carbon molecule bonds to the four-carbon molecule oxaloacetate to form citrate, the namesake of this pathway. This pathway is also known as the tricarboxylic acid cycle, because citrate contains three carboxylic acid groups, or the Krebs cycle after the physician/biochemist Sir Hans Krebs, who earned a Nobel Prize for his work on elucidating this sequence of reactions. Further enzyme reactions remove carbon molecules in the form of carbon dioxide (CO_2) and H_2O. By doing so, the cycle produces three molecules of NADH, one molecule of $FADH_2$, and one molecule of guanosine triphosphate (GTP), all of which may be used in energy production and storage by oxidative phosphorylation (see below) and conversion to ATP.

The regulation of the citric acid cycle is mainly via the availability of each enzyme's substrate as well as inhibition when the concentration of any enzyme's product gets too high. One prime example is citrate, the product of the first reaction, which inhibits not only the cycle but also phosphofructokinase activity in glycolysis. That citrate is the first intermediate of the cycle and is important in the control of the body's carbohydrate resources. As with glycolysis, the same general rules also apply: if energy levels (in this case NADH or $FADH_2$, which are converted to ATP via the later oxidative phosphorylation pathway) are high, the citric acid cycle is slowed or intermediates are diverted to other purposes.

Red Blood Cells (Erythrocytes) and Glycolysis: Red blood cells are unique in the human body in that they lose their nuclei and organelles, including mitochondria, early in their development (Chapter 14). As a result, red blood cells are unable to utilize the citric acid cycle or oxidative phosphorylation and are solely reliant on ATP production from glycolysis. NADH produced from glycolysis is used to produce lactate from pyruvate; the resulting lactate diffuses from the red blood cells and is transported into and used by tissues such as heart muscle and liver. In addition, red blood cells do not respond to insulin like other tissues in the human body and are, therefore, not regulated by this important hormonal signal that controls carbohydrate metabolism.

OXIDATIVE PHOSPHORYLATION

The third and final pathway of conversion of carbohydrates to energy is oxidative phosphorylation (Figure 6-5). This metabolic pathway takes place exclusively inside the inner mitochondrial membrane, taking advantage of differences in concentration of the products of the citric acid cycle to drive the formation of ATP.

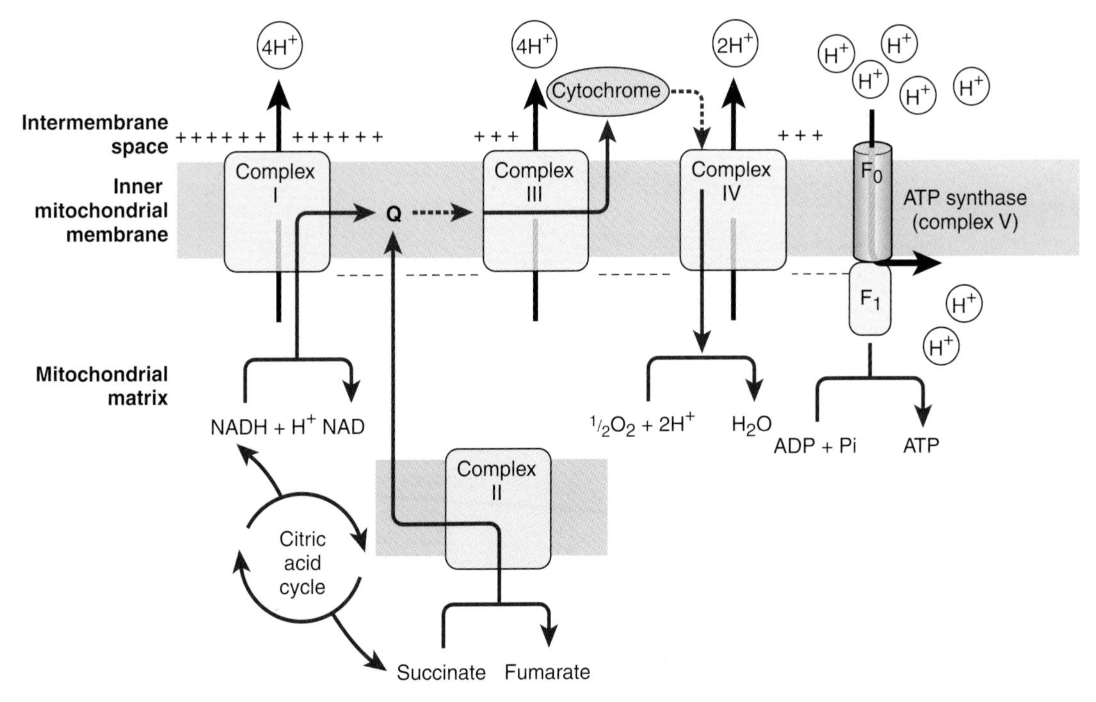

Figure 6-5. Overview of Oxidative Phosphorylation. Electrons are accepted from NADH (Complex I), succinate (Complex II), reduced Q protein via Complex II-generated FADH2 (Complex III), and from cytochrome *c* (Complex IV). The resulting excess of hydrogen drives ATPase synthase (Complex V) to produce ATP. The process uses molecular oxygen (O_2) and the final product of this series of reactions is water (H_2O). [Adapted with permission from Murray RA, et al.: Harper's Illustrated Biochemistry, 28th edition, McGraw-Hill, 2009.]

Specifically, oxidative phosphorylation includes both the respiratory chain (i.e., oxidative) and ATP synthesis via ATP synthase [i.e., phosphorylation of adenosine diphosphate (ADP)]. The respiratory chain portion involves four complexes (complexes I–IV) and a Q protein that help to transport H^+ from the inside of the mitochondria to the intermembrane space between the inner and outer membranes. These H^+ ions can then readily diffuse to the cytoplasm so that the pH in the intermembrane space and the cytoplasm become equivalent. Complex II is actually the same enzyme seen in the citric acid cycle (succinate dehydrogenase; Figure 6-4) and is responsible for the generation of one $FADH_2$ and a molecule of fumarate for each molecule of succinate. **Complex I** accepts electrons from NADH, **complex II** accepts electrons from succinate, and **complex III** accepts electrons from reduced **Q protein** that itself receives electrons from $FADH_2$ generated from complex II. **Complex IV** accepts electrons from **cytochrome** *c* molecules (see Figure 6-5). The process uses molecular oxygen (O_2) and the final product of this series of reactions is H_2O.

The four complexes are thought to be associated into a "super" complex, allowing the H^+ to be channeled from one complex to another; if this concept is correct, oxidative phosphorylation illustrates how association of several enzymes together can increase their efficiency, a theme that will be repeated several times in human biochemistry. ATP synthase (also called complex V) then transports the H^+ back through the inner membrane into the matrix, utilizing the energy from the H^+ concentration gradient to form ATP (Figure 6-5).

Toxins and Poisons—Blocking Oxidative Phosphorylation: Because oxidative phosphorylation represents a key step in the production of biological energy, it is also the target of various chemical agents. **Cyanide** and **carbon monoxide**, both toxic substances, interfere with hemoglobin molecules but also specifically inhibit complex IV, blocking the respiratory chain. Tissues highly dependent on energy production (e.g., nervous and heart muscles) are particularly affected, resulting in headache, dizziness, seizures, coma, and cardiac arrest leading to death. Antidotes (e.g., nitrites and sodium thiosulfate) work by causing the release of complex IV from the cyanide molecule followed by metabolism of the nontoxic product. Some insecticides (e.g., rotenone) also work via blocking specific complexes of oxidative phosphorylation.

NADH is often produced outside the **mitochondrial matrix** (e.g., glycolysis, conversion of lactate to pyruvate) but cannot cross the inner mitochondrial membrane for conversion to ATP. To address this problem, the body utilizes two molecular "shuttle" mechanisms. The first relies on the enzyme **glycerol-3-phosphate**

ATP Synthase: The enzyme responsible for production of ATP offers unique insights into protein conformational changes in a membrane-bound protein. **ATP synthase** is composed of two subunits, a cylindrical-shaped $\mathbf{F_0}$ protein with a hollow central channel and a stalk and ball-shaped $\mathbf{F_1}$ **headpiece** subunit (see the figure below). The entire complex appears similar to and, in fact, works much like an electrical generator with H^+ replacing the flow of H_2O. The biochemical generator is complete with a third rod-shaped "stator" subunit, which holds the F_0 and F_1 subunits together (see the figure).

The F_0 subunit is composed of several smaller proteins, with their hydrophilic amino acids on the interior and hydrophobic amino acids adjacent to the membrane. When H^+ flow back into the interior mitochondrial membrane space, they cause rotation of the F_0 subunit within the surrounding membrane. Studies indicate that the H^+ may influence particular amino acid R-groups of the internal F_0 structure to create the rotation. The F_1 headpiece stalk, connected directly to the F_0 subunit, spins inside the stationary external ball portion (held in place by the stator protein subunit), producing a conformational change that binds ADP, catalyzes the reaction to form ATP, and then releases ATP. At the end of the rotating cycle, the ball-shaped subunit is ready for a repeat round of ATP synthesis.*

ATP Synthase F_0–F_1 Complex

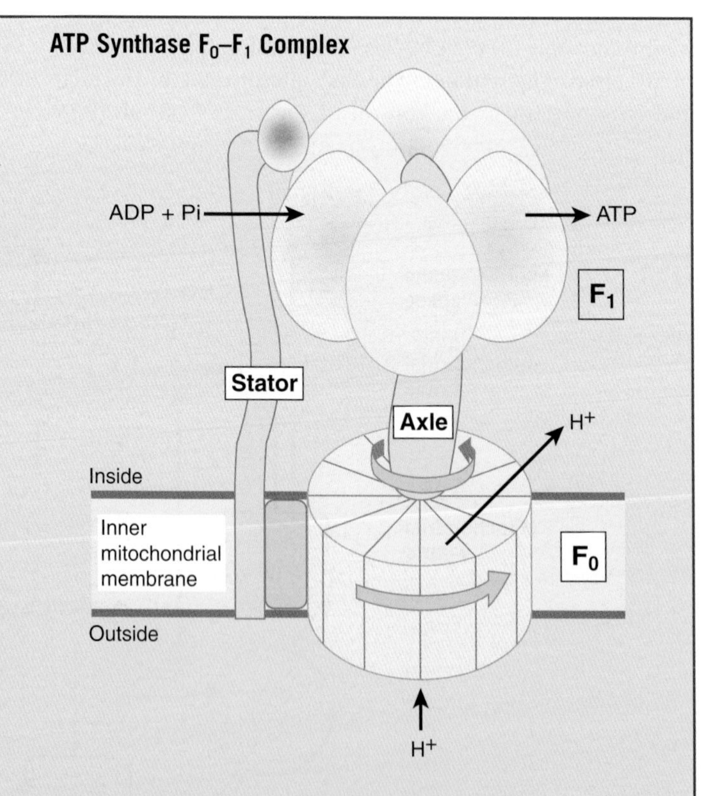

[Adapted with permission from Murray RA, et al.: Harper's Illustrated Biochemistry, 28th edition, McGraw-Hill, 2009.]

*Controversy surrounds the actual number of ATP molecules generated by oxidative phosphorylation with suggestions ranging from two to three ATP per NADH (and numbers in between). $FADH_2$ produces only about two ATP molecules because of its later entry at complex III into oxidative phosphorylation. Some scientists feel that the number may actually change depending on cellular conditions. This book has adopted the convention of exactly three ATP per NADH and exactly two ATP per $FADH_2$.

dehydrogenase, which has both cytoplasmic and inner mitochondrial membrane isoforms. The cytosolic enzyme converts glycerol-3-phosphate into dihydroxyacetone phosphate by oxidizing NADH to NAD^+. The mitochondrial enzyme captures the energy by catalyzing the oxidation of dihydroxyacetone phosphate via the conversion of flavin adenine dinucleotide to $FADH_2$. $FADH_2$ then continues through the oxidative phosphorylation pathway as above to yield two ATP molecules. A more efficient shuttle involves conversion of oxaloacetate to malate by reduction with NADH in the cytosol; this reaction is catalyzed by an isozyme of malate dehydrogenase from the citric acid cycle except in reverse (Figure 6-4). The malate is transported into the mitochondrial matrix where the reverse reaction takes place, resulting in NADH that yields three ATP molecules. Malate is ultimately converted to aspartate, which returns the carbons to the cytosol.

GLUCONEOGENESIS

Gluconeogenesis, the creation of new glucose, consists of 11 steps (some simply the reverse of reactions found in glycolysis) and is the first anabolic or "synthetic" carbohydrate pathway to be discussed. The process takes place almost exclusively in the liver, although some production of new glucose occurs in the kidney. Other tissues (e.g., muscles) contain many of the gluconeogenic enzymes with the exception of the final step that actually produces glucose. In muscles, these reactions allow for the conversion of pyruvate to glycogen.

Gluconeogenesis occurs when the body has low energy stores (e.g., starvation, lengthy periods of exercise) and elects to divert non-carbohydrate molecules [e.g., lactic acid, glycerol, and amino acids (except lysine and leucine)] into ATP molecules (see Chapter 10). Points of entrance by these molecules are shown in Figure 6-6. Any of the intermediates of the citric acid cycle and molecules that feed into the cycle can also be diverted to gluconeogenesis via oxaloacetate. The basic gluconeogenesis pathway (Figure 6-6) shows the conversion of pyruvate to glucose. Referring back to the glycolysis pathway above (Figure 6-2), two pyruvate molecules are formed by the enzymatic breakdown of each glucose molecule. Therefore, two pyruvate molecules are required by gluconeogenesis to form one glucose molecule.

However, gluconeogenesis is not simply a reversal of glycolysis. First, gluconeogenesis is a far more energy-expensive pathway, utilizing four ATP (two at each of two steps), two GTP, and two NADH (the equivalent of three additional ATP molecules per NADH) per glucose produced; a simple reversal of glycolysis would use two fewer ATP and GTP molecules each. This energy investment does not include that used by the substrate molecule to enter gluconeogenesis. Second, four instead of three different enzymes are seen in gluconeogenesis at reactions important to drive formation and regulation of glucose production. In order to drive the energetically favorable glycolysis pathway in reverse, the conversion of each pyruvate to phosphoenolpyruvate requires one ATP and one GTP molecule per pyruvate (i.e., four total high-energy phosphate molecules per glucose molecule produced) as well as two enzymes specific to the gluconeogenic pathway, the enzymes pyruvate carboxylase and phosphoenolpyruvate carboxykinase. Third, the next steps to fructose-1,6-bisphosphate are glycolysis in reverse and, hence, occur in the cytoplasm. Although glycolysis generates NADH and ATP within this portion of the pathway, gluconeogenesis requires the net consumption of two ATP and two NADH molecules per glucose molecule produced. Fourth, because gluconeogenesis cannot regenerate the ATP used in the reverse direction (phosphofructokinase-1) during glycolysis, the conversion of fructose-1,6-bisphosphate back to fructose-6-phosphate requires a third new gluconeogenic enzyme, fructose-1,6-bisphosphatase. This reaction is driven by the energy from cleavage and release of an inorganic phosphate (Pi). The next to last step is a simple reversal of the glycolysis reaction. Lastly, the final conversion of glucose-6-phosphate to glucose occurs in the lumen of the endoplasmic reticulum, with glucose being transported into the cytoplasm and, subsequently, into the blood to maintain its concentration during starvation or stress. Again, energy release of Pi drives the reaction forward.

As the body does not want to "waste" energy-converting molecules into glucose; gluconeogenesis is highly regulated, in particular with respect to glycolysis (see Chapter 10). Not surprisingly, gluconeogenesis is regulated at the exact points important for glycolysis but by different enzymes. **Pyruvate carboxylase, phosphoenolpyruvate carboxykinase, fructose-1,6-bisphosphatase**, and **glucose-6-phosphatase** replace the glycolysis enzymes pyruvate kinase, phosphofructokinase-1, and hexokinase or glucokinase. High acetyl-CoA and ATP, representative of high-energy stores in the body, activate pyruvate carboxylase, the first step in gluconeogenesis, whereas the analogous glycolytic step, pyruvate kinase, the final step in glycolysis is inhibited, insuring that only one pathway—glycolysis or gluconeogenesis—is functional at any one time. Primary regulation of gluconeogenesis in humans occurs at phosphoenolpyruvate carboxykinase via cortisol and glucagon effects on DNA transcription levels of the enzyme. High concentrations of citrate, evidence of plentiful ATP production and carbon availability, and low amounts of fructose-2,6 bisphosphate cause increased activity of the enzyme, fructose-1,6-bisphosphatase, helping to regulate gluconeogenesis in mammals; in non-mammal species, this enzyme is, in fact, the primary regulator of gluconeogenesis. Further details of regulation are discussed in Chapter 10.

Gluconeogenesis and Metformin: Most patients with **type 2 diabetes mellitus** have gluconeogenesis rates in the liver three times that of persons without this disease. **Metformin**, one of the family of **biguanide** medications, works in the liver by specifically inhibiting gluconeogenesis. Although several mechanisms of biguanide action may be at work, evidence points to an increase in cytosolic adenosine monophosphate, indicating to the body a low ATP concentration and prompting use of glucose via glycolysis. Metformin also increases the body's sensitivity to insulin, increases cellular uptake of circulating glucose, the breakdown of fatty acids (all which promote glucose use instead of new production), and decreases glucose absorption by the intestines.

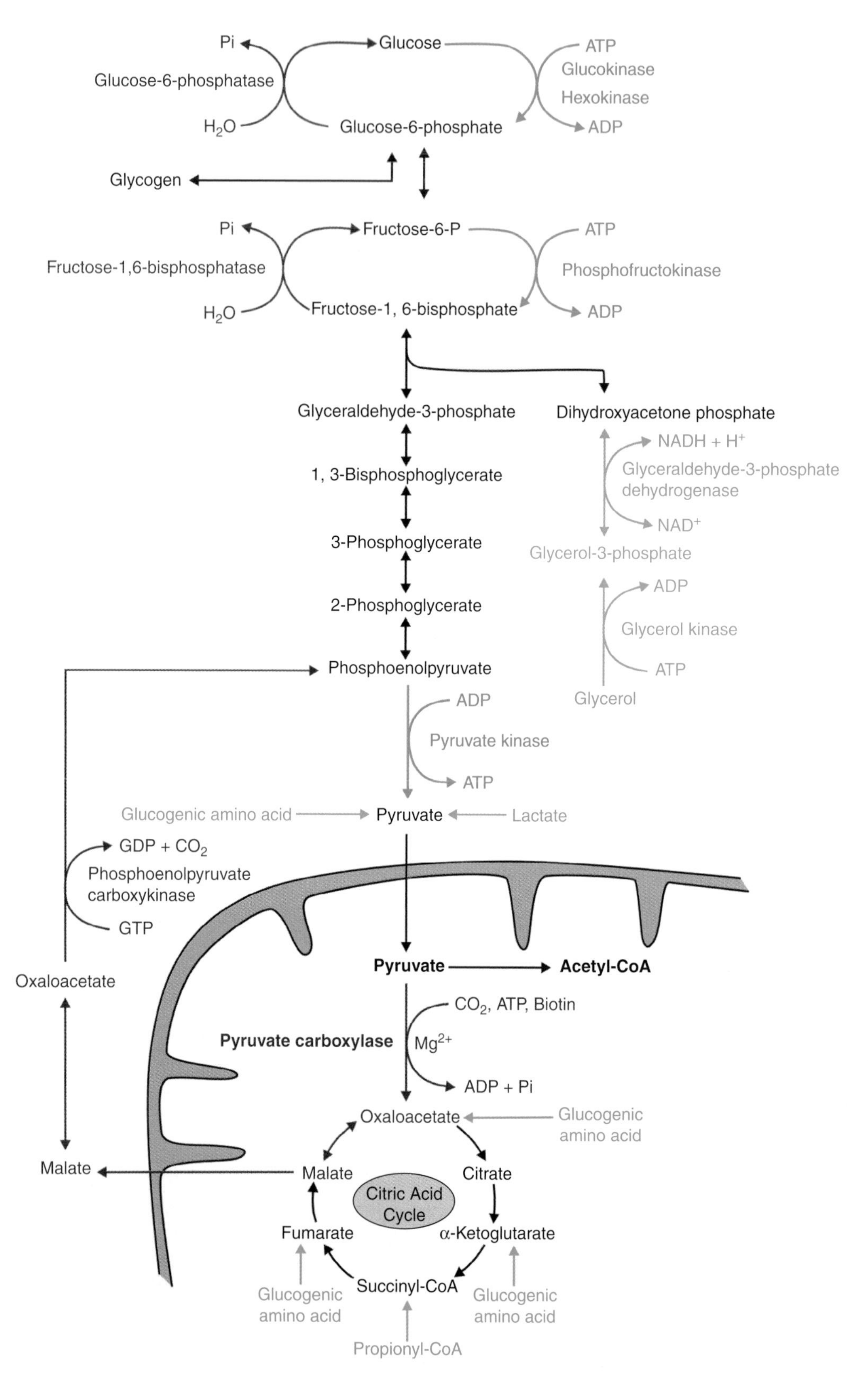

Figure 6-6. Gluconeogenesis. The pathway of gluconeogenesis includes 11 enzymatic steps, which form one molecule of glucose from two molecules of pyruvate. Reactions and enzymes specific only to gluconeogenesis are shown in red. Irreversible reactions specific for glycolysis are shown in green. Additional substrates for gluconeogenesis are shown in blue and may also enter gluconeogenesis as denoted by the associated arrows. [Adapted with permission from Naik P: Biochemistry, 3rd edition, Jaypee Brothers Medical Publishers (P) Ltd., 2009.]

THE PENTOSE PHOSPHATE PATHWAY

The second "synthetic" carbohydrate pathway to be discussed is the **pentose phosphate pathway**, sometimes referred to as the pentose phosphate shunt. This pathway takes place in the cytoplasm and produces five-carbon, pentose carbohydrate molecules for nucleotide and nucleic acid synthesis. It is also the major pathway for the production of the important reducing molecule **nicotinamide adenine dinucleotide phosphate (NADPH)**, a similar molecule to NADH but with a very different function. NADPH has an additional phosphate group on the carbon 2 position of the adenine ribose molecule. It is the primary molecule indirectly responsible for eliminating toxic oxygen "radicals" (e.g., hydrogen peroxide)—highly reactive oxygen molecules produced by some reactions in the body, which can cause significant damage to tissues. NADPH is also required for certain reactions in the synthesis of nucleic acids (Chapter 3) as well as fatty acids, cholesterol and steroids, and nucleic acids (Chapter 7), which require a reducing molecule.

The pentose phosphate pathway occurs in two phases. The first, called the **oxidative phase**, produces NADPH and ribulose-5-phosphate, a pentose sugar molecule (Figure 6-7A). The second, called the **nonoxidative phase**, produces products for glycolysis (Figure 6-7B) or for nucleotide synthesis (Figure 6-7C). Regulation of this pathway is via NADPH concentration—when $NADP^+$ concentration is high, the oxidative pathway is accelerated; when NADPH concentration is high, the oxidative pathway is inhibited. Normal concentrations of NADPH in the cytoplasm are 100 times that of NADP, so the pathway will only be turned on when NADPH stores are used.

Glucose-6-phosphate Dehydrogenase (G6PDH) Deficiency: Abnormally low levels of **G6PDH** is the most common inherited disease of enzyme deficiency, affecting an estimated 400 million persons worldwide, mainly those of African, Middle Eastern, or South Asian descent. The disease is inherited via the X chromosome in a recessive manner and, therefore, affects males much more often than females. However, females with two affected X chromosomes suffer from an immune disorder called **chronic granulomatous disease**, which blocks the action of immune cells that utilize the oxygen radicals to kill bacteria, often leading to death due to infection. Multiple mutations can affect G6PDH activity with five different classes of the disease clinically defined.

This pathway is the only way that red blood cells may eliminate reactive oxygen molecules, which can seriously damage the red blood cell membrane and cell wall and lead to red blood cell death. As a result, any problems with the pentose phosphate pathway, including G6PDH deficiency, can be detrimental to red blood cells and the body. Although most individuals with G6PDH deficiency are asymptomatic, severely affected patients display symptoms of red blood cell breakdown resulting in liver and kidney problems. In newborn babies, this may result in damage to the brain, termed **kernicterus**, and death.

Diagnosis is by signs of red blood cell breakdown and liver problems and is confirmed by a direct assay of G6PDH activity. Individuals diagnosed with the deficiency must avoid use of a large number of **sulfa drugs**, including various antibiotics; a burn medication; several anticonvulsants; thiazide and loop diuretics (used to treat high blood pressure), a class of medications used to treat diabetes; some glaucoma medications; certain pain medications; and, finally, several antimalarial medications. Treatment of G6PDH deficiency is avoidance of agents, which cause red blood cell destruction and, in some cases, blood transfusions and/or removal of the spleen, the organ responsible for removal of damaged and destroyed red blood cells.

However, one advantage of moderate G6PDH deficiency has been noted in resistance to malarial infection by *Plasmodium falciparum*. In certain individuals, usually of African or Mediterranean descent, a decreased level of G6PDH activity leads to weakening of the red blood cell membrane, which does not affect the person but does stop reproduction and further infection by the malaria parasite. Persons with sickle cell disease also enjoy a heightened resistance to malaria infections. A suggested mechanism is the selected clearance by the spleen of the red blood cells partially damaged by the malarial parasite, whereas uninfected red blood cells are preserved.

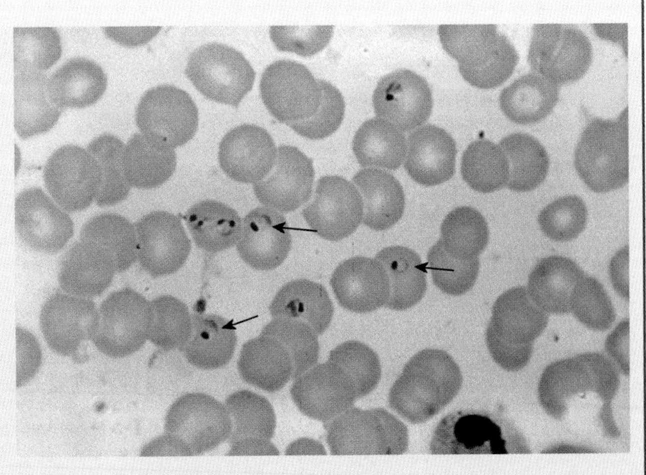

Arrows indicate ring phase of *Plasmodium falciparum* in infected cells.
[Reproduced with permission from Chamberlain NR: The Big Picture: Medical Microbiology, 1st edition, McGraw-Hill, 2009.]

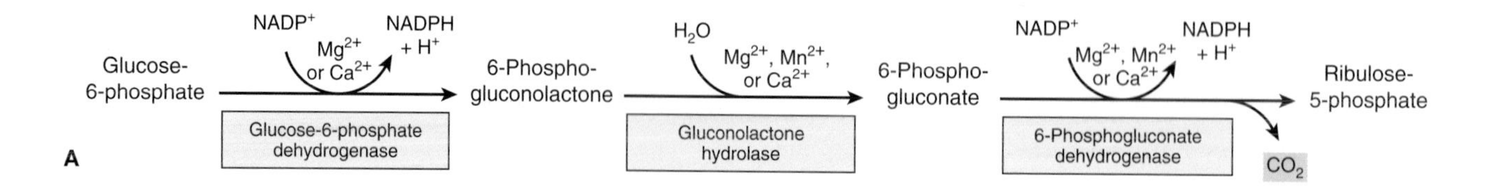

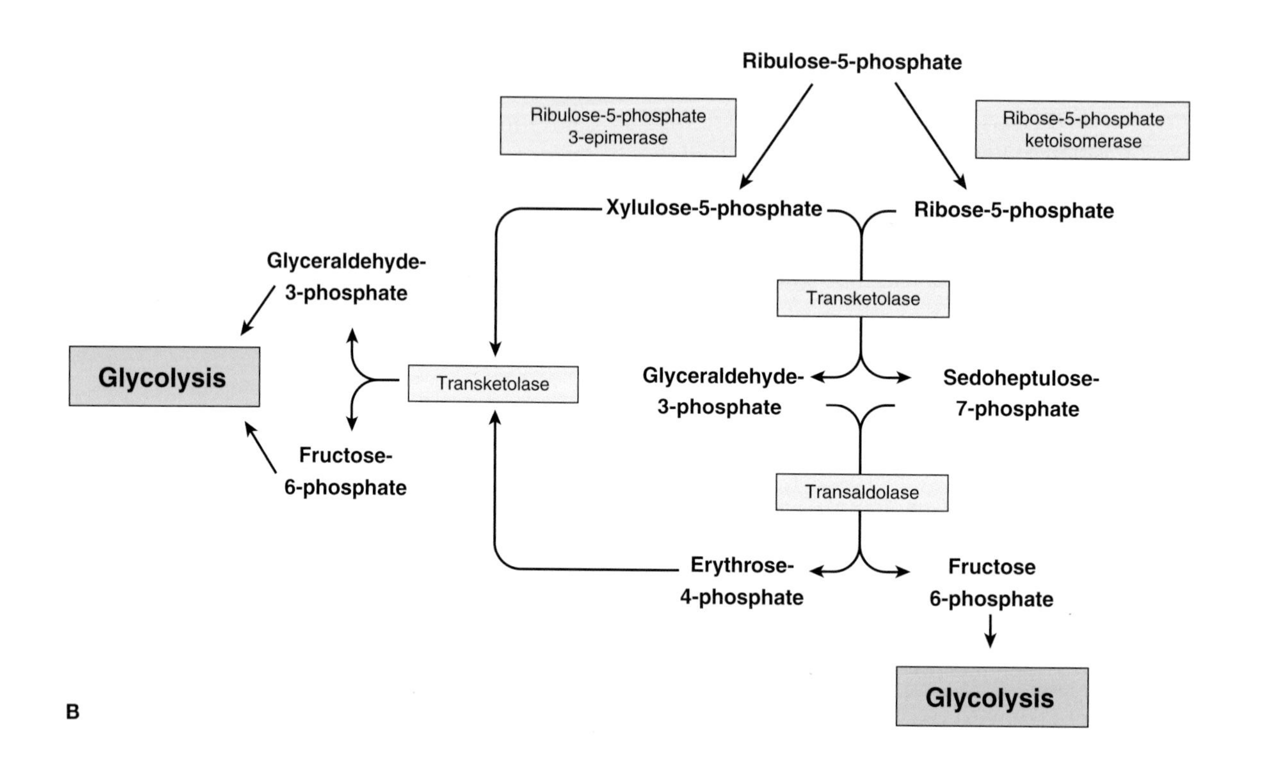

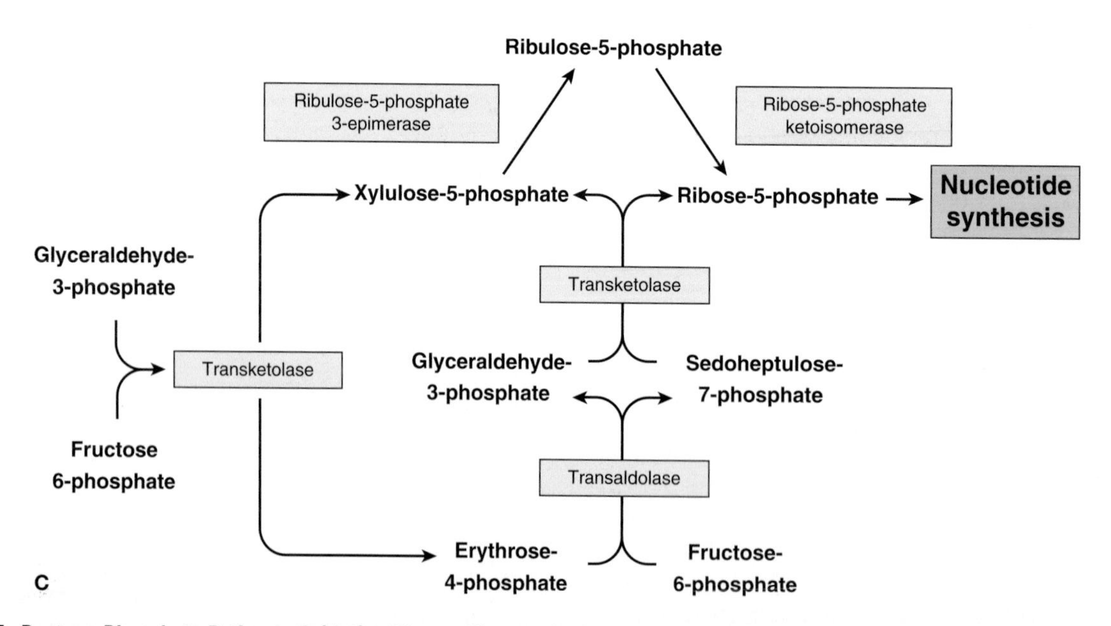

Figure 6-7. A. Pentose Phosphate Pathway, Oxidative Phase. Glucose-6-phosphate is converted into ribulose-5-phosphate with generation of two molecules of NADPH and one molecule of CO_2 (blue shading). Various cations (Mg^{2+}, Mn^{2+}, and Ca^{2+}) play roles as cofactors in these reactions. The overall reaction for this process is: glucose-6-phosphate + 2 $NADP^+$ + H_2O $\rightarrow$ ribulose-5-phosphate + 2 NADPH + 2 H^+ + CO_2. **B–C. Pentose Phosphate Pathway, Nonoxidative Phase**. Ribulose-5-phosphate is further converted to 3-, 4-, 5-, 6-, and 7-carbon carbohydrates, which can be used for glycolysis (**B**), or in a reversal of reactions, glycolytic intermediates can be used for nucleotide synthesis (**C**). Enzymes involved in the reactions are indicated. [Adapted with permission from Murray RA, et al.: Harper's Illustrated Biochemistry, 28th edition, McGraw-Hill, 2009.]

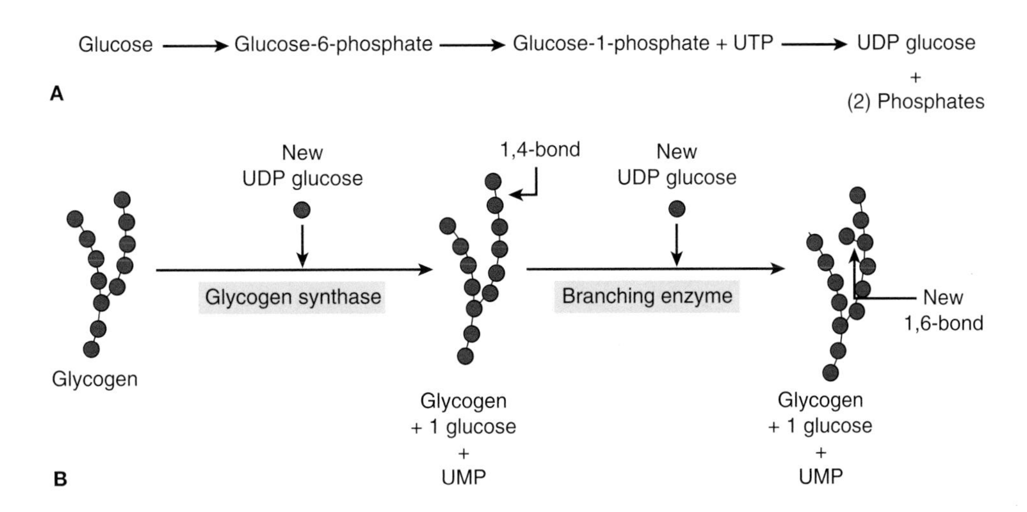

Figure 6-8. A–B. Glycogen Synthesis. A. The synthesis of glycogen begins with the conversion of glucose to glucose-1-phosphate and subsequent bonding to uridine triphosphate (UTP) to form uridine diphosphate–glucose (UDP glucose) and two phosphate groups. **B.** UDP glucose serves as the source for the addition of new glucose molecules to an existing glycogen molecule either via a 1,4-bond or a 1,6-bond with a resulting uridine monophosphate (UMP) molecule. Enzymes are shown in blue-shaded boxes. [Adapted with permission from Naik P: Biochemistry, 3rd edition, Jaypee Brothers Medical Publishers (P) Ltd., 2009.]

GLYCOGEN SYNTHESIS

Step 1 of glycolysis, production of **glucose-6-phosphate** catalyzed by **hexokinase**, involves an investment of one ATP molecule and is, therefore, a committing step toward utilization of the glucose molecule. If energy (ATP) stores are high, the body may decide to store the glucose molecule in the form of **glycogen** (Chapters 2 and 10; Figure 6-8). The first step of this process is the conversion of glucose-6-phosphate to glucose-1-phosphate. Glucose-1-phosphate reacts with the nucleotide uridine triphosphate (one of the nucleotides in RNA) to form uridine diphosphate glucose (UDP glucose), the reaction depends on the energy released from the two phosphate bonds hydrolyzed during this reaction (Figure 6-8A). Glucose from UDP glucose is then attached to a preexisting glycogen molecule (Figure 6-8B), which includes a protein called **glycogenin** as the core of the growing glycogen chain. The new bond is formed between carbon 1 of the glycogen molecule and carbon 4 of the new glucose. A separate "branching" enzyme forms bonds between carbons 1 and 6, approximately every 10 glucose molecules.

Glycosyltransferase Inhibitors: The enzymes that link carbohydrate molecules together to form glycogen and similar complex sugar molecules are collectively known as **glycosyltransferases**. This family of enzymes is essential to the efficient storage of carbohydrate energy, as well as the synthesis of important structural proteins. As a result, inhibitors of glycosyltransferases have been developed for many medical and other uses. The drug **caspofungin** kills **fungi** by inhibiting bonds between carbons 1 and 3 of glucose molecules in fungal cell walls, leading to their death. **Ethambutol** blocks growth of the bacteria that cause tuberculosis by blocking carbon 5 on the carbohydrates of the bacteria's cell wall. Other glycosyltransferase inhibitors are also being developed as agents against cancer and viral infections, including against the human immunodeficiency virus (HIV).

GLYCOGEN BREAKDOWN

Glycogenolysis, the breakdown of glycogen to release glucose, is an important pathway in energy use and reactions that utilize carbohydrates. Glucose-6-phosphate is the entry point of sugars liberated from glycogen catabolism into glycolysis, thus bypassing the first regulatory step of hexokinase. Glucose from the digestion of starch molecules, a major part of the human diet, also enters glycolysis at this point.

Glycogenolysis involves three steps (Figure 6-9). First, the repetitive removal of glucose residues by **phosphorylase** breaking bonds between carbons 1 and 4, producing a glucose-1-phosphate molecule. However, this process stops when four glucose residues are left before the carbons 1 and 6 branch points. Second, a **glucan transferase** enzyme moves three residues to the linear part of the glycogen chain, leaving only the final branched glucose residue. Third, a **debranching enzyme** cleaves the carbons 1 and 6 bond. A final enzyme converts the resulting glucose-1-phosphate molecules into glucose-6-phosphate (not shown) for entry into glycolysis.

Removing glucose residues from the linear chain and branch point residues continues as long as the body requires them. The process is controlled by the addition of a phosphate to (leading to activation) or removal of a phosphate from (leading to inhibition) the first enzyme in the process. Phosphorylation of this enzyme is, in turn, under the control of the hormones **epinephrine** and **glucagon**, both of which promote activation of glycogenolysis to produce new glucose in liver. The liver is the primary organ involved in glucose utilization or storage for the body (see Chapters 10 and 11 for further discussion). As a result, the liver contains an enzyme, glucose-6-phosphatase, which removes the phosphate from glucose, allowing it to leave the liver cell and be transported to other tissues. Muscles, too, can store glucose in the form of glycogen but do not contain glucose-6-phosphatase. Glucose released during glycogenolysis is, therefore, used by the muscle cells under control by epinephrine but not by glucagon.

Glycogenolysis Inhibitors: As with glycosyltransferases, the enzymes responsible for breakdown of glycogen and other complex carbohydrate molecules are also the targets of several medications. **Acarbose** and **miglitol**, two medications used in the treatment of diabetes, inhibit the release of new glucose residues in the small intestine and pancreas, thereby reducing the availability of these glucose molecules to the body. **Zanamivir** (trade name **Relenza**) and **oseltamivir** (trade name **Tamiflu**) are both inhibitors of a reaction involving cleavage of carbohydrate bonds that allows new influenza A and B virus particles to be released from an infected cell. As a result of these medications, the new viral particles are trapped and soon die. Because of the mechanism involved, these medications must be taken at the very beginning of a suspected influenza infection before replication and release of new viruses takes place.

Because of the importance to the proper use of carbohydrate and other energy sources, defects in the pathways of glycogen synthesis and breakdown within the liver and muscles, collectively known as **glycogen storage diseases** (GSDs), can cause major medical problems. As further research has identified specific enzyme deficits, several GSDs have been combined. Ten GSDs have been identified and are listed in Table 6-1.

MODIFIED CARBOHYDRATES (GLYCOPROTEINS, GAGS)

Glycoproteins and **GAGs**, discussed in Chapter 2, are proteins with a chain of carbohydrates, known as an **oligosaccharide**, attached to a core protein after protein synthesis. These carbohydrate molecules can be attached to the amino (NH_3) group of an asparagine amino acid (known as "***N*-glycosylation**") or to the hydroxyl (OH) group of amino acids serine, threonine, or specially modified hydroxylysine or hydroxyproline residues (known as "**O-glycosylation**"). Enzymes from the family of glycosyltransferases, also seen in glycogen synthesis (see above), catalyze the additions of carbohydrates to the protein molecules. Only eight carbohydrate molecules are usually seen in glycoproteins—glucose, galactose, mannose, fructose, xylose, *N*-acetylgalactosamine, *N*-acetylglucosamine, and *N*-acetylneuraminic acid. GAG carbohydrates can vary widely but usually include galactose, glucosamine, glucuronic acid, and/or iduronic acid. Specific glycosyltransferases are responsible for addition of a particular carbohydrate molecule to a specific amino acid residue or to the growing oligosaccharide chain.

The resulting glycoproteins/GAGs perform important functions both in the cytoplasm and in the membrane (glycosylated portions of the protein usually point out into the extracellular environment), the latter is important as cell–cell signaling molecules. Functions of GAGs are listed in Table 2-2; glycoprotein functions are reviewed in Table 6-2.

Glucosamine/Chondroitin and Osteoarthritis: Certain GAGs serve structural roles in joint cartilage, possibly acting as molecular "shock absorbers." **Osteoarthritis** is a disease characterized by decreased cartilage in weight-bearing joints such as knees, hips, and back due to "wear-and-tear" on these joints. One therapy for osteoarthritis involves the oral replacement of **glucosamine** and **chondroitin** by tablets. Although evidence is lacking for true effectiveness, studies have shown that these tablets pose no health threat to patients and many note marked improvement in osteoarthritis pain.

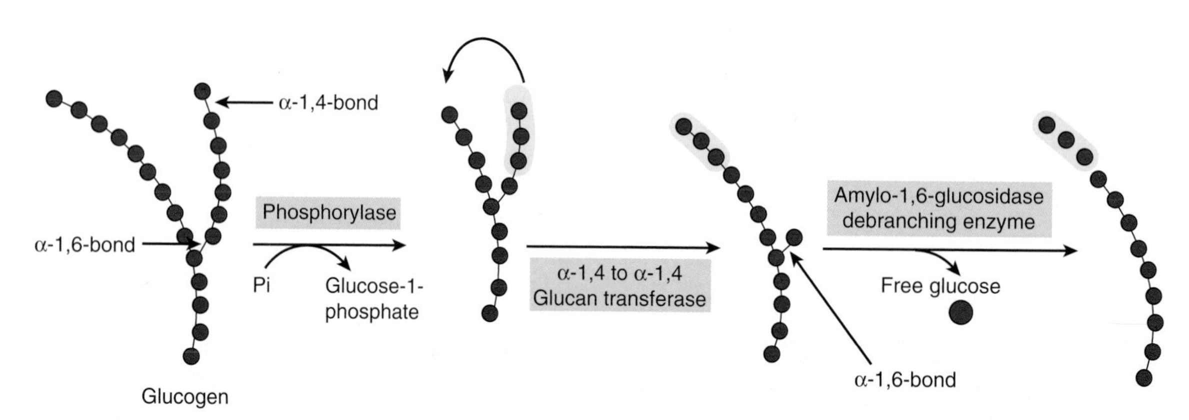

Figure 6-9. Glycogen Breakdown. The breakdown of glycogen proceeds by glycogen phosphorylase cleavage of the α 1,4-bonds until four residues remain before the α 1,6-bond. Glucan transferase then moves three residues to the other chain. Finally, the remaining α 1,6-bound glucose is released as free glucose by a debranching enzyme. Resulting glucose-1-phosphate molecules are subsequently converted to glucose-6-phosphate for reentry into glycolysis (not shown). [Adapted with permission from Naik P: Biochemistry, 3rd edition, Jaypee Brothers Medical Publishers (P) Ltd., 2009.]

TABLE 6-1. Glycogen Storage Diseases

Name	Enzyme Deficiency	Symptoms	Treatment
GSD type 0	Glycogen synthase (affects only liver)	Fasting low sugar, lactate, and alanine as well as high ketones and fatty acids. During feeding, high sugar and lactate. Can have muscle cramping.	Diet changes to avoid low and high blood sugar levels (avoid processed carbohydrates to prevent glucose conversion to lactate, include high-protein meals and nighttime consumption of uncooked cornstarch [slow-release form of glucose]).
Von Gierke's disease (GSD type I)	Glucose-6-phosphatase	Low sugar, enlarged liver and kidney (due to glycogen accumulation) and liver problems, increased lactate/uric acid/triglycerides, and growth failure and developmental delay.	Diet changes to avoid low and high blood sugar levels. Frequent high glucose or cornstarch feeds every 3–4 h with approximately one-third of total daily carbohydrates at night. Avoid high lactate, uric acid, and triglyceride levels.
Pompe's disease (GSD type II)—infant, juvenile, and adult-onset forms	Acid maltase (also affects lysozymes)	Glycogen accumulation leads to progressive muscle weakness (affecting primarily heart and skeletal muscles) and enlarged heart and liver. Enlarged tongue may cause swallowing problems. If untreated, infant form can lead to death usually by the age of 2 years.	Enzyme replacement using myozyme (alglucosidase alfa). Gene replacement therapy is also being developed.
Cori's or Forbe's Disease (GSD Type III)	Glycogen debranching enzyme	Glycogen accumulation in liver, skeletal muscle and/or heart, leading to associated medical problems. Fasting low sugar.	High protein diet to promote gluconeogenesis.
Andersen's Disease (GSD type IV)	Glycogen branching enzyme	Accumulation of long and unbranched glycogen molecules (i.e., amylose), which precipitate in the liver and heart, leading to failure and death usually by the age of 1 year.	Symptomatic treatment of liver and heart failure. Transplantation as required.
McArdle's disease (GSD V)	Muscle glycogen phosphorylase	Inability to perform glycogenolysis leads to muscle cramping and breakdown from exercise. Can result in kidney failure due to muscle breakdown products.	Limited exercise to avoid muscle breakdown. Treatment of kidney disease as required.
Hers' disease (GSD types VI, VIII, and X)	Liver glycogen phosphorylase	Glycogen accumulation in liver, leading to enlargement. Growth retardation. Low blood sugar, high ketones, cholesterol, and fatty acids are possible.	Usually mild effect on overall health. Diet management as required.
Tarui's disease (GSD type VII)	Muscle phosphofructokinase	Inability to perform glycogenolysis leads to muscle cramping and breakdown from exercise. Low red blood cell levels due to breakdown. Can result in kidney failure due to breakdown products.	Often worse effect on health than GSD V. Limited exercise to avoid muscle breakdown. Treatment of kidney disease as required.
GSD Type IX	Phosphorylase kinase	Glycogen accumulation in liver, muscle, and rarely heart, leading to enlargement and disease. Growth retardation and delayed muscle development/function. Low blood sugar, high ketones, cholesterol, and fatty acids are possible.	Symptomatic treatment. Diet changes to avoid low and high blood sugar levels, including high intake of starches and protein and avoidance of fasting.
Fanconi–Bickel syndrome (GSD type XI)	Glucose transporter (GLUT2; primary transport protein for glucose entrance into cells)	Low sugar, enlarged liver and kidney (due to glycogen accumulation) and liver problems, increased lactate/uric acid/triglycerides, and growth failure and developmental delay.	Regulate glucose intake and avoid low and high blood sugar levels. Frequent feeds high in glucose or cornstarch, occurring every 3–4 h with approximately one-third of total daily carbohydrates at night. Avoid high lactate, uric acid, and triglyceride levels.

TABLE 6-2. Biochemical Roles of Glycoproteins

Glycoprotein(s)	Location(s) in Human Body	Function(s)
Mucins ("mucus glycoproteins")	Lungs, saliva, digestive tract, connective tissue	Lubrication (lung, connective tissue) and protection from bacterial invasion (digestive tract). Increased in cancers of the pancreas, lung, breast, ovary, and colon and lung diseases such as asthma, bronchitis, chronic obstructive pulmonary disease, and cystic fibrosis. Decreased levels may cause development of ulcerative colitis.
Antibodies, major histocompatability complex	White blood cells (immune system)	Specific glycosylated proteins serve as recognition/activation molecules for various cells of the immune system.
Glycoprotein IIb/IIIa	Platelets	Platelet aggregation and clot formation. Deficient or altered glycoprotein IIb/IIIa can lead to serious bruising/bleeding problems (Glanzmann's thrombasthenia, idiopathic thrombocytopenic purpura).
Hormones	Various tissues and functions	Follicle-stimulating hormone, luteinizing hormone, thyroid-stimulating hormone, human chorionic gonadotropin, alpha-fetoprotein, erythropoietin.
Transferrin, ceruloplasmin	Blood	Iron and copper transport throughout the body.
Calnexin and calreticulin	Endoplasmic reticulum	Ensure proper folding and glycosylation of newly synthesized glycoproteins.
Notch glycoproteins (receptors)	Nerve, arteries, heart, pancreas, intestines, bone, mammary glands, and cytoskeleton	Involved in cell development and differentiation in a variety of tissues. Altered Notch activity may lead to certain leukemias, multiple sclerosis, Tetralogy of Fallot, Alagille syndrome, and other diseases.
Various glycoprotein receptors and membrane-bound molecules	Sperm–egg, virus (e.g., HIV) and bacterial coats, lectins, selectins, hormones, drugs	Involved in cell recognition, attachment, and initiation of biological processes via signaling.

REVIEW QUESTIONS

1. What are the key features of glycolysis and gluconeogenesis and how are they similar and different?
2. What are the key features of the citric acid cycle?
3. What are the key features of oxidative phosphorylation and its relationship to the citric acid cycle and glycolysis?
4. What are the key features of the pentose phosphate pathway?
5. What are the key features of glycogen metabolism and how does it relate to glycolysis and gluconeogenesis?
6. How is carbohydrate metabolism regulated and how does this regulation relate to synthesis and degradation of the various carbohydrates important to humans and choices between other metabolic pathways?

CHAPTER 7

LIPID METABOLISM

OVERVIEW

Lipids perform several essential functions, including forming biological membranes, efficient storage of energy, and as components of several important structural and functional molecules. Lipid metabolism includes both the synthesis and degradation of fatty acids and/or more complex lipid molecules. The choice between synthesis and degradation represents an important regulatory step in human biology and reflects the level of food and, therefore, energy stores available to the body. Several processes are involved in this decision of producing fatty acids/lipids or, instead, directing their precursors to energy production via carbohydrate metabolic pathways. Separate but still dependent on this process, the production of cholesterol and several lipid-derived hormones and signaling molecules is essential for multiple functions in the human body. As in most metabolic pathways, deficiencies or problems can lead to serious disease states. As seen in other chapters, understanding of the exact mechanism of these problematic metabolic steps also allows treatment of these diseases.

FATTY ACID METABOLISM

FATTY ACID SYNTHESIS

Although most fatty acids needed by humans are supplied in the diet, fatty acid synthesis plays a role in certain tissues to convert any excess of sugar molecules into the more efficient storage form of fatty acid/lipid molecules and/or to produce specialized lipid molecules. The link between sugar and fatty acid/lipid metabolism is seen at acetyl coenzyme A (CoA), the intermediate between glycolysis and the citric acid cycle. As will be seen below, acetyl-CoA is also the initial molecule of fatty acid metabolism. Therefore, acetyl-CoA is an important branch point of human metabolism and its use is highly controlled to reflect the body's nutritional state and needs. As an illustration of this fact, the first and committing step of fatty acid synthesis is the addition of a carboxyl (CO_2^-) group, donated by HCO_3^-, to acetyl-CoA to produce **malonyl-CoA**.

$$\underset{\text{Acetyl-CoA}}{CH_3-\overset{O}{\overset{\|}{C}}-S-CoA} + \underset{\text{Bicarbonate}}{HCO_3^-} \xrightarrow[\text{Acetyl-CoA carboxylase}]{\text{ATP} \;\; \text{Biotin} \;\; \text{ADP + Pi}} \underset{\text{Malonyl-CoA}}{COO^- - CH_2 - \overset{O}{\overset{\|}{C}} - S - CoA} + H^+$$

Although the reaction is simple, the choice to commit the potential energy production from carbohydrates to produce lipid molecules reflects several important biological principles in human metabolism. First, the enzyme that catalyzes this reaction, **acetyl-CoA carboxylase**, is produced as smaller, inactive protein complexes. When the concentration of **citrate** is high (the first intermediate of the citric acid cycle and a sign of abundant sugar resources), these smaller complexes join to form active enzymatic polymers (Figure 7-1). Therefore, citrate increases fatty acid synthesis when the body has plentiful energy and needs to store this energy in an efficient fashion. If lipid stores are high, though, increased concentration of **palmitoyl-CoA** (the final product of fatty acid synthesis) decreases fatty acid synthesis by depolymerizing the active enzyme complex into the original smaller, inactive complexes. The balance between citrate and palmitoyl-CoA or, more simply, a measure of sugar versus fat levels (e.g., high sugar/fat indicates plentiful food supplies; low sugar/fat indicates low food and/or starvation conditions) controls the activity of this committing step and, therefore, utilizes available energy supplies most efficiently.

There is a second, longer term, hormonal regulation that allows the human body even further control of energy sources. **Glucagon**, the hormone that promotes the breakdown of glycogen to glucose, and/or **epinephrine**, released in times of

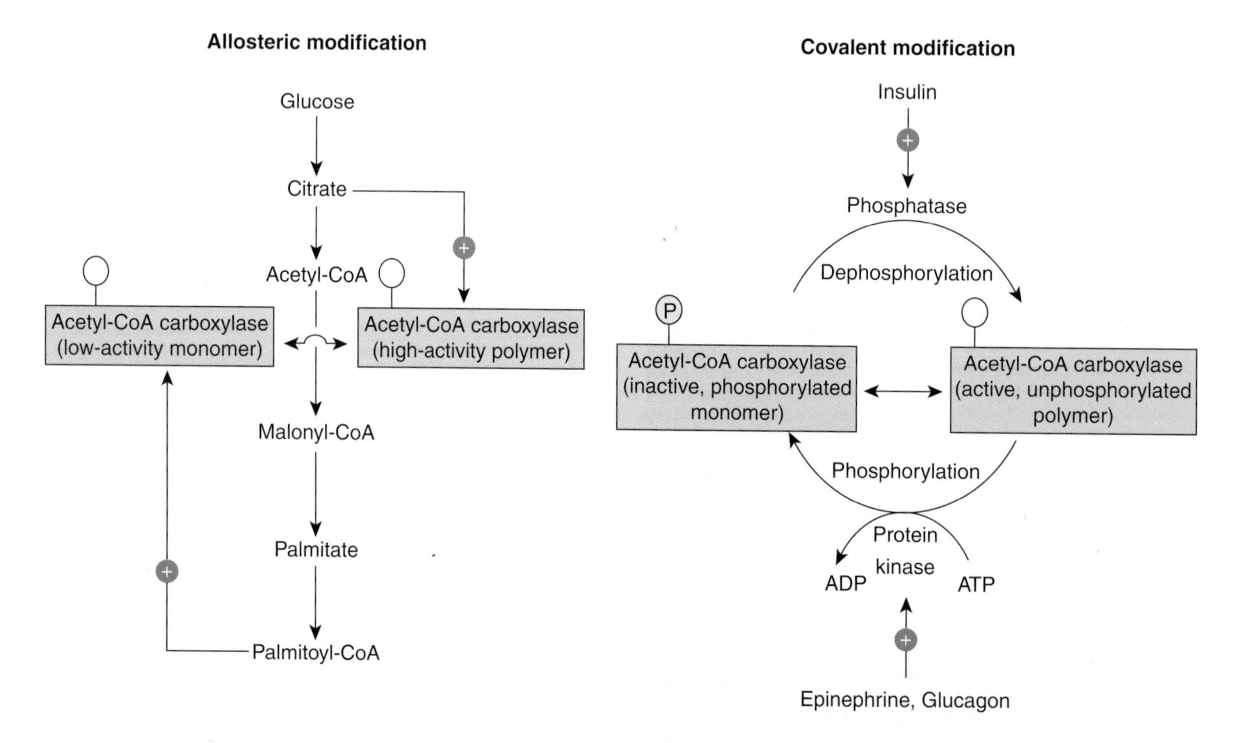

Figure 7-1. Regulation of the Acetyl-CoA Carboxylase and Production of Malonyl-CoA by Hormones. Palmitate and citrate regulate malonyl-CoA production allosterically by enhancing the formation of the low-activity monomer or the highly active polymer form of acetyl-CoA carboxylase, respectively. Both forms are unphosphorylated (♀). Insulin, epinephrine, and glucagon regulate acetyl-coA carboxylase activity by influencing the phosphorylation (inactivation, Ⓟ) or dephosphorylation (activation, ♀) of the monomer. [Adapted with permission from Naik P: Biochemistry, 3rd edition, Jaypee Brothers Medical Publishers (P) Ltd., 2009.]

stress when energy requirements are higher, activate phosphorylation of acetyl-CoA carboxylase, causing inhibition of fatty acid synthesis (Figure 7-1). Alternatively, **insulin**, the hormone that promotes uptake and storage of sugars (Chapter 10), causes dephosphorylation, which increases production of malonyl-CoA and, therefore, fatty acid synthesis. Therefore, glucagon and epinephrine drive acetyl-CoA molecules away from lipid storage and toward the production of energy via the citric acid cycle. Additionally, food deprivation or chronic high-fat diet decreases the amount of the enzyme that produces malonyl-CoA.

Following the committing step of malonyl-CoA production, fatty acid synthesis is a relatively simple pathway that takes place on a multienzyme–protein complex in the cytoplasm. This enzyme complex, called **fatty acid synthase**, contains seven separate subunits, catalyzing linked enzymatic reactions, and an acyl carrier protein "arm," which carries the growing fatty acid chain through to palmitate, the 16-carbon-long fatty acid with single carbon–carbon bonds. The overall reaction for production of one palmitate molecule, including the seven adenosine triphosphate (ATP) molecules to generate malonyl-CoA via acetyl-CoA carboxylase, is as follows:

$$\text{Acetyl-CoA} + 7\ \text{Malonyl-CoA} + 14\ \text{NADPH} + \text{H}^+ \xrightarrow{\text{Fatty acid synthase}} \text{Palmitate} + 7\ CO_2 + 14\ \text{NADP}^+ + 8\ \text{CoA}$$

Other fatty acids are derived from palmitate by specific enzymatic reactions. Production of fatty acids with an *odd* number of carbon atoms relies on the substitution of the three-carbon, **propionyl-CoA**, molecule for acetyl-CoA to initiate the fatty acid synthase reaction. The basic process of fatty acid production, which relies on the cofactor nicotinamide adenine dinucleotide phosphate (NADPH) to drive the reaction, is shown in Figure 7-2. Similar to the malonyl-CoA reaction, the quantity of the enzyme complex is positively regulated by higher sugar levels and negatively regulated by high fatty acid/fat levels in the diet.

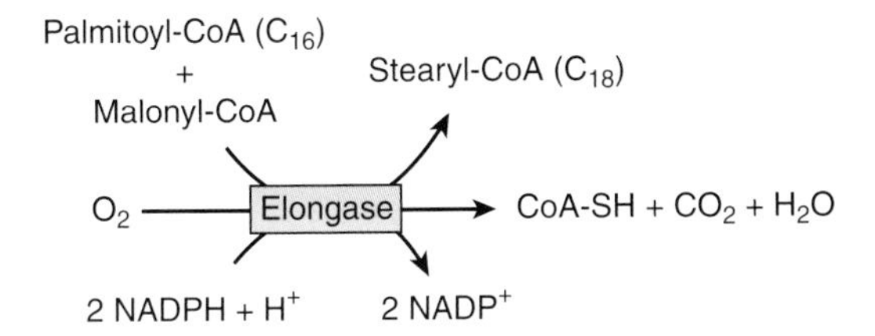

[Adapted with permission from Naik P: Biochemistry, 3rd edition, Jaypee Brothers Medical Publishers (P) Ltd., 2009.

Palmitate is used to produce longer fatty acid chains by the addition of more two-carbon units from malonyl-CoA as indicated by the equation below. This reaction mainly occurs in the endoplasmic reticulum by a four-step process including the enzyme **fatty acid elongase** with continued use of NADPH as a source of both energy and reducing power. Fatty acid elongation can also occur in the mitochondria (NADH dependent) and in selected tissues (see Sidebar) where particular fatty acid types are required.

Finally, the production of carbon–carbon double bonds in fatty acids takes place via a **desaturase** enzyme, utilizing NADH, in the endoplasmic reticulum.

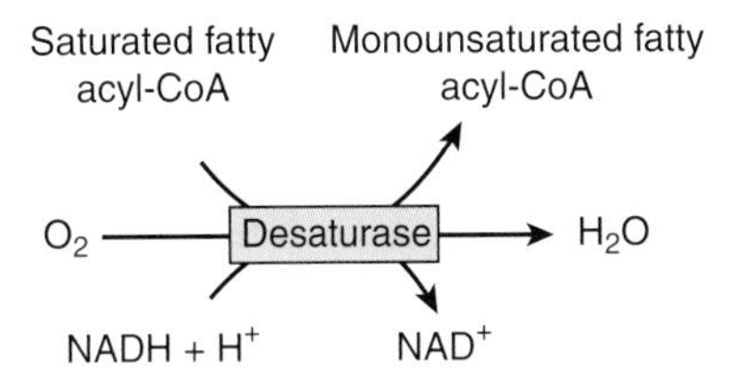

[Adapted with permission from Naik P: Biochemistry, 3rd edition, Jaypee Brothers Medical Publishers (P) Ltd., 2009.]

Lipid Synthesis in Selected Tissues: New synthesis of fatty acids in the human body takes place in liver, adipose (fat) tissue, kidney, brain, and the mammary glands. The liver is intimately involved in maintaining the balance of lipids, including cholesterol (Chapter 11). Adipose tissue, commonly referred to as fat cells, plays a key role in the storage of lipids in the form of triacylglycerol molecules. **Triacylglycerol** can store 9 kilocalories of energy per gram (kcal/g) versus only 4 kcal/g for **carbohydrates**, making triacylglycerol the most efficient form of energy storage in the body. Fatty acid, triacylglycerol, and cholesterol syntheses are also found in kidneys, and increased levels of lipids are seen in various kidney diseases (Chapter 18). The brain is one of only a few tissues that use ketone bodies (reviewed below) as an energy source. Because the brain uses approximately 25% of the body's total energy production, this alternative energy source to carbohydrates most certainly illustrates an essential adaptation for survival during periods of reduced food intake. The brain also synthesizes lipid molecules essential in the growth and maturation of the nervous system from infancy to adulthood. Lack of these synthetic functions leads to a number of neurological diseases. Mammary glands also respond during pregnancy by producing triacylglycerol containing medium-chain fatty acids found in human milk. These lipids, which make up 3%–5% of human breast milk, are essential for the development of the newborn.

Essential Unsaturated Fatty Acids: Humans cannot produce desaturated bonds beyond the ninth and tenth carbons and, therefore, need to acquire fatty acids with more distal double bonds from dietary sources (Chapter 3). These essential, unsaturated fatty acids are required for production of molecules such as prostaglandins (important in inflammatory reactions, various aspects of pregnancy, the spread of some cancers, control of blood flow to the kidney, regulation of ion flow across membranes, the conduction of nerve impulses, and modulation of sleep), thromboxanes (required for blood clotting and possibly involved in constriction of blood vessels, e.g., Prinzmetal's angina), and leukotrienes (important molecules in immune reactions including asthmatic and allergic reactions as well as certain cardiovascular and neuropsychiatric diseases).

FATTY ACID DEGRADATION

The process of the breakdown or degradation of fatty acids starts in the cytoplasm with the linkage of a fatty acid to CoA forming a **fatty acyl-CoA** molecule. Following this linkage, fatty acids *less* than 12 carbons in length cross the mitochondrial membranes without assistance. However, fatty acids longer than 12 carbons are transported into the mitochondria via a unique process involving the molecule carnitine. In the first of three steps, the fatty acyl-CoA links to carnitine by the enzyme **carnitine palmitoyltransferase I** (**CPT I**), located on the outer mitochondrial membrane (Figure 7-3). The new molecule moves across the mitochondrial inner membrane via a specific membrane protein called translocase. Inside the mitochondria, **CPT II**, located on the inside of the inner mitochondrial membrane (Figure 7-3), removes the carnitine molecule. This carni-

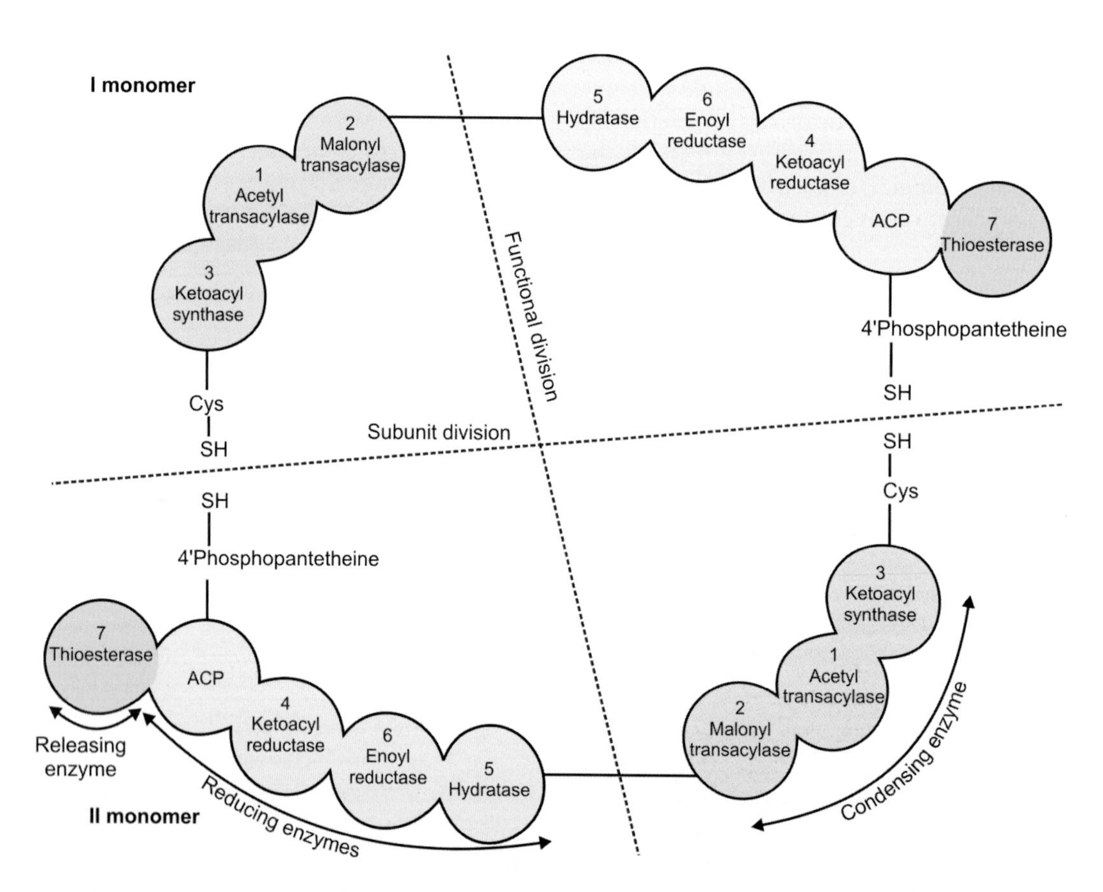

Figure 7-2. Fatty Acid Synthase. The human fatty acid synthase enzyme complex is composed of two identical monomers (I and II, "subunit division"), which are, themselves, split into condensing and reducing enzyme halves ("functional division"). The enzymes responsible for the growing fatty acid chain are lined up (although not in the linear order of reactions), starting at ketoacyl synthase. As one reaction concludes, the product is carried to the next enzyme by an "acyl carrier protein" (ACP), a section of the enzyme complex, which can flip back and forth between the successive enzyme reactions. Building blocks of acyl or malonyl subunits are attached to the ACP by acetyltransacylase or malonyltransacylase, respectively. Reducing enzymes creates the final carbon–carbon bond of the growing fatty acid chain. The new fatty acid product either returns to the start of the process or, once a 16-carbon, palmitate molecule is achieved, is released by thioesterase. The exact tertiary and quaternary structure of fatty acid synthase is still in question but all models agree with the successive enzymatic processes and the role of the ACP. [Reproduced with permission from Naik P: Biochemistry, 3rd edition, Jaypee Brothers Medical Publishers (P) Ltd., 2009.]

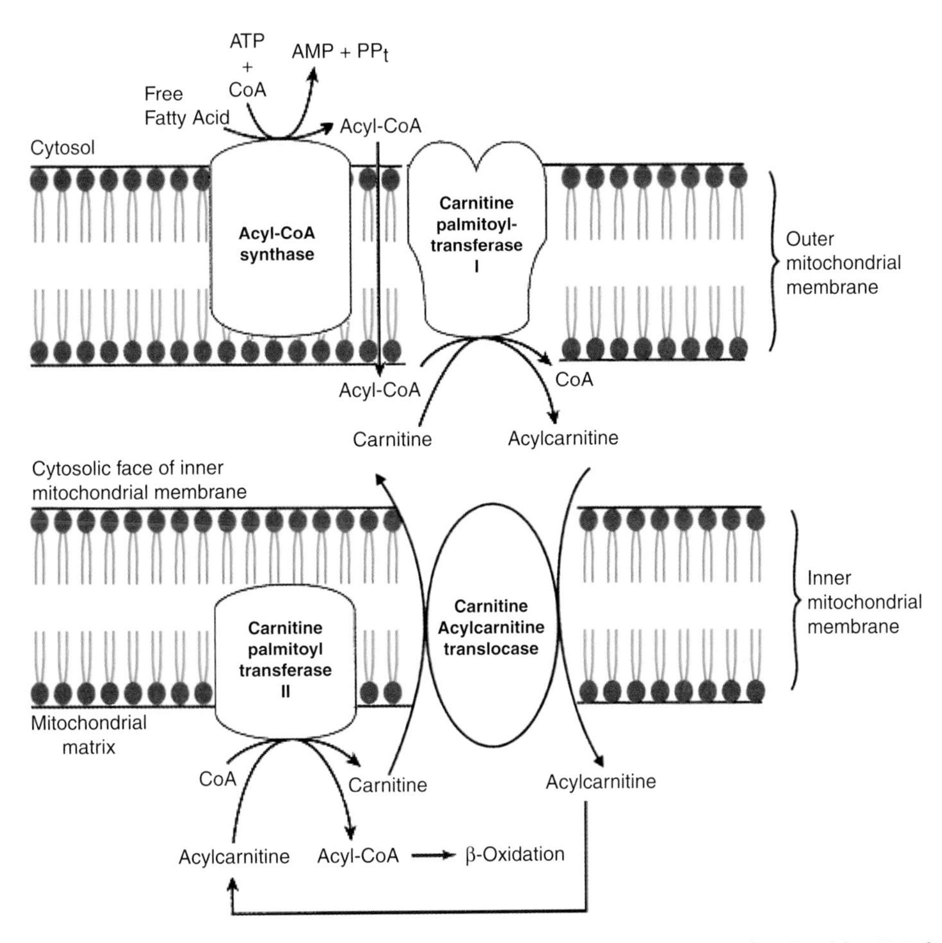

Figure 7-3. Transport of Fatty Acids Longer than 12 Carbons Cross the Mitochondrial Membrane Using Carnitine Palmitoyltransferase I and II. [Adapted with permission from Naik P: Biochemistry, 3rd edition, Jaypee Brothers Medical Publishers (P) Ltd., 2009.]

tine molecule moving out of the mitochondria exchanges with a new molecule of palmitoylcarnitine moving into the mitochondria. High concentration of malonyl-CoA, the molecule representing commitment to the synthesis of fatty acids, prevents the transport of fatty acids into mitochondria by inhibiting CPT I. Increased amounts of fatty acyl-CoA reverse this inhibition and stimulates fatty acid degradation.

CPT Diseases: Not surprisingly, deficiencies in **CPT I or II** lead to severe disease states. CPT I deficiency is an autosomal recessive disorder that usually strikes in infancy or early childhood and causes low sugar, low ketone levels, and increased ammonia levels in association with fasting or illness (i.e., when the body attempts to use fatty acids as an energy source). Patients with CPT I deficiency can develop damaged and enlarged livers (partly due to the buildup of fatty acids), kidney problems, muscle breakdown, seizures, coma, and impaired growth. Without a diet of only medium-chain triglycerides (whose fatty acids can cross the mitochondrial membranes independent of carnitine) and the avoidance of fasting, death is common.

CPT II deficiency occurs in three different forms and is the most common inherited disorder of mitochondrial long-chain fatty acid oxidation. Inheritance can be autosomal recessive or due to multiple forms of heterozygous mutations that affect CPT II activity. Overall, CPT II deficiency occurs more often in males than in females. The three forms include a lethal neonatal form, a severe infantile hepatocardiomuscular form, and a milder form, which may occur from infancy to adulthood. The first two forms cause low sugar, low ketone levels, liver failure, heart damage, fatigue, seizures, and, often, early death. The second and third forms involve breakdown of muscle during periods of exercise, fasting, or infection; reddish-brown urine usually results from the resulting muscle breakdown products. Treatment of CPT II deficiency includes a high-carbohydrate/low-fat diet to encourage only carbohydrate metabolism (including infusion of glucose during infections, use of medium-chain fatty acids in the diet, maintaining constant feeding so as to avoid fasting states, and restricting lengthy periods of exercise), adequate hydration to help flush breakdown products through the kidneys without damage, and the addition of carnitine to the diet to convert potentially toxic, long-chain fatty acids to molecules, which can be eliminated via other metabolic pathways.

Once inside the mitochondria, fatty acid breakdown continues in a four-step process called the **β-oxidation cycle**, which removes two carbons from the fatty acid with each cycle (Figure 7-4). The first step involves the **acyl-CoA dehydrogenase** enzyme, which removes electrons from the two end carbons to form a carbon–carbon double bond. The electrons are transferred to a flavin adenine dinucleotide (FAD) cofactor molecule associated with the enzyme that later produces ATP via the electron transport chain. In the second step, **enoyl-CoA hydratase** resaturates the double bond with hydrogen (H^+) and a hydroxyl (OH^-) group from a water molecule. Next, the bond is oxidized to a ketone by **L-3-hydroxyacyl-CoA dehydrogenase**, with the transfer of another electron to the cofactor NAD^+, which later produces ATP via the electron transport chain. In the fourth and final step, the potentially reactive ketone and another CoA molecule react via the enzyme **β-ketothiolase** (sometimes called acetyl-CoA acyltransferase) to form a two-carbon acetyl-CoA fragment and a new fatty acyl-CoA that is now two carbons shorter. The acetyl-CoA fragment is then used for other metabolic pathways for energy production or synthesis of new molecules. The four-step cycle repeats until the entire fatty acid chain is degraded into acetyl-CoA units. In the final step for an even-chain fatty acid, two acetyl-CoA molecules are formed from the remaining four-carbon fragment.

Fatty acid chains with an odd number of carbon atoms are broken down similarly (Figure 7-5), although the final cycle produces a three-carbon propionyl-CoA molecule and an acetyl-CoA molecule. A carboxyl (CO_2^-) group is then added to the propionyl-CoA molecule in a two-step process to produce succinyl-CoA, one of the major molecules of the citric acid cycle (Chapter 6).

Mitochondrial Trifunctional Protein (MTP): Three of the enzymes involved in β-oxidation, namely **2,3-enoyl-CoA hydrase**, **3-hydroxyacyl-CoA dehydrogenase**, and **3-ketoacyl-CoA thiolase** (see inside box in figure), are part of the MTP. This enzyme complex catalyzes the final three steps of fatty acid degradation (see the figure below).

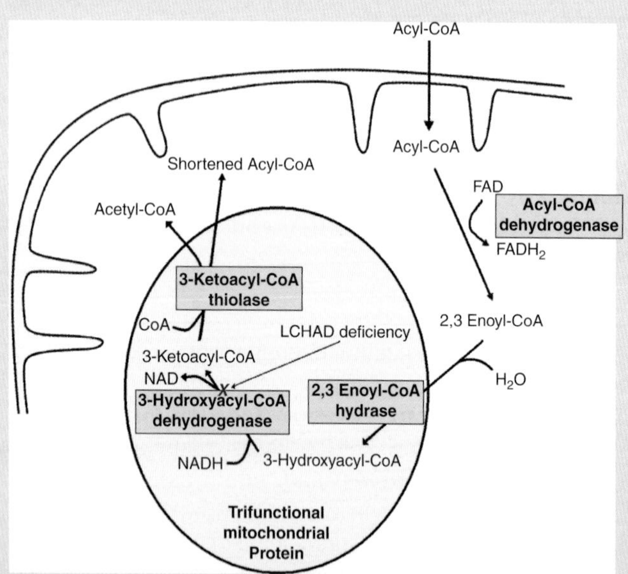

[Adapted with permission from Naik P: Biochemistry, 3rd edition, Jaypee Brothers Medical Publishers (P) Ltd., 2009.]

MTP deficiency results when all three enzyme activities are deficient. Several mutations have been found to cause this disease, which stops the conversion of lipids/fats to energy. Symptoms include low sugar, progressive destruction of nerves in the limbs and digits, breakdown of muscle tissue, and liver or heart damage, which can result in early death.

Mutations also affect isolated enzyme activities of the MTP complex. Low levels of the first enzyme enoyl-CoA hydrase result in abnormally small body, arms, fingers, and head. Deficiencies of the second enzyme 3-hydroxyacyl-CoA dehydrogenase actually involve two separate enzymes, which degrade fatty acids of a particular length. Medium- and short-chain 3-hydroxyacyl-CoA dehydrogenase deficiency or long-chain 3-hydroxyacyl-CoA dehydrogenase (LCHAD) deficiency involves loss of β-oxidation of medium- and short-chain fatty acids or long-chain fatty acids, respectively. As a result, the body cannot convert these particular fatty acid chains to energy and increased levels build up in particular tissues and organs causing damage. These diseases are present in infancy or later in life. Symptoms of the infant disease include poor feeding, vomiting/diarrhea, fatigue, low blood sugar, muscle weakness, heart and lung problems, changes in the retina, coma, and death. The disease that presents in later life is usually less severe with symptoms including poor muscle tone and weakness, increased breakdown of muscles, and problems with nerves in the arms and legs. Mothers carrying a fetus with LCHAD deficiency may also have liver disease as well as **HELLP syndrome**, which includes red blood cell destruction (**H**emolysis), **E**levated **L**iver enzymes (indicative of liver injury), and **L**ow **P**latelets. Diseases involving isolated deficiencies of 3-ketoacyl-CoA thiolase have not been described.

Specific attacks of all these disease states can be brought on by decreased food intake and stresses such as viral infections. Therefore, treatment involves diet restrictions to avoid the affected fatty acid chain(s), addition of carnitine to the diet to promote alternative metabolic pathways, and the avoidance of fasting and infective stresses.

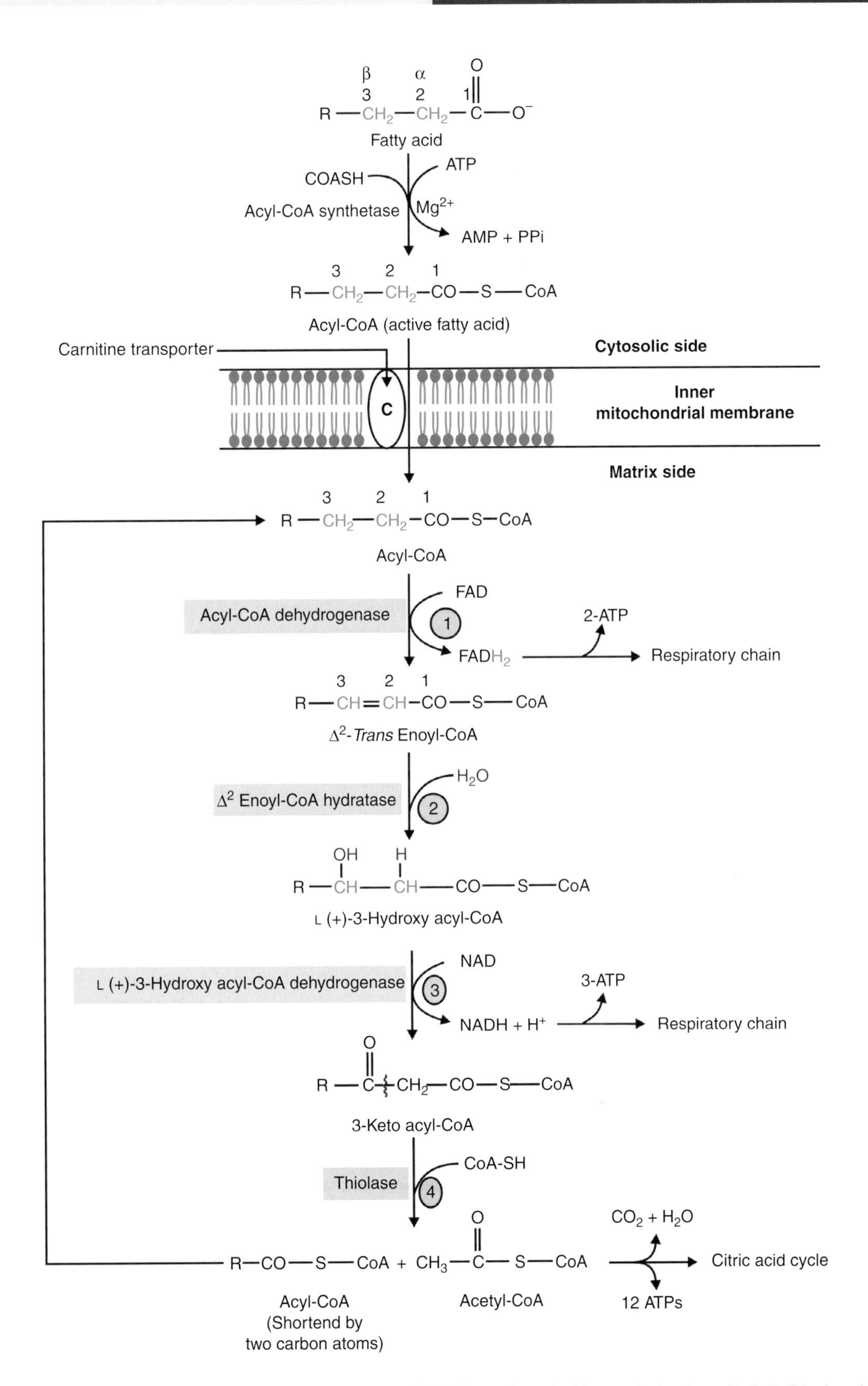

Figure 7-4. Fatty Acid Degradation (β-Oxidation, see the text for details). [Reproduced with permission from Naik P: Biochemistry, 3rd edition, Jaypee Brothers Medical Publishers (P) Ltd., 2009.]

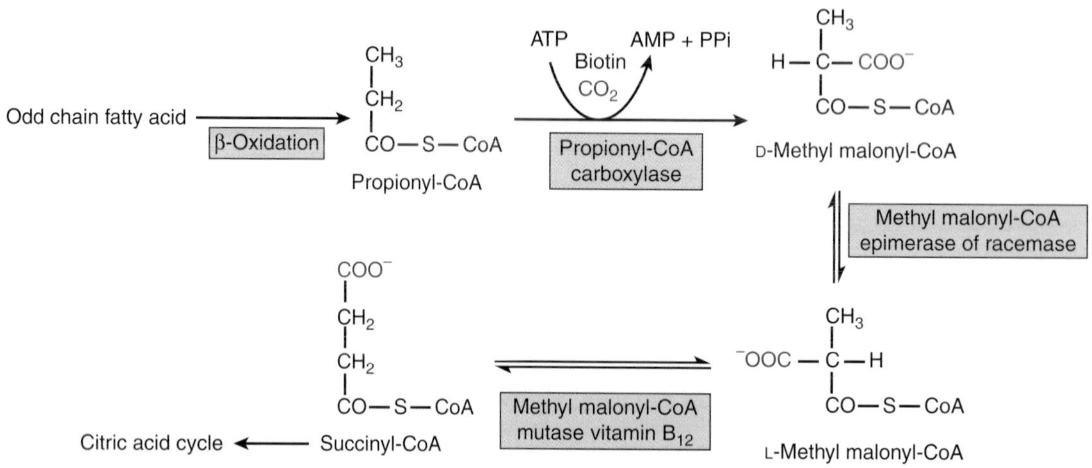

Odd chain fatty acid → β-Oxidation → CH_3–CH_2–CO—S—CoA Propionyl-CoA

ATP, Biotin, CO_2, AMP + PPi — Propionyl-CoA carboxylase → H—C(CH_3)—COO^-, CO—S—CoA D-Methyl malonyl-CoA

Methyl malonyl-CoA epimerase of racemase ⇌ $^-$OOC—C(CH_3)—H, CO—S—CoA L-Methyl malonyl-CoA

Methyl malonyl-CoA mutase vitamin B_{12} ⇌ COO^-–CH_2–CH_2–CO—S—CoA Succinyl-CoA → Citric acid cycle

Figure 7-5. Odd Number Fatty Acid Degradation (β-Oxidation, see the text for details). [Adapted with permission from Naik P: Biochemistry, 3rd edition, Jaypee Brothers Medical Publishers (P) Ltd., 2009.]

Degradation of unsaturated fatty acids with carbon–carbon double bonds depends on the exact conformation of the double bond. Because part of the process of producing the acetyl-CoA fragment involves the creation of a double bond (Step 1), some unsaturated fatty acids can simply be metabolized by the same enzymatic process. However, if the double bond is not in the correct conformation or at the wrong carbon for the acyl-CoA dehydrogenase or the enoyl-CoA hydratase, two other enzymes provide assistance (Figure 7-6). If the double bond is not in the proper conformation, **enoyl-CoA isomerase** converts it to the correct structure, followed by the normal reaction catalyzed by **enoyl-CoA hydratase.** If the double bond is in the wrong location, **2,4-dienoyl-CoA reductase** changes the position of the double bond utilizing NADPH and, if needed, the enoyl-CoA isomerase insures it is the correct conformation.

Two points are worth emphasizing about the breakdown of fatty acids. First, for every two-carbon acetyl-CoA fragment produced, one $FADH_2$ and one NADH are produced, resulting in the later production of two and three ATP molecules, respectively. Thus, one molecule of palmitate converted to eight acetyl-CoA molecules can be used to produce 35 ATP molecules. Each acetyl-CoA molecule can then potentially generate an additional 12 ATP molecules via oxidation by the citric acid cycle. Second, this process illustrates the essential roles of FAD^+ and NAD^+ and their associated vitamins riboflavin (vitamin B_2) and nicotinic acid (niacin) in the temporary storage and subsequent transfer of biological energy.

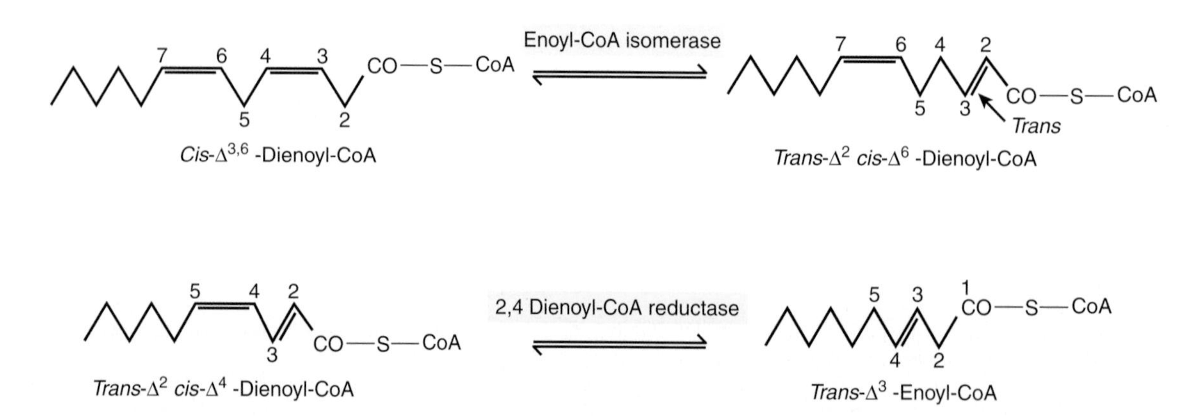

Figure 7-6. Alteration of Carbon–Carbon Double Bonds. Enoyl-CoA isomerase changes a *trans*-double bond at carbon 3 to a *cis*-double bond at carbon 2, and 2,4-dienoyl reductase changes the double bond at carbon 4 to a single bond. These alterations allow normal fatty acid degradation by enoyl-CoA hydratase to continue. [Adapted with permission from Naik P: Biochemistry, 3rd edition, Jaypee Brothers Medical Publishers (P) Ltd., 2009.]

Multiple Acyl-CoA Dehydrogenase Deficiency (MADD): The first enzymatic step in the breakdown of fatty acids is actually via activity of one of three separate enzymes, depending on the initial length of the molecule. **Long-chain acyl-CoA dehydrogenase (LCAD)** helps to degrade fatty acids longer than 12 carbons. **Medium-chain acyl-CoA dehydrogenase (MCAD)** degrades fatty acids between 6 and 12 carbons in length. Finally, **short-chain acyl-CoA dehydrogenase (SCAD)** helps to break down fatty acids shorter than six carbons.

MADD is the generic term referring to the deficiency of any of the three enzymes (LCAD, MCAD, or SCAD) catalyzing the first step in the degradation of fatty acids. This is an autosomal recessive disease with a varying occurrence rate depending on the specific enzyme involved (MCAD deficiency is the most common disorder of the three enzymes, occurring in approximately 1 in 17,000 people). An enzyme deficiency leads to a halt in the breakdown of fatty acids and a requirement for energy production from carbohydrate stores. In addition, tissues that utilize ketone bodies are unable to produce this alternative energy source. As a result, levels of sugar (hypoglycemia) and ketone (hypoketonuria) decrease and abnormal by-products of the "stuck" fatty acid degradation increase. Some of these abnormal fatty acids impede normal ammonia metabolism, leading to toxic ammonia levels, and dramatically decrease the body's ability for gluconeogenesis—the production of new carbohydrate molecules from alternative sources such as amino acids.

Patients with MADD have a number of clinical symptoms including sweaty feet and stale breath odor (from the abnormal fatty acids), nausea and vomiting, increased liver size (hepatomegaly) and liver damage (sometimes resulting in yellowing of the skin also known as jaundice), distortions of the skull and face (e.g., forehead, nose, lips, and ears), muscle weakness, breathing problems and lung damage, and defective kidneys. Death in infancy is common.

Finally, mitochondria are unable to degrade fatty acids greater than 22 carbons. In the case of **very-long-chain fatty acids** (**VLCFA**), breakdown occurs in peroxisomes, organelles found in all eukaryotes that provide specialized lipid metabolism as well as processing of toxic substances. Peroxisomes metabolize VLCFA down to an eight-carbon octanyl-CoA, which is then further processed by mitochondria as described above. Peroxisomal oxidation of fatty acids is driven not by ATP but rather by

Diseases of the Peroxisome: Peroxisomes are often-forgotten, intracellular organelles that eliminate toxic substances, break down special types of fatty acids, synthesize bile acids and plasmalogen (an important type of phospholipid in the myelin membrane surrounding nerve cells), and perform posttranslational processing of proteins. Deficiencies in peroxisomes or their various functions lead to serious diseases. One such disorder, **Zellweger syndrome**, leads to decreased numbers of peroxisomes and/or deficiencies in any of a variety of peroxisomal enzymes. The syndrome usually results from mutations in genes coding for any of several **peroxin** proteins, which recognize specific amino acid sequences on newly translated proteins and label them for transport to peroxisomes.

Two closely related diseases of peroxisome function, **adrenoleukodystrophy (ALD)** (also called Addison–Schilder Disease or Siemerling–Creutzfeldt Disease) and **Refsum disease**, combine to form a trio called the "Zellweger spectrum." ALD is caused by the deficiency of a membrane protein that transports VLCFA into the peroxisome. ALD is usually recessively inherited via the X chromosome and, therefore, strikes only young males (symptoms usually become evident between the age of 4 years and 10 years) at a rate of 1 in 20,000. However, women carrying a single mutation may develop less serious symptoms in adulthood (called adrenomyeloneuropathy). Refsum disease is caused by a specific mutation on either chromosome 6 or 10, which decreases the breakdown of phytanic acid, a 16-carbon fatty acid found normally in the human diet. Patients with Refsum disease usually present with symptoms in childhood or early adolescence.

The three diseases of the Zellweger spectrum share many similar symptoms. Absent β-oxidation of VLCFA leads to damaging increases in levels of 24- to 30-carbon fatty acids, which result in an enlarged liver, jaundice, and intestinal bleeding. Without production of plasmalogen, the insulating myelin membrane is compromised, leading to progressive brain and nerve damage with seizures, loss of vision and hearing, decreasing muscle tone and strength culminating in the inability to move (again due to decreased production of the myelin sheath), and, in infants, poor or absent suckling or swallowing ability. ALD patients often develop adrenal gland failure secondary to buildup of the VLCFAs in these organs. A variant of ALD predominately strikes the spinal cord with symptoms including weakness and numbness of the limbs and problems with urination and defecation. Death usually occurs in childhood or early adolescence due to problems occurring in the affected organs.

Treatment for Zellweger syndrome is mainly supportive, including prevention of infections; however, death usually occurs before the first birthday. Some success in the treatment of ALD has been reported with bone marrow transplantation and a diet with low intake of VLCFA and inclusion of **Lorenzo's oil**, a mixture of 18- and 22-carbon triglycerides. Further research is attempting to provide further support for this treatment and to elucidate the mechanism of Lorenzo's oil. Finally, patients with Refsum disease are maintained on a diet with no phytanic acid (found in beef, lamb, tuna, cod, and haddock); attempts to find alternative/natural therapies are also ongoing.

the production of hydrogen peroxide (H_2O_2), a highly energized molecule, which is converted to water and oxygen by the enzyme catalase found only in peroxisomes. Similarly, though, the fatty acid substrate is transported into the peroxisome by a carnitine acyltransferase and the final step in the process is via a similar peroxisomal β-ketothiolase.

METABOLISM OF COMPLEX LIPIDS

TRIACYLGLYCEROL SYNTHESIS

Triacylglycerols, commonly referred to as triglycerides, are the predominant storage form of lipids (Chapter 3, Figure 3-2B. In fact, the term "fat" is actually a name for triacylglycerol stores. As noted previously, triacylglycerol stores 9 kcal/g versus only 4 kcal/g for carbohydrates. Synthesis of triacylglycerols mainly takes place on the smooth endoplasmic reticulum of the liver but can also be generated in adipose (fat) cells. Regardless of the location of synthesis, the starting molecule is glycerol-3-phosphate produced in liver from glycerol stores or in adipose cells from dihydroxyacetone phosphate, the product of the fourth step of glycolysis (Chapter 6). Triacylglycerol molecules form via a simple series of enzymatic reactions catalyzed by a **triacylglycerol synthase enzyme complex**. This collection of enzymes sequentially attaches fatty acid chains to glycerol-3-phosphate to produce a triacylglycerol molecule. As always, the choice to divert stored liver glycerol or the potential energy from carbohydrate molecules depends on whether or not carbohydrate levels are high (fed) or low (fasting).

Once synthesized, triacylglycerols produced in the liver combine with cholesterol and phospholipids together with a lipoprotein called apolipoprotein B to form a complex referred to as **very-low-density lipoprotein** (**VLDL**), which is released into the blood for transport to other tissues. VLDL plays an important role in triacylglycerol transport and low-density lipoprotein (LDL) levels and is also one of the lipids measured when assessing a patient's cholesterol and lipid status. VLDL will be examined in much more detail in Chapters 11 and 16.

PHOSPHOGLYCERIDE SYNTHESIS

Phosphoglycerides are lipid molecules with two fatty acid chains and a specific "head group," which defines that particular phosphoglyceride molecule. The various phosphoglycerides and their functions are discussed in Chapters 3 and 8. The synthesis of serine-, inositol-, and ethanolamine-phosphoglycerides follows the same enzymatic pathway as triacylglycerols up to and including the addition of the second fatty acid molecule to form **phosphatidic acid** (Figure 7-7). Following this reaction, an activated cytidine diphosphoglycerol intermediate formed from the nucleotide cytidine triphosphate (CTP) binds at the phosphate group and then is replaced by the serine or inositol head group (Figure 7-7A). Phosphatidyl ethanolamine is formed by the loss of the COO^- group from the serine (Figure 7-7B). Phosphatidyl choline is from dietary sources by ATP-driven phosphorylation of the head group, attachment to CTP, and final attachment to the diacylglycerol molecule. This is also a second and alternative synthetic pathway for phosphatidyl ethanolamine (Figure 7-7C).

KETONE BODY SYNTHESIS

If carbohydrate levels are low, the human body will often revert to an alternative source of energy in the form of ketone bodies. **Ketone bodies**, produced in the liver from acetyl-CoA, are a collective term for the molecules **acetoacetate** and **D-3-hydroxybutyrate**, as well as **acetone** that is produced by the spontaneous degradation of acetoacetate (Figure 7-8A).

The body can convert stored ketone bodies back into oxidative metabolism by a series of enzymatic reactions that mimic the reverse of ketone body synthesis. Acetoacetate and D-3-hydroxybutyrate are produced by the liver from acetyl units. High concentrations especially of D-3-hydroxybutyrate and, therefore, acetyl units available for the citric acid cycle decrease this release of adipose-derived fatty acids. In the utilization process, which occurs in tissues other than the liver and red cells, ketone bodies are converted back to two acetyl-CoA molecules by the addition of succinyl-CoA and then directed into the citric acid cycle (Figure 7-8B). In times of low carbohydrate levels and starvation, this series of reactions is used as the primary source of energy by the heart, kidneys, muscles, and brain.

Ketone body synthesis/degradation is regulated by the concentration of oxaloacetate, the last molecule of the citric acid cycle (Chapter 6), which combines with acetyl-CoA to continue the process of production of energy from carbohydrates. Therefore, if oxaloacetate concentration is low, indicative of low carbohydrate levels, the body can direct its resources to the production of new glucose molecules (gluconeogenesis, Chapter 6) and ketone bodies to produce energy. At the hormonal level, glucagon and epinephrine also increase breakdown of triacylglycerols and the production of fatty acids, acetyl-CoA, and, as a result, ketone bodies. The hormone cortisol has the same but longer lasting effects. If oxaloacetate concentration is high, entry into the citric acid cycle will increase because the acetyl units will condense with the oxaloacetate. Insulin has the opposite effect, promoting storage of fatty acids in the form of triacylglycerols and inhibiting ketone body synthesis; the effect of insulin is decreased or negated by high cortisol levels seen in Cushing's syndrome. Not surprisingly, high levels of ketone bodies are seen in forms of poorly treated diabetes (Chapter 10), low-cholesterol/high-fat diets, and fasting/starvation.

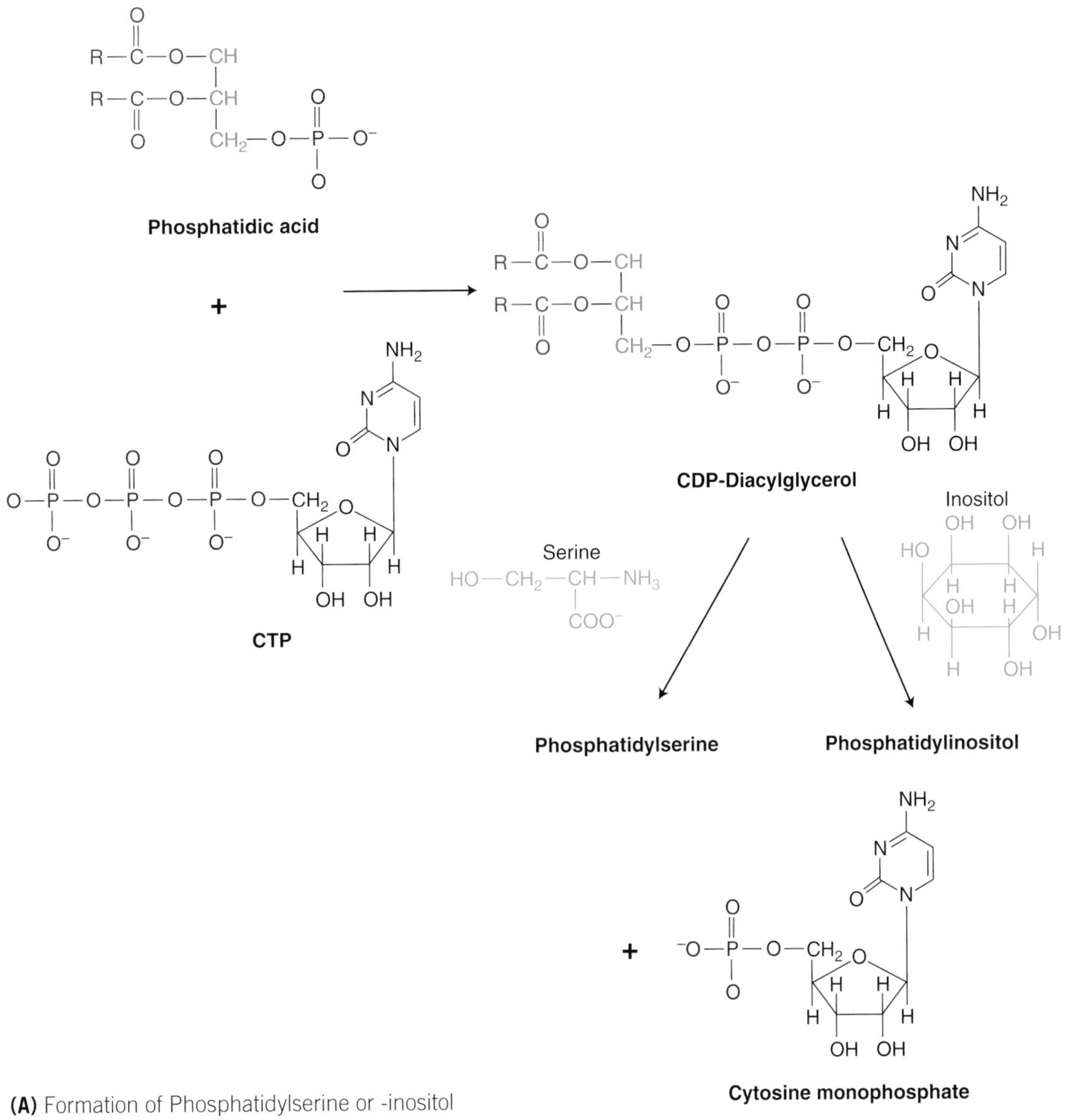

Phosphatidic acid

CTP

CDP-Diacylglycerol

Serine

Inositol

Phosphatidylserine

Phosphatidylinositol

Cytosine monophosphate

(A) Formation of Phosphatidylserine or -inositol

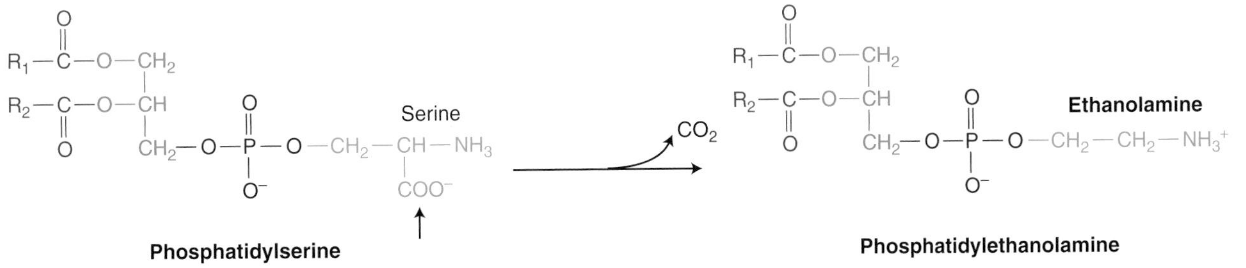

Phosphatidylserine

CO_2

Ethanolamine

Phosphatidylethanolamine

(B) Formation of Phosphatidyl Ethanolamine from Phosphatidyl Serine

Figure 7-7. A–C. Synthetic Pathways of Phosphatidyl Serine, Inositol, Ethanolamine, and Choline. Phosphatidyl serine and inositol are produced from phosphatidic acid, utilizing cytidine triphosphate (CTP). Loss of a carbon dioxide molecule from phosphatidyl serine produces phosphatidyl ethanolamine. Phosphatidyl choline and ethanolamine synthesis also utilize CTP following ATP phosphorylation of the choline or ethanolamine head group. CDP, cytidine 5′-diphosphate. [Adapted with permission from Naik P: Biochemistry, 3rd edition, Jaypee Brothers Medical Publishers (P) Ltd., 2009.]

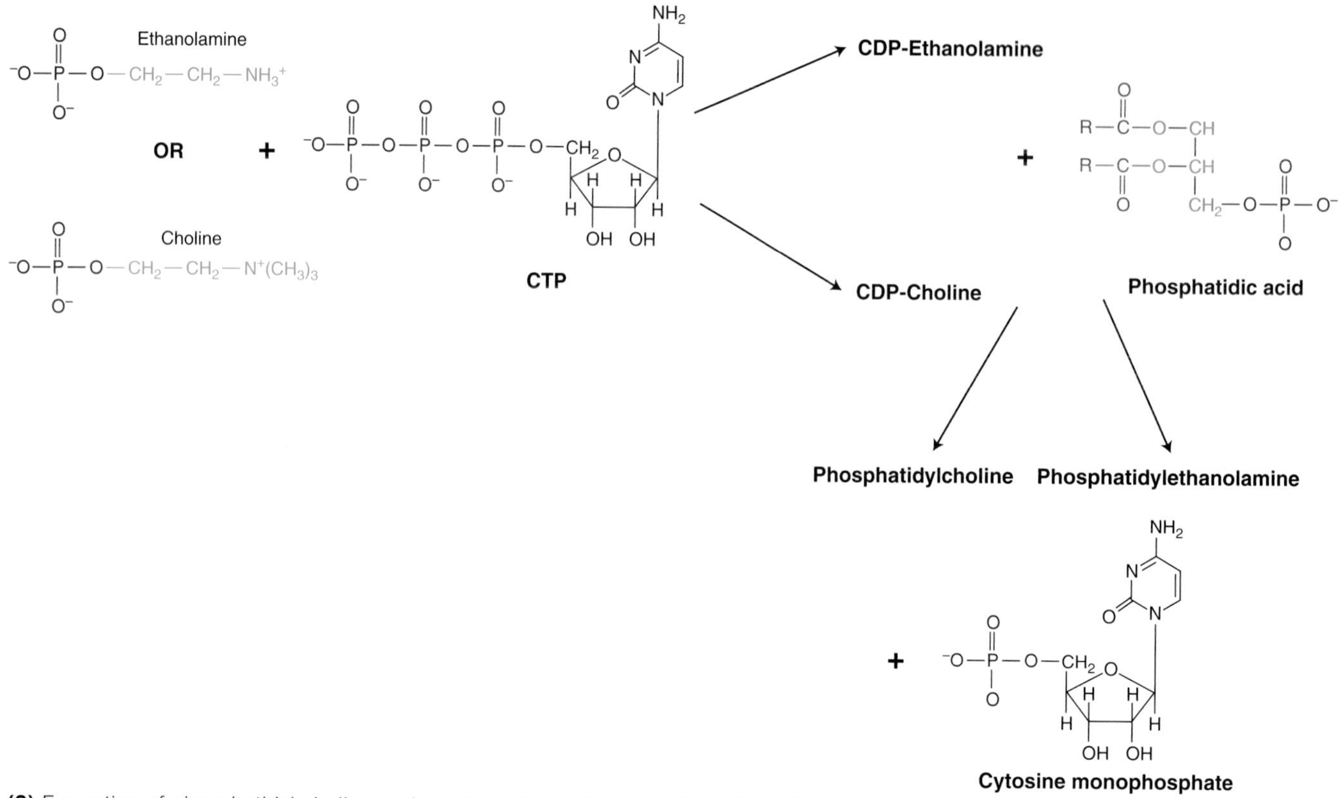

Ethanolamine

OR +

Choline

CTP

CDP-Ethanolamine

CDP-Choline

+

Phosphatidic acid

Phosphatidylcholine Phosphatidylethanolamine

+

Cytosine monophosphate

(C) Formation of phosphatidyl choline and an alternative pathway for formation of phosphatidyl ethanolamine

Figure 7-7. (*Continued*)

Cushing's syndrome: Uncontrolled and excessive production of the hormone cortisol by the adrenal glands, a condition known as **Cushing's syndrome**, is caused by a tumor in the cortex portion of the adrenal glands. Excessive use of some medications can cause a similar condition known as hyperadrenocorticism or hypercorticism. Patients with these diseases exhibit a "Cushingoid" appearance from increased and uncontrolled deposition of fat, resulting in rapid weight gain mainly in the trunk of the body, rounding of the face known as "moon face," a large fat pad in the back of the neck known as a "buffalo hump," and stretch marks in the skin from the rapid weight increase. Patients may also experience other hormonal symptoms including excess sweating, male-pattern hair growth in females (hirsutism), impotence, infertility, and cessation of menses and various psychiatric problems such as depression, anxiety, and psychosis. Treatment is usually by surgical removal of the tumor. Interestingly, other mammals also often suffer from Cushing's syndrome, including dogs, horses, and ferrets.

CERAMIDE/SPHINGOLIPIDS SYNTHESIS

Sphingosine is the precursor molecule for ceramide, which replaces glycerol as the "backbone" of sphingolipid molecules, including cerebrosides, sulfatides, globosides, and gangliosides (Chapter 3), essential in many neurological and muscle functions and in cell–cell recognition and interactions. Sphingosine is produced from a palmitoyl-CoA and a serine molecule in a reaction that involves NADPH and FAD (Figure 3-4A), again showing the importance of these molecules and their respective vitamins in lipid production. Sphingosine, in turn, links to a long-chain fatty acid activated by CoA to form the molecule **ceramide**. Ceramide is then utilized by several independent enzymatic reactions to produce a variety of sphingolipids molecules of varying function.

Gangliosides are the most complex sphingolipids, being built by the successive addition of carbohydrate molecules via specific glycosyltransferase enzymes. These enzymes add glucose, galactose, and other modified sugars in a defined order and precise positions (Figure 3-4F) to create one of more than 15 ganglioside molecules. The function of these molecules is reviewed in Chapter 3 and will be discussed in more detail in later chapters.

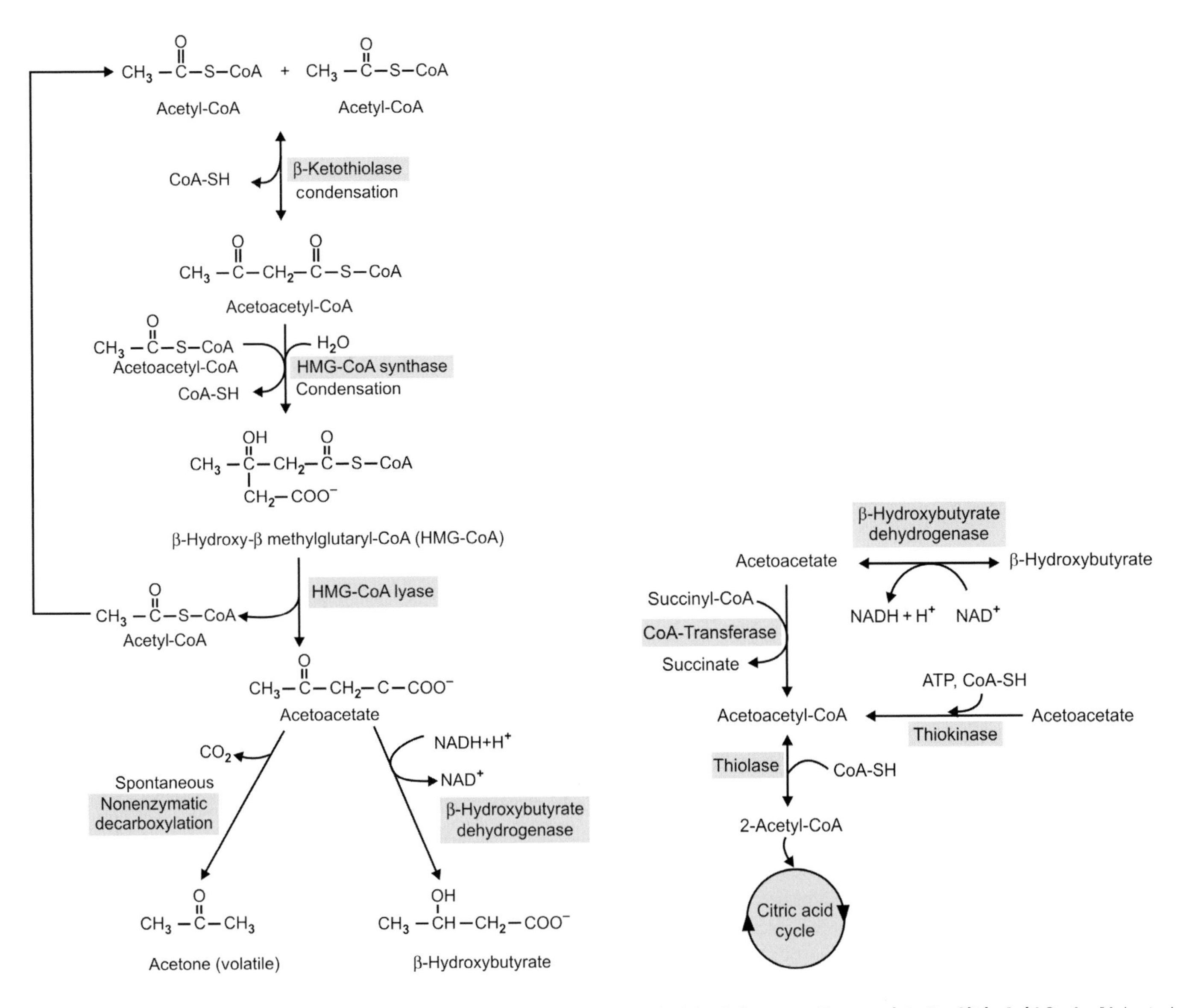

Figure 7-8. A. Ketone Body Synthesis. B. Conversion of Ketone Bodies to Acetyl-CoA for Subsequent Entrance into the Citric Acid Cycle. [Adapted with permission from Naik P: Biochemistry, 3rd edition, Jaypee Brothers Medical Publishers (P) Ltd., 2009.]

Tay–Sachs disease: As discussed in Chapter 3, gangliosides are found in highest concentration in cells of the nervous system and are normally broken down inside lysosomes by orderly removal of the sugar residues, partly by the action of the enzyme **β-*N*-acetylhexosaminidase**. When this enzyme is missing or deficient, ganglioside GM2 cannot be broken down and **Tay–Sachs disease** results. Symptoms of weakness and delayed development of muscle skills are usually evident before the age of 1 year with subsequent blindness and death, usually by the age of 3 years. Neurons of affected patients are literally swollen with lysosomes filled with GM2. Tay–Sachs disease occurs 100 times more often in Ashkenazi (Eastern European) Jews.

CHOLESTEROL SYNTHESIS

Although about half of the body's requirement for cholesterol is usually obtained from the diet, the remainder must be synthesized. Any deficiencies in this synthesis are, therefore, important in normal human physiology and disease. The main location of cholesterol synthesis in humans is the liver/intestinal tract. Cholesterol receives all of its carbon molecules from the two-carbon molecule acetyl-CoA, once again showing the importance of this intermediate in carbohydrate and lipid metabolism (Figure 7-9). The first step involves the formation of **3-hydroxy-3-methyl-glutaryl-CoA (3-HMG-CoA)** from one molecule of acetyl-CoA, one molecule of acetoacetyl-CoA, and a water molecule. Interestingly, any 3-HMG-CoA made in the mitochondria is used to produce

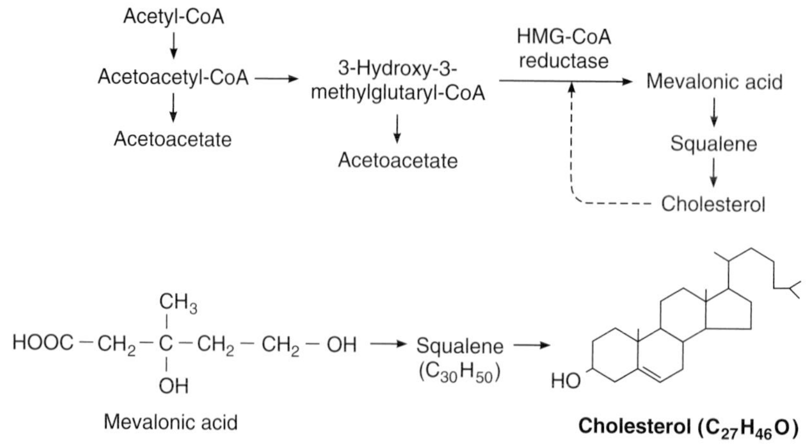

Acetyl-CoA
Acetoacetyl-CoA → 3-Hydroxy-3-methylglutaryl-CoA → (HMG-CoA reductase) → Mevalonic acid
Acetoacetate
Acetoacetate
Squalene
Cholesterol

$HOOC-CH_2-C(CH_3)(OH)-CH_2-CH_2-OH$ → Squalene ($C_{30}H_{50}$) →
Mevalonic acid
HO
Cholesterol ($C_{27}H_{46}O$)

Figure 7-9. Cholesterol Synthesis. Summary of steps in the synthesis of cholesterol (top). The molecular structures of mevalonic acid and the final cholesterol molecule are shown in the bottom panel. [Reproduced with permission from Barrett KE, et al.: Ganong's Review of Medical Physiology, 23rd edition, McGraw-Hill, 2010.]

ketone bodies (see above), whereas 3-HMG-CoA made in the cytoplasm forms cholesterol. 3-HMG-CoA is converted to the six-carbon mevalonate in a reaction catalyzed by the HMG-CoA reductase enzyme. This step in cholesterol synthesis is irreversible and is the committing step in the pathway. This step is also the primary target of medical therapy for high cholesterol diseases (see Sidebar). The six-carbon **mevalonate** is converted to the five-carbon **isopentenyl pyrophosphate** through a series of three reactions that involve two molecules of NADPH and one of ATP. Isopentenyl pyrophosphate molecules then join successively together to form a 10-carbon, 15-carbon (**farnesyl pyrophosphate**), and finally a 30-carbon **squalene** molecule. This step again involves NADPH. At this point, the growing molecule is still a linear chain of carbon molecules. For the final stage in cholesterol synthesis, termed "cyclization," three separate steps utilize NADPH and one atom of molecular oxygen (O_2) to form new single- and double-carbon bonds, remove three methyl (CH_3) groups, and create a hydroxyl (OH^-) group (Figure 7-9). The first cyclic intermediate is lanosterol, and this is converted to cholesterol via a complex series of reactions.

Regulation of new cholesterol synthesis and, therefore, levels in the human body rely on **HMG-CoA reductase**, the enzyme catalyzing the first and committing step of producing mevalonate (Figure 7-9). Specifically, the activity of this enzyme is decreased by high concentrations of mevalonate, cholesterol, and its products. Decreased HMG-CoA reductase activity results from (a) decreased DNA to messenger RNA (mRNA) transcription, (b) lowered mRNA for protein translation, and (c) increased degradation of existing HMG-CoA reductase enzyme complexes. The enzyme is also under direct hormonal control with **glucagon** and **epinephrine** decreasing activity via phosphorylation, whereas insulin increases activity by dephosphorylation. In relation to the hormonal control, low concentrations of ATP decrease cholesterol synthesis as the body directs its resources toward energy production rather than lipid synthesis.

Statins: Statins are an important class of medications, which reduce cholesterol in patients with high cholesterol and heart disease. Statins work by inhibiting the critical and committing step in cholesterol synthesis, the enzyme **HMG-CoA reductase**, responsible for the conversion of HMG-CoA into mevalonate. Statins, whose structure is similar to HMG-CoA, mimic the substrate and competitively inhibit the enzyme. Decreased production of mevalonate and, therefore, cholesterol also causes increased production of receptors for LDL cholesterol, the most harmful cholesterol-transporting lipoprotein, increasing clearance of LDL from the bloodstream.

REVIEW QUESTIONS

1. What are the key features of fatty acid synthesis and how is it regulated?
2. What is the role of fatty acid synthase in fatty acid synthesis?
3. How are fatty acids with an odd number of carbons synthesized?
4. How are unsaturated fatty acids produced?
5. What are the functions of carnitine palmitoyltransferase I and II (CPT I and II) and mitochondrial trifunctional protein (MTP) in the degradation of fatty acids?
6. What are the key features of ketone body synthesis and how do ketone bodies contribute to meeting energy demands in the body?
7. What are the key features and regulation of triacylglycerol, phosphoglycerides, and ceramide synthesis?
8. What are the key features and regulation of cholesterol synthesis?
9. What role do HMG-CoA and HMG-CoA reductase play in cholesterol synthesis?

MEMBRANES

OVERVIEW

Lipids, driven by their hydrophobic and hydrophilic portions, form the biological membranes found in all living creatures. These include the cell and nuclear membranes as well as membranes that are part of organelles such as mitochondria and the endoplasmic reticulum (ER). Membranes provide separation of different environments to permit a variety of biological functions. Membranes are not static structures, though, rather they are dynamic and fluid and allow selective movement of ions, energy sources, vitamins and cofactors, and waste. Complex lipids such as cholesterol and sphingolipids both affect the structure of membranes where they are found and are also involved in specific functions.

The variety of lipid and protein molecules, which make up membranes, are responsible for essential functions such as channels and transport across the membrane as well as signaling. Membrane receptors, along with their cytoplasmic partners, transmit signals via steroid hormones. An important type of membrane receptors are G-proteins with intrinsic enzyme activity. The resulting secondary messenger molecules amplify and transmit this signal to various parts of the cytoplasm and nucleus. This ability to transmit a message across a membrane is paramount to not only the normal functions of cells but also their ability to perform specialized functions, which define the human body.

Not surprisingly, lipids and membrane structure and function are important in disease processes as well as treatments. Any deficiencies or problems with lipid synthesis or breakdown lead to a serious disease state and/or death. Modulation of membrane fluidity is, itself, important in membrane functions and, as a result, a variety of diseases. For example, multiple bacteria and viruses are infective and several medications work simply because their hydrophobic nature affects and directly targets lipids and membranes.

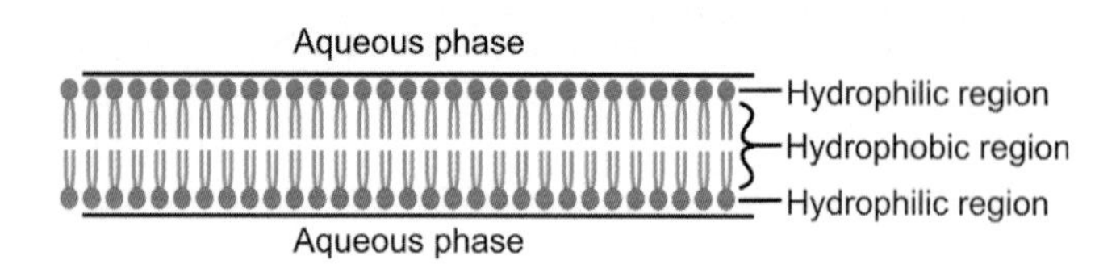

Figure 8-1. Simplified Representation of Lipid Bilayer/Membrane. The lipid bilayer is composed of lipid molecules, with hydrophilic head groups (e.g., phosphate) forming the outer surfaces and hydrophobic tails grouped together in the hydrophobic center. See Chapter 3 for more detailed discussion of lipid molecule structures. [Adapted with permission from Naik P: Biochemistry, 3rd edition, Jaypee Brothers Medical Publishers (P) Ltd., 2009.]

MEMBRANE STRUCTURE

LIPIDS

Membranes are composed of lipids arranged in a **lipid bilayer**, with the hydrophilic glycerol and phosphate "head" groups of the lipid molecules forming the two outside layers and the hydrophobic "tail" groups arranged inside (Figure 8-1).

A majority of membrane lipids are **phospholipids**, usually with 16- or 18-carbon tails, some saturated and some unsaturated. Other types of lipids, including cholesterol, modulate the structure and, therefore, the fluidity of the membrane and, in turn, affect some membrane functions.

Particular lipids are found in specific membranes and are responsible for specific function as summarized below:

- **Cardiolipin (Diphosphatidylglycerol)**—Several locations including the inner mitochondrial membrane. It has a negative charge and may be involved in (a) decreasing blood clots, (b) stabilizing important respiratory chain enzymes in the mitochondria, (c) moving proteins and cholesterol from the outer to the inner mitochondrial membrane, (d) assisting in the proper folding of mitochondrial proteins (Chapter 9) as a chaperone, and (e) possibly regulating deoxyribonucleic acid (DNA) synthesis.
- **Phosphatidylserine (PS)**—Located in the inner platelet membrane and, during activation, the exterior platelet membrane. It carries a negative charge and is the primary phospholipid that promotes the anticoagulant protein C pathway, providing feedback inhibition of thrombin formation.
- **Phosphatidylethanolamine (PE)**—Located in both the interior and exterior of cell membranes. It carries a neutral charge and promotes the anticoagulant protein C pathway, but to a lesser degree than PS.
- **Phosphatidylcholine (PC)**—Located in the interior and exterior of cell membranes. It carries a neutral charge and promotes the anticoagulant protein C pathway, but to a lesser degree than PS.
- **Phosphatidylinositol**—Located in the interior and exterior of cell membranes as well as the nuclear membrane. It carries a positive charge and promotes the anticoagulant protein C pathway, but to a lesser degree than PS, and is important in signaling processes for initiation of DNA replication.

Cardiolipin and Disease: Cardiolipin has been suggested to play several important roles in a variety of tissues, and deficiencies in the production and modification of this lipid, not surprisingly, lead to several human disorders. **Barth syndrome** is an X-linked disease resulting from defects in an enzyme that remodels cardiolipin. The disease results in decreased phosphorylation of adenosine diphosphate molecules, presumably related to cardiolipin's proposed role in stabilizing respiratory chain enzymes. Changes in the amount or composition of cardiolipin in heart and brain mitochondria may also play a role in diseases such as **heart failure**, **diabetes**, **Alzheimer's disease**, and **Parkinson's disease**. Finally, *Trepenoma pallidum*, the bacterium responsible for the disease **syphilis**, produces antibodies against cardiolipin; the biological and disease implications of this are still not well understood.

The smaller head groups of phospholipids such as **PS** and **PE** are preferred on the inner side of the membrane bilayer. On the other hand, PS is often located on the outside of the bilayer where it can increase adherence to other cells and tissues. A membrane-bound enzyme called "**flippase**" catalyzes the process of moving particular phospholipid molecules from one side of the bilayer to the other when required. Other phospholipids, because of their head group size and charges as well as the number and location of their tail double bonds, create a more fluid membrane more amenable to areas of cell curvature or cell–cell connections (Figure 8-2).

Composition of the plasma membrane and, therefore, fluidity has been implicated in several disease processes, including the following:

1. During stages of infection with malaria, the composition of infected red blood cells is markedly changed to mimic that of

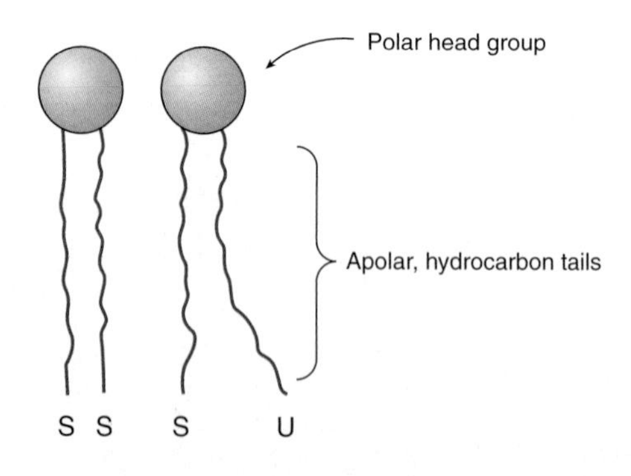

Figure 8-2. Membrane Packing of Saturated and Unsaturated Lipid Tails. The lipid bilayer is composed of lipid molecules whose tail groups can be saturated ("S") or unsaturated ("U"). Saturated tails pack close together, whereas unsaturated tails have a looser packing structure, leading to a more fluid lipid membrane. Increased amounts of saturated or unsaturated lipids can markedly affect membrane fluidity and, therefore, the function of the membrane at that location. [Adapted with permission from Murray RA, et al.: Harper's Illustrated Biochemistry, 28th edition, McGraw-Hill, 2009.]

the infecting parasite. This may assist in the rupture of red blood cells and the subsequent spread of the parasites that is a hallmark of this disease.

2. Various bacteria and viruses are believed to change the plasma membrane of target cells to allow the bacteria/virus to fuse and, therefore, infect the modified cells.
3. Human immunodeficiency virus (HIV) fuses with specific parts of an infected cell's membrane similar to its own viral membrane in composition and fluidity, allowing new viral particles to emerge and spread to other cells. Research has shown that changing membrane composition of target cells may allow protection from HIV infection.
4. The parasite *Entamoeba histolytica* produces proteins that bind to and form holes in human cell membranes, killing these target cells. However, the parasite membrane has a markedly different membrane composition, including different phospholipids and a high level of cholesterol, which protects itself from these deadly proteins.
5. Membrane composition is markedly changed in the platelets of patients with frequent migraines. Although the mechanism is not known, these changes may be part of the cause of migraine headaches and a better understanding could, therefore, lead to a novel treatment.
6. Virgin olive oil alters the composition of plasma membranes and affects regulatory proteins in the lipid bilayer, which results in beneficial changes in carbohydrate and lipid metabolism.

The presence and amount of cholesterol also influence the fluidity of membranes. Cholesterol's hydroxyl group readily interacts with phospholipid head groups or exterior water molecules and the remaining hydrophobic tetracyclic structure and "tail" also naturally insert inside the lipid bilayer (Figure 8-3). Hydrogen bonding stabilizes these interactions. Depending on the adjacent lipids, cholesterol either increases or decreases the lipid tail packing. Localized membrane "rafts" containing high levels of cholesterol have been noted by researchers and are felt to directly influence not only membrane functions but also the functions of the membrane proteins within them.

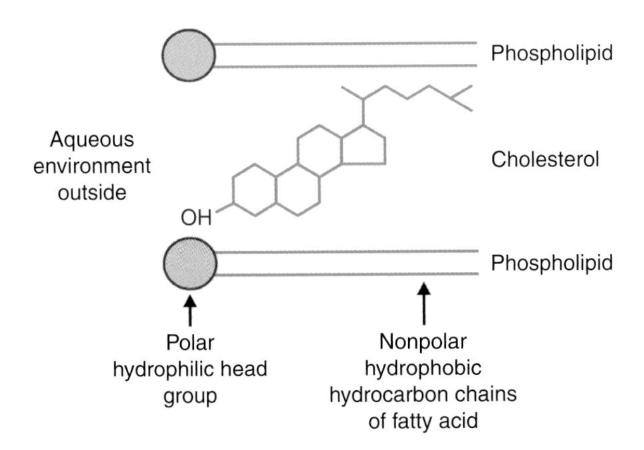

Figure 8-3. Cholesterol Interaction with Membrane Lipids. Illustration of the positioning of cholesterol within the lipid membrane. The hydroxyl group on one end of the molecule prefers to be near the polar, lipid head groups of the external aqueous environment, whereas the remainder of the cholesterol molecule prefers to be within the hydrophobic lipid tails. [Reproduced with permission from Naik P: Biochemistry, 3rd edition, Jaypee Brothers Medical Publishers (P) Ltd., 2009.]

Polyene Antifungals: Antifungal medications including **Nystatin** and **Amphotericin B*** (see the figures below) share the structural basis of the polyene molecule (i.e., multiple carbon=carbon double bonds) opposite an additional carbon chain containing multiple hydroxyl groups (OH). This structure enhances binding to **ergosterol**, a cholesterol-like, complex lipid found in fungal cell membranes. Binding of these antifungals to ergosterol decreases membrane fluidity and leads to leakage of the inner contents and fungal cell death. Human cell plasma membranes are not affected beacuse they do not contain ergosterol, and the antifungal structure does not bind to cholesterol.

Nystatin

Amphotericin B

[Adapted with permission from Katzung BG, et al.: Basic and Clinical Pharmacology, 11th edition, McGraw-Hill, 2009.]

Ergosterol

[Adapted with permission from Murray RA, et al.: Harper's Illustrated Biochemistry, 28th edition, McGraw-Hill, 2009.]

*Amphotericin B is well known for a variety of serious and even fatal side effects, which are believed to be due to mechanisms other than its effects on fungal membrane fluidity.

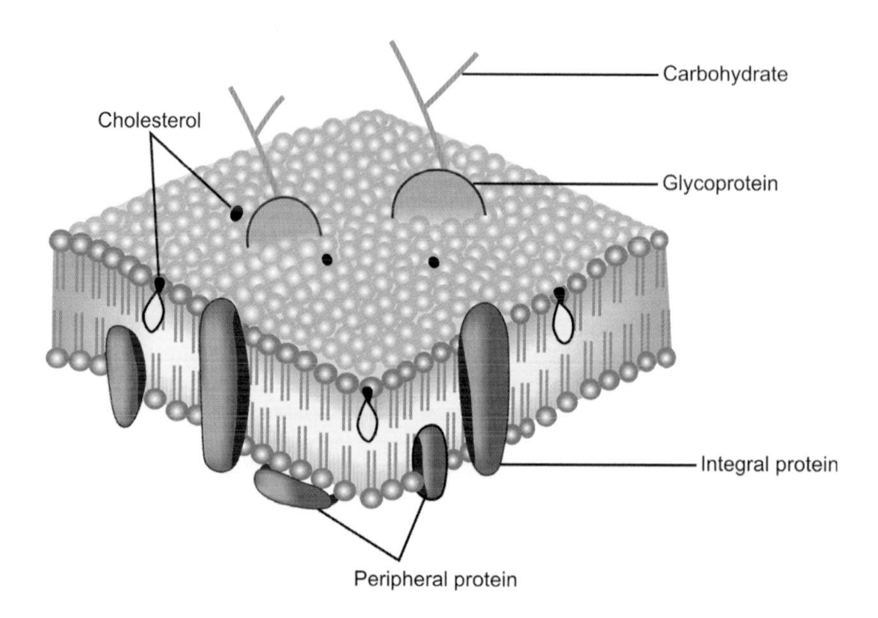

Figure 8-4. Peripheral and Integral Membrane Proteins. The relative positions of peripheral and integral membrane proteins in the lipid bilayer are illustrated. Membrane proteins can also contain carbohydrates resulting in membrane glycoproteins important in membrane signaling and functions. The presence and positioning of cholesterol are also illustrated. [Adapted with permission from Naik P: Biochemistry, 3rd edition, Jaypee Brothers Medical Publishers (P) Ltd., 2009.]

PROTEINS

Proteins are the second major part of biological membranes and make up approximately 20%–80% of both the structural and functional components of these membranes. Membrane proteins are classified as **peripheral**—predominately on the inner or the outer surface of the membrane bilayer—and **integral**—predominately within the membrane (Figure 8-4). Peripheral membrane proteins are believed to mainly act as anchor points for attachment to external structures (e.g., the extracellular matrix) and internal points (e.g., the actin, microtubule, and intermediate filament cytomatrix; Chapter 12). Integral proteins are usually composed of uncharged, hydrophobic amino acids so they can enter and stay in the hydrophobic environment of the lipid bilayer. Integral proteins serve several functions including channels, "carrier proteins" (transporting molecules through the membrane), or as a signaling protein (changing a portion of their internal cytoplasmic structure in response to a signal at an exposed external part of their structure to activate additional, closely associated molecules, leading to a change of function, e.g., turning on a gene).

As already illustrated, the membrane content of cholesterol and saturated/unsaturated lipids impacts membrane fluidity. Membrane proteins form connections to the extracellular matrix, the internal cytomatrix, and even other cells that can allow particular proteins and lipids to collect at special areas of a biological membrane. Thus, a cell can orchestrate a particular "area" or "side" where it can receive signals and/or transport particular molecules both in and out and/or other essential functions. These specialized "domains" within membranes are called "lipid rafts" (Figure 8-5). Various examples of these membrane protein functions are discussed below.

Hemoglobinopathies: Diseases of red blood cells, known collectively as "**hemoglobinopathies**," often result from changes in lipid composition and/or defects in proteins associated with the red cell membrane. **Elliptocytosis** and **spherocytosis** (see the Figure) are two such diseases that result from defects in spectrin and ankyrin and other red cell membrane proteins important in the stabilization of the normal biconcave shape of red blood cells.

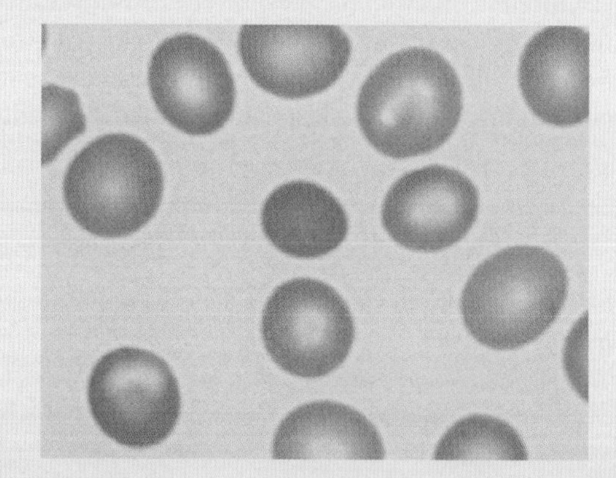

Normal erythrocytes surrounding elliptocyte/spherocyte shown in the center
[Courtesy of Dr. Walter Kemp, Montana State Department of Justice]

Hemoglobinopathies involving changes in lipid composition of the red cell membrane are also common. Red blood cells carefully maintain a specific, asymmetric distribution of hundreds of phospholipids and proteins in the lipid bilayer. Specific defects in either number or bilayer composition of PS, PC, sphingomyelin, and cholesterol are seen in diseases such as sickle cell, thalassemias, anemias, and liver disease. These changes also lead to destruction or removal of these blood cells as part of the disease processes.

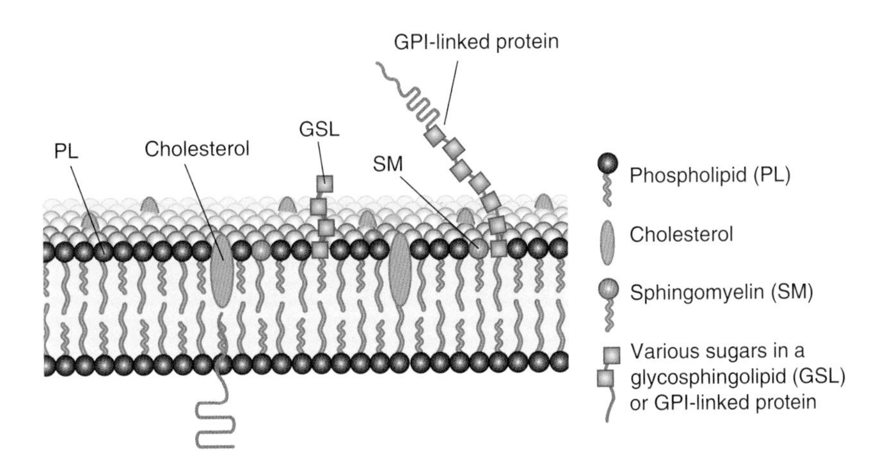

Figure 8-5. Diagram of Lipid Raft. Lipid rafts are somewhat thicker and contain higher amounts of specialty lipids [e.g., sphingomyelin, gangliosides (not shown), saturated phospholipids (non zig-zag tails), and cholesterol]. Membrane proteins can also contain carbohydrates resulting in membrane glycoproteins important in membrane signaling and functions. [Adapted with permission from Murray RA, et al.: Harper's Illustrated Biochemistry, 28th edition, McGraw-Hill, 2009.]

MEMBRANE FUNCTIONS

A primary function of biological membranes is to maintain the structural integrity and the individual functions of cells, the nucleus, and organelles. Separation of environments allows differences in concentration of ions and molecules that literally allow life to exist. As examples, if biological molecules such as carbohydrates passed easily through membranes, cells could not perform any significant biological functions. If cells were unable to selectively increase and maintain the concentration of certain ions on the inside and outside of the membrane, they could not maintain their required internal environment or perform functions such as conduction of a nerve impulse. However, the fact that molecules such as oxygen, carbon dioxide, nitrogen, and urea are able to move through the lipid bilayer without any specialized transport proteins is actually beneficial to the human body (Figure 8-6A). Special circumstances of nonselective transport of molecules have also been developed. For example, because of its reliance on ketones as an energy source in times of low carbohydrate levels, brain membranes allow free flow of ketone bodies without the need of a transport protein.

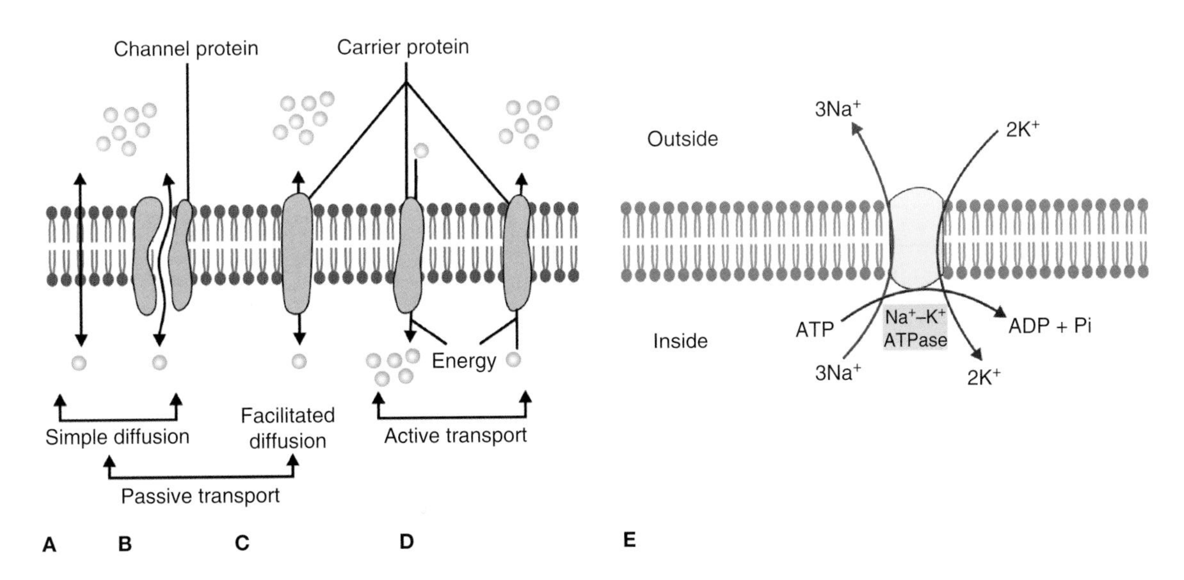

Figure 8-6. A-D. Illustration of Membrane Transport Proteins. Simple diffusion through (**A**) lipid bilayer or (**B**) simple channel. Facilitated diffusion via a (**C**) channel and (**D**) active transport either into or out of the cell. See the text for further description. **E. Na^+–K^+ ATPase Membrane Channel.** Three sodium ions (Na^+) inside the cell bind to the Na^+–K^+ ATPase pump, which promotes phosphorylation of the protein by a bound adenosine triphosphate (ATP) molecule. This phosphorylation causes a conformational change of the pump, which carries and releases the Na^+ to the outside of the cell. Adenosine diphosphate (ADP) is released from the protein and two potassium ions (K^+) bind to the new conformation of the pump. Binding of K^+ promote dephosphorylation of the protein, causing a reverse conformational change, which carries and releases the K^+ to the inside of the cell. ATP rebinds to the pump, which also helps accelerate release of the K^+, returning the pump to its original state. [Reproduced with permission from Naik P: Biochemistry, 3rd edition, Jaypee Brothers Medical Publishers (P) Ltd., 2009.]

MEMBRANE CHANNELS

Proteins that create **membrane channels** often have multiple α-helical and/or β-strand secondary structures that form a tube-like channel through the membrane. Hydrophobic amino acids on the outside of the protein's tertiary and quaternary structure (Chapter 1) form an exterior that readily inserts itself into the membrane and stays within the hydrophobic lipid tails. Hydrophilic or charged amino acids on the inside of the channel form a suitable passageway that allows charged and/or uncharged molecules to move through the previously impeding lipid membrane.

Simple channels (Figure 8-6B) have no active control over the molecules that can enter through them except by the diameter of the tube and, sometimes, the exact hydrophobic/hydrophilic environment in the interior of the tube. Simple channels, therefore, only work via a concentration gradient that provides the drive for molecules to cross the membrane. An example of a simple channel is the **gap junction** that creates a passageway in the gap between two cells allowing movement of ions, sugars, amino acids, and nucleotides between the cells. Gap junctions are important in organs such as the heart, bone and eye lens, and in developing tissues.

Facilitated protein channels (Figure 8-6C) are similar to simple channels (no energy is used, driven by a concentration gradient) but the transport is sped up or facilitated from particular amino acids that help to form the channel. These channels can also have well-positioned amino acids (often charged) in the tube that serve as a "gate" to regulate the movement of specific molecules through the channel. These **gated channels** (not shown) have amino acids that block transport but a signal (e.g., binding of a specific molecule or a change in voltage) causes the protein to change conformation, moving the gatekeeper amino acid(s) out of the way. These amino acids can be specifically hydrophobic or hydrophilic or even charged (Chapter 1), depending on the molecule to be transported.

Ionophore Antibiotics/Antifungals: The important function of membrane channels in maintaining the correct balance of various ions inside and outside of cells has been utilized in the development of an important class of antibacterial and antifungal medications—the "**ionophores.**" This class of drugs includes peptide- or lipid-based molecules that insert into biological membranes. Although mechanisms differ slightly, all disrupt the ionic balance of the target cell by allowing unregulated or abnormal movement or permeability of ions, leading to cell death. The structure of certain ionophores lends itself to selective insertion into only bacterial or yeast cells with limited or no effect on human cells, thereby allowing treatment of the infection without adverse effects on the human patient.

Active transport (Figure 8-6D) involves a carrier protein that undergoes a conformational change via the release of energy or phosphorylation from nucleotide molecules. Active transport can move molecules into or out of a cell. A prime example of active transport is the **sodium–potassium ATPase (Na^+–K^+-ATPase) pump** (Figure 8-6E). The Na^+–K^+ ATPase pump helps to establish a high concentration of Na^+ outside and a high concentration of K^+ inside a cell. This Na^+–K^+ gradient is required for nerve and muscle function as well as to drive other channels to transport carbohydrates and amino acids and nutrients into cells. Proper Na^+ and K^+ cellular concentrations also help to maintain the correct cell volume because these ions affect the movement of water molecules into and out of cells. Other molecular pumps will be discussed in later chapters.

Membrane Channels and Channel Blockers: Membrane channels of all types are found throughout the human body performing a wide variety of functions. These channels are prime targets for the development of various medications to treat diseases associated with membrane channel function and/or dysfunction.

One example is calcium (Ca^{2+}) channels, which are important in the rate and force of heart contraction as well as contraction/relaxation of smooth muscle in blood vessels. Patients with high blood pressure are often treated by a class of medications known as **calcium channel blockers (CCBs)**, easily identified by the suffix "-dipine." CCBs decrease the total transport of Ca^{2+} into muscle cells, thereby decreasing heart contraction (rate and force) and increasing the diameter of blood vessels leading to lower blood pressure.

Histamine (H_2) receptors and the **H^+–K^+ ATPase proton pump** are both responsible for the transport of acidic hydrogen ions (H^+) into the stomach to aid digestion. In diseases such as heartburn, gastritis, peptic ulcer disease, reflux disease, Barrett's esophagitis, gastrinomas, and Zollinger–Ellison syndrome, it is helpful to lower the amount of acid content in the stomach. In each of these cases, specific **H_2 blockers** with the suffix "tidine" and **proton pump inhibitors**, ending with the suffix "prazole," block H^+ transport.

A "carrier protein" can also facilitate the movement of large or charged molecules through lipid membranes. In this instance, the charged molecule to be transported binds to the membrane protein and, via a conformational change, is surrounded by the carrier protein's hydrophobic exterior. This carrier protein can then cross from one side of the lipid bilayer to the other with its molecular passenger safely shielded from the lipid tail groups only to emerge on the other side where it can release the molecule into a friendlier hydrophilic environment.

Aquaporins: Aquaporins are a family of integral membrane proteins found in the kidney and other organs, which help to regulate water flow into and out of cells. Defects of one of these channel proteins **aquaporin-2 (AQP2)** cause excessive loss of water in some patients with **nephrogenic diabetes insipidus** or increased water retention in **pregnancy** and **congestive heart failure**. Use of the psychiatric drug lithium can induce problems by decreasing the amount of AQP2, whereas the peptide hormone vasopressin works via increasing AQP2 concentrations in membranes.

MEMBRANE SIGNALING

Many integral membrane proteins do not form a channel or physically move through the membrane but, instead, transmit a message (signal) from one side of the lipid bilayer to the other. The **signaling** function of membrane proteins is seen frequently and is vastly important in human biology in multiple cell types. These membrane proteins function by a change in their own conformation from external binding of a molecule, phosphorylation of one of its amino acids, and/or then interaction with other proteins, and so on. This conformational change affects the internal portion of the signaling protein and can then activate internal proteins, leading to a selected function(s). There are several different types of membrane protein signaling, summarized in Table 8-1.

The first example of signaling molecules in Table 8-1 are the **steroid hormones** (see Chapter 3) and can be further divided into five categories, depending on their particular receptor—**androgens**, **estrogens**, **glucocorticoids**, **mineralocorticoids**, and **progestagens** (precursors of a variety of progesterones). Vitamin D, although not technically a steroid, transduces signals such as steroid hormones and is, therefore, usually included with this group. The steroid hormones work in two different ways (Figure 8-7). First, the hormone may bind to a membrane surface receptor, which then either activates other signaling pathways inside the cell or moves from the membrane into the nucleus to activate DNA transcription factors. Second, and more commonly, steroids are lipid soluble and can, therefore, pass easily through the cell membrane without the aid of a membrane receptor. Once inside the cell, the hormone can bind with cytoplasmic receptors to activate signaling or may continue to receptors in the nucleus to activate transcription factors and DNA synthesis. Steroid hormones that affect DNA synthesis are called **genomic** (i.e., affect *genes*), whereas those that do not affect transcription are referred to as **nongenomic**.

One of the most predominant and best understood membrane signaling mechanisms is via the **G-protein** family (see Table 8-1). All of these receptors rely on the conformational change resulting from the exchange/conversion of the nucleotides guanosine diphosphate (GDP) and guanosine triphosphate (GTP) to convey an external signal to the inside of the cell. G-proteins are all composed of three internal peripheral protein subunits α-, β-, and γ-subunits, which associate with an integral membrane protein receptor. The β- and γ-subunits are closely bound and are represented as the dimer β/γ. G-proteins can be divided into five classes, $\mathbf{G_s}$, $\mathbf{G_i}$, $\mathbf{G_q}$, $\mathbf{G_{12/13}}$, and $\mathbf{G_t}$, depending on differences in the α-subunit, which affects interaction with the external signaling molecule (Table 8-2).

Despite the differences in the α-subunit and effects among the five classes, the general signaling mechanism is almost identical and is summarized in Figure 8-8 and as follows:

1. The α-subunit with a bound GDP interacts with the internal portion of the receptor as well as the β/γ-subunits.
2. Upon binding of the signaling protein to the exterior of the receptor, a GTP molecule displaces the GDP on the α-subunit, causing the dissociation of the α-subunit from the β/γ-subunits.

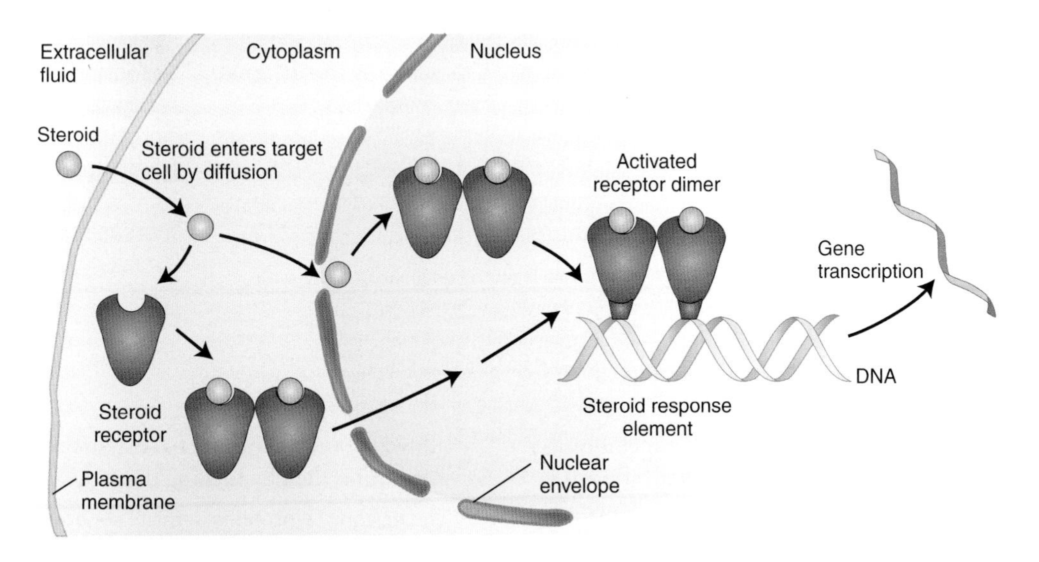

Figure 8-7. Activation by Steroid Hormone Receptor. A lipid-derived steroid hormone traverses the membrane to bind to a cytoplasmic receptor (bottom left), and the hormone–receptor complex enters the nucleus as a homodimer (upper). Alternatively, the unbound hormone may traverse through the nuclear membrane with subsequent binding to a hormone receptor to form a homodimer in the nucleus. Either mechanism leads to binding to specific parts of deoxyribonucleic acid (DNA) known as steroid response elements and activation of transcription of selected DNA. [Adapted with permission from Kibble JD and Halsey CR: The Big Picture: Medical Physiology, 1st edition, McGraw-Hill, 2009.]

TABLE 8-1. Summary of Membrane Signaling Receptors

	Group	Ligand Type	Hormone Examples	Activity
Intracellular receptor (crosses membrane)	I	Cholesterol-derived hormones	Steroids—androgens (testosterone), estrogens, glucocorticoids (cortisol), mineralocorticoids (aldosterone), progesterone; non-steroid—vitamin D_3	Bind internal transcription factors/initiates DNA synthesis
	I	Small, hydrophobic, signaling molecules	Retinoic acid, thyroxine (T4), triiodothyronine (T3)	Bind to specific protein factors that initiate synthesis of specific genes; increase various metabolic functions
	–	Ion channel	Some neurotransmitters, ions, and nucleotides	Various effects
Cell surface receptor (stays in/at membrane)	IIA (stimulatory)	Polypeptide hormones (G_s receptor)	Adrenocorticotropic hormone, antidiuretic hormone (ADH or vasopressin) (in kidney), corticotrophin-releasing hormone, calcitonin, epinephrine (β-adrenergic), follicle-stimulating hormone, glucagon, luteinizing hormone, parathyroid hormone, prostaglandins, thyroid-stimulating hormone	Increased cAMP (adenyl cyclase)
	IIA (inhibitory)	Polypeptide hormones (G_i receptor)	Epinephrine (α_2- adrenergic), somatostatin	Decreased cAMP (adenyl cyclase)
	IIB	Polypeptide hormones (G_q receptor)	Epinephrine (α_1-adrenergic), angiotensin II, ADH (arterioles), gonadotropin-releasing hormone, thyrotropin-releasing hormone	Increased IP_3/Ca^{2+} and diacylglycerol (phospholipase C)
	IIC	Integral receptor tyrosine kinase activity (some rely on G_{13} receptor)	Epidermal growth factor, fibroblast growth factor, insulin, insulin-like growth factor I, platelet-derived growth factor, and other cytokines and growth factors	Autophosphorylation of specific residues on receptor leads to activation of other signal pathways (e.g., phospholipase C, MAP kinase)
	IIC′	Soluble receptor-associated tyrosine kinase (e.g., Janus kinase) activity	Colony-stimulating factor, growth hormone, leptin, prolactin, many cytokines (interferon-γ)	Autophosphorylation of receptor leads to phosphorylation/activation of transcription factors
	IID	Integral guanyl cyclase activity	Atrial natriuretic peptide	Increased cGMP (guanyl cyclase)

DNA, deoxyribonucleic acid; cAMP, cyclic adenosine monophosphate; MAP, mitogen-activated protein.

3. The freed α-subunit can then interact with other membrane-bound proteins (**effectors**), leading to their activation (G_s, G_q, $G_{12/13}$, and G_t) or inhibition (G_i). Increasing evidence also suggests that the β/γ-subunits can activate different effector protein signaling pathways (e.g., L-type Ca^{2+} channels involved in various CCB medications).
4. The α-subunit's inherent GTPase activity, often enhanced by an accessory GTPase activating protein or sometimes by the effector proteins, eventually converts GTP back to GDP, allowing this subunit to reassociate with the β- and γ-subunits, thereby turning the activation off.

The various G-proteins activate several important membrane proteins, which lead to the conveyance of the signal via **second messenger molecules**. Examples include **cyclic adenosine monophosphate (cAMP)** (Figure 8-9A), generated by **adenyl cyclase** (also known as adenylyl or adenylate cyclase), which leads to phosphorylation of any of several molecules by **protein kinase A** and (Figure 8-9B) cleavage of the membrane lipid

TABLE 8-2. G-Protein Receptor Classes

Class Name	Signaling Effects
G_s	cAMP production via adenyl cyclase and protein kinase A signaling (multiple targets)
G_i	Inhibits cAMP production; minor role in stimulation of phospholipase C
G_q	Stimulates phospholipase C
$G_{12/13}$	Activation leads to changes in the actin cytoskeleton and, therefore, regulation of cell cycle and motility
G_t	"Transducin" molecules found in rods and cones couple visual signals between rhodopsin and cGMP phosphodiesterase

cAMP, cyclic adenosine monophosphate; cGMP, cyclic guanosine monophosphate.

phosphatidylinositol 4,5-bisphosphate by **phospholipase C** to form **diacylglycerol** and **inositol triphosphate (IP_3)**. IP_3 subsequently leads to the release of Ca^{2+} generated from internal cellular stores (usually inside the ER) by subsequently activated ion channels and/or enzymes. These second messengers lead to multiple membrane, cytoplasmic, and nuclear effects. For example, the calmodulin kinases activate myosin molecules causing muscle contraction, help to regulate secretion of neurotransmitter molecules, regulate a variety of transcription factors, modulate carbohydrate storage and use, and may be essential to brain function. **Protein kinase C** is known to change membrane structure, regulate transcription and cell growth, assist in immune responses, and provide key activation of proteins involved in learning and memory. Importantly, the multistep, signaling process allows selected amplification of a small signal at the exterior of the cell membrane into a potentially large response within the cell.

The final group of membrane receptors (see Table 8-1) is the "soluble receptor-associated tyrosine kinases," including the **receptor tyrosine kinases**, **Janus kinases**, and **integral guanyl cyclase** (also known as guanylyl or guanylate cyclase) receptors (Figure 8-9C).

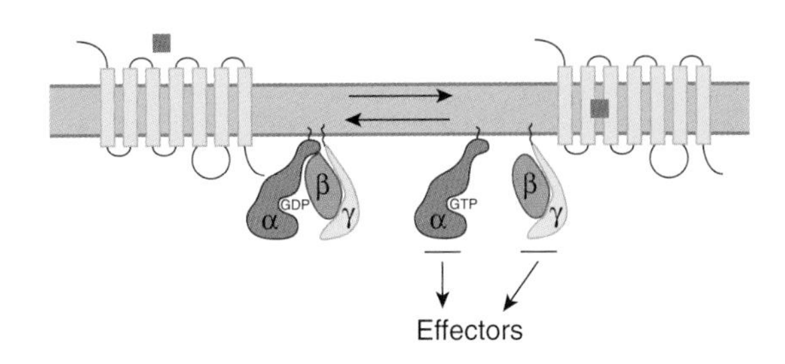

Figure 8-8. Function of G-proteins. The ligand (■) binds to its receptor, which interacts with the inactive α/β/γ complex with bound GDP (left) and causes an exchange of GTP for the GDP on the α-subunit with dissociation of the β- and γ-subunits (right). These subunits can then move within the membrane and serve as "effectors" to activate other protein signaling pathways. GTP can be hydrolyzed back to GDP via a GTPase causing re-association of the α–β–γ complex and termination of the signal. [Reproduced with permission from Barrett KE, et al.: Ganong's Review of Medical Physiology, 23rd edition, McGraw-Hill, 2010.]

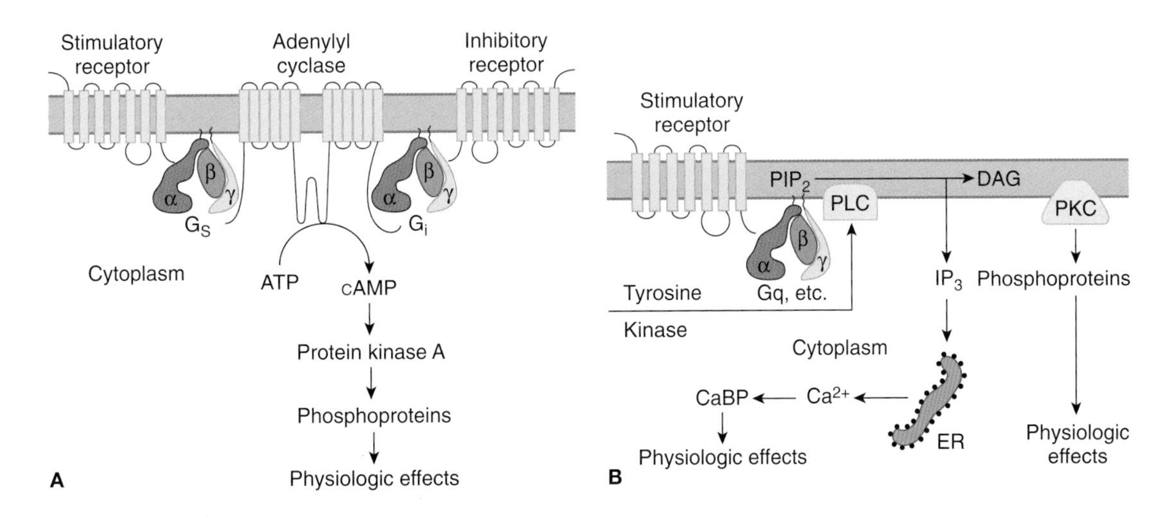

Figure 8-9. A. Cyclic adenosine monophosphate (cAMP) Signaling. Activation of adenylyl cyclase by the α-subunit results in the conversion of adenosine triphosphate (ATP) to cAMP which then activates protein kinase A and phosphorylation of various signaling proteins that elicit physiological effects such as the expression of specific genes. Stimulatory ligands activate via G_s proteins, whereas inhibitory ligands act via G_i proteins. **B. Phospholipase C–Protein Kinase C (PKC) Signaling.** Activation of phospholipase C, usually via the α-subunit of the G-protein G_q, results in the cleavage of the membrane lipid phosphatidylinositol 4,5-bisphosphate (PIP_2) into diacylglycerol (DAG) and inositol triphosphate (IP_3). Hydrophilic IP_3 leaves the membrane and enters the cytoplasm to release calcium (Ca^{2+}) from the endoplasmic reticulum (ER). Ca^{2+} subsequently activates Ca^{2+} binding proteins (CaBP) leading to physiological effects such as activation of enzymes and/or expression of specific gene products. Hydrophobic DAG remains in the membrane and can activate PKC leading to separate phosphorylation of proteins and resulting physiological effects. Ca^{2+} is also required to synergistically maximize this effect of DAG on PKC. *(continued)*

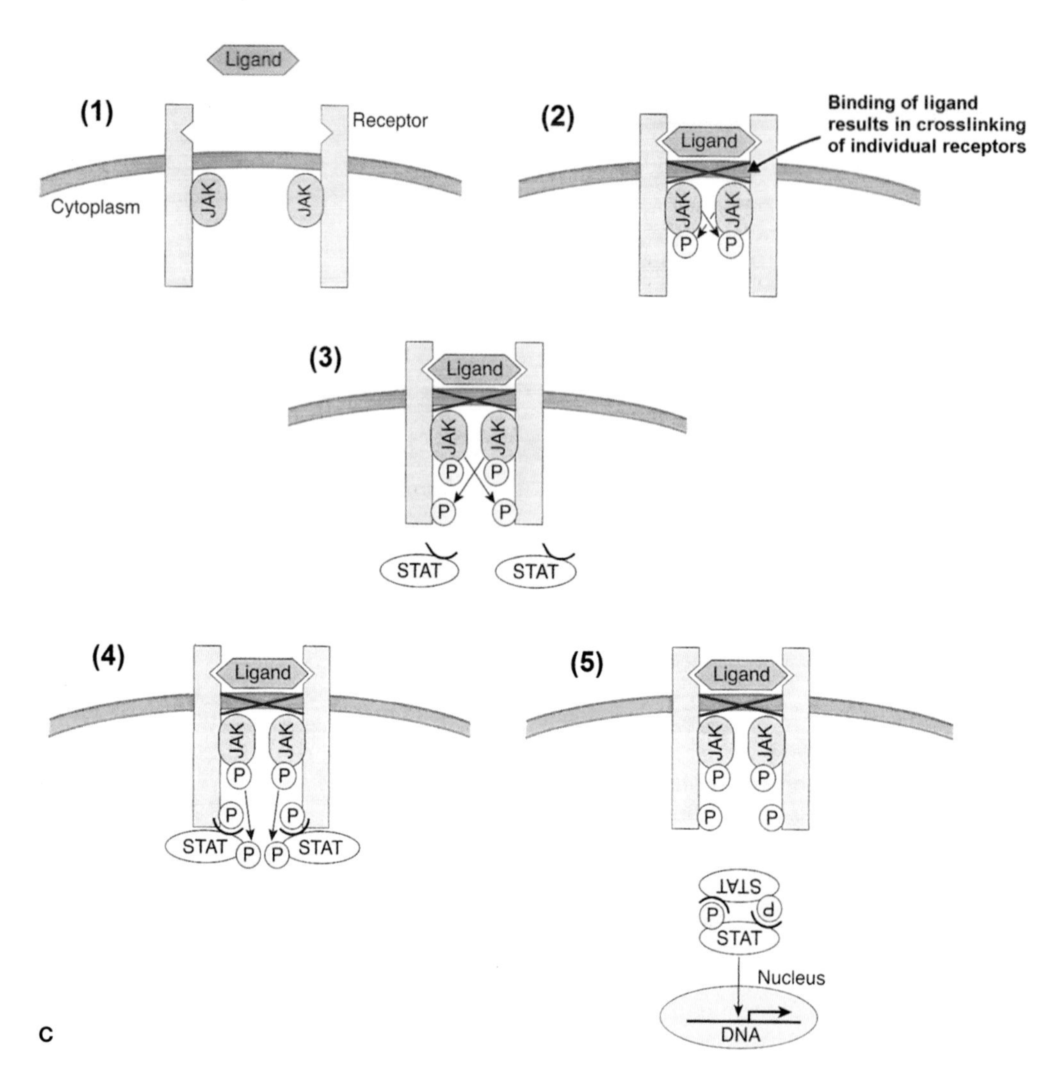

Figure 8-9. (*Continued*) **C. Janus Kinase and Signal Transducer and Activator of Transcription (JAK–STAT) Signaling.** The ligand binds to the monomeric form of the receptor (1), leading to dimerization and cross-linking of the two receptors. Each JAK cross-phosphorylates (see arrows) the tyrosine residue of other JAK, leading to their activation (2). The activated, dimeric JAKs cross-phosphorylate (see arrows) tyrosine residues on the other receptor (3). The phosphorylated receptors now contain an appropriate site for binding with a number of other signaling molecules such as STAT (note semicircular binding pockets on STAT molecules), leading to STAT phosphorylation (4). The phosphorylated, dimeric STAT molecules subsequently may travel to the nucleus and bind to response elements in deoxyribonucleic acid (DNA) with propagation of the initial signal by the expression of certain genes (5). [Adapted with permission from Barrett KE, et al.: Ganong's Review of Medical Physiology, 23rd edition, McGraw-Hill, 2010.]

All three receptors have the ability, after activation by binding of the signaling molecule, to carry out a specific enzymatic function (i.e., phosphorylation of its own selected tyrosine residues, production of cyclic guanosine monophosphate). These enzymatic changes then lead to activation of a signaling pathway specific for the signaling molecule.

REVIEW QUESTIONS

1. What are the basic structure, key features, and important components of the lipid bilayer and biological membranes?
2. How do specific phospholipids and cholesterol molecules affect the functioning of biological membranes?
3. What is meant by peripheral and integral membrane proteins and what functions can they play?
4. How would you describe simple, gated, and ATPase-dependent membrane channels and what is their significance?
5. What is the significance of signaling proteins, carrier proteins, and second messenger molecules?
6. What are the major classes of receptors and second messenger molecules?

CHAPTER 9

DNA/RNA FUNCTION AND PROTEIN SYNTHESIS

OVERVIEW

The nucleus is often represented as a relatively empty structure, containing only deoxyribonucleic acid (DNA) being replicated and transcribed along with a few accessory molecules to help in the process. To the contrary, the nucleus is actually a highly organized, membrane-bound structure that is literally filled with proteins, nucleotides, carbohydrates, and lipids with multiple functions. Various proteins are involved, along with the nuclear membrane, in the organization of chromosomes, which also helps to regulate the processes of DNA replication and transcription, and, subsequently, protein synthesis. Other proteins directly influence the expression of genes via direct interactions with specific nucleotide sequences. Posttranslational modifications affect both protein function and direct particular proteins to intracellular and/or extracellular destinations.

STRUCTURE OF THE NUCLEUS

The nucleus is surrounded by a double membrane called the **nuclear membrane** (also known as the nuclear envelope), with the outer layer being continuous with the endoplasmic reticulum (ER) in the cytoplasm and, like the ER, containing ribosomes and newly synthesized proteins. The inner and outer nuclear membranes are also continuous at the sites of **nuclear pores**. Approximately 2000 **nuclear pores** are contained within the nuclear membrane, and each pore can allow movement of about 1000 molecules in and out of the nucleus per second. These nuclear pores, composed of proteins called **nucleoporins**, are directly analogous to the membrane channels discussed in Chapter 8 and transport several types of molecules, including ribonucleic acid (RNA) and ribosomes, proteins, carbohydrates, and lipids. Smaller molecules and ions pass through the nuclear pore by simple diffusion, but larger proteins and RNA molecules are blocked by a spoke-like gate inside the channel and must be actively assisted by carrier proteins called "**importins**" or "**exportins**" by a process that requires two guanosine triphosphate (GTP) molecules. Each type of RNA molecule [messenger RNA (mRNA) and transfer RNA (tRNA)] that must be transported into the cytoplasm has an exportin and a specific amino acid (AA), **nuclear export sequence**, which directs them to a specialized nuclear pore for their selected transport out of the nucleus.

> **Leptomycins: Leptomycins A and B,** originally developed as antifungal drugs, specifically alkylate an importin, which results in inhibition of the nuclear export of several RNAs and transcription regulators in cell cycle control. An additional leptomycin is **HIV-1 "regulator, which allows HIV to take over host protein synthesis."** The leptomycins also stabilize **p53**, known to suppress tumor development/growth. Because of their cell cycle effects (see below), leptomycins are now being considered for cancer therapy.

HISTONES

The most important structure inside the nucleus is **chromatin**, consisting, in humans, of the 46 chromosomes and their associated proteins. These proteins enable not only efficient packing of over 12 billion nucleotides in human DNA, but also selective unwinding of these chromosomes to expose genes for DNA replication, DNA to RNA transcription and mRNA processing. **Histones** are one major example of these associated proteins, and are separated into six classes: H1, H2A, H2B, H3, H4, and H5. Two each of histones H2A, H2B, H3, and H4 assemble into an eight-subunit **nucleosome** and wrap 146 or 147 DNA base pairs of the simple, double helix structure (Figure 9-1) around the complex.

Several of these nucleosomes, linked together by about 50 DNA base pair sequences (Figure 9-2A), create an approximately 11-nm "beads on a string" chromatin structure (Figure 9-2B). Addition of the H1 histone (Figure 9-1) covers the entry and exit

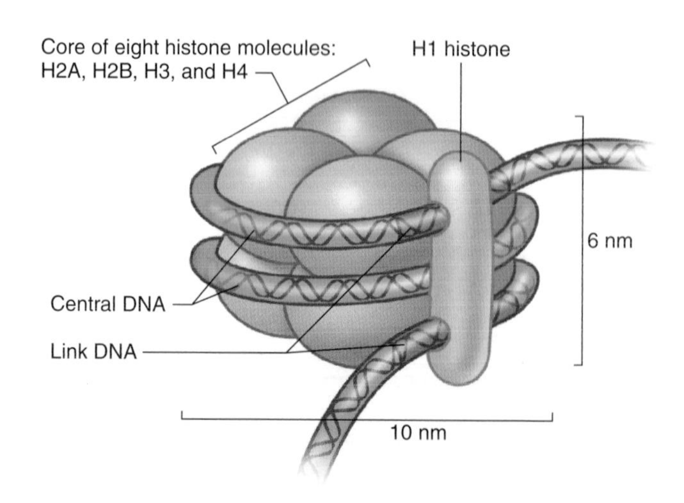

Figure 9-1. Histone Structure. The histone contains a core of histone molecules, including pairs of H2A, H2B, H3, and H4, wrapped by double-helix DNA and held together by histone H1. See text for further details. DNA, deoxyribonucleic acid. [Reproduced with permission from Mescher AL: Junqueira's Basic Histology Text and Atlas, 12th edition, McGraw-Hill, 2010.]

points of the DNA molecule and allows DNA to form coiled-coil structures, including the 30-nm chromatin fiber (Figure 9-2C). The 30-nm fiber can readily unwind into its 11-nm component to allow DNA replication and transcription. Analogous proteins called **protamines** are found in sperm and allow even denser packing of DNA in the sperm head than histones.

Important protein attributes (Chapter 1) include positive/negative/uncharged AAs (primary structure), alpha helices (secondary structure), and grooves formed between separate histone proteins that provide and optimize essential interaction points between the histones and the DNA molecules. The properties of nucleotides also contribute to this binding because increased numbers of adenosine (A) and thymine (T) residues at the minor groove of DNA allow improved nucleosome binding. Not surprisingly, histone AA sequences and, therefore, higher order structures are highly conserved in biology. However, histones do much more than just wind DNA to a manageable size. Several histones are modified on their N-terminal tails and globular areas (see posttranslational modifications below). These modifications include the addition of methyl, acetyl, phosphate, ubiquitin, and adenosine diphosphate groups among others. The modifications change how the proteins interact with the DNA molecule, to expose (e.g., acetylation) and conceal (e.g., methylation) particular parts of the DNA strand as a part of activation and inhibition of gene transcription (gene regulation), as well as the repair of DNA errors from both DNA replication and ongoing processes that lead to nucleotide mutations.

NUCLEAR MATRIX/SCAFFOLD

Further compaction of the structure is provided by the **nuclear matrix** or "**scaffold**" proteins, which help in maintaining the structure and integrity of the DNA molecules. The exact structure and function of this structural element of the nucleus are

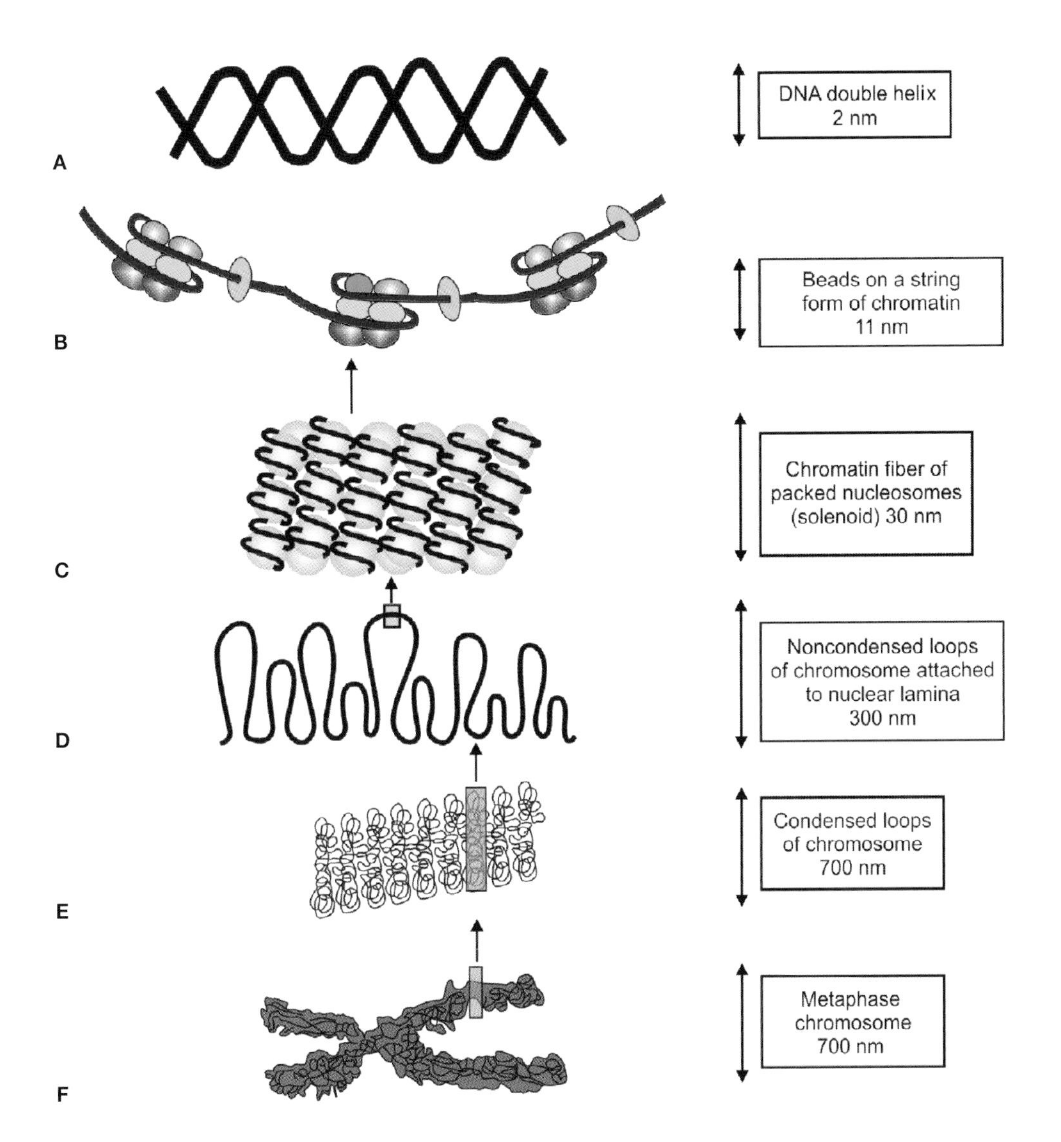

Figure 9-2. A-F. DNA Structure from Chromosome to Double-Helix. Initial DNA structure beyond the simple, double helix (**A**) results from histones H2A, H2B, H3, and H4 interactions with DNA to create (**B**) 11-nm "beads on a string" structure. (**C**) The subsequent addition of histone H1 creates the 30-nm fiber. Further tertiary and quaternary structures of DNA are created by the addition of various scaffolding proteins, which condense DNA into varying compact chromosome structures. The level of structure varies depending on the cell cycle stage and, as a result, the requirement for DNA transcription or replication [e.g, (**D**) noncondensed, interphase structures (300 nm) and (**E**) condensed, metaphase structure (700 nm), ending with (**F**) the highly compacted structure of the metaphase chromosome]. DNA, deoxyribonucleic acid. [Adapted with permission from Naik P: Biochemistry, 3rd edition, Jaypee Brothers Medical Publishers (P) Ltd., 2009.]

still highly disputed. However, the existence is generally accepted of a supportive, highly dynamic network that produces both chromatin structure and, thereby, its functions by influencing the molecules with which it interacts. In corroboration of the concept of a nuclear matrix, **"scaffold/matrix attachment elements" (S/MARs)** have been proposed/identified in which specific DNA sequences would be able to connect to the scaffolding proteins. The existence and function of S/MARs is also now considered in biotechnology involved in gene therapy and modulation of DNA expression to modify disease states.

The nuclear matrix or **lamina** is a structure thought to be a three-dimensional network of **intermediate filaments**, a third type of structural protein (Chapter 12), as well as peripheral and integral membrane proteins attached to the inner lipid bilayer. Within the nucleus, additional chromatin structure is created from the scaffolding proteins by the formation of further supercoiled structures. These higher orders of structure include a 30-nm zigzag formation and a 100-nm fiber (not shown), as well as a 300-nm fiber (Figure 9-2D) with noncondensed and, therefore, accessible loops and a highly condensed 700-nm fiber (Figure 9-2E) that is part of the final metaphase chromatin structure (Figure 9-2F) found in cells. Other "scaffolding" proteins link the chromatin to the nuclear matrix (essential for chromosome movement during cell replication) and the cytoplasmic matrix to the nuclear membrane and internal nuclear structures. Of course, chromatin structure changes dramatically from interphase throughout DNA replication.

Laminopathies: More than 2000 variants of medical disorders of the nuclear membrane are known, many of which are caused by mutations that affect the proteins of the nuclear membrane or lamina. Each affects an otherwise important role of proteins involved in transport or structure of the membrane bilayer or its membrane proteins, which directly affect DNA replication and translation and the processes of protein translation and transport. These "**laminopathies**" have far-reaching effects, often resulting in childhood or adolescent disorders of skeletal and cardiac muscles, lipid, skin, nerves, and white blood cells, as well as cancers, diabetes, premature aging, and even early death.

The nuclear matrix is not only actively involved in chromosome separation and cell division but also helps to maintain the required structure for each of these functions. The nuclear matrix creates a chromatin structure in the nucleus that is ordered and constrained in discrete territories that reflect specific functions, but which, at times of need such as mitosis, can alter dramatically while still maintaining control and organization. The result of the primary, secondary, tertiary, and quaternary folding of the DNA genome is the compaction of about 1.8 m of human DNA about 40,000-fold so it can fit inside a microscopic nucleus. Of course, this level of compaction changes throughout the cell cycle, being maximally compacted during mitosis and minimally compacted during interphase. Finally, changes in the DNA structure, including the marked condensation and expansion of chromatin, occur during the normal cell cycle.

Changing Nuclear Matrix in Cancer: Recent research has shown that early changes in many human cancers include a notable and measurable change in the proteins that make up the nuclear matrix. Although evaluation of these changes is still ongoing, the ability to detect these changes in the nuclear matrix may offer a reliable way to detect cancers at very early stages of the disease. Reversal of these changes may also offer clinicians, insights into prognosis and cures.

NUCLEOLUS AND RIBOSOME SYNTHESIS

Also found within the nucleus is the **nucleolus**, made of proteins and nucleic acids, where **ribosomal RNA (rRNA)** is produced and **ribosomes** are assembled for export into the cytoplasm. The protein components of ribosomes, often called "**r-proteins**," are made in the cytoplasm and transported to the nucleolus via a connected network of nuclear membrane pores and nucleolar channels. rRNA is transcribed in the nucleus by RNA **polymerases pol I, II, and III**, and often methylated or shortened. The resulting rRNA sequences are targeted to the nucleolus where they join with the r-proteins to make the 40S and 60S subunits of mammalian ribosome, which is subsequently exported through a nuclear pore to the cytoplasm.

DNA REPLICATION AND TRANSCRIPTION

The conversion of information in the genetic code to functioning proteins and nucleic acids relies on the ability of the DNA to replicate itself and to transcribe its code to mRNA for use in protein synthesis. DNA replication results in very few mismatched nucleic acid pairs, with an approximate error rate of only one mismatched nucleic acid per 10 million nucleotides. The process of transcription also has mechanisms that reduce errors, but these are far less efficient than that seen in DNA replication. Both processes involve several proteins essential to the process as functional or regulatory elements.

DNA REPLICATION

The process of uncoiling double-stranded DNA, faithfully copying each DNA strand and then separating the two, new, double-stranded copies, is called **replication**. The process starts at an **origin of replication (ori)**, a particular sequence of nucleic acids at which a **pre-replication complex (pre-RC)** can bind and the replication can start. Approximately 100,000 origins of replication can be found in each human cell, allowing the copying of DNA to proceed in a parallel fashion from all of these points, thereby speeding the process of DNA replication. The pre-RC is composed of the following four proteins: (a) a six-subunit **origin recognition complex** binds first to the origin of replication; (b) two cell cycle regulatory proteins, **Cdc6** and **Cdt1**, ensure that the cell is prepared for DNA replication; and (c) the **minichromosome maintenance complex**, which is believed to contain proteins essential for the establishment of a **replication fork**. The replication fork is the point where two DNA strands, one termed the **leading strand** and the other the **lagging strand**, are separated and DNA copying occurs. The coiled-coil, double-helical DNA structure examined in Chapter 4 is initially unwound by the enzyme **DNA helicase** (possibly part of the minichromosome maintenance complex) by breaking the hydrogen bonds between complementary nucleic acids. **Single-stranded binding proteins** attach to the new DNA strands to keep them separated. An enzyme termed **primase** then produces a short strand of RNA (sometimes with DNA) to serve as a primer for the remainder of the process. The enzyme **DNA polymerase** replicates each DNA strand in the 5′ to 3′ direction by adding the correct, matching nucleotide triphosphate to the 3′-hydroxyl end of the primer strand. As each new nucleic acid is added, a new phosphodiester bond is formed, utilizing the energy contained in the remaining diphosphate group. This process is continuous on the leading strand but, as DNA polymerase can only add in the 5′ to 3′ direction, short chains of newly added nucleic acids, called **Okazaki fragments**, are generated on the lagging strand. The enzyme **DNA ligase** joins the Okazaki fragments together as lagging strand replication proceeds. The process of replication along the coiled-coil structure of DNA soon leads to an unfavorable DNA conformation that is wound about itself. To relieve this problem, a **DNA topoisomerase** efficiently cuts the phosphate backbone, "untangles" the DNA strands, and then repairs the cut, leaving the DNA otherwise unaltered. This entire replication process is depicted in Figure 9-3.

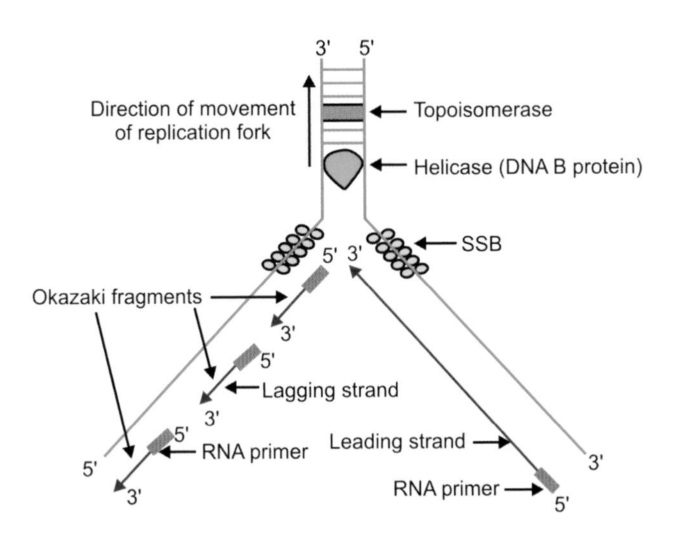

Figure 9-3. Overview of DNA Replication. Summary of DNA replication, illustrating the replication fork, leading and lagging strand synthesis, and the various proteins involved in replication and unwinding. A detailed description is offered in the text. DNA, deoxyribonucleic acid; RNA, ribonucleic acid. [Reproduced with permission from Naik P: Biochemistry, 3rd edition, Jaypee Brothers Medical Publishers (P) Ltd., 2009.]

TRANSCRIPTION

Transcription is the process whereby genetic information from a DNA sequence is utilized to create an equivalent RNA—mRNA if the gene codes for a protein, tRNA for a tRNA, rRNA for assembly of a ribosome, or even catalytic RNA, termed **ribozyme**. This DNA sequence contains not only the gene for the RNA (**coding sequence**) but also regulatory sequences that dictate when and how the particular RNA will be produced. Again, the DNA is transcribed or "read" from 3′ to 5′ and occurs only on one of the DNA strands, known as the **template strand**.

Transcription starts with binding of the enzyme **RNA polymerase** to a **promoter** sequence on the DNA, usually located from 10 to 35 bases before the start of the actual gene (Figure 9-4). RNA polymerase binding depends on a variety of protein **transcription factors**. Transcription factors vary in type and activity, but all interact in some way with the DNA (either directly or indirectly) and the RNA polymerase to either enhance or block binding to the DNA molecule. The variation of transcription factors is part of the regulatory mechanism that allows differential expression of genes. Once bound, the RNA polymerase travels down the DNA from 3′ to 5′ while matching the appropriate RNA to its DNA counterpart, utilizing uracil matched with adenine instead of thymine. As in DNA replication, energy for the formation of the phosphodiester bond is derived from hydrolysis of the two terminal phosphate bonds of the nucleoside triphosphate. Unlike replication, multiple RNA polymerases can transcribe on a single DNA gene sequence, allowing rapid production of the RNA product. The termination of transcription is still not well understood but is followed by the addition of several adenine units to the 3′-end of the new RNA (**polyadenylation**, also known as a poly-A tail). Next, the new RNA is usually spliced and reannealed to remove noncoding regions, and a 7-methyl guanosine nucleic acid is added to the 5′-end (known as a **5′ cap**) to protect the RNA from various enzymes (exonucleases) during its subsequent activities. If the RNA is destined for protein synthesis, the molecule is exported from the nucleus to the cytoplasm or ER, depending on the final destination of the protein (see Protein Trafficking below). The 5′ cap also aids in the recognition of mRNA by the ribosomes.

Reverse Transcriptase and Retroviral Therapy: Several types of viruses, including those that cause human immunodeficiency virus (HIV)/acquired immune deficiency syndrome, have the ability to transcribe their RNA into DNA as part of the process of infecting a host cell and utilizing that cell's replication machinery. The viral enzyme that performs this RNA to DNA replication is known as **reverse transcriptase**. Clinicians can utilize this unique ability of viruses against them by using a class of medications known as **reverse transcriptase inhibitors**, commonly called **retrovirals**. One such retroviral is **Zidovudine** or **Azidothymidine (AZT)**, an analog of thymidine (see figure below). Because reverse transcriptase enzymes are only found in these viruses and the thymidine analog is 100 times less specific for the patient's human DNA polymerases, the medication preferentially blocks virus replication. The azido group also has a hydrophobic quality that allows AZT to cross through the plasma membrane for easy delivery to infected cells, after which internal cellular enzymes convert the AZT molecule into an active 5′-triphosphate form that can incorporate into and terminate reverse transcription.

AZT molecule

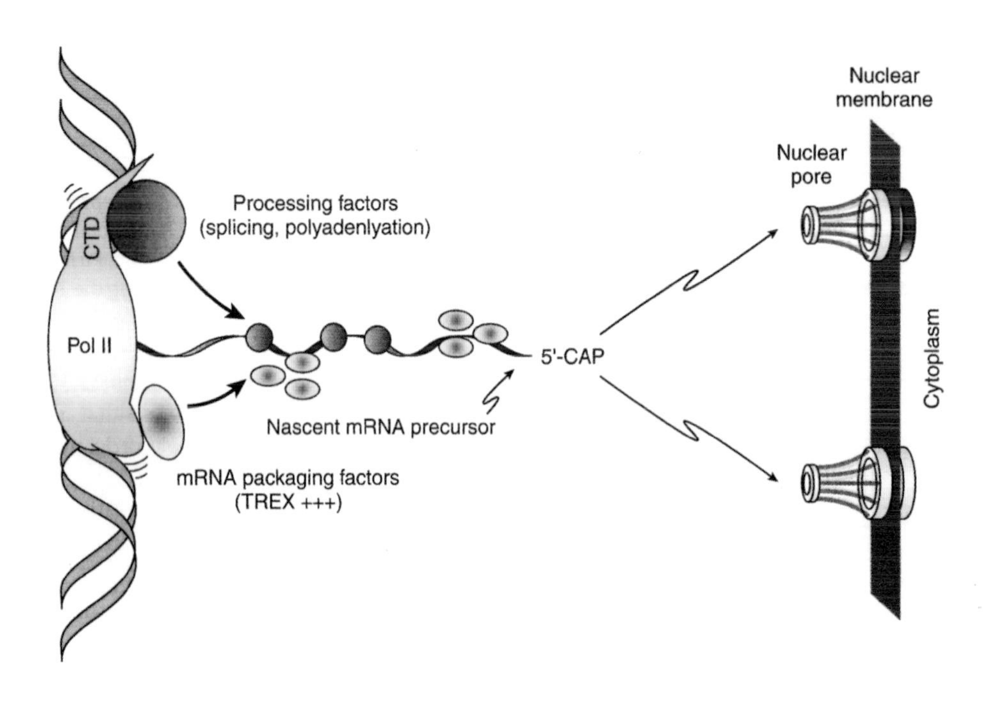

Figure 9-4. mRNA Processing and Nuclear Export. Following transcription of DNA to produce the initial complementary mRNA molecule, subsequent processing occurs, including splicing, polyadenylation, and addition of the 5 cap. The mRNA molecule bound for protein synthesis is then exported through nuclear pores into the cytoplasm. See the text for further details. CTD, carboxy terminal domain; DNA, deoxyribonucleic acid; mRNA, messenger ribonucleic acid. [Adapted with permission from Murray RA, et al.: Harper's Illustrated Biochemistry, 28th edition, McGraw-Hill, 2009.]

PROTEIN SYNTHESIS

Proteins are chains of AAs connected by peptide bonds formed between the carboxylic acid (COOH) group of one AA and the amino group (NH_3) of the second AA (Chapter 1, Figure 1-4A). Because each protein has a specific AA sequence, mRNA copied from the DNA gene template for that specific protein provides the nucleotide sequence/codon (Chapter 4) instructions to link the proper AAs together. Enzymes bring together the RNA and the necessary substrates along with an energy source, ATP, for efficient production of proteins. This process, called "**translation**," is briefly described below and summarized in Figure 9-5.

Protein synthesis requires the concerted interaction of mRNA, tRNA, several accessory proteins, called **initiation factor** (**IF**) and **elongation (EF) factor**, and **ribosomes**. The mRNA molecule is copied from the DNA gene and contains the copy of the nucleotide sequence for the chosen protein. Individual tRNA molecules can bind to only one specific type of AA. This **tRNA–AA** binds to the mRNA molecule via **anticodons** using the same binding rules as DNA double strands (e.g., a tRNA that binds the starting mRNA codon AUG has an anticodon sequence of UAC), insuring the specific order of AAs required for proper production of the protein. The IF and EF accessory proteins serve a number of roles, including enabling binding of the mRNA molecule to the ribosome, movement of the mRNA along the ribosome to the start point of the synthesis, docking of the tRNA–AA, joining of the 60S subunit to the 40S–mRNA complex (Figure 9-5A), and movement of the mRNA and growing peptide chain (Figure 9-5B), as well as accuracy assurance. IF and EF roles will not be discussed further.

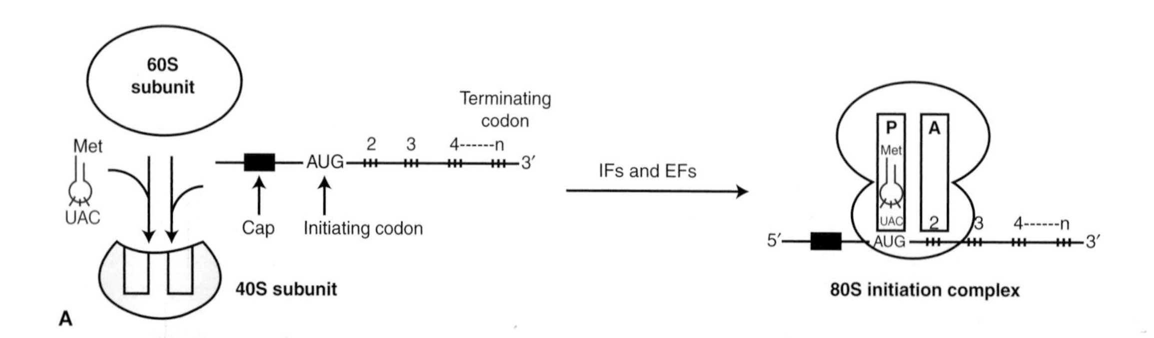

Figure 9-5. A. Formation of the Ribosome Complex for "Translation." The components of protein synthesis include the 60S and 40S ribosomal subunits, mRNA, and tRNA–AA molecules, the first of which codes for methionine (AUG mRNA sequence with analogous UAC tRNA sequence). Several initiation factors (IFs) and elongation factors (EFs) are involved in the formation of the active 80S ribosomal complex. (*continued*)

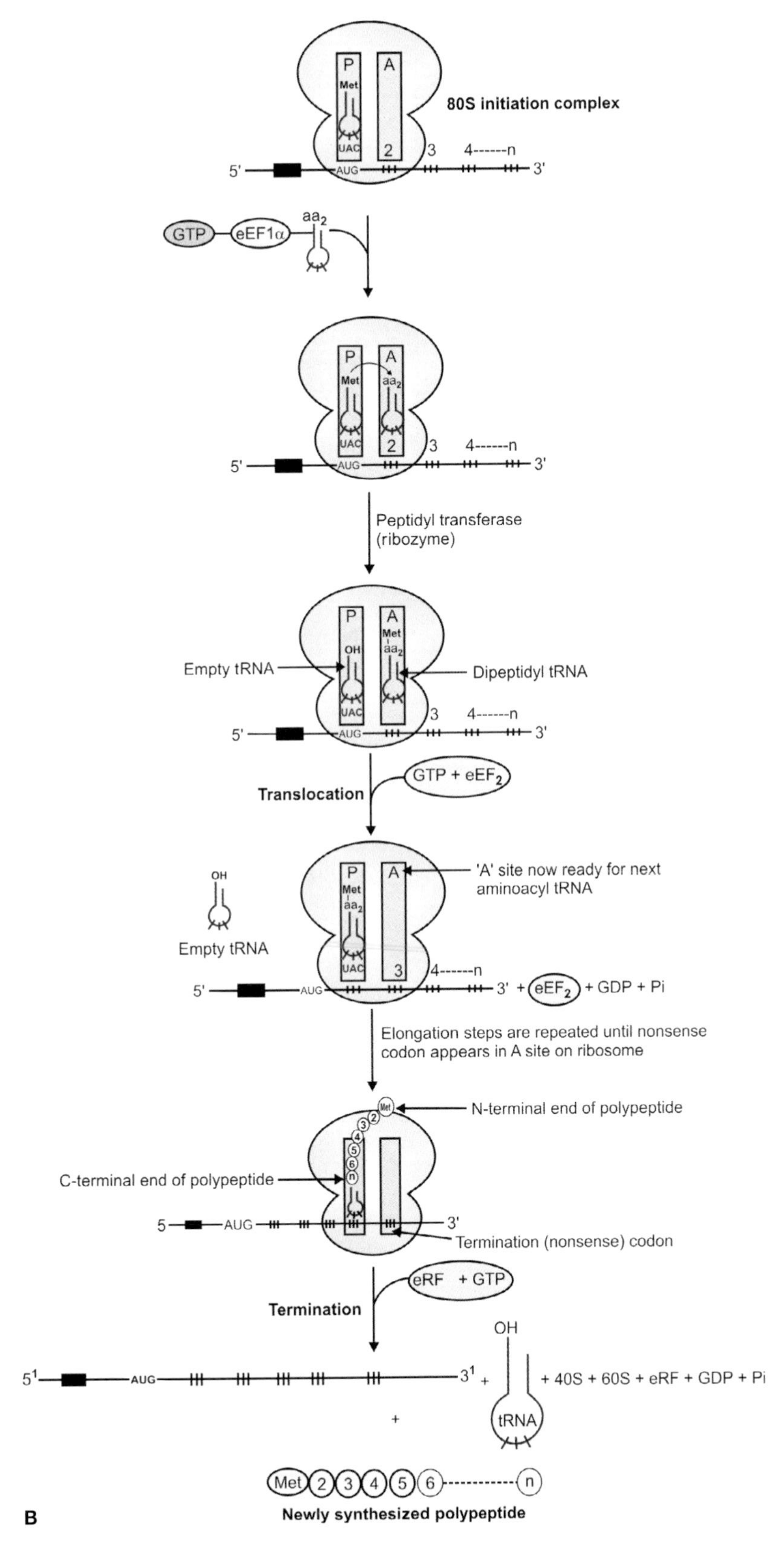

Figure 9-5. *(Continued)* **B. Protein Synthesis ("Translation").** Simplified model of protein synthesis. The first tRNA–AA is bound to the P-site followed by binding of the second tRNA–AA to the A-site [aided by elongation factor 1α ($eEF_{1\alpha}$)]. A peptide bond is formed between the two AAs, and the mRNA moves down the ribosome [aided by elongation factor 2 (eEF_2)], releasing the first tRNA molecule and freeing the A-site, so another tRNA–AA may bind. This process is repeated until a nonsense codon is reached in the mRNA sequence, signaling the end of the protein. Termination is followed by dissociation of the ribosomal complex (aided by eRF) and release of the new peptide. AA, amino acid; GTP, guanosine triphosphate; mRNA, messenger ribonucleic acid; tRNA, transfer ribonucleic acid. [Adapted with permission from Naik P: Biochemistry, 3rd edition, Jaypee Brothers Medical Publishers (P) Ltd., 2009.]

Elongation Factors, Bacteria, and Disease: *Corynebacteria diptheriae* causes the disease **diphtheria**, which, until effective immunization was available, caused up to 15,000 deaths a year, mainly in children and older adults. The bacteria *Pseudomonas aeruginosa* causes frequent infections of burns, wounds, and medical devices (catheters, ventilators, etc.), as well as serious problems in immunocompromised persons. **Rotaviruses** cause millions of childhood deaths each year from serious diarrhea. All three of these infective agents work by mimicking human accessory proteins and binding to elongation factors. Although the two bacteria inhibit an elongation factor, rotaviruses actually "hijack" the factor and use the host's protein synthesis machinery for reproduction and further infection.

The ribosome is a two-part structure composed of proteins and four strands of rRNA. The first part of the ribosome is a 40S subunit ("S" stands for "Svedberg," a unit of molecular size) along with one rRNA, which helps in establishing the proper AUG codon start site on the mRNA. The 40S subunit also contains the binding sites for the mRNA strand and two binding sites for incoming AAs—the "**P-site**" (named for the peptide chain that grows at this site) and the "**A-site**" (named because new AAs bind here). The second part of the ribosome is a 60S subunit with three rRNA molecules, which help in stabilization of the entire ribosome, peptide bond formation, and movement of the ribosome, along the mRNA strand. The 60S subunit binds to the 40S subunit only after the mRNA is in place, the first tRNA–AA is bound at the P-site, and the AUG start codon is found and correctly aligned. Assembly of this complex relies on the energy obtained from GTP. Following the formation of the mRNA–first tRNA–AA–ribosome complex, a second tRNA with its specific, bound AA binds to the A-site. A peptide bond is formed between the two AAs and the ribosome moves down the mRNA (a process that utilizes ATP energy), the P-site tRNA is released, and the process repeats itself until the end of the protein is reached.

Macrolides and the Machinery of Protein Synthesis: The class of drugs called macrolides includes the antibiotics **azithromycin**, **clarithromycin**, **erythromycin**, and **roxithromycin** (used for many respiratory, urinary, and soft-tissue infections), and the immunosuppressant drugs **tacrolimus** and **sirolimus** (used after organ transplants to reduce organ rejection). These medications block the movement of the ribosome along the mRNA molecule. Macrolides take advantage of the different 50S ribosomal subunit found in bacteria and do not affect the human 60S subunit. Macrolides also tend to concentrate in white blood cells and are, therefore, conveniently transported to the specific sites of infection. As a result, macrolides selectively and efficiently treat bacterial infections without harm to the patient. Many other classes of antibiotics take advantage of the differences in human and bacterial protein translation for their selective actions.

POSTTRANSLATIONAL TRAFFICKING/ MODIFICATION

Inherent in the process of protein synthesis/translation is the production of a string of AAs, the primary protein structure, which must fold into the correct secondary, tertiary, and quaternary structures to produce a functioning final protein and/or protein complex (Chapter 1). Although some smaller proteins are able to simply fold as translation occurs and the AA sequence is generated, most of the larger and/or more complex proteins require specialized proteins, known as **chaperones**. These chaperones assist in the resulting folding and unfolding of single, sequence proteins (i.e., secondary and tertiary structures) and the assembly of multiple proteins (quaternary structure).

Some chaperone proteins are known as **"heat shock" proteins** because they are sometimes preferentially utilized to stabilize protein folding at times of cellular stresses such as an increase in temperature. In fact, a major role of chaperones is to temporarily stabilize the newly synthesized AA chain, especially in a cytoplasm already filled with other proteins, so it does not randomly aggregate into a nonfunctional structure. Other chaperones are directly involved in the actual folding, dissociating from their associated protein after the final structure is achieved. Still other chaperone proteins are involved in some membrane transport and even the breakdown of certain proteins. Because of their close association with protein translation, chaperones are often found in or near the ER.

Protein trafficking or **targeting** directs a protein made for a particular location (e.g., nucleus, cytoplasm, membranes, organelles, or extracellular) to its correct destination. This process relies on **signal sequences** normally contained in the first 20–30 AAs of the protein. Exceptions include proteins bound for the nucleus, which have signals throughout their entire sequence, some peroxisome-bound proteins with a signal sequence in the last three AAs, and mitochondrial proteins, which have two signal sequences to direct them to either the mitochondrial matrix or the intermembrane space. In addition, protein synthesis for cytoplasmic-, nucleus-, and mitochondria-bound proteins usually takes place on **free ribosomes**, whereas extracellular-, peroxisome-, and membrane-bound proteins are usually synthesized on **ER-bound ribosomes**. Accessory proteins called **sequence recognition particles (SRPs)** and **SRP docking protein** also help to direct proteins targeted for secretion/peroxisome/membranes to the ER and then the Golgi apparatus for further sorting and routing.

The final step in the synthesis of many proteins is **posttranslational modification**, the addition or removal of AAs (e.g., signal sequence and initiation methionine), carbohydrates (e.g., specific carbohydrate sequences for glycoproteins/glycosaminoglycans), or chemical modifications (e.g., addition of lipids, phosphate, hydroxyl or methyl groups, and/or cysteine–cysteine disulfide bonding). These changes usually involve extracellular or membrane-bound proteins and, thus, generally occur in the ER and/or Golgi apparatus. Proenzymes (Chapter 5) can also be converted to their active form at this time.

Protein Folding, Chaperones, and Mad Cow Disease: Mad cow disease/Bovine spongioform encephalopathy and its human equivalent **Creutzfeldt–Jakob disease** result from infectious, misfolded proteins known as **prions** (***pr****otein-aceous and* ***in****fectious viri****ons***). Although current research is still ongoing and is often very controversial, prions are felt to infect their host as a normally folded protein, which then replicates and misfolds into a tightly packed structure of β-sheets, referred to as an "amyloid fold." These proteins aggregate into structures known as amyloid plaques, which create "holes" in the normal brain tissue and a resulting "spongioform" appearance that is typical of the disease (see picture below). Prions affect the brain/nervous tissue leading to detrimental and fatal neurological symptoms, including dementia, memory/speech/movement/balance problems, marked personality changes, hallucinations, and seizures. Some researchers are looking into the possible role of chaperone proteins in the prevention and/or treatment of this disease.

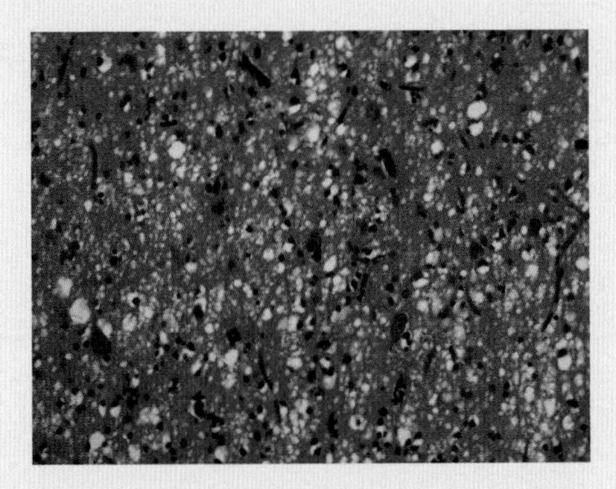

Courtesy of Dr. Dennis K. Burns, Department of Pathology, University of Texas Southwestern Medical Center

CONTROL OF GENE EXPRESSION

The ability of the body to control the expression of genes and their resulting mRNAs and proteins allows the selective initiation and cessation of the various biological reactions that allow life. Because conservation of resources is important, most regulation occurs at the level of gene transcription, although translational control as well as posttranslational activation via protein modifications also plays an important part in the overall regulation.

Transcriptional control (DNA → RNA) relies on **promoter** regions of DNA, usually, but not always, located close to the starting point of the transcription. These promoter regions allow binding of the RNA polymerase and regulatory **transcription factor** and **enhancer** proteins. DNA sequences that appear in several different promoter regions have been identified, such as the **TATA box**, an eight base-pair sequence often exclusively consisting of adenine and thymine nucleotides. Many transcription factors also share conserved secondary and tertiary protein structures that allow them to bind to particular parts of the DNA sequence; examples are shown in Table 9-1). The binding of different transcription factors and/or enhancers to these sequences are essential for transcription and can increase or decrease the transcription of a specific gene. **Steroid hormones** (Chapter 3) are specific type of transcription factors that bind often to unique receptors (cytoplasmic or nuclear). These steroid receptors subsequently bind to specific DNA sequences called **steroid response elements** for steroid activation of particular genes. Steroid receptors always contain an area for steroid binding, a modified zinc finger structure for DNA binding, a sequence that activates transcription, and an area that allows receptor dimerization. Examples of steroids and their action will be covered in Section III.

Iron Response Element (IRE): IRE is a specific span of approximately 25–30 nucleotides found in the mRNA for proteins such as **ferritin** (responsible for controlled iron storage) and **transferrin** (responsible for iron transport). These nucleotides bind to form a secondary hairpin structure, which can then bind proteins involved in regulation of iron concentration in the human body. During times of low iron concentration, these proteins bind to the IRE of ferritin and decrease its translation to increase the available iron by decreasing its storage. These same proteins bind to response elements found in transferrin, which stabilizes the mRNA leading to increased translation and, therefore, increased iron acquisition. IREs, therefore, allow regulation at the RNA level of concentrations of free and stored iron.

Gene expression is also controlled by the reversible acetylation of histone, lysine AA residues in nucleosomes, carried out by specific enzymes known as **histone acetyl transferases (HATs)** and **histone deacetylases**. The addition of acetyl groups allows access of RNA polymerases to the DNA, whereas the removal of the acetyl groups blocks this access. Some transcription factors or their associated proteins actually contain HAT activity to allow them access to their target genes. In a similar way, the addition of methyl groups to cytosine nucleotide residues usually results in decreased transcriptional activity, and the removal of the methyl group is usually required for expression of that gene. Finally, several alternative forms of gene expression are also seen. In some instances, promoters only occur in particular tissue or cell types. Alternative splicing/removal/reannealing can also occur, affecting mRNA molecules that contain introns that create several different related mRNAs/proteins with differing regulation (up to 30% of human genes may be affected by this process). Deactivation of one allele of a gene or even an entire chromosome can also occur as is seen by the inactivation of one of the X chromosomes in females during initial embryo development.

TABLE 9-1. Examples of Some Common DNA Binding Motifs

Name	Structure	DNA Binding
Helix–turn–helix		The second helix provides majority of binding to the major groove of DNA promoter region, whereas the first helix stabilizes this interaction.
Helix–loop–helix	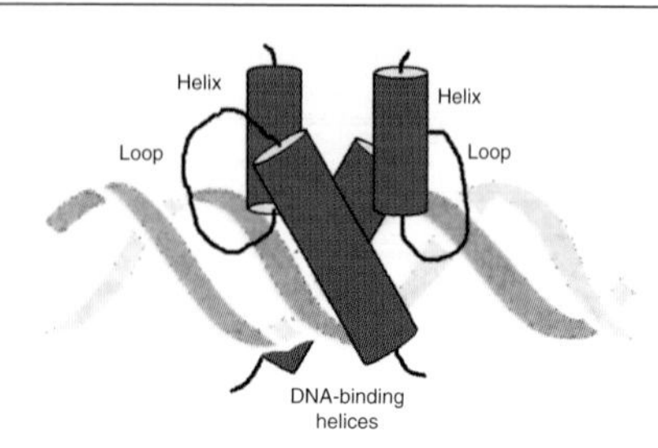	Basic AAs in helices facilitate DNA binding. One helix binds to an "E-box" nucleotide sequence of C–A–X–X–T–G, where "X" is any nucleotide.
Leucine zipper	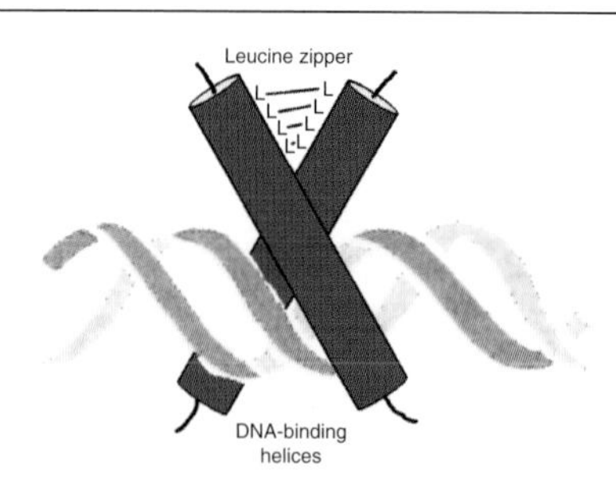	Protein dimer that binds at multiple leucine residues and inserts into and binds with the major groove of DNA via hydrogen bonding and hydrophobic forces.
Zinc finger	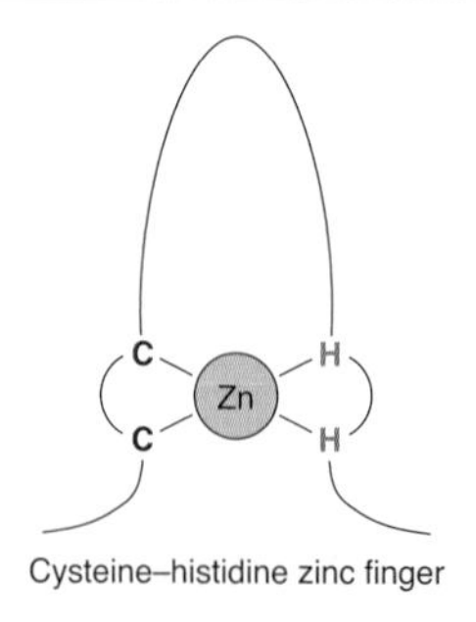 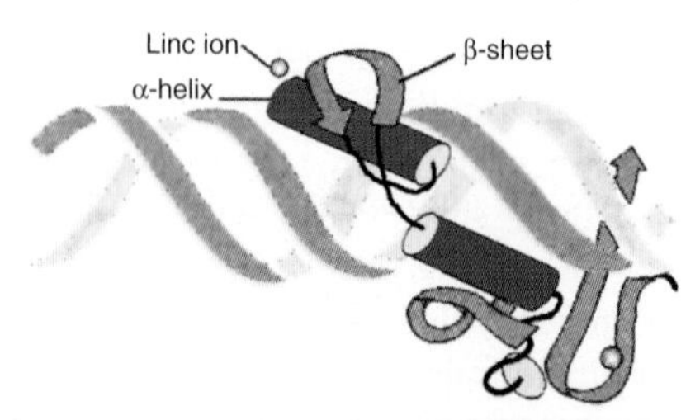 Reproduced with permission from Murray RA, et al.: Harper's Illustrated Biochemistry, 28th edition, McGraw-Hill, 2009.	Combination of α-helices and β-sheets stabilized by cysteine and histidine AA binding of zinc. Usually two to four repeated zinc fingers bind to the major grooves of DNA via the α-helices.

AA, amino acid; DNA, deoxyribonucleic acid.

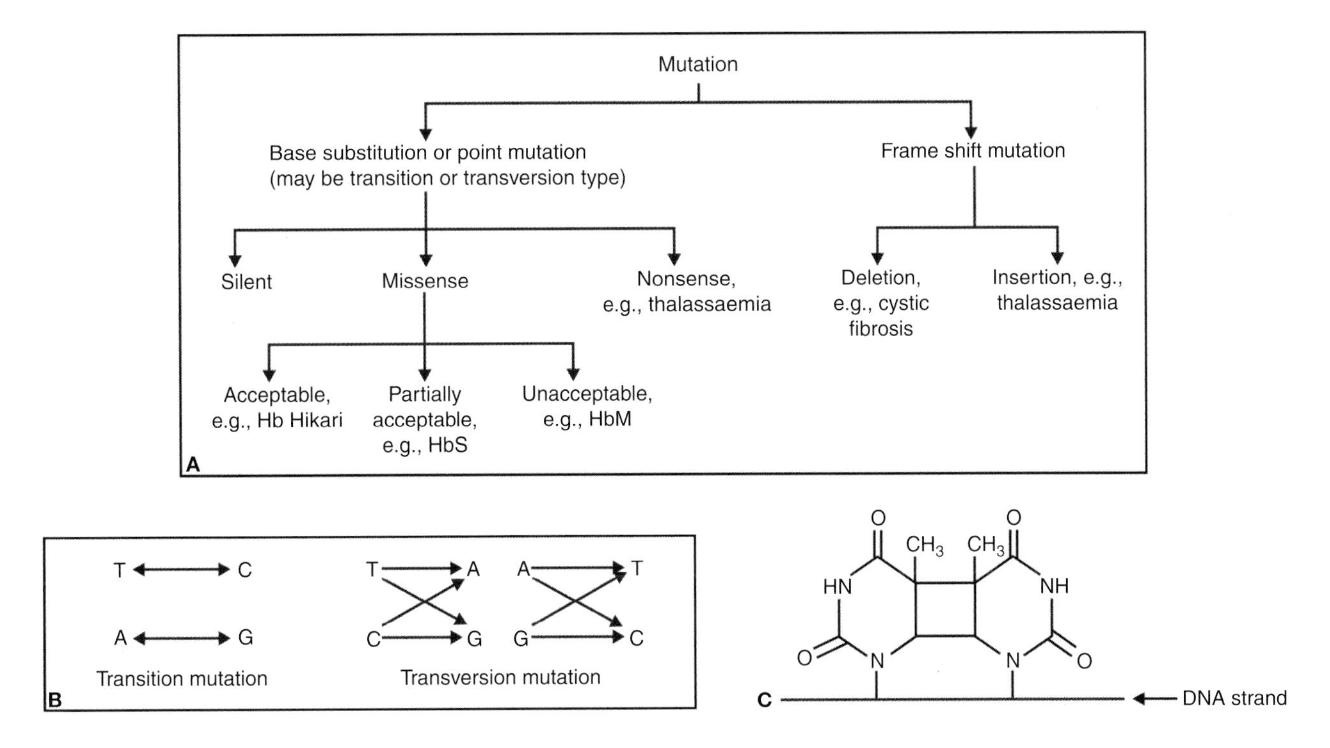

Figure 9-6. A-C. DNA Mutations. Overview of mutations and some of their clinical implications (**A** and **B**) and example thymidine dimer (**C**). See text for further discussion. DNA, deoxyribonucleic acid. [Reproduced with permission from Naik P: Biochemistry, 3rd edition, Jaypee Brothers Medical Publishers (P) Ltd., 2009.]

MUTATIONS AND REPAIR MECHANISMS

Accidental changes or "**mutations**" in DNA or RNA sequence (Figure 9-6) are estimated to occur at an approximate rate of 1000–1,000,000 per cell per day in the human genome, and every new cell is believed to contain approximately 120 new mutations. One kind of mutation is **point mutation**, which includes two different types (Figure 9-6B): **transition** occurs when a single purine nucleotide is changed to a different purine (A ↔ G) or a pyrimidine nucleotide is changed to a different pyrimidine nucleotide [C ↔ T(U)] and **transversion** occurs when the orientation of a single purine and pyrimidine nucleotide is reversed [A/G ↔ C/T(U)]. Point mutations can be further classified as **silent** (when the same AA is coded), **missense** (when a different AA is coded), **neutral** (when an AA change occurs but does not affect the protein's structure or function), or **nonsense** (when a stop codon results, terminating translation and shortening the resulting protein). **Insertion** and **deletion** mutations involve more than one nucleotide and can not only change the translated protein, but, if not occurring in multiples of three nucleotides, can also change the entire reading of the mRNA sequence (**frameshift** mutation). Chemical-, radiation-, and even viral-caused mutations can also cause single- or double-**strand breaks** and/or nucleotide–nucleotide or nucleotide–protein cross-linking. A common example of cross-linking occurs between adjacent thymidine dimers whose purine bases can form a cyclobutane ring when exposed to ultraviolet light (Figure 9-6A–C).

Although some of these mutations are beneficial, offering resistance to disease or improved structure and/or function, they can often lead to disease and/or death of the cell or organism. The human body has, therefore, developed a variety of repair mechanisms to attempt to restore the altered nucleotide sequence. In the case of single-nucleotide mutations, particular **endonuclease** enzymes sense and remove (**excision**) the incorrect base and the correct nucleotide is replaced by a polymerase and ligase. A particular example is the endonuclease specific for the repair of thymine dimers. Strand breaks are repaired by ligases or, when complex, by excision and subsequent retranslation/ligation. Cell cycle check points (see below) also offer the body the chance to verify the accuracy of the DNA sequence prior to commitment to cell replication and division.

Thymidine Dimer Repair and Xeroderma Pigmentosum: Although a relatively uncommon disorder **Xeroderma pigmentosa** is a direct example of how an autosomal recessive mutation in the enzyme(s) required for the repair of thymidine dimers can lead to a disease. The loss of this important repair mechanism results in diseases such as **basal cell carcinoma**, **squamous cell carcinoma**, and **malignant melanoma**—all as a result of unrepaired, ultraviolet light damage. The only treatment for affected individuals is reducing the exposure to sunlight and supportive care, although most patients die by the age of 20 years.

REGULATION OF CELL GROWTH AND DIFFERENTIATION

Cell cycle is the ordered pathway undertaken by all mammalian cells, which defines not only times of both cell division and replication, but also those times of cell growth and functional activities. The cell cycle is divided into **G_1 (gap 1), S (synthesis), G_2 (gap 2/interphase)**, and **M (mitosis)** phases. Cells can also enter a **G_0 ("resting")** phase. The progression through the cell cycle depends on a variety of proteins, which insure that a cell is ready for the next cell cycle phase ("checkpoints"). These proteins include two types, the **cyclins** and the **cyclin-dependent kinase (CDKs)**, which join in different heterodimer combinations that regulate various parts of the cell cycle. The cyclin component of this protein complex regulates the function of the phosphorylation activity of the bound CDK. Different concentrations of cyclins during the cell cycle, as determined by regulation of their protein synthesis in response to various molecular signals, lead to activation or inhibition of a CDK. An overview of the cell cycle and the role of cyclins and CDKs are shown in Table 9-2 and Figure 9-7.

Microtubules in the Cell Cycle and Cancer Treatment Strategies: The growth of microtubules during G_2 and their subsequent breakdown during M phases of the cell cycle have led to varying strategies in the treatment of cancers. One such cancer agent is **vincristine**, which binds to tubulin protein and blocks microtubule polymerization, stopping mitosis at metaphase. Vincristine is used in **non-Hodgkin's lymphoma**, **Wilm's tumor**, and **acute lymphoblastic leukemia** treatments among others. **Taxol** is another important cancer treatment medication obtained from the Pacific yew tree and is used for patients suffering from cancers of the breast, lung, ovaries, head and neck, and for Kaposi's sarcoma. Taxol works by stabilizing microtubules, thereby inhibiting the progression of cancer cells undergoing sister chromatid separation. Vincristine and taxol also affect a patient's healthy cells, but to a lesser extent than the much faster replicating and dividing cancer cells.

TABLE 9-2. Overview of the Cell Cycle

	Phase	Description
	G_0	Period of no cell replication/division, usually entered from G_1 restriction checkpoint. Cell cycle progression is halted by an inhibitory protein **cyclin-dependent kinase inhibitor 2A** that blocks the interaction of **CDK4** with cyclin D. G_0 cells lose cell cycle proteins, including cyclins and CDKs.
Interphase	G_1	Period of high rate of cell growth/protein production in preparation for DNA synthesis. Entry into G_1 relies on the G_1 restriction checkpoint in which binding of **CDK4** to **cyclin D** leads to phosphorylation of the tumor suppressor **retinoblastoma (Rb)** protein. Phosphorylated Rb dissociates from the E2F transcription promoter, leading to the expression of cyclin E. Cyclin E activates CDK2, allowing progression into S-phase. These checks prepare the cell for DNA synthesis and cell cycle continuation.
Interphase	S	Synthesis phase during which DNA replication occurs. RNA transcription is very low except for that involved in histone production.
Interphase	G_2	Second period of cell growth between DNA synthesis and cell division (mitosis), mainly involving synthesis of microtubules for the mitotic spindle and chromosome condensation. The G_2 checkpoint prepares the cell for cell division and cell cycle continuation. This process is controlled by dephosphorylation of tyrosine residues in the **maturation/mitosis-promoting factor** complex, made of cyclin B (or A) and CDK1. CDK1 is activated, causing irreversible progression to mitosis. When the cell is unprepared for mitosis, Cdc25 can be inactivated by phosphorylation by a protein kinase.
Cell Division	M	No cell growth. Alignment of the replicated chromosomes and the tension created by the protein attachment of the mitotic spindle to the centromeres and the mitotic plate initiate mitosis. When the tension is sufficient, cyclin B is degraded, allowing activity of the **anaphase-promoting complex (APC).** Activated APC leads to degradation of the kinetichore structural protein **securin**, releasing its inhibition on **separase**, the protein that leads to the separation of the sister chromatids.

CDK, cyclin-dependent kinase; DNA, deoxyribonucleic acid.

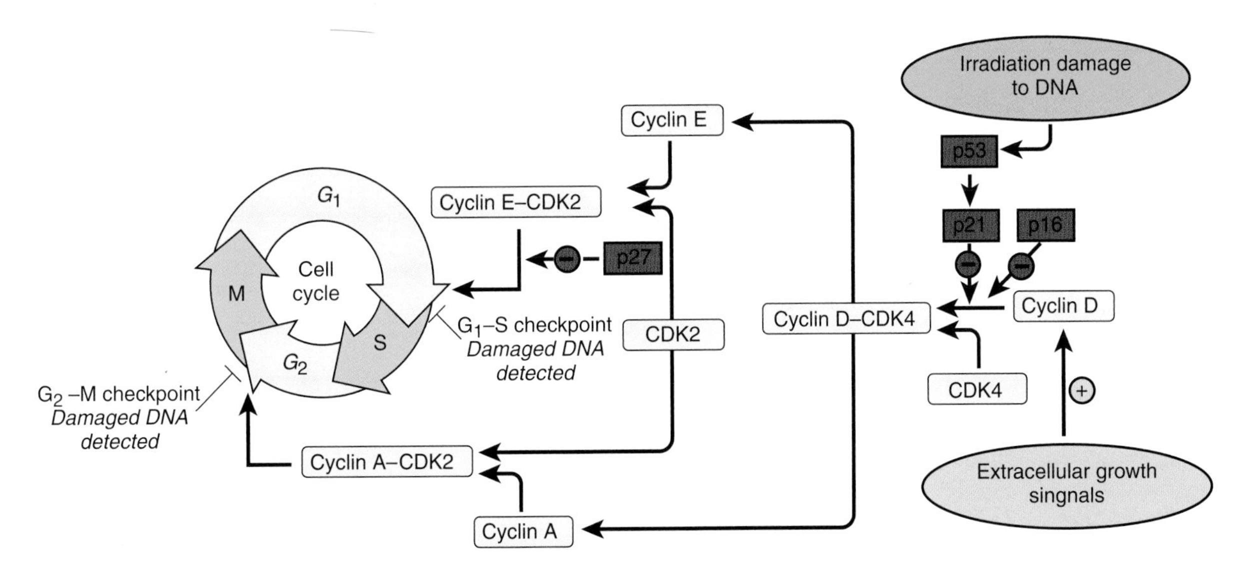

Figure 9-7. Overview of Cell Cycle Control. Review of various signals that regulate progression through the cell cycle. Checkpoints include G_1–S and G_2–M transition points where DNA damage is detected and repaired, S-phase where completion of replication is assessed, and M-phase where the integrity of the spindle apparatus is verified. Cyclins A, D, E, CDK2, and CDK4 and the protein regulators p16, p 27, and p53, as well as DNA irradiation damage and extracellular signals play roles as illustrated. Transforming growth factor-β also increases p27 inhibitory activity (not shown). CDK, cyclin-dependent kinase; DNA, deoxyribonucleic acid. [Adapted with permission from Murray RA, et al.: Harper's Illustrated Biochemistry, 28th edition, McGraw-Hill, 2009.]

Various inhibitors can stop the cell cycle by inactivating the cyclin–CDK heterodimer complex. When DNA damage is present, the tumor suppressor protein **p53** senses the problem and activates the inhibitor **p21**, which blocks the cyclin D–CDK4 complex and, subsequently, the G_1 to S transition. The inhibitors **p16** and **p27**, the latter activated by the growth inhibitor transforming growth factor-β (not shown in figure), can also inhibit the cell cycle by inhibiting the cyclin D–CDK4 or cyclin E–CDK2 complex, respectively. The overall control of the cell cycle by cyclins and CDKs is shown in Figure 9-7.

REVIEW QUESTIONS

1. What are the basic structural components of the nucleus and their roles?
2. What are the key events involved in DNA replication and the role of relevant proteins and enzymes?
3. What are the key events involved in transcription and the role of relevant proteins and enzymes?
4. What are the key events involved in protein synthesis including the roles of mRNA, tRNA, and accessory proteins?
5. What is the role of chaperones?
6. What is the significance of protein trafficking?
7. What are the key aspects in the control of gene expression?
8. What are the various types of mutations and how do repair mechanisms relate to these mutations?
9. What are the various phases of the cell cycle and their functions?

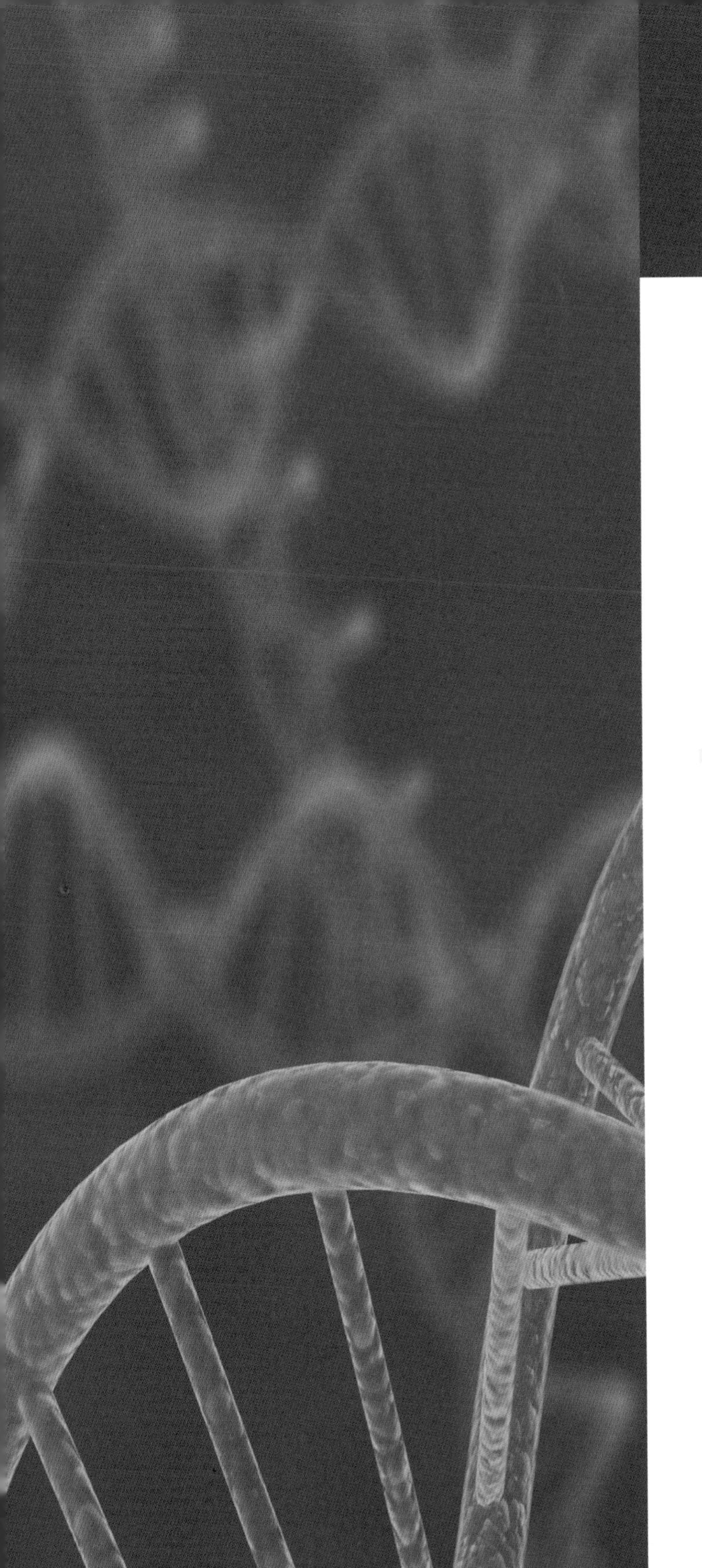

SECTION II

INTEGRATED USMLE-STYLE QUESTIONS AND ANSWERS

QUESTIONS

II-1. A previously normal 2-month-old female presents with jittery spells several hours after meals. She has low blood glucose. Physical exam reveals a liver edge 4 cm below the right costal margin. Percussion of the right chest and abdomen confirms hepatomegaly. The infant increases her blood glucose after breastfeeding but it is not maintained at normal levels upon fasting. Which of the following is the most likely diagnosis of this patient?

A. Fructosemia with inability to liberate sucrose from glucose

B. Galactosemia with inability to convert lactose to glucose

C. Glycogen storage disease

D. Growth hormone deficiency with inability to maintain glucose

E. Intestinal malabsorption of lactose

II-2. A well, 2-year-old girl contracted a viral illness at day care with vomiting, diarrhea, and progressive lethargy. She presents to the clinic with disorientation, a barely rousable sensorium, cracked lips, sunken eyes, lack of tears, flaccid skin with "tenting" weak pulse with low blood pressure, and increased deep tendon reflexes. Laboratory tests show low blood glucose, normal electrolytes, elevated liver enzymes, and (on chest X-ray) a dilated heart. Urinalysis reveals no infection and no ketones. The child is hospitalized and stabilized with 10% glucose infusion. Admission laboratories show elevated medium-chain fatty acylcarnitines in blood and 6–8 carbon dicarboxylic acids in the urine. Which of the following abnormalities is the most likely diagnosis in this child?

A. Carnitine deficiency

B. Defect of medium-chain coenzyme A dehydrogenase

C. Defect of medium-chain fatty acyl synthetase

D. Mitochondrial defect in fatty acid transport

E. Mitochondrial defect in the electron transport chain

II-3. Which of the following is the correct order of the following steps in protein synthesis?

1. A peptide bond is formed.
2. The small ribosomal subunit is loaded with initiation factors, messenger ribonucleic acid (mRNA), and initiation aminoacyl–transfer RNA (tRNA).
3. The intact ribosome slides forward three bases to read a new codon.
4. The primed small ribosomal subunit binds with the large ribosomal subunit.
5. Elongation factors deliver aminoacyl–tRNA to bind to the A site.

A. 1, 2, 5, 4, 3

B. 2, 3, 4, 5, 1

C. 2, 4, 5, 1, 3

D. 3, 2, 4, 5, 1

E. 4, 5, 1, 2, 3

II-4. A middle-Eastern family presents for evaluation because their infant son died in the nursery with severe hemolysis and jaundice. The couple had two prior female infants who are alive and well, and the wife relates that she lost a brother in infancy with severe hemolysis induced after a viral infection. The physician suspects glucose-6-phosphate dehydrogenase deficiency, implying defective synthesis of which of the following compounds?

A. Deoxyribose and nicotinamide adenine dinucleotide phosphate (NADP)

B. Glucose and lactate

C. Lactose and NADPH

D. Ribose and NADPH

E. Sucrose and nicotinamide adenine dinucleotide

II-5. Which of the following is an important intermediate in the biosynthesis of unesterified fatty acids?

A. Carnitine

B. Cholesterol

C. Fatty acyl-coenzyme A (CoA)

D. Glucose

E. Malonyl-CoA

II-6. Several disorders, such as one form of α_1-antitrypsin deficiency, can result from mistargeting of proteins into the wrong cellular compartments. New proteins destined for secretion are synthesized in which of the following?

A. Free polysomes

B. Golgi apparatus

C. Nucleus

D. Rough endoplasmic reticulum

E. Smooth endoplasmic reticulum

II-7. Children with urea cycle disorders present with elevated serum ammonia and consequent neurologic symptoms including altered respiration, lethargy, and coma. Several amino acids are intermediates of the urea cycle, having side ammonia groups that join with free carbon dioxide (CO_2) and ammonia to produce net excretion of ammonia as urea (NH_2CONH_2). Which of the following amino acids has an ammonia group in its side chain and is thus likely to be an intermediate of the urea cycle?

A. Arginine

B. Aspartate

C. Glutamate

D. Methionine

E. Phenylalanine

II-8. Cholera toxin causes massive and often fatal diarrhea by causing the continual synthesis of cyclic adenosine monophosphate (cAMP). Which of the following mechanisms would account for this effect?

A. Activation of G_i protein

B. Irreversibly activation of adenylyl cyclase

C. Locking of G_s protein into an inactive form

D. Prevention of guanosine triphosphate (GTP) from interacting with G_q protein

E. Rapid hydrolysis of G protein GTP to guanosine diphosphate (GDP)

II-9. Which of the following events occur during the formation of phosphoenolpyruvate from pyruvate during gluconeogenesis?

A. Acetyl-CoA is utilized

B. Adenosine triphosphate (ATP) is generated

C. CO_2 is required

D. GTP is generated

E. Inorganic phosphate is consumed

II-10. Regulation of which of the following enzymes is most important in controlling lipogenesis?

A. Acetyl-CoA carboxylase

B. Acyl-CoA synthetase

C. Carnitine–acylcarnitine translocase

D. Carnitine–palmitoyl transferase

E. Fatty acid synthase

II-11. The sequence of the template deoxyribonucleic acid (DNA) strand is 5′-GATATCCATTAGTGAC-3′. What is the sequence of the RNA produced?

A. 5′-CAGUGAUUACCUAUAG-3′

B. 5′-CTATAGGTAATCACTG-3′

C. 5′-CUAUAGGUAAUCACUG-3′

D. 5′-GTCACTAATGGATATC-3′

E. 5′-GUCACUAAUGGAUAUC-3′

II-12. Which of the following is an enzyme that is activated by hydrolysis of a proenzyme form?

A. Heparin

B. Keratin

C. Lactase

D. Pepsin

E. Phenylalanine hydroxylase

II-13. Which of the following is an accurate description of signal transduction in response to a peptide hormone?

A. Calcium activation of calmodulin-dependent kinase.

B. Diacylglycerol (DAG) causing an increase of intracellular Ca^{2+}.

C. Inositol trisphosphate (IP_3) binding to and activating protein kinase B.

D. Phospholipase A_2 catalyzing the cleavage of membrane phospholipids to release DAG.

E. Release of potassium ions from the endoplasmic reticulum.

II-14. Which of the following is an energy-requiring step of glycolysis?

A. Hexokinase

B. Phosphoenolpyruvate carboxykinase

C. Phosphoglycerate kinase

D. Pyruvate carboxylase

E. Pyruvate kinase (PK)

II-15. Which of the following processes generates the most ATP?

A. Citric acid cycle

B. Fatty acid oxidation

C. Glycogenolysis

D. Glycolysis

E. Pentose phosphate pathway

II-16. Which of the following statements accurately describes features of ribosomes?

A. An integral part of transcription

B. Bound together so tightly they cannot dissociate under physiologic conditions

C. Composed of RNA, DNA, and protein

D. Composed of three subunits of unequal size

E. Found both free in the cytoplasm and bound to membranes

II-17. Children with cystinosis have growth delay, photosensitivity with crystals in the lens of their eyes, and progressive renal failure because of accumulation of cystine in cellular lysosomes. The defect involves a specific lysosomal membrane receptor that facilitates cystine egress, and an effective therapy has been found using oral cysteamine, a compound similar in structure to cystine. This therapy reflects the general principle that competitive inhibitors typically resemble the structure of which of the following?

A. An allosteric regulator of enzyme/receptor activity

B. Enzyme or receptor protein

C. Enzyme reaction products

D. Substrates or ligands that bind the enzyme/receptor

E. The cofactor

II-18. Which of the following kinases plays a role in signal transduction for growth hormone, leptin, and prolactin?

A. Adenylate kinase

B. Janus kinase

C. Mitogen-activated protein (MAP) kinase

D. Protein kinase A

E. Pyruvate kinase (PK)

II-19. A 2-month-old boy is brought to the emergency department in a coma after sleeping through the night and failing to awaken in the morning. He is given intravenous glucose and awakens. Serum levels of pyruvate, lactate, and alanine are elevated, whereas aspartic acid levels are reduced. A muscle biopsy shows no abnormalities, and vitamin supplementation is ineffective. Which of the following is the most likely diagnosis in this patient?

A. Isocitrate dehydrogenase deficiency

B. Phosphofructokinase (PFK) deficiency

C. Pyruvate carboxylase deficiency

D. Pyruvate dehydrogenase (PDH) complex deficiency

E. Pyruvate kinase deficiency

II-20. Which of the following best explains why statin therapy is effective for individuals with hypercholesterolemia?

A. Bind to low-density lipoprotein (LDL) receptor, displacing cholesterol and inhibiting cholesterol synthesis.

B. Inhibit 3-hydroxy-3-methylglutaryl (HMG)-CoA reductase, a key regulator of cholesterol synthesis.

C. Inhibit HMG-CoA synthase, key step for synthesis of mevalonate that inhibits fatty acid synthesis.

D. Stimulate synthesis of *trans*-unsaturated fatty acids.

E. Stimulate thiolase, thus making more malonyl-CoA for inhibition of the tricarboxylic acid cycle.

II-21. GTP is required by which of the following steps in protein synthesis?

A. Aminoacyl–tRNA synthetase activation of amino acids

B. Attachment of mRNA to ribosomes

C. Attachment of ribosomes to endoplasmic reticulum

D. Attachment of signal recognition protein to ribosomes

E. Translocation of tRNA–nascent protein complex from A to P sites

II-22. Inherited deficiency of the enzyme methylmalonyl-CoA (MMA-CoA) mutase causes serum and urine accumulation of methylmalonic acid. Recognition that pernicious anemia (due to deficiency of vitamin B_{12}) can involve accumulation of methylmalonic acid led to successful treatment of some patients with MMA-CoA mutase deficiency using excess B_{12}. Studies of purified MMA-CoA mutase enzyme from normal individuals then showed enhanced mutase activity when B_{12} was added to the reaction mixture. These facts are best reconciled by which of the following roles for Vitamin B_{12}?

A. Cofactor for the MMA- CoA mutase enzyme

B. Competitive inhibitor of MMA-CoA mutase enzyme

C. Covalently attached group for the enzyme methylmalonic acid-CoA mutase

D. Feedback inhibitor of MMA-CoA mutase enzyme

E. Precursor for methylmalonic acid synthesis

II-23. Which of the following statements describes where integral proteins are located?

A. Associated with DNA in the nucleus

B. In the lumen of the endoplasmic reticulum

C. In the mitochondrial matrix

D. Mostly in the blood

E. Predominantly within cell membranes

II-24. A 7-year-old boy arrives at the emergency department asleep in his father's arms. The boy's mother explains that the boy spent the night throwing up and experiencing severe diarrhea. She is concerned about the vomiting and his inability to stay awake. History indicates the boy was healthy yesterday, but became ill at dinnertime after spending time playing in the basement of their apartment complex that afternoon. Further inquiry reveals that an exterminator had been hired to take care of a rat problem in the apartment. He had used a poison (Rotenone) that blocks complex I in the respiratory chain of oxidative phosphorylation. The boy is pale and not cyanotic. An analysis of this patient's metabolism would likely indicate impaired function of which of the following enzymes?

A. Glucose-6-phosphate dehydrogenase

B. PFK

C. Pyruvate carboxylase

D. PDH complex

E. Succinate dehydrogenase

II-25. A prime diagnostic indicator for the first presentation of diabetes mellitus type 1 is the presence of ketonuria and severe acidosis; the acidosis produces exaggerated attempts at respiratory compensation known as Kussmaul breathing. Which of the following are "ketone bodies" that would be elevated in serum and urine from a child with diabetic ketoacidosis?

A. Acetone and ethanol

B. Fumarate and succinate

C. Oxaloacetate and pyruvate

D. Pyruvate and lactate

E. β-Hydroxybutyrate and acetoacetate

II-26. Aminoacyl–tRNA synthetases must be capable of recognizing which of the following?

A. A specific amino acid and the 40S ribosomal subunit

B. A specific amino acid and the 60S ribosomal subunit

C. A specific ribosomal RNA (rRNA) and a specific amino acid

D. A specific tRNA and a specific amino acid

E. A specific tRNA and the 40S ribosomal subunit

II-27. Your patient presents with a deficiency of the enzyme that catalyzes synthesis of N-acetylglutamic acid. Which of the following would be a consequence of this deficiency in the patient?

A. Decreased digestion of dietary protein

B. Decreased excretion of uric acid

C. Increased amino acids in the blood

D. Increased fatty acids in the blood

E. Increased glucose in the blood

II-28. A patient presents with anemia and hyperbilirubinemia, reflecting excessive red blood cell hemolysis. Her symptoms include fatigue, pallor, and jaundice. Further evaluation shows a defect in spectrin of her red blood cell membranes. Which of the following is the likely diagnosis?

A. Essential fatty acid (EFA) deficiency

B. Pyruvate kinase deficiency

C. Spherocytosis

D. Tarui's disease

E. Zellweger syndrome

II-29. Which of the following describes the malate shuttle system?

A. Carries NADH from the cytoplasm directly into the mitochondrial matrix.

B. Generates three molecules of ATP in mitochondria per each NADH from glycolysis.

C. Moves NADH from the mitochondria to the cytoplasm.

D. Relies on malate dehydrogenase in the inner mitochondrial membrane.

E. Requires reduction of pyruvate to lactate for oxidizing NADH.

II-30. Which of the following is the final primary product of the fatty acid synthase reaction in adipose tissue?

A. Acetyl-CoA

B. Malonyl-CoA

C. Palmitic acid

D. Palmitoyl-CoA

E. Propionic acid

II-31. A mutation that results in a valine replacement for glutamic acid at position 6 of the β-chain of hemoglobin S hinders normal hemoglobin function and results in sickle cell anemia when the patient is homozygous for this mutation. This is an example of which of the following types of mutation?

A. Deletion

B. Frameshift

C. Insertion

D. Missense

E. Nonsense

II-32. In liver, which of the following inhibitory effects is the key regulatory event that ensures newly synthesized palmitoyl-CoA is not immediately oxidized?

A. Acyl-CoA synthetase by malonyl-CoA.

B. Carnitine–acylcarnitine translocase (CAT) by palmitoyl-CoA.

C. Carnitine–palmitoyl transferase-I (CPT-I) by malonyl-CoA.

D. CPT-I by palmitoylcarnitine.

E. Carnitine–palmitoyl transferase-II (CPT-II) by acetyl-CoA.

II-33. Some patients with familial hypercholesterolemia produce a truncated form of the LDL receptor, termed the "Lebanese" allele, which lacks three of the five domains of the protein and causes it to be retained in the endoplasmic reticulum. Analysis of the mutant gene indicated that the sequence of the protein was normal up to the point where it terminated. The genetic change that produced the mutant LDL receptor in these cases can be classified as which type of mutation?

A. Deletion

B. Insertion

C. Missense

D. Nonsense

E. Silent

ANSWERS

II-1. The answer is C. Important carbohydrates include the disaccharides maltose (glucose–glucose), sucrose (glucose–fructose), and lactose (galactose–glucose), and the glucose polymers starch (cereals, potatoes, and vegetables) and glycogen (animal tissues). Humans must convert dietary carbohydrates to simple sugars (mainly glucose) for fuel, employing intestinal enzymes and transport systems for enzymatic digestion and absorption. Simple sugars (galactose and fructose) are converted to glucose by liver enzymes, and the glucose is reversibly stored as glycogen. Enzymatic deficiencies in intestinal digestion (e.g., lactase deficiency in those with lactose intolerance), in sugar to glucose conversion (e.g., galactose to glucose conversion in galactosemia), or in glycogenesis/glycogenolysis (e.g., in those glycogen storage diseases) result in glucose deficiencies (low blood glucose or hypoglycemia) with potential accumulation and toxicity to hepatic tissues. The infant had been normal during breastfeeding, excluding low glucose due to growth hormone deficiency, and could readily digest breast milk lactose with absorption and conversion to glucose. Low glucose during fasting and liver enlargement implies altered regulation of glycogen synthesis/release due to one of the enzyme deficiencies within the category of glycogen storage disease. The hepatomegaly results from the accumulation of excessive amounts of glycogen (Table 6-1; Figure 6-9).

II-2. The answer is B. Fatty acid oxidation is a major source of energy after glycogen is depleted during fasting. Fatty acids are first coupled with CoA, transferred for mitochondrial import as acylcarnitines, and degraded in steps that remove two carbons. The fatty acyl-CoA dehydrogenases, enoyl hydratases, hydroxyacyl-CoA dehydrogenases, and thiolases that carry out each oxidation step are present in three groups with specificities for very long-/long-, medium-, and short-chain fatty acyl esters. As would be expected, deficiencies of long-chain oxidizing enzymes have more severe consequences than those for short chains because they impair many more cycles of two-carbon removal. Long-chain deficiencies may be lethal in the newborn period, whereas medium- or short-chain deficiencies may be undetected until a child goes without food for a prolonged time and must resort to extensive fatty acid oxidation for energy. Medium-chain CoA dehydrogenase deficiency can be fatal if not recognized, and sometimes presents as sudden unexplained death syndrome (usually at older ages than sudden infant death syndrome, that is, mostly from respiratory problems). The deficit of acetyl-CoA from fatty acid oxidation impacts gluconeogenesis with hypoglycemia, and the energy deficit leads to heart, liver, and muscle disease that may be lethal. Unlike most causes of hypoglycemia, the impaired fatty acid oxidation does not produce ketones (nonketotic hypoglycemia). Carnitine is tied up as medium-chain acylcarnitines and is secondarily deficient in fatty acid oxidation disorders. Rare primary carnitine deficiencies [as in answer option (e)] impair oxidation of all fatty acids because they cannot be imported into mitochondria.

II-3. The answer is C. Despite some differences, protein synthesis in prokaryotes and eukaryotes is quite similar. The small ribosomal subunit is 30S in prokaryotes and 40S in eukaryotes. The large ribosomal subunit is 50S in prokaryotes and 60S in eukaryotes. The intact ribosome is consequently larger in eukaryotes (80S) and smaller in prokaryotes (70S). At the start of translation, initiation factors, mRNA, and initiation aminoacyl–tRNA bind to the dissociated small ribosomal subunit. The initiation tRNA in prokaryotes is N-formyl methionine in prokaryotes and simply methionine in eukaryotes. Only after the small ribosomal subunit is primed with mRNA and initiation aminoacyl–tRNA does the large ribosomal subunit bind to it. Once this happens, elongation factors bring the first aminoacyl–tRNA of the nascent protein to the A site. Then peptidyl transferase forges a peptide bond between the initiation amino acid and the first amino acid of the forming peptide. Now uncharged initiation tRNA leaves the P site and the peptidyl–tRNA from the A site moves to the now vacant P site with the two amino acids attached. The ribosome advances three bases to read the next codon and the process repeats. When the stop signal is reached after the complete polypeptide has been synthesized, releasing factors bind to the stop signal, causing peptidyl transferase to hydrolyze the bond that joins the polypeptide at the A site to the tRNA. Factors prevent the reassociation of ribosomal subunits in the absence of new initiation complex (Figure 9-5).

II-4. The answer is D. Glucose-6-phosphate dehydrogenase (G6PD) is the first enzyme of the pentose phosphate pathway, a side pathway for glucose metabolism whose primary purpose is to produce ribose and NADPH. Its deficiency is the most common enzymopathy, affecting 400 million people worldwide. It contrasts with glycolysis in its use of NADP rather than NAD for oxidation, its production of CO_2, its production of pentoses (ribose, ribulose, and xylulose), and its production of the high-energy compound (5-phosphoribosyl-1-pyrophosphate) rather than ATP. Production of NADPH by the pentose phosphate pathway is crucial for reduction of glutathione, which in turn removes hydrogen peroxide via glutathione peroxidase. Erythrocytes are particularly susceptible to hydrogen peroxide accumulation, which oxidizes red blood cell membranes and produces hemolysis. Stresses such as newborn adjustment, infection, or certain drugs can increase red blood cell hemolysis in G6PD-deficient individuals, leading to severe anemia, jaundice, plugging of renal tubules with released hemoglobin, renal failure, heart failure, and death. Because the locus encoding G6PD is on the X chromosome, the deficiency exhibits X-linked recessive inheritance with severe affliction in males and transmission through asymptomatic female carriers. Ribose-5-phosphate produced by the pentose phosphate pathway is an important precursor for ribonucleotide synthesis, but alternative routes from fructose-6-phosphate allow ribose synthesis in tissues without the complete cohort of pentose phosphate enzymes or with G6PD deficiency. The complete pentose phosphate pathway is active in liver, adipose tissue, adrenal cortex, thyroid, erythrocytes, testis, and lactating mammary gland. Skeletal muscle has only low levels of some of the enzymes of the pathway but is still able to synthesize ribose through fructose-6-phosphate (Figure 6-7).

II-5. The answer is E. Acetyl-CoA is carboxylated to form malonyl-CoA through the addition of CO_2 by acetyl-CoA carboxylase. The acetyl- and malonyl-CoA groups are added to sulfydryl groups of fatty acid synthase multienzyme complex (one on each subunit) through transacylation reactions. Condensation forms acetoacetyl-S-enzyme on one subunit and a free sulfhydryl group of the other subunit—a sequence of enzyme reactions then converts the acetoacetyl-S-enzyme to acyl (acetyl) enzyme. A second round of two-carbon addition begins, as another malonyl-CoA residue displaces the acyl-S-enzyme to the other sulfhydryl group, and then condenses to extend the acyl group by two carbons.

Fatty acid synthesis then proceeds by successive addition of malonyl-CoA residues with condensation, causing the acyl chain to grow by two carbons with each cycle (Figure 7-1).

II-6. The answer is D. Protein synthesis occurs in the cytoplasm, on groups of free ribosomes called polysomes, and on ribosomes associated with membranes, termed the rough endoplasmic reticulum. However, proteins destined for secretion are only synthesized on ribosomes of the endoplasmic reticulum and are synthesized in such a manner that they end up inside the lumen of the endoplasmic reticulum. From there, the secretory proteins are packaged in vesicles. The Golgi apparatus is involved in the O-glycosylation and packaging of macromolecules into membranes for secretion.

II-7. The answer is A. Arginine is an amino acid used in proteins that is also part of the urea cycle. Citrulline and ornithine are amino acids not used in proteins but important as urea cycle intermediates. Aspartate is condensed with citrulline to form argininosuccinate in the urea cycle, and acetylglutamate is a cofactor in the joining of CO_2 with ammonia to form carbamoyl phosphate at the beginning of the urea cycle (Figure 5-9A; also see Table 1-1).

II-8. The answer is B. Cholera toxin is an 87-kDa protein produced by *Vibrio cholerae*, a Gram negative bacterium. The toxin enters intestinal mucosal cells by binding to G_{M1} ganglioside. It interacts with G_s protein, which stimulates adenyl cyclase. By ADP-ribosylation of G_s, the toxin blocks its capacity to hydrolyze bound GTP to GDP. Thus, the G protein is locked in an active form, and adenyl cyclase stays irreversibly activated. Under normal conditions, inactivated G protein contains GDP, which is produced by a phosphatase catalyzing the hydrolysis of GTP to GDP. When GDP is so bound to the G protein, the adenyl cyclase is inactive. Upon hormone binding to the receptor, GTP is exchanged for GDP and the G protein is in an active state, allowing adenyl cyclase to produce cAMP. Because cholera toxin prevents the hydrolysis of GTP to GDP, the adenyl cyclase remains in an irreversibly active state, continuously producing cAMP in the intestinal mucosal cells. This leads to a massive loss of body fluid into the intestine within a few hours (Figures 8-8, 8-9A; Table 8-2).

II-9. The answer is C. In the formation of phosphoenolpyruvate during gluconeogenesis, oxaloacetate is an intermediate. In the first step, catalyzed by pyruvate carboxylase, pyruvate is carboxylated with the utilization of one high-energy ATP phosphate bond:

$$\text{Pyruvate} + \text{ATP} + CO_2 \rightarrow \text{Oxaloacetate} + \text{ADP} + P_i$$

The pyruvate in gluconeogenesis is derived mostly from lactate and to a large extent from alanine, as well as to a lesser extent from some other amino acids.

In the second step, catalyzed by phosphoenolpyruvate carboxykinase, a high-energy phosphate bond of GTP drives the decarboxylation of oxaloacetate:

$$\text{Oxaloacetate} + \text{GTP} \rightarrow \text{Phosphoenolpyruvate} + \text{GDP} + CO_2$$

In contrast to gluconeogenesis, the formation of pyruvate from phosphoenolpyruvate during glycolysis requires only PK, and ATP is produced (Figure 6-6).

II-10. The answer is A. Acetyl-CoA carboxylase catalyzes the first step of lipogenesis in which acetyl-CoA is linked to malonyl-CoA. This enzyme is activated by citrate via polymerization and inactivated by palmitoyl-CoA that causes depolymerization to the monomeric form. The monomeric form can undergo phosphorylation by epinephrine or glucagon to put it into an inactive conformation that cannot readily polymerize. Insulin activation of a phosphatase reverses this covalent modification. Acetyl-CoA does not readily cross the mitochondrial membrane. Instead, citrate translocates to the cytosol where it is cleaved to acetyl-CoA and oxaloacetate by ATP-citrate lyase. Citrate increases in the fed state and indicates an abundant supply of acetyl-CoA for lipogenesis. Although fatty acid synthase is in the pathway, its regulation is far less important. Acyl-CoA synthetase is not in the lipogenesis pathway per se because it activates fatty acids to the fatty acyl-CoA form regardless of their source. Carnitine–acylcarnitine translocase and carnitine–palmitoyl transferase are both involved in the process of fatty acid degradation (Figure 7-1).

II-11. The answer is E. The template strand refers to the DNA strand that is transcribed into mRNA. As for DNA, mRNA is synthesized in the 5′ to 3′ direction. The template strand is always read in the 3′ to 5′ direction. The opposite DNA strand is known as the coding strand and has the same sequence as the mRNA transcript, except that U replaces T in mRNA. Choices b and d are DNAs, as they contain T instead of U.

II-12. The answer is D. Pepsin is secreted in a proenzyme form (pepsinogen) in the stomach. Unlike the majority of proenzymes, it is not activated by protease hydrolysis but instead by spontaneous acid hydrolysis. Hydrochloric acid secreted by the stomach lining creates the acid environment. All the enzymes secreted by the pancreas exist also in a proenzyme form including trypsinogen, chymotrypsinogen, procarboxypeptidase, and proelastase. Lactase and phenylalanine hydroxylase are active in their native forms. Heparin and keratin are not enzymes.

II-13. The answer is A. Hormone signal transduction ultimately leads to the activation of a variety of kinases. Some of these depend on calcium for their activity. One of these involves calcium first attaching to a calcium-binding protein, calmodulin. Hence, this kinase is referred to as calmodulin-dependent kinase. Protein kinase C depends on interaction with both calcium and DAG for its full activity. The DAG is generated by the cleavage of membrane phospholipids catalyzed by the enzyme phospholipase C. The other product of this cleavage is IP_3 that causes the release of calcium from the endoplasmic reticulum where it is stored (Figure 8-9B).

II-14. The answer is A. Hexokinase catalyzes the conversion of glucose to glucose-6-phosphate in the energy-requiring first step of glycolysis. ATP is also required in the conversion of fructose-6-phosphate to fructose 1,6-bisphosphate by PFK. ATP is generated in the conversion of 1,3-bisphosphoglycerate to 3-phosphoglycerate by phosphoglycerate kinase and in the conversion of phosphoenolpyruvate to pyruvate by PK. Both phosphoenolpyruvate carboxykinase and pyruvate carboxylase are energy-requiring reactions except that these occur in the gluconeogenesis pathway (Figure 6-2).

II-15. The answer is B. The pentose phosphate pathway does not generate any ATP but instead forms NADPH and ribose phosphate. Glycolysis produces a net two ATP molecules per glucose. The citric acid cycle produces a net 12 ATP per turn of the cycle. Fatty acid oxidation of palmitate results in a total of 129 ATP. Electron transport in the respiratory chain results in five ATP for each of the first seven acetyl-CoA produced by the oxidation of palmitate for a total of 35 ATP. Each of the eight acetyl-CoA molecules produced from palmitate results in 12 ATP from the citric acid cycle for 96 total ATP. This gives a total of 131 ATP per palmitate oxidized, minus two ATP for the initial activation of palmitate for a grand total of 129 ATP per palmitate.

II-16. The answer is E. The two subunits of ribosomes are composed of proteins and rRNA. Ribosomes are found in the cytoplasm, mitochondria, and bound to the endoplasmic reticulum. Transcription refers to the synthesis of RNA complementary to a DNA template and has nothing immediately to do with ribosomes.

II-17. The answer is D. Ligand–receptor and substrate–enzyme reactions are both saturable processes with similar dependence of reaction rate on ligand/substrate and receptor/substrate concentrations. Competitive inhibitors function by binding to the substrate or ligand-binding portion of the active site and thereby block access to the substrate (Figure 5-3). Thus, the structures of competitive inhibitors tend to resemble the structures of the substrate and are often called substrate or ligand analogs. The effects of competitive inhibitors can be overcome by raising the concentration of the substrate. The amount the substrate must be increased is dependent on the concentration of the inhibitor, the affinity of the inhibitor for the enzyme, and the affinity of the substrate for the enzyme. For membrane receptors such as that in the lysosome that is defective in cystinosis, the reaction may be one of membrane transport such that internal substrate/ligand is in equilibrium with external substrate/ligand. Thus, lysosome–internal cystine is substrate and lysosomal–external cystine a product, in a sense, such that lysosomal–external cysteamine will effectively decrease external cystine concentration and lead to egress of lysosomal cystine through its defective transporter.

II-18. The answer is B. Janus kinase is a soluble receptor-associated tyrosine kinase that helps fully activate the receptor after binding of the appropriate hormone (e.g., growth hormone, prolactin, and leptin). Adenylate kinase catalyzes the reversible interconversion of 2 ADP $\leftrightarrow$ ATP + AMP. MAP kinase is associated with transduction of signals from growth factors (e.g., epidermal growth factor, fibroblast growth factor, insulin-like growth factor, and other cytokines and growth factors) but not from growth hormone, which is more similar to the metabolic hormones than the growth factors. Protein kinase A is involved in transducing signals from hormones that act via G_s protein (e.g., catecholamines, glucagon, follicle-stimulating hormone, parathyroid hormone, among others). PK is the last enzyme of the glycolytic pathway (Figure 8-9C; Table 8-1).

II-19. The answer is C. Deficiencies of PK and PFK are ruled out by the elevated serum lactate levels, as these are glycolytic enzymes. The coma is associated with a fasting hypoglycemia, which is indicative of pyruvate carboxylase deficiency. The elevated lactate and alanine occurs because the pyruvate required for pyruvate carboxylase in gluconeogenesis is derived mostly from lactate and to a large extent also from alanine, as well as to a lesser extent some other amino acids. The reduction of aspartic acid levels occurs because there is reduced formation of oxaloacetate, the product of the pyruvate carboxylase reaction and oxaloacetate provides the carbon backbone for synthesis of aspartate (Figure 6-6).

II-20. The answer is B. Cholesterol is formed in five steps. The first step, biosynthesis of mevalonate, is catalyzed by three enzymes—acetyl-CoA thiolase, HMG-CoA synthase, and HMG-CoA reductase. Thiolase catalyzes the condensation of two molecules of acetyl-CoA to form acetoacetyl-CoA. HMG-CoA synthase catalyzes the addition of a third molecule of acetyl-CoA to form HMG-CoA. This compound is reduced to mevalonate by HMG-CoA reductase. This enzyme is the principal regulatory step in the pathway. In the second step of cholesterol synthesis, mevalonate is phosphorylated and decarboxylated to produce isopentyl diphosphate. Six of these isoprenoid units are condensed to form squalene in the third step. Lanosterol is formed in the fourth step and is subsequently converted to cholesterol (Figure 7-9).

II-21. The answer is E. Two molecules of GTP are used in the formation of each peptide bond on the ribosome. In the elongation cycle, binding of aminoacyl–tRNA delivered by EF-2 to the A site requires hydrolysis of one GTP. Peptide bond formation then occurs. Translocation of the nascent peptide chain on tRNA to the P site requires hydrolysis of a second GTP. The activation of amino acids with aminoacyl–tRNA synthetase requires hydrolysis of ATP to AMP plus PP_i.

II-22. The answer is A. Small molecules may be integral (covalently attached) parts of enzymes (prosthetic groups) or cofactors that participate in enzyme–substrate interaction or conversion. Prosthetic groups cannot be dissociated from the enzyme by dilution and thus will not be obvious components of the enzyme reaction when reconstituted in the test tube. Cofactors, such as vitamin B_{12} for MMA-CoA mutase, associate reversibly with enzymes or substrates and can be added in vitro to obtain enhancement of the catalyzed reaction(s). Competitive or feedback inhibitors interact at substrate or allosteric binding sites of the enzyme, reducing effective substrate concentration and reaction rate or converting the enzyme to a less active conformation. Vitamin B_{12} (cyanocobalamin) is a cofactor for MMA-CoA mutase, accelerating the conversion of methylmalonic acid to succinyl-CoA through activity of its cobalt group. Certain defects in MMA-CoA mutase can be ameliorated by intramuscular B_{12} injections so that effective B_{12} concentration and mutase activity are increased.

II-23. The answer is E. Membrane proteins are classified as **extrinsic**—predominantly on the inner or the outer surface of the membrane bilayer—and **intrinsic**—predominantly within the membrane. Intrinsic proteins are usually composed of uncharged, hydrophobic amino acids so they can enter and stay in the hydrophobic environment of the lipid bilayer. Intrinsic proteins serve several functions including channels, "carrier proteins"—transporting molecules through the membrane as a "carrier protein" or as a signaling protein—changing a portion of their internal cytoplasmic structure in response to a signal at an exposed external part of their structure to activate additional, closely associated molecules leading to a resulting function (e.g., turning on a gene) (Figure 8-4).

II-24. The answer is D. This patient exhibits several signs of acute complex I poisoning. Because complex I of oxidative phosphorylation is at least partially inhibited by the child consuming this poison, the ability of mitochondrial NADH to be oxidized is impaired. Consequently, NAD will become limiting for the PDH complex (Chapter 6). Although glycolysis also requires regenerating NAD, enzymes in this pathway can operate normally because NADH can be oxidized by the conversion of pyruvate to lactate. Glucose-6-phosphate dehydrogenase in the pentose phosphate pathway will be unaffected because it uses NADP not NAD (Figure 6-7A). Likely pyruvate carboxylase will have increased activity because pyruvate not used by the PDH complex could be processed to oxaloacetate. Finally, succinate dehydrogenase will be unaffected because it produces $FADH_2$ rather than NADH. Hence, it feeds electrons into complex II instead of complex I of the respiratory chain.

II-25. The answer is E. The ketone bodies, β-hydroxybutyrate and acetoacetate, are synthesized in liver mitochondria from acetyl-CoA. The liver produces ketone bodies under conditions of fasting associated with high rates of fatty acid oxidation. The inability to get glucose into extrahepatic cells because of insulin deficiency in diabetes also increases fatty acid oxidation and ketogenesis. The acid groups of β-hydroxybutyrate and acetoacetate cause acidosis and an anion gap (sum of serum sodium and potassium minus the sum of chloride and bicarbonate) that is greater than normal (over 8–15). In the case of diabetes, the "hidden anions" that add to bicarbonate in balancing the cations can be recognized as ketones through urine ketostix testing. In metabolic disorders such as methylmalonic aciduria or fatty acid oxidation defects, there are scanty abnormal or no ketones (if fat cannot be oxidized) so the hidden anions must be identified by plasma acylcarnitine or urine organic acid profiles. Acetone is a ketone body produced in diabetes that produces an acid breath during ketoacidosis (Figure 7-8A).

II-26. The answer is D. Aminoacyl–tRNA synthetases are responsible for charging a tRNA with the appropriate amino acid for translation. Charging a tRNA is a two-step reaction. In the first step, the enzyme forms an aminoacyl–AMP enzyme complex in a reaction that requires one ATP. In the second step, the activated amino acid is attached to the appropriate tRNA and the enzyme and AMP are released.

II-27. The answer is C. N-acetylglutamic acid is a stimulant of the urea cycle. A patient with a deficiency of the enzyme that catalyzes its formation would have a decreased activity of the urea cycle. Because the nitrogen for the cycle largely comes from amino acids, the blood concentration of amino acids would increase. A decreased ability to process these amino acids in fasting/starvation for glucose production could potentially lead to decreased blood glucose. Fatty acid usage would be unaffected. Although the urea cycle processes the nitrogen from amino acids derived from dietary proteins, their digestion would be unaffected. Instead, blood amino acids would also rise after a meal. Although secretion of urea would be diminished, uric acid is derived from the degradation of purine nucleotides and that would be unaffected.

II-28. The answer is C. Diseases of red blood cells often result from changes in lipid composition and/or defects in proteins associated with the red blood cell membrane. Spherocytosis is one such disease that results from defects in spectrin, ankyrin, or other red blood cell membrane proteins important in the stabilization of the normal biconcave shape of red blood cells. This unique shape is required for red blood cells to travel easily and undamaged through blood vessels. As the name implies, red blood cells appear as spheres and patients often suffer from low red blood cell numbers (anemia) due to the resulting destruction of the affected cells. EFA deficiency, PK deficiency, and Tarui's disease (PFK deficiency) can all lead to anemia for various reasons. EFA deficiency causes instability of the red blood cell membrane and hence hemolysis. Both PK and PFK deficiencies result in insufficient energy production to maintain red blood cell membrane integrity. Although Zellweger syndrome can lead to jaundice, this is related to diminished liver function and not to red blood cell issues.

II-29. The answer is B. The role of the malate shuttle is, under aerobic conditions, to oxidize cytosolic NADH produced during glycolysis to regenerate NAD to keep the glycolytic pathway operating. Because there is no transporter to carry NADH into the mitochondria, the electrons retrieved during the cytosolic oxidation of NADH must be carried across the mitochondrial inner membrane to the matrix as malate. In the matrix, malate dehydrogenase in the citric acid cycle (Figure 6-4) uses the malate to produce NADH that is subsequently processed via oxidative phosphorylation (Figure 6-5) to produce three molecules of ATP. Under anaerobic conditions, NAD is regenerated for glycolysis by the reduction of pyruvate to lactate.

II-30. The answer is C. The primary fatty acid that is stored for energy purposes is the 16-carbon palmitic acid. The product of fatty acid synthase is always the free (unesterified) fatty acid. In a subsequent step catalyzed by acyl-CoA synthetase, the fatty acid is activated to its fatty acyl-CoA form. Acetyl-CoA is a precursor for the synthesis of fatty acids, and malonyl-CoA is an intermediate in the process. Propionic acid is a product of odd-chain fatty acid degradation (Figure 7-2).

II-31. The answer is D. Missense mutations are those in which a single base change (point mutation) results in a codon that encodes for a different amino acid residue. The effects of these types of mutations can range from very minor or even undetectable to major, depending on the importance of the altered residue to protein folding and function. Nonsense mutations are also point mutations in which the affected codon is altered to a stop (nonsense) codon, resulting in a truncated protein. Frameshift mutations are due to one or two base pair insertions or deletions such that the reading frame is altered. These mutations generally lead to truncated proteins as well because, in most protein coding regions, the unused reading frames contain numerous stop codons.

II-32. The answer is C. Malonyl-CoA is a unique intermediate in the pathway for fatty acid synthesis and will only be high in concentration in the cell when fatty acid synthesis is active. When the cell is synthesizing fatty acids, it would be energetically illogical to immediately oxidize the newly formed fatty acid. The simplest means of slowing fatty acid oxidation is to prevent the formation of the fatty acylcarnitine derivative that must be formed for transport of the fatty acid into the mitochondrial matrix for oxidation. Inhibition of acyl-CoA synthetase would not work because the fatty acyl-CoA form must be produced for esterification of glycerol to form triacylglycerol for storage. Any control by palmitoyl-CoA would be ineffective because this is the substrate for fatty acid oxidation. Inhibition of CAT would be illogical because the fatty acylcarnitine derivatives would accumulate and be useless for any other function. Inhibition of carnitine–palmitoyl transferase-I by palmitoylcarnitine would be feedback inhibition by its product and would not be effective. Inhibition of carnitine–palmitoyl transferase-II by acetyl-CoA would cause accumulation of palmitoylcarnitine in the mitochondrial matrix, leading to a depletion of carnitine (Figure 7-3).

II-33. The answer is D. Production of a truncated protein indicates that a mutation has occurred, but this phenomenon may have arisen from a frameshift mutation (insertion or deletion) or by a nonsense mutation. The most likely possibility is a nonsense mutation because sequence analysis of the truncated protein showed that it had normal (wild-type) sequence. Insertion and deletion events often produce a stretch of garbled or abnormal protein sequence at the C-terminal end of the truncated protein arising from out-of-frame translation of the mRNA downstream of the mutation until a stop codon is encountered.

SECTION III

APPLIED BIOCHEMISTRY

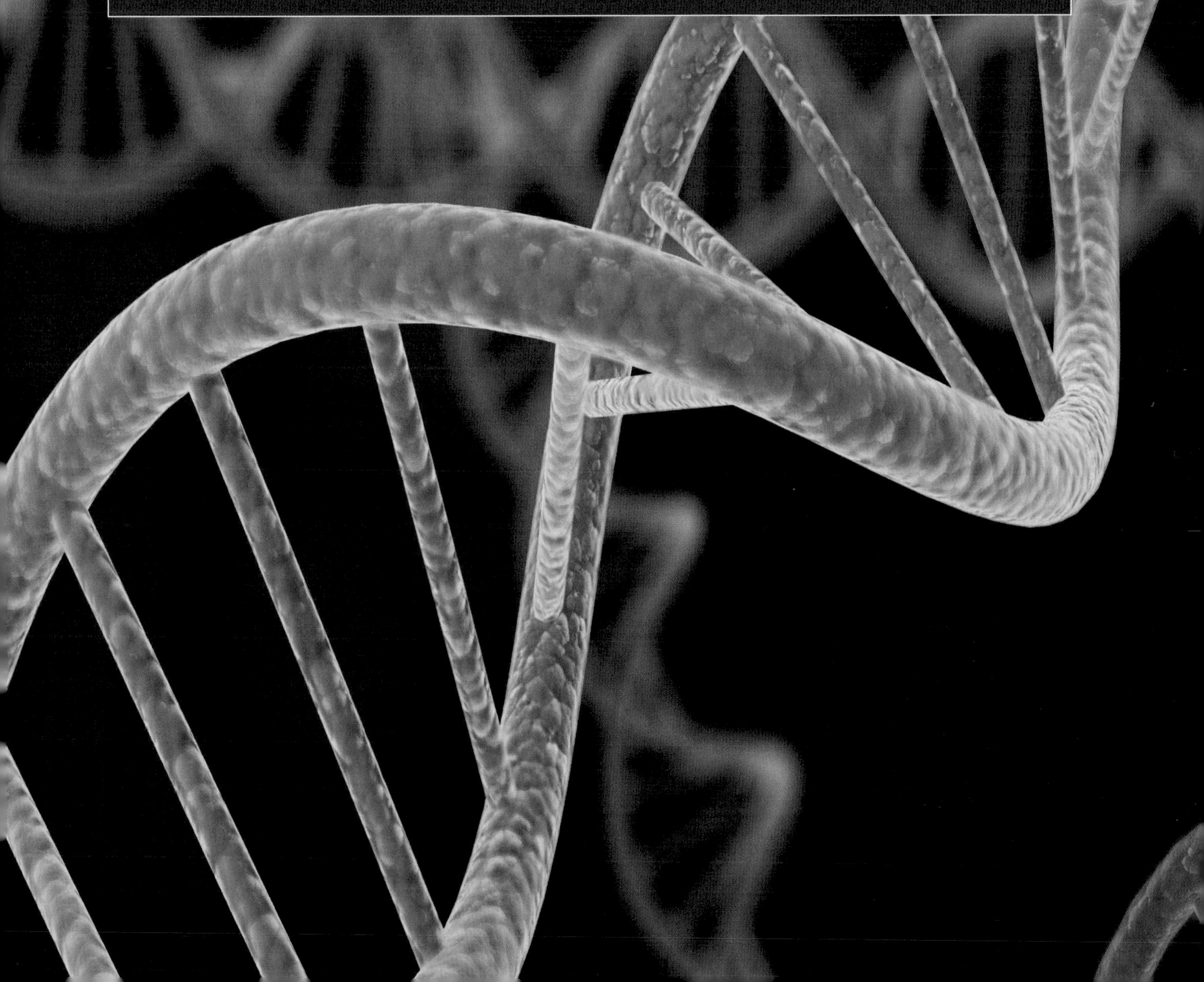

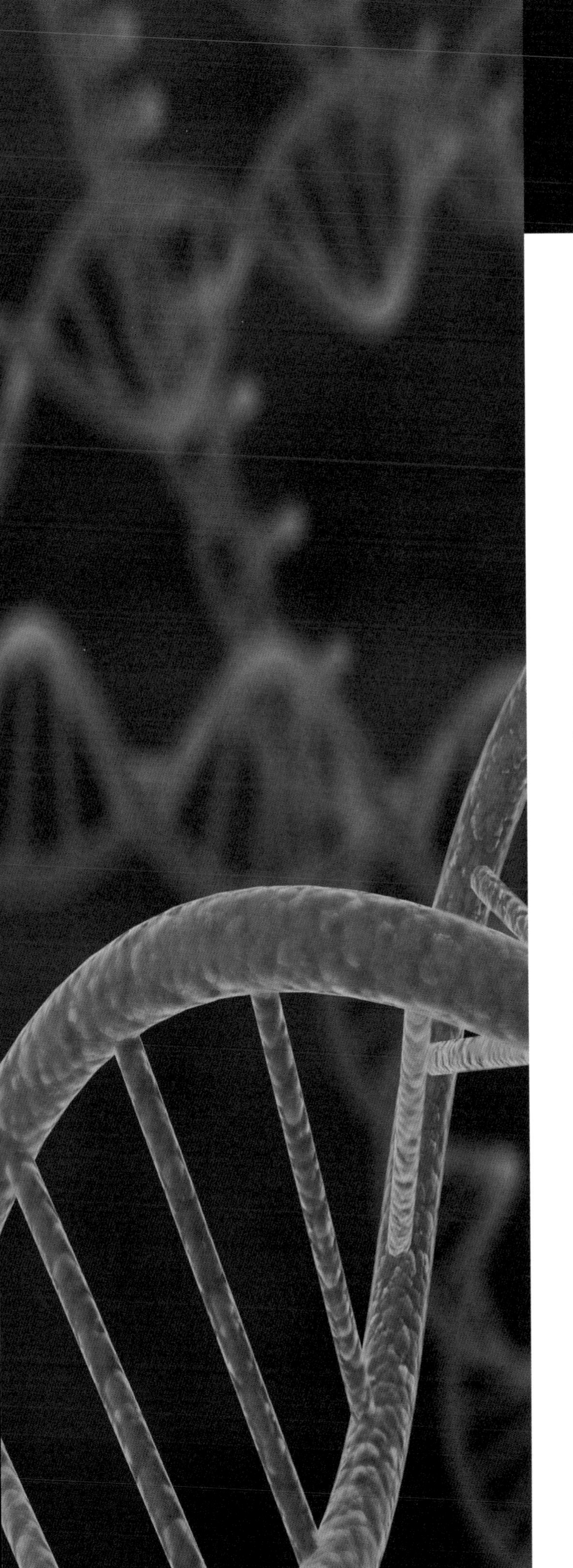

CHAPTER 10

METABOLISM AND VITAMINS/MINERALS

Co-authors/Editors: Maria L. Valencik and Cynthia C. Mastick

University of Nevada School of Medicine, Department of Biochemistry, Reno, NV

OVERVIEW

The integration of metabolism is a story of supply and demand. Food is ingested to supply energy but must be converted to the carbohydrate, lipid, and amino acid forms the body can use, primarily glucose and fatty acids. Individual cells then convert the fuels to usable energy, adenosine triphosphate (ATP) and nicotinamide adenine dinucleotide (NADH). The body demands energy to function but individual organs and tissues require particular sources of energy under varying conditions.

To convert consumed food into the needed energy, the body uses a variety of organs, each with unique metabolic profiles, to integrate and regulate the use and storage of energy. Specific regulatory points of biochemical pathways provide immediate control of the usage, conversion, or storage of food energy. Various hormones can also regulate these biochemical pathways to provide longer term control of food conversion and energy usage. Essential to both of these processes is the maintenance of glucose homeostasis. Finally, vitamins and minerals serve important functions as cofactors in many of these metabolic reactions. Their deficiency or excess can lead to numerous disease states.

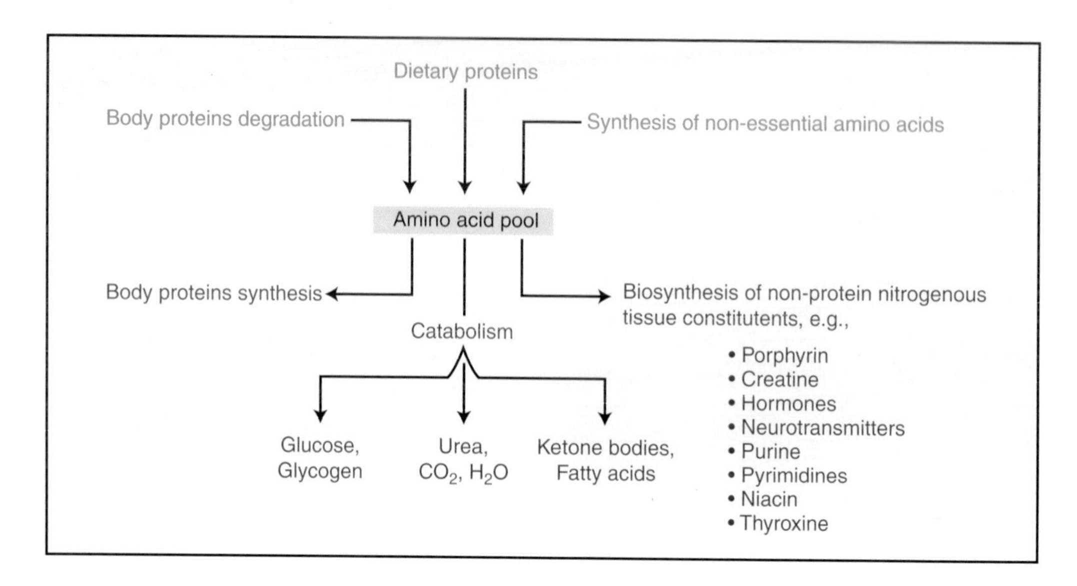

Figure 10-1. Summary of Amino Acid Metabolism. [Reproduced with permission from Naik P: Biochemistry, 3rd edition, Jaypee Brothers Medical Publishers (P) Ltd., 2009.]

METABOLIC ROLES OF MAJOR BIOCHEMICAL MOLECULES

The first consideration is the major sources of energy that can be used by the body, the nutrients required for their metabolism, and the biochemical pathways that integrate them.

Amino acids (Chapter 1, Figure 10-1) provide several major biochemical functions, including serving as (1) the building blocks of proteins; (2) the precursors of hormones, neurotransmitters, and other important signaling molecules (such as nitrous oxide); and (3) contributors to the purine and pyrimidine components of nucleic acids, co-enzymes [NADH and flavin adenine dinucleotide ($FADH_2$)], and other fundamental biological molecules. Additionally, excess amino acids can enter the citric acid cycle and can be used to generate or store biological energy (Chapter 5). Furthermore, the metabolism of some amino acids can be funneled into glucose synthesis (gluconeogenesis) during food deprivation.

Carbohydrates (Chapter 2, Figure 10-2) perform a fundamental role as the primary energy-production source for the human body. Glycolysis and the subsequent metabolic pathways form the primary energy molecules ATP, NADH, and $FADH_2$ via the oxidation of glucose and other carbohydrates (Chapter 6). Storage of carbohydrates as glycogen offers a readily available source of energy when dietary carbohydrate intake is low (Chapter 2). Carbohydrates are also important in the synthesis of nicotinamide adenine dinucleotide phosphate (NADPH) (Chapter 6) and nucleic acids (Chapter 4).

Lipids (Chapter 3, Figure 10-3) are nonpolar biomolecules. In most tissues, they serve a primary structural role as the components of biological membranes, creating a lipid bilayer via their hydrophobic and hydrophilic entities (Chapters 7 and 8). Their roles in membranes as well as in pathological processes such as atherosclerosis (Chapter 16) have raised the awareness of saturated, mono-unsaturated, and poly-unsaturated forms with regard to their role in diet. However, in adipose tissue, triglycerides are the major storage form of biological energy and their oxidation yields more energy per carbon than carbohydrates (Chapter 7). Lipolysis of triglycerides mobilizes fatty acids that generate energy through β-oxidation and produces the substrates necessary for ketone body (acetoacetate and β-hydroxybutyrate) synthesis, an essential fuel source during prolonged starvation. Oxidation of both fatty acids and ketone bodies spares glucose by preventing its oxidation. The consumption of dietary cholesterol and fats has a large impact on lipid metabolism through the generation of plasma lipoproteins [chylomicrons and low-density lipoprotein (LDL) via very-low-density lipoprotein (VLDL)]. The resultant elevation of harmful lipids/lipoproteins (dyslipidemia) has negative metabolic consequences that directly impact health and disease throughout all socioeconomic classes of modern society.

Vitamins, both lipid and nonlipid derived, serve important roles as cofactors in metabolic pathways and reactions (see end of this chapter). Several diseases, including scurvy, rickets, and Wernicke–Korsakoff syndrome, result directly from deficiencies of vitamins or, as in pernicious anemia, from the body's inability to properly absorb them. **Minerals**, including sodium, potassium, chloride, calcium, phosphate, iron, and others, play major roles in the regulation of metabolic enzymes involved in digestion, in the use and/or storage of food metabolites, and in the elimination of waste products.

Even more important is the integration of metabolism of these molecules in the human body and how regulation can be maintained by interrelationships between their anabolic and catabolic metabolism. In this regard, the body's ability to sense energy levels, respond to hormone signaling, and upregulate and downregulate particular metabolic pathways is paramount for the body to maintain the proper and controlled level of metabolic function and for the myriad of structural and functional processes to occur, which allow life.

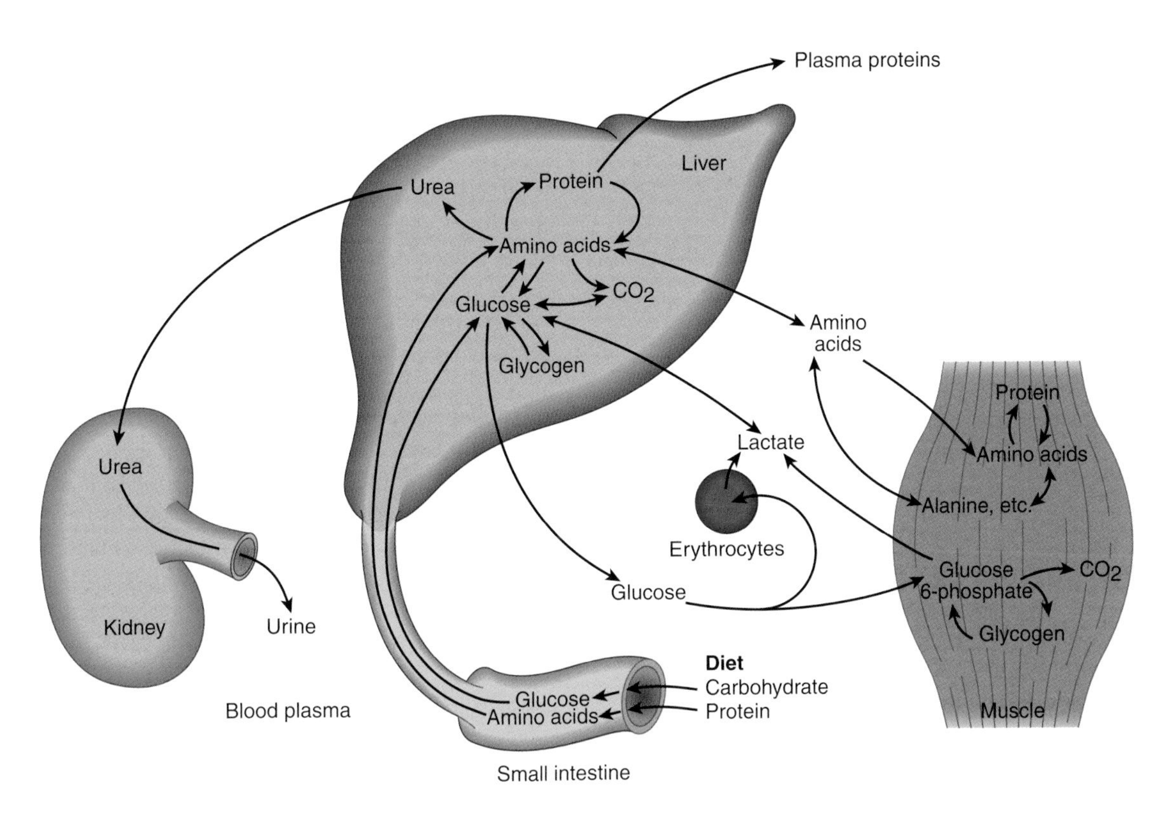

Figure 10-2. Transport and Fate of Major Carbohydrates and Amino Acids. [Reproduced with permission from Murray RA, et al.: Harper's Illustrated Biochemistry, 28th edition, McGraw-Hill, 2009.]

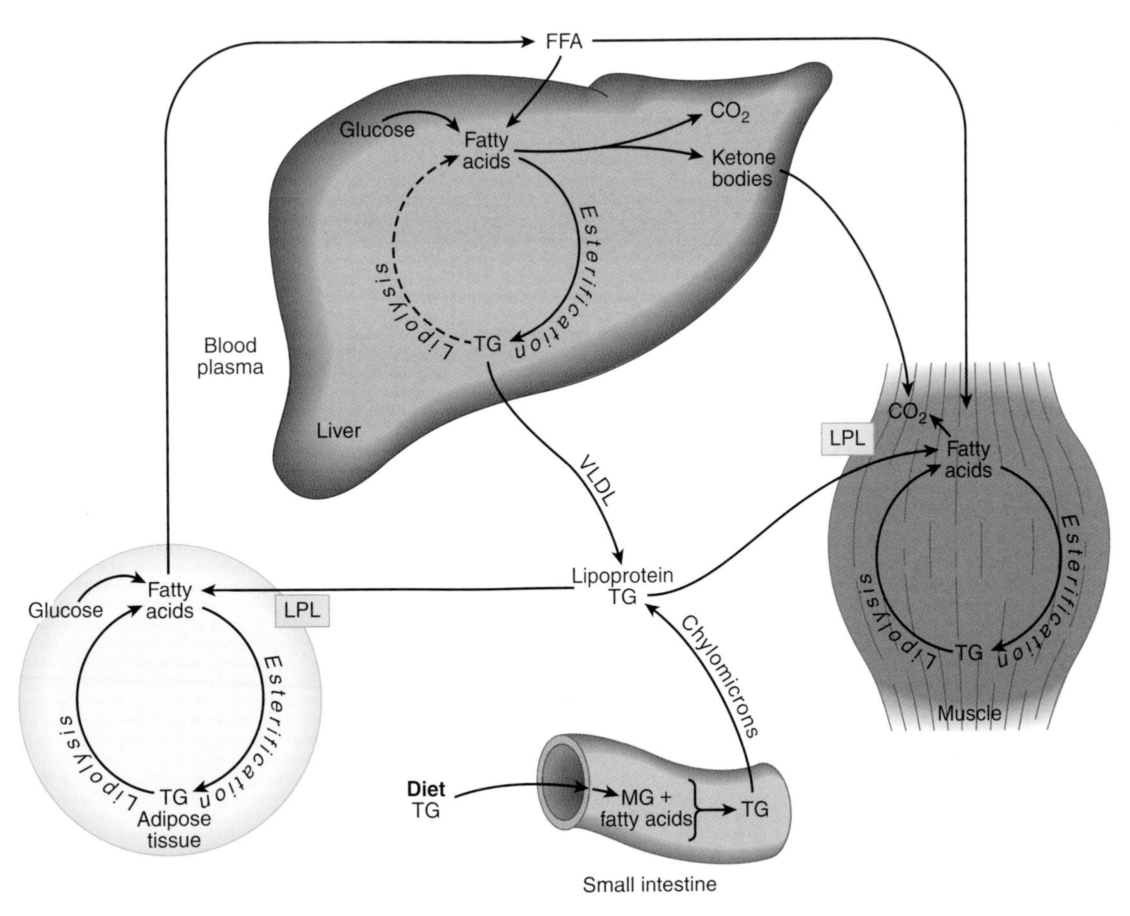

Figure 10-3. Transport and Fate of Major Lipid Substrates and Metabolites. FFA, free fatty acids; LPL, lipoprotein lipase; MG, monoacylglycerol; TG, triacylglycerol; and VLDL, very-low-density lipoprotein. [Reproduced with permission from Murray RA, et al.: Harper's Illustrated Biochemistry, 28th edition, McGraw-Hill, 2009.]

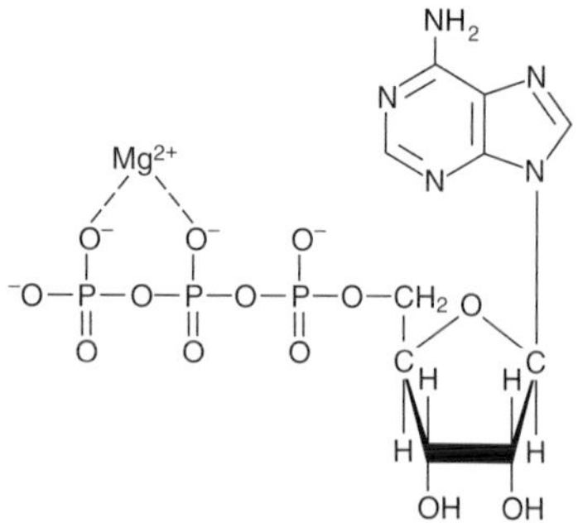

Figure 10-4. Adenosine Triphosphate Structure with Its Magnesium Cofactor. [Reproduced with permission from Murray RA, et al.: Harper's Illustrated Biochemistry, 28th edition, McGraw-Hill, 2009.]

INTEGRATION AND REGULATION OF METABOLISM

ATP, associated with magnesium (Mg^{2+}) for stability, is the primary form of biological energy utilized by the human body (Figure 10-4).

As such, the catabolic oxidation of carbohydrates (glycolysis, citric acid cycle, and oxidative phosphorylation), fatty acids/lipids/ketone bodies (fatty acid degradation), and amino acids all lead eventually to the production of ATP. In contrast, anabolic metabolic processes (gluconeogenesis, glycogen synthesis, lipid synthesis, triglyceride synthesis, and amino acid synthesis) consume ATP, NADH, and/or NADPH to store energy (glucose), to store energy, or to build essential biomolecules. Coupled to all of these processes is the need to eliminate waste products, including CO_2 (exhalation, acid–base balance), reactive and/or free-radical species (antioxidants), and urea (urea cycle). These concepts are summarized in Figure 10-5.

These metabolic pathways are intimately linked at several points in biochemical pathways, but are also separated into distinct compartments and/or organelles (e.g., cytoplasm versus mitochondria versus nucleus, etc.) to allow the necessary regulation and control. Additionally, each organ has unique metabolic needs and functions as summarized in Figure 10-6. These functions and needs must be coordinated in a variety of organs to maintain a constant supply of energy while preserving some energy for the future. The body accomplishes this goal by using the nervous system and hormonal signals to differentially stimulate and inhibit biochemical pathways within various organs in response to supply and demand. The main signals used to regulate metabolism are insulin, glucagon, catecholamines, glucocorticoids, and growth hormone (in children).

The remainder of this chapter will focus on the metabolism in three major tissues, the liver, adipose tissue, and skeletal muscle (Figure 10-6). The liver actively provides the quick fuel (glucose) your body needs, whereas adipose tissue provides

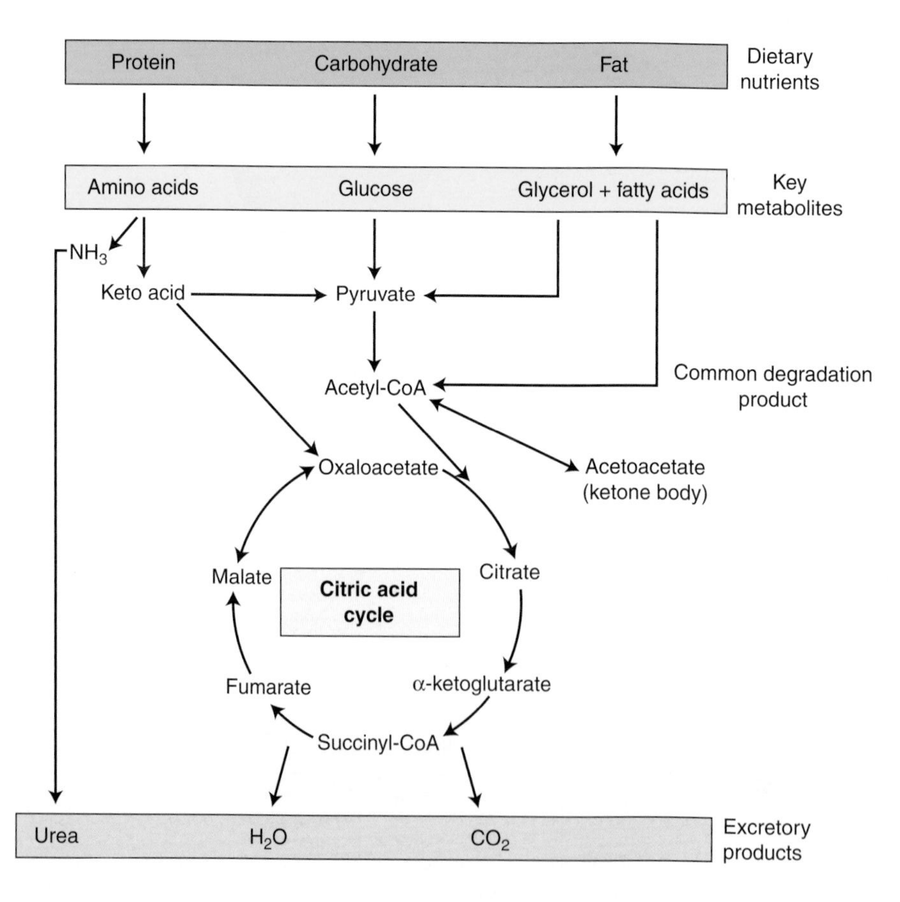

Figure 10-5. Interrelationship Between Proteins, Carbohydrates, and Fats. ATP, adenosine triphosphate; CoA, coenzyme A. [Adapted with permission from Naik P: Biochemistry, 3rd edition, Jaypee Brothers Medical Publishers (P) Ltd., 2009.]

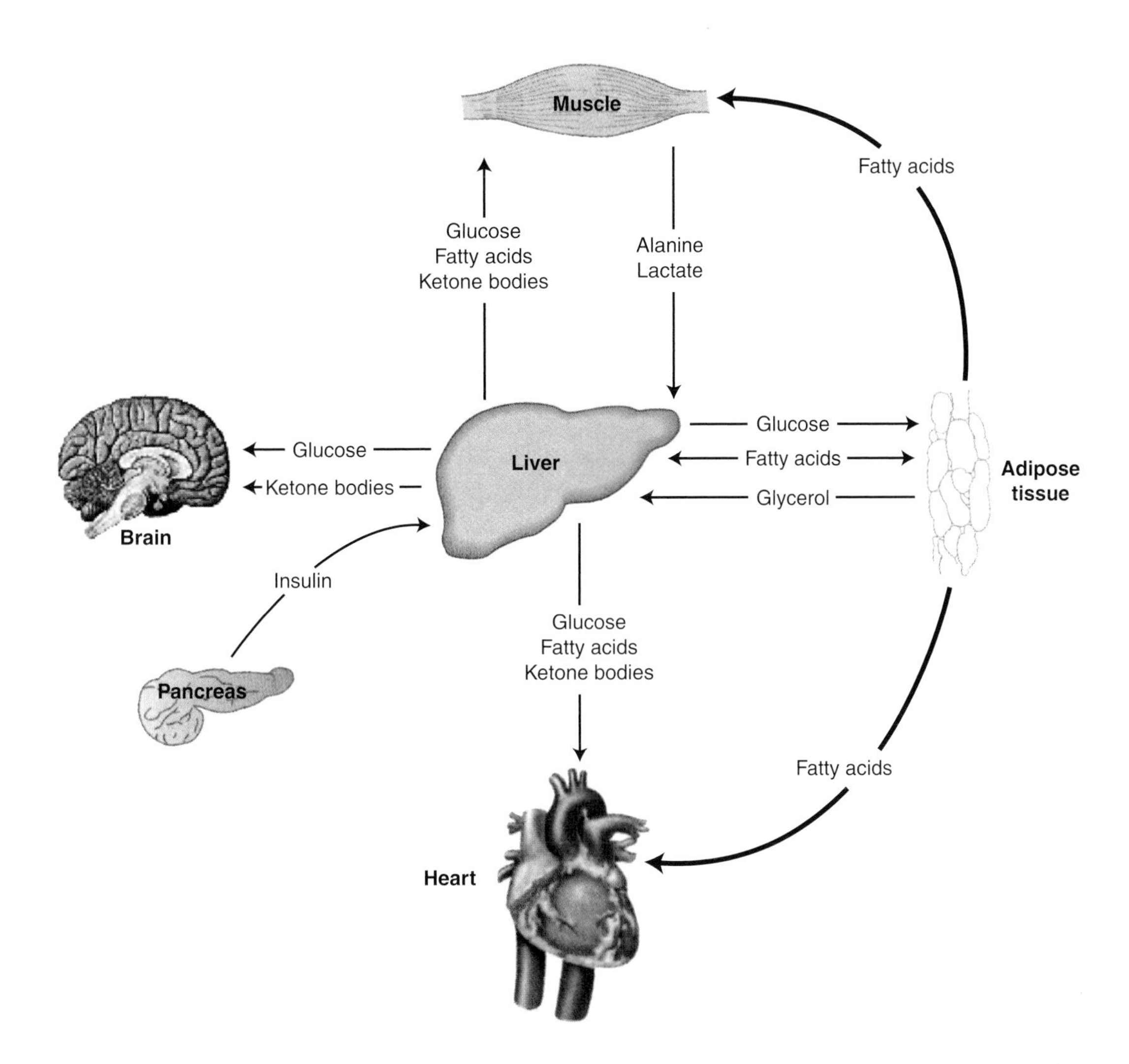

Figure 10-6. Integration of Metabolism Among Major Organs. [Adapted with permission from Naik P: Biochemistry, 3rd edition, Jaypee Brothers Medical Publishers (P) Ltd., 2009.]

long-term energy storage. Finally, skeletal muscle and the rest of your body constantly demand this energy. For example, the brain consumes approximately 90 g of glucose in a day, 20% of the average diet.

The supply and demand of energy must be continuously provided via dietary intake or breakdown of stores to balance with the energy requirements of respiration, transport, motility, and synthesis of cells and tissues (Figure 10-7). Overall, the average adult uses approximately 24 kcal of energy per kilogram of body mass to insure proper health and to maintain proper weight.

Several key biomolecules (glucose-6-phosphate or G6-P, pyruvate, and acetyl coenzyme A or acetyl-CoA) link the biochemical pathways for carbohydrates, lipids, and amino acids/proteins and the pathways they funnel into are tightly regulated and tissue specific (Figure 10-8). G6-P, pyruvate, and acetyl-CoA link the anabolic and catabolic pathways of carbohydrate metabolism to maintain a constant supply of energy to maintain homeostasis under constantly changing conditions. The particular pathways and regulation also depend on the specific functions and needs of each tissue type.

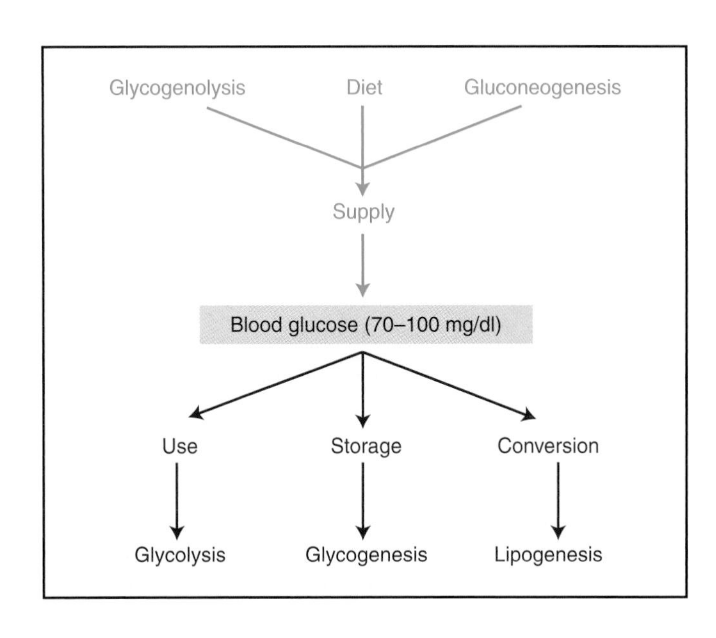

Figure 10-7. Factors Affecting Blood Glucose. [Reproduced with permission from Naik P: Biochemistry, 3rd edition, Jaypee Brothers Medical Publishers (P) Ltd., 2009.]

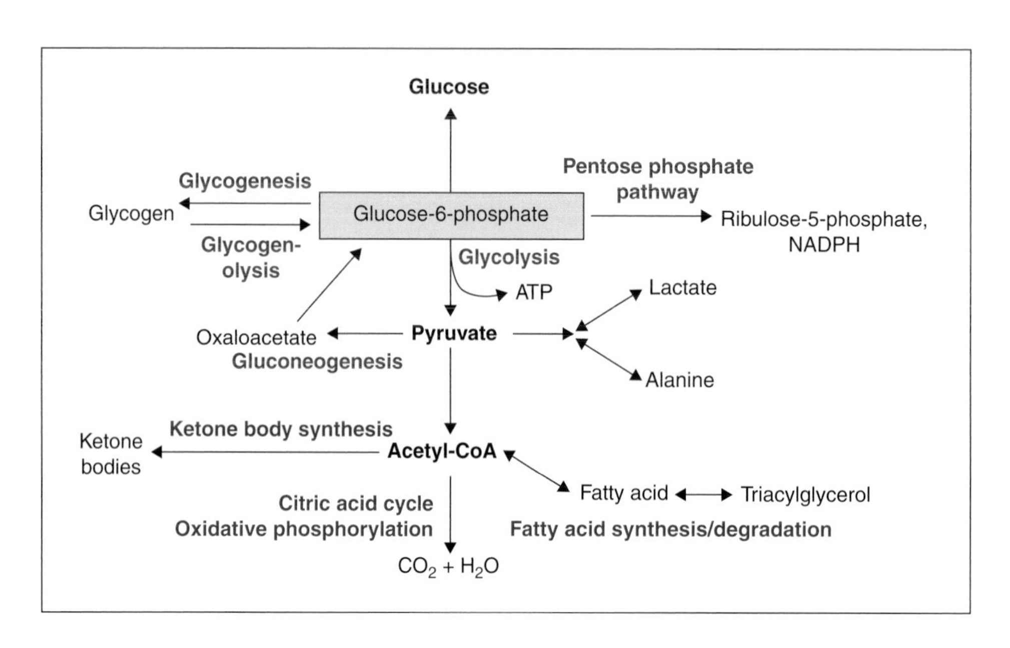

Figure 10-8. Summary of Important Control Points of Metabolism. The three important intermediaries, glucose-6-phosphate, pyruvate, and acetyl-CoA are indicated. Metabolic pathways of importance are indicated in red. See text for full discussion. ATP, adenosine triphosphate; CoA, coenzyme A; NADPH, nicotinamide adenine dinucleotide phosphate. [Adapted with permission from Naik P: Biochemistry, 3rd edition, Jaypee Brothers Medical Publishers (P) Ltd., 2009.]

GLUCOSE-6-PHOSPHATE

Metabolic regulation at this first branch point, **G6-P**, is clearly illustrated in the liver (Figure 10-8). After the ingestion of carbohydrates, glucose taken up by the liver is converted to G6-P by glucokinase. This phosphorylation uses one ATP molecule and traps the glucose within liver cells (hepatocytes). Subsequently, G6-P is metabolized via one of the following three pathways: (a) **glycogenesis**—the storage of carbohydrates as glycogen, (b) **glycolysis**—the production of ATP, or (c) the **pentose phosphate pathway**—the production of NADPH and/or five-carbon (pentose) sugars (Chapter 6, Figure 10-8). The pathway chosen depends upon the activation state of key enzymes (**glycogen synthase** and **phosphofructokinase-1**), substrate availability (G6-P, ATP, and $NADP^+$), and allosteric effectors [ATP, adenosine monophosphate (AMP), fructose 2,6-bisphosphate (F2,6BP), hydrogen ions (H^+), and citrate]. The key enzymes in glycogenesis and glycolysis are predominantly regulated by hormone-stimulated, covalent modification (**phosphorylation**), whereas the allosteric effectors fine-tune the flow of carbons through these pathways. In contrast, the pentose phosphate pathway is primarily regulated by the availability of G6-P and $NADP^+$ (Chapter 6).

In the well-fed state, when ATP and citrate concentrations are high, **phosphofructokinase-1** is allosterically inhibited, slowing down the committed step of glycolysis (the production of **fructose 1,6-bisphosphate**) leading to increased concentrations of G6-P. The increased concentration of G6-P can stimulate carbohydrate storage in two ways. First, G6-P is a positive allosteric effector of **glycogen synthase**, leading to the formation of glycogen. Second, G6-P indirectly inhibits **glycogen phosphorylase** thereby inhibiting **glycogenolysis** (glycogen degradation). Alternatively, when the ratio of $NADP^+$ to NADPH is high, G6-P can be shuttled into the pentose phosphate pathway to generate NADPH (reductive energy). This reducing power is used to synthesize a variety of biomolecules such as, fatty acids, cholesterol, nucleotides and other cofactors as needed. Under conditions where the ratio of $NADP^+$ to NADPH is low, the pentose pathway will not operate regardless of the concentration of G6-P.

The production of glycogen in the liver is further controlled by the hormones, insulin and glucagon (see below), and the resulting phosphorylation or dephosphorylation of glycogen synthase. Degradation of glycogen is decreased concomitantly by counterregulatory dephosphorylation or phosphorylation of glycogen phosphorylase. Eating breakfast, after an overnight fast, stimulates glycogen synthesis (and inhibits glycogen breakdown) in preparation for the next period of fasting. This replenishment is controlled by the favorable high ratio of insulin to glucagon, leading to activation of glycogen synthase activity and decreased glycogen phosphorylase activity. Under these conditions, the demand for de novo synthesis of lipid will rise after the glycogen is replaced, using carbons from excess dietary carbohydrate (to synthesize fatty acids). Additionally, if cholesterol biosynthesis is active, excess acetyl Co A from fatty acid catabolism can be synthesized into cholesterol. Once lipid biosynthesis commences, the utilization of NADPH increases the $NADP^+$/NADPH ratio favoring flux through the pentose phosphate pathway.

After a meal, the key regulator that restarts glycolysis in liver is **F2,6BP**. F2,6BP concentration is controlled by a **bifunctional enzyme** that includes both kinase and phosphatase active sites. Under conditions of a high ratio of insulin to glucagon, the bifunctional enzyme (**phosphofructokinase-2/fructose 2,6-bisphosphatase**) is dephosphorylated, leading to stimulation of phosphofructokinase-2. The resulting F2,6BP formed allosterically activates phosphofructokinase-1 and hence increases glycolysis while simultaneously inhibiting fructose 1,6 bisphosphatase, therefore shutting down gluconeogenesis. Following food deprivation, these events are reversed with a high ratio of glucagon to insulin, favoring phosphorylation of the bifunctional enzyme stimulating the fructose 2,6-bisphosphatase activity and leading to decreased phosphofructokinase-1 activity.

PYRUVATE

The second major branch point in the integration of metabolism is at **pyruvate** (Chapter 6, Figure 10-8). Pyruvate can be converted into four different substrates: lactate, alanine, oxaloacetate, and acetyl-CoA, depending upon the energy needs of a cell. Therefore, it is an important integration point where carbons are shuttled between energy storage, energy generation, and/or biosynthetic reactions. In the liver, pyruvate can undergo oxidative decarboxylation to enter the citric acid cycle and ultimately generate ATP when energy levels are low. Specifically, low energy levels inhibit the activity of an important regulatory enzyme, **pyruvate dehydrogenase kinase**. This inhibition prevents phosphorylation of the pyruvate dehydrogenase complex to an inactive state. Furthermore, this kinase is inhibited by NAD^+, pyruvate, and sulfhydryl form of CoA (non-acetylated), substrates of pyruvate dehydrogenase. Therefore, when substrates are plentiful, pyruvate is oxidatively decarboxylated to acetyl-CoA. In the liver, pyruvate is also the point where lactate and alanine (see below) can be actively funneled into either the citric acid cycle or gluconeogenesis via pyruvate carboxylation to oxaloacetic acid (Chapter 6) when liver glycogen or blood glucose levels are low. During starvation, gluconeogenesis can produce up to 160 g of glucose in a day, half of this from amino acids. Half of this glucose will be used by the brain. As blood glucose levels stabilize and gluconeogenesis is no longer required, oxaloacetate can re-enter the glycolytic pathway at phosphoenolpyruvate or shuttle back into the mitochondria, as malate, to be used in the citric acid cycle. If energy is abundant, high NADH and acetyl-CoA concentrations activate pyruvate dehydrogenase kinase and also serve as allosteric inhibitors of enzymatic activities within the PDH complex. This effectively turns off the pyruvate dehydrogenase complex by phosphorylation and allosteric control and shuts down the citric acid cycle. High ATP and acetyl-CoA concentrations also stimulate **pyruvate carboxylase**, the first step of gluconeogenesis (hormone regulation of gluconeogenesis is even more important; see below).

Skeletal muscle illustrates another important way that pyruvate can be metabolized (Figure 10-9). If oxygen levels are low and **anaerobic respiration** becomes important (such as during a quick sprint), pyruvate can be converted to **lactate** by **lactate dehydrogenase** with an oxidation of one NADH to NAD^+, the latter being essential for sustaining glycolysis. In this scenario, ATP is solely derived from anaerobic glycolysis. Lactate can subsequently be converted back to glucose for energy production via the **Cori cycle** (in the liver); when lactate concentrations get too high, feedback inhibition blocks further conversion of pyruvate to lactate. High lactate concentrations also create the sensation of "burning" in muscles, which serves as a signal to the body to limit further use of these muscles. Furthermore, pyruvate can also be converted in muscle tissue to the amino acid **alanine** via the **alanine transaminase** reaction. In a manner analogous to the Cori cycle, the **alanine cycle** then converts this alanine back to pyruvate in the liver where it is used to produce new glucose via gluconeogenesis as a source of energy for anaerobic glycolysis in muscle. Once oxygen levels are restored in skeletal muscle, production of ATP via citric acid cycle/oxidative phosphorylation resumes.

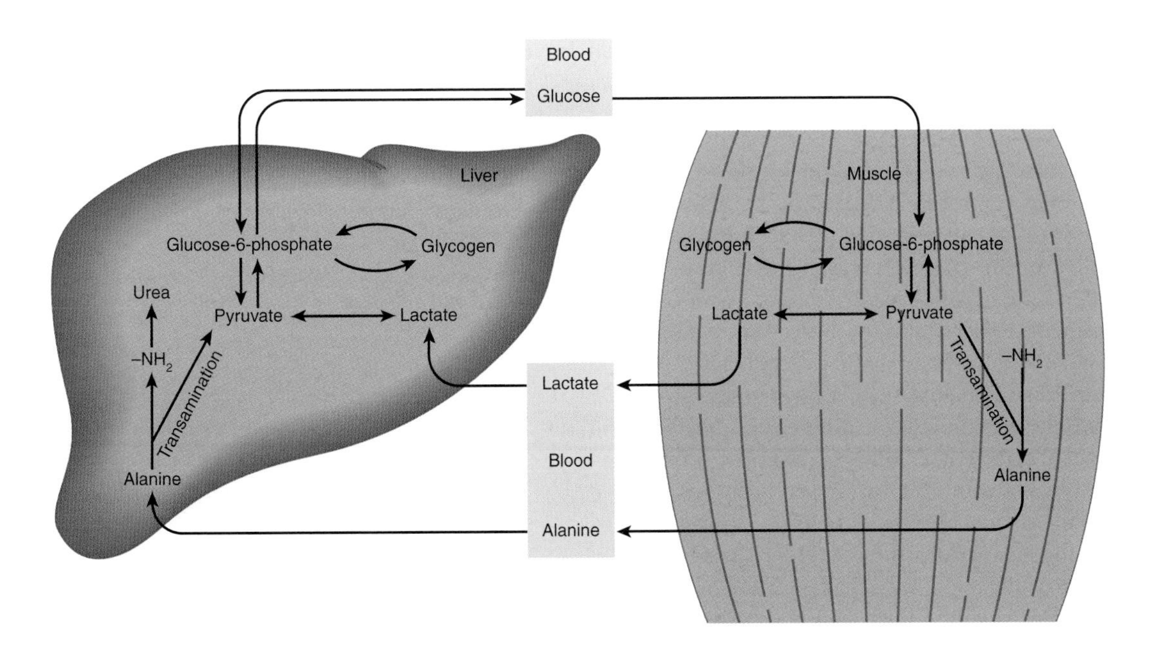

Figure 10-9. The Lactic Acid (Cori) and Glucose–Alanine Cycles. Carbons from glucose metabolism in muscle are recycled to the liver either as lactate or alanine for reconversion to glucose. Hence, when these cycles operate glucose carbons are spared. [Adapted with permission from Murray RA, et al.: Harper's Illustrated Biochemistry, 28th edition, McGraw-Hill, 2009.]

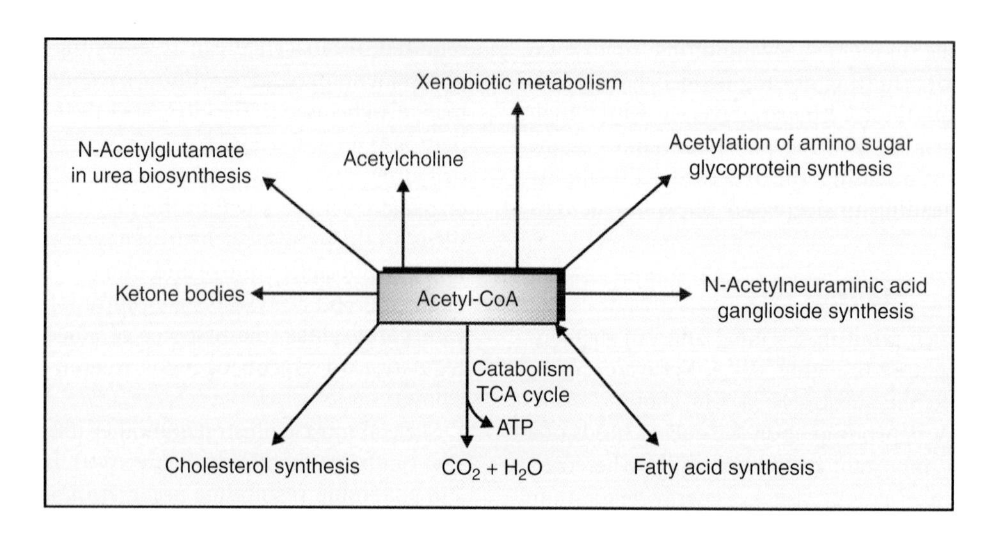

Figure 10-10. Overview of Acetyl-CoA Metabolism. ATP, adenosine triphosphate; CoA, coenzyme A; TCA, tricarboxylic acid. [Adapted with permission from Naik P: Biochemistry, 3rd edition, Jaypee Brothers Medical Publishers (P) Ltd., 2009.]

ACETYL-COA

Acetyl-CoA is the third branch point of primary metabolic control, and coordinates carbohydrate, ketone, and fat/lipid pathways (Chapter 6, Figures 10-8 and 10-10). Acetyl-CoA is a substrate for the citric acid cycle and can be oxidized to generate energy. However, when energy levels are high (high NADH/NAD^+ ratio), NADH inhibits the citric acid cycle at the isocitrate dehydrogenase and α-ketoglutarate dehydrogenase steps. Accumulation of $FADH_2$ also occurs, leading to an increase in succinyl-CoA that inhibits the cycle as well. Hormones also play a key, longer term role in the regulation of fatty acid synthesis and degradation (see below). Acetyl-CoA is also required for production of the neurotransmitter acetylcholine (see below and Chapter 19).

In the fed state, excess acetyl-CoA can be directed toward synthesis of cholesterol and/or fats/triacylglycerols in the liver. During starvation, fatty acid oxidation supplies energy for hepatocytes to drive gluconeogenesis. Furthermore, any excess acetyl-CoA generated will be used for the synthesis of ketone bodies. The ketones cannot be oxidized by the liver and are exported and used as an alternate fuel for the brain, heart, and muscles. In fact, during fasting/starvation, the brain will be heavily reliant on ketone bodies, using them for up to 70% of its energy requirements, especially in prolonged starvation. Hormones also regulate ketone body synthesis (see below).

In both nutritional circumstances, acetyl-CoA may activate **pyruvate carboxylase,** although for different purposes. In the fed state, pyruvate carboxylase converts pyruvate to oxaloacetate, which condenses with acetyl-CoA producing citrate, the first product of the citric acid cycle. The citrate is transported to the cytoplasm for fatty acid synthesis. High citrate activates **acetyl-CoA carboxylase** to promote the formation of fatty acids (Chapter 7). Citrate, when too high, inhibits phosphofructokinase-1, thus blocking glycolysis to prevent unnecessary metabolism of more glucose to pyruvate. The G6-P that backs up can be cycled through the pentose pathway to provide NADPH for fatty acid synthesis, as described above, or may be directed toward glycogen synthesis. In the starvation state, high acetyl-CoA from oxidation of fatty acids stimulates pyruvate carboxylase to promote gluconeogenesis.

Low concentrations of **citrate** and the other intermediates of the citric acid cycle as well as low **ATP/NADH/$FADH_2$** promote continuation of the citric acid cycle and oxidative phosphorylation. The citric acid cycle intermediates can also be used for the production of amino acids or as an energy source (Chapter 5). Low citrate/high palmitoyl-CoA (from lipolysis) concentrations prevent fatty acid synthesis. The resultant decrease of **malonyl-CoA**, an allosteric inhibitor of carnitine palmitoyltransferase 1 (CPT1), favors formation of palmitoyl carnitine by CPT1, with subsequent transport across the mitochondrial membrane and oxidation in the mitochondria.

HORMONAL CONTROL OF METABOLISM

The coordination of metabolic pathways to achieve this essential balance primarily depends on hormone; nerve and signaling pathways, including insulin, glucagon, catecholamines (Chapter 19), glucocorticoids (slower, stress-related changes); and cytokines. Errant control leads to disease states if glucose levels are high (diabetes mellitus or DM) or low (hypoglycemia) and, if too low, even death due to coma.

INSULIN

Insulin (Figure 10-11) is the anabolic hormone of the well-fed state and an important signal to stimulate storage of excess nutrients as glycogen and triglycerides (fat in adipose tissue).

The action of insulin (Figure 10-12) is experienced by three main targets, the liver, adipose tissue, and striated muscle. The synthesis and release of insulin is stimulated by glucose and potentiated by amino acids. In the liver, insulin stimulates glycogenesis (glycogen synthesis), fatty acid synthesis, glycolysis, and the pentose phosphate pathway. In the adipose tissue, it stimulates glucose and fatty acid uptake and triglyceride synthesis (energy

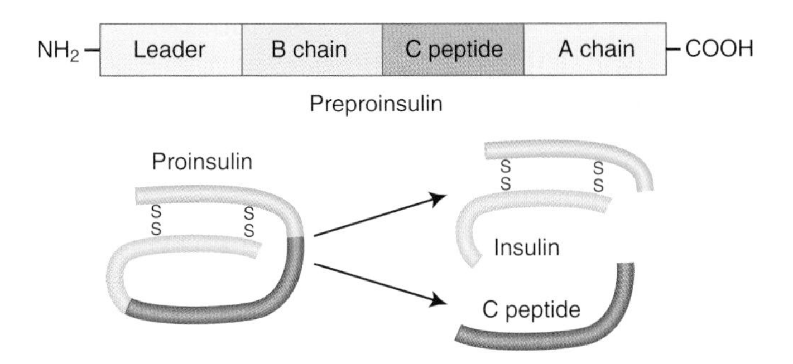

Figure 10-11. Preproinsulin Processing. Preproinsulin (top) is composed of a leader sequence (blue), A (green) and B (yellow) insulin chains, and C (red) peptide. Removal of the leader sequence produces proinsulin (bottom left). Cleavage of C-peptide from proinsulin leads to the production of active insulin. Because C-peptide is produced in equal amounts to insulin, it has become an important measure of insulin production. [Adapted with permission from Kibble JD and Halsey CR: The Big Picture: Medical Physiology, 1st edition, McGraw-Hill, 2009.]

storage). Similarly, in skeletal muscle, it stimulates glucose uptake, glycogenesis, and protein synthesis. It is noteworthy that insulin does not influence glucose metabolism in either the brain or red blood cells.

The release of insulin from the pancreatic β-cells (Figure 10-13) is the result of increased blood glucose concentrations. Glucose enters the β-cells via the glucose transporter 2 (GLUT2) (passive transport). The GLUT2 has a weak affinity for glucose so that it favors glucose uptake only after a meal, when blood glucose levels are high, rather than in the fasted state. Following glucose oxidation, the increased ATP concentration stimulates K^+ channels and depolarizes the cell membrane. This depolarization opens voltage-gated Ca^{2+} channels. Other signals related to production of inositol trisphosphate, a second messenger, stimulate Ca^{2+} release from the endoplasmic reticulum, resulting in high intercellular Ca^{2+} concentration and triggering the release of insulin.

Insulin affects the metabolism of cells that have insulin receptors: liver cells (hepatocytes), fat cells (adipocytes), and muscle cells (Table 10-1). The brain and red blood cells are not affected by insulin. Insulin works via a tyrosine kinase receptor, which phosphorylates target proteins that lead to a number of metabolic effects. One effect is the rapid translocation of a glucose transporter 4, GLUT4, from vesicles to the cell surface of skeletal and cardiac muscle and fat cells, increasing glucose transport into these cells. Insulin also regulates metabolic enzymes such as glycogen synthase and phosphorylase through activation of type I phosphatase and dephosphorylation.

GLUCAGON

Glucagon (Figure 10-14) is the hormone of fasting produced by pancreatic α-cells, adjacent to the insulin-producing β-cells. Glucagon signals via G-protein coupled receptors and the secondary messenger molecule cyclic AMP. In contrast to many mammals, glucagon acts almost exclusively on the liver in humans (Table 10-2). Primarily, it stimulates glycogenolysis, gluconeogenesis, and fatty acid oxidation. Glucagon levels increase two-to threefold in response to hypoglycemia, and the liver begins production of glucose from glycogen. During times of high blood glucose, glucagon is reduced to half of its normal level. Glucagon also stimulates the release of insulin, thereby allowing insulin-sensitive cells to take up the released glucose. The delicate balance of glucagon and insulin levels is how the body maintains glucose homeostasis under varying conditions.

CATECHOLAMINES

Catecholamines, including norepinephrine and epinephrine, the latter being primarily the hormone responsible for the "fight or flight" response to external stresses, can provide almost immediate (within seconds) regulation of metabolism (Figure 10-15). Specifically, they stimulate glycogenolysis and glycolysis for the production of ATP in the muscle. At the same time, they inhibit glycolysis in the liver and stimulate glycogenolysis to provide glucose for the blood. More recently, synaptically released catecholamines have emerged as the main physiological pathway for the activation of lipolysis under conditions of

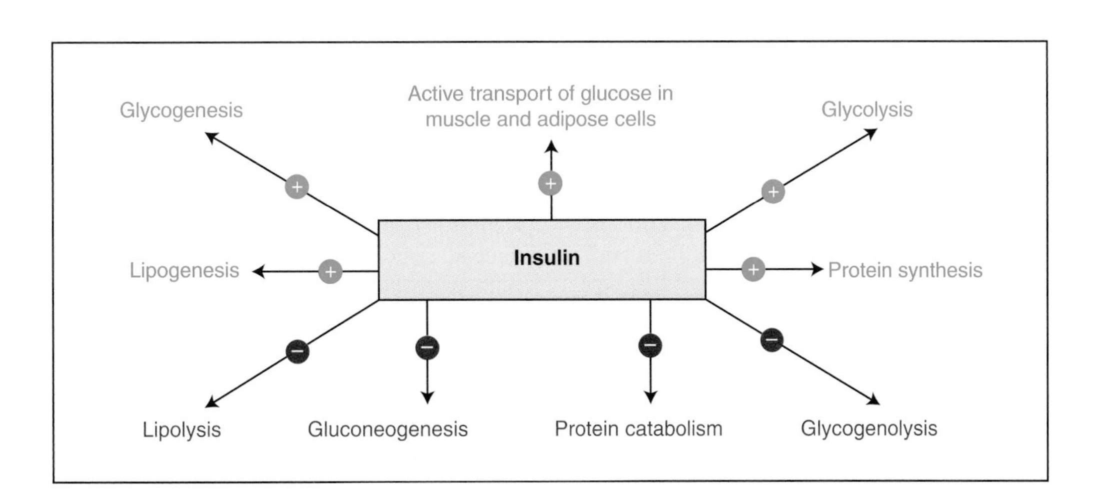

Figure 10-12. Metabolic Systems Affected by Insulin. [Adapted with permission from Naik P: Biochemistry, 3rd edition, Jaypee Brothers Medical Publishers (P) Ltd., 2009.]

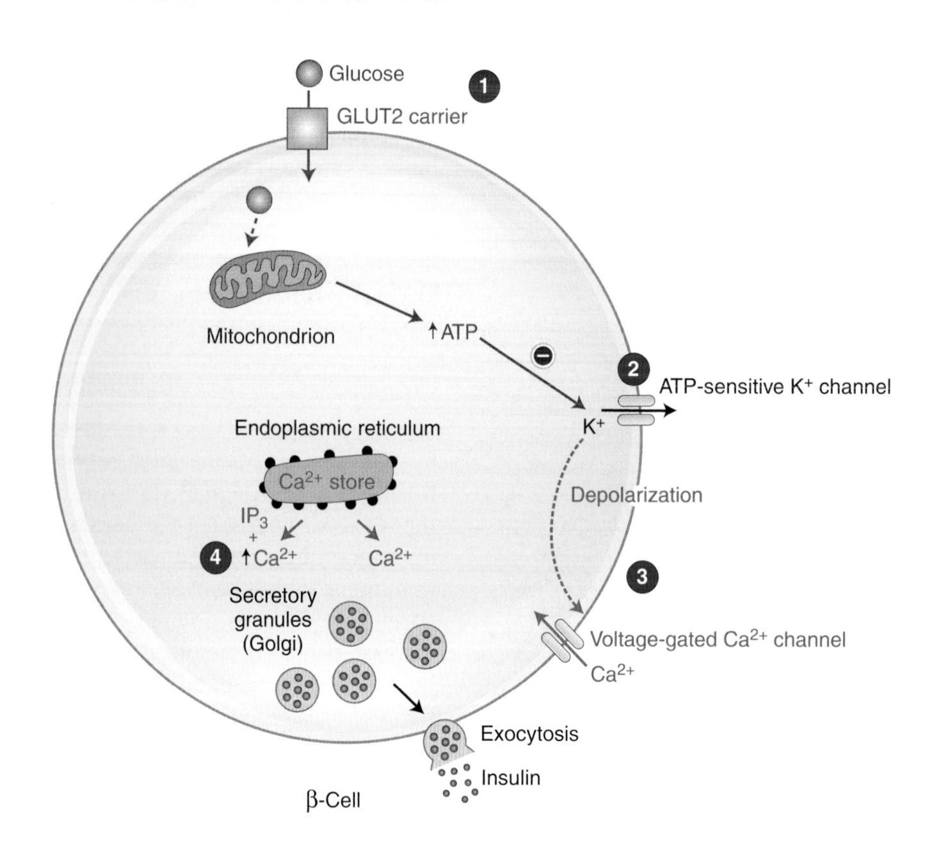

Figure 10-13. Regulation of Secretion of Insulin via Glucose. Pancreatic β-cells are induced to secrete insulin by (1) the uptake of glucose and its oxidative metabolism in mitochondria, which produces increased concentration of ATP. (2) Increased ATP causes closure of the ATP-sensitive K^+ channels, leading to depolarization. (3) The depolarization of the cell causes in influx of Ca^{2+} from voltage-gated Ca^{2+} channels. (4) The increased Ca^{2+} along with IP_3 and other signaling induces further release of Ca^{2+} from the endoplasmic reticulum (ER), which prompts exocytosis of insulin, produced in the ER, and subsequently processed and released from secretory granules of the Golgi apparatus. ATP, adenosine triphosphate; GLUT2, glucose transporter 2; IP_3, inositol trisphosphate. [Adapted with permission from Kibble JD and Halsey CR: The Big Picture: Medical Physiology, 1st edition, McGraw-Hill, 2009.]

fasting (a condition of chronic stress). The effects of epinephrine on metabolism are summarized in Table 10-3.

GLUCOCORTICOIDS

Cortisol, a **glucocorticoid**, is a chronic stress hormone that also regulates metabolism but in the time frame of hours to days. With prolonged stress, the hypothalamus increases secretion of corticotrophin-releasing factor, which subsequently leads to production and secretion of adrenocorticotropic hormone from the anterior pituitary gland and then cortisol from the adrenal glands. Cortisol has much of the same influence on metabolism as epinephrine but functions via activation of transcription and translation of genes rather than modulation of enzyme activity. Under conditions where insulin declines and/or cortisol levels rise, cortisol stimulates transcription of lipases involved in lipogenesis (glucose sparing), enzymes involved in gluconeogenesis and glycogenesis in the liver, and in the breakdown of muscle protein. The net effect is restored blood glucose and larger glycogen stores in the liver. However, this increase is at the expense of muscle and bone and ultimately impairs immunological function.

DIABETES MELLITUS (DM)

DM is a condition characterized by either the total lack of insulin (Type 1) or resistance of peripheral tissues to the effects of insulin (Type 2). Both diseases lack the signaling effect of insulin in the presence of normal or high glucagon and other metabolic signals (Figure 10-16). The disease of DM is due to the imbalance in carbohydrate metabolism and its effects on other metabolic pathways.

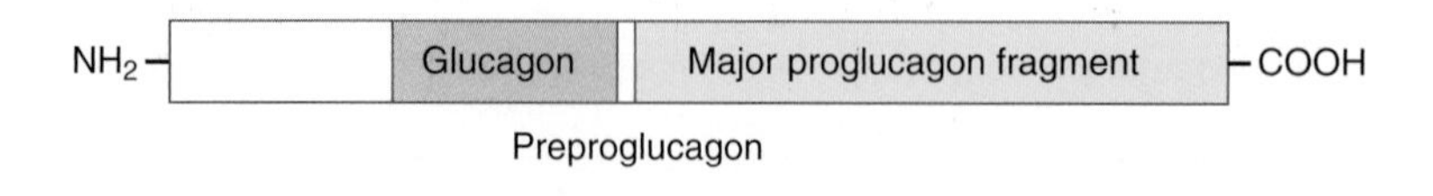

Figure 10-14. Preproglucagon. Preproglucagon, produced by pancreatic α-cells, is processed to active glucagon (orange). [Adapted with permission from Kibble JD and Halsey CR: The Big Picture: Medical Physiology, 1st edition, McGraw-Hill, 2009.]

TABLE 10-1. Insulin Effects on Metabolism

Organ	Pathway	Effect	Enzyme Activity/Metabolite Altered
Liver (100 μU/ml)	Glucose phosphorylation	↑	Glucokinase
	Glycolysis	↑	Phosphofructokinase-1 (increases concentration of fructose 2,6-bisphosphate), pyruvate kinase, phosphofructokinase-2
	Gluconeogenesis	↓	Phosphoenolpyruvate carboxykinase, fructose 1,6-bisphosphatase, glucose-6-phosphatase, fructose 2,6-bisphosphatase
	Glycogen synthesis	↑	Glycogen synthase
	Glycogenolysis	↓	Glycogen phosphorylase
	Fatty acid synthesis	↑	Acetyl-CoA carboxylase, citrate lyase, and malic enzyme
	Pentose phosphate pathway	↑	Glucose-6-phosphate (G6-P) dehydrogenase
Adipose Tissue	Glucose uptake (@ 10 μU/ml)	↑	GLUT4*
	Glycolysis	↑	Phosphofructokinase-1
	Pentose phosphate pathway	↑	G6-P dehydrogenase
	Pyruvate oxidation	↑	Pyruvate dehydrogenase
	Triglyceride lipolysis	↑	Lipoprotein lipase
	Fatty acid/triglyceride synthesis	↑	Acetyl-CoA carboxylase
	Lipolysis	↓	Hormone-sensitive lipase
Skeletal Muscle	Glucose uptake (@ 50 μU/ml)	↑	GLUT4*
	Glycolysis	↑	Phosphofructokinase-1 (increases concentration of fructose 6-phosphate)
	Glycogen synthesis	↑	Glycogen synthase
	Glycogenolysis	↓	Glycogen phosphorylase
	Protein synthesis	↑	Translation (nonselective)

*GLUT4 transport of glucose into cells is the rate-limiting step of glucose metabolism and is 10–20× increased in the plasma membranes of adipose and skeletal muscle cells in response to insulin.
CoA, coenzyme A; GLUT4, glucose transporter 4.

In **type 1 DM**, autoimmune destruction of the pancreatic β-cells leads to a complete loss of insulin production. Although the liver can make glucose, glycogen synthesis is impeded. In the absence of insulin, gluconeogenesis is unrestrained, elevating blood glucose (Figure 10-17). However, muscle and fat cells cannot take up available blood glucose via the GLUT4. Thus, the body is unable to clear the elevated blood glucose, and the peripheral tissues (muscle and fat) are starved for glucose even when present at very high levels in the blood. Furthermore, in the absence of insulin, glucagon secretion is uncoupled from the blood glucose levels (insulin is an important physiological regulator of glucagon secretion). Unopposed glucagon, together with the other counter regulatory hormones (catecholamines, cortisol, and growth hormone), inhibits glycogen synthesis and stimulates gluconeogenesis, glycogenolysis, and lipolysis. Increased lipolysis leads to elevation of **free fatty acids** in the blood stream. These fatty acid molecules are partly taken up by liver and incorporated into **lipoproteins** to increase VLDL and LDL levels, a risk factor for heart disease. **Ketone bodies** are also produced because of the excess of lipolysis, which cannot be

Diabetes and the Polyol Pathway. DM is a disease with the hallmark of elevated blood glucose. Although the mechanism is not completely agreed, the high glucose levels are specifically detrimental to the **kidneys, retina**, and **nerves** because of their ability to transport glucose without the aid of insulin. In these tissues, excess glucose enters the **sorbitol–aldolase reductase pathway** (also known as the **polyol pathway**) where it is reduced to sorbitol and then fructose, oxidizing **NADPH** to $NADP^+$ and reducing NAD^+ to **NADH**, during the enzymatic reactions.

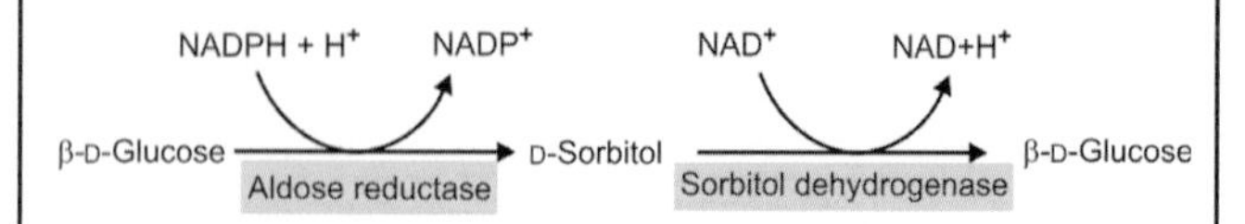

Reproduced with permission from Naik P: Biochemistry, 3rd edition, Jaypee Brothers Medical Publishers (P) Ltd., 2009.

When glucose concentrations are normal, this pathway is minimally active because of low affinity of glucose for the enzyme **aldolase reductase**. However, when glucose levels are high, this reaction is more prominent. The resulting decrease in NADPH and increase in NADH affects other enzymatic reactions that use these molecules as cofactors. NADPH is required for production of reduced **glutathione** and nitric oxide required for detoxification of **reactive oxygen species**. Lowered NAD^+ also leads to additional reactive oxygen species, and production of **inositol** (signaling, including insulin receptor) is also decreased. The effects of lowered NADPH and NAD^+ result in continual damage to those tissues where the polyol pathway is most prominent, causing kidney, eye, and nerve problems seen in many diabetic patients.

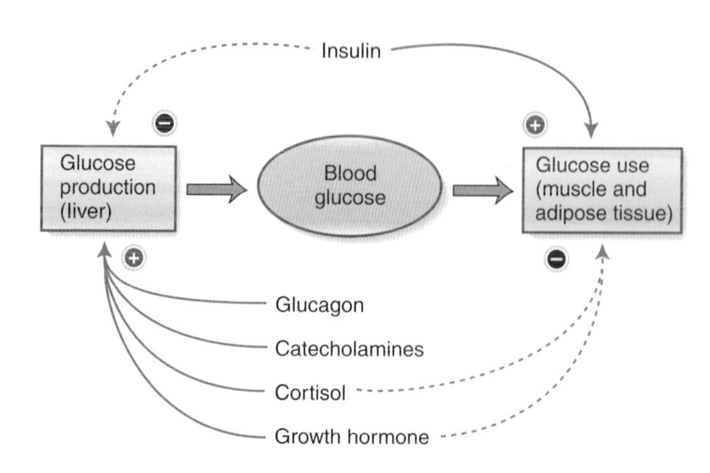

Figure 10-15. Integrated Control of Blood Glucose Concentration. [Adapted with permission from Kibble JD and Halsey CR: The Big Picture: Medical Physiology, 1st edition, McGraw-Hill, 2009.]

inhibited in the absence of insulin. This can result in the dangerous condition ketoacidosis, if the ketone body level becomes too elevated. The only available treatment is the injection of exogenous insulin into the body. However, even with optimal control, the damaging effects of elevated glucose and lipids eventually lead to medical complications.

Type 2 DM is characterized by the production of insulin but resistance of its effects on target tissues. As a result of this resistance, the human body acts as if there is a relative deficiency of insulin, even when present at high levels (Figure 10-18). The disease shares many traits with type 1 DM. As in type 1, gluconeogenesis is unrestrained, and muscle and fat cells do not take up glucose via the GLUT4. As a result, high levels of blood glucose are present. However, the liver still can make glycogen, and lipolysis is kept in check because of decreased but present insulin. However, plasma lipoproteins are typically elevated, often as a consequence of obesity and poor nutrition. Ketoacidosis is not a common sequela to type 2 DM. However, it can occur in

TABLE 10-2. Glucagon Effects on Metabolism

Organ	Pathway	Effect	Enzyme Activity/Metabolite Altered
Liver	Glycolysis	↓	Phosphofructokinase-1, pyruvate kinase, phosphofructokinase-2
	Gluconeogenesis	↑	Phosphoenolpyruvate carboxykinase, fructose 1,6-bisphosphatase, glucose-6-phosphatase, and fructose 2,6-bisphosphatase
	Glycogen synthesis	↓	Glycogen synthase
	Glycogenolysis	↑	Glycogen phosphorylase
	Fatty acid synthesis	↓	Acetyl-CoA carboxylase
	Fatty acid oxidation	↑	Expression of carnitine palmitoyltransferase 1

CoA, coenzyme A.

TABLE 10-3. Epinephrine Effects on Metabolism

Organ	Pathway	Effect	Enzyme Activity/Metabolite Altered
Liver	Glycolysis	↓	Phosphofructokinase-1, decreased F2,6-bisphosphate
	Glycogen synthesis	↓↓	Glycogen synthase
	Gluconeogenesis	↑	Pyruvate kinase, fructose 1,6-bisphosphatase, fructose 2,6-bisphosphatase
	Glycogenolysis	↑↑↑	Glycogen phosphorylase
	Fatty acid synthesis	↓	Acetyl-CoA carboxylase, citrate lyase, and malate dehydrogenase
Adipose Tissue	Lipolysis	↑↑↑	Hormone-sensitive lipase
	Triglyceride uptake from lipoproteins	↓	Lipoprotein lipase
Skeletal muscle	Glycolysis	↑↑↑	Changes concentration of fructose 6-phosphate
	Glycogen synthesis	↓↓	Glycogen synthase
	Glycogenolysis	↑↑↑	Glycogen phosphorylase
	Triglyceride uptake from lipoproteins	↑	Lipoprotein lipase

CoA, coenzyme A; GLUT4, glucose transporter 4.

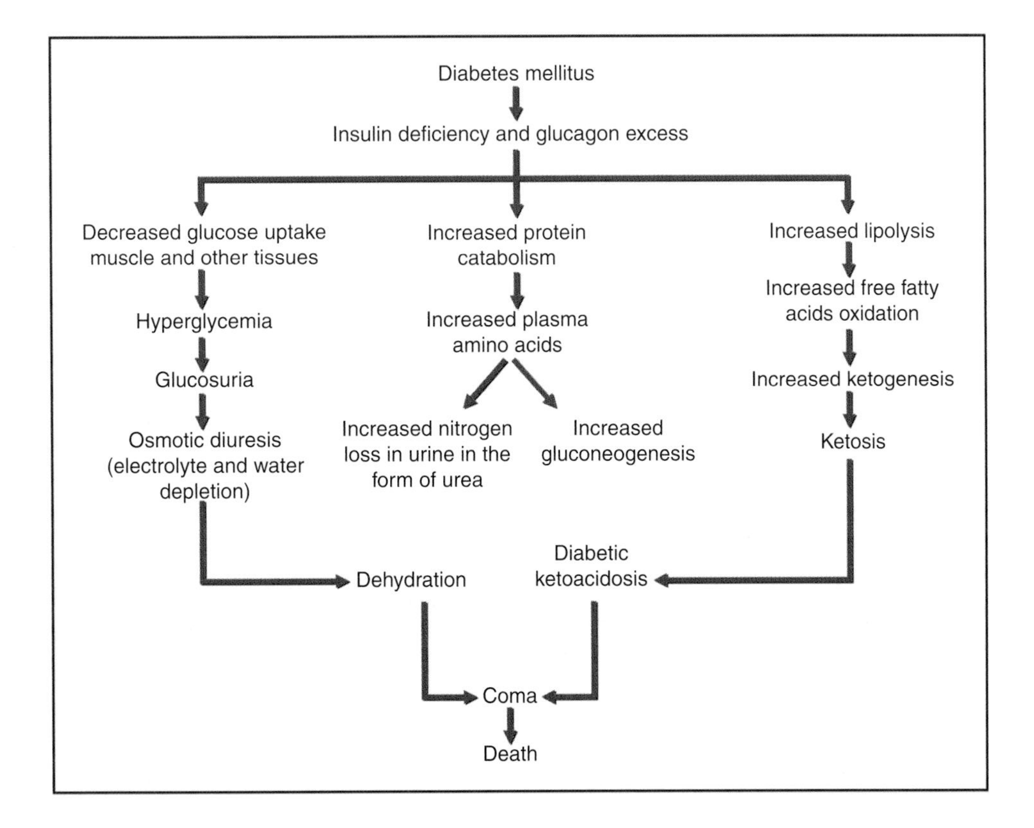

Figure 10-16. Metabolic Events Occurring in Diabetes Mellitus. Overview of effects incurred by the deficiency of insulin and excess glucagon, including those on glucose, proteins/amino acids, and lipids. All effects lead to dehydration and the condition of diabetic ketoacidosis, which can be fatal. [Reproduced with permission from Naik P: Biochemistry, 3rd edition, Jaypee Brothers Medical Publishers (P) Ltd., 2009.]

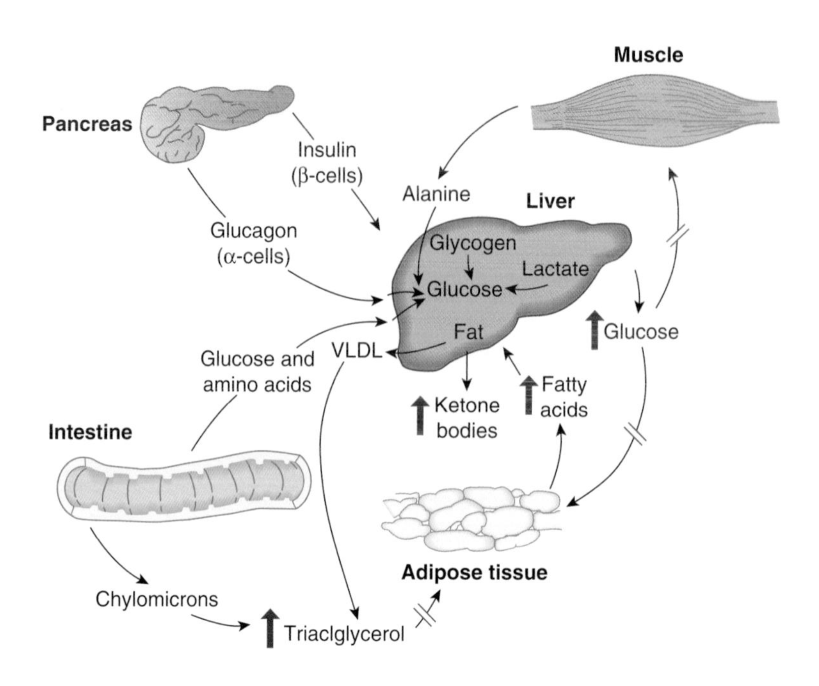

Figure 10-17. Type 1 Diabetes Mellitus. The effect of this disease on organs and major metabolic pathways is illustrated. The absence of insulin in type 1 diabetes inhibits (red bars) the uptake/conversion of glucose by muscle and adipose tissue, leading to an increase in glucose (red arrow) synonymous with the disease. Uptake of fatty acids from triacylglycerols by adipose tissue is also inhibited (red bars), leading to an increase in its levels (red arrow). Resulting changes in metabolism lead to increased fatty acids and ketone bodies (red arrows), the latter of which contributes to diabetic ketoacidosis seen in type 1 diabetic patients. See the text for further discussion. VLDL, very-low-density lipoprotein. [Adapted with permission from Katzung BG, et al.: Basic and Clinical Pharmacology, 11th edition, McGraw-Hill, 2009.]

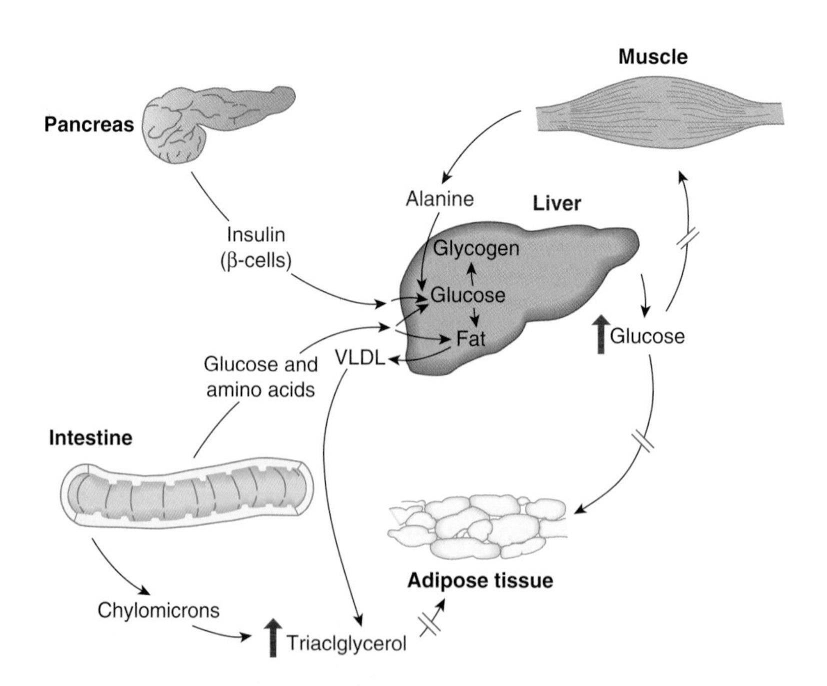

Figure 10-18. Type 2 Diabetes Mellitus. The effect of this disease on organs and major metabolic pathways is illustrated. Insulin resistance in type 2 diabetes inhibits (red bars) the uptake/conversion of glucose by muscle and adipose tissue, leading to an increase in glucose (red arrow) synonymous with the disease. Uptake of fatty acids from triacylglycerols by adipose tissue is also inhibited (red bars), leading to increase in its levels (red arrow). Unlike type 1 diabetes, fatty acids and ketone bodies are not increased (not shown) and diabetic ketoacidosis is, therefore, rare. See the text for further discussion. VLDL, very-low-density lipoprotein. [Adapted with permission from Katzung BG, et al.: Basic and Clinical Pharmacology, 11th edition, McGraw-Hill, 2009.]

type 2 DM patients under conditions of additional metabolic stress and after pancreatic failure leads to decreased production and secretion of insulin. Some older people with type 2 DM may experience a different serious condition called **Hyperosmolar hyperglycemic nonketotic syndrome (HHNS)**, a condition in which the body tries to get rid of excess sugar by passing it into the urine. HHNS is usually brought on by an illness, infection, or other factors.

DM and Advanced Glycation End Products: Increased levels of sorbitol and fructose in diabetic patients lead to nonspecific attachment onto proteins of carbohydrate molecules via carbohydrate–nitrogen links (see the figure below), for example, lysine and arginine amino acids, proportional to the level of glucose in the body. These erroneously modified proteins are referred to as **advanced glycated (also known as glycosylation) endproducts (AGEs)**. The best known AGE is hemoglobin A_{1C} in circulating red blood cells, which forms the basis for testing of diabetic control. AGE proteins and their breakdown products cause oxidant damage to the kidney and also increase the production of cytokines (e.g., tumor necrosis factor-β), which damages the glomerulus. In addition, they increase the **permeability of blood vessels**, increase **oxidized LDL** levels and increase cytokine-related **oxidative stress**.

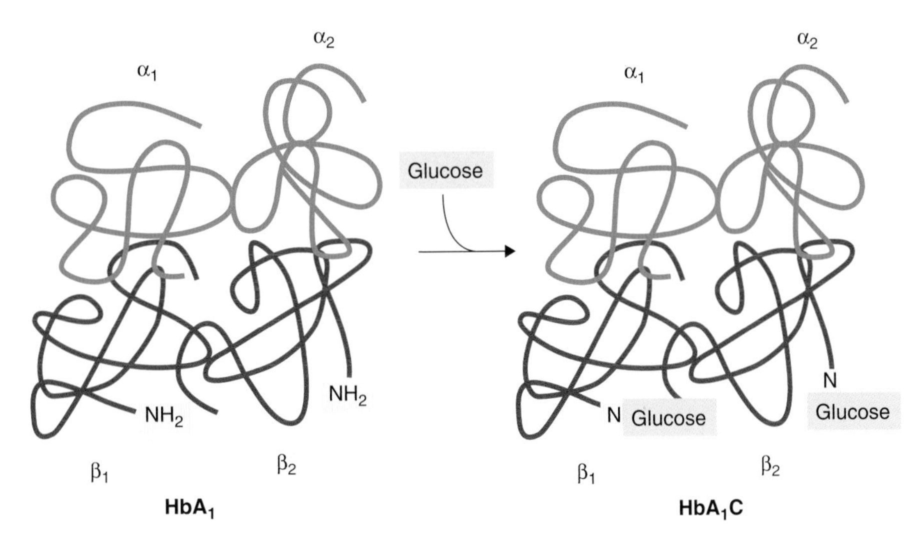

Adapted with permission from Naik P: Biochemistry, 3rd edition, Jaypee Brothers Medical Publishers (P) Ltd., 2009.

AGE proteins also bind to a **receptor for advanced glycation endproducts (RAGE).** Activation of RAGE leads to **nuclear factor kappa B** induction of several inflammatory gene products resulting in a chronic inflammatory environment. This long-term inflammation promotes such varied disease states as atherosclerosis with accompanying heart attacks and strokes. Damage to the nerves and retina is also prominent, leading to **diabetic neuropathy** and **retinopathy**, the latter of which often leads to avoidable blindness.

VITAMINS AND MINERALS

VITAMINS

Vitamins and minerals serve as important cofactors in several, essential enzymatic reactions, including protein, carbohydrate, and fat metabolism as well as the formation of tissues. Vitamins are usually divided into fat soluble (vitamins A, D, E, and K) and water soluble (vitamins B and C).

Although the human body or resident, benevolent bacteria can produce some vitamins, most, along with the required minerals, are obtained from the diet. Some vitamins can also be stored in the body for different times. Deficiencies of particular vitamins in the diet, therefore, often do not manifest themselves for a lengthy period of time. Vitamins A, D, and B_{12}, for example, are stored in large enough quantities so that absence in the diet will not be noticed for months or years. Deficiencies in vitamins and minerals may lead to tissue damage and/or a number of diseases (e.g., beriberi, scurvy, night blindness, pellagra, and rickets), many of them being fatal. At the other end of the spectrum, excess amounts of lipid-soluble vitamins can be toxic. The contribution of vitamins during fetal development is especially important; excess maternal vitamin A during pregnancy can lead to significant birth defects. Vitamins required by humans are summarized in Table 10-4.

MINERALS

Minerals, specifically those essential atoms in the human diet, are simple chemical elements required for the existence and subsistence of life. The basic chemical elements include carbon, hydrogen, nitrogen, oxygen, phosphate, and sulfur. Additional minerals that play a smaller but still important role are sodium, potassium, calcium, magnesium, iodine, and zinc. Other minerals are required by the human body, although the exact number of essential minerals is still controversial. A summary of minerals important to biochemical functions of the human body is included in Table 10-5.

TABLE 10-4. Summary of Vitamins

Vitamin	Function	Diseases
A (retinol)	Forms visual pigment (retinal) for reception of light impulses in the eye; possible antioxidant role	Deficiency—Night blindness, dry eyes/cornea (keratomalacia). Excess—Usually from vitamin supplements. Skin discoloration and dryness (e.g., at angle of mouth), hair loss, bone thinning and abnormal growth, birth defects. Acute signs include headache, vision blurring, and nausea/vomiting.
B_1 (thiamine)	Forms thiamine. Essential cofactor for dehydrogenase, decarboxylation, and transketolase reactions	Deficiency—Wet beriberi (cardiovascular), dry beriberi (neurological), and Wernicke–Korsakoff syndrome.
B_2 (riboflavin)	Required for electron transfer to and from $FAD(H_2)$ in carbohydrate and fatty acid metabolism; flavin mononucleotide in the respiratory chain	Deficiency—Sore, red throat and mouth, dry and cracked lips, mouth with inflamed tongue, scaly skin (e.g., scrotum, labia and edges of nose, and anemia).
B_3 (niacin)	Converted to nicotinamide, required for electron transfer to and from NAD(H) or NADP(H) in carbohydrate, and lipid metabolism	Deficiency—Pellagra (the four Ds—diarrhea, dermatitis, dementia, and death). Other symptoms include sunlight sensitivity, cardiovascular, and neurological disorders. Excess—Flushing/itching of skin, stomach upset, and liver problems up to full failure, retinal damage that may result in blindness, and worsening of diabetes or gout.
B_5 (pantothenic acid)	Required for synthesis of coenzyme A (CoA), an important molecule in carbohydrate, protein, and lipid metabolism and biosynthesis	Deficiency—"Pins and needles" nerve sensations (parathesias).
B_6 (pyridoxine, pyridoxal, pyridoxamine)	Converted to pyridoxal-5-phosphate, required for aspects of amino acid, carbohydrate, and fatty acid metabolism as well as production of the vitamin nicotinic acid (niacin) and catecholamine neurotransmitters (dopamine, norepinephrine, and epinephrine) and serotonin	Deficiency—Sideroblastic (atypical nucleated erythrocytes) anemia, skin and mouth sores, and peripheral nerve dysfunction (neuropathy).
B_7 (biotin)	Required coenzyme for enzymes that add a carboxyl group in degradation of leucine and fatty acid metabolism as well as gluconeogenesis	Deficiency—Hair loss; chronic inflammation of the eye (conjunctiva), skin, and small intestine; and nerve dysfunction.
B_9 (folic acid)	Required coenzyme for nucleosides/nucleotide synthesis and repair of DNA and methylation reactions (e.g., DNA) as well as replication and growth of red blood cells; formation of methyl-B_{12}	Deficiency—Fetal birth defects, especially neurological ("neural tube defects"), megaloblastic (enlarged red blood cells) anemia.
B_{12} (cyanocobalamin)	Required for synthesis of amino acid methionine, important in methylation reactions (e.g., DNA). Also, important in synthesis of hemoglobin, protein and fat metabolism, and maintenance of nerve cells	Deficiency—Megaloblastic (enlarged red blood cells) anemia and pernicious (due to B_{12} malabsorption) anemia; increased homocysteine (homocystinuria) that may lead to cardiovascular disease; methylmalonicaciduria.

TABLE 10-4. Summary of Vitamins (Continued)

Vitamin	Function	Diseases
C (ascorbic acid)	Important antioxidant. Also, coenzyme for collagen/bone, carnitine, and norepinephrine synthesis as well as tyrosine and peptide hormone function, blood vessel function, and wound healing	Deficiency—Scurvy (weak collagen leading to defective and weak bone and connective tissue formation). Poor wound healing. Excess—Skin flushing and rashes, headache, diarrhea, nausea/vomiting, indigestion.
D (calciferol)	Important in bone growth and remodeling as well as absorption/regulation of blood calcium and phosphate concentration	Deficiency—Rickets (children) and osteomalacia (adults). Acute overdose—Symptoms due to resulting high calcium level include nausea/vomiting weakness, increased urination and fluid intake, and possible, irreversible kidney damage.
E (tocopherols/tocotrienols)	Important antioxidant that protects lipids/cell membranes. Possible role in protecting neurons and reducing cholesterol	Deficiency—Possible neurological problems (ataxia), increase in cardiovascular disease, anemia, congenital heart disease, and prostate cancer. Excess—May increase chance of death.
K (phyloquinone, menaquinone)	Cofactor for carboxylation of glutamate, required for several blood coagulation factors, osteocalcin activity (bone metabolism), and blood vessel health	Deficiency—Bleeding and poor clot formation.

DNA, deoxyribonucleic acid; $FADH_2$, flavin adenine dinucleotide.

TABLE 10-5. Summary of Important Minerals

Mineral	Function	Disease(s)
Sodium (Na^+)	Involved in maintenance of fluid volume and osmotic pressure per the kidneys and associated hormones (e.g., renin, aldosterone, antidiuretic hormone, atrial natriuretic peptide). Essential for generation and maintenance of electric or transport potential across membranes (e.g., nerve conduction, muscle contraction, and membrane pumps)	Hyponatremia—Neurological symptoms secondary to cell swelling and electrolyte imbalance; potentially fatal. Hypernatremia—Deficit in free water in the body. Variable symptoms, including neurological, potentially fatal.
Potassium (K^+)	Usually the partner to sodium, essential for generation and maintenance of electric and transport potential across membranes (e.g., nerve conduction, muscle contraction, and membrane pumps), as well as potassium-specific pumps.	Hypokalemia and hyperkalemia—Muscle and neurological symptoms; both may lead to fatal abnormal heart rhythm, especially hyperkalemia.
Chloride (Cl^-)	Involved in conjunction with sodium in maintenance of fluid volume and osmotic pressure per the kidney. Essential role in neurological functions (e.g., glycine and GABA neurotransmitters) and acid–base balance via transport of bicarbonate.	Hypochloremia and hyperchloremia—Often secondary to vomiting and/or diarrhea; usually asymptomatic but may have respiratory symptoms.

(continued)

TABLE 10-5. Summary of Important Minerals (Continued)

Mineral	Function	Disease(s)
Calcium (Ca^{2+})	Required for bone formation and remodeling; important cofactor for several enzymes and signal for signaling pathways (i.e., diacylglycerol/IP_3), including blood clotting and muscle contraction; neurotransmitter for some neuron signals and plays a prominent role in maintaining a potential difference across membranes	Hypocalcemia—Neurological symptoms; may be followed by potentially fatal spasms of larynx and abnormal heart rhythm. Hypercalcemia—Constipation (*groans*), psychotic episodes (*moans*), pain in *bones*, kidney *stones*, and depression, etc. (*psychiatric overtones*); abnormal heart rhythm can also develop.
Magnesium (Mg^{2+})	Magnesium stabilizes phosphate groups, including those in ATP; cofactor in several enzymatic processes	Hypomagnesemia—Muscle weakness, nerve problems/tremors, psychiatric episodes/epileptic fits; may lead to heart failure. Hypermagnesemia—Weakness, breathing problems, and potentially fatal heart rhythms.
Phosphorous (P, usually found in the form of PO_4^{-3})	Essential structural and functional element for nucleic acids, bone/teeth, and phospholipid component of membranes; addition or removal of phosphate to/from a protein/enzyme serves as a key regulator of enzymes	Hypophosphatemia—Nerve, bone, red and white blood cells, membrane, and muscle functional problems. Hyperphosphatemia—(Interference with other minerals, promotes calcification of soft-tissue organs).
Iron (Fe^{2+} or Fe^{3+})	Essential cofactor in numerous enzymes and proteins (e.g., heme); essential for oxidation processes or oxygen transport	Iron deficiency (anemia). Iron excess (hemachromatosis).
Iodine (I_2)	Essential element for thyroid hormones; can act as antioxidant outside of thyroid, may play a role in the development of breast and/or stomach cancer, and affects immune system and salivary gland health	Iodine deficiency (goiter; cretinism).
Zinc (Zn^{2+})	Cofactor in almost 100 enzymes, serving a multitude of roles in metabolism, transcription and translation, acid–base balance, immune function, and protein synthesis; part of unique, tertiary protein structures (e.g., zinc fingers); part of nerve response of glutamate and essential for learning	Zinc deficiency—Directly impacts the enzymatic processes that rely on it; initial signs may be seen in skin, hair, and nails. Zinc excess—Can impair the absorption of other ions (e.g., iron and copper); corrosive damage to soft tissues.
Manganese (Mn^{2+})	Essential cofactor for several types of enzymes involved in numerous biological functions as well as several specific types of peptides	Manganese deficiency—Possible association with inflammatory diseases, diabetes, and some neurological and psychiatric problems. Manganese excess (manganism)—Progressive neurological/psychiatric symptoms.
Copper (Cu^{2+})	Cofactor in several enzymes involved in electron transport or oxidation–reduction reactions (e.g., cytochrome *c* oxidase). Also, used for electron transport	Copper deficiency (anemia symptoms, decreased metabolism, and psychiatric manifestations). Copper excess/Wilson's disease (neurological and psychological effects).

TABLE 10-5. Summary of Important Minerals (Continued)

Mineral	Function	Disease(s)
Sulfur (S, usually joined with H, O, and/or C)	As part of cysteine and methionine amino acids, plays an essential role in component of primary and tertiary protein structure via disulfide bond as well as role in sulphur-containing enzymes (e.g., cytochrome *c* oxidase, coenzyme A (CoA); reduction of reactive species via glutathione	NA
Cobalt (Co^{2+})	Component of cobalt-containing cofactors/enzymes, the most prominent of which is vitamin B_{12}	Cobalt excess—Potentially fatal. Cobalt deficit (pernicious anemia).
Nickel (Ni^{2+})	Important cofactor in some enzymes (e.g., urease), especially those involved in reduction reactions	Nickel deficiency—Potential impact on involved enzymes, although not manifested as symptoms. Nickel excess—Skin irritant and potential cancer-causing agent.
Chromium (Cr^{3+} and Cr^{6+})	Possible role in carbohydrate and/or lipid metabolism	Chromium deficiency (extremely rare; effects controversial). Chromium excess (Cr^{3+}—damage to DNA; Cr^{6+}—can act as a cancer-causing agent and damages internal organs).
Fluoride (F^-)	Role in strengthening of teeth and bone and, as such, used for prevention of cavities and treatment of osteoporosis	Fluoride deficiency—Possible connection to weakened teeth and bones. Fluoride excess—Neuromuscular and other symptoms that can result in death.
Selenium (Se)	Essential cofactor for certain antioxidant enzymes (e.g., glutathione peroxidase), which remove reactive oxygen species; believed to be cofactor in thyroid hormone conversion of T_4 to T_3	Selenium deficiency—Rarely seen but may contribute to destruction of heart or connective tissue; also affects thyroid hormone synthesis Selenium excess (selenosis)—Affects liver and lungs; potentially fatal.
Molybdenum (MoO_4^{2-})	Cofactor in several enzymes, including oxidizing enzymes (e.g., xanthine oxidase)	Affects enzymes requiring cofactor; neurological symptoms may result; possible association with development of esophageal cancer.

ATP, adenosine triphosphate; GABA, gamma-aminobutyric acid; IP_3, inositol trisphosphate.

REVIEW QUESTIONS

1. What are the general metabolic roles of the major biochemical molecules—carbohydrates, lipids, and amino acids?
2. What are the primary metabolic contributions of small intestine, liver, adipose tissue, pancreas, and muscle?
3. Why are glucose-6-phosphate, pyruvate, and acetyl-CoA considered primary intermediates in metabolism of the major biochemical molecules?
4. How does the concentration of glucose-6-phosphate determine the fate of glucose after entry into the liver cell or muscle cell?
5. What is the role of fructose 2,6-bisphosphate in the control of liver carbohydrate metabolism?
6. What factors in the fed and starved states determine the fate of pyruvate in terms of its oxidation, use for glucose synthesis, or role in fatty acid formation? How do these factors balance each other?

7. What factors in the fed and starved states determine the fate of acetyl CoA in liver in terms of its oxidation, use for fatty acid synthesis, or conversion to ketone bodies? How do these factors balance each other?

8. How does the ratio of insulin to glucagon determine the balance of carbohydrate metabolism in the body in terms of effects on metabolic pathways in specific tissues?

9. How does the ratio of insulin to glucagon determine the balance of fat metabolism in the body in terms of effects on metabolic pathways in specific tissues?

10. What are the key effects of epinephrine on metabolism and how do these relate to the body's needs in the fight or flight response?

11. What are the roles of glucocorticoids in stress or starvation?

12. What are the key metabolic disturbances associated with type 1 and type 2 diabetes?

13. What are the fat-soluble vitamins, their functions, and diseases of deficiency?

14. What are the B vitamins, their functions, and diseases of deficiency?

15. What are the functions of vitamin C and diseases of deficiency?

16. What are the functions of the following minerals and the consequences of their deficiency and/or excess—sodium, potassium, calcium, magnesium, phosphorus, iron, and iodine?

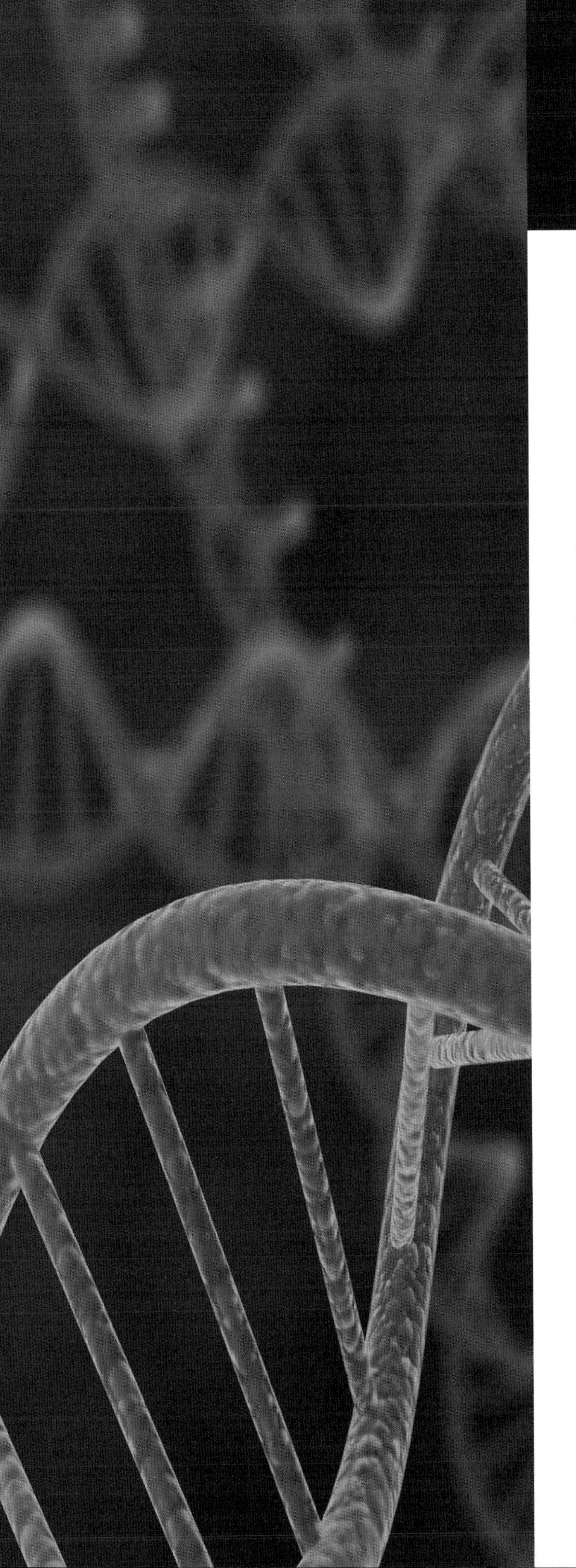

CHAPTER 11

THE DIGESTIVE SYSTEM

Editor: Kshama Jaiswal
Department of Surgery, Denver Health Medical Center, Denver, Colorado

OVERVIEW

The digestive or gastrointestinal system is roughly defined as the anatomical component from the mouth to the anus, including organs responsible for transit, mechanical breakdown, digestion and absorption of foodstuffs, as well as the efficient elimination of solid waste. Included are the mouth and dentitia, pharynx and esophagus, stomach, small intestine, liver, gall bladder, pancreas, large intestine, rectum, and anus. As with the complex integration and control of metabolism, the digestive system is, itself, under the influence of neurological and hormonal regulation that both activates and inhibits many of its complex actions. Most of these complex actions are, themselves, simple biochemical processes of proteins, carbohydrates, lipids, and nucleosides/nucleotides to include enzyme reactions with associated activating and inhibitory molecules, membrane-spanning protein channels, and pumps all leading to the production and storage of energy and essential building blocks for current or future use.

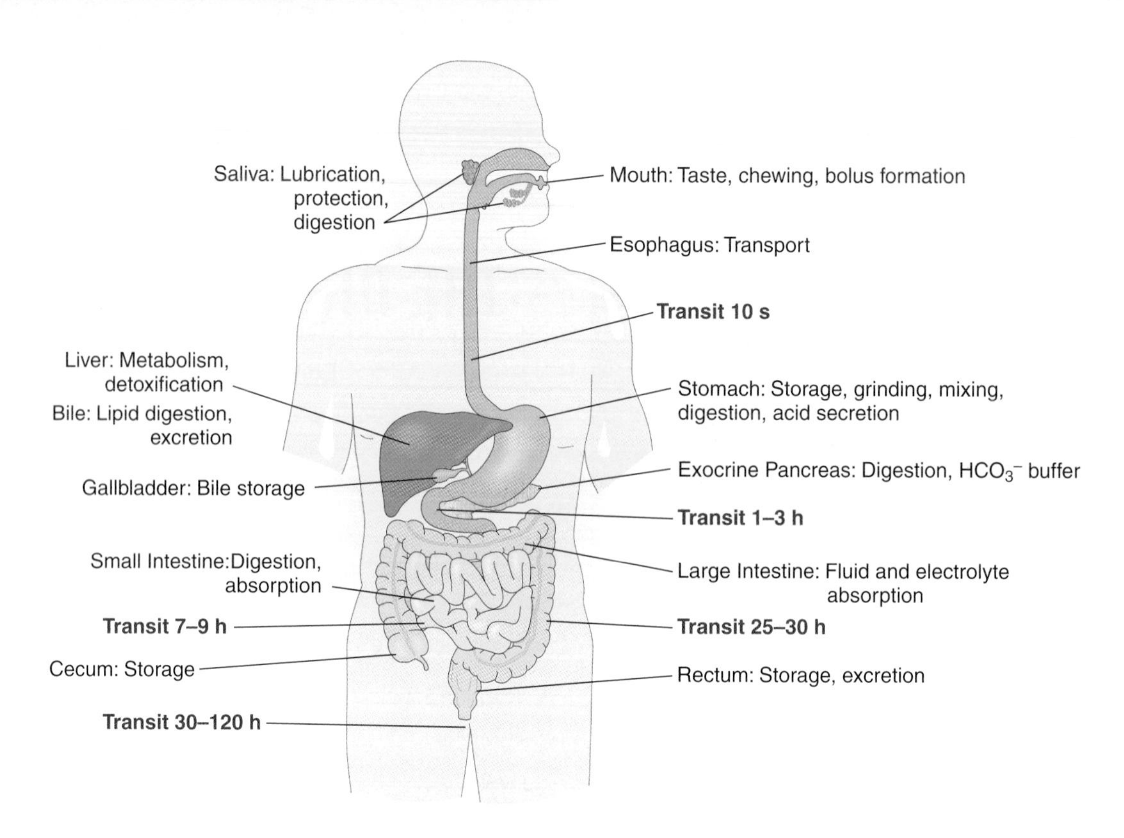

Figure 11-1. Overview of the Digestive System. Components of digestion and transport of food from the mouth to the rectum/anus are shown, including a summary of their contributions and the average amount of time for food to reach their location after ingestion. [Reproduced with permission from Kibble JD and Halsey CR: The Big Picture: Medical Physiology, 1st edition, McGraw-Hill, 2009.]

SUMMARY OF THE DIGESTIVE SYSTEM

The digestive system is the collection of organs responsible for the digestion of ingested foods and liquids (Figure 11-1). This system is classically considered to start at the mouth and continue via the esophagus, stomach, small and large intestines, and end at the rectum/anus. Besides their mechanical and enzymatic breakdown of foodstuffs, considerable contributions toward digestions are supplied from the liver and pancreas.

MOUTH

Teeth provide the initial mechanical breakdown of food via chewing or mastication and, therefore, the formation and health of teeth is important. Proteins serve as an important structural component of teeth enamel. Like bone, teeth require the collagen-like proteins **amelogenin** (over 90% composed of proline, glutamine, and histidine amino acids), **ameloblastin**, **enamelin**, and **tuftelin** (a phosphorylated glycoprotein), which organize, initiate, and direct calcium phosphate crystal formation and help anchor teeth to the gums.

The possible breakdown of enamel is equally important and is also dependent on biochemical processes. Dental plaque formation and tooth decay both rely on enzymatic processes for formation as well as prevention. Plaque is caused mainly by the normal oral bacteria *Streptococcus mutans*, *Lactobacillus acidophilus*, *Fusobacterium nucleatum*, *Actinomyces viscosus*, and *Nocardia spp*. When these organisms form a layer on teeth, those closest to the teeth exist in an oxygen-deficient environment and convert to anaerobic respiration for energy production. This enzymatic process turns carbohydrates into lactic acid from pyruvate (Chapter 6) and results in a pH of below 5.5, leading to tooth decay, the demineralization process that causes cavities. Glucose, fructose, and especially sucrose (common table sugar) are the main carbohydrate sources.

The mouth is also the location for the start of the processes of carbohydrate, lipid, and protein digestion via important enzymes contained in saliva. Saliva production, along with feelings of hunger and satiety, is controlled by a variety of neurobiochemical processes, which start with the initial thoughts or **cephalic phase** of eating. The physical presence and act of chewing and tasting food, known as the orosensory or **gustatory phase**, elicits further signals that enhance saliva formation and expression. The various factors affecting hunger and satiety are discussed in Chapter 19. Saliva also traps molecules produced by normal oral bacteria that adds to the taste sensation of otherwise odorless and tasteless food compounds. The hormone **gustin**, produced in saliva and an activator of a calmodulin-dependent cyclic adenosine monophosphate (cAMP) phosphodiesterase (Chapter 8), is also thought to play an important part in taste bud formation. Digestive enzymes contained in saliva and their roles in digestion are listed in Table 11-1.

TABLE 11-1. Composition of Saliva

Item	Function/Notes
Water	98%–99% of total
Electrolytes	1%–2% of total: sodium (2–21 mM), potassium (35 mM), calcium (1.2–2.8 mM), magnesium (0.08–0.5 mM), chloride (5–40 mM), bicarbonate (25 mM), and phosphate (1.4–39 mM)
Major Digestive Enzymes	
α-amylase	Initiates random digestion of amylose and amylopectin chains, producing maltotriose, maltose, amylose, glucose, and oligosaccharides.
Lingual lipase (secreted by tongue)	Initiates hydrolysis of long-chain triglycerides into diacylglycerol and free fatty acids, which continues into and through stomach. Optimal activation requires acidic (~pH 4) environment, so vast majority of activity is in stomach.
Minor Digestive Enzymes	
Salivary acid phosphatases A + B	Removes phosphate group from proteins by reaction orthophosphoric monoester + $H_2O \rightarrow$ alcohol + H_3PO_4.
N-acetylmuramoyl-L-alanine amidase	Removes *N*-acetylmuramoyl from amino acid residue in certain glycoproteins.
NAD(P)H dehydrogenase (quinone)	Catalyzes the following reaction: $NAD(P)H + H^+ + \text{a quinone} \rightleftharpoons NAD(P)^+ + \text{a hydroquinone}$
Superoxide dismutase	Changes highly reactive superoxide molecules into oxygen and hydrogen peroxide (H_2O_2) by oxidation of a metal ion (e.g., copper, zinc, manganese, iron, or nickel).
Glutathione transferase	Detoxifies several types of molecules containing the sulfur-containing, tripeptide glutathione, produced from the cysteine, glutamic acid, and glycine amino acids, via reduction and conjugation.
Aldehyde dehydrogenase (class 3)	Catalyzes the oxidation (dehydrogenation) reaction $RCHO + NAD^+ + H_2O \rightarrow RCOOH + NADH + H^+$
Glucose-6-phosphate isomerase	Changes glucose-6-phosphate into fructose 6-phosphate.
Antibacterial Compounds	
H_2O_2	Highly reactive oxygen-containing molecule.
Immunoglobulin A	Main immunoglobulin in saliva; plays a critical role in mucosal immunity.
Antimicrobial Enzymes	
Lactoferrin	Produces small peptides, called lactoferricin and kaliocin-1, which, coupled with iron, inhibit bacterial and viral binding to cell membranes. Also exhibits antifungal properties.
Lactoperoxidase (salivary)	Kills bacteria via formation of reactive bromine and iodine species.
Lysozyme	Hydrolyzes peptidoglycan component of bacterial cell wall at 1,4-β-linkages between *N*-acetylmuramic acid and *N*-acetyl-D-glucosamine residues.

(continued)

TABLE 11-1. Composition of Saliva (Continued)

Item	Function/Notes
Thiocyanate $(SCN)^-$	Reactive sulfur-containing molecule
Mucus	Secretion offering lubrication and protection to teeth, tongue, and epithelial cells of the gums/inner mouth as well as the remainder of the digestive tract.
Mucopolysaccharides	Long and unbranched polysaccharides consisting of a repeating disaccharide unit, often referred to as Glycosaminoglycans. See Chapter 2.
Glycoproteins	Proteins with large, attached carbohydrate chains. See Chapter 2.
Cells	~8 million human/ml; ~500 million bacterial/ml. Bacterial metabolism leads to production of thiols, amines, and organic acids causing bad breath (halitosis).

"Meth Mouth": Patients who abuse **methamphetamines** are prone to marked dental decay, known colloquially as "Meth Mouth." Methamphetamine acts on the **α-adrenergic receptors** of the vasculature of the salivary glands, causing vasoconstriction and reducing salivary flow, depriving the oral environment of saliva's buffering activity to counteract acidity and prevent demineralization of enamel. Methamphetamine-induced vomiting also exposes teeth to acids. In addition, methamphetamine overstimulates the sympathetic nervous system, eventually depleting norepinephrine and dopamine and altering concentrations of other central nervous system (CNS) neurotransmitters such as serotonin, acetylcholine, and glutamate. This reduction increases the demand for adenosine triphosphate (ATP); methamphetamine users may compensate by consuming more carbohydrates in the form of sugars and starches. Observers specifically report a high intake of carbonated soft drinks among meth users. At the same time, users typically abandon oral hygiene. In short, methamphetamine use encourages an environment that maximizes caries risk—decreased saliva, frequent exposure to sugar, and lack of plaque control.

STOMACH

Stomach is the location of continued mechanical breakdown of food via the actions of its smooth muscle layers but, perhaps more importantly, the site of the activation and activity of a number of regulated enzymes and exposure to acid (pH 1–2) that initiates metabolism. The other main function of the stomach is the regulated and coordinated secretion of hydrogen atoms and various digestive enzymes under the control of the autonomic nervous system and several hormones. These molecules are produced from a variety of cell types found in various parts of the stomach as shown in Figure 11-2 and summarized below. Hormones that affect the stomach are described in Table 11-2.

1. **Mucus (Neck) Cells.** Found in all parts of the stomach, these cells produce a viscous mixture of protective enzymes and **mucins**, large glycoproteins (Chapter 1), via stimulation of a **myristylated alanine-rich C kinase**. Mucins have low glycosylation on the amino and carboxy terminal ends but very high levels of glycosylation (via serine, threonine, and asparagine amino acids) in the central part of its amino acid sequence. These mucin glycoproteins are also interlinked by cysteine–cysteine disulfide bonds (Chapter 1) that form large aggregate gels filled with water and protected from enzymes by their dense carbohydrate coating. The secreted mucus provides a protective barrier against the

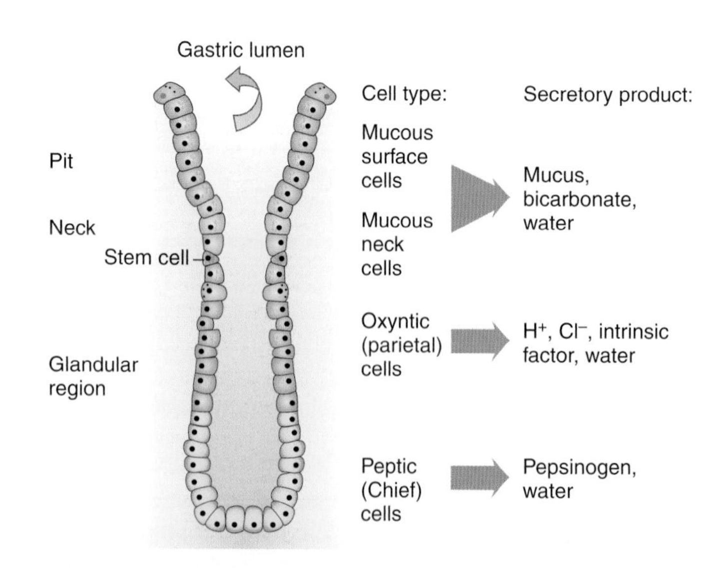

Figure 11-2. Hormone Production from the Stomach. Specific portions of the stomach are responsible for production of hormones as illustrated above (see text for further description). [Reproduced with permission from Kibble JD and Halsey CR: The Big Picture: Medical Physiology, 1st edition, McGraw-Hill, 2009.]

TABLE 11-2. Hormones Affecting the Stomach

Molecule	Location Produced	Action
Cholecystokinin (CCK)	Peptide hormone produced by cells in the duodenum	Main effect is the contraction of smooth muscle of the gall bladder and simultaneous secretion of pancreatic enzymes to increase digestion. However, emptying of the stomach and gastric acid secretion is also decreased by CCK as digestion progresses beyond the stomach.
Enteroglucagon	Mainly in terminal ileum and colon from the prohormone preproglucagon	Decreases the production of gastric acid by parietal cells and smooth muscle contraction (motility) of the stomach, thereby decreasing gastric emptying.
Gastrin	Produced by G-cells in response to presence of undigested proteins and/or distension of the antrum of the stomach	Inhibited by pH <4 and by the hormone somatostatin (see below). Gastrin stimulates hydrochloric acid, pepsinogen, and intrinsic factor secretion from parietal cells, pepsinogen/rennin by Chief cells, as well as histamine release from enterochromaffin-like cells. Smooth muscle contraction (i.e., motility) of the stomach is also enhanced by gastrin.
Gastric inhibitory peptide (GIP)	Peptide hormone produced in the duodenum/jejunum	Decreases gastric acid release by parietal cells as well as smooth muscle contraction (motility) of the stomach. GIP may also increase insulin secretion and fatty acid metabolism (e.g., milk digestion) by activating lipoprotein lipase.
Motilin	Peptide hormone made mainly in the duodenum/jejunum	Increases smooth muscle contraction (fundus, antrum, and gall bladder). Also, stimulates secretion of somatostatin, pancreatic peptide, and pepsinogen.
Secretin	Produced in the duodenum	Several effects including increased secretion of water and bicarbonate as well as insulin (following glucose intake) from the pancreas and bile from the liver. In the stomach, secretin inhibits G-cell production of gastrin, thereby reducing the acidity (pH) of digested food leaving the stomach and entering the duodenum. Lowered pH allows maximal activation of pancreatic enzymes being secreted into this part of the small intestine. Secretin also enhances secretion of pepsin as well as the hormones glucagon, pancreatic polypeptide, and somatostatin.
Somatostatin/ growth inhibiting hormone	Produced in the stomach, intestines, and pancreas	Decreases release of gastrin, CCK, secretin, motilin, vasoactive intestinal peptide (VIP), GIP, and enteroglucagon, decreasing stomach secretion and contraction.
VIP	Peptide hormone produced in digestive tract and brain	Causes relaxation of smooth muscle (lower esophageal sphincter, stomach, and gallbladder), stimulates pepsinogen secretion, dilutes bile and pancreatic juice, increases bicarbonate production in the pancreas, decreases gastrin-induced gastric acid secretion, and increases water secretion in intestine.

digestive enzymes and highly acidic conditions found in the stomach.

2. **Parietal (Oxyntic) Cells.** Found in all parts of the stomach, parietal cells are stimulated by the hormones **histamine** via the H_2 receptor (G_s—adenyl cyclase/cAMP) and **gastrin** via a CCK2 receptor [G_q—phospholipase C/inositol triphosphate (IP_3)/Ca^{2+}] as well as by the vagus (parasympathetic) nerve via acetylcholine and the M_3 receptor (G_q—phospholipase C/IP_3/Ca^{2+}). Stimulation of adenyl cyclase is known to specifically activate the H^+/K^+ ATPase active transport channel and gastric acid production. These cells produce three main components that are as follows:
   a. **Gastric acid** [mainly hydrogen (H^+) and chloride (Cl^-) ions] via a unique **H^+/K^+ ATPase active transport channel** that pumps hydrogen ions into the stomach against a very high concentration gradient (approx. 3 million to 1).

Gastric acid functions in protein denaturization, pepsinogen activation (see below), and inhibition of bacterial growth.

b. Bicarbonate ion (HCO_3^-). Excreted into the blood as part of the overall H^+ pumping mechanism.

c. Intrinsic factor. Required for intestinal absorption of vitamin B_{12} (see Chapter 10).

Achlorydia: The destruction or damage of **parietal cells** leads to a marked reduction in the production of gastric acid, essential for the initial breakdown of food and activation of stomach enzymes (e.g., pepsin). Although several disease states and/or surgical interventions can affect parietal cells, immune destruction of the cells leads to the condition of **achlorydia/hypochloridia** in which decreased amounts of hydrochloric acid are produced. The resulting higher stomach pH leads to symptoms of gastroesophageal reflux, pain and fullness from inadequate digestion, and increased growth of bacteria, normally limited by low pH, which can lead to diarrhea and decreased absorption of essential ions (e.g., magnesium and zinc) and vitamins (e.g., C, K, B-complex), which themselves lead to other disease states. Treatment is via supplementation with **Betaine HCl**, a form of hydrochloric acid, which survives into the stomach, and any required vitamins and minerals.

3. **Chief (Zymogenic) Cells.** Produce **pepsinogen**, the proenzyme form of the important enzyme **pepsin**, which cleaves peptide bonds, preferably at hydrophobic and aromatic [phenylalanine (Phe), tryptophan, and tyrosine] amino acids. Stimulation of chief cell secretions is by the vagus nerve, acidic conditions per gastric acid, or by the hormones gastrin or **secretin** (produced in the duodenum). In infancy, Chief cells also produce the enzyme **rennin**, which aids milk absorption by breaking the Phe–methionine peptide bond in milk protein kappa-casein. Secretion of rennin is stimulated by ingestion of milk by the human infant but the gene product is turned off past this stage.
4. **Enterochromaffin-like Cells (ELCs).** Found in the gastric glands. These cells secreted histamine, which activates parietal cells and gastric acid production. ELCs are, themselves, activated by the hormone gastrin, pituitary adenyl cyclase-activating peptide, and vagus nerve. ELCs are inhibited by **somatostatin**.
5. **G Cells.** Found in the antrum. Secrete the hormone **gastrin** (see Table 11-2), which both increases the secretion of and works along with histamine to stimulate parietal cells to produce hydrochloric acid and Chief cells to produce pepsin. Secretion of gastrin is increased by parasympathetic vagus nerve activity via release of **gastrin-releasing peptide** or by the presence of amino acids in the stomach.
6. **Prostaglandin E_2.** Binding to its receptor stimulates smooth muscle contraction of the gastrointestinal tract and decreases parietal cell secretion of gastric acid while increasing mucus production. The action is via the G_i protein receptor, which inhibits the production of cAMP by adenyl cyclase and, therefore, parietal cell H^+/K^+ ATPase pump activity.

Misoprostol and Gastric Ulcers. Misoprostol (see the figure), a synthetically produced **prostaglandin E_1**, is sometimes used to prevent gastric ulcers because of its ability to inhibit parietal cell production of gastric acid. Misoprostol is normally used only for treatment of or prophylaxis against nonsteroidal anti-inflammatory drug-induced peptic ulcers because other medication classes (H_2-receptor blocker and protein pump inhibitors, PPIs) are more effective for long-term care of acid reflux and similar disorders.

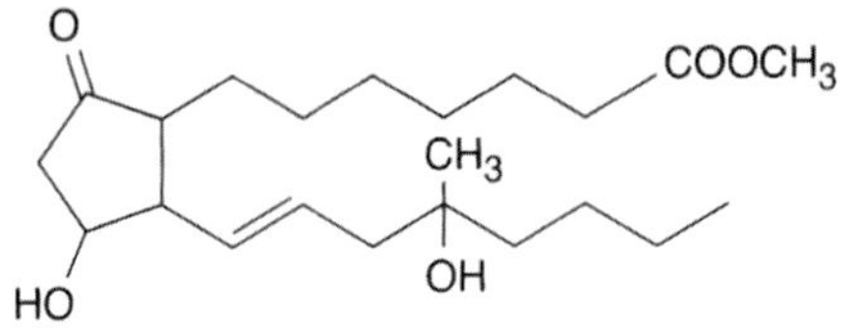

Misoprostol (prostaglandin E_1 analog)

Reproduced with permission from Katzung BG, et al.: Basic and Clinical Pharmacology, 11th edition, McGraw-Hill, 2009.

The actions of each of the above cell types and, therefore, the environment of the stomach are controlled by vagus nerve signals as well as the hormones listed below. Of note is the fact that several hormones that affect the stomach are produced and also act on organs outside the stomach (e.g., small intestine) to decrease stomach motility and/or digestion (Figure 11-3). In this manner, the body not only "turns on" the stomach when food is present but also turns it "off" when food has traveled further along the digestive tract.

Acid Reflux (Gastroesophageal Reflux Disease or GERD), Barrett's esophagus, and the Vagus Nerve. GERD affects a large percentage of the population either acutely or chronically. Untreated GERD can sometimes lead to precancerous changes in the lower part of the esophagus called **Barrett's esophagus**, which can progress to potentially fatal **adenocarcinoma** of the esophagus. The surgical severing of the vagus nerve (**vagal nervectomy** or **vagotomy**) was once used for treatment due to this nerve's prominent role in parietal and G cell (producers of gastrin), secretory activity. The procedure, although usually effective for GERD and peptic ulcer disease, also carried a number of unwanted side effects because of the innervation of other organs by the vagus nerve unless careful and selective surgery was performed. Many treatment methods have since been developed, including the **H_2 blocker** class of medications, which inhibit the promotion of acid secretion by histamine and **PPIs**, which block the unique H^+/K^+ ATPase pump found in the stomach. As a result, a vagotomy is now rarely performed for GERD treatment.

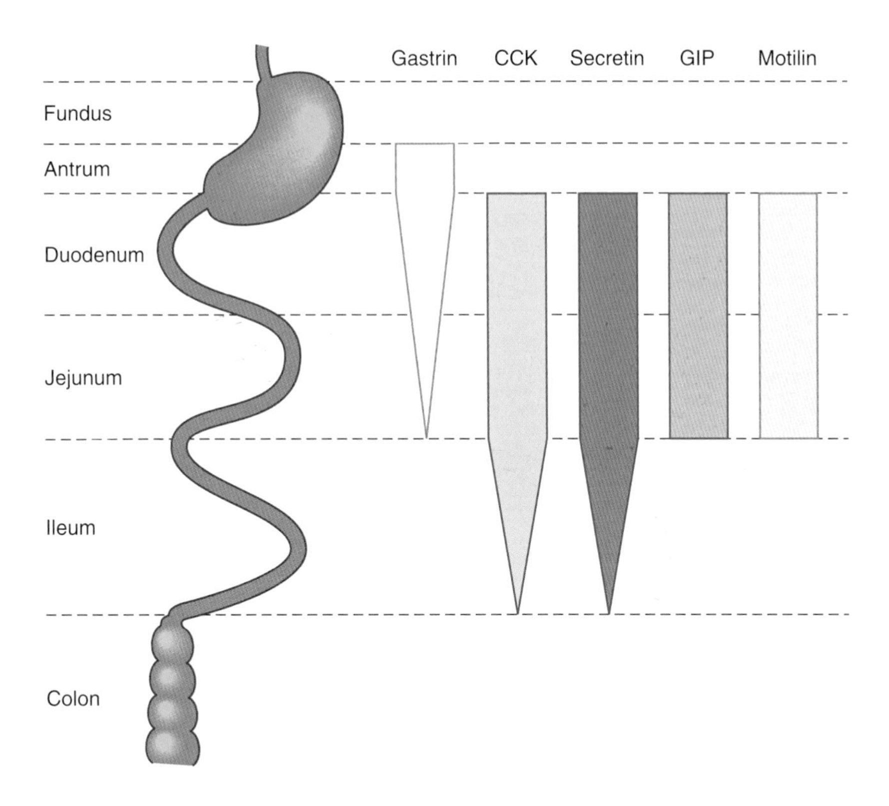

Figure 11-3. Hormone Production by the Digestive System. Sites of production of the five major gastrointestinal hormones along the length of the gastrointestinal tract. The width of the bars indicates the relative abundance at each location. CCK, cholecystokinin; GIP, gastric inhibitory peptide. [Reproduced with permission from Barrett KE, et al.: Ganong's Review of Medical Physiology, 23rd edition, McGraw-Hill, 2010.]

LIVER

The liver plays multiple, critical roles (see Table 11-3) in such diverse functions as a major part in carbohydrate, protein, and lipid metabolism/regulation; synthesis and secretion of several blood clotting (coagulation) factors; synthesis of bile; the degradation of hemoglobin/bilirubin from decomposing red blood cells (Figure 11-4); synthesis and clearance of cholesterol; storage of glycogen, vitamins, A, D, B_{12}, iron, and copper; breakdown and elimination of a variety of toxic substances (including alcohol); and the synthesis of a variety of molecules from albumin to insulin-like growth factor 1 to angiotensin (see Chapters 12 and 16). These functions are carried out by hepatocytes, the functional cells of the liver.

Bilirubin Measurement and Disease: Laboratory measurement of **bilirubin** is always reported as the **conjugated (direct)**, **unconjugated (indirect)**, and **total bilirubin**. A dye used to detect bilirubin quickly produces a red-violet azobilirubin compound with conjugated bilirubins; these are, therefore, referred to as "direct" bilirubin. The addition of ethanol to the sample makes all bilirubins (conjugated and unconjugated) react quickly with the dye and yields "total bilirubins." Unconjugated bilirubin levels are calculated by subtracting direct bilirubin from total bilirubin and are, therefore, called "indirect" bilirubin. **Urobilinogen** is measured by the addition of a reagent containing *p*-dimethylaminobenzaldehyde, which turns a dark pink/red proportional to the urobilinogen in the test sample. Urobilin does not react with this compound and, therefore, usually only urobilinogen is measured. Stercoblin, found exclusively in feces, is usually not measured.

Diseases of the liver, gallbladder, and red blood cells all affect the amount of unconjugated/conjugated bilirubin as well as the amounts of resultant urobilinogen and stercobilin. Simple observations and laboratory measurement of these molecules can greatly assist in determining these disease states. For example, gallstones, chronic liver disease with cirrhosis, or other diseases that block or decrease the secretion of conjugated bilirubin into the bile will lead to an increased leakage of bilirubin from liver cells and a resulting rise in urine bilirubin. This rise can both be measured and seen in urine by the resulting dark amber color. The lack of bilirubin secretion into the intestine will also decrease the amount of stercobilin creating white or pale feces. Diseases that cause increased breakdown of red blood cells (e.g., hemolytic anemias) will increase the amount of unconjugated bilirubin in blood tests; however, as the unconjugated form is not soluble in water, urine bilirubin will not increase and urine color will remain the same. However, the excess heme from the destroyed red blood cells will increase the amount of urobilinogen/urobilin that can be measured.

In newborns, the initial lack of liver enzymes responsible for conjugation and intestinal bacteria, which convert bilirubin to urobilinogen, may lead to the diseases of **hyperbilirubinemia** and potentially lethal **kernicterus**. These same conditions also lead to paler feces in newborns and elevated levels of bilirubin create a yellow skin color (jaundice).

TABLE 11-3. Summary of Liver Functions

Function	Molecule(s)	Description
Amino acid/ protein metabolism	Amino acid synthesis	Liver enzymes are responsible for most of amino acid synthesis (Chapters 5 and 10).
	Protein degradation	Liver enzymes are responsible for most of protein degradation (Chapters 5 and 10).
Carbohydrate metabolism	Gluconeogenesis	Liver enzymes account for a significant amount of gluconeogenesis (Chapters 6 and 10).
	Glycogenesis	Liver enzymes account for a significant amount of glycogen synthesis (Chapters 6 and 10).
	Glycogenolysis	Liver enzymes account for a significant amount of glycogen breakdown (Chapters 6 and 10).
Lipid metabolism	Lipogenesis	Liver enzymes are responsible for much of lipid (triglyceride) synthesis (Chapters 7 and 10).
	Cholesterol/ lipoproteins	Liver enzymes are responsible for endogenous cholesterol/lipoprotein metabolism. These reactions and pathways are discussed in detail in the text below.
	Apolipoproteins	Major site of **apolipoprotein** synthesis, proteins responsible for increasing the solubility and transporting dietary fats in the blood, lipoproteins components involved in cholesterol metabolism. Apolipoproteins can also serve as cofactors and can bind to receptors as part of their function.
Coagulation factor synthesis/ clot formation and breakdown	Coagulation factors	Synthesis of coagulation factors I (fibrinogen), II (Prothrombin), V, VII, IX, X, and XI (Chapter 14).
	Fibronectin (soluble)	Soluble **fibronectin**, which differs from the insoluble, extracellular matrix form, is a glycoprotein, which, along with **fibrin**, helps to form the initial blood clot following injury. The soluble fibrin/fibronectin clot is replaced by other matrix proteins, including the insoluble form of fibronectin, as part of the process of wound healing.
	α2-macroglobulin	Functions as an inhibitor of **thrombin** coagulation and **plasmin/kallikrein** fibrinolysis (Chapter 14).
	α1-antitrypsin	Serine protease inhibitor, which covalently binds to **trypsin** and inactivates its function, including cleavage of lung elastase. Deficient α1-antitrypsin leads to a variety of diseases in the lungs, including **cystic fibrosis**, and **congenital, panacinar emphysema/ Chronic obstructive pulmonary disease (COPD)**.
	Antithrombin III	Glycoprotein serine protease in blood, which inactivates **thrombin** (coagulation factor IIa) as well as **kallikrein** and **plasmin** molecules, thereby inhibiting clot formation. Antithrombin III activity is increased by the binding of heparin. Antithrombin III also inactivates trypsin and other serine protease enzymes of the classical complement pathway (Chapter 15).
	Plasminogen/ plasmin	Serine protease, produced as **plasminogen** in the liver and subsequently converted to **plasmin** in the blood by **coagulation factor XII (Hageman)**, **tissue plasminogen activator**, and/ or **urokinase plasminogen activator**. Activated plasmin breaks down fibrin/fibronectin clots (**fibrinolysis**). Plasmin activates parts of the complement system and collagen-cleaving enzymes known as **collagenases** (Chapter 13). Self-cleavage of plasminogen produces the molecule **angiostatin**, a potent inhibitor of the formation of new blood vessels. Plasmin also breaks down the wall of Graafian follicles to allow **ovulation** (Chapter 20).

TABLE 11-3. Summary of Liver Functions (Continued)

Function	Molecule(s)	Description
	α2-antiplasmin	Serine protease inhibitor, which inactivates **plasmin** and, thereby, inhibits fibrinolysis, the breakdown of the initial fibrin clot formed upon injury.
Bile synthesis	Bile	Utilized in small intestine for digestion and absorption of lipids. Includes water, bile acids (normally conjugated to taurine or glycine), bile pigments, including bilirubin, from breakdown of hemoglobin porphyrin molecules, cholesterol, phospholipids, and bicarbonate.
Breakdown of hemoglobin	Bilirubin	Breakdown of heme from degenerating red blood cells starts in the spleen with reduction of the heme by reduced nicotinamide adenine dinucleotide phosphate (NADPH) to biliverdin/free Fe^{3+} ion and then a further reduction with NADPH to bilirubin. This form is known as **unconjugated bilirubin.** Bilirubin is transported by albumin (see below) to the liver where it is conjugated to glucuronic acid by the enzyme **UDP-glucuronosyltransferase**, a process that makes the molecule more soluble. This form is known as **conjugated bilirubin** and is excreted in bile into the intestine where it is converted by bacteria into **urobilinogen**. Urobilinogens are partly absorbed in the intestine and finally excreted in urine as **urobilin** or are converted to **stercobilin** for excretion in feces. Urobilinogens are responsible for the yellow color of urine; stercobilin is responsible for the brown color of feces.
Urea cycle		Main site of conversion of amino acid nitrogen to urea via the urea cycle (Chapter 18). Urea synthesis also occurs, although to a lesser extent, in the kidney.
Detoxification	Various toxins, drugs, and alcohol	Breakdown and elimination of a variety of toxic substances, including **toxins**, **medications**, and **alcohol**. Many toxic molecules are normally metabolized in two overall steps (phases I and II). Drug detoxification also occurs to a lesser extent in the digestive system, lungs, kidneys, and skin. **Phase I** normally occurs prior to phase II reactions and usually involves reactions that increase the polar nature of the molecule (e.g., reduction/oxidation, hydrolysis, and cyclization/decyclization). Many drugs are designed to be activated, inactivated, or modified for elimination in urine or feces (via bile conjugation) by phase I reactions. Phase I enzymes in the liver include the **cytochrome p450 system** (oxidation and reduction reactions) and alcohol dehydrogenase (converts alcohol to acetaldehyde) and acetaldehyde dehydrogenase (converts acetaldehyde to acetic acid). Alcohol metabolism can also occur in other tissues, including stomach epithelium (men only) and the brain. **Phase II** reactions normally involve the addition of biochemical groups (e.g., glucuronic acid, sulfonates, glutathione, methylation, acetylation, and/or amino acid residues) to polar groups added in phase I, including carboxyl (COOH), hydroxyl (OH), amino (NH_2), and sulfhydryl (SH) groups. Phase II reactions normally permanently inactivate the toxin or drug.
Storage	Albumin	Important carrier protein of multiple molecules in the blood, including thyroid hormones and other fat-soluble hormones (see additional hormone transport proteins below). Also, transports fatty acids on their way to the liver for storage, or oxidation for energy generation, unconjugated bilirubin (see above), and several medications. Other globulins produced by the liver serve the same role(s), although to a lesser extent than albumin.
	Glycogen	The liver stores glucose in the form of **glycogen** (Chapter 6).
	Vitamin A (retinol)	Stores vitamin A transported from the intestine esterified with palmitate via chylomicrons. The liver can store up to a 2-year supply of vitamin A. See below for release, transport, and use.

(continued)

TABLE 11-3. Summary of Liver Functions (Continued)

Function	Molecule(s)	Description
	Vitamin B_{12} (cobalamin)	Storage of vitamin B_{12} (approximately 50% of body's total). Because of the efficient recirculation and restorage by the liver, years worth of vitamin B_{12} can be stored.
	Vitamin D (as calcidiol)	Vitamin D in the liver is converted by carbon 25-hydroxylation of vitamin D_3 (**cholecalciferol**) by cholecalciferol 25-hydroxylase into a prohormone form called "**calcidiol**" (25-hydroxy vitamin D_3). When released for use, calcidiol is then converted to its active form by a second hydroxylation at the 1 position to form "**calcitriol**" (1, 25-dihydroxy vitamin D_3) by the kidney. The liver can store up to a 4-month supply.
	Vitamin E	Ingested vitamin E is taken up by the liver but only the α-tocopherol form is stored. Other forms (β-, γ-, and δ-tocopherols and α-, β-, γ-, and δ-tocotrienols) are metabolized and excreted.
Transport/ "carrier" proteins	α-fetoprotein	Binds to calcium ions affecting the total, available calcium concentration (Chapter 13 and Appendix II). Serves as a pH buffer and as an osmotic molecule to maintain colloid osmotic pressure (oncotic pressure) that influences the movement of water from and to the blood. Serves a similar role in the developing fetus and, as such, is used as part of a prenatal screen for Down syndrome, neural tube defects, and abdominal wall defects (omphalocoele). Serves as a tumor marker for cancer of liver cells and some germ cell and testicular cancers.
	Ceruloplasmin	Enzyme that contains six copper atoms and is responsible for carrying approximately 90% of the body's total copper (additional 10% is contained in albumin).
	Haptoglobulin	Transports free hemoglobin molecules released from degenerating red blood cells. Also produced by several other tissues, including kidney, skin, and lung.
	Hemopexin	Transports free heme porphyrin molecule released from degrading hemoglobin. In doing so, it preserves iron and protects the body from the damaging oxidative effects of the free heme group.
	Insulin-like growth factor 1 (IGF-1)-binding protein	Transports IGF-1 (see below).
	Retinol-binding protein	Binds to de-esterified, alcohol form of retinol (vitamin A) released from storage in the liver (see above) and transports to tissues in the body.
	Sex hormone-binding protein	Transports testosterone and estradiol. Also produced in the placenta, testes, uterus, and brain.
	Thyroxin-binding globulin	Transports thyroxine (T_4) and 3,5,3′-triiodothyronine (T_3).
	Transcortin	Transports cortisol, aldosterone, and progesterone.
	Transferrin	Important carrier protein of iron (Fe^{3+}) as well as the primary store of iron in the body. Transferrin is composed of two, identical monomers, linked by disulfide bonds, each of which can bind and carry one or two Fe^{3+} ions. Primarily produced in liver but also made in other tissues (e.g., brain).

TABLE 11-3. Summary of Liver Functions (Continued)

Function	Molecule(s)	Description
	Transthyretin	Transports thyroxine (T_4). Also produced in the choroid plexus and retinal pigment epithelium.
	Vitamin D-binding protein	Transports vitamin D to tissues in the body. Also, has actin binding activity that may serve as a scavenger role for actin monomers released from injured cells or tissues.
Miscellaneous	Angiotensinogen	Peptide hormone that, when converted to angiotensin I by the enzyme renin (and subsequently to angiotensin II by angiotensin-converting enzyme), raises blood pressure via a number of mechanisms (see Chapter 18).
	C-reactive protein (CRP)	Protein whose liver production is increased because of inflammation, specifically the release of interleukin 6 (IL-6) by macrophages and adipocytes. CRP binds to phosphocholine molecules on degenerating cells to activate the complement system (see above and Chapter 14), leading to their phagocytosis by macrophages; CRP may play other roles in the immune system. CRP is used as a marker for inflammation. Its use for risk assessment for heart attack, high blood pressure, high cholesterol/lipids, and diabetes is still being investigated but has been shown not to be as useful as once thought. Investigations of CRP measurement for cancer screening are also ongoing.
	Complement proteins	Synthesis of complement proteins C 1–9, including the complement component 3 utilized in both the classical and alternative complement pathways (Chapter 15).
	Insulin	The majority of insulin is degraded in liver cells. Other cells are also able to breakdown insulin as well as other hormones.
	IGF-1	Polypeptide hormone mainly responsible for growth in early childhood. Also produced as an autocrine hormone in several target tissues (see binding protein above).
	Fetal red blood cell (erythrocyte) production	Site of production of fetal red blood cells (erythrocytes), containing hemoglobin F (Chapter 14). The liver is the sole site of production during the first trimester and is gradually replaced by the developing bone marrow.
	Kupffer cells (reticulendothelial system)	Monocyte/macrophage-type cells of the reticuloendothelial system, which serve as antigen monitors sampling circulating antigens to determine whether an immune response should be mounted.
	Thrombopoietin	Glycoprotein hormone that promotes the production of platelets by bone marrow from precursor megakaryocyte cells. Also produced in the kidney.

LIPID METABOLISM IN THE LIVER

The liver plays a central role in lipid metabolism. It is the only organ capable of the disposal of significant quantities of **cholesterol**, either via excretion into the bile or by metabolism to bile acids, both of which are lost to some degree in the feces. Dietary cholesterol, which is packaged into chylomicrons in the intestine for transport in the blood, ultimately ends up in the liver where it combines with the pool of cholesterol synthesized from acetyl coenzyme A derived from β-oxidation of saturated fatty acids and is either excreted via the bile as bile salts or distributed to other tissues via **low-density lipoprotein (LDL)**. In the fed state, the liver converts excess dietary glucose to fatty acids that are esterified to glycerol (Chapter 7). The resulting **triglycerides** are packaged, with cholesterol, into **very-low-density lipoprotein (VLDL)**, which is secreted into the blood to deliver fatty acids primarily to adipose cells and muscle. Finally, bile salts made in the liver are needed for the absorption of dietary lipids (e.g., triglycerides and cholesterol) and fat-soluble vitamins.

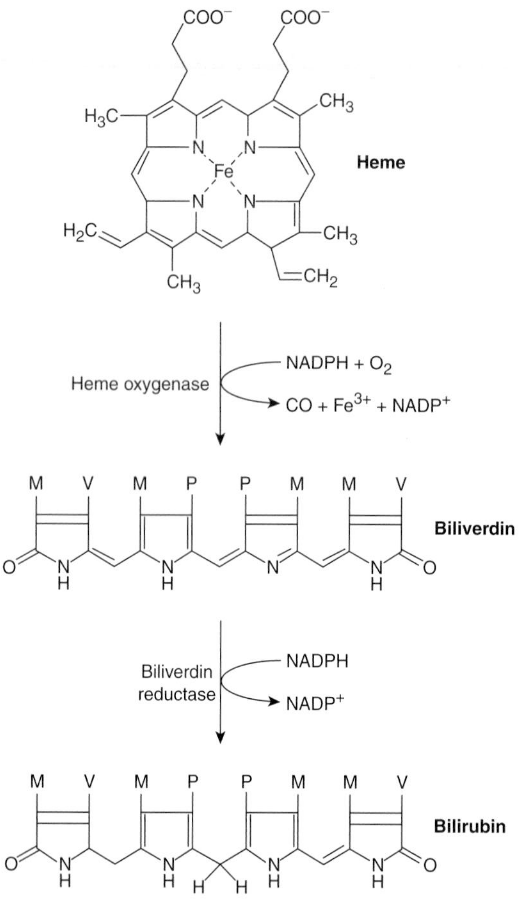

Figure 11-4. Degradation of Hemoglobin to Bilirubin. Overview of breakdown in the reticuloendothelial system of heme molecules that are released from decomposing red blood cells. Steps producing biliverdin and bilirubin are shown. The gradual breakdown of heme results in several intermediate products that create the colors seen in an evolving and healing bruise. NADPH, nicotinamide adenine dinucleotide phosphate. [Reproduced with permission from Barrett KE, et al.: Ganong's Review of Medical Physiology, 23rd edition, McGraw-Hill, 2010.]

The liver is the source of the lipoprotein VLDL and most apolipoproteins (apo) and is actively involved in the endogenous transport/metabolic pathway of cholesterol and associated lipoproteins (high-density lipoprotein, LDL, and VLDL; see Chapter 3). Its major function is to transport the endogenously synthesized triacylglycerol from the liver to extrahepatic tissues. VLDL secreted into the circulation also contains apo B100 (also synthesized in liver) that is required for the proper assembly and export of VLDL particles. The second major function of VLDL is to carry liver-generated cholesterol esters to peripheral cells after its conversion to LDL (see Chapter 16). The further steps in cholesterol transport and delivery to peripheral tissues are described in Chapter 16.

GALL BLADDER

The gall bladder serves to store and concentrate **bile** produced in the liver. Bile is released into the **duodenum** at the **ampulla of Vater** by smooth muscle contraction of the *muscularis externa* layer and relaxation of the **Sphincter of Oddi**. Release is in response to secretion of **cholecystokinin (CCK)**, the name of a group of related peptide hormones secreted from I-cells in the duodenum with similar structure to gastrin (see above). CCK secretion is increased by entrance of fat- or protein-containing food into the duodenum but decreased by the actions of somatostatin (see above). Bile is composed of water, a variety of ions [Na^+, K^+, Ca^{2+}, Cl^-, and HCO_3^- (bicarbonate)], lipids (fatty acids, phospholipids, and cholesterol), proteins, and, most importantly for digestion, more than 30 g/l of anion forms of bile acids (see Chapter 3). As noted in Chapter 3, glycine- and taurine-conjugated bile acids are the major forms. These bile acids have detergent properties enabling them to surround and break up triglycerides and phospholipids fat particles in food (Figure 11-5), allowing lipid-degrading enzymes (e.g., pancreatic lipase) to act. Bile acids also promote improved absorption

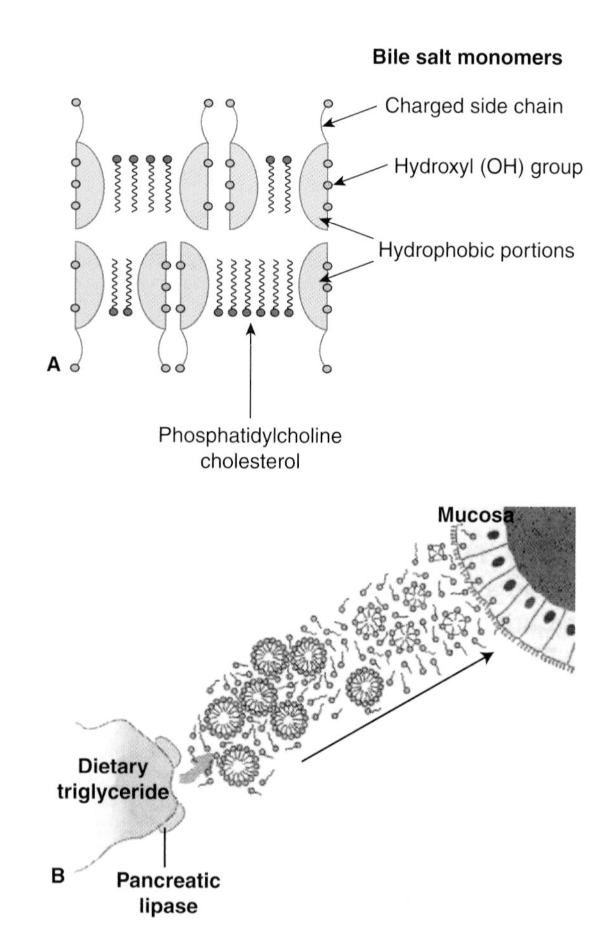

Figure 11-5. A–B. Role of Bile Salts in Triglycerides Metabolism. A. Bile salts form micelles with fatty acids released from dietary triglycerides by pancreatic lipase. The hydrophobic, charged side chain and hydroxyl (OH) groups of bile salt monomers (see Chapter 3) are well suited for micelle formation. [Adapted with permission from Barrett KE, et al.: Ganong's Review of Medical Physiology, 23rd edition, McGraw-Hill, 2010.] **B.** These micelles promote increased degradation of the triacylglycerol, by the lipase/colipase complex, to two fatty acids and monoacylglycerol followed by transport of these products to the mucosa, as part of new micelles, where they are absorbed for reassembly of triacylglycerols.

of fats, including the fat-soluble vitamins A, D, E, and K (Chapters 3 and 10).

Besides expression of bile from the gall bladder, CCK serves an important role in stimulating the pancreas to release digestive enzymes, including **trypsin, chymotrypsin, pancreatic lipase**, and **pancreatic amylase**, to aid in fat and protein digestion. CCK also slows down stomach emptying to allow proper digestion of these food particles and decreases the production of gastric acid (see above). Finally, CCK serves as a neuropeptide regulating hunger/satiety working via CCK receptors found in the CNS. As discussed above, the CCK receptor also binds gastrin and functions via a G_q protein-coupled signaling process, which activates phospholipase C with subsequent production of IP_3 and Ca^{2+}. In the CNS, CCK works via regulation of the activity of dopamine and possibly GABA (Chapter 19). CCK activity is also involved in the activity of opioids in the CNS.

Cholesterol Gallstones: The formation of cholesterol gallstones, one type of cholelithiasis, occurs most often in women older than 40 years who have had several pregnancies. Obesity increases the risk. Gallstones are also increased in women of Caucasian and Hispanic origin. These risk factors are sometimes described by the five "Fs"—female, forty, fertile, fat, and fair—but the mechanism appears to be far more complicated than these simple factors. High cholesterol and relatively low bile salt levels are known risk factors as well as bile stasis in the gallbladder due to infrequent, weak, and incomplete contractions and emptying of its contents. Particular proteins found in bile have also been noted to enhance or inhibit the precipitation of cholesterol into stones. Interestingly, levels of the dietary supplement melatonin may also play a role in gallstone formation as, among many suggested functions, it also promotes lower levels of cholesterol and reduces oxidation levels in the gall bladder.

PANCREAS

The pancreas plays a predominant role in the digestion of fats and proteins because of the pancreatic juice and various digestive enzymes that it produces and secretes into the duodenum (**exocrine functions**), sharing the ampulla of Vater with the common bile duct. In addition, the pancreas produces several hormones (**endocrine functions**). The exocrine functions occur in the **pancreatic acini**, small clusters of cells and ducts. One function of these acini is the production of bicarbonate ions (HCO_3^-), under control of the hormone **secretin**, for alkalinization of the acid contents leaving the stomach (Figure 11-6).

A second function, mainly under the control of CCK, is the synthesis and secretion of enzyme proteases, including the proenzymes trypsinogen and chymotrypsinogen as well as pancreatic lipase, pancreatic amylase, phospholipase A2, lysophospholipase, and cholesterol esterase (Table 11-4).

As an endocrine organ, the pancreas produces and secretes several hormones from five different cell types forming the Islets of Langerhans, summarized in Table 11-5.

The endocrine pancreas is regulated not only by other hormones but also by the autonomic nervous system (Chapter 19). Sympathetic (adrenergic) innervation specifically increases α-cell secretions while decreasing that from β-cells. Parasympathetic (muscarinic) innervation increases secretions from both α- and β-cells.

SMALL INTESTINE (DUODENUM, JEJUNUM, AND ILEUM)

The **small intestine** is the approximately 16–20 ft. long section of the digestive tract in which a majority of the digestion of food and absorption of nutrients occurs. The effective functional length of the small intestine is increased by a factor of about 500 by folds/invaginations (*plicae circulares* and rugae) of the intestinal wall and also the projections (villi) of the **enterocyte** cell borders which

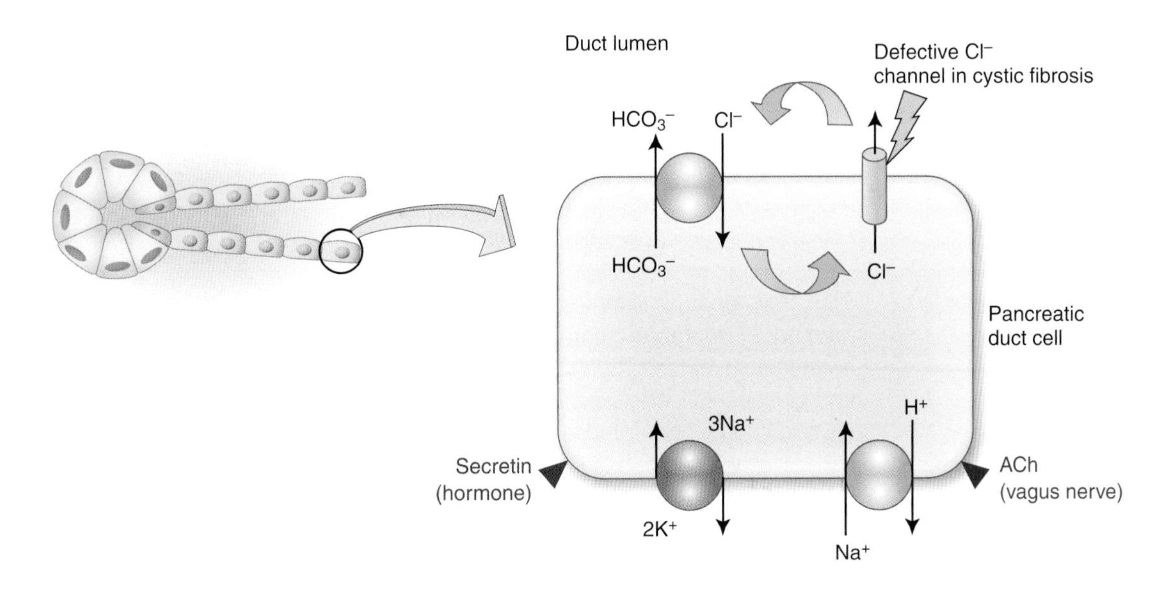

Figure 11-6. Mechanism for Production of Bicarbonate Ions (HCO_3^-) by the Pancreas. Ach, acetylcholine. [Reproduced with permission from Kibble JD and Halsey CR: The Big Picture: Medical Physiology, 1st edition, McGraw-Hill, 2009.]

TABLE 11-4. Exocrine Pancreatic Enzymes and Their Functions

Pancreatic Enzyme	Function
Elastase	Cleaves carboxyl side of peptide bond when first amino acid is a small, hydrophobic amino acid (e.g., alanine, glycine, and/or valine).
Trypsinogen/trypsin	Cleaves lysine–arginine peptide bond in the middle of a peptide chain when arginine is on the carboxy terminal side.
Chymotrypsinogen/chymotrypsin	Cleaves carboxyl side of peptide bond when first amino acid is tyrosine, tryptophan, or phenylalanine.
Pancreatic carboxypeptidases A1, A2, and B	Remove one amino acid from carboxy terminus of a peptide chain.
Pancreatic lipase	Cleaves triacylglycerol "fat" molecules into glycerol, two free fatty acids, and monoglyceride.
Pancreatic amylase	Cleaves starches at α-1,4-glycosidic bonds.
Phospholipase A2	Cleaves phospholipid molecules (e.g., phosphatidylethanolamine, phosphatidylcholine, and phosphatidylinositides) at the second glycerol carbon to release the signaling molecules arachidonic acid (Chapter 3) and lysophospholipids.
Lysophospholipase	Cleaves 2-lysophosphatidylcholine (one of the family of phosphatidylcholines) to produce glycerophosphocholine and a carboxylate molecule. Glycerophosphocholine is a precursor for choline production, an essential nutrient and precursor of the neurotransmitter acetylcholine.
Cholesterol esterase	Cleaves the fatty acid, R-group from a cholesterol ester molecule.

TABLE 11-5. Endocrine Pancreatic Hormones, Cell source, and Functions

Cell Type	Pancreatic Hormone	Function
Alpha	Glucagon	When blood sugar levels are low, promotes production of glucose from glycogen and from gluconeogenic precursors in the liver; moderately increases insulin release.
Beta	Insulin	Promotes uptake and storage of glucose by liver and muscle, and uptake of glucose for fatty acid production in fat cells. Inhibits conversion of lipid molecules to energy; inhibits glucagon secretion.
	Amylin	Works in cooperation with insulin at the brain stem level to control glucose levels by slowing stomach emptying, inhibiting secretion of stomach and pancreatic digestive enzymes as well as gall bladder bile, and inhibiting the secretion of glucagon. Affects calcium turnover in bones.
Delta	Somatostatin	Inhibits anterior pituitary gland release of growth hormone and thyroid-stimulating hormone. Suppresses secretion of gastrin, CCK, secretin, motilin, VIP, GIP, enteroglucagon (see Table 11-2 above), and the pancreatic hormones, especially insulin and glucagon. Decreases stomach emptying and intestinal peristalsis by reducing smooth muscle contraction. Inhibits adenyl cyclase activity in parietal cells of the stomach.
Epsilon	Ghrelin	"Hunger hormone" that activates various parts of the hypothalamus to encourage increased food and fat intake.
Pancreatic polypeptide cells	Pancreatic polypeptide	Regulates pancreatic secretion (endocrine and exocrine). Increased levels are seen with protein intake, fasting, and exercise. Decreased levels are caused by somatostatin and increased glucose.

CCK, cholecystokinin; GIP, gastric inhibitory peptide; VIP, vasoactive intestinal peptide.

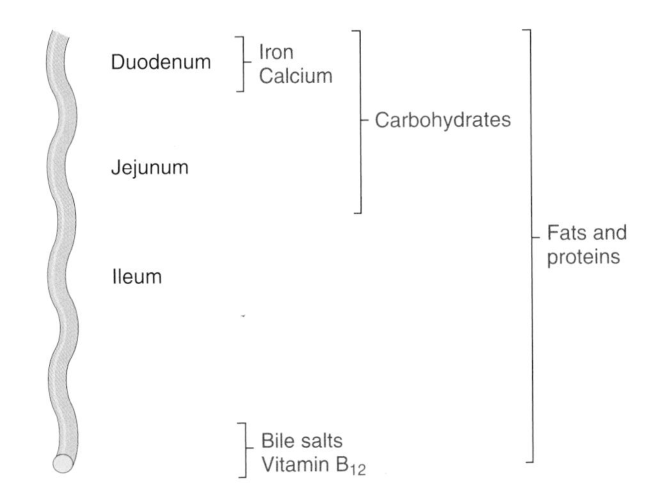

Figure 11-7. Summary of Absorption by Small Intestine. Absorption of various nutrients is noted for the duodenum, jejunum, and ileum. [Reproduced with permission from Kibble JD and Halsey CR: The Big Picture: Medical Physiology, 1st edition, McGraw-Hill, 2009.]

line it. The three different parts of the small intestine—the **duodenum**, the **jejunum**, and the **ileum**—perform different functions, depending on the presence of digestive enzymes and absorptive capabilities of their particular cell types (Figure 11-7).

Short Gut Syndrome: Short gut (bowel) syndrome results from either congenital or surgically caused shortening of the small intestine with resulting loss of the essential digestive and absorptive functions. Surgery for Crohn's disease, cancers, trauma, and bariatric (obesity) treatments are the primary causes. Symptoms which include diarrhea, pain, and secondary disorders from limited vitamin and nutrient absorption usually do not present unless more than two-thirds of the small intestine (i.e., less than 7 ft. remaining) is affected. Some patients can improve via enlargement and increased functioning of the remaining intestinal length and/or slowing of movement of food through the intestine to optimize digestion and absorption. Treatment is symptomatic and by supplementation of required nutrients, although surgical attempts at intestinal lengthening or transplant are being attempted.

Infectious/Inflammatory Diseases of the Small Intestine: The constant exposure of the digestive tract and, in particular, the small intestine to ingested microorganisms leads to multiple occurrences of **gastroenteritis** or inflammation of the stomach and/or small intestine. Although a majority of causes are viral (e.g., rotavirus), various bacteria (e.g., *Escherichia coli, Shigella, Salmonella, Campylobacter, Vibrio cholerae,* and *Clostridium*), protozoan (e.g., Giardia), and parasites (e.g., *Ascaris lumbricoides,* flatworms, and tapeworms) can also infect the small intestine. All infections lead to acute changes in the ability of the intestinal lining to digest and absorb nutrients and water, leading to acute diarrhea and nutrient deficiencies.

Digestion and absorption of each of the three parts of the small intestine are summarized in Table 11-6.

Pernicious Anemia: Pernicious anemia is a decrease in red blood cells count caused by a decrease of absorbed **vitamin B_{12}** due to the lack of **intrinsic factor**. Intrinsic factor is normally produced by the parietal cells of the stomach and is essential in allowing the absorption of this essential vitamin in the ileum. The cause of pernicious anemia is normally by **atrophic gastritis** and resulting immune attack against intrinsic factor and these cells. Gastric bypass surgery can also cause an artificially produced form of nonfunctional parietal cells. Symptoms include loss or alteration of nerve sensation (**paresthesia**) usually in the fingers and toes, inflammation of the tongue (**glossitis**), and weakness (secondary to the anemia) among others. Pernicious anemia is one of the **megaloblastic anemias**, one cause of which is folate deficiency, leading to defects in red blood cells DNA synthesis and abnormally enlarged red blood cells. Treatment is by oral and sublingual supplementations, both of which allow absorption of vitamin B_{12} via locations other than the ileum, or by injections or other absorptive methods.

LARGE INTESTINE/ANUS

The final part of the gastrointestinal tract is the approximately 5 ft. long, large intestine and anus/rectum, which mainly serves as a location for continued, passive water reabsorption, vitamin absorption, and transport of indigestible food for elimination as feces (Figure 11-8). Many of these functions are dependent on "gut flora," a variety of from 300 to 1000 different, symbiotic bacterial species and four or more fungal species that live in the intestine and allow a large number of biochemical reactions that are essential to life. The *Bacteroides* genus of bacteria appears to be most abundant and play an especially important role.

One role of gut flora is the final digestion of **dietary fiber**, starches, and carbohydrates that the human body cannot metabolize, including lactose in lactose-intolerant individuals. These polysaccharide chains are converted to **short-chain fatty acids** (e.g., acetic acid, propionic acid, butyric acid, lactic acid, and isovaleric acid) by bacterial **fermentation** and passively absorbed into the blood stream. Indigestible proteins (e.g., collagen and elastin) are also broken down by bacterial fermentation pathways. Bicarbonate ions secreted by the large intestine help to reduce acidity produced by the fermentation reactions. In addition, symbiotic, intestinal bacteria also increase the absorption of remaining lipids and minerals such as calcium, magnesium, and iron. These bacterial processes not only allow utilization of these energy sources but also increase water absorption and reduce levels of dangerous intestinal bacteria while increasing growth of beneficial bacteria.

Gut flora also augment levels of **vitamin K** and **biotin (vitamin B_7)**, which are absorbed by the large intestine, as a normal by-product of their metabolism. Gut flora are also responsible for production of some **modified (secondary) bile salts**. In addition, these bacteria are important in the development

TABLE 11-6. Overview of Functions of the Small Intestine

Section	Molecule/Gland or Transporter	Digestion	Absorption	Other
Duodenum—alkalinization of acidic mixture from stomach and addition of pancreatic and intestinal enzymes, iron absorption	Mucus/alkaline solution (Brunner's glands)	Alkalinization allows activity of intestinal enzymes	Activation of enzymes aids digestion of food and absorption of the products. However, deactivation of pepsin at pH > 5.0.	Protects and lubricates duodenal wall by coating mucosa and neutralizing the acid solution from the stomach.
	Prosecretin/secretin (S cells in the crypts of Lieberkühn)	Increases bicarbonate ion production by Brunner's glands and the pancreas (G_s/cAMP receptor) to neutralize stomach acid and activate intestinal enzymes.		Increased and activated by acidic (pH 4–4.5) secretions from the stomach. Inhibited by H_2 blockers, which increase pH above these levels.
		Stimulates release of the enzyme pepsin in the stomach as well as insulin, glucagon, pancreatic polypeptide, and somatostatin by the pancreas.		Along with cholecystokinin (CCK), regulates rate of stomach emptying and inhibits gastrin release from G cells in the stomach.
		Increases bile/bile salt secretion from the liver/gall bladder.		Presence of bile and fatty acids increases level of secretin.
	CCK	Increases production/secretion of pancreatic juices and enzymes.	Iron absorption (active)	Increased by fats and proteins entering the duodenum. Release inhibited by somatostatin.
		Increased production of bile/bile salts and release from gall bladder by stimulating smooth muscle contraction and relaxation of Sphincter of Oddi.		Along with secretin, regulates rate of stomach emptying and inhibits gastrin release from G cells in the stomach.

TABLE 11-6. Overview of Functions of the Small Intestine (Continued)

Section	Molecule/Gland or Transporter	Digestion	Absorption	Other
Jejunum—large numbers of folds and cellular villi and microvilli allow a majority of nutrient absorption	Few Brunners' glands			Jejunum pH is usually 7–8, negating need for further alkalinization and allowing full activity of enzymes.
	Glucose and galactose [sodium/glucose transporter 1 (SGLT1)]	Continued digestion of carbohydrates by pancreatic enzymes (e.g., amylase), intestinal enzymes (e.g., dextrinase, glucoamylase, maltase, sucrase, and lactase) and normal, intestinal bacteria (over 500 different species).	SGLT1 provides active transport via Na^+ gradient.	Glucose or galactose are transported from the enterocytes (basal side) to blood stream via GLUT2.
	Fructose (GLUT5)		Facilitated transport of fructose from high concentration in intestine to low concentration inside intestinal absorptive cells via GLUT5 (enterocytes, apical side).	Fructose is transported from the enterocytes (basal side) to blood stream via GLUT2.
	Amino acids	Continued digestion by pancreatic enzymes (e.g., chymotrypsin, trypsin, and carboxypeptidase) and intestinal enzymes (e.g., pepsin, dipeptidase, and aminopeptidase) amino acid products.	Cotransport with Na^+ using the Na^+ gradient.	
	Lipids	Continued solubilization and digestion by bile/bile salts and pancreatic lipase.	Free fatty acids and monoglycerides (glycerol) are absorbed either by passive diffusion or by protein-dependent transport. Triglycerides are resynthesized inside the cells from monoglycerides and packaged into chylomicrons, which are then secreted into lacteals (lymphatic capillaries).	
Ileum—continued digestion and absorption. Immunoprotection		Continued digestion of carbohydrates, peptides, and lipid molecules by digestive enzymes as in the jejunum.	Absorption of vitamin B_{12}/intrinsic factor and bile salts, which enter enterohepatic circulation. Continued absorption of carbohydrates, amino acids, and lipids as in jejunum.	Peyer's patches containing concentrated numbers of macrophages, B- and T-lymphocytes, dendritic cells and specialized antigen sensing/presenting cells (e.g., M cells). Offer immunological protection to the intestine from the exposure to ingested microorganisms.

cAMP, cyclic adenosine monophosphate.

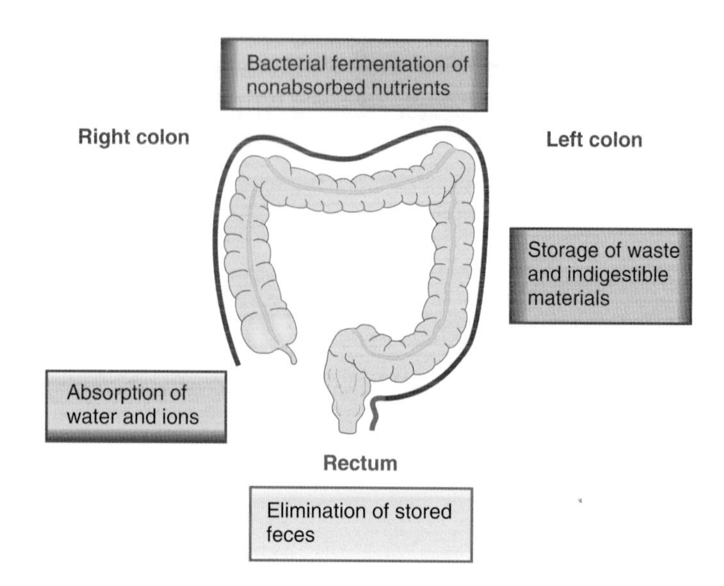

Figure 11-8. Summary of Functions of the Large Intestine. Absorption of water and ions by the ascending and proximal transverse colon and storage and transport of waste materials by the distal transverse, descending, and sigmoid colon and the rectum is illustrated. [Reproduced with permission from Kibble JD and Halsey CR: The Big Picture: Medical Physiology, 1st edition, McGraw-Hill, 2009.]

and growth of essential, intestinal **lymph tissue**; the development of **antibodies** to harmful pathogens; metabolism and elimination of ingested **carcinogens**; the continued replication and growth of the cells lining the intestine; and protective changes in the expression of cell-surface molecules on these cells.

Antibiotic Use and *Clostridium difficile*: Gut flora provides a number of essential functions, which includes keeping the growth of unwanted bacteria and yeast reduced. This ability to prevent harmful species from overproducing in the human intestine is sometimes called the "barrier effect." Antibiotics used in the treatment of diseases can sometimes adversely affect this balance by inadvertently reducing the number of helpful gut flora. The increasing use of broad-spectrum antibiotics that eliminate multiple bacterial species, especially the family of **fluoroquinolones**, has augmented this problem.

One common bacterium that grows in the large intestine in the wake of overaggressive antibiotic use is ***C. difficile***. Overgrowth by this organism and its release of harmful toxins leads to chronic diarrhea, bloating, and abdominal pain. These toxins are believed to inactivate a family of G-proteins/GTPase receptors by altering essential recognition and binding sugar residues on their surface. Continued growth of *C. difficile* can lead to the serious condition of **pseudomembranous colitis**, a severe infection of the colon. This disease results from marked inflammation of the intestinal lining and a resulting membrane-like structure composed of fibrin, lymphocytes and monocytes, and dead and dying lining cells. Patients who have been in the hospital or nursing homes are increasingly susceptible to this problem due to increased numbers of background *Clostridium* in their intestines. Treatment of the condition is either by cessation of antibiotic use or with the oral use of the antibiotics **metronidazole** or **vancomycin**.

REVIEW QUESTIONS

1. What are the general roles of the mouth, stomach, liver, gallbladder, pancreas, and small intestine in digestion?
2. What are the major digestive enzymes of saliva and the specific function of each?
3. What is the role of mucus?
4. What are the functions of the parietal (oxyntic) and Chief cells in the stomach?
5. What are the functions of the major digestive hormones and where is each produced?
6. What are the primary metabolic roles of liver?
7. What factors related to clotting and clot dissolution are secreted by the liver and what are their functions?
8. What transport/carrier proteins are secreted by liver?
9. What roles does the liver play in lipid transport and metabolism, especially with regard to cholesterol and associated lipoproteins?
10. What are the functions of each of the exocrine pancreatic hormones?
11. What are the functions of each of the endocrine pancreatic hormones and in which cell type is each produced?
12. What are the main functional molecules/glands of each section of the small intestine?
13. What is intrinsic factor and the consequence of its deficiency?
14. What are the main functions of the large intestine?

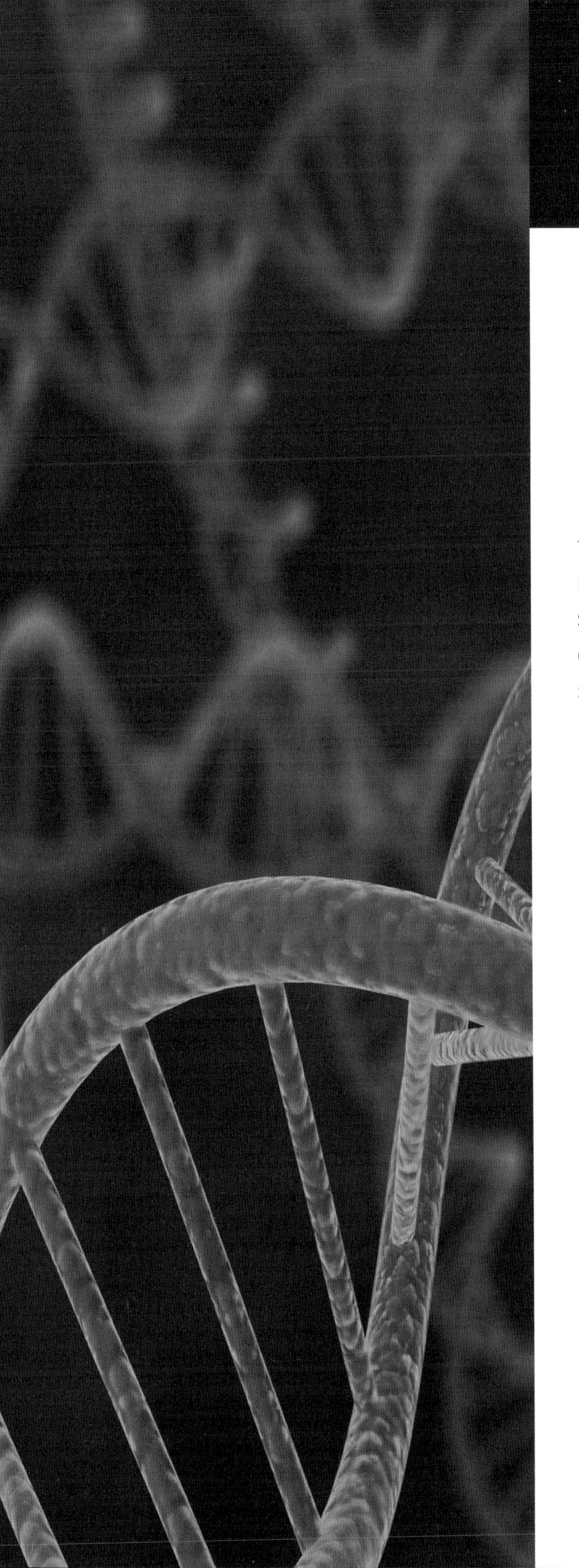

CHAPTER 12

MUSCLES AND MOTILITY

Co-author/Editor: Darren Campbell
Division of Sports Medicine,
U.S. Air Force Academy

OVERVIEW

Movement and its regulation, on a microscopic and/or macroscopic scale, are essential for human life. Muscle, composed of actin, myosin, and a variety of structural and regulatory proteins, is one of the main cell/tissue types involved in this movement. Skeletal, cardiac, and smooth muscles offer a coordinated and regulated means to move the human body from place to place; interact with its surroundings; keep nutrients flowing to and waste products flowing from various cells; or move nutrients, blood, lymph, and other molecules. Specialized proteins and the means to provide energy for these processes have evolved for particular tissues and functions.

Microtubules with tubulin, dynein, and kinesin molecules and associated structures such as cilia, flagella, centrioles, basal bodies, centromeres, and mitotic spindles provide another important mechanism for a variety of cell movements and internal cell functions that affect all types of cells and allow the division of cells. Intermediate filaments (IFs) also serve several essential roles of cell motility. Nonmuscle cells utilize all of these mechanisms—actin/myosin, microtubule/dynein/kinesin, and IFs—to achieve a wide array of functions throughout the human body.

THE BASIC COMPONENTS OF MUSCLE

Muscle is an organ that specializes in transforming chemical energy into mechanical work or movement. The muscular system comprises all the individual anatomic muscles. Muscle tissue is derived from the mesodermal layer of embryologic germ cells and is divided into three main types: skeletal muscle, cardiac muscle, and smooth muscle (Figure 12-1). The structure of these different types of muscles is similar but the architecture and regulation are often very different, making their functions unique. Nonmuscle cells are a fourth cell type that utilizes many of the same proteins and processes for motility.

There are two major proteins involved in contraction in all types of muscle: **actin** and **myosin**. These proteins convert chemical energy into mechanical work through an interaction with **adenosine triphosphate (ATP)**. The triggering and control of this process are different for each muscle type. Skeletal, cardiac, and smooth muscles have a repeating unit called the **sarcomere**, containing the muscle fibers composed of actin and myosin and accessory proteins (see below). Surrounding these muscle fibers is the equivalent of a plasma membrane called the **sarcolemma**. Within the sarcolemma is the sarcoplasm containing mitochondria and a **sarcoplasmic reticulum (SR)**, the equivalent of a smooth endoplasmic reticulum (SER). Like the SER in other cell types, the SR contains calcium ions (Ca^{2+}) essential for the initiation of muscle activation. More specifically, specialized invaginations of the SR, called **transverse tubules (T-tubules)**, interdigitate between the muscle fibers to provide efficient delivery of calcium.

ACTIN

The term **actin** describes both the globular, single amino acid (monomer) chain (**G-actin**) and the **thin filament** (**F-actin**) structure formed from two parallel strands of multiple G-actin molecules, which wind to form a double helix structure (Figure 12-2A). Three major classes of actin exist, including α-actin (found in skeletal, cardiac, and smooth muscles) and β- and γ-actins (found in nonmuscle cells). These F-actin filaments provide structure as part of the cell's **cytoskeleton**. The cytoskeleton is also directly linked to the plasma membrane via integral membrane proteins and intracellular junctions (Chapter 8) and, as such, is important in many signaling functions.

Actin thin filaments interact with myosin thick filaments (Figure 12-2C and see below) and are integral in a multitude of cell motility functions, including muscle contraction, the movement of intracellular vesicles, cell division and cytokinesis, immune response, phagocytosis, wound healing, and others. Actin is also required in the nucleus for RNA polymerase I, II, and III complex formation and function, export of RNA and proteins from the nucleus, and some aspects of chromatin remodeling.

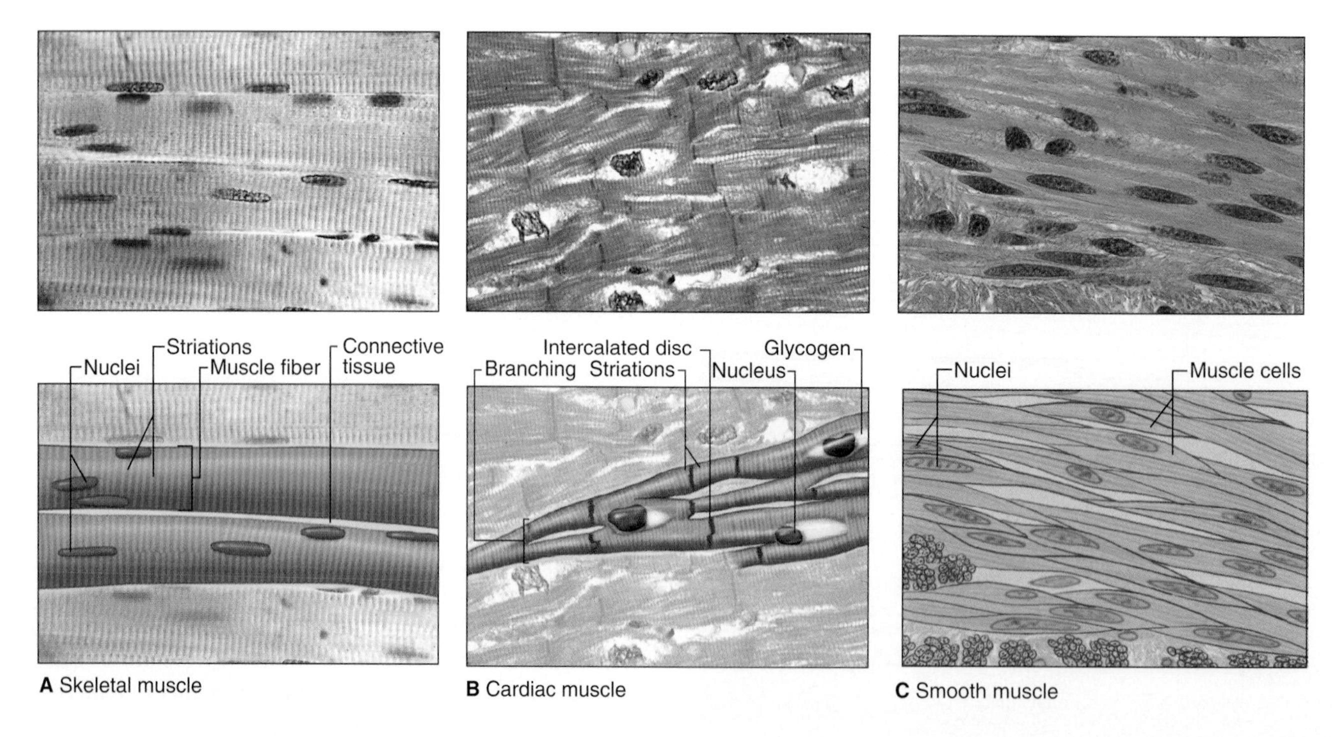

Figure 12-1. A–C. The Three Types of Muscle. Light micrographs of each type, accompanied by labeled drawings. **A.** Skeletal muscle is composed of large, elongated, multinucleated fibers that show strong, quick, voluntary contractions. **B.** Cardiac muscle is composed of irregular branched cells bound together longitudinally by intercalated disks and shows strong, involuntary contractions. **C.** Smooth muscle is composed of grouped, spindle-shaped cells with weak, involuntary contractions. The density of intercellular packing seen reflects the small amount of extracellular connective tissue present. [Reproduced with permission from Mescher AL: Junqueira's Basic Histology Text and Atlas, 12th edition, McGraw-Hill, 2010.]

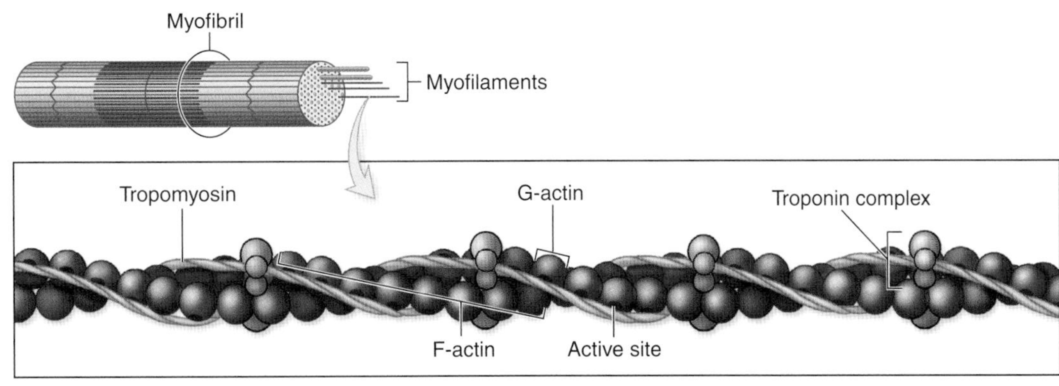

A Thin filament

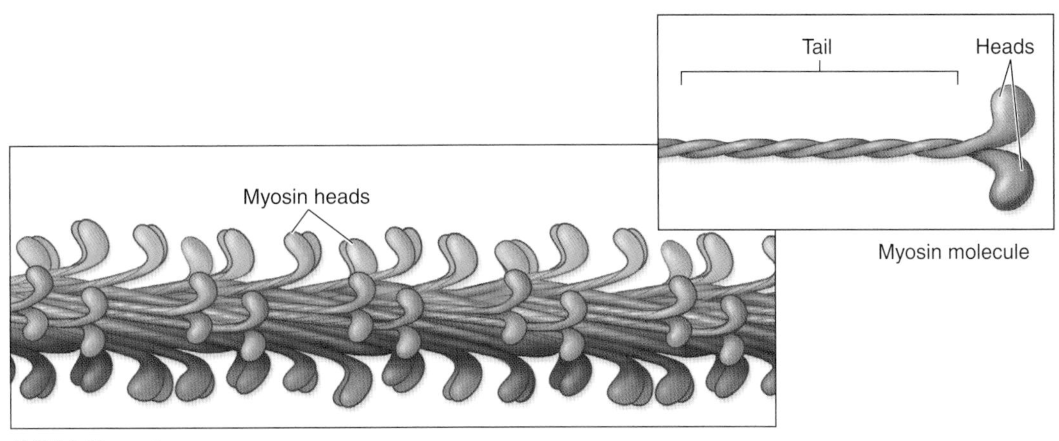

B Thick filament

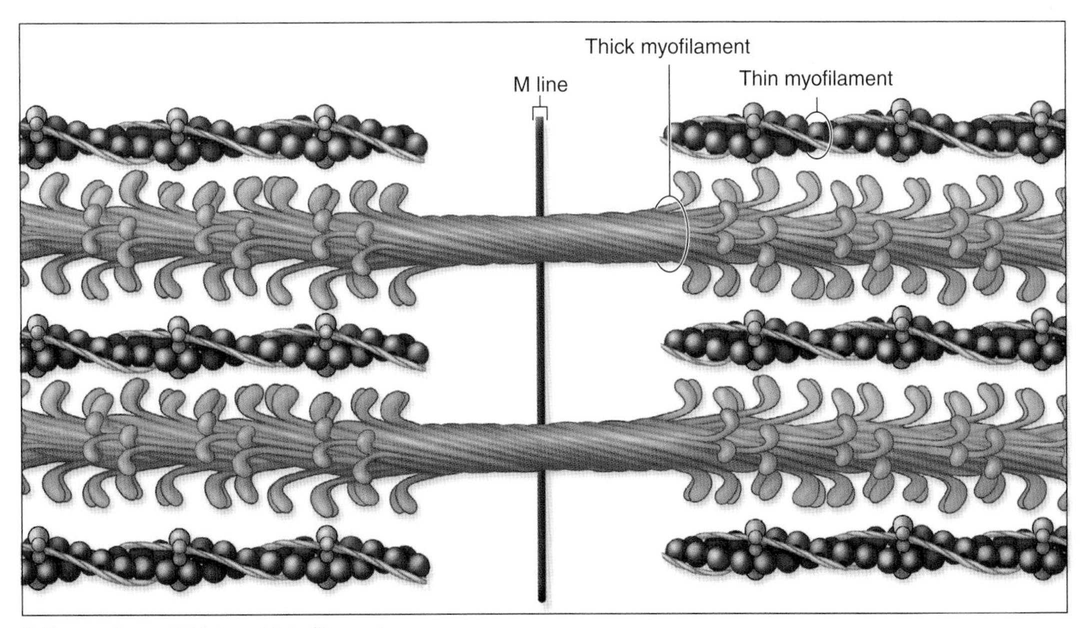

C Comparison of thick and thin filaments

Figure 12-2. A–C. Molecules Composing Thin and Thick Filaments. The contractile proteins are the thin and thick myofilaments within myofibrils. **A.** Each thin filament is composed of F-actin, tropomyosin, and troponin complexes. **B.** Each thick filament consists of many myosin heavy chain molecules bundled together along their rod-like tails, with their heads exposed and directed toward neighboring thin filaments. **C.** Besides interacting with the neighboring thin filaments, thick myofilament bundles are held in place by less well-characterized, myosin-binding proteins within the M-line. [Reproduced with permission from Mescher AL: Junqueira's Basic Histology Text and Atlas, 12th edition, McGraw-Hill, 2010.]

TROPOMYOSIN–TROPONINS

Tropomyosin is a regulatory protein that binds to F-actin thin filaments and reversibly blocks myosin head binding sites (Figure 12-2A–C). In noncontracting muscle, **troponin T and troponin I** molecules (part of a **troponin** complex, Figure 12-2A) hold tropomyosin over the myosin-binding site. Upon exposure to Ca^{2+} released from the SR, **troponin C**, the third protein of the troponin complex, creates a conformational change of the tropomyosin–troponin complex exposing the site. When calcium concentrations fall, the change reverses and the block is again present. Smooth muscle and nonmuscle cells do not contain troponin and rely on other regulatory processes to control actin–myosin contraction/force generation. Troponin C contains a specific amino acid sequence named an EF hand. In this tertiary structure motif (Chapter 1), an α-helix (termed E for its sequence in a series of α-helices) is joined to an F α-helix by a loop of approximately 12 amino acids. The helix–loop–helix forms a structure reminiscent of an extended forefinger (E helix) and thumb (F helix) with the curve between representing the loop. Among the 12 amino acids of the loop are positions, 1, 3, 5, 7, 9, and 12, which bind the Ca^{2+}. Amino acid 12 is always glutamate or aspartate whose R-group provides two partially negatively charged oxygen molecules (Chapter 1) for binding the positively charged calcium. The small size of a highly conserved glycine amino acid at residue 6 provides space for this structure to form. The other amino acids are mainly hydrophobic that stabilize the E and F helices and, therefore, the helix–loop–helix structure.

MYOSIN

The human genome holds more than 40 different types of **myosin** genes. These different genes produce variations of myosin shapes, which in turn affect the speed at which the filaments move. To date, there are 12 classes of myosin defined in the human genome. All myosin molecules have three domains, including a **head**, **neck**, and **tail** portion of the amino acid sequence (Figure 12-3A). The myosin head binds actin thin filaments and uses hydrolysis of ATP to adenosine diphosphate (ADP) to generate force. The tail domain varies depending on the type of myosin but, in general, can bind to transport vesicles or combine with other myosin molecules. Some myosin tail domains can also regulate the proteins' activity by folding over and effectively blocking the myosin head domain. The neck domain serves as a link or lever arm for the transfer of this force between the head and tail. The varying types of myosin proteins produce different speeds of movement, depending on the length of the neck region. Four **myosin light chains**: two **essential light chains**, and two **regulatory light chains** are also bound to the neck regions (Figure 12-3A). The function of these light chains is discussed below.

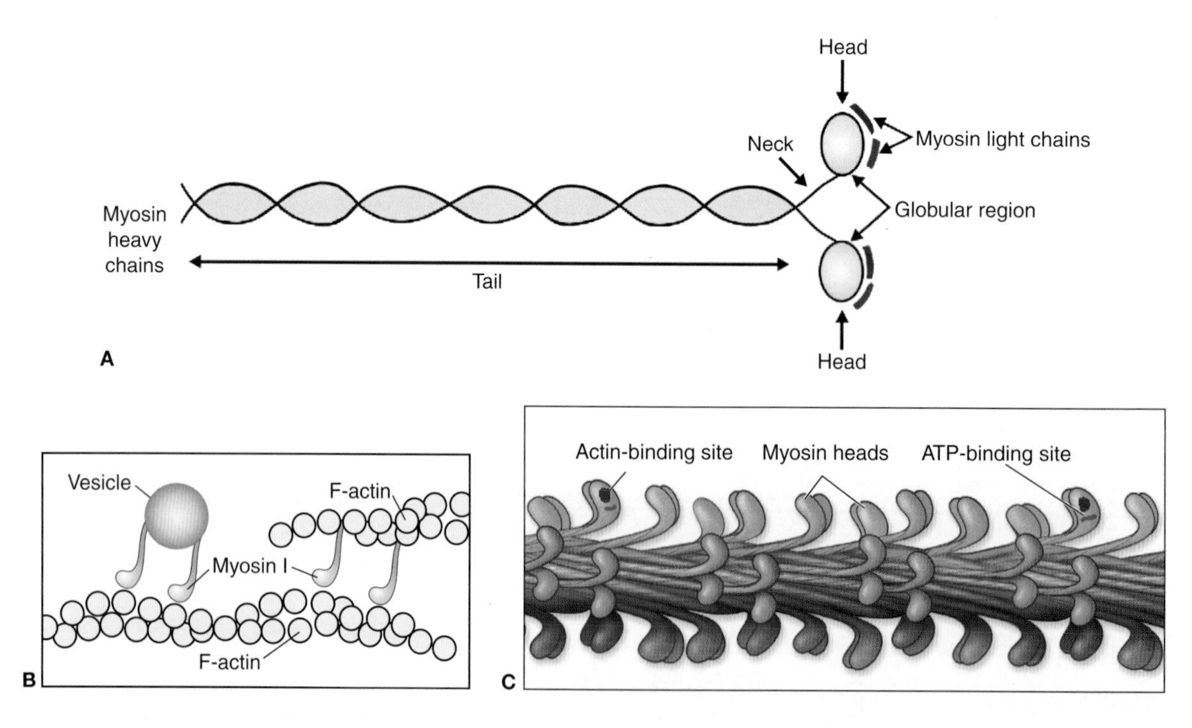

Figure 12-3. A–C. Myosin I and II. A. Basic Myosin Molecule. The heavy and light chains (essential–red, regulatory–dark blue) form the basic structure of all myosin molecules, including the long tail composed of two coiled α-helices and the globular head region. **B. Myosin I.** Vesicles for transport are attached by the myosin I tail section. The single head of myosin I moves down the actin filament in a similar manner to other myosin heads. Myosin I molecules can also link between F-actin filaments, resulting in limited motility similar to myosin II molecules. [Adapted with permission from Naik P: Biochemistry, 3rd edition, Jaypee Brothers Medical Publishers (P) Ltd., 2009.] **C. Myosin II.** The tail section of the two heavy chains of myosin II forms a coiled-coil arrangement. This structure can interact with several other myosin II tails to form thick filaments. The two head regions of each myosin molecule contain an actin-binding site (red circle) and an ATP-binding site (blue line). Coordinated movement of the myosin heads produces force and, therefore, movement when interacting with actin filaments. See text for further details. [Adapted with permission from Mescher AL: Junqueira's Basic Histology Text and Atlas, 12th edition, McGraw-Hill, 2010.]

Myosin I is the simplest form consisting of a basic myosin head, neck, and a tail of varying lengths (Figure 12-3A–B). The basic function of myosin I is most often described as vesicle transport, although other functions are emerging. **Myosin II** is composed of two identical amino acid chains, termed the "heavy chains" (Figure 12-3A). The tail section of each chain coils around the other, making a dimer myosin filament with two heads. Multiple myosin II dimers bind at the tail domain, forming a structure called a **thick filament** (Figures 12-2B–C and 12-3C). Additional myosin subtypes have structures similar to either myosin I or myosin II and are responsible for diverse types of cell motility. These functions include intracellular transport (both to and from the nucleus to the cell periphery), maintenance and movement of vesicles and organelles in subregions of the cytoplasm, and, possibly, the perception of light in the retina and sound waves in the inner ear.

MYOSIN LIGHT CHAINS

Each myosin head contains two **essential myosin light chains** and two **regulatory myosin light chains** bound to the neck region (Figure 12-3A). Different forms of myosin light chain are found on different myosin types and influence their activity and regulation. Myosin light chains serve an important regulatory role for myosin activity in smooth and nonmuscle cell contraction where the tropomyosin–troponin mechanism is absent. In skeletal muscle, the myosin light chains influence the speed of contraction but are not essential for myosin activity. In cardiac muscle, myosin light chains affect myosin head ATPase and are believed to have some role in the development of heart failure. Activity of myosin in smooth/nonmuscle cells relies on phosphorylation of serine 19 on the regulatory chains by **myosin light chain kinase (MLCK)**. MLCK activity is regulated by Ca^{2+} that binds to the regulatory protein calmodulin, also by an EF hand-binding motif (see above). Calmodulin subsequently activates MLCK. When calcium concentrations fall, **myosin light chain phosphatase (MLCP)** removes the phosphate group and inactivates the regulatory light chains and, therefore, myosin. Calcium also inhibits MLCP by the actions of another serine/threonine enzyme called **Rho kinase**; other signaling pathways and enzymes have also been implicated in myosin light chain phosphorylation and dephosphorylation. One enzyme, sphingosine-1-phosphatase, may play an important role in modulating contraction and relaxation of blood vessels and, therefore, maintenance of blood flow in response to changes in blood pressure.

ACTIN-BINDING PROTEINS

Actin-binding proteins (ABPs) are a varied group of proteins that regulate actin filament formation and length as well as actin–myosin interactions. Although more prominent in nonmuscle cell and smooth muscle contraction, some, including tropomyosin and the troponins (see above), are found in skeletal and cardiac muscles. One such ABP, **α-actinin** binds actin thin filaments to skeletal muscle Z-lines and smooth muscle dense bodies (Figures 12-4 and 12-8A). α-Actinin is also believed to bind to actin filament bundles and separates each thin filament by approximately 35 nm. This separation allows myosin thick filaments the proper spacing for optimal contraction. Additional ABPs cross-link thin filaments (**filamin**), sever thin filaments (**gelsolin**, **tropomodulin**, and **villin**) and cap them to restrict their length (**capZ**), regulate actin–tropomyosin activity in smooth muscle cells where troponins are absent (**caldesmon**), regulate myosin ATPase activity (**calponin**), and connect microfilaments to the cell membrane (**dystrophin**, **vinculin**, and **integrins**). Many other ABPs also exist and perform other functions in a variety of cell types.

EXCITATION–CONTRACTION COUPLING

Muscle achieves its major function of generating movement by contraction through a series of binding events and enzymatic reactions. This process is called **excitation–contraction coupling** (Figure 12-4). Certain parts of this mechanism are conserved in all types of actin–myosin contraction/motility, whereas alternative forms of initiation and/or regulation are found in particular muscle types. Many of the components are similar in nonmuscle cells, but many differences exist because of the variations in organization and additional proteins and their functions. Nonmuscle cell contraction will be discussed separately below.

Two basic processes must happen in all muscle types for contraction to occur: (1) binding of myosin heads to the actin thin filament and (2) stimulation of myosin to generate force. Both of these processes rely, at least in part, on Ca^{2+}. In all muscle types, a signal from a nerve or pacemaker cells causes an initial increase in Ca^{2+} concentrations. This relatively small influx of Ca^{2+} leads to a larger release from the SR. This concept of a small calcium signal leading to higher calcium concentrations and contraction is termed **calcium-induced calcium release (CICR)**. The release from the SR varies depending on the type of muscle and is discussed below. T-tubules allow delivery of this calcium to all areas of a muscle unit to permit synchronous contraction. The released calcium interacts with an ABP—troponin/tropomyosin in skeletal and cardiac muscle and caldesmon in smooth muscle—to allow myosin head binding to actin thin filaments.

Myosin force generation is the second basic function required for contraction. In all muscle types, ATP binds to the myosin molecule head. An ATPase on the myosin head cleaves ATP into ADP and a phosphate molecule (PO_4^{3-}), producing a charged form of the myosin protein that binds to the now open F-actin filaments at an angle. Release of the phosphate molecule from the myosin head causes it to swivel to a more acute angle, resulting in a ratcheting movement between it and actin. This ratcheting movement slides the thick and thin filaments past each other, converting the chemical energy of the ATP into the mechanical energy moving the filaments. Release of the ADP molecule and binding of a new ATP molecule breaks the bond between myosin head and F-actin thin filament. As this process repeats many times, the overall length of the sarcomere shortens. Calcium also plays a prominent role in smooth muscle myosin activity via activation of regulatory myosin light chains (see above).

The cycle continues as long as the concentration of calcium in the sarcoplasm of the cell remains elevated. At the end of this

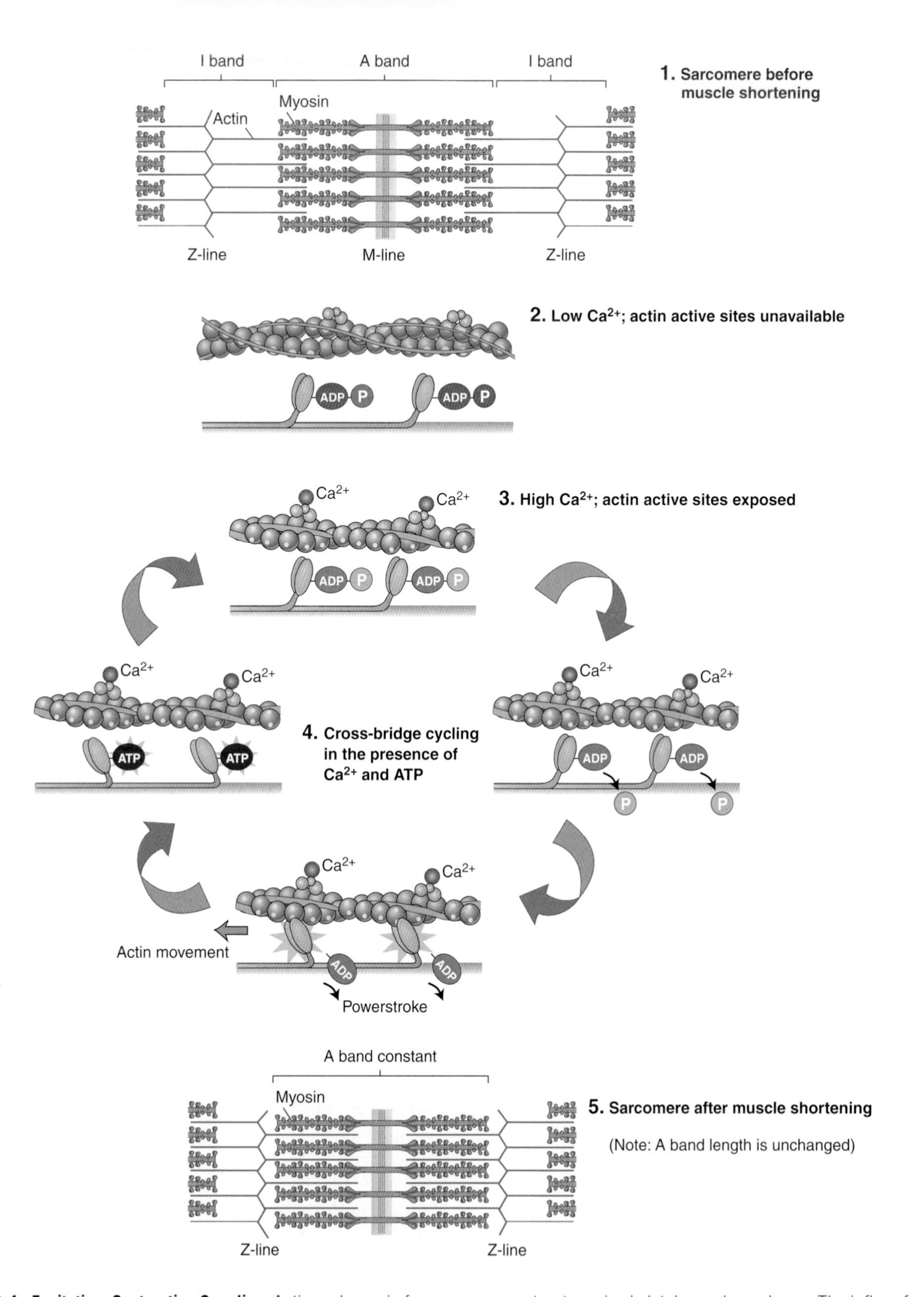

Figure 12-4. Excitation–Contraction Coupling. Actin and myosin form sarcomere structures in skeletal muscle as shown. The influx of calcium ions (Ca^{2+}) allows myosin heads from thick filaments to bind to the actin filaments. Ca^{2+} also activates myosin head cross-bridge cycling, leading to force generation and, ultimately, shortening of the sarcomere. Shortening of multiple sarcomeres leads to muscle contraction. The process is similar in cardiac and smooth muscle contraction, although structural organization and regulation differ. ADP, adenosine diphosphate; ATP, adenosine triphosphate. [Reproduced with permission from Kibble JD and Halsey CR: The Big Picture: Medical Physiology, 1st edition, McGraw-Hill, 2009.]

process, the calcium entry into the cells slows and begins to be collected again by the SR by **a sarco/endoplasmic reticulum Ca^{2+}-ATPase** pump, also powered by ATP. The pump can produce an approximately 10,000-fold higher concentration inside the SR versus the sarcoplasm. An additional protein **calsequestrin** also binds Ca^{2+} inside the SR to reduce the effective, free Ca^{2+} concentration that the pump must work against. Calsequestrin can bind over 40 Ca^{2+} per protein molecule not by an EF hand structure but, instead, by using charged amino acids residues and changes of secondary structure to α-helices. Once the calcium concentration lowers, calcium is released from troponin C, the previous conformational change is reversed and troponin T and I again block the binding of myosin to the actin binding site. As the cycle ends, a new molecule of ATP then binds to the myosin head, which displaces the ADP and the initial sarcomere length is restored.

SKELETAL MUSCLE

STRUCTURE AND GENERAL OVERVIEW

Skeletal muscle comprises the bulk of the muscle mass found in our bodies (Figure 12-5). The average adult male is made up of approximately 42% skeletal muscle and an average adult female consists of 36% skeletal muscle. Each muscle has an origin and insertion on the skeleton, with the individual muscle fibers running parallel or oblique to the long axis of the muscle. Skeletal muscles are capable of producing a forceful contraction and moving the elements of the muscle over a relatively large distance. With the shortening of length during the contraction, this muscle will often change in diameter, like the swell produced when contracting the biceps muscle. In skeletal muscle, the F-actin thin filaments are bound end to end at a structure called the Z-line (Figure 12-4). The repeated Z-lines and alternating thin and thick filaments give the muscle the striated appearance seen on light microscopy.

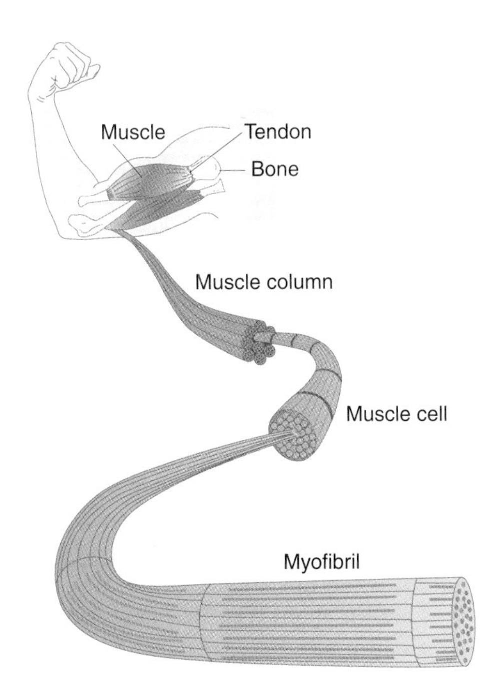

Figure 12-5. Skeletal Muscle. Skeletal muscle is highly organized. At the molecular level, repetitive sarcomeres (see text above) form a myofibril. Multiple myofibrils form muscle cells, columns, and, ultimately, muscle. [Reproduced with permission from Kibble JD and Halsey CR: The Big Picture: Medical Physiology, 1st edition, McGraw-Hill, 2009.]

Centronuclear/Myotubular Myopathies: Centronuclear/Myotubular myopathies (CNMs) are a rare series of inherited disease states in which skeletal muscle nuclei are located in the center of the muscle cells instead of at the peripheral edges. The exact causes have not been found but errors in the development of embryological muscle cells appear to be common. A suspect enzyme called **myotubularin**, which has dephosphorylating activity, may be part of the X-linked form, properly termed myotubular myopathy. X-linked CNM presents at birth with severe loss of muscle tone (hypotonia), affects muscles required for breathing, and often results in death before a cause can be determined. Patients also often have an associated narrow and long head with palate, finger, and chest structural abnormalities. Autosomal dominant and recessive forms also exist, properly termed centronuclear myopathies. The autosomal dominant version may be because of mutations affecting the protein **dynamin**, a guanosine triphosphate-dependent transport protein. Symptoms of CNM may never present (e.g., autosomal recessive form) or may include variable ranges of increasing muscular weakness, which often presents in the teens or 20s. There is no cure for CNM diseases and treatment is only supportive.

Growth, regeneration, and activity of skeletal muscle is under the control of several signaling pathways, which activate specific gene transcription/translation, leading to the production of muscle proteins as well as regulatory proteins, including those involved in energy production for muscle contraction. Some are dependent on motor neuron activity and/or contraction [e.g., more glucose transporter type 4 (GLUT4) receptors are brought to the surface in highly contracting muscles to increase glucose uptake] and, therefore, are directly influenced by specific muscle activity.

Skeletal muscle is under voluntary control and the contraction is driven by motor neurons in the spinal cord and the neurotransmitter **acetylcholine** (Chapter 19). Skeletal muscle neurons are directly associated with ryanodine receptors in the SR, which receive the nerve signal and depolarize L-type voltage-dependent calcium channels in the T-tubules. This depolarization causes a change in the foot structures that bridge the T-tubules and SR, allowing release of calcium into the sarcoplasm. The free calcium binds with the regulatory molecule **calmodulin** and activates troponin C, unmasking the myosin-binding site by displacing troponin-I and tropomyosin.

Acetylcholine and Skeletal Muscle Disease: Skeletal muscle relies on the neurotransmitter **acetylcholine** to initiate muscle contraction. Disruption of acetylcholine's activity leads to disease states including myasthenia gravis and Lambert–Eaton syndrome. The major form of **myasthenia gravis (MG)** is caused by antibodies against the body's own receptor for acetylcholine at the postsynaptic junction (**nicotinic–acetylcholine receptor**). The antibodies block or destroy the receptor negating the action of acetylcholine. The body's natural **cholinesterase** activity degrades the neurotransmitter before activation of muscle contraction can be achieved. The cumulative effect of both leads to weakening of muscles after extended periods of activity (**fatigability**). Muscles often affected include eyes and eyelids, facial/neck muscles responsible for expression, speech and swallowing, muscles involved in breathing, and arm/leg muscles. Examination can often be normal unless measurement to fatigue of muscles is included. Blood tests for the antibodies are not always conclusive, leading to other neurological or muscle biopsy tests. Although rarely performed, intravenous administration of the cholinesterase inhibitors **edrophonium** or **neostigmine** can relieve symptoms and assist in diagnosis. Patients diagnosed with MG are treated with similar cholinesterase inhibitors (neostigmine or **pyridostigmine**) and/or immune suppressants (steroids and other medications). With proper treatment, life expectancy is normal. **Lambert–Easton syndrome** affects patients in a similar manner to MG but is caused by antibodies against a presynaptic calcium channel that stops the release of acetylcholine. Like MG, Lambert–Eaton syndrome is rare and often difficult to diagnose.

SKELETAL MUSCLE TYPES

Skeletal muscle can be divided into different types, depending on the myosin fibers present. There are two general categories: slow twitch (type I) and fast twitch (type II) with type II divided into three additional subtypes (type IIa, IIb, and IIx). The types of muscle fibers have many differences including contraction time, ability to be fatigued, and the maximum duration of use.

Slow-twitch (type I) fibers fire more slowly than the fast-twitch fibers. They contain the heme-containing, oxygen-binding protein **myoglobin**, which serves as an important oxygen storage protein capable of releasing oxygen during periods of hypoxia. Myoglobin is also used clinically as a marker for damaged muscle and can be elevated during periods of muscle cell damage such as **rhabdomyolysis** (see sidebar). Type I fibers are sometimes referred to as "red" because of the presence of **heme/Fe^{2+}** in the myoglobin as well as increased numbers of mitochondria and small blood vessels/capillaries. These fibers, therefore, are able to primarily use oxygen for more efficient **oxidative phosphorylation** (Chapter 6) to produce energy, mainly from fatty acids mobilized from **triglycerides**. As a result, these fibers can participate in slow force generation and contraction for long periods of time without fatigue, especially when compared with fast-twitch fibers. This muscle fiber type helps endurance athletes, who often have a greater percentage of slow-twitch muscle fibers than nonendurance athletes, compete over long distances for an extended period of time. Type I myosin is also found in muscles involved in posture.

Rhabdomyolysis: Rhabdomyolysis or "**rhabdo**" is the medical term for the abnormal breakdown of skeletal muscle caused by trauma (e.g., major crush injuries), disease states, and/or medication side effects. Rhabdo is marked by the appearance of abnormally high levels of muscle components in the blood stream. Myoglobin is a specific laboratory marker used by clinicians who suspect rhabdomyolysis in their patients. The sudden release of proteins and various ions can cause kidney failure, neurological problems, and irregular heart rhythms, which can be fatal. Inflammation caused by the damage to muscle can also lead to serious conditions such as compartment syndrome where portions of the body can lose their blood supply because of swelling. The sudden swelling can also lead to shock. Treatment is by intravenous fluids and supportive care often including dialysis to preserve kidney function and to correct ion imbalances. Compartment syndrome may require surgical opening of muscle connective tissue (fasciotomy) to relieve the pressure from swelling. Prompt and professional care of anyone in danger of or showing signs of rhabdomyolysis usually results in a good outcome for the patient.

Type IIa fast-twitch fibers are also known as intermediate fast-twitch fibers and contain high amounts of myoglobin, mitochondria, and capillaries. To some extent, they bridge the gap between the slow-twitch and fast-twitch fibers in that they can use both aerobic and anaerobic metabolism to create energy using **creatine phosphate** and **glycogen**. These fibers do not generate the same force as type IIb fibers nor are they as fatigue resistant as the slow-twitch fibers. **Type IIb fast-twitch fibers** are the classic fast-twitch muscle fibers. These fibers are capable of producing quick and powerful muscle contractions necessary for the sudden explosive need for strength of a power lifter or the burst of speed needed from a sprinter. This muscle fiber has the highest rate of firing of all the muscle fiber types. Type IIb fibers do not contain myoglobin and have fewer capillaries than type I. Therefore, they primarily use anaerobic metabolism also using **creatine phosphate** and **glycogen**. As a result, they are sometimes called white fibers and are also more sensitive to fatigue. Less is known about **type IIx fibers**, which may have some of the oxidative metabolism of slow-twitch fibers along with the biophysical properties of fast-twitch fibers.

In individual muscles, there is usually a combination of different types of fibers (Figure 12-6). Each individual motor unit, however, contains a single α-motor neuron and the muscle fibers it innervates will contain the same type of muscle fibers. These fibers all have the same contractile characteristics and fire as an all or none phenomenon. The distribution of these

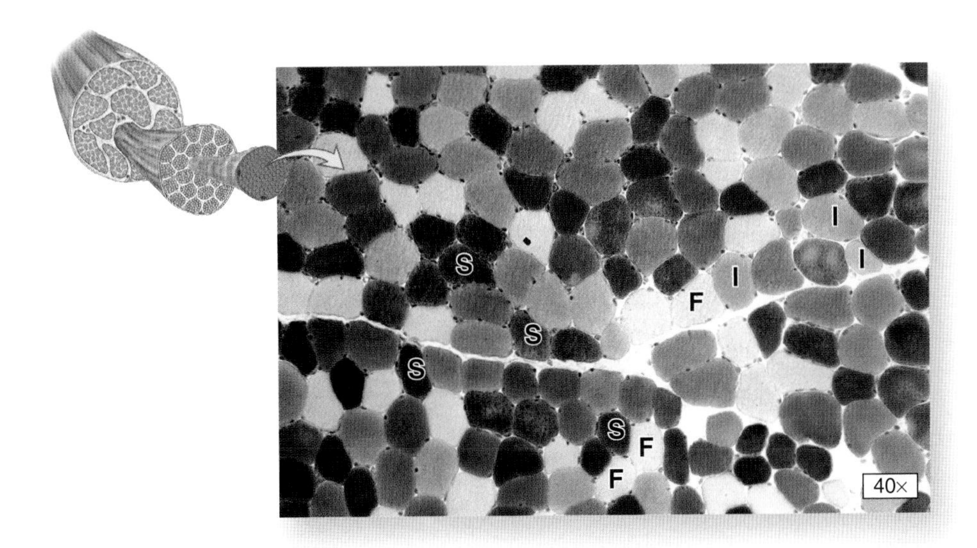

Figure 12-6. Skeletal Muscle Fiber Types. Cross section of skeletal muscle stained histochemically to detect the density of myofibrillar myosin ATPase can be used to demonstrate the distribution of slow (S)-type I fibers, intermediate (I)-type IIa fibers, and fast (F)-type IIb fibers. [Reproduced with permission from Mescher AL: Junqueira's Basic Histology Text and Atlas, 12th edition, McGraw-Hill, 2010.]

different types of motor units within a single skeletal muscle is determined genetically and by the specific function of the muscle. It is this mix that determines the function and fatigability of the individual muscle. A muscle with more fast-twitch motor units will have quicker and more powerful contractions but be more susceptible to fatigue, whereas a muscle with fatigue-resistant slow-twitch motor units will generate less powerful contractions but be able to sustain the contractions for longer periods of time. There is some evidence that muscles are able to adapt and switch muscle fiber types with specific training.

Muscle Mass Loss from Sarcopenia to Astronauts: Several causative factors can lead to the loss of skeletal muscle mass. One common cause is **sarcopenia**, the loss of skeletal muscle tissue due to aging, estimated at 0.5%–1% per year from age 25 years on. Although decreased exercise and use may be part of the reason, biochemical changes also result in decreased muscle fiber growth, including smaller muscle fiber circumference and replacement of muscle by fat or fibrous tissue. Part of the mechanism may be the increasing failure during aging of muscle satellite cells that help in muscle growth and repair. Skeletal muscle also appears to exhibit decreased response to growth hormone and testosterone as part of its slow decline.

Loss of muscle mass is also seen in astronauts exposed to a period of microgravity. During space flight, the weightless environment means that muscles normally used on an almost constant basis for posture and movement will no longer be bearing weight, resulting in loss of muscle tissue called **atrophy**. Loss of bone accompanies muscle loss and adds to a general state of deconditioning when the astronauts return to the earth's gravity. Often the most significant bone loss is at sites of muscle attachment. Attempts to counteract this problem include regular exercise on specially constructed exercise equipment and even a space suit with bungee cords adapted to provide counterforce for major muscle groups. The problem of muscle mass loss for both ground-based patients and space travelers is often studied using lengthy (days to weeks) research studies wherein individuals are restricted to horizontal bed rest. Animal models in which muscles are made nonweight bearing have been developed to also study this problem.

CARDIAC MUSCLE

Cardiac cells, called **cardiomyocytes**, share some of the same structures as skeletal muscle, including having several nuclei in one cell and a similar organization of thin and thick filaments, giving the resulting muscle a striated appearance (Figure 12-7). Like some types of skeletal muscle, they have a high content of myoglobin and a large number of mitochondria to drive oxidative phosphorylation. However, cardiac cells may be branched instead of the normally straight and uniform structure found in skeletal muscle.

Cardiac muscle normally derives energy aerobically from **triglycerides/fat** (65%), **glucose** (30%), and **proteins/ketone bodies** (5%) or alternatively from **lactic acid** transported from skeletal muscle. In fact, cardiac cells are almost totally reliant on aerobic respiration (only about 10% of the energy needs can ever be derived from anaerobic respiration) and oxygen delivery is, therefore, key to cardiac cell function and survival. Decreased

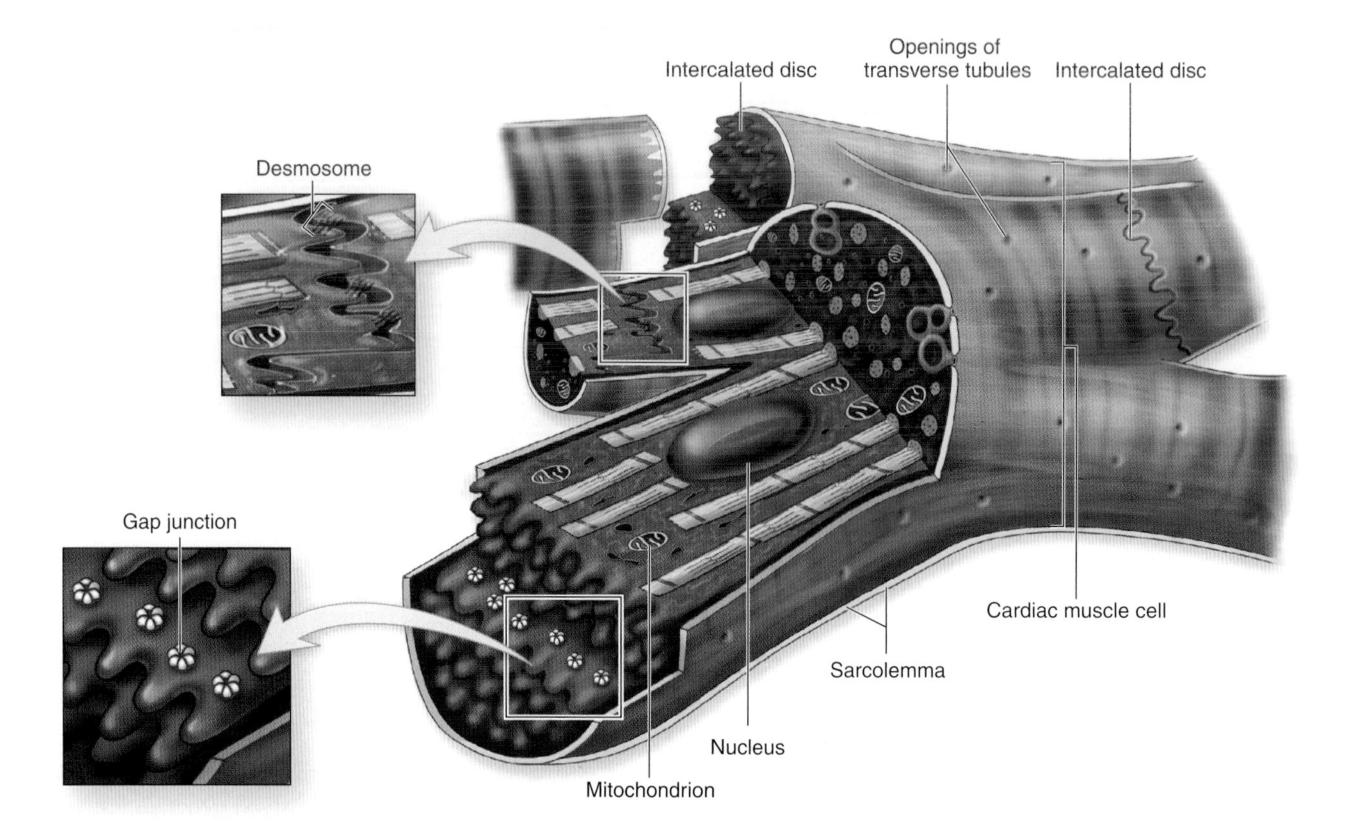

Figure 12-7. Cardiac Muscle Structure. Diagram of cardiac muscle cells indicates characteristic features of this muscle type. The fibers consist of separate cells with interdigitating processes wherein they are held together. These regions of contact are called the intercalated disks (IDs), which cross an entire fiber between two cells. The transverse regions of the step-like ID have abundant desmosomes and other adherent junctions, which hold the cells firmly together. Longitudinal regions of these disks contain abundant gap junctions, which form "electrical synapses" allowing contraction signals to pass from cell to cell as a single wave. Cardiac muscle cells have central nuclei and myofibrils that are less dense and organized than those of skeletal muscle. Also the cells are often branched, allowing the muscle fibers to interweave in a more complicated arrangement within fascicles that produces an efficient contraction mechanism for emptying the heart. [Reproduced with permission from Mescher AL: Junqueira's Basic Histology Text and Atlas, 12th edition, McGraw-Hill, 2010.]

blood supply and oxygen as well as removal of waste and carbon dioxide (CO_2) via the coronary arteries lead to heart-related chest pain (angina) and heart attacks.

Cardiac muscle produces a strong and forceful contraction but is not under voluntary control. Instead of nerve input, special internal pacemaker cells contract on a regular basis and then propagate this contractile impulse to the remainder of the heart. Contractions are further controlled by special cardiac muscle fibers that are innervated by the autonomic nervous system. T-tubules are fewer but larger and run only along the Z-lines or discs (Figure 12-7). Cardiac cells are also joined to each other by intercalated discs (IDs), which serve both structural and impulse transmission roles (Figure 12-7). The IDs serve as cytoskeletal anchors between individual cells to allow the coordination of the contraction into a multicell composite. IDs also contain gap junctions (Figure 12-7; Chapter 8), which allow the rapid propagation of flow of ions (Chapter 19) from one cell to the next. Both functions are essential for the rapid and synchronized contraction required by the heart.

CICR is the predominate mechanism in cardiac muscle. Voltage triggering of a **dihydropyridine receptor** induces calcium to flow into the sarcoplasm. Once inside the sarcolemma, the initial influx of calcium binds to a **ryanodine receptor** on the SR and releases much larger stores of Ca^{2+} in a manner similar to inositol triphosphate. Ca^{2+} flow through **L-type calcium channels** to activate the cardiomyocytes. These L ("long") channels stay open for an extended period of time to sustain the influx of calcium. This rise in calcium concentration causes troponin–tropomyosin conformation changes as previously described, actin–myosin binding and contraction. The release of temporal and spatial calcium wave and the integrated actions resulting from the IDs lead to the organized contraction of various parts of the heart muscle to produce the force to propel blood throughout the body. See Chapter 16 for additional details.

SMOOTH MUSCLE

Smooth muscle not only shares several qualities with skeletal and/or cardiac muscle but also differs in others. Smooth muscle activity relies on the interaction of actin thin filaments and myosin thick filaments. Although smooth muscle contains actin and myosin filaments and ABPs; in contrast to skeletal and cardiac muscle, there is much less structural organization. Smooth muscle, therefore, has a smooth appearance (i.e., not striated) under a light microscope. However, closer examination does show local zones of contractile elements that somewhat mimic the sarcomere structure. This structure allows smooth muscle cells to perform specific and directed functions, both temporally and spatially.

Unlike skeletal and cardiac muscle cells, smooth muscle cells have only one nucleus and lack troponins to regulate actin–myosin head binding (Figure 12-8A–B). Smooth muscle cells also do not contain T-tubules, and the SR is organized in only a loose network around the cell. Instead, calmodulin, caldesmon, and calponin regulate contraction via calcium (see Figure 12-9 and associated text below). Like cardiac muscle, the cytoskeletal elements of individual smooth muscle cells are anchored together as **dense bodies** (Figure 12-8A), which mimic the Z-line of skeletal muscle. Dense bodies, composed of the intermediate filament (IF) protein **desmin**, anchor the actin thin filaments. Other IFs (e.g., vimentin) and structures known as **adherens junctions** are also prominent in the organization of smooth muscle. These cytoskeletal/membrane connections allow transduction of force generation between individual smooth muscle cells. Smooth muscle cells also contain gap junctions, which allow the quick spread of Ca^{2+} fluxes between cells. Together, adherens and gap junctions allow coordinated, multicellular contractions similar to cardiac muscle.

Like cardiac muscle, smooth muscle contractions can be initiated by pacemaker cells, which contract on a regular basis and are innervated by the autonomic nervous system. However, nerve impulses, hormone signals, and even stretching can also initiate contraction. Smooth muscle cells do not produce the forceful contraction of skeletal muscle or cardiac muscle but they have the ability to undergo either a slow contraction for long periods of time with little energy use (**tonic contraction**) or a rapid contraction and relaxation (**phasic contraction**). Tonic contractions are seen in blood vessel walls. Phasic contractions are important for portions of the gastrointestinal tract involved in peristalsis (e.g., esophagus or small intestine).

Paramount to any contraction by smooth muscle is an influx of Ca^{2+}. Smooth muscle SR contains specific and specialized areas where ion channels and signaling pathway receptors are found together. Examples of these receptors include those for G proteins, neurohormones, activation of individual kinases, L-type calcium channels (see above), and calcium-sensitive and insensitive potassium channels. Areas of the SR are also

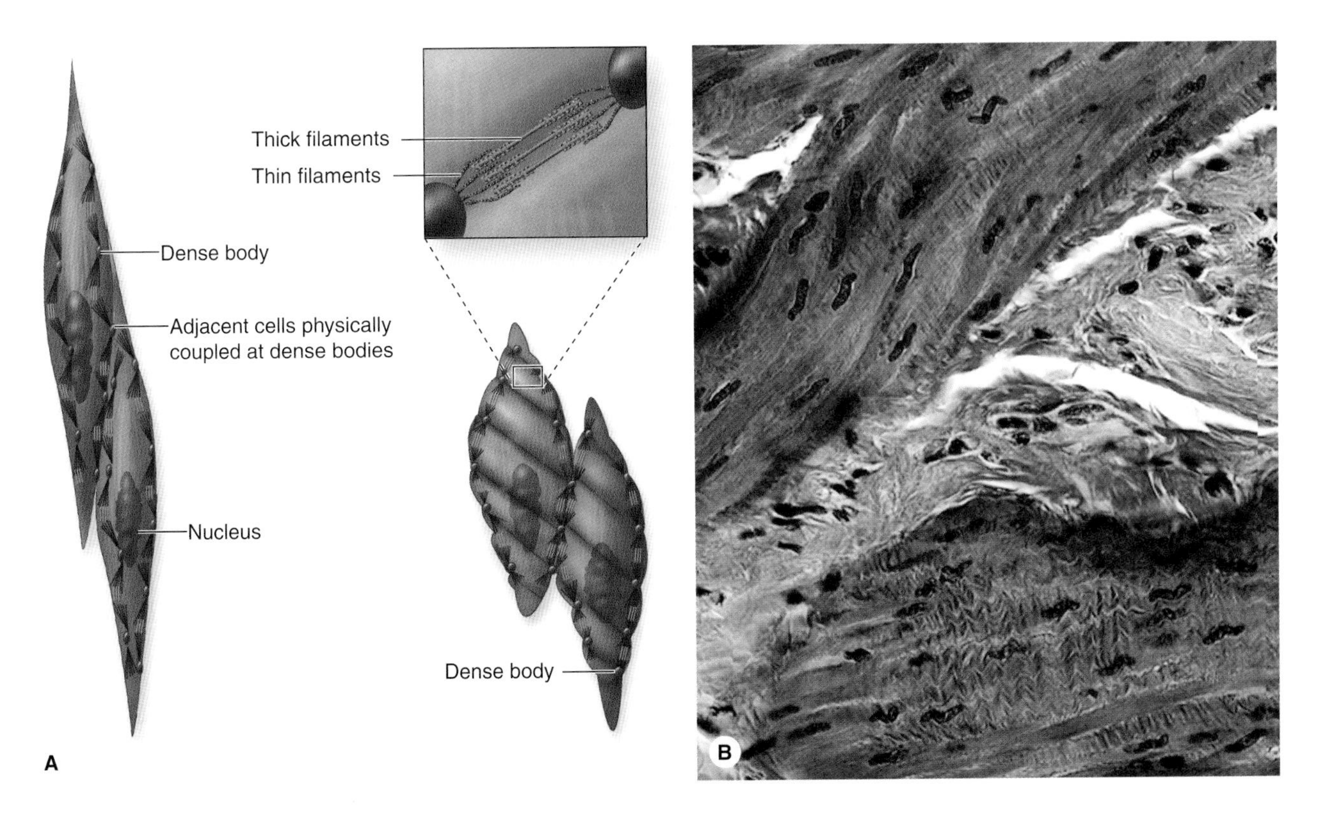

Figure 12-8. A–B. Smooth Muscle Contraction. Most molecules that allow contraction are similar in the three types of muscle, but the filaments of smooth muscle are arranged differently and appear less organized. **A.** The diagram shows thin filaments attach to dense bodies located in the cell membrane and deep in the cytoplasm. Dense bodies contain α-actinin for thin filament attachment. Dense bodies at the membrane are also attachment sites for intermediate filaments and for adhesive junctions between cells. This arrangement of both the cytoskeleton and contractile apparatus allows the multicellular tissue to contract as a unit, providing better efficiency and force. **B.** Contraction decreases the length of the cell, deforming the nucleus and promoting contraction of the whole muscle. The micrograph shows a region of contracted tissue in the wall of a urinary bladder. The long nuclei of individual fibers assume a cork-screw shape when the fibers contract, reflecting the reduced cell length at this time. [Reproduced with permission from Mescher AL: Junqueira's Basic Histology Text and Atlas, 12th edition, McGraw-Hill, 2010.]

near the external cell membrane allowing extracellular signals (e.g., hormones) and calcium influxes to also activate contraction. Signals for contraction (signal from an adjacent smooth muscle cell via a gap junction, nerve, hormone, and stretching) result in release of calcium from the SR (via the ryanodine receptor, see above) as well as possible influx of external calcium through the L-type channels. The resulting increase in Ca^{2+} concentration initiates actin–myosin interactions and force generation.

Regulation of contraction by calcium differs in many ways from skeletal and cardiac muscle. Smooth muscle cells do not contain troponin, and tropomyosin only serves an accessory role in actin–myosin force generation. Instead, the proteins caldesmon and calponin inhibit myosin ATPase enzymatic function and, possibly, actin–myosin binding through mechanisms that are still being studied. A general overview is that an influx of calcium activates the protein calmodulin, which leads to caldesmon and calponin phosphorylation and loss of inhibitory effects. Calmodulin activation also activates MLCK, which leads to the phosphorylation of the regulatory myosin light chain (Figure 12-9). MLCP removes the phosphate group when calcium concentrations fall, inactivating myosin and stopping contraction.

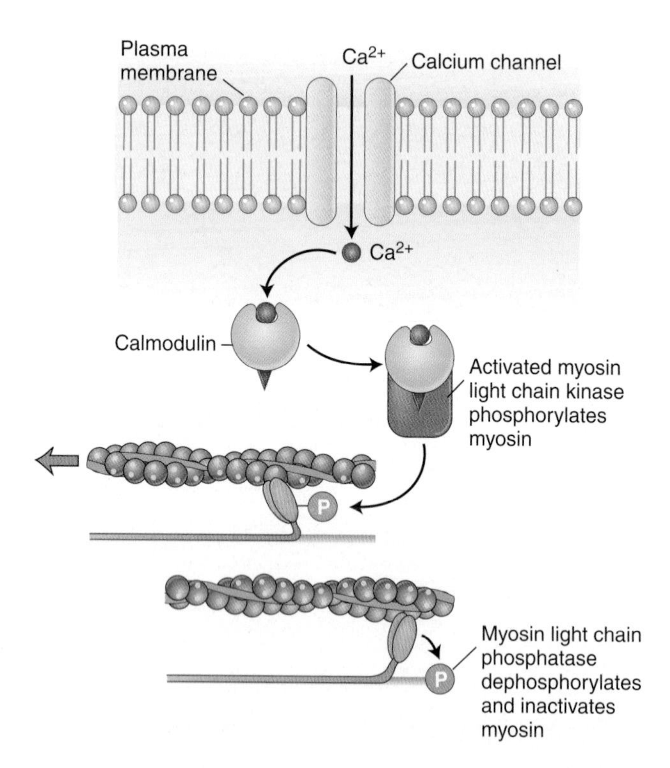

Figure 12-9. Regulation of Smooth Muscle Myosin Force Generation by Calcium Ions. The influx of calcium ions (Ca^{2+}) leads to activation of calmodulin and then myosin light chain kinase (MLCK). MLCK phosphorylates the regulatory light chains on the myosin molecule and activates its ATPase and force-generating capabilities. Myosin light chain phosphatase (MLCP) can remove the phosphate group, inactivating myosin activity. See text for more details. [Reproduced with permission from Kibble JD and Halsey CR: The Big Picture: Medical Physiology, 1st edition, McGraw-Hill, 2009.]

The lack of sarcomere structure in smooth muscle also enables **polymerization** of G-actin to F-actin thin filaments to partly regulate contraction. This process can form new thin filaments at local areas in a smooth muscle where contraction or force generation is required. Breakdown of existing F-actin thin filaments can stop contraction. The exact role of actin polymerization and **depolymerization** in smooth and in nonmuscle cells is still being determined. Several ABPs involved in these changes are affected by calcium as well.

Smooth muscle cell types are found throughout the human body and vary in function but all rely on the basic actin–myosin contractile mechanism. Examples and their associated functions are shown in Table 12-1.

ENERGY PRODUCTION AND USE IN MUSCLES

Humans use energy in the form of **ATP** to drive the necessary muscle contractions. During exercise, both aerobic and anaerobic energy-producing systems need to function because different types of muscle fibers are activated, relying on one or both energy sources. The body is able to move the energy-producing molecules back and forth between the aerobic and anaerobic pathways, depending on the type of exercise/muscle contraction and duration of exercise. Adaptation between aerobic and anaerobic metabolism is also regulated by increased transcription and translation of factors such as **hypoxia-inducible factor-1α**. Muscles mainly use one of three different energy sources (Figure 12-10) to produce this ATP—**muscle glycogen (glucose)**, **creatine phosphate**, or **triglycerols/fatty acids** from adipose tissue.

Creatine phosphate, in essence a storage form of energy for muscle, can use its high energy phosphate bond to readily produce ATP in a reaction catalyzed by the enzyme **creatine kinase**. **Muscle glycogen** is a major storage depot from which glucose-6-phosphate for the glycolytic pathway can be produced. Some types of muscle fibers rely primarily on **fatty acids** from triacylglycerols (triglycerides) as fuel. Both glycogen and fatty acids can be metabolized via aerobic respiration by types I and IIa muscle fibers and produce ATP by oxidative phosphorylation. Under aerobic conditions, the metabolism of glucose yields **pyruvate** and energy in the form of ATP. The pyruvate is then completely oxidized to CO_2 with the formation of additional 15 ATP for each pyruvate molecule. Additionally, the nicotinamide adenine dinucleotide (NADH) generated from glycolysis can be processed by the malate–aspartate shuttle to generate additional three ATP per NADH. Hence, under aerobic conditions, each glucose molecule can yield up to 38 ATP.

$$\text{Glucose} + 2\ \text{ADP} + 2\ \text{NAD}^+ + 2\ \text{P}_i \rightarrow 2\ \text{Pyruvate} + 2\ \text{ATP} + 2\ \text{NADH} + 2\ \text{H}^+$$

$$2\ \text{Pyruvate} + 8\ \text{NAD}^+ + 2\ \text{FAD} + 2\ \text{ADP} + 2\ \text{P}_i \rightarrow 6\text{CO}_2 + 8\ \text{NADH} + 2\ \text{FADH}_2 + 2\ \text{ATP} + 8\ \text{H}^+$$

In periods of low oxygen, the citric acid cycle and oxidative phosphorylation are limited and anaerobic respiration is used. Types IIb and IIx muscle fibers rely mainly on anaerobic ATP production. When this happens, pyruvate accumulates and is then converted to **lactic acid** by the action of the enzyme

TABLE 12-1. Types and Functions of Smooth Muscle Cells

Smooth Muscle Type	Function	Comments
Blood vessels	Control of vessel diameter and resistance. Forms capillary sphincters.	Directly affects blood pressure and controls backflow of blood in capillary beds.
Lymph vessels	Movement of lymph	Propels lymph throughout lymphatics.
Digestive tract	Control of movement of food	Rhythmic and coordinated contraction (peristalsis) serves to move digesting food through the digestive tract.
Respiratory tract	Control of diameter of bronchial airways	Spasms can cause asthmatic reactions/wheezing.
Urinary tract	Urinary bladder contraction, renin production, mesangial cells in kidney glomeruli	Coordinated contraction moves and expels urine. Sense changes in blood to regulate excretion of renin. Also affects glomerular filtration rate.
Reproductive tract	Uterine contraction, vaginal and penis (urethral) contractions	Forms the myometrium and birth canal. Rhythmic contractions are part of female and male orgasms.
Skin	Arrector pili that cause hair to stand vertically	Part of involuntary thermal regulation (although fear can also cause contraction), the efficacy in humans is limited.
Eye	Ciliary muscles and iris, optimization of vision	Changes lens focus and iris diameter.
Connective tissue	Secretion of collagen, elastin, and other connective tissue proteins and glycoproteins	Interact with extracellular matrix as part of viscoelasticity of tissues.

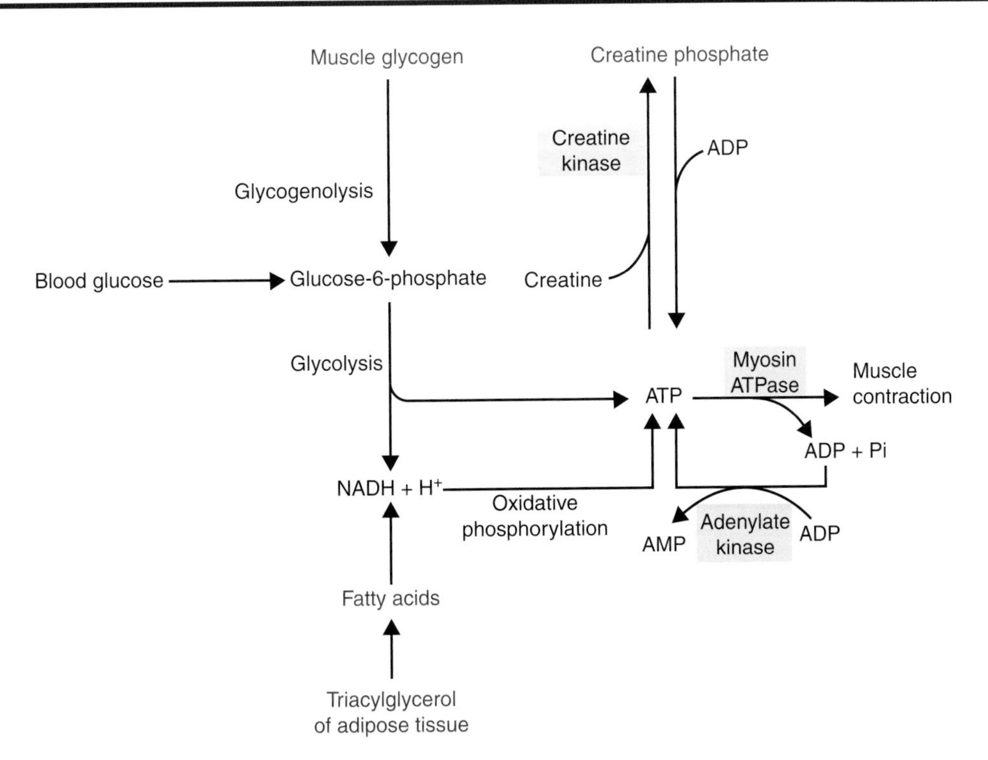

Figure 12-10. Sources of ATP in Muscles. ADP, adenosine diphosphate; AMP, adenosine monophosphate; ATP, adenosine triphosphate. [Reproduced with permission from Naik P: Biochemistry, 3rd edition, Jaypee Brothers Medical Publishers (P) Ltd., 2009.]

lactate dehydrogenase and use of an **NADH** molecule. In this way, NAD^+ is regenerated to sustain flux through glycolysis.

$$\text{Pyruvate} + H^+ + \text{NADH} \rightarrow \text{Lactic acid} + NAD^+$$

This point when lactate production is greater than lactate utilization is known as the **lactate threshold**. This threshold usually occurs at around 65% of a person's maximal oxygen consumption. If muscle activation continues, then lactic acid accumulates in the muscle tissue and can affect important metabolic functions, giving the muscle pain and the sensation of muscle fatigue. The surplus of lactic acid is ultimately transported to the liver or heart muscle where it is converted back into pyruvate and into the citric acid cycle.

McArdle's disease: Glycogen storage diseases are a series of disorders that impact on the body's ability to produce, store, and/or breakdown this important energy source. One type of glycogen storage disease, type V or **McArdle's disease**, results from the absence of **myophosphorylase**, the muscle form of glycogen phosphorylase. Patients usually present in childhood with muscle pain, fatigue, cramps, and weakness with excessive myoglobin in the urine, indicative of rhabdomyolysis (see above) during prolonged periods of exercise. Progressive symptoms and muscle mass loss and weakness are usually evident as the patient ages. Diagnosis may be complicated by a "**second wind**" felt by patients. Although once thought to be simply due to the use of alternative energy sources such as fatty acids and proteins, evidence now suggests a possible role of increased numbers of GLUT4 receptors on the plasma membrane, allowing increased glucose uptake for glycolysis instead of relying on glycogen stores.

MICROTUBULE-BASED MOTILITY

Microtubules function as a structural cytoskeleton but they also serve as a "track" system for transport within the cell. With the use of motor proteins, intracellular components such as organelles including mitochondria, vesicles, or granules can be moved within the cell. Chromosomes can also be moved with unique attachment proteins. Joined together, microtubules form the structure of more complex cellular motility components such as cilia or flagella. Centrioles, such as cilia and flagella, are also made of microtubules. Centrioles come in pairs that are oriented at right angles to each other and are responsible for setting up the spindle that moves the chromosomes during mitosis.

Microtubules form the backbone of all of these structures. Microtubules are made of linear polymers of a globular protein called **tubulin** (Chapter 1) and have a polar arrangement within the cell with the positive end directed toward the periphery and the negative end toward the center. Like actin, there are slightly different forms of tubulin (α to ε), which form different microtubule structures in different parts of the cell and body. α- and β-tubulin form cytoplasmic microtubules, γ-tubulin forms centrosomes, and δ- and ε-tubulin are involved in centrioles and the mitotic spindle apparatus.

Microtubule Polymerization and Cancer Therapy: Microtubule function is essential for a variety of cellular functions including several aspects of mitosis, which involves **polymerization** and **depolymerization** of **tubulin** into microtubule polymers. The medications **colchicine**, **nocodazole**, **vincristine**, and **colcemid** all inhibit polymerization by interfering with the tubulin monomer. The **taxane** family of drugs, including **taxol**, break down existing microtubules. All lead to the disruption of mitosis and eventual death of the affected cell. Although these effects are used in the treatment of **gout** (inhibition of microtubule-dependent, neutrophil motility, as well as uric acid crystallization) and various other inflammatory diseases, these agents have also emerged as important **cancer therapies**. Because cancerous cells often rapidly divide, agents that inhibit microtubules have emerged in the treatment of cancer as well as other diseases. The inhibition of mitosis affects cancer cells more than normal cells, leading to a partly selective treatment agent. Unfortunately, side effects of these microtubule/mitosis inhibitors are still apparent, limiting their use and/or causing additional problems for patients.

The polar alignment of microtubules plays a role as part of the transport system within a cell, using a conversion of chemical energy to kinetic energy via the motor proteins dynein and kinesin, very similar in mechanism to the interaction of actin and myosin. With the arrangement of the microtubule with the negative end toward the center and the positive end toward the periphery, the dynein molecules are responsible for transport toward the nucleus and kinesins are responsible for transport toward the plasma membrane.

Dynein, like myosin, is a large ATPase with two heads and uses the hydrolysis of ATP to move the components of a cell by literally walking along the microtubule network structure (Figure 12-11A). The polarity of the microtubule plays a role in the direction of the movement of the dynein to the negative end of the microtubule or toward the nucleus. Two forms of dynein exist. **Cytoplasmic dynein** organizes intracellular organelles and moves vesicles within cells as well as chromosomes and the mitotic spindle during mitosis. **Axonemal dynein** functions in cilia or flagella motility. The two heads of cytoplasmic dynein are joined together at a tail with intermediate and light chains, the latter of which attaches the dynein molecule to the organelle or cargo being transported.

Kinesins are also microtubule-based motor proteins that function via the hydrolysis of ATP and, like dynein, are responsible for moving substances around the inside of a cell (Figure 12-11B). Like dynein, kinesin molecules are also involved in

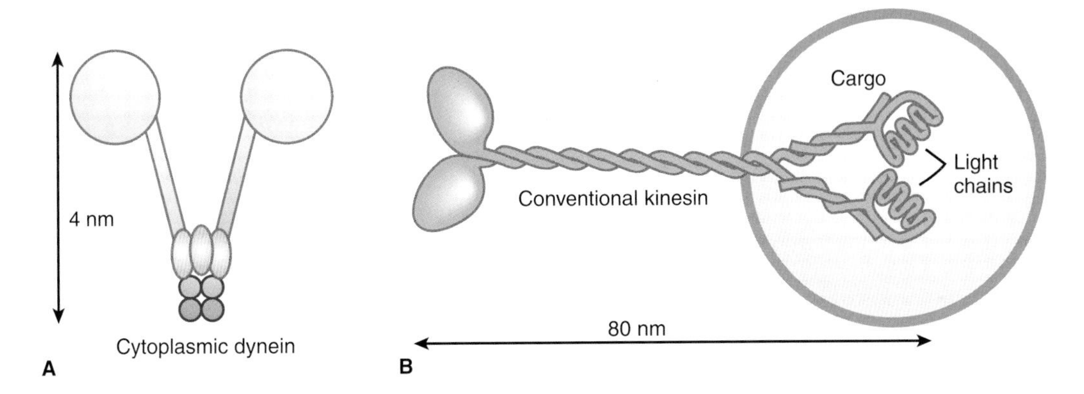

Figure 12-11. A–B. Microtubule Motors—Dynein and Kinesin. A. Structure and size of cytoplasmic dynein. **B.** Structure and size of kinesin, including its associated regulatory light chains and connection to its cargo vesicle. [Adapted with permission from Barrett KE, et al.: Ganong's Review of Medical Physiology, 23rd edition, McGraw-Hill, 2010.]

mitotic spindle formation and chromosome movement during mitosis. The first kinesin was discovered in 1985, with many more subfamilies of kinesins discovered since then. All human kinesins are responsible for intracellular transport toward the positive end of the microtubule network or toward the plasma membrane. Kinesin molecules resemble myosin II molecules with two heavy chains and two light chains. The heavy chains have two heads at one end, which bind to and transport along the microtubules; the tail portion of the heavy chain attaches to the cargo organelle or vesicle via the light chains.

In **cilia** and **flagella**, the microtubules are arranged in an organized structure (Figure 12-12). In cilia and flagella, two microtubules are joined together to form a doublet. The IF protein **nexin** (see Table 12-2) forms links along the microtubules to hold them together. Nine of these doublets form a circle around two microtubules in the center. Extending from

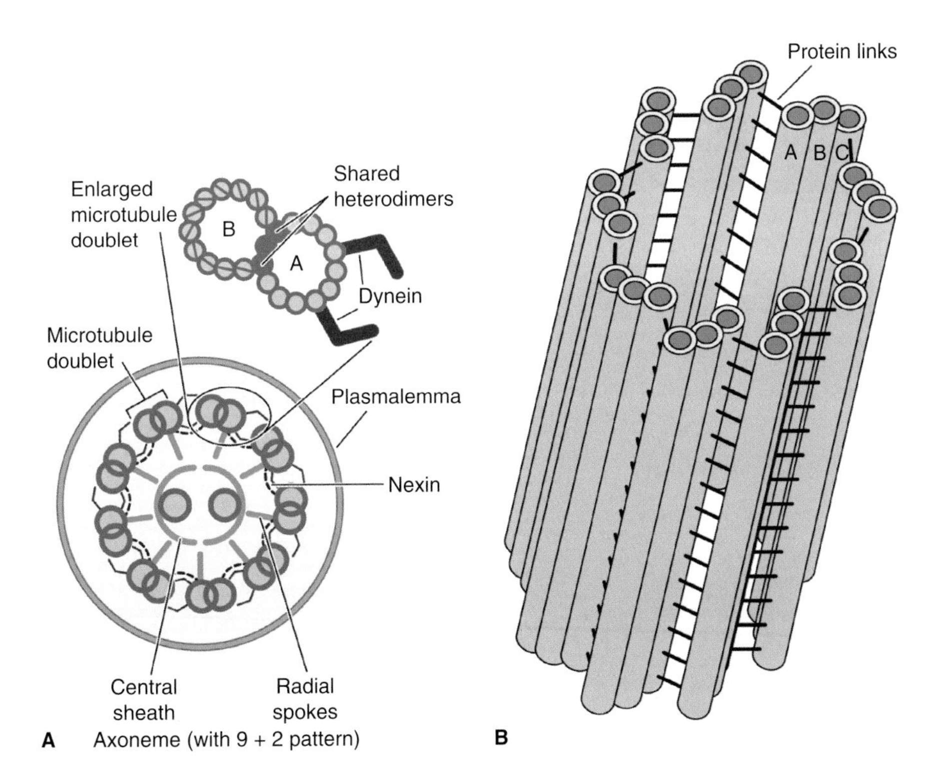

Figure 12-12. A–B. Structure of Cilia and Flagella. The structure of cilia and flagella is composed of (**A**) a central microtubule doublet surrounded by nine outer doublets with dynein arms between them. The dynein creates movement of the cilia or flagella. Radial spokes and nexin provide structure and limits to this movement. The base of the main cilia or flagella structure includes (**B**) a centriole/basal body with a triplet microtubule structure, which serves as an anchor. See text for further details. [Reproduced with permission from Mescher AL: Junqueira's Basic Histology Text and Atlas, 12th edition, McGraw-Hill, 2010.]

the doublets are "arms" that connects neighboring doublets, composed of axonemal dynein. **Radial spokes** also provide structure between the inner and outer microtubule doublets. The organization of the microtubules allows the dynein to slide along one tubule, while attached with its tail to another microtubule creating a bend in the cilium or flagella. However, the nexin linkages allow for only limited lengthwise movement and, therefore, create a bend in the cilium. The dynein bridges are also regulated so the sliding can lead to synchronized bending. A **centriole/basal body** structure at the base of the cilia or flagella and composed of triplets of tubulin monomers helps to form and anchor these longer structures.

Microtubules, Cilia, and Flagella—Roles in Disease Processes: Although often rare, defects in cilia/flagella, known as ciliopathies, lead to several diseases/syndromes including the following:

Kartagener Syndrome/Primary Ciliary Dyskinesia—Defective cilia in the respiratory tract, Eustachian tube, and Fallopian tubes leading to chronic lung infections, ear infections and hearing loss, and infertility. Possible association with "situs inversus," a condition in which major internal organs are "flipped" left to right.

Senior–Loken Syndrome/Nephronopthisis—Eye disease and formation of cysts in the kidneys, leading to renal failure.

Bardet–Biedl Syndrome—Dysfunction of cilia throughout the body leading to obesity because of inability to sense satiation, loss of eye pigment/visual loss and/or blindness, extra digits and/or webbing of fingers and toes, mental and growth retardation and behavioral/social problems, small and/or misshaped genitalia (male and female), enlarged and damaged heart muscle, and kidney failure.

Alstrom Syndrome—Childhood obesity, breakdown of the retina leading to blindness, hearing loss, and type 2 diabetes.

Meckel–Gruber Syndrome—Formation of cysts in kidneys and brain leading to renal failure and neurological deficits, extra digits, and bowing/shortening of the limbs.

Increased Ectopic (Tubal) Pregnancies/Male Infertility—Deficient cilia in Fallopian tubes or flagella/sperm tail motility.

Autosomal Recessive Polycystic Kidney Disease—Much rarer than the autosomal dominant form, dysfunction of basal bodies and cilia in renal cells leads to alterations of the lung and kidneys, resulting in a variety of secondary medical conditions and often death.

Parkinson's and Alzheimer's Diseases—Although work is still ongoing, researchers now feel that some forms of Parkinson's and Alzheimer's diseases may result, in part, from damage to microtubules and associated proteins. Treatments aimed at stabilizing microtubules may help sufferers from these specific proteins.

INTERMEDIATE FILAMENTS

The **IF** is the name for a single protein member of a family of related cytoskeletal linear proteins (monomer forms) or the resulting filaments that these proteins form. The filaments forms are about 10 nm in diameter and, as the name implies, are between the size of 6 nm actin microfilaments and either 15–20 nm myosin thick filaments or 25 nm microtubules (Figure 12-13). All IF monomer proteins have two head regions for binding and lengthy, connecting α-helical portion. IFs are formed from the intertwining of the α-helical section of pairs of the same (homodimer) or different (heterodimer) monomers or pairs of pairs (homo- or heterotetramer) IFs can stretch and

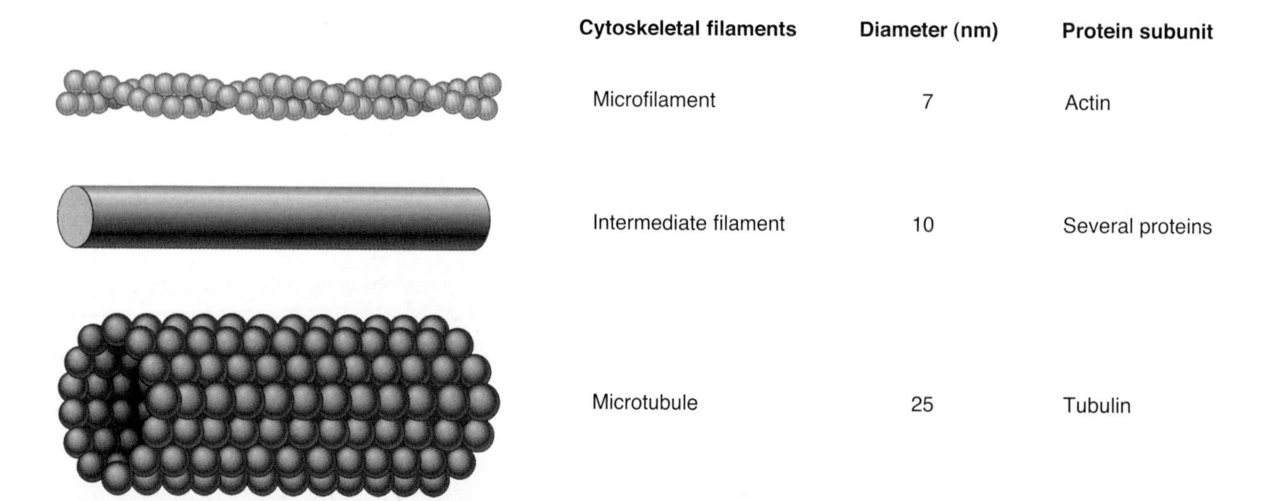

Figure 12-13. Comparison of Cytoskeletal Proteins. Relative sizes of actin microfilaments, intermediate filaments, and microtubules are shown. [Reproduced with permission from Barrett KE, et al.: Ganong's Review of Medical Physiology, 23rd edition, McGraw-Hill, 2010.]

TABLE 12-2. Intermediate Filament Types and Functions

Type	Function
I	Acidic Keratins—Form a variety of tissue-specific keratin filaments. Each type I IF protein pairs with a type II proteins (heterodimers) in the cytoplasm of epithelial cells, providing strength to skin, hair, and nails.
II	Basic Keratins—Form a variety of tissue-specific keratin filaments. Each type II IF protein pairs with a type I proteins (heterodimers) in the cytoplasm of epithelial cells, providing strength to skin, hair, and nails.
III	Desmin—Full function(s) not fully identified. Contributes to Z-disk structure in muscle cells as well as cell–cell adhesion.
	Glial Fibrillary Acidic Protein—Found in central nervous system cells as well as numerous other tissue types where it helps the mechanical strength of the cytoskeletal structure and helps determine cell shape. Form four-protein (tetramer) structures.
	Peripherin—Found in neurons where it may function in cell growth and development as well as regeneration.
	Vimentin—Found in skin cells (fibroblasts), white blood cells (leukocytes), and cells lining blood vessels. Plays a structural/anchoring roll for the cytoskeleton as well as the nucleus, endoplasmic reticulum, and mitochondria. Made of four proteins (tetramer).
IV	α-Internexin—Found in neural cells and involved mainly in linear nerve axon growth.
	Neurofilaments—Found in nerve axons and responsible for determining the diameter of an axon and, therefore, its speed of nerve signal transduction.
	Synemin (Desmuslin)—Found in all cell types, although its only function as yet determined is linking contracting actin–myosin fibers to the Z-disk of skeletal muscle cells, thereby helping to transmit the force generated.
	Syncoilin—Coupled with desmin, plays an unknown role in the Z-disk structure of muscle cells.
V	Nuclear Lamins—Part of the nuclear lamina on the inner surface of the nuclear membrane and provide general structure to the nucleus. Are phosphorylated by mitosis promoting factor and broken down during mitosis/dissolution of the nucleus and nuclear membrane.
VI	Nexin—Found mainly in nerve cells but also in other cell types, where it plays a still elusive role in the changing structure and organization of these cells. This function appears to be modified by phosphorylation. May form two (dimer) or four (tetramer) protein structures.

IF, intermediate filament.

compact because of their structure and, thus, can serve to transmit force or absorb mechanical stress. There are six basic types based on structural similarities (Table 12-2).

NONMUSCLE CELLS

Nonmuscle cells include skin fibroblasts; platelets; immune system lymphocytes and monocytes/macrophages (Chapter 15); neurons, including photoreceptor cells of the retina (Chapter 19); chondrocytes and osteocytes (Chapter 13); various cells of the kidney, pancreas, and intestines (Chapter 11); as well as the thyroid and blood vessels. As such, nonmuscle cells are responsible for diverse tasks such as wound healing, immune response, specialized nerve functions, connective tissue and bone formation and health, digestion, regulation of metabolism, cardiovascular disease, and even embryological development. As noted previously, the contractile apparatus in nonmuscle cells is far less structured than skeletal, cardiac, or smooth muscle types but does preserve many of the same functional and regulatory elements. Actin thin filaments and myosin II thick filaments are major parts of nonmuscle cell motility and interact in an analogous way. Other types of myosin (e.g., myosin I, myosin V, and others) are also prominent in nonmuscle cell motility that includes intracellular transport (Figures 12-3B and 12-11B). Regulation of contraction is much like smooth muscle where Ca^{2+} regulate actin filament stability and length, aspects of actin–myosin head binding, and, finally, myosin force generation.

REVIEW QUESTIONS

1. What are the roles of the basic muscle components including how they may interact with each other to facilitate muscle function?
2. What are the three domains of myosin and how is each involved in muscle contraction?
3. From where and how is calcium released for skeletal muscle contraction and what is its role in the process?
4. What is the role of myosin ATPase in muscle contraction?
5. How is calcium sequestered and concentrated in the sarcoplasmic reticulum after contraction cycles?
6. What is the role and mechanism of action of acetylcholine in muscle contraction?
7. What are the types and metabolic characteristics of the different muscle fiber types?
8. How does structure and contraction of cardiac muscle differ from skeletal muscle?
9. What are the types and functions of smooth muscle cells?
10. How do muscles generate energy for function in anaerobic versus aerobic conditions?
11. What are the functions of microtubules and the associated roles of dynein and kinesin in motility?
12. What are the types and functions of intermediate filaments?
13. How is nonmuscle cell contraction alike and different from contraction by other muscle types?

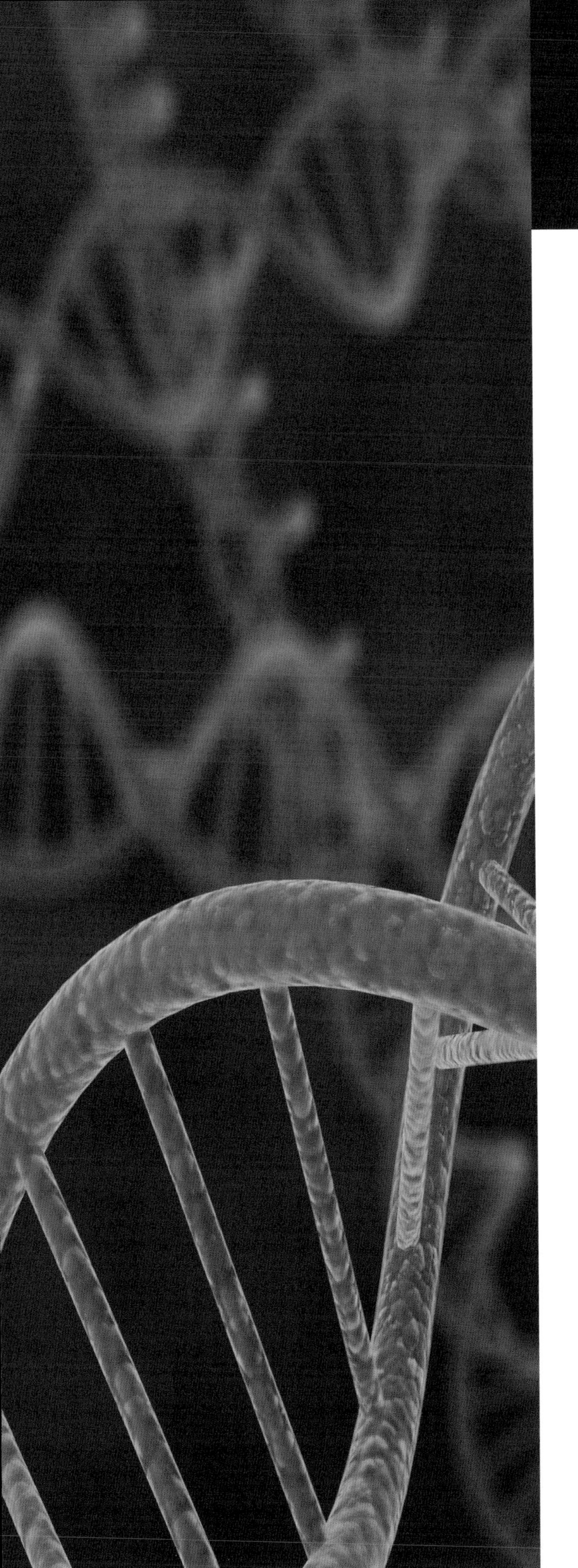

CHAPTER 13

CONNECTIVE TISSUE AND BONE

Editor: Jacques Kerr, BSc, MB, BS, FRCS, FCEM
Lead Consultant in Emergency Medicine, Borders General Hospital, Melrose, Scotland, United Kingdom

OVERVIEW

Connective tissue provides a framework and support for a large variety of structures, including organs, blood vessel walls, as well as the better known functions of connecting muscle to bone and bone to bone. Chondrocytes are responsible for the production of connective tissue components, including the principal protein, collagen. The three types of connective tissues offer a wide variety of functions to fulfill the many roles needed by the body. A multitude of diseases result from abnormalities, deficiencies, or overproduction of connective tissues or their components.

Bones provide a mechanical structure for the human body and, in that role, also allow effective muscle contraction and, therefore, movement. Bones offer protection for the body's internal organs, especially brain, heart, lungs, liver, stomach, and spleen. In the ear, bones transduct sound waves from the ear drum to the inner ear. Osteoblasts, osteocytes, and osteoclasts produce, break down, remodel, and repair both the organic and inorganic matrix that make up bones. In doing so, these cells, as well as several regulatory molecules, help to regulate calcium and phosphate metabolisms and levels in the body. Bones also have a synthetic function. Within their marrow, bones produce red and white blood cells as well as growth factors; and store fatty acids as yellow marrow. Finally, bones provide for the storage of certain minerals, including calcium and phosphorus and, to a lesser extent, zinc, copper, and sodium. In an analogous role, bone can temporarily absorb and store toxic heavy metals to reduce their effects on the body.

CONNECTIVE TISSUE

Connective tissue is a fibrous tissue made mainly of collagen (Chapter 1) and proteoglycans (Chapter 2) that forms, supports, and/or connects various organs in the body, attaches muscles to bones (e.g., tendons) and bones to bones (e.g., ligaments), forms the supportive matrix during bone formation (see below), and makes up various structures such as parts of blood vessels and intestinal walls. One major example of connective tissue is collagen, which is found in various forms throughout the body (Figure 13-1A–D).

Formation of connective tissue relies on **chondrocytes**, which both produce and maintain the collagen matrix. Chondrocytes differentiate from **osteochondrogenic** cells, which can alternatively develop into osteoblasts (see below). Although the different structures vary widely, three major fiber forms of connective tissue fibers are prominent—collagenous fibers, elastic fibers, and reticular fibers.

Collagen is the principal component of **collagenous fibers**, and the type of collagen determines the structural and functional qualities of that particular form. Collagen is believed to represent about a quarter of the total protein in human body. Although 29 types of collagen have been discovered, each varying in its structure, location in parts of the body, and functions, four (types I–IV) make up a vast majority of connective tissue. Structure and formation of a type I collagen fiber is illustrated in Figure 13-2A–B.

Varying combinations of the different types of collagen can alter this basic structure for particular cell functions. Most collagenous fibers follow the previously described, left-handed, triple-helix of type I (Figure 13-3A), with multiple glycine, proline/hydroxyproline, and hydroxylysine amino acids contributing to its structure.

The second type of connective tissue is the **reticular fiber**, composed of type III collagen and forming an ordered, "reticulum" meshwork instead of a linear structure (Figure 13-3B).

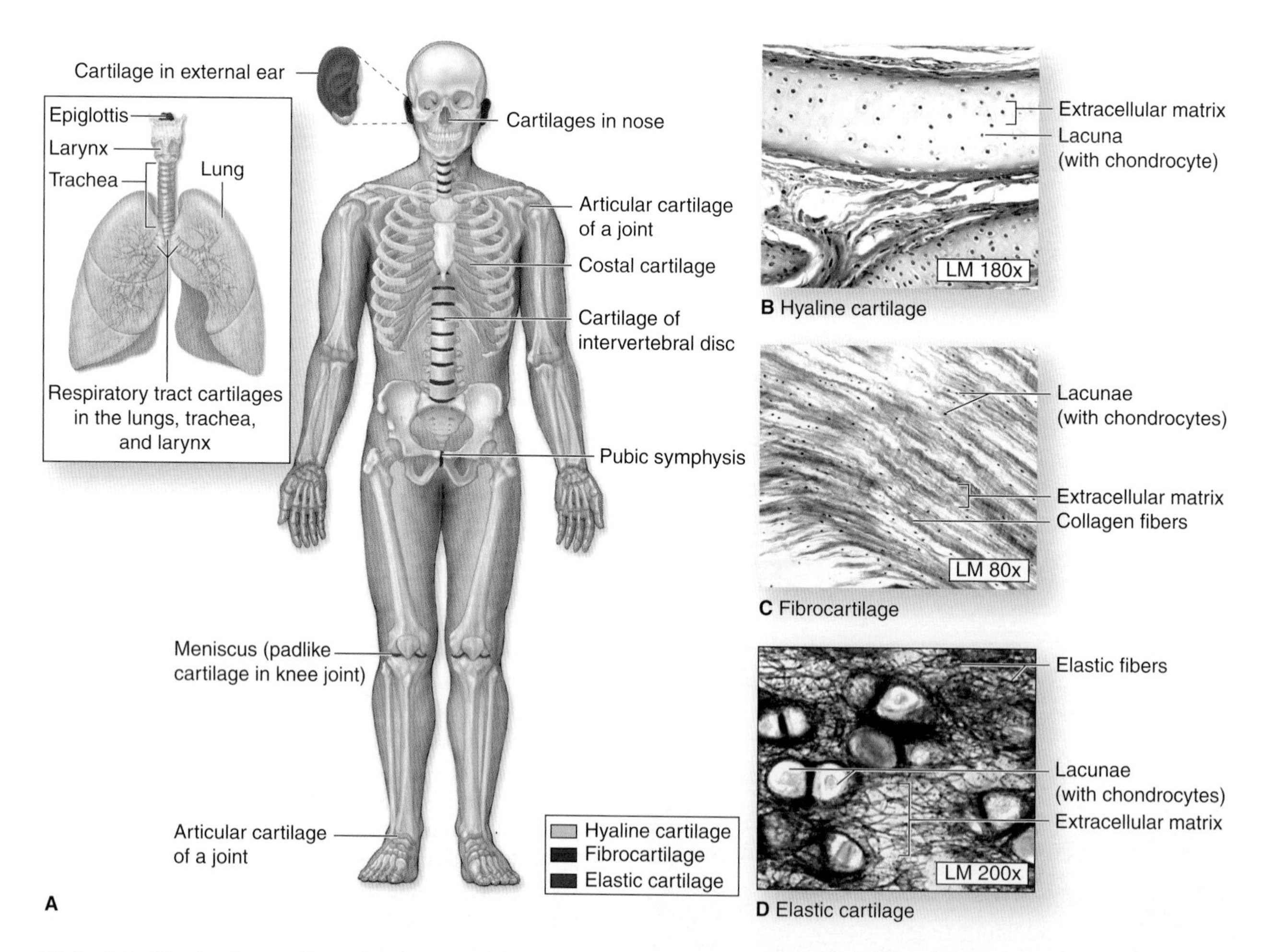

Figure 13-1. A–D. Distribution of Cartilage in Adults. (**A**) There are three types of adult cartilage distributed in many areas of the skeleton, particularly in joints and where pliable support is useful, as in the ribs, ears, and nose. Cartilage support of other tissues throughout the respiratory system is also prominent. The photomicrographs show the main features of (**B**) hyaline cartilage, (**C**) fibrocartilage, and (**D**) elastic cartilage. [Reproduced with permission from Mescher AL: Junqueira's Basic Histology Text and Atlas, 12th edition, McGraw-Hill, 2010.]

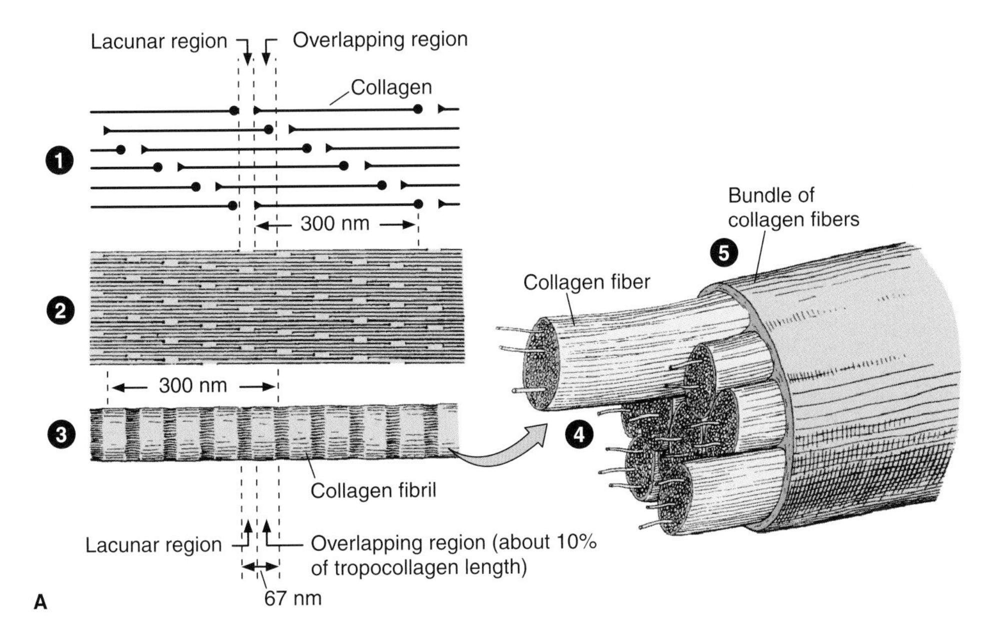

Figure 13-2. A. Structure of Collagen. Illustration of structural arrangement of individual, overlapping collagen molecules (1), collagen fibrils, collagen fibers (2 and 3), and collagen fiber bundles (4 and 5). The overlapping nature of this structural organization produces the required strength and flexibility of this molecule. [Reproduced with permission from Mescher AL: Junqueira's Basic Histology Text and Atlas, 12th edition, McGraw-Hill, 2010.] (*continued*)

The reticular network links these proteins to carbohydrates, including glucose, galactose, mannose, and fucose (Chapter 3). Reticular fibers are found mainly in liver, muscle, bone marrow, lymph system, and various other tissues and organs, where it offers a supportive framework.

The final type of connective tissue fiber is the **elastic fiber** (Figure 13-3C). Elastic fibers are formed from smaller **microfibrils**, made mainly of the protein **tropoelastin/elastin** and the glycoprotein **fibrillin**, and cross-linking polypeptides (Figure 13-4A–B). The microfibrils are in the form of irregular, random coils, with glycine, valine, alanine, proline, and lysine amino acids contributing to the stable structure.

Elastic fibers are found in several tissue types, but mainly in skin, the outer ear, larynx and epiglottis, blood vessels, lungs, bladder, some intervertebral discs, and as an attachment between teeth and the underlying bones. As the name suggests, elastic fibers can be easily stretched (up to 1.5× their original length). Diseases related to defects in elastin include Menkes disease, Hurler disease, William's syndrome, cutis laxa, Buschke–Ollendorf syndrome, and pseudoxanthoma elasticum. Changes in the elastic fibers in the heart and/or arteries may also play a role in high blood pressure, as the ability to absorb the force of heart contraction with appropriate blood vessel stretch and rebound is diminished.

Diseases of connective tissues are usually associated with defects in the component collagen molecule or deficiencies in essential nutrients required for their synthesis (e.g., vitamin C) and secretion. A large group of autoimmune diseases of connective tissue also exists, in which the body's immune system mistakenly attacks parts of the body. If too much collagen is produced, scleroderma may result, an autoimmune disease characterized by thickening and hardening of the skin and deleterious changes in blood vessels. Most of the connective tissue diseases are diagnosed by physical examination, looking for abnormalities in the organ(s) involved that reflect changes in the connective tissue. DNA studies and biopsies can help confirm the condition(s). Autoimmune connective tissue diseases are normally diagnosed by symptoms and by blood tests revealing diagnostic antibodies or reactive molecules. The collagen types, functions, and related diseases are reviewed in Table 13-1.

COMPONENTS OF BONE

Bone is made up of the inorganic mineral **hydroxyapatite**, known more formally as calcium apatite [$Ca_5(PO_4)_3(OH)$ or, in its crystal-structure form, $Ca_{10}(PO_4)_6(OH)_2$]. Carbonated hydroxyapatite is also the main component of the enamel in teeth, and variable chemical forms can cause natural variations in teeth color. In addition, bone is composed of an organic matrix, primarily composed of type I collagen and three primary cell types—osteoblasts, osteoclasts, and osteocytes (Figure 13-5A-B).

Osteoblasts, a specialized form of a fibroblast cell, which produces bone, arise from **osteoprogenitor cells**, whose growth and development are influenced by fibroblast growth factor (FGF), platelet-derived growth factor (PDGF), transforming growth factor-β (TGF-β), and bone morphogenetic proteins (BMPs). FGF signaling molecules bind to a receptor with tyrosine kinase (phosphorylation) activity, which initiates a number of signaling pathways inducing the expressions of selected genes.

TGF-β signaling usually occurs by first combining two TGF-β proteins via a unique "cysteine knot" structure, in

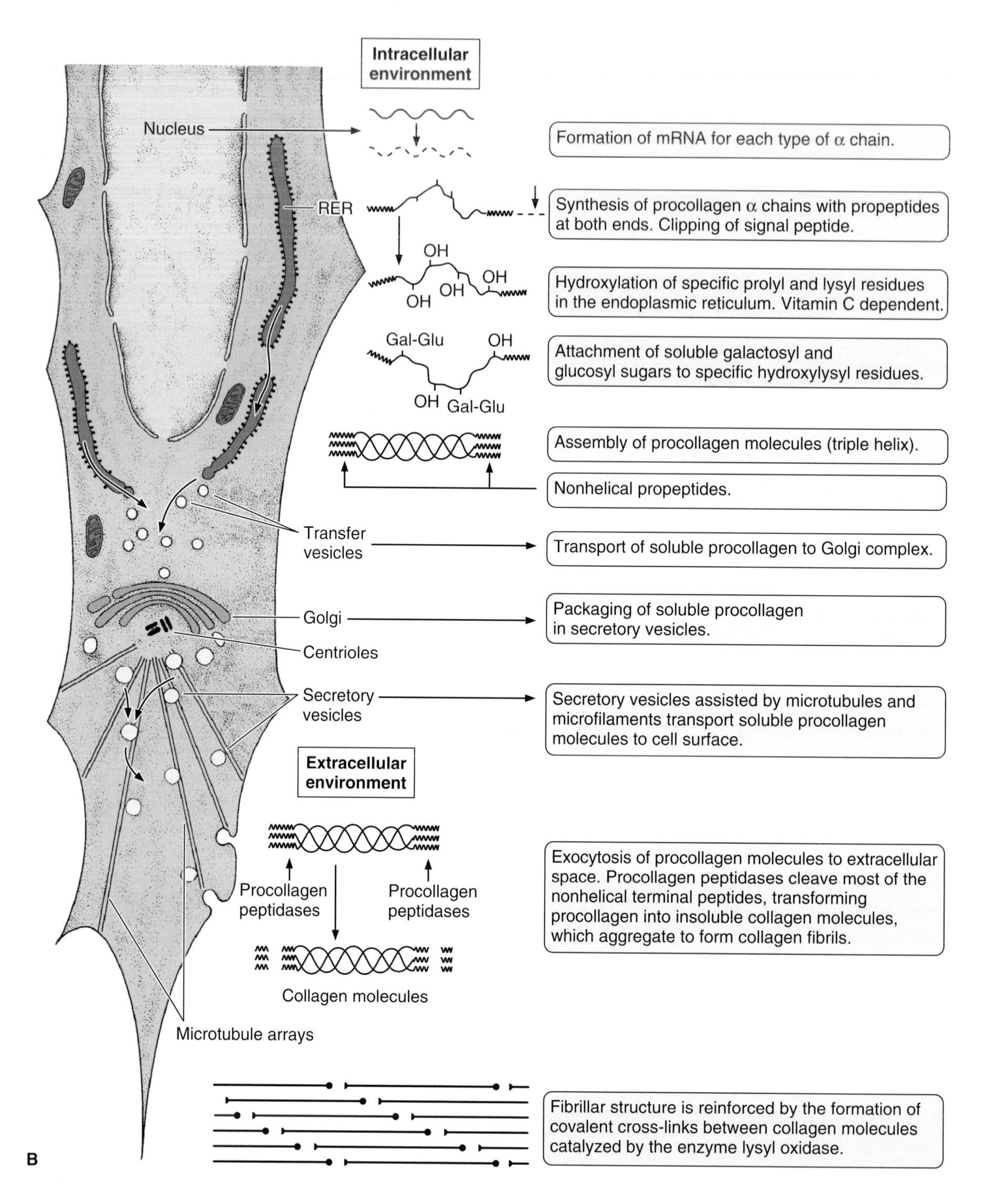

Figure 13-2. (*Continued*) **B. Formation of Collagen Fibers.** Collagen peptides are synthesized by ribosomes in the lumen of rough endoplasmic reticulum (RER) as "preprocollagen" molecules (not shown) with N-terminal signal peptides. Signal peptides are removed to produce "procollagen" molecules. Proline and lysine amino acid residues are hydroxylated via prolyl hydroxylase and lysyl hydroxylase enzymes, which depend on vitamin C as a cofactor. Some amino acid residues are also glycosylated. The procollagen molecules form a left-handed triple helix within the RER lumen. The helical molecules are transported to the Golgi apparatus, where they are processed and secreted via exocytosis to the cell exterior. The amino- and carboxy-terminal ends are removed by the enzyme procollagen peptidase to form collagen fibrils with further cross-linking of hydroxylysine and lysine residues on different tropocollagen molecules by the enzyme lysyl oxidase. Cross-linked collagen fibrils form collagen fibers. Because there are many slightly different genes for procollagen α chains and collagen production depends on several posttranslational events involving several other enzymes, many diseases involving defective collagen synthesis have been described. [Reproduced with permission from Mescher AL: Junqueira's Basic Histology Text and Atlas, 12th edition, McGraw-Hill, 2010.]

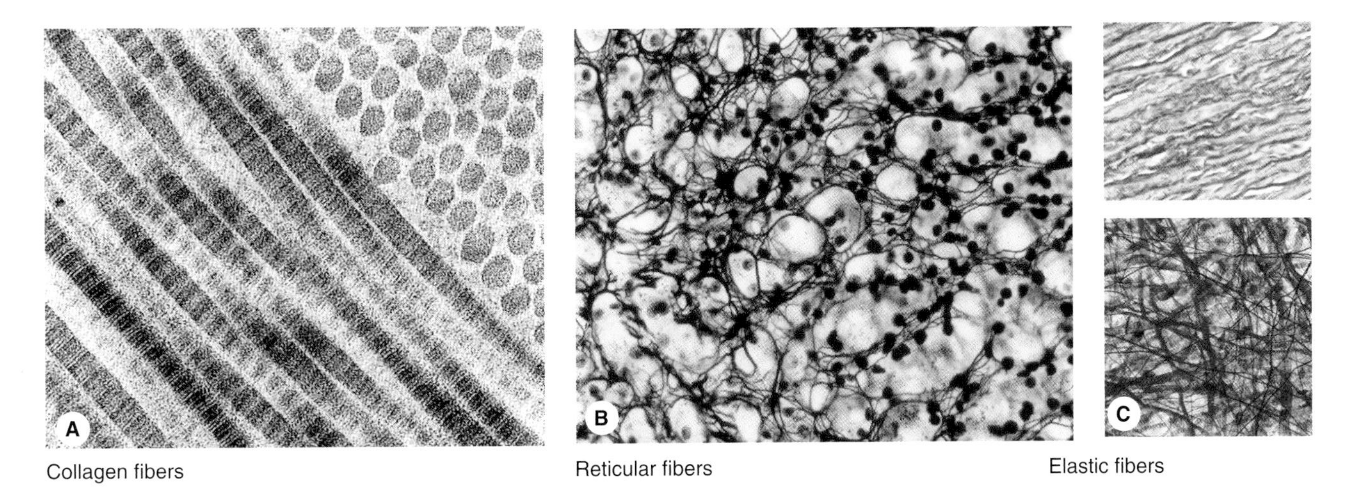

Figure 13-3. A–C. Representative Types of Collagen-Based Connective Tissue. (**A**) Molecules of type I collagen, the most abundant type, assemble to form much larger structures. Transmission electron microscopy shows that fibrils cut longitudinally and transversely. In longitudinal sections, the fibrils display alternating dark and light bands, which are further divided by cross-striations, and in cross section, the cut ends of individual collagen molecules can be seen. (**B**) Silver-stained sections of adrenal cortex illustrate a network of reticular fibers, which provides a framework for cell attachment. Reticular fibers contain type III collagen, which is heavily glycosylated. (**C**) Elastic fibers are composed of a third type of connective tissue, formed from microfibrils of elastin and fibrillin and add resiliency to the connective tissue. They can be seen between layers of smooth muscles in the wall of elastic arteries such as the aorta (pink, upper panel) and in structures such as mesentery (dark magenta, lower panel). [Reproduced with permission from Mescher AL: Junqueira's Basic Histology Text and Atlas, 12th edition, McGraw-Hill, 2010.]

3 Allysine + Lysine $\xrightarrow{\text{Condensation}}$ Desmosine

Desmosine

A

Stretch
Relax
Cross-link
Single elastin molecule

B

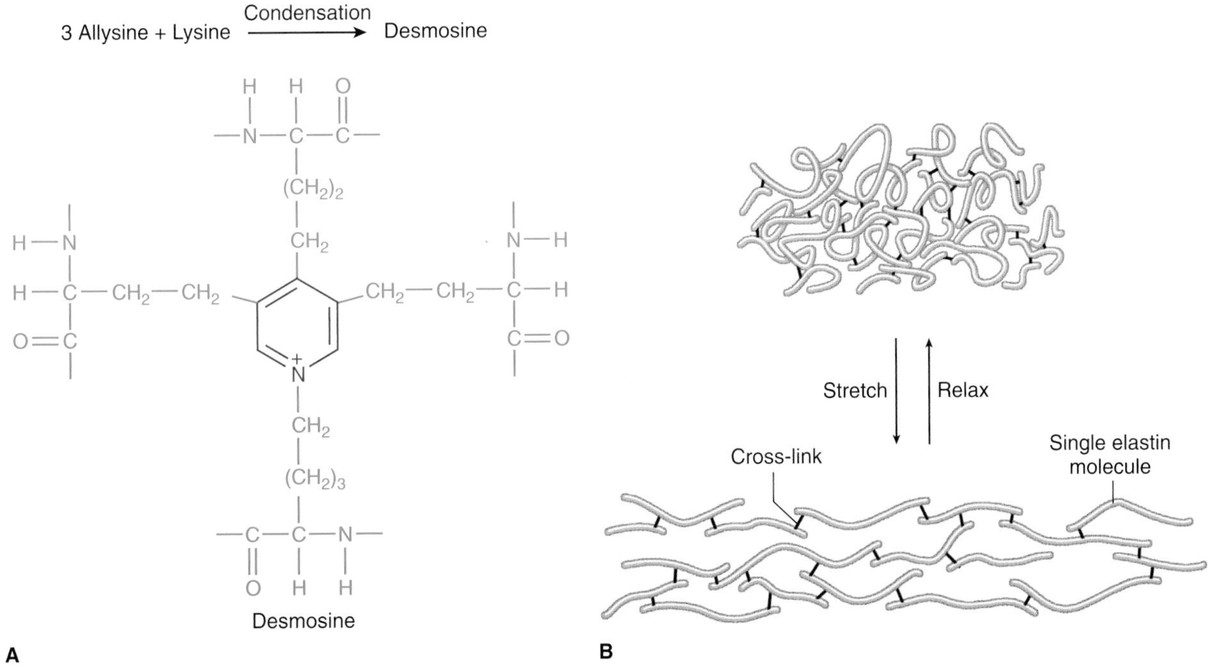

Figure 13-4. A–B. Quaternary Structure of Elastic Fibers. Elastin microfibrils may form internal and external (between different elastin fibers) cross-links to form the cross-link structure "desmosine," which provides both strength and elasticity. Illustration of function of elastin fibers provides strength and elasticity to a number of tissues (see text). 13-4A. [Reproduced with permission from Naik P: Biochemistry, 3rd edition, Jaypee Brothers Medical Publishers (P) Ltd., 2009.] 13-4B. [Reproduced with permission from Mescher AL: Junqueira's Basic Histology Text and Atlas, 12th edition, McGraw-Hill, 2010.]

TABLE 13-1. Summary of Collagen Types

Type	Location/Functions	Associated Diseases
I (two subtypes)	Found in skin, tendons, ligaments, blood vessels, muscles, various organs, bones, and teeth. Forms scar tissue during wound healing. Most abundant form of collagen.	Ehlers–Danlos syndrome (types I, II, and VII A and B), osteogenesis imperfecta (types I–IV), Caffey's disease (infantile cortical hyperostosis), atypical Marfan syndrome, possible association with osteoporosis
II (one subtype)	Hyaline cartilage (major component of this type), vitreous humor (eye)	Types II and XI collagenopathies [achondrogenesis type 2, hypochondrogenesis, Kniest dysplasia, otospondylomegaepiphyseal dysplasia, spondyloepimetaphyseal dysplasia (Strudwick type), spondyloepiphyseal dysplasia congenita, spondyloperipheral dysplasia, Stickler syndrome, Weissenbacher–Zweymüller syndrome, platyspondylic lethal skeletal dysplasia (Torrance type)]
III (one subtype)	Skin, muscle, intestines, liver, lung, bone marrow, blood vessels, lymphatic system, lens of eyes, and uterus. Also, main component of granulation tissue and reticular fibers.	Ehlers–Danlos syndrome (types III and IV), aneurysms (aortic and arterial)
IV (six subtypes)	Forms sheets versus fibers. Major component of all cell basement membranes (basal lamina)	Alport syndrome, HANAC (**h**ereditary **a**ngiopathy with **n**ephropathy, **a**neurysms, and muscle **c**ramps), familial benign hematuria, and Goodpasture's syndrome
V (three subtypes)	Surface of cells, most interstitial tissues, placenta and other fetal tissues, and hair. Often found with type I.	Ehlers–Danlos syndrome (types I–III)
VI (three subtypes)	"Fibril-associated collagens" with regularly appearing globular domains giving a "beaded filament" appearance. Often associated with type I in extracellular matrix of most interstitial tissues and microfibrils.	Bethlehem myopathy, Ullrich scleroatonic muscular dystrophy
VII (one subtype)	Component of basement membrane (basal lamina). Anchors stratified squamous epithelial to underlying supportive framework tissue (stroma). Also found in retina.	Dystrophic epidermolysis bullosa and epidermolysis bullosa acquisita
VIII (two subtypes)	Component of basement membrane (basal lamina) of the corneal endothelium.	Posterior polymorphous dystrophy, type 2
IX (three subtypes)	"Fibril-associated collagens" with regularly appearing globular domains giving a "beaded filament" appearance. Along with type II, major component of hyaline cartilage. Also, associates with type XI. Also found in vitreous humor.	Early-onset arthritis, epiphyseal dysplasia, Stickler syndrome (recessive variant)
X (one subtype)	Product of endochondral ossification (see below)	Schmid-type metaphyseal chondrodysplasia (SMCD) and Japanese-type spondylometaphyseal dysplasia (SMD)

TABLE 13-1. Summary of Collagen Types (Continued)

Type	Location/Functions	Associated Diseases
XI (two subtypes)	Muscle, joints, various organs, skin, nose, inner ear and earlobe, vitreous humor, and nucleus pulposus of vertebral discs	Inherited deafness, types II and XI collagenopathies (see a full list under type II), and Marshall syndrome
XII (one subtype)	"Fibril-associated collagens" with regularly appearing globular domains giving a "beaded filament" appearance. Often associated with type I. Found in extracellular matrix of interstitial tissues, embryonic tissue, skin, and growth plate	May play a role in the development of atherogenesis
XIII (one subtype)	Found in low levels in various connective tissues and, unlike most collagens, appears to be a plasma membrane-bound form. Function unknown, but binds with integrins, fibronectin, and basement membrane (basal lamina).	Unknown
XIV (one subtype)	"Fibril-associated collagens" with regularly appearing globular domains giving a "beaded filament" appearance. Believed to be associated with the extracellular matrix.	Unknown
XV (one subtype)	"Fibril-associated collagens" with regularly appearing globular domains giving a "beaded filament" appearance. Found in multiple tissues, but mainly adheres basement membrane (basal lamina) to underlying structures (stroma).	Defects may cause breakdown of muscle and/or small blood vessels.
XVI (one subtype)	"Fibril-associated collagens" with regularly appearing globular domains giving a "beaded filament" appearance. Often associated with types I and II in extracellular matrix of smooth muscles and amniotic sac membrane (amnion). Also found in fibroblasts and keratinocytes.	Unknown
XVII (one subtype)	Unlike most collagens, appears to be a plasma membrane-bound form, playing an important part in the structure of hemidesmosomes, adhering epidermal keratinocytes to the underlying basement membrane. Reportedly binds to keratin, α-actinin, and dystonin among others.	Dysfunction leads to condition of junctional epidermolysis bullosa with easy skin blistering, skin, mucus membrane and nail breakdown, hair loss, and adverse changes in teeth structure. May also cause bullous pemphigoid with intermittently formed blisters (bullae) of skin and mucus membranes.
XVIII (one subtype)	Unlike most collagens, contains collagen-like and non–collagen-like domains. Found in extracellular matrix. Cleavage of the protein to a 20 kDa, C-terminal fragment produces the protein **endostatin**, which is a strong inhibitor of the migration and proliferation of endothelial cells for the growth of new blood vessels. Mechanism may be by inhibition of growth factors.	Knobloch syndrome with retinal and neural tube structural abnormalities. Endostatin derivative has shown marked effectiveness in cancer treatments (e.g., endocrine, carcinoid, and non-small cell lung carcinoma).

(continued)

TABLE 13-1. Summary of Collagen Types (Continued)

Type	Location/Functions	Associated Diseases
XIX (one subtype)	"Fibril-associated collagens" with regularly appearing globular domains giving a "beaded filament" appearance. Function unknown, but known to associate with types I and II as part of extracellular matrix structure.	Unknown
XX (one subtype)	Function unknown, but believed to be involved in adhesion in the extracellular matrix.	Unknown
XXI (one subtype)	"Fibril-associated collagens" with regularly appearing globular domains giving a "beaded filament" appearance. May assist in assembly of extracellular matrix of smooth muscle cells in heart (right side > left side), blood vessels (e.g., aorta), trachea, stomach, jejunum, colon, liver, pancreas, kidney, testis, uterus and placenta, and lymph nodes.	Unknown
XXII (one subtype)	Found in skeletal muscle and heart, where it appears to play a yet unknown function in junctions between heart cells and between muscles and tendons.	Unknown
XXIII (one subtype)	Unlike most collagens, is found in the plasma membrane of lungs, cornea, brain, skin, tendons, and kidney, and contains collagen-like and non-collagen-like domains. Appears to be involved in binding of epithelial cells to basement membrane (basal lamina) possibly via type IV.	Unknown
XXIV (one subtype)	Found in developing bone and eye usually with type I. Function unknown, but may be involved in collagen expression during fetal development.	Unknown
XXV (one subtype)	Unlike most collagens, is found in the plasma membrane of brain. Function unknown, but may regulate collagen fibril elongation.	Found in senile plaques in Alzheimer disease
XXVI (one subtype)	Found in myoid and pre-theca cells in testis and ovary. Function unknown.	Unknown
XXVII (one subtype)	Found in cartilage in eye and ear and developing skeletons (e.g., growth plate). May be involved in calcification of cartilage during bone growth and development.	Unknown
XXVIII (one subtype)	"Fibril-associated collagens" with regularly appearing globular domains giving a "beaded filament" appearance. Appears to be present in dorsal root ganglia, sciatic nerve, and peripheral nerves, although it has mainly been studied in mouse models.	Unknown
XXIX (one subtype)	Found in skin, lungs, and intestines. Appears to function in cell adhesion, but also may be associated with the development of allergic responses.	Atopic dermatitis

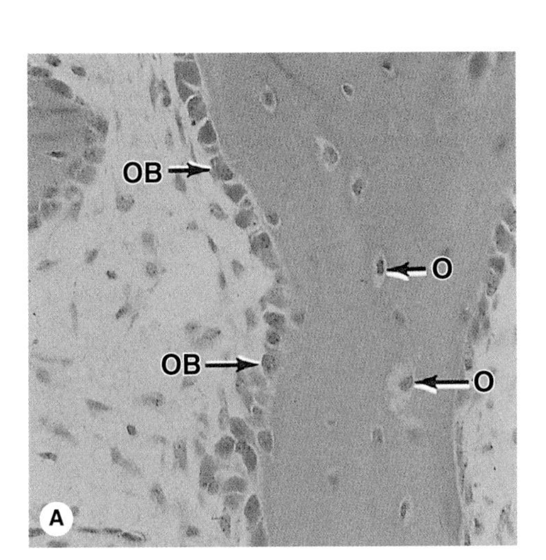

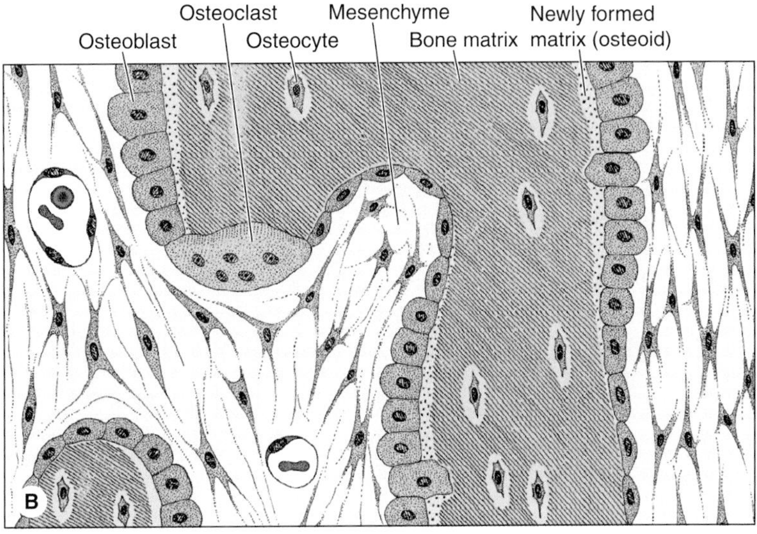

Figure 13-5. A–B. Osteoblasts and Osteocytes. (**A**) The photomicrograph of developing bone shows the location and morphological differences between osteoblasts (OB) and osteocytes (O). Rounded osteoblasts, derived from the mesenchymal cells nearby, appear as a simple row of cells adjacent to a very thin layer of lightly stained matrix covering the more heavily stained matrix. The lightly stained matrix is osteoid. Osteocytes are less rounded and are located within lacunae. In thin spicules of bone such as those seen here, canaliculi are usually not present. (**B**) Schematic diagram shows the relationship of osteoblasts with osteoid, bone matrix, and osteocytes. [Reproduced with permission from Mescher AL: Junqueira's Basic Histology Text and Atlas, 12th edition, McGraw-Hill, 2010.]

Osteoarthritis: Osteoarthritis (OA), sometimes called "degenerative" or "wear and tear" arthritis, results from the breakdown/loss of cartilage and bones in joints, resulting in pain, stiffness, and decreased movement of the affected areas. Although a number of infective, inflammatory, and miscellaneous secondary causes are known, most cases have no identifiable primary cause. However, the resulting effects are believed to be due to decreasing water content of the cartilage, causing a reduction in the **proteoglycan** content. With less proteoglycans, the cartilage is more prone to degradation that eventually leads to the gradual loss of the joint components, inflammation from these breakdown products, and, occasionally, reactive growth of bone "spurs."

Treatment is often multifactorial, including pain medications, moderate exercise, and, occasionally, surgery, although these often have less than optimal effectiveness. A variety of alternative treatments have thus been developed, including acupuncture, herbs, supplements, fish oils, and, probably most prominent, use of glucosamine and chondroitin sulphate. **Glucosamine**, simply glucose with an amino (NH_2) group at the second carbon, is a known component of glycosaminoglycans (GAGs) and proteoglycans. **Chondroitin**, a GAG with sulfate (SO_3) groups, is also a part of the proteoglycan structure of cartilage. As a result, some believe that the oral intake of these two supplements can help to restore the proteoglycan content of cartilage, leading to relief in OA symptoms. To date, research studies and analyses have not shown any objective improvement of OA, although they have also shown that no adverse effects result from their use and, more importantly, that a strong placebo effect may actually help many patients. Further studies are ongoing.

which nine, highly conserved, cysteine amino acids on each monomer form disulfide bonds with the analogous cysteine residue. The TGF-β homodimer structure then interacts with a dimer of type II receptors (Figure 13-6). The bound TGF-β dimer/type II receptor dimer then recruits two type I receptors, creating a four-protein (tetramer) receptor along with two-protein (dimer) substrate complex. Binding of these six proteins causes a conformational change of the type II receptor dimer, which leads to **serine kinase** activity and phosphorylation of serine amino acids on the type 1 receptor. These phosphorylated serine residues can then bind to a variety of specific signaling proteins (known as SMADs and SARAs, the latter of which utilize a zinc-finger binding motif; see Table 9-1). These signaling proteins combine and enter the nucleus, where they act as transcription-activating factors for a variety of gene products. BMP substrates signal in a similar manner to TGF-β, utilizing the SMAD proteins as transcription factors.

Fibrodysplasia Ossificans Progressiva (FOP) and BMPs: **FOP** is a very rare, but potentially, extremely debilitating disease, in which the repair process of fibrous tissues such as muscles, tendons, and ligaments is changed to bone repair, resulting in mineral deposition/bone matrix formation in the normally flexible tissue. Progression of the disease literally locks the patient's body parts in place, a process sometimes described by the expression "turned to stone." The disease is believed to be caused by an autosomal dominant mutation of **BMP4**. Recent research indicates that BMP4 is actually the protein **activin**, a protein dimer better known for its role in the secretion of follicle-stimulating hormone (Chapter 20). Mutated activin erroneously affects its **BMP type I receptor**, promoting the activation of SMAD proteins, gene transcription, and bone matrix growth. Lymphocytes, which respond to the initial damage to the fibrous tissue, are believed to be the carriers of the mutated protein. There is no known cure for FOP, although it is possible that shark squalamine, a protein under trial, which prevents bone growth in cartilage, may provide some treatment.

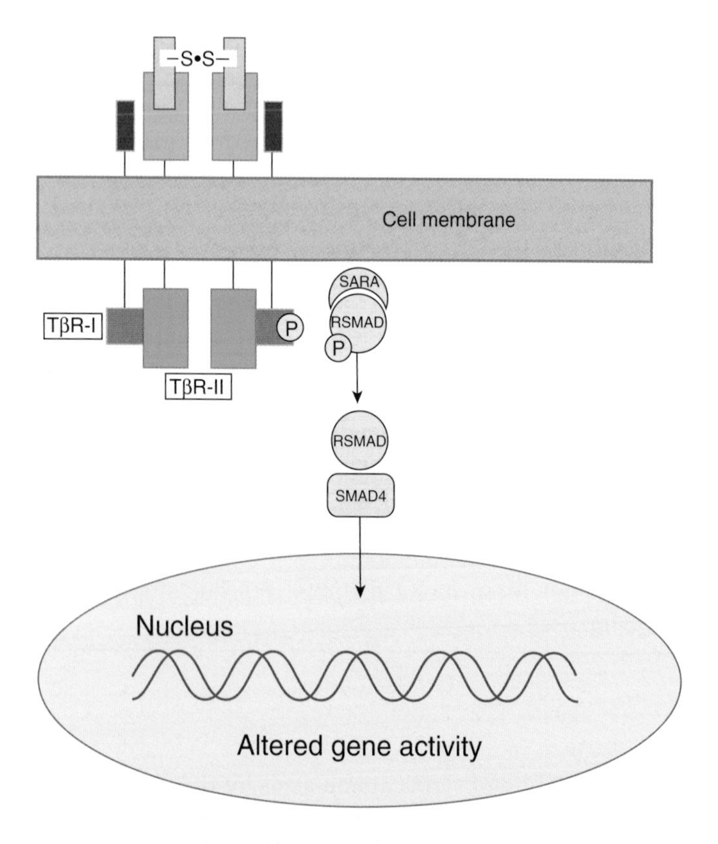

Figure 13-6. Mechanism of TGF-β Signaling. The TGF-β protein (light blue) forms a dimer via a series of eight disulfide bonds (represented by –S•S–). This dimer recruits two type I (TβR-1) and two type II (TβR-II) TGF-β receptors. Formation of tetramer receptor structure leads to serine amino acid phosphorylation on the type I cytoplasmic tail and subsequent signaling via the SARA, RSMAD, and SMAD4 proteins, leading to increased transcription in the nucleus. [Adapted with permission from Barrett KE, et al.: Ganong's Review of Medical Physiology, 23rd edition, McGraw-Hill, 2010.]

Osteoblasts produce the protein **osteoid**, made primarily of type I collagen, and are also responsible for laying down the hard mineral matrix (Figure 13-7). Osteoid is first synthesized as approximately 300-nm long, **tropocollagen** "microfibrils," which interdigitate and bind via hydrogen and covalent bonding with neighboring microfibrils to form a sheet-like collagen matrix of parallel fibrils. Gaps between tropocollagen molecules offer sites for hydroxyapatite mineralization.

In addition to osteoid, osteoblasts also produce "**ground substance**," composed primarily of the GAG **chondroitin sulphate** (Chapter 2) and several osteoblast-derived glycoproteins, including **osteocalcin** (function not completely understood, but believed to regulate bone formation), **osteonectin** (binds calcium and collagen and initiates mineralization process), **osteopontin**, also known as **bone sialoprotein** [production increased by calcitriol (Vitamin D_3), and thought to help anchor osteoclasts to the bone matrix for bone resorption, but may also stimulate formation of first hydroxyapatite crystals for new mineralization]. Newly formed osteoblasts also secrete the cytokine (Chapter 15) macrophage colony-stimulating factor (M-CSF), another disulfide homodimer molecule, which binds to its receptor, leading to tyrosine phosphorylation. This activity promotes the growth and differentiation of osteoclasts (see below) to help maintain the balance between bone formation and resorption, as well as calcium levels (see below).

Following laying down of the osteoid and ground substance matrix, osteoblasts secrete vesicles with the enzyme **alkaline phosphatase**, which removes phosphate groups from hydroxyapatite. The modified hydroxyapatite and remnant vesicles then act as centers for deposit of calcium and phosphate crystals and, as a result, newly mineralized bones.

Collagen Synthesis and Scurvy: The proper production of collagen is essential for osteoid/bone formation, as well as other connective tissues discussed below. Part of collagen's unique structure is the presence of hydroxylated proline and lysine residues, which stabilize the triple-helical structure essential for collagen's structure and function. **Vitamin C (ascorbic acid)** is a required cofactor for these hydroxylation reactions, and the deficiency of this vitamin causes the disease **scurvy**. Symptoms of scurvy include abnormal bleeding (from affected capillaries), skin discoloration, nonhealing wounds and gum deterioration/loss of teeth (from connective tissue), and weakening of bones (from adverse effects on bone formation/remodeling). Untreated scurvy can be fatal. The treatment for scurvy is simple supplementation of vitamin C in the diet. In Latin, "ascorbic" literally means "no scurvy."

Osteoclasts are modified forms of monocyte/macrophage cells that break down bones and are regulated in differentiation and growth by the binding of **receptor activator of nuclear factor kappa B (RANK)** ligand and by the activity of the cytokine M-CSF (noted above). Binding of the RANK ligand to its receptor [a member of the tumor necrosis factor (TNF) receptor

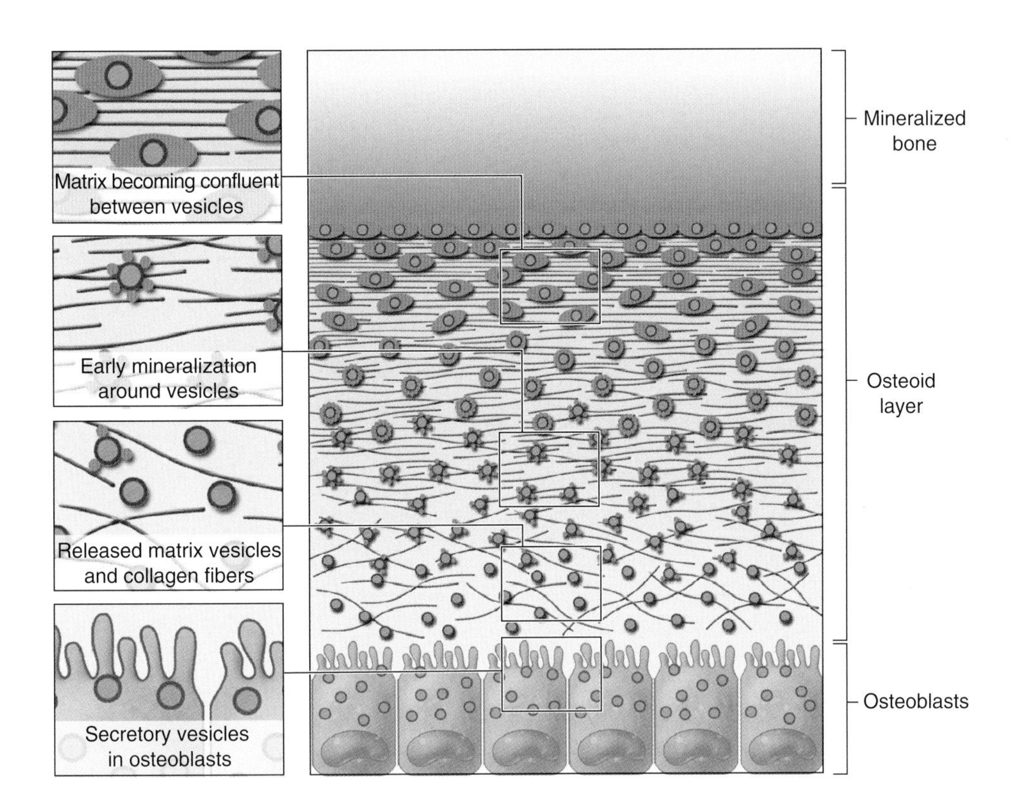

Figure 13-7. Mineralization in Bone Matrix. From their ends adjacent to the matrix, osteoblasts secrete type I collagen, several glycoproteins, and proteoglycans. Some of these factors, notably osteocalcin and certain glycoproteins, bind Ca^{2+} with high affinity, thus raising the local concentration of these ions. Osteoblasts also release very small membrane-enclosed matrix vesicles with which alkaline phosphatase and other enzymes are associated. These enzymes hydrolyze PO_4^{2-} ions from various macromolecules, creating a high concentration of these ions locally. The high ion concentrations cause crystals of $CaPO_4$ to form on the matrix vesicles. The crystals grow and mineralize further with the formation of small growing masses of hydroxyapatite $[Ca_{10}(PO_4)_6(OH)_2]$, which surround the collagen fibers and all other macromolecules. Eventually, the masses of hydroxyapatite merge as a confluent solid bony matrix as calcification of the matrix is completed. [Reproduced with permission from Mescher AL: Junqueira's Basic Histology Text and Atlas, 12th edition, McGraw-Hill, 2010.]

family] activates nuclear factor kappa B (NF-κB), an important transcriptional activator protein. M-CSF binds to a tyrosine kinase receptor, leading to a complicated and not yet fully understood signaling pathway that leads to osteoclast differentiation. These osteoclast effector molecules are produced by osteoblasts and surrounding cells, and both RANK and M-CSF are required for osteoclast production. Differentiation of osteoclasts is inhibited by the molecule osteoprotegerin (**OPG**) (see below), which inhibits RANK ligand binding to its receptor.

Osteoclasts perform the opposite function to osteoblasts, namely resorption and/or remodeling of the mineral matrix. Bone remodeling (see below) allows the body to change existing bone structure during growth and to repair microfractures that result from normal mechanical stress, as well as from pathological fractures due to injury. Remodeling also provides a source of calcium as part of the regulation of calcium levels in the body (see below).

Osteoclast resorption/bone remodeling is accomplished by several mechanisms. First, osteoclasts are able to attach to bone via a unique **podosome** structure, in which intracellular actin filaments attach to membrane **integrin receptor** molecules (Figure 13-8), which subsequently bind to a specific arginine–glycine–aspartate amino acid sequence on the osteopontin protein (discussed earlier). Once bound, an osteoclast-derived **carbonic anhydrase** produces hydrogen ions via the reaction $H_2O + CO_2 \rightarrow HCO_3^- + H^+$ and pumps them out into the bone matrix by a specialized ATPase pump. Osteoclasts also produce an Fe^{2+}-containing glycoprotein, **tartrate resistant acid phosphatase (TRAP)**, present in abundance in osteoclast lysosomes. TRAP's Fe^{2+} binds with a phosphate in hydroxyapatite and cleaves the phosphate ester bond by nucleophilic attack of a hydroxyl group. Free phosphate and calcium ions that are released are initially taken up by vesicles, but then subsequently released into the blood stream by exocytosis. TRAP is also believed to reduce osteopontin/bone sialoprotein activity (see above) via dephosphorylation and also influences growth, differentiation, migration, and activity of osteoblasts, as well as the migration of osteoclasts. TRAP also generates reactive oxygen species, which help to resorb bone. Several **cathepsin** enzymes, especially the **cathepsin K enzyme**, are also produced by osteoclasts, which degrade the collagen matrix via selective cleavage of collagen and other proteins. These cathepsins have no apparent specificity for particular peptide bonds, although some preference for amino acids with large hydrophobic side chains has been noted. Other enzymes such as **aspartate protease** and **matrix metalloprotease** also aid in the breakdown of the inorganic and organic matrices. Osteoclast activity is regulated by several hormones, which are discussed below.

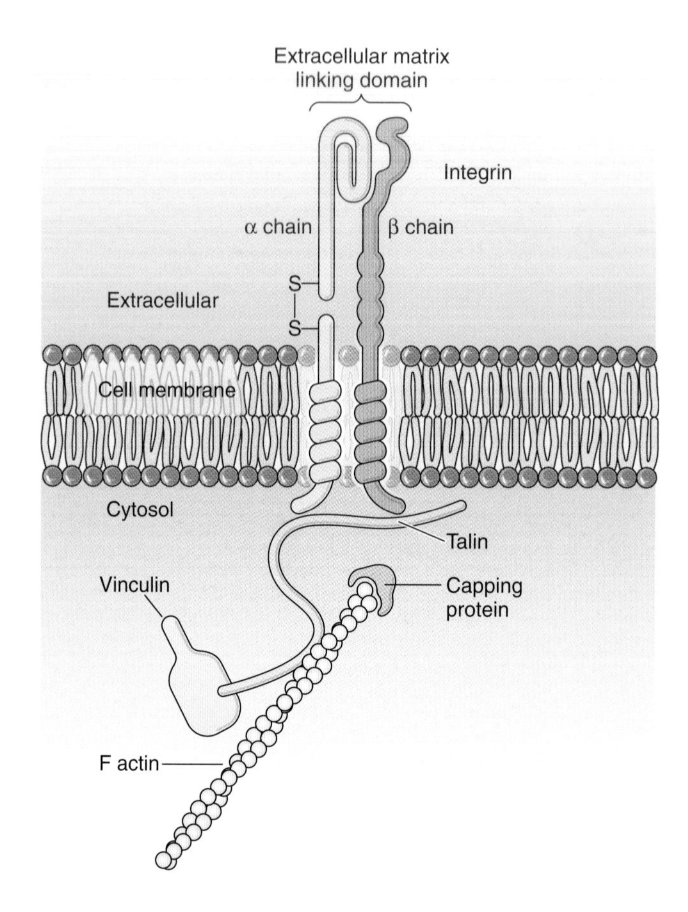

Figure 13-8. Integrin Cell Surface Matrix Receptor. By binding to a matrix protein and to the actin cytoskeleton (via talin) inside the cell, integrins serve as transmembrane links by which cells adhere to components of the extracellular matrix (ECM). The molecule is a heterodimer, with α and β chains. The head portion may protrude some 20 nm from the surface of the cell membrane into the ECM, where it interacts with fibronectin, laminin, or collagens. [Reproduced with permission from Mescher AL: Junqueira's Basic Histology Text and Atlas, 12th edition, McGraw-Hill, 2010.]

Osteoid Mineralization and Osteomalacia/Rickets: The process of mineralization of osteoid by osteoblasts is critical for the proper development of bones. When osteoblasts are unable to form hydroxyapatite or when sufficient calcium and/or phosphate are not available, it results in **osteomalacia**. In children, this condition is known as **rickets**. Osteomalacia, literally meaning "bone softness," exhibits normal amounts of organic collagen matrix, but deficient mineralization. This is different from osteoporosis (see below) in which the bone matrix is normally mineralized but decreased. As a result, patients suffering from osteomalacia have weak and easily fractured bones. Most often, osteomalacia/rickets is caused by deficiency of vitamin D either in the diet (i.e., poor intake or poor intestinal absorption) or secondary to low sun exposure/absorption. Other causes may include kidney or liver disease (or other disorders that affect vitamin D metabolism and/or absorption), decreased phosphate levels, cancers, and medication side effects (e.g., anticonvulsant medications). Symptoms and signs of osteomalacia include bone pain (often starting in the lumbar region of the spine, pelvis, and legs) and related muscle and nerve weakness/numbness. Laboratory tests show low calcium levels in serum and urine (often accompanied by low serum phosphate), high alkaline phosphatase (see above) and parathyroid hormone (PTH) levels (see below), and radiological evidence of pseudofractures and/or bone loss. Treatment involves correction of the problem and replacement of calcium and vitamin D; full recovery often takes 6 months or longer.

Osteocytes, the third and most abundant type of cell contained in the organic bone matrix, are actually altered forms of osteoblasts. Osteoblasts change into osteocytes when they become trapped inside the growing bone matrix. As part of this change, osteocytes link together via specialized "**canalicular**" extensions containing gap junctions (Chapter 8), which allow the interchange of nutrients and waste products. Like osteoblasts, osteocytes are involved in bone formation, although their exact role and function(s) are still unknown. Emerging research indicates a unique ability of osteocytes to "sense" strain and flow of fluid through the canaliculi channels. The mechanical and fluid-derived strain appears to activate osteocytes to regulate bone remodeling and growth via osteoclast-directed signaling to and between osteoblasts and osteocytes.

Osteoporosis/Mechanisms of Bisphosphonate Action: **Osteoporosis** is a bone condition in which the amount of bone mineral is significantly lowered, leading to an altered and weakened bone matrix and a markedly increased risk of fracture. These fractures are often seen in vertebrae, leading to the stooped-over posture of **lordosis** and hip fractures.

The cause of osteoporosis can be multifactorial but can be generalized as a condition in which bone resorption outweighs bone formation. Decreased estrogen stimulation resulting in reduced bone formation as well as deficiency of calcium and vitamin D are often the principal causes, especially in postmenopausal women of European or Asian descent. Heavy use of alcohol, smoking, and some medical conditions (e.g., endocrine, hypogonadal, or certain blood disorders) are also known risk factors for the development of osteoporosis. Chronic use of medications including glucocorticoids (e.g., adrenal insufficiency, severe asthmatics, transplant recipients), levothyroxine, lithium, certain barbiturates and anti-seizure drugs, heparin and warfarin, the newer diabetes medication class of thiazolidinediones, and proton pump inhibitors can also contribute to osteoporosis by decreasing calcium and/or vitamin D metabolism and/or osteoblast activity.

Although increased calcium and vitamin D intake along with exercise and estrogen replacement can help promote bone formation, the development of **bisphosphonates** has allowed the directed treatment of this condition (see below). Bisphosphonates are taken up by and kill osteoclasts by either replacing the terminal phosphate in adenosine triphosphate (ATP), rendering the molecule inoperable (nonnitrogenous bisphosphonates), or inhibiting 3-hydroxy-3-methylglutaryl-coenzyme A reductase enzyme farnesyl diphosphate synthesis (nitrogenous bisphosphonates) (Chapter 7). Although required for cholesterol biosynthesis, this enzyme is also essential for modification and correct trafficking of some membrane proteins and for particular cytoskeletal functions of osteoclasts that enable them to resorb bones. Alternative medications include **calcitonin** (inhibition of osteoclasts), **PTH** (stimulation of osteoclasts), and **OPG** (RANK receptor inhibition) analogues. Emerging treatments, including **sodium fluoride** and **strontium**, may offer additional treatment modalities.

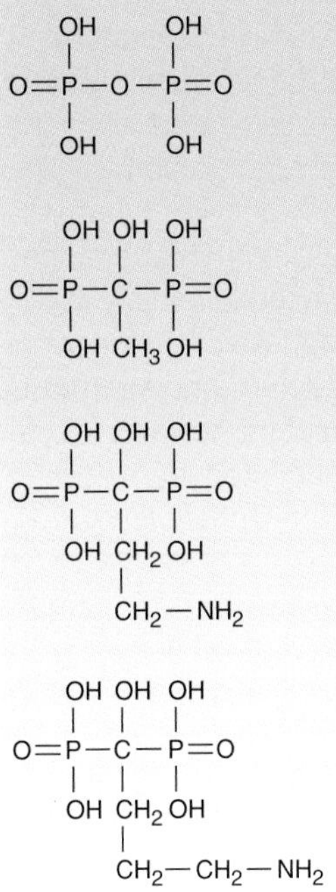

```
   OH    OH
   |     |
O=P--O--P=O
   |     |
   OH    OH
```
Inorganic pyrophosphoric acid

```
   OH OH  OH
   |  |   |
O=P--C--P=O
   |  |   |
   OH CH3 OH
```
Etidronate: ethane-1-hydroxy-1, 1-bisphosphonate

```
   OH OH  OH
   |  |   |
O=P--C--P=O
   |  |   |
   OH CH2 OH
      |
      CH2—NH2
```
Pamidronate: 3-Amino-1-hydroxy-propylidene bisphosphonate

```
   OH OH  OH
   |  |   |
O=P--C--P=O
   |  |   |
   OH CH2 OH
      |
      CH2—CH2—NH2
```
Alendronate: 4-Amino-1-hydroxy-butylidene bisphosphonate

Reproduced with permission from Katzung BG, et al.: Basic and Clinical Pharmacology, 11th edition, McGraw-Hill, 2009.

BONE GROWTH AND REMODELING

The growth of bone and its constant remodeling in response to normal growth, microfractures (due to normal or abnormal mechanical stresses), and fractures due to trauma is highly regulated and balanced between new bone formation via osteoblasts and bone resorption via osteoclasts. The amount of "normal" remodeling decreases markedly from complete bone turnover in infancy to only approximately 10% of bone remodeling in adults in any given year.

The balance of osteoblast and osteoclast activity is dependent on several factors. Apart from FGF, PDGF, TGF-β, and BMPs noted above, osteoblast function is stimulated by the pituitary-derived, polypeptide growth hormone, which serves both to augment bone matrix formation and to increase the retention of calcium in the body. Androgens and estrogens also promote osteoblast activity/bone growth. Estrogens function via promoting the increased secretion of the cytokine (Chapter 15) glycoprotein **OPG**, which competitively binds to the RANK ligand TNF family receptor (discussed earlier). By blocking RANK ligand activity, OPG/osteoclastogenesis inhibitory factor effectively stops the development of osteoclasts from their monocyte/macrophage precursor cells. PTH and vitamin D also affect bone growth by increasing calcium levels (see below).

Osteoblasts can also stimulate the activity of osteoclasts. Increased PTH and vitamin D from low serum calcium levels along with signals from osteocytes increase the production of RANK ligand and cytokine **interleukin 6 (IL-6)**. Although IL-6 is better known for its role in fever and inflammation, when secreted by osteoblasts, it also serves, together with RANK receptor activation, to promote the development and growth of osteoclasts. IL-6's receptor signals via the Janus kinase–signal transducer and activator of transcription pathway, leading to phosphorylation of signal molecules and gene transcription (Chapter 8 and Figure 8-9C). Although estrogen promotes osteoblast growth, it can also block IL-6 activity, thereby inhibiting osteoclast growth. As noted above, M-CSF is also involved in osteoclast activation by promoting the development and growth of osteoclasts from the monocytes/macrophage precursor cells. M-CSF also decreases the secretion of the inhibitory hormone, osteoprotegerin (see above).

Paget's Disease: **Paget's disease** of the bone, also known as **osteitis deformans**, is a condition of excessive bone turnover (breakdown and reformation), resulting in bone deformities, pain, decreased strength, and arthritis and fractures. A genetic linkage has been suggested as the possible role of paramyxovirus, although no convincing evidence has been found. An emerging belief suggests involvement of an abnormal response to **vitamin D** and an increased response of the **RANK ligand**, possibly to the actions of **IL-6**. The disease starts with increased, localized resorption by osteoclasts in long bones and the skull. A resulting increase of osteoblast activity results in a quickly laid-down and disorganized bone matrix, fibrous connective tissue, and new blood vessels rather than the organized and sturdy structure of regular bones. The increased osteoclast and osteoblast activity eventually diminishes, leaving permanent and detrimental changes in the bone. Although no longer used clinically, skull X-rays show these changes as diagnostic "cotton wool spots." Treatment is normally via calcium and vitamin D supplementation along with bisphosphonates and, occasionally, calcitonin. Patients are monitored by following alkaline phosphatase, calcium, and phosphate levels.

Bone growth in a fetus and a child varies slightly from an adult and occurs via two slightly different processes. **Intramembranous ossification** occurs in the skull and requires the development of localized areas of bone growth, known as ossification centers. Initially, newly differentiated osteoblasts aggregate and secrete bone matrix around themselves to form **bone spicules**. As more bone matrix is secreted, the spicules join together to form an open bone network called **trabeculae**. The trabeculae join together, centers of bone marrow and a periosteum develops, and a more compact bone is laid down. Intramembranous ossification is also essential during the repair of bone fractures (see below). The alternative method of early bone formation is **endochondral ossification**, responsible for long bone formation, as well as some flat and irregular bones. Endochondral ossification is also responsible for continued growth during childhood and parts of fracture healing. This mode of bone formation starts with the formation of a cartilage matrix, followed by the development of ossification centers where the mineral matrix is laid down. A primary ossification center occurs in the center of the newly forming bone to form the **diaphysis**; a secondary ossification center is established near the ends of the growing bone as part of the **epiphyseal plate**, which is completely replaced by inorganic bone matrix near the end of the second decade of life (**epiphyseal closure**).

Bone Fractures and Healing: A **fracture** of a bone initiates a series of three main phases of healing-**reactive**, **reparative,** and **remodeling**, which are responsible for both quick stabilization of the fracture and a slower restoration of the damaged bone to its near-original state. Immediately following a fracture, the **reactive phase** begins, in which blood vessels initially constrict and a clot forms, which, along with fibroblasts, form an initial matrix of fibrous connective tissue with a blood vessel supply. A few days after the fracture, the **reparative phase** begins, with development of chondroblasts and osteoblasts from fibroblast cells within the periosteum. These cells produce an initial **woven bone "callus,"** composed of quickly and haphazardly organized hyaline cartilage (type II collagen and chondroitin sulphate), which unites the broken bone fragments and provides some support for the fracture. This hyaline cartilage matrix is replaced by a regular and parallel array of type I collagen **lamellar bone** matrix via endochondral ossification. The growth of a more permanent blood vessel supply modifies the lamellar bone to **trabecular bone**, with almost all of the original strength of the bone restored. The final phase, **remodeling**, can take from several weeks to months to complete, with selected resorption by osteoclasts and redeposition by osteoblasts until a final **compact bone** matrix is produced, closely mimicking the original bone.

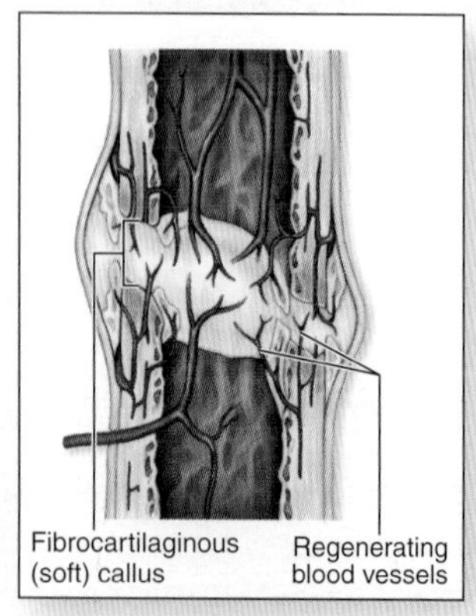

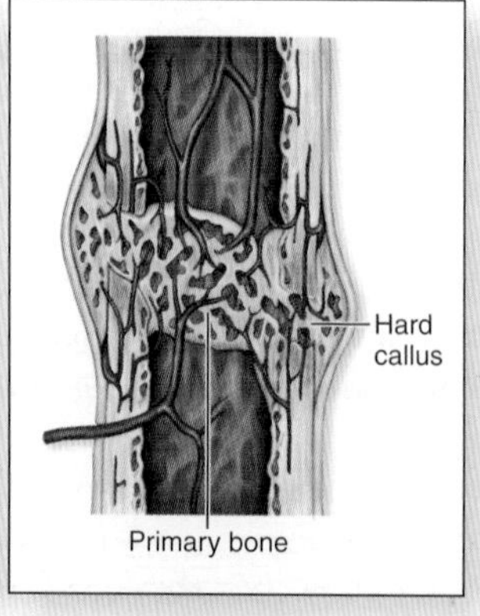

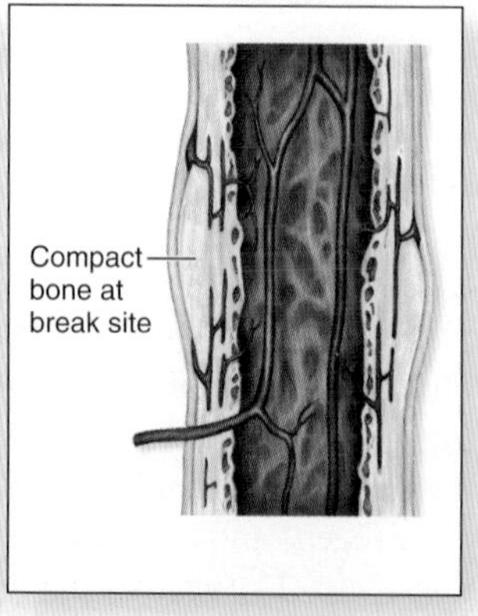

Adapted with permission from Mescher AL: Junqueira's Basic Histology Text and Atlas, 12th edition, McGraw-Hill, 2010.

REGULATION OF CALCIUM LEVELS

Because bone represents a large storage depot of readily available calcium, the regulation of calcium levels in the body influence the balance of bone formation and breakdown. Thus, a multiorgan, regulated process involving the intestines, kidneys, and bones works in synchrony to absorb new calcium, store new and old calcium, and, when necessary, free stored calcium for the body's use. The normal diet allows an addition of approximately 5 mmol of calcium per day. The kidneys excrete a net of 5 mmol of calcium, thereby negating any dietary gains. Bone, which contains approximately 99% of the body's total calcium, therefore, serves as both a storage depot and a source for calcium. Intestinal calcium absorption is strongly influenced by vitamin D, which has the important role of increasing the number of calcium-binding proteins, known as **calbindin**, on the cells of the small intestine as well as the kidney. Increased amounts of calbindin present in the intestine directly increase the amount of calcium absorbed and may also stimulate an ATP-dependent calcium pump, which transports calcium into the blood stream. Kidney reabsorption of calcium is also increased by vitamin D; the kidney is also responsible for converting 25(OH) vitamin D_3 into the active form of calcitriol, 1,25$(OH)_2$ vitamin D_3 (Chapter 3). In addition, **PTH** and **calcitonin** play prominent roles in the regulation of intestinal absorption, kidney excretion/reabsorption, and bone turnover of calcium. PTH is secreted by the parathyroid glands in response to low serum calcium levels (sensed by special receptors on the parathyroid cells) and/or increased serum phosphate, which binds to and decreases the level of free calcium ions. Increased calcium ions lead to a reduction in PTH secretion. PTH is responsible for increasing Ca^{2+} concentration via the action of a G_q protein, which stimulates phospholipase C to produce inositol 1,4,5-trisphosphate (IP_3) and, therefore, increase calcium release (Chapter 8). Additionally PTH acts via G_s to increase protein kinase A activity.

Calcitonin is a polypeptide hormone produced by the parafollicular C-cells of the thyroid gland and is responsible for decreasing calcium levels. Calcitonin levels are increased when calcium levels are high, as well as by gastrin secretion (Chapter 11). Although the exact mechanism of calcitonin function is still being determined, it is known to act not only via a G_q-protein receptor mechanism but also via a G_s-protein receptor (adenyl cyclase/cyclic adenosine monophosphate) mechanism. Interaction with these G proteins is known to inhibit osteoclast resorption by quiescence (Q-effect) and retraction (R-effect) of the osteocyte membrane. Calcitonin receptors are mainly found on osteoclasts and also on ovaries and testes. The actual importance of calcitonin in the regulation of calcium in humans is still in question. The effects of PTH and calcitonin are illustrated in Table 13-2.

PTH-Related Peptide and Cancer: A common consequence of some cancers can be increased levels of calcium (hypercalcemia) with secretion of **PTH-related peptide (PTH-rP)**. PTH-rP has significant homology with PTH, but only in the amino-terminal 13 amino acids, which make up the receptor-binding region. Because PTH-rP is not secreted by the parathyroid glands, hypercalcemia produces no feedback regulation of its production. The rise in calcium levels due to PTH-rP is termed **humoral hypercalcemia of malignancy (HHM)**. Cancers usually associated with HHM include breast and lung cancers and multiple myeloma.

TABLE 13-2. Summary of PTH and Calcitonin Effects on Target Organs

Organ	PTH Effect	Calcitonin Effect
Intestine	PTH increases levels of 1,25$(OH)_2$ vitamin D_3 (calcitriol), which increases the levels of the intestinal protein calbindin. As a result, increased amounts of dietary calcium are absorbed.	Inhibits intestinal absorption of calcium. Mechanism poorly understood, but believed to involve inhibition of vitamin D effects.
Kidney	PTH directly increases reabsorption of calcium in the distal tubules and thick ascending limb, while also reducing reabsorption of phosphate in the proximal tubule, thereby increasing the excretion of phosphate and, thus, the amount of available free calcium. PTH also increases the activity of the enzyme 25(OH) vitamin D_3 1-α-hydroxylase in the kidney, leading to higher levels of active 1,25$(OH)_2$ vitamin D_3 (calcitriol).	Increases calcium reabsorption in the loop of Henle. Decreases phosphate reabsorption, primarily in the proximal convoluted tubule; decreased phosphate increases free calcium levels.
Bone	PTH binds to osteoblast and increases RANK receptor numbers. Increased RANK stimulation leads to growth of osteoclasts, leading to bone resorption and the release of calcium.	Inhibits the activity of osteoclasts and, therefore, breakdown/reabsorption of bones. Exact mechanism not completely understood.

PTH, parathyroid hormone.

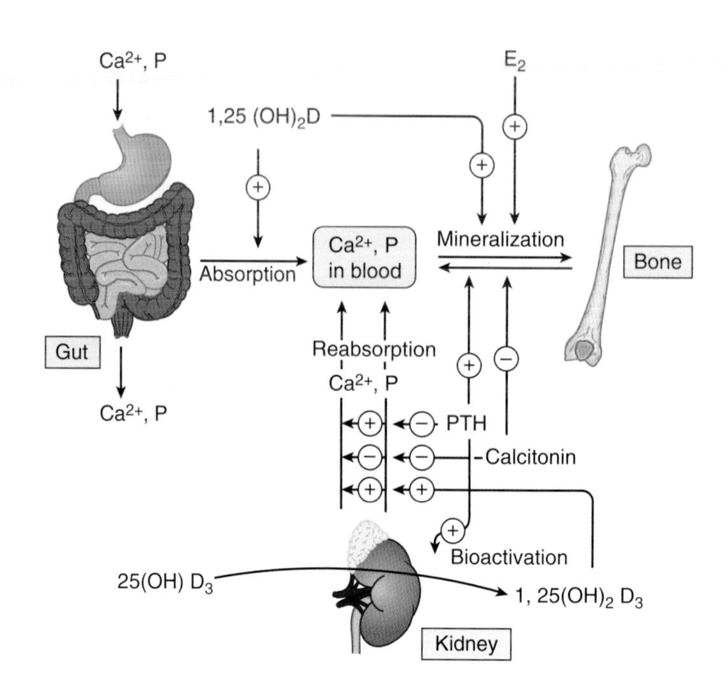

Figure 13-9. Control of Blood Calcium and Phosphate Levels. Overview of the regulation of calcium (Ca^{2+}) and phosphate (P) levels by the coordinated actions of active vitamin D [1,25$(OH)_2D_3$], parathyroid hormone (PTH), calcitonin, and estradiol (E_2). See the text above and Table 13-2 for further details. [Adapted with permission from Katzung BG, et al.: Basic and Clinical Pharmacology, 11th edition, McGraw-Hill, 2009.]

A summary of calcium regulation in the body is shown in Figure 13-9.

MARKERS OF BONE FORMATION AND RESORPTION

On the basis of various mechanisms involved in the formation of bone matrix, a number of chemical markers have been characterized, which allow clinicians to measure and monitor bone formation and breakdown. Calcium can be measured as either **total calcium** or **ionized calcium**, with calcium bound to serum proteins (e.g., serum albumin) affecting the measurement of calcium freely available to the body. Abnormal levels of albumin (hypoalbuminemia or hyperalbuminemia) will alter the amount of bound calcium, and therefore, albumin levels are often tested as well, so a **corrected calcium** level can be calculated. Vitamin D levels, specifically the active 1,25$(OH)_2$ D_3 (calcitriol) form, will also impact on calcium metabolism and may provide insight into a number of related disease states. Phosphate levels are also important in relation to the binding and excretion of calcium and are, thus, often measured. The important role of PTH in calcium, phosphate, and vitamin D regulation and particular diseases may also prompt a clinician to order measurement of its level. Calcitonin's importance in calcium regulation usually precludes its measurement, except as a tumor marker for medullary thyroid adenocarcinoma. Finally, alkaline phosphatase, the enzyme that promotes mineralization of newly forming bone, is sometimes measured in patients suffering from bone disorders. As alkaline phosphatase has three different isoforms [intestinal, placental, and nonspecific (e.g., liver/bone/kidney)] and is found in all tissues in the human body, particular isoenzymes need to be isolated and their separated level(s) carefully considered. Alkaline phosphatase is often measured in conditions such as Paget's disease of bone, osteosarcoma, cancers that have metastasized to bone, bone fractures, osteomalacia (rickets), achondroplasia, congenital hypothyroidism (previously known as cretinism), renal osteodystrophy, hypophosphatasia, osteoporosis/estrogen use, and/or vitamin D deficiency. However, alkaline phosphatase levels are more often used to measure blockage of liver bile ducts than for general bone and calcium studies.

REVIEW QUESTIONS

1. How would you describe the three main types of connective tissue and how do they differ?
2. What are the functions of chondrocytes?
3. What is the basic pathway for the formation of collagen?
4. What are the major components of bone (inorganic and organic)?
5. How would you describe the three types of cells in the organic matrix, including their function(s) and regulation?
6. How would you describe the processes of bone formation and resorption, including their regulation?
7. How would you describe the mechanisms regulating calcium levels and bone formation, including the roles of calcium, vitamin D, parathyroid hormone, and calcitonin?
8. What are the major markers of bone formation/resorption?

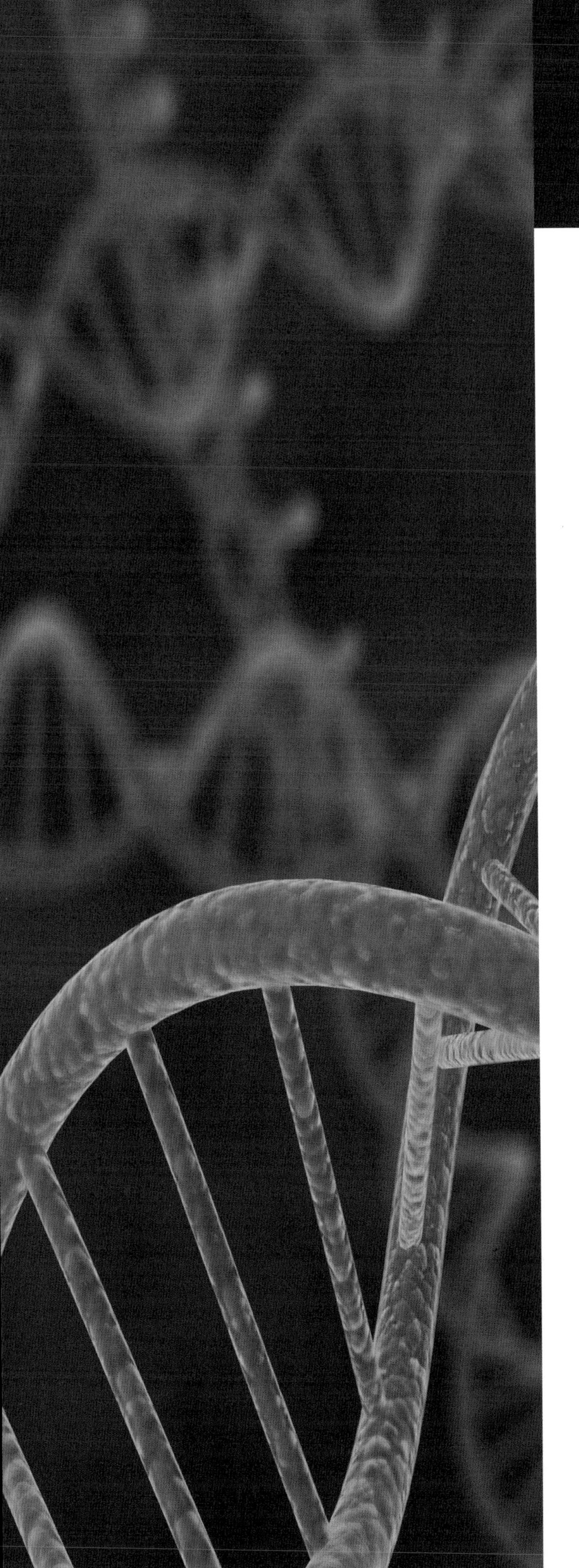

CHAPTER 14

BLOOD

Co-Authors/Editors: Matthew Porteus, MD, PhD
Divisions of Cancer Biology, Hematology/Oncology, and Human Gene Therapy, Stanford University, Stanford CA

Tina Mantanona
University of Texas Southwestern Medical Center, Dallas, TX

OVERVIEW

Blood, including red cells, white cells, platelets, and a collection of specialized proteins, serves an essential physiologic function as it carries molecules from one part of the body to another. This chapter discusses the biochemical underpinnings of how blood transports oxygen from the lungs and iron to the tissues that need them without causing damage to the tissues that do not. This transport system also insures adequate waste removal from the human body. In addition, the biochemical properties of clotting, the system that maintains the integrity of the blood vasculature, are discussed. In describing the normal way in which blood carries out these functions, the discussion will illustrate, from a biochemical perspective, the ways diseases may result when these processes go awry.

These three important roles, however, are just a small fraction of the critical functions that blood carries out to maintain the proper performance of the human body. The core biochemical principles of these three systems—the role of allostery in regulating ligand binding in the oxygen–hemoglobin system, the specialized functions of proteins for iron transport and storage, and the importance of biochemical cascades as illustrated by the clotting system—apply broadly to other functions of the blood and other organ systems in general. Thus, by understanding these fundamental biochemical mechanisms, a general foundation for understanding the normal physiology and pathophysiology of other systems can be established.

BASIC COMPONENTS OF BLOOD

Hematology, the study of blood, has been of central importance in medicine throughout history. For much of history, health and illness were considered a reflection of different "humors" in the body: black bile, yellow bile, phlegm, and blood. Good health was attained when all the humors were in harmonic balance and illness was considered as the result of an imbalance in the humors. Thus, the goal of healers was to restore the balance of the humors in those who were ill. To manipulate blood, healers employed methods such as applying leeches, cupping, and bloodletting in hope of achieving health through creating proper balance of humors.

A human adult contains approximately 5 liters (l) of blood, with the entire volume circulating through the body every 1–2 min. Centrifugation separates blood into a cellular layer, which collects at the bottom end of the collection tube, and a noncellular layer, located at the top of the tube (Figure 14-1). If whole blood clots before centrifugation, the remaining noncellular component is called **serum**. On the other hand, if anticlotting additives are added to the test tube, the noncellular layer is called **plasma**. The cellular and noncellular layer each account for approximately 50% of the total blood volume.

The cellular compartment consists of red blood cells (RBCs), the most abundant cell in the body, white blood cells, and platelets. The generation of new RBCs (**erythrocytes**) from bone marrow is termed **erythropoiesis**, and it relies on the kidney-secreted hormone **erythropoietin (Epo)** (see below). White blood cells, which primarily serve to fight infections, can be further divided into specific subtypes and are discussed in Chapter 15. Platelets are small cells that primarily participate in clotting. The clotting system consists of both cellular and noncellular components and will be discussed below. Similar to the cellular layer, the noncellular layer can be divided into various components, each of which carries out essential functions for maintaining health. For instance, within the noncellular component are carrier proteins, which transport other proteins or small molecules. Important examples include **hemoglobin (Hgb)**, which carries oxygen (O_2) and carbon dioxide (CO_2), and transferrin, which transports iron and regulates iron metabolism. Other carrier proteins transport small peptides, lipophilic hormones (Chapter 3), and even drugs.

The most abundant noncellular protein in the blood is **albumin** (3.5–5 g/dl). Albumin is secreted by liver cells (hepatocytes, discussed in Chapter 11), serves as a carrier protein, and, most importantly, serves to maintain **oncotic pressure**. Oncotic pressure is the term used to describe the force that keeps fluid from leaking out of a container through a diffusible barrier. When albumin levels become low (**hypoalbuminemia**), such as with liver failure (inadequate albumin is made) or in protein-losing diseases (where too much albumin is lost through either urine or stool), the oncotic pressure of the blood falls. This change in pressure causes noncellular fluid to leak from the

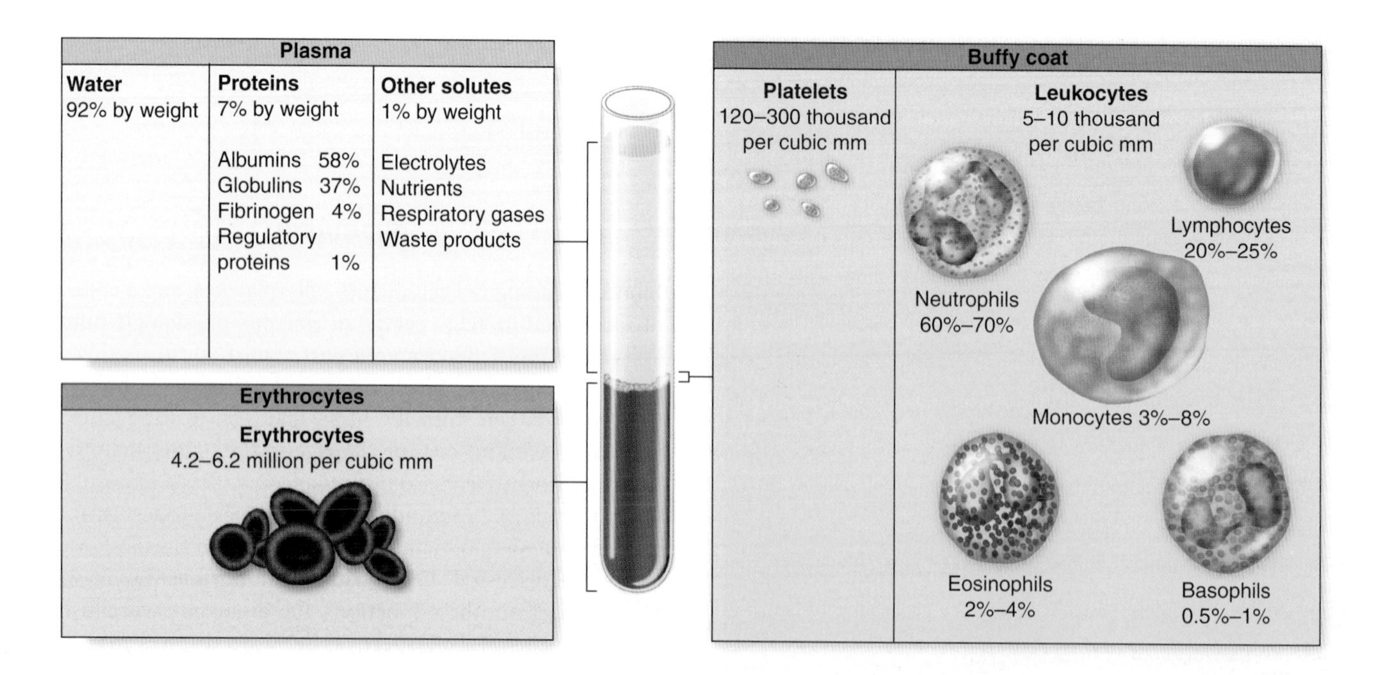

Figure 14-1. Separation of Blood into its Basic Components. Upon centrifugation, blood separates into a noncellular "plasma" layer at the top of the tube; a middle "buffy coat," composed of platelets and white blood cells; and a bottom layer of mostly red blood cells. Because this sample contains anticlotting additives, the noncellular layer is called plasma. [Reproduced with permission from Mescher AL: Junqueira's Basic Histology Text and Atlas, 12th edition, McGraw-Hill, 2010.]

blood into the tissues, resulting in **edema** (swelling). Edema can be life-threatening, particularly pulmonary edema, where lungs become fluid filled and O_2 exchange is impaired. Pulmonary edema from pure hypoalbuminemia is extremely rare, and the most common causes of pulmonary edema are heart failure and inflammation. **Immunoglobulins** (antibodies) form the second most common type of protein in the blood (2.3–3.5 g/dl). Although immunoglobulins play a minor role in maintaining blood oncotic pressure, the most serious complication from low immunoglobulin levels is an increased risk of infection (Chapter 15). The blood also transports **hormones**, small bioactive molecules that affect the function of one or multiple target organs and tissues (Chapters 1 and 3). In addition to protein components, serum contains an incredible variety of nonprotein components, including minerals, electrolytes, lipoprotein particles, and essential nutrients (such as glucose and glutamine).

Anemia: Anemia occurs when the quantity of RBCs is lower than normal. In severe forms of anemia, the capacity of the blood to carry adequate O_2 to the peripheral tissues is compromised. Anemia can be caused by either the failure to produce enough RBCs (e.g., **iron deficiency** or **aplastic anemia**), a loss of RBCs (**bleeding**), an increased destruction of RBCs that cannot be compensated for by the bone marrow (e.g., **sickle cell anemia** or **autoimmune hemolytic anemia**), or by sequestration, in which RBCs are hidden away (usually in the spleen). The most common cause of anemia, and the most common nutritional deficiency in the world, is iron deficiency anemia. In this condition, there is insufficient dietary iron for erythropoiesis. The failure to meet ongoing, persistent demands of the bone marrow to create new RBCs results in vague symptoms of weakness and fatigue. Serious cases can manifest as shortness of breath and even death.

RED BLOOD CELL (RBC) FUNCTIONS

RBCs are small cells (6–8 μm in diameter) whose primary function is to transport O_2 from the lungs to the peripheral tissues. A secondary function of RBCs is to carry CO_2, generated by the peripheral tissues, back to the lungs for elimination by exhalation (Chapter 17) through the mouth and nose. Hgb is the O_2-carrying molecule in RBCs. Hgb consists of four *heme* molecules with four *globin* chains, giving the molecule its name. Adult Hgb is a tetramer of four globin chains: two **α-globin** subunits and two **β-globin** subunits (Figure 14-2A–C and 14-3A–B).

Each heme molecule consists of two parts: a **protoporphyrin molecule** and **iron**. The local environment created by each globin subunit maintains iron in the ferrous (Fe^{2+}) state. In this form, iron has six "coordination sites" that interact with the protoporphyrin ring and other molecules. The first four of the coordination sites bind to the protoporphyrin ring along its plane, suspending the iron in the middle of the ring (Figure 14-4A). The fifth site forms a strong, covalent linkage to Hgb through histidine (called the "proximal histidine") of the globin chain (Figure 14-4A). Sites 1 through 5 are fixed. The sixth site, however, is not fixed and accounts for the versatility of Hgb binding. It is the sixth site that binds and releases O_2 as well as other gases. The deoxygenated protoporphyrin ring has a dome-like structure because of steric hindrances and other forces from the surrounding amino acids (Figure 14-4B).

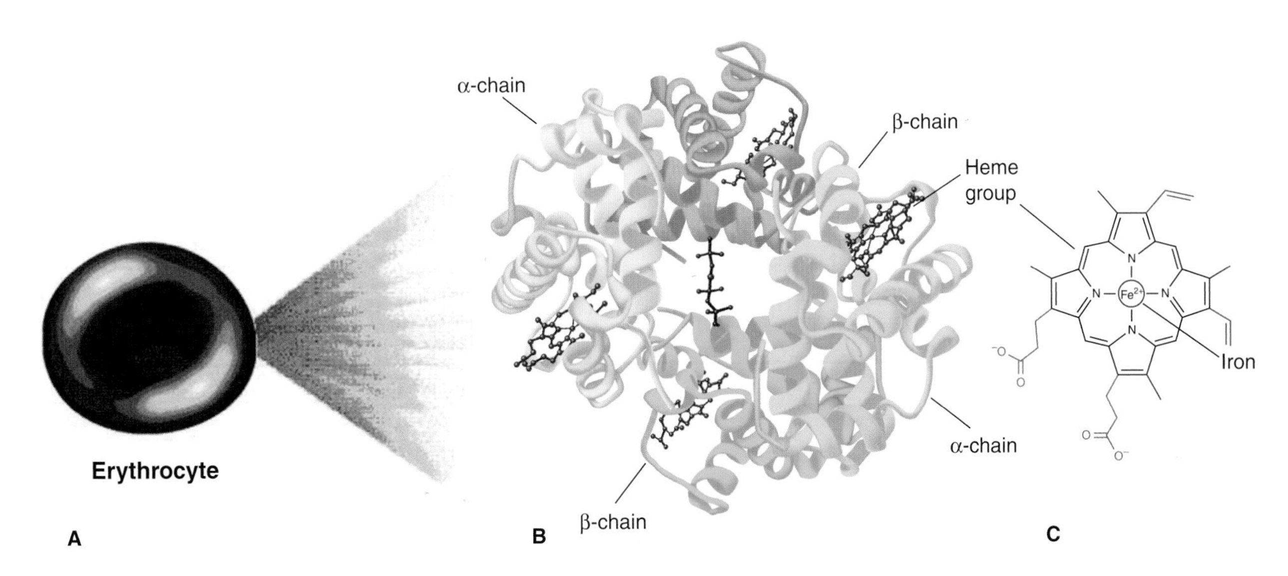

Figure 14-2. A–C. Basic Components of a Red Blood Cell. (**A**) Each erythrocyte (red blood cell) contains many molecules of (**B**) hemoglobin (Hgb), a four-subunit molecule. Each subunit of Hgb contains (**C**) one heme molecule. [Adapted with permission from Murray RA, et al.: Harper's Illustrated Biochemistry, 28th edition, McGraw-Hill, 2009.]

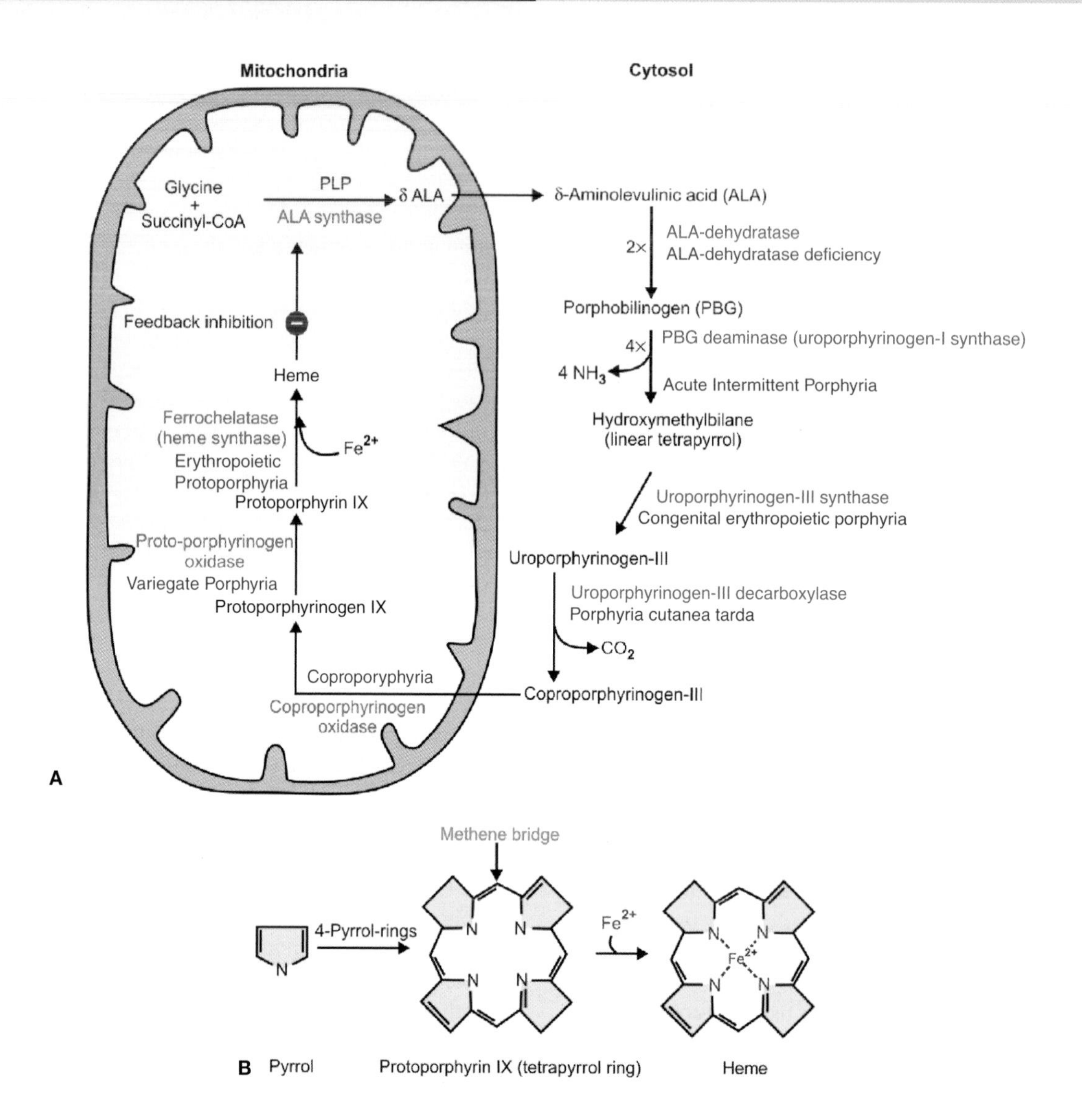

Figure 14-3. A–B. Synthesis and Structure of Hemoglobin (Hgb). (**A**) Heme is synthesized via multiple enzymatic steps (indicated in blue), which occur in both the mitochondria and cytosol. Increased levels of heme negatively regulate ALA synthase, the first enzyme in the synthetic pathway. The final step, catalyzed by ferrochelatase, is responsible for the addition of Fe^{2+} to the finished heme molecule. Defective enzymes in all of the remaining synthetic steps lead to a series of diseases noted in red. (**B**) A heme molecule consists of a protoporphyrin ring with a centrally suspended iron molecule (see the text below). This iron atom binds oxygen (O_2); therefore, one Hgb molecule can bind a maximum of four O_2 molecules. CoA, coenzyme A; PLP, pyridoxal phosphate. [Adapted with permission from Naik P: Biochemistry, 3rd edition, Jaypee Brothers Medical Publishers (P) Ltd., 2009.]

Carbon Monoxide (CO) Poisoning: CO is a dangerous gas not only because it is undetected by our sense of smell but also because it has an affinity 200 times greater than that of O_2 for Hgb's sixth coordination position. As a result, CO will displace O_2 for binding to Hgb. The displacement of O_2 by CO severely decreases O_2 delivery to the peripheral tissues and decreases the cell's ability to carry out oxidative phosphorylation. The detection of CO poisoning can be difficult. Notably, a patient does not have to appear blue to be in dire need of O_2; patients are just as pink when Hgb is saturated with either O_2 or CO. The treatment for CO poisoning is to first remove the patient from the source of CO exposure and, in extreme cases, to put the patient in a hyperbaric O_2 chamber. In a hyperbaric O_2 chamber, the equilibrium between Hgb binding O_2 and CO is changed and the high-pressure O_2 displaces the CO molecule from the sixth coordination position of the Hgb molecule.

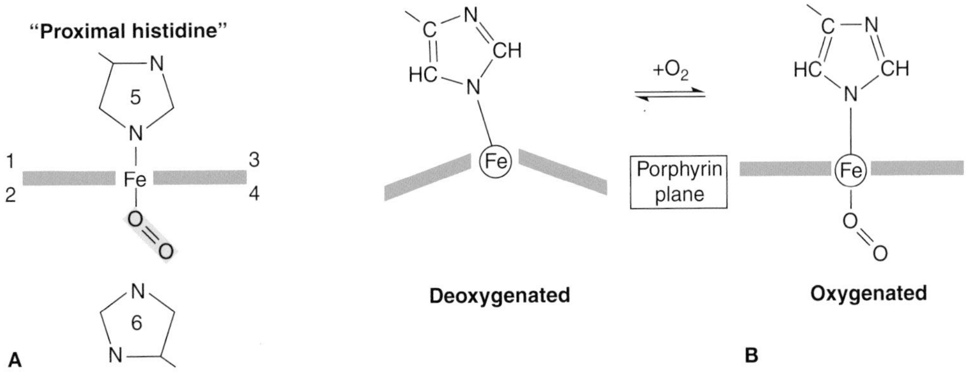

Figure 14-4. A–B. Binding of Oxygen (O_2) to Hemoglobin (Hgb) Molecule. (**A**) Binding of O_2 by the Fe^{2+} atoms' fifth (5) coordination site histidine with additional regulation by the sixth coordinate site (6) histidine. Coordination sites 1–4 are symbolically indicated in the plane of the porphyrin ring. (**B**) When the heme molecule's sixth coordination site is empty, the iron molecule projects outward from the center of the protoporphyrin ring, causing the molecule to assume a domed formation. Upon O_2 binding, the heme molecule assumes a planar conformation (right). [Adapted with permission from Naik P: Biochemistry, 3rd edition, Jaypee Brothers Medical Publishers (P) Ltd., 2009.]

DISEASES ASSOCIATED WITH INADEQUATE SYNTHESIS OF HEMOGLOBIN COMPONENTS

Several disease states are associated with the inadequate synthesis of any of the final Hgb components. Of the four components of Hgb (the **protoporphyrin ring**, α-globin, β-globin, and iron), cells must synthesize all but iron, which it absorbs from the intestinal tract. These diseases are summarized in Table 14-1.

OXYGEN BINDING

TENSE AND RELAXED HGB

Each Hgb unit is in equilibrium between two different structural states: **tense (T)** or **relaxed (R)** (Figure 14-5). Hgb in the T state has a low affinity for O_2 and is more likely to surrender O_2 molecules it is holding. Hgb in the "R" state has an approximately 100-fold increase in affinity for O_2 as compared with the T state; R state Hgb is less likely to surrender O_2 to tissues. A mutation in a globin gene or a small molecule that stabilizes the T state (or destabilizes the R state) will result in a higher proportion of O_2-binding units being in the T state. An overall shift to the right in the O_2 dissociation curve will occur (lower affinity for O_2) (Figure 14-6). Conversely, mutations or small molecules that destabilize the T state (or stabilize the R state) will result in a higher proportion of O_2-binding units being in the R state. An overall shift in the O_2 dissociation curve to the left (higher affinity) will occur.

ALLOSTERIC BINDING OF O_2 BY Hgb

Each of the four subunits of Hgb contains one protoporphyrin ring, thereby allowing a total of four O_2 molecules to bind per Hgb molecule. However, O_2 binding is not a simple, linear process. The R and T states exist in equilibrium with each other and this equilibrium is what determines the overall **sigmoid** or **"S" shape** of the O_2 dissociation curve (Figure 14-6). The binding of the first O_2 to a single Hgb site sets in motion a series of conformational changes of the globin, polypeptide chains that increase the probability that the other three sites will adopt the "R" state (the high affinity state). Binding of subsequent O_2 molecules also increases the binding of the other sites (although much less than the binding of the first O_2). In addition, more relative R state than T state shifts the curve to the left. Conversely, more T than R will shift the curve to the right.

The role of Hgb is to transport O_2 from higher partial pressure of O_2 (pO_2) environments (such as lungs) to lower pO_2

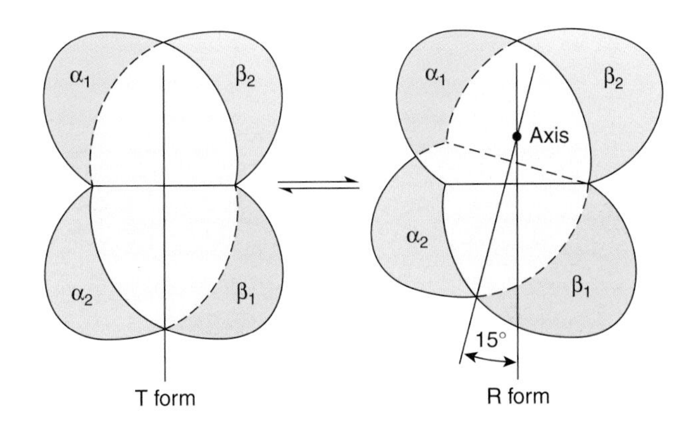

Figure 14-5. Hemoglobin Binding of Oxygen (O_2) Causes a Change in the Shape of the Hemoglobin (Hgb) Molecule. Hgb molecules exist in an equilibrium of deoxy/tense and oxy/relaxed states. Deoxy Hgb (left) has a low affinity for O_2 and is associated with the tense (T) state and the release of O_2 to the tissues. Binding of O_2 causes an approximately 15° rotational change in the quaternary structure to form oxy Hgb (right). Oxy Hgb has a high affinity for O_2 and is associated with the R state. Oxy Hgb will bind O_2 more than release O_2. [Reproduced with permission from Murray RA, et al.: Harper's Illustrated Biochemistry, 28th edition, McGraw-Hill, 2009.]

environments (such as tissues). Because the ligand affinity of Hgb is increased by successive binding, O_2 is a **positive allosteric effector** (Chapter 5). The S shape of the O_2–Hgb dissociation curve depicts a **positive cooperativity** relationship between the percent saturation of the Hgb molecule and pO_2 (Figure 14-6). The positive cooperativity of O_2 binding is an essential characteristic of Hgb that makes it both an efficient O_2 carrying and delivery molecule. This characteristic also allows more O_2 delivery to tissues with high metabolic demands (low pO_2). Correspondingly, tissues that need less O_2 because of fewer metabolic demands have higher pO_2 and extract, appropriately, less O_2.

TABLE 14-1. Summary of Hemoglobin Diseases

Condition/Cause	Types	Clinical Manifestations
Porphyrias—inability to efficiently synthesize the protoporphyrin ring.	NA	These rare diseases display a spectrum of manifestations: acute abdominal pain, skin and tooth discoloration, and hemolysis [destruction of red blood cells (RBCs)]. An inciting agent such as an environmental exposure (e.g., sunlight) or drug (e.g., barbiturate) often precipitates clinical manifestations.
Thalassemias*—defect in the amount of globin chains that are synthesized. There are four α-globin genes and two β-globin genes.	**α-Thalassemia**—underproduced α-globin subunit because of mutated α-globin genes.	**Carrier state—**A single α-globin gene is mutated. No clinical manifestations.
		α-Thalassemia trait—Two α-globin genes are mutated. There is a mild microcytic anemia (small RBCs). This anemia is usually not clinically relevant but must be distinguished from iron deficiency, discussed below. In contrast to iron deficiency, in α-thalassemia trait, blood iron studies are normal.
		Hemoglobin (Hgb) H—Three of four α-globin genes are mutated. There is not enough α-globin made to match the demands of the body. This is treated either with chronic RBC transfusions or by bone marrow transplantation.
		Hgb "Barts"—All four α-globin genes are mutated. This condition is not compatible with life, and results in hydrops fetalis and intrauterine death (α-globin is required for normal fetal development).
	β-Thalassemia—underproduced β-globin subunit because of mutations in the β-globin genes.	**β-Thalassemia trait**—One β-globin gene is mutated. Like α-thalassemia trait, β-thalassemia trait causes a mild microcytic anemia that must be distinguished from iron deficiency.
		β-Thalassemia major and β-thalassemia intermedia— Both β-globin genes are mutated. Patients requiring chronic blood transfusions are classified as major; those who do not require are classified as intermediate.
Iron deficiency anemia—inadequate iron supply to sustain erythropoiesis.	NA	Believed to affect one billion people. Symptoms include pallor, weakness, and lethargy. Severe and prolonged iron deficiency can result in neuropsychological problems. Treatment is by addition of iron to the diet via tablets.

*The only curative treatment for thalassemias is to completely replace the bone marrow (bone marrow makes all the cellular components of blood). Bone marrow transplantation involves finding an immunologically matched donor. This insures that the recipient's immune system does not attack the donor's and vice versa. The recipient's endogenous bone marrow is ablated using chemotherapy and/or radiation, and the donor bone marrow is infused to replace it.

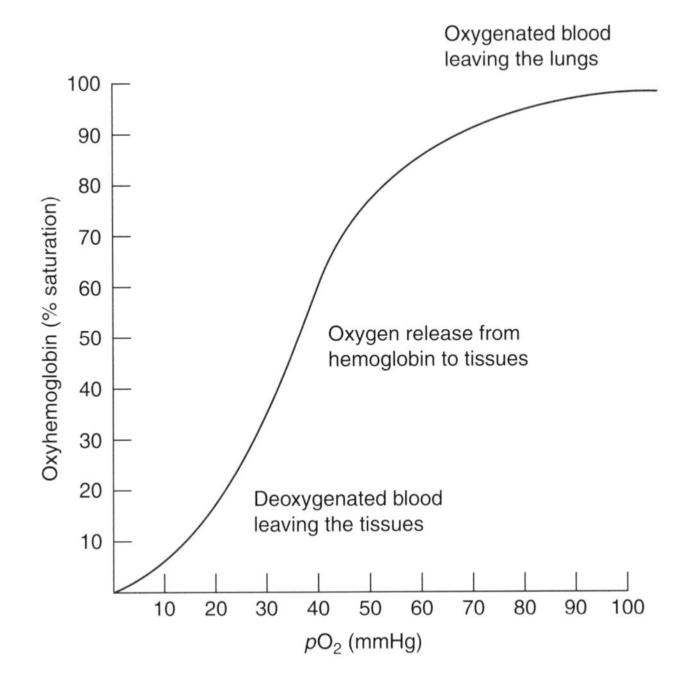

Figure 14-6. The Oxygen (O_2)–Hemoglobin (Hgb) Dissociation Curve. The O_2–Hgb dissociation curve depicts the saturation of Hgb with O_2 as a function of the partial pressure of O_2 (pO_2). pO_2 is a measure of the concentration of O_2 in the tissues. The pO_2 of metabolically active tissues is approximately 40 mmHg. At pO_2 = 40 mmHg, the affinity for O_2 steeply declines and Hgb surrenders O_2 to the tissues. [Adapted with permission from Kibble JD and Halsey CR: The Big Picture: Medical Physiology, 1st edition, McGraw-Hill, 2009.]

REGULATION OF O_2 BINDING

An important concept in understanding Hgb function is that Hgb must be able to carry out two seemingly opposing functions: (1) to become fully saturated with O_2 in the lungs and (2) to be able to efficiently unload O_2 in the tissues. **That is, O_2 binding does not guarantee O_2 delivery.** In fact, if Hgb binds O_2 too tightly, tissues would not be able to extract O_2 from the Hgb molecule, and functionally it would be equivalent to there being no O_2 at all. It is the *release* of O_2 from Hgb that drives tissue oxygenation. **Negative allosteric regulators** such as increased **temperature**, increased **CO_2**, and decreased **pH** (increased H^+ concentration) will cause the curve to shift to the right and help Hgb unload O_2 in the tissue (Figure 14-7).

2,3-bisphosphoglycerate (2,3-BPG) (also known as 2,3-diphosphoglycerate) is a small molecule created as a side product of glycolysis from 1,2-BPG (Chapter 6). It is present in the same concentration as Hgb in the RBC (~2 mM) and is the key molecule in assuring that Hgb does not bind O_2 too tightly (insuring delivery of O_2 to the tissues). The biochemical mechanism of 2,3-BPG action is that it specifically fits into a cleft present in the T state, which is not present in the R state (Figure 14-8A). This stabilizes the T state through a series of noncovalent interactions in the globin chain, thus shifting the O_2 dissociation curve to the right (Figure 14-8B). Because 2,3-BPG binds at a site separate from the ligand-binding site (in this case, the ligand is O_2) and through that binding affects the affinity of ligand (O_2) binding, 2,3-BPG is an **allosteric effector** (Chapter 5) of Hgb function.

Acid–base balance in the body is regulated by an integrated combination of mechanisms found in the lungs (Chapter 17), kidney (Chapter 18), and blood (Figure 14-9). The regulation of Hgb binding of O_2 by acid–base status, in particular **H^+** and **CO_2** (both regarded as acids), is called the **Bohr effect.** H^+ directly facilitates the formation of salt bridges that stabilize the T state (the low-affinity state). CO_2 directly stabilizes the T state by binding to the amino terminal ends of the globin chains to generate a T state stabilizing **carbamate** moiety. In addition, CO_2 indirectly stabilizes the T state when it is converted by **carbonic anhydrase** into **carbonic acid.** Carbonic acid generates extra H^+, which then act as described above. Metabolism creates H^+ and CO_2, and the more demanding the tissue, the more H^+ and CO_2 are created. When more acid is created (a decrease in pH), this causes a right-shift in the O_2–Hgb dissociation curve.

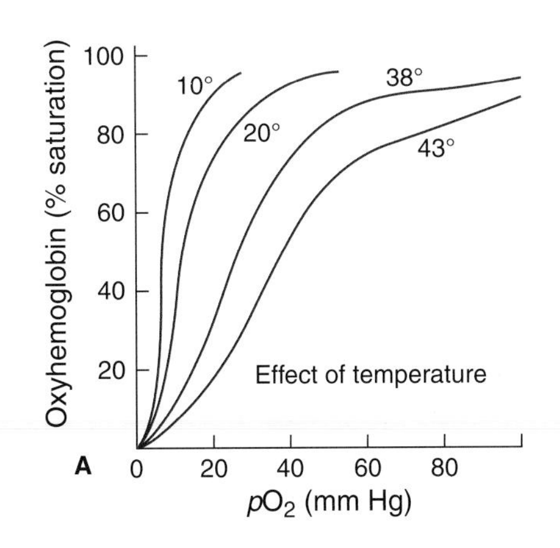

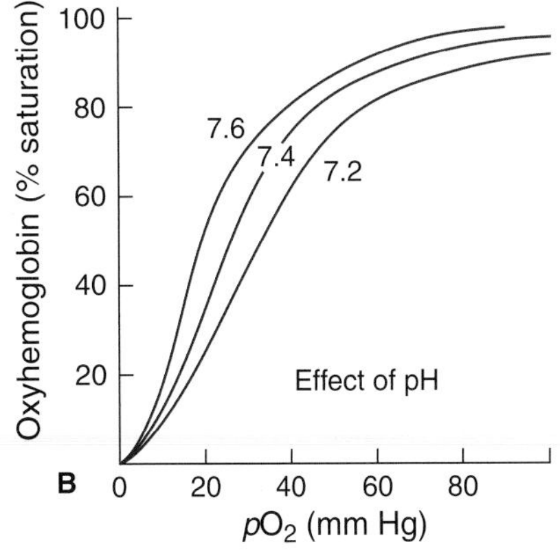

Figure 14-7. Effects of Temperature and pH on Hemoglobin (Hgb) Binding of Oxygen (O_2). (**A**) Increasing temperature (e.g., 10, 20, 38, and 43°C) causes reduced binding of O_2 by Hgb, reflecting reduced affinity for O_2 (T state > R state). In this higher temperature environment, Hgb will more readily release O_2 as noted by the right shift of the curves. (**B**) Increasing pH (e.g., 7.2, 7.4, and 7.6) causes increased binding of O_2 by Hgb. As the curve shifts left, Hgb has an increased affinity for O_2 (R state > T state). [Adapted with permission from Kibble JD and Halsey CR: The Big Picture: Medical Physiology, 1st edition, McGraw-Hill, 2009.]

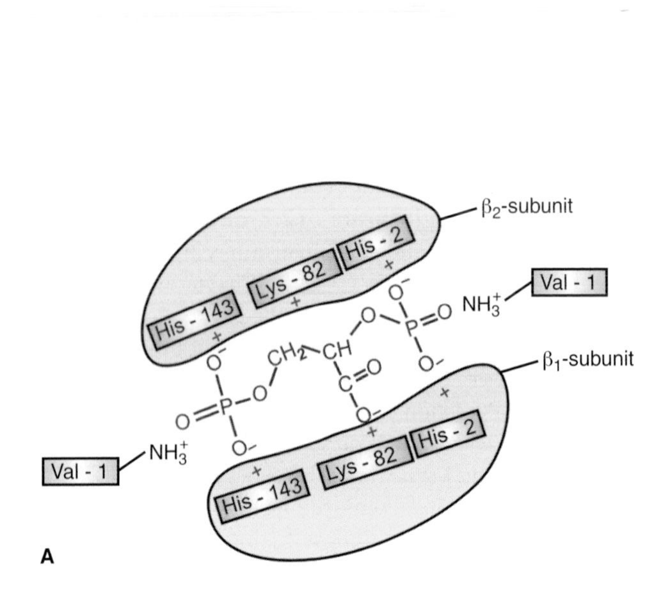

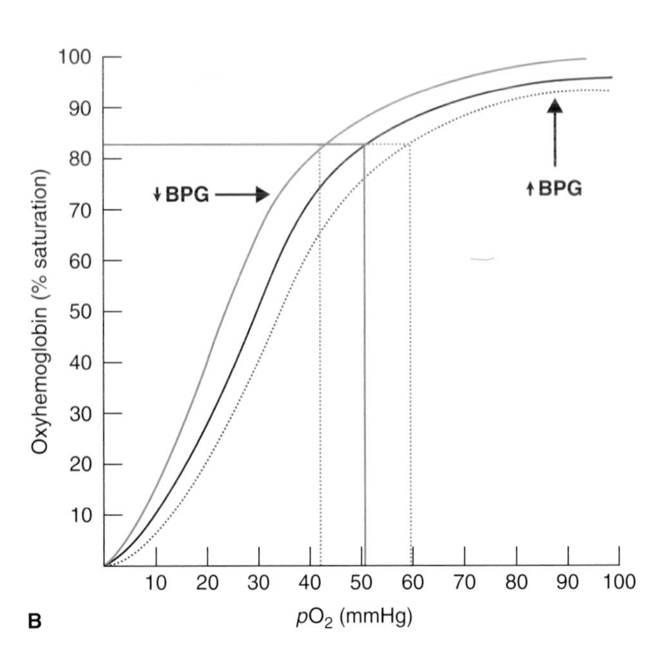

Figure 14-8. A–B. Binding of 2,3-BPG to Deoxyhemoglobin. (**A**) 2,3-BPG (in red) fits into a cleft and interacts with the amino terminal valine (Val)'s NH_3^+ group plus the positively charged R-group sidechains of lysine (Lys) and histidine (His) on both β-globin chains. These same interactions do not occur in the relaxed (R) state of oxy Hgb. [Adapted with permission from Naik P: Biochemistry, 3rd edition, Jaypee Brothers Medical Publishers (P) Ltd., 2009.] (**B**) These noncovalent bonds stabilize the tense (T) [low oxygen (O_2) affinity] state and change the normal binding (blue, solid line), promoting the release of any bound O_2 molecules to tissues. Thus, increasing BPG (red, dotted line) results in a right shift (lower binding of O_2) and decreasing BPG (green line) results in a left shift (higher binding of O_2) of the binding curve. 2,3-BPG, 2,3-bisphosphoglycerate. [Adapted with permission from Kibble JD and Halsey CR: The Big Picture: Medical Physiology, 1st edition, McGraw-Hill, 2009.]

A right-shifted O_2 binding curve means that, for a given pO_2, the affinity of Hgb for O_2 will be decreased and the probability that O_2 will be delivered to the tissue will be increased.

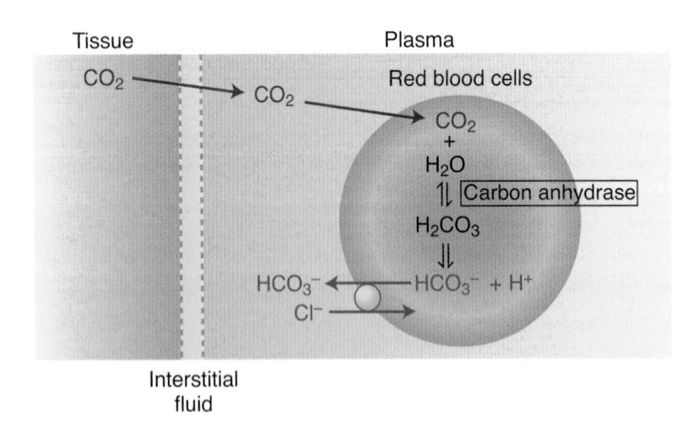

Figure 14-9. Fate of Carbon Dioxide (CO_2) in the Red Blood Cell (RBC). Upon entering the RBC, CO_2 is rapidly hydrated to H_2CO_3 by carbonic anhydrase. H_2CO_3 is in equilibrium with H^+ and its conjugate base HCO_3^-. H^+ can interact with deoxy hemoglobin, whereas HCO_3^- can be transported outside of the cell via band 3, an HCO_3^-/Cl^- anion exchanger in the RBC membrane (green circle). In effect, for each CO_2 molecule that enters the RBC, there is an additional HCO_3^- or Cl^- in the cell. [Adapted with permission from Kibble JD and Halsey CR: The Big Picture: Medical Physiology, 1st edition, McGraw-Hill, 2009.]

Hypoxia, Altitude Sickness, and Acetazolamide: Altitude sickness can occur when people move from low altitudes to high altitudes, resulting in inadequate O_2 to the peripheral tissues, especially the brain. The main symptom of altitude sickness is headache and nausea that usually resolve without any intervention. In certain circumstances, altitude sickness can cause life-threatening **pulmonary edema** (fluid filling the lungs, causing severe shortness of breath and difficulty in breathing) and cerebral edema (swelling of the brain, causing severe headaches, disorientation, confusion, coma, and possibly death). The only treatment is to move the patient to lower elevations as rapidly as possible. To prevent the onset of symptoms, **acetazolamide**, a **carbonic anhydrase inhibitor**, is prescribed prophylactically to those prone to altitude sickness. Inhibiting carbonic anhydrase before moving to higher altitude causes an alkalinization of the urine and consequent acidification of the blood. The increased H^+ in the blood shifts the O_2 dissociation curve to the right through the Bohr effect and increases O_2 unloading in the peripheral tissues, including the brain.

In summary, the relative stabilization of the T state by three factors (H^+, CO_2, and carbonic acid) in the tissue immediately shifts the O_2 dissociation curve to the right locally, thus increasing the amount of O_2 delivered. A lower affinity means that Hgb will more readily **release O_2** to tissues with high H^+ (low pH) and/or high CO_2 levels. Metabolically active tissues (e.g., muscle cells) produce high amounts of these molecules. By decreasing the affinity of Hgb for O_2, the Bohr effect insures the ability of Hgb to surrender O_2 to the tissues where it is needed most. In addition, metabolically active tissues generate heat, raising the temperature of the tissue locally. This increase in temperature also decreases the affinity of Hgb for O_2 and facilitates the unloading of O_2 in tissues that are generating heat through their metabolic activities. When the RBCs return to the normal pH and temperature of the lung capillary bed, the excess CO_2 is exhaled (Chapter 17) and the normal equilibrium between the R and T states is restored.

How Does the Fetus Get Enough O_2? The sigmoid shape of the O_2 dissociation curve explains how Hgb can bind O_2 in the lungs and deliver O_2 to the tissues. But how does the developing fetus, a period of high macromolecular synthesis, cell growth, and consequent O_2 requirement, get enough O_2 to grow? The fetus does not breathe air. Instead, it must extract O_2 from the mother's RBCs across the **placental vasculature** and then deliver sufficient O_2 to its own developing tissues. The fetus solves this problem by using a different set of globin genes as part of its Hgb molecule. In contrast to the mother, whose Hgb molecule consists of two α- and two β-globin chains creating **HgbA**, the fetus uses two α- and two γ-globin chains creating **fetal** Hgb **(HgbF)**. HgbF has a lower affinity for **2,3-BPG** than HgbA because of a serine at position 143 in the γ-globin gene instead of histidine (see Figure 14-8A for histidine position). This decreased affinity for 2,3-BPG results in HgbF having an increased affinity for O_2 and a shift in the O_2 dissociation curve to the left. Thus, HgbF is able to extract O_2 from HgbA in the placenta and then deliver the O_2 to the fetal tissues. The shift to the left is precise, however, because although it allows the fetal Hgb to "steal" O_2 from the maternal Hgb, it is not so significant that the fetal tissues cannot remove the O_2 for their own use.

PHYSIOLOGIC RESPONSE TO INADEQUATE O_2 DELIVERY

The inadequate delivery of O_2 to tissues (**hypoxia**) induces a complex variety of normal and pathophysiologic compensatory mechanisms. Cells must turn to anaerobic respiration (Chapter 6), which produces **lactic acid** and lowers the pH. As described above, a low pH has an immediate effect on O_2 delivery and extraction through the Bohr effect. The next level of compensation to decreased O_2 delivery is to increase the concentration of the allosteric molecule, 2,3-BPG. The presence of 2,3-BPG lowers the O_2 affinity of Hgb, resulting in a rightward shift of the O_2 binding curve and increased O_2 delivery to tissues in low O_2 conditions. The compensatory increase in 2,3-BPG concentration takes about 24 hrs to occur.

A slow compensatory mechanism of inadequate O_2 delivery is via the production and secretion of erythropoietin (Epo) to create more RBCs. Kidney cells have high O_2 demands because they need abundant adenosine triphosphate to power multiple channels and pumps that regulate electrolyte balance. Thus, these cells are ideally suited to be sensors of O_2 delivery. When there is inadequate delivery of O_2 to kidney cells, a transcription factor called **hypoxia-inducible factor** is not degraded and is able to activate the transcription of multiple genes, including the blood hormone **Epo**, leading to an increased level of Epo in the blood stream. Epo migrates through the blood to the bone marrow (where RBCs are made), binds to the Epo receptor, and activates a **Janus kinase (JAK) and Signal Transducer and Activator of Transcription (STAT)** intracellular signaling cascade (Chapter 8). Activation of this cascade does not stimulate RBC precursor proliferation; rather, the JAK/STAT pathway protects RBC precursors from their normally programmed cell death (apoptosis). Increased RBC survival accounts for an increase in hematocrit (the amount of RBCs in the blood, see Appendix II). Ultimately, the increased numbers of mature RBCs augment the O_2-carrying capacity of the blood. It takes approximately 1 week for an RBC to be made and, thus, the full effect of this slow response requires about a month to manifest itself; however, beneficial effects can be found in just a few days. Patients with kidney disease do not secrete enough Epo to maintain an adequate quantity of RBCs and, thus, are given supplemental Epo as treatment to prevent severe anemia.

SICKLE CELL DISEASE (SCD)

Normal RBCs have a biconcave shape that maximizes their ability to pack Hgb and deliver O_2. RBCs are also flexible and deformable so that they can squeeze through the capillaries. **SCD** is caused by a recessive, single-nucleotide mutation in the β-globin gene of Hgb, inherited from both parents (i.e., homozygous). This mutation results in the substitution of the normal glutamic acid (negative charge) with a valine (hydrophobic) at amino acid in position 6. SCD is associated with early mortality (the average lifespan is 40–45 years in the United States and much shorter in developing nations) and lifelong medical problems.

Sickle RBCs bind O_2 normally (Figure 14-10A, left panel), and the oxygenated sickle RBCs have a normal biconcave shape. However, when the sickle Hgb unloads its O_2, the normal conformational change exposes the valine at position 6 to the surface creating a "hydrophobic (sticky) patch" (Figure 14-10A, middle panel). The hydrophobic patch on one deoxygenated Hgb molecule can interact with the hydrophobic patch on a second Hgb molecule (Figure 14-10A, right panel), creating stiff Hgb polymers (Figure 14-10B). These internal polymers cause the RBC to stiffen and adopt abnormal shapes, including a crescent or "sickle" shape that gives the disease its name (Figure 14-10C). These stiffened RBCs lose

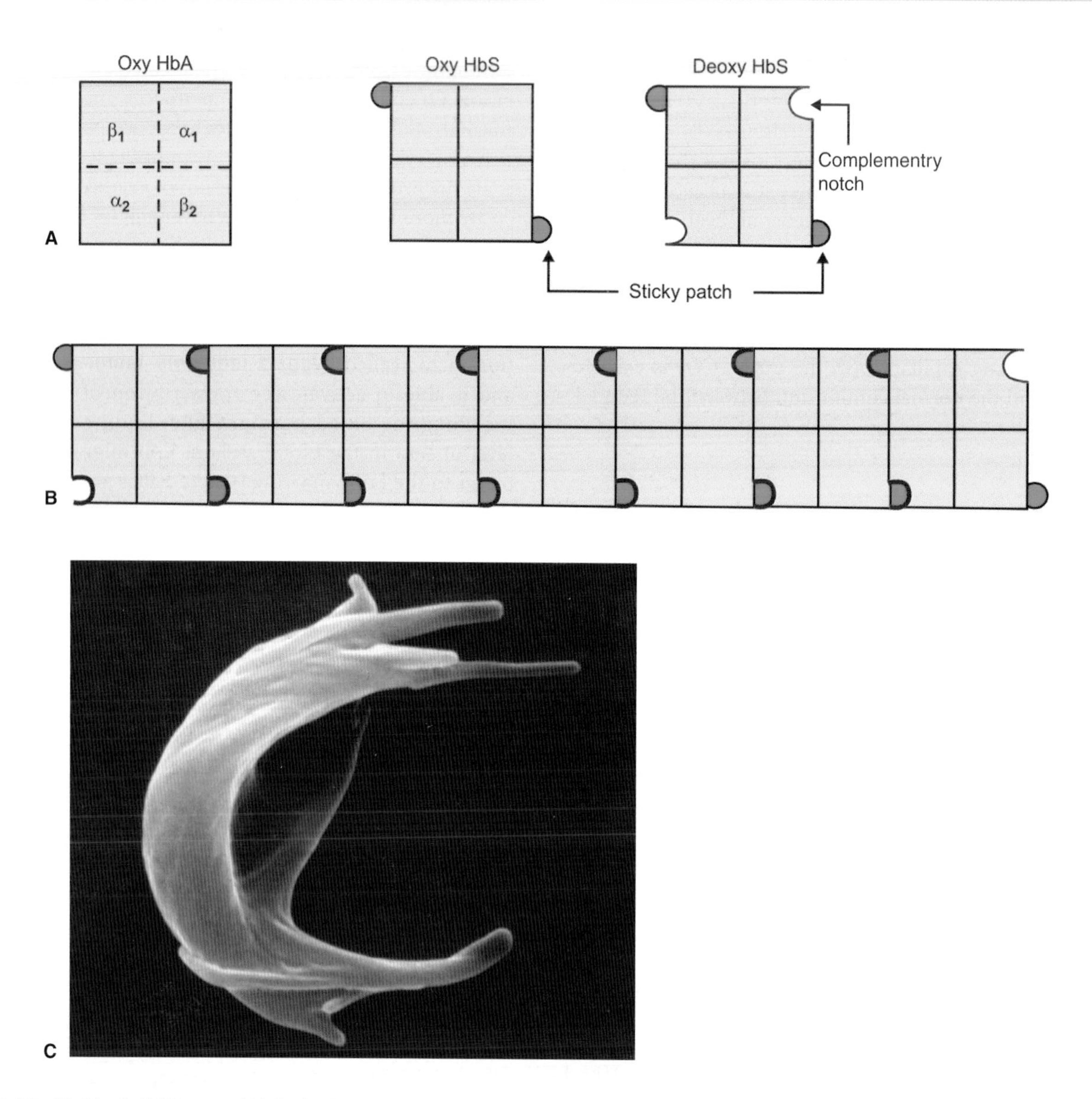

Figure 14-10. Sickle Cell Disease. (**A**) A single nucleic acid change in the β-globin subunit gene (oxy HbA, left panel) causes a substitution of the amino acid valine for the normal glutamate, creating a hydrophobic "sticky patch" on the exterior (Oxy HbS, middle panel). Release of O_2 from the sickle cell hemoglobin (Hgb) (deoxy HbS, right panel) creates a complimentary conformational change on the α-subunit that can bind to the sticky patch and lead to (**B**) polymerization. [Adapted with permission from Naik P: Biochemistry, 3rd edition, Jaypee Brothers Medical Publishers (P) Ltd., 2009.] (**C**) Polymers of sickle Hgb deform the natural biconcave shape of the red blood cell into a sickle shape. [Reproduced with permission from Mescher AL: Junqueira's Basic Histology Text and Atlas, 12th edition, McGraw-Hill, 2010.]

their pliability and cannot easily squeeze through normal capillaries, thereby forming blockages. Dehydration can worsen this effect because small changes in the hydration status of the RBC can have dramatic effects on sickle cell Hgb polymerization. When Hgb rebinds O_2, the resulting conformational change hides the hydrophobic patch, allowing separation into monomers and resumption of a normal, biconcave form. After multiple rounds of excessive Hgb polymerization (sickling) and recovery, erythrocyte membranes lose their elasticity, adopt a permanently stiffened form, and are ultimately destroyed and broken down (hemolysis).

Inheritance of the mutation from only one parent (heterozygous) is termed **sickle cell trait**. Patients with sickle cell trait are typically asymptomatic, unless placed under extreme conditions (low O_2 or severe dehydration). Sickle cell trait also provides relative resistance to diseases such as malaria and is believed to have been a positive evolutionary trait in areas of the world where malaria is endemic. For example, in Sub-Saharan Africa, there is such severe selective pressure that nucleotide change has been selected for multiple times. Other less common variants of SCD all have at least one mutated copy of the β-globin gene.

SCD and Hydroxyurea: SCD leads to a number of lifelong, clinical consequences due to alterations in the biophysical properties of the RBC. Veno-occlusive crises (also known as "painful crises") result from obstructions of blood vessels due to the inability of tough, inflexible RBCs to squeeze through capillaries. This obstruction leads to organ ischemia (diminished blood flow) and may result in irreversible damage and necrosis of vital brain regions (stroke) and chronic injury to organs such as the spleen and kidney. The lifespan of a sickle RBC is only about 10 days, significantly shorter than the approximately 120-day lifespan of a normal RBC, because they are destroyed by the spleen and liver because of their abnormal shapes (hemolysis). This markedly shortened lifespan leads to significant anemia (the Hgb concentration in the blood is ~50% of normal). The bone marrow is unable to compensate for the increased RBC destruction (hemolysis). The persistent severe anemia and the persistent microvasculature blockages can ultimately lead to chronic end-organ damage in essentially every organ of the body over time. Increased risk of infections by encapsulated organisms such as *Streptococcus* (causing sepsis) or *Salmonella* (causing osteomyelitis) are also a common problem for SCD patients.

One form of treatment is the use of **hydroxyurea**, which causes an increase in the expression of γ-globin, a variant globin gene that can replace β-globin in the Hgb molecule. Hgb molecules that contain the γ-globin gene do not have a valine at position 6 and they block the formation of Hgb polymers, reducing the RBC deformity problems.

IRON

Blood uses carrier proteins to transfer essential nutrients as part of each person's metabolism. One important example is iron that is involved in a vast array of important biologic reactions:

1. Binds O_2 as part of Hgb molecule. An adult human has approximately 4 g of iron, of which about two-thirds is employed in the O_2-carrying role of Hgb.
2. Mediates a wide variety of oxidative–reductive reactions by serving as an essential cofactor for many proteins via oxidation between ferric (Fe^{3+}) and reduced Fe^{2+} states.
3. Iron is important for many microorganisms. Keeping iron sequestered from these invaders is an important part of the immune system.

IRON METABOLISM

In most well-rounded Western diets, meats and green, leafy vegetables provide adequate iron to prevent iron deficiency. Certain iron-poor diets, such as vegan diets, however, can result in iron deficiency if not supplemented. Once iron is ingested, it is converted from the Fe^{3+} state to the Fe^{2+} state by intestinal **ferric reductase**. Fe^{2+}, not Fe^{3+}, is transported into the intestinal epithelial cell through the **divalent metal transporter (DMT1)** protein. Fe^{2+} is transported out of the intestinal epithelial cells into blood through a second transporter, **ferroportin** (Figure 14-11).

TRANSFERRIN

Once in the blood, the Fe^{2+} is quickly oxidized back to $Fe^{3+}$and bound by the iron carrier protein, **transferrin**. The affinity of transferrin for Fe^{3+} at pH 7.4 is 10^{-23}, which means that transferrin will bind Fe^{3+} even when its concentration is 10^{-23} (10 yoctomolar or 0.01 zeptomolar). This affinity suggests that in the entire 5 l blood volume of an adult, there would be only approximately five free molecules of Fe^{3+} at a time. This exceedingly high affinity is the biochemical mechanism that the human body has adapted to prevent any free iron from existing in the blood stream (Figure 14-12).

The iron bound transferrin circulates through the blood stream until it binds to a transferrin receptor on the surface of a cell. Cells that express transferrin receptors have high iron demands. These include developing RBCs, dividing cells, and microorganisms. The transferrin–iron–transferrin–receptor complex is endocytosed through the classic clathrin-coated pit pathway. Following endocytosis, pH of the endocytic vesicle is reduced to 5, liberating iron from its carrier protein and making iron available for biologic reactions.

FERRITIN

Although Hgb is the most abundant protein that uses iron, ferritin is the most important protein for iron storage. Ferritin consists of a 24-unit multimer of **heavy (H)** and **light (L)** chains that create a hollow shell. Iron is imported into the shell as Fe^{2+} but is converted to Fe^{3+} within the ferritin core by the H chain. A fully loaded molecule of ferritin contains 4500 atoms of iron. It is for this reason that the ferritin molecule is metaphorically referred to as a "bag of rust." A pathologic form of iron deposition is called **hemosiderin**. This is a nonstructured conglomerate of intracellular iron and is often a pathologic consequence of prolonged inflammation. Prolonged deposition of hemosiderin can cause fibrosis (scarring) of tissues.

The importance of iron is highlighted by the fact that the human body has no natural way to excrete it. Unlike other ions, such as sodium, potassium, and calcium, kidneys do not excrete excess iron in the urine. Indeed, the physiological default is to conserve iron. Iron deficiency is much easier to treat than iron overload. Accumulated iron is toxic to a variety of tissues, especially the heart and liver. Infusion of an iron chelator, such as **deferoxamine** or **deferasirox**, provides the only means of reducing some of the iron accumulation. Chelation works by the drug-binding free iron in the blood followed by excretion of the iron–drug complex. The body does lose a small amount of iron each day by the sloughing of skin and epithelial cells and in women through their menstrual flow. However, the same reactivity that makes it a valuable cofactor for proteins also means that free iron can be dangerous because

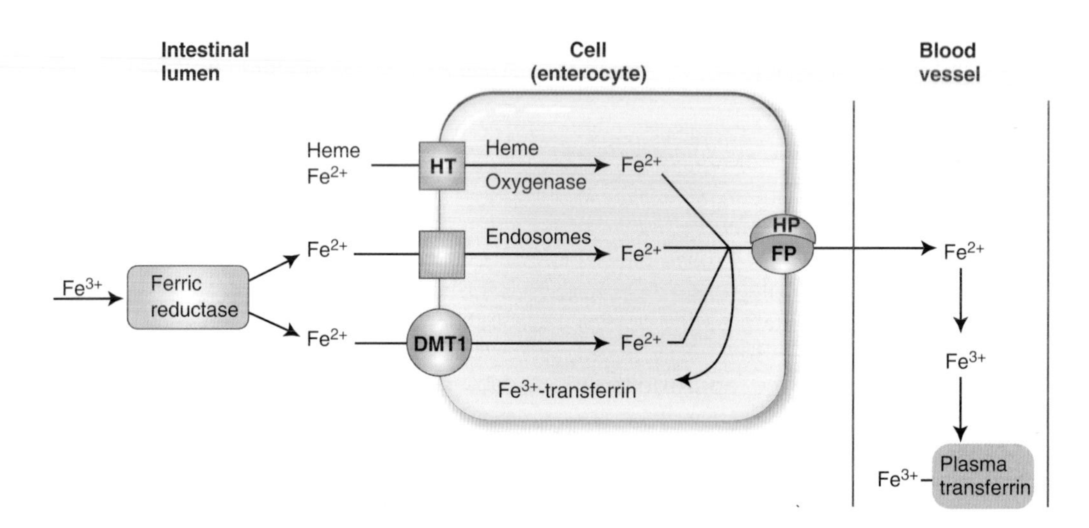

Figure 14-11. Overview of Iron Transport. Intestinal lumen ferric (Fe^{3+}) reductase reduces Fe^{3+} to Fe^{2+}. Fe^{2+} is transported from the lumen into the intestinal epithelial cell through heme transporter (HT), endosomes, and/or divalent metal transporter 1 (DMT1). Fe^{2+} can be converted back to Fe^{3+} and bound to transferrin within the intestinal cell or can be transported into the blood by ferroportin (FP) and hephaestin (HP). The Fe^{2+} oxidized to Fe^{3+}, which binds to plasma transferrin, is carried through the circulation to the tissues. [Adapted with permission from Kibble JD and Halsey CR: The Big Picture: Medical Physiology, 1st edition, McGraw-Hill, 2009.]

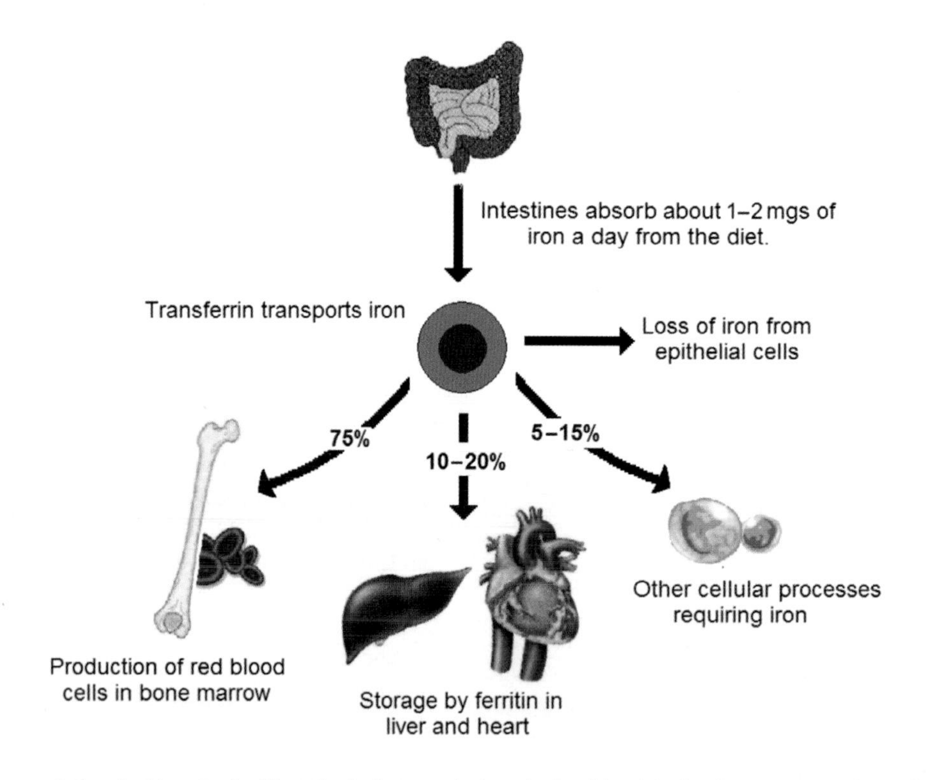

Figure 14-12. Iron Transportation by Transferrin. Transferrin transports iron in the blood to the bone marrow to make hemoglobin and red blood cells (erythropoiesis). Transferrin also carries iron to the liver and heart for storage in ferritin molecules, as well as to other parts of the body for various enzymatic and other functions.

it can easily generate damaging O_2 free radicals. Thus, the body has a powerful system to control the metabolism of this precious and perilous metal.

REGULATION OF IRON AVAILABILITY BY HEPCIDIN

Although the human body has no natural way of excreting excess iron, it is able to regulate the uptake and availability of iron through the protein **hepcidin**. Hepcidin is synthesized as an 84-amino acid precursor, which is then processed to the 25-amino acid active form. Hepcidin acts as a negative regulator of **ferroportin**, blocking the ability of cells to export iron from the cytoplasm into the blood. In the intestinal cell, this inhibition results in decreased iron absorption from the diet. In **reticuloendothelial cells**, the primary storage depot for iron, this results in a decreased ability of iron to be mobilized from storage pools to cells that need it.

Hepcidin is made in the liver, and the levels of hepcidin in the blood are controlled by a variety of different stimuli, including the total iron stores in the body, the erythropoietic demands of making RBCs, hypoxia, and inflammation. When iron stores are high, the level of hepcidin is increased and the amount of iron absorbed is decreased. When the demand for making RBCs is increased, for example, in response to acute blood loss, the level of hepcidin decreases and the amount of bioavailable iron is increased. Similarly, when tissues do not receive sufficient O_2 (hypoxia), it signals that more RBCs are needed, and hepcidin production is decreased. Finally, inflammation, such as signaled through the inflammatory cytokine interleukin-6 (IL-6), causes an increase in hepcidin production (Figure 14-13). This inflammatory regulation is thought to reflect a host defense mechanism because it sequesters iron away from infectious organisms, limiting their growth. If the inflammatory state persists, however, hepcidin sequesters iron away from both the microorganism and the human cells. There is insufficient iron to support the demands for RBC production, and a state of **anemia of inflammation** or **anemia of chronic disease** develops.

Iron Deficiency Anemia: Iron deficiency anemia is the most common nutritional disorder in the world, believed to affect 1 billion people. In children, in the developed world, the most common cause of iron deficiency anemia is the excess consumption of **cow's milk**. Excess cow's milk can cause inflammation damaging the intestinal lining, resulting in blood loss as well as a diminished capacity to absorb iron.

The first manifestation of iron deficiency is an anemia that stems from an inadequate iron supply to sustain erythropoiesis. The symptoms of iron deficiency anemia include pallor, weakness, and lethargy. Severe and prolonged iron deficiency can result in neuropsychological problems. The diagnosis of iron deficiency is usually made by laboratory studies that demonstrate a microcytic anemia (smaller, pale RBCs), reflective of poor Hgb production. Blood studies also show low **ferritin** level and low **transferrin saturation** (only 10% contain iron as compared with the normal 30–40%). Because ferritin is an acute phase reactant that increases in times of stress and illness, sometimes it can be paradoxically high even in iron-deficient states.

The treatment for iron deficiency is to give iron. The most bioavailable dietary iron is in red meat, but often the patient is unable or unwilling to pursue this method. In this situation, oral elemental iron is given. Because iron is transported by DMT1 as Fe^{2+}, not Fe^{3+}, some physicians will advise their patients to simultaneously drink orange juice or take **ascorbic acid (vitamin C)**. Ascorbic acid reduces Fe^{3+} iron to the Fe^{2+} state, facilitating the absorption of elemental iron. In treating iron deficiency anemia in children, caused by excessive milk consumption, it is important to both administer elemental iron and dramatically decrease milk consumption. In rare cases of adult and child anemia, such as poor compliance or anatomic or genetic defects in iron absorption, iron is administered intravenously. Supplemental iron is given until the microcytic anemia is resolved and normal levels of ferritin and transferrin saturation are achieved.

CLOTTING

Blood vessels are subjected to occasional trauma (e.g., cuts and lacerations) and also to routine microtraumas by seemingly benign activities such as running or knocking on a door. Without clotting, these events would cause **hemorrhage (bleeding)**, which is the continuous flow of RBCs from the intravascular space into the extravascular space and tissues. Running would result in bleeding into the knee **joint (hemarthroses)**, and knocking on a door would leave **ecchymoses** (bruises) over the knuckles. To prevent hemorrhage, it is essential to form clots at trauma sites in the vessels. The clots must be formed quickly to minimize blood loss. Clotting also needs to occur in a controlled and localized fashion, neither occluding the vessel nor causing clots to form at remote sites that have no need for clot formation. The key principles of clotting are as follows:

1. Clots consist of platelets, the protein **fibrin**, and RBCs. Fibrin assumes a netlike matrix that spans the full thickness of the clot, ensnaring RBCs throughout the thickness of the meshwork. Ensnaring the RBCs in the fibrin network creates the clot that stops hemorrhage.
2. The formation of a clot occurs in two phases:
   a. **Platelet plug formation**—*Temporary* repair until a proper clot can form.
   b. **Clot formation**—*Longer* term and stronger than a platelet plug. Clot formation stops hemorrhage long enough for tissues to repair.

PLATELET PLUG FORMATION CONSISTS OF ADHESION, AGGREGATION, AND ACTIVATION OF PLATELETS

Damage to the endothelial lining of blood vessels exposes collagen on the basement membrane (Figure 14-14A). Damage also exposes **von Willebrand Factor (vWF)**, normally located between the endothelium and the basement membrane (Figure 14-14B).

1. **Adhesion** of platelets to the endothelium: Collagen is highly thrombogenic and platelets will adhere to it. vWF, contained within platelets and endothelial cells, enhances platelet adhesion by increasing the number of links between platelets and collagen fibrils.
2. **Aggregation** of platelets to one another: Platelets stick to one another by fibrin linking the **glycoprotein IIb/IIIa (Gp IIb/IIIa)** of one platelet to the Gp IIb/IIIa of another.

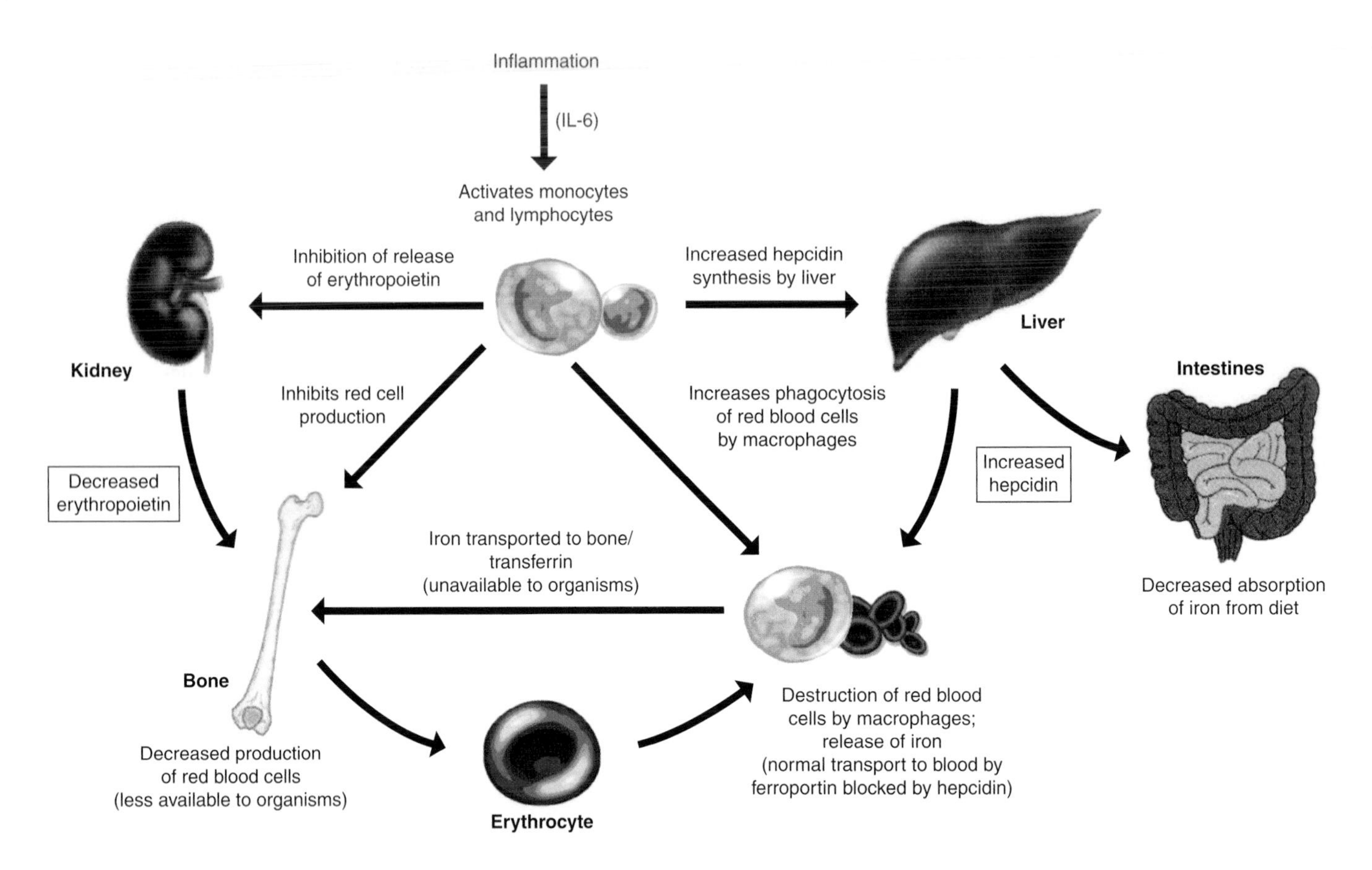

Figure 14-13. Action of Hepcidin in an Inflammatory Response. Response to inflammation (e.g., an infection) leads to activation of monocytes and lymphocytes, often by IL-6. The resulting immunological signals elicit responses from several organs, including increased hepcidin synthesis in the liver, inhibition of erythropoietin from the kidneys, inhibition of red blood cell (RBC) production in the bone marrow, and increased macrophage phagocytosis of RBCs. Increased hepcidin also decreases iron absorption from dietary sources. All of these effects decrease the availability of iron for infective organisms, which aids in their destruction and clearance. IL-6, interleukin-6.

3. **Activation:** The adhesion step causes platelets to release their stored granules (**activation**) that contain the following, which all act to enhance plug formation and limit bleeding:
   a. **Adenosine diphosphate (ADP)**—increases expression of Gp IIb/IIIa on platelets and causes them to swell.
   b. **Prostaglandin Thromboxane A2 (TXA_2)** (Chapter 3) activates a G-coupled protein receptor (G_q). TXA_2 increases vasoconstriction (to decrease blood flow and limit hemorrhage) and the aggregation of platelets.
   c. **PLA2** increases platelet adhesion to fibrin via Gp IIb/IIIa.
   d. ADP and Ca^{2+} lead to additional fibrin deposition.

THE CLOTTING CASCADE

The Clotting cascade consists of a series of ordered enzymatic steps whose end result is the formation of a sturdy clot. There are two limbs of the clotting cascade: the **extrinsic pathway** and the **intrinsic pathway** (Figure 14-15). **Factors II–XII** (named in order of discovery, not in order of activation) are inactive **serine proteases** that are sequentially activated by enzymatic cleavage of the serine protease that immediately precedes it (the "a" following a factor number designates the active form).

Aspirin (Acetylsalicylic Acid) Prevents Platelet Aggregation: Nonsteroidal anti-inflammatory drugs (NSAIDs) act by inactivating or inhibiting **cyclooxygenase (COX)** activity (Chapter 3). Aspirin noncompetitively and irreversibly inhibits the COX enzyme via acetylation. Other NSAIDs such as indomethacin and ibuprofen are reversible, competitive inhibitors of COX. In general, NSAIDs inhibit the synthesis of molecules responsible for various aspects of pain and inflammation: prostaglandins, prostacyclins, and thromboxanes. Aspirin is the only NSAID that blocks **TXA_2** production, limiting **platelet plug** formation (see the text). Thus, low-dose aspirin is used to treat potential prothrombotic diseases and to reduce the risk of heart attack and stroke (which are in part caused by thrombus formation in the wrong vessels) in susceptible patients. Because aspirin inhibits platelet plug formation, people taking aspirin are more prone to bruising and prolonged bleeding after injury.

The intrinsic pathway is triggered when Factor XII, an *inactive* serine protease, encounters exposed **collagen and other polyanions** from an injured blood vessel. A protein complex is formed, which transforms Factor XII into an activated serine

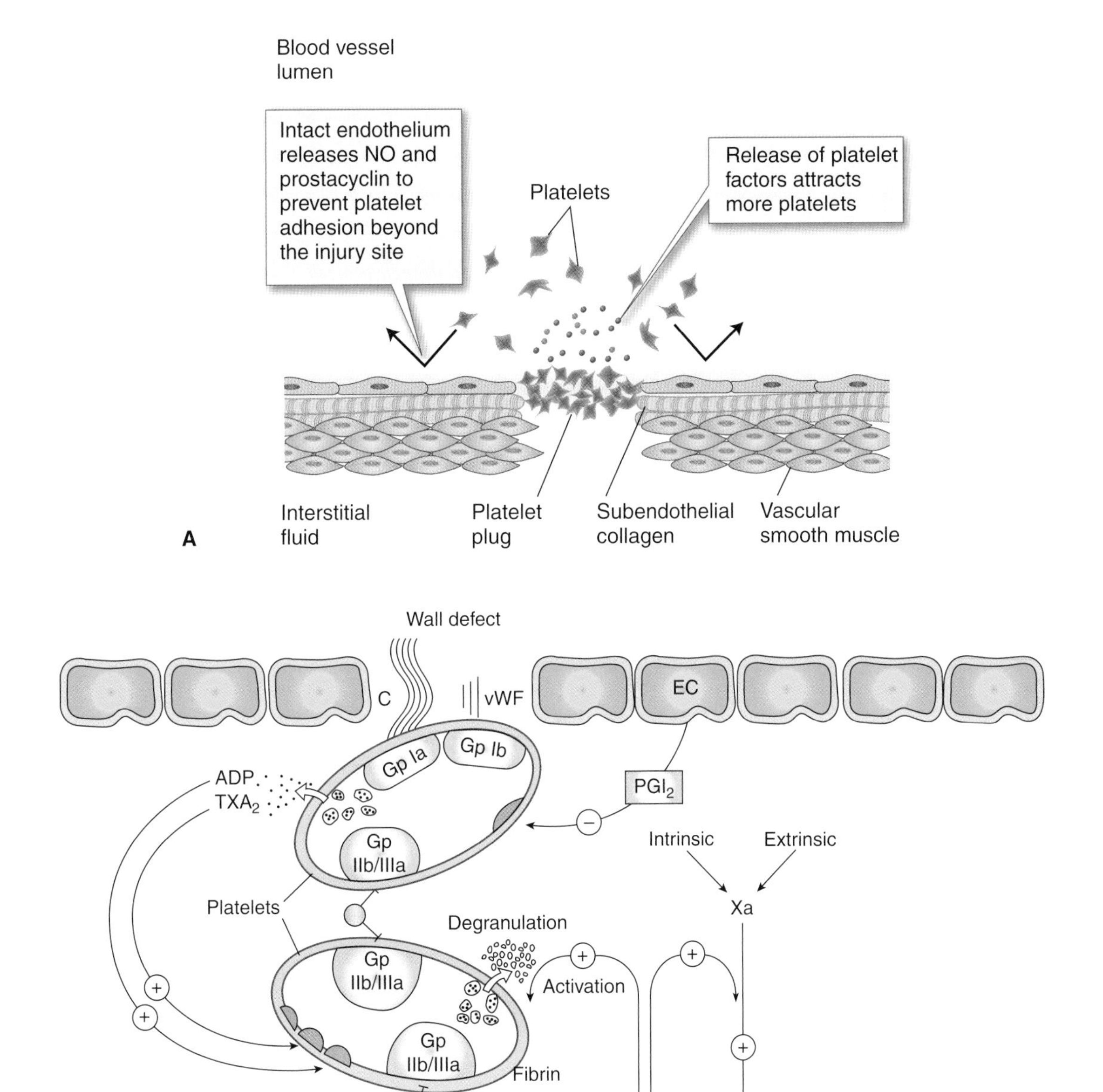

Figure 14-14. A. Formation of the Platelet Plug. Platelet plug formation results from the initial response of platelets to a site of blood vessel wall injury. Control of formation of the initial plug is via nitrous oxide and prostacyclins, as noted in the text. NO, nitric oxide. [Reproduced with permission from Kibble JD and Halsey CR: The Big Picture: Medical Physiology, 1st edition, McGraw-Hill, 2009.] **B. Mechanism of Clot Formation by Platelets.** Following the initial formation of a platelet plug, partly aided by exposed von Willebrand factor (vWF) at the injury site (see the figure), platelets continue to aggregate via fibrin linking of their Gp IIb/IIIa receptors. Adhesions by these receptors lead to activation of the platelets and degranulation, which releases adenosine diphosphate (ADP), thromboxane A2 (TXA_2), and other factors, all of which further increase the formation of initial platelet plug/clot. As noted in Figure 14-14A, nitric oxide and antiplatelet prostacyclin (PGI_2) limit plug and clot formation beyond the injury. The clotting cascade noted in the figure is described further below. EC, endothelial cell. [Reproduced with permission from Simmons ML and Decker JW: Br Heart J, McGraw-Hill, 1995.]

protease, XIIa. XIIa cleaves inactive Factor XI into activated Factor XIa. Factor XIa is a serine protease that cleaves the inactive form of Factor IX into IXa. Factor IXa cleaves inactive Factor VIII to VIIIa. **High-molecular-weight kininogen** and **prekallikrein** also trigger the intrinsic pathway by activating Factor XII. Injury triggers the activation of Factor III and sets off the extrinsic pathway. Factor III cleaves inactive Factor VII (the most abundant of the clotting factors) to active VIIa.

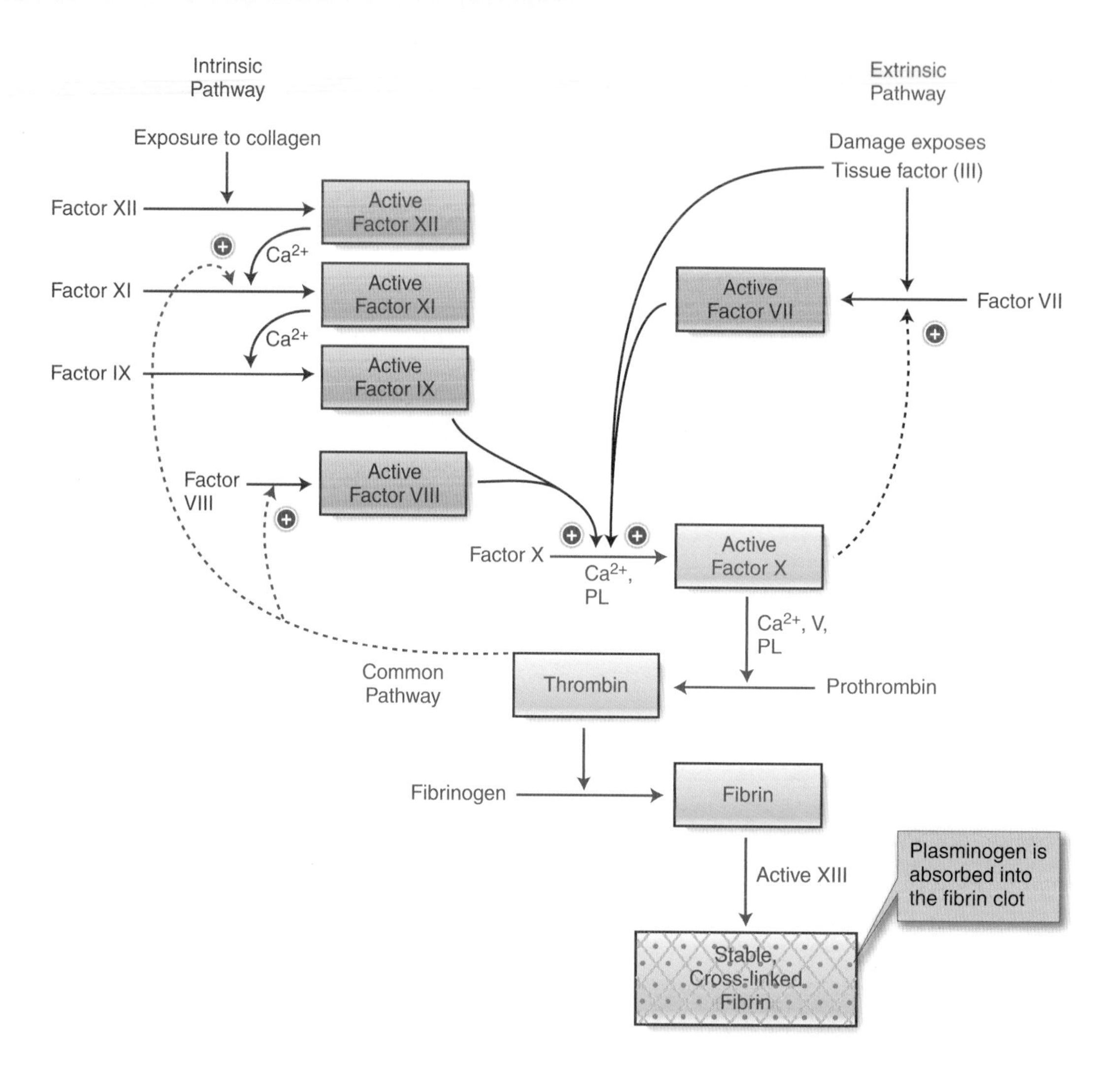

Figure 14-15. Extrinsic and Intrinsic Pathways of the Clotting Cascade. See the text for details of both extrinsic and intrinsic pathways. [Reproduced with permission from Kibble JD and Halsey CR: The Big Picture: Medical Physiology, 1st edition, McGraw-Hill, 2009.]

The extrinsic and intrinsic limbs of the clotting cascade operate independent of one another and converge into a common pathway, beginning with the cleavage of Factor X into Xa. Subsequent activity of Factor Xa and Va convert **prothrombin** into **thrombin**. (Factor Xa also positively regulates Factor VII of the extrinsic pathway.)

Although it is the aggregation and cross-linking of fibrin molecules that directly result in clot formation (see below), **thrombin** formation is the committed step in clot formation. Thrombin positively regulates the intrinsic pathway factors Factor XI as well as Factor VIII, the terminal step of the intrinsic limb of the clotting cascade. The activation of Factor VIII perpetuates coagulation, causing more and more clot to form. It is this widespread availability of fibrinogen and the thrombogenic versatility of thrombin that allow for a prompt response to hemorrhage, no matter where it occurs.

THE FIBRIN MESHWORK

Unlike other clotting factors, proteolysis does not transform **fibrinogen** into an active serine protease. Rather, the cleavage of fibrinogen results in the formation of **fibrin** monomers. Fibrin monomers cross-link with one another, forming a netlike fibrin mesh that ensnares RBCs, putting an end to the spilling of RBCs by forming a clot.

Fibrinogen, a large (340 kD), soluble hexamer (six subunits), is formed by two trimers made of α, β, and γ chains (Figure 14-16A). The two heterotrimers are linked through disulfide bonds. The secondary and tertiary structures (Chapter 1) form three globular domains: one between the two heterotrimers and one on each end of the fibrinogen molecules. The N termini of the α and β chains project from the center globular domain and are the sites where thrombin cleaves off cysteines (Figure 14-16B).

Following proteolysis, fibrin aggregates to form a soft clot, which connects the central globular domains to the C-terminal domains of other (different) fibrin molecules. Factor XIIIa stabilizes the soft clot into a hard clot. Factor XIIIa catalyzes the formation of a cross-link between one end of a globular domain and a different fibrin's globular domain, stabilizing the soft clot into a hard clot (Figure 14-17A–B).

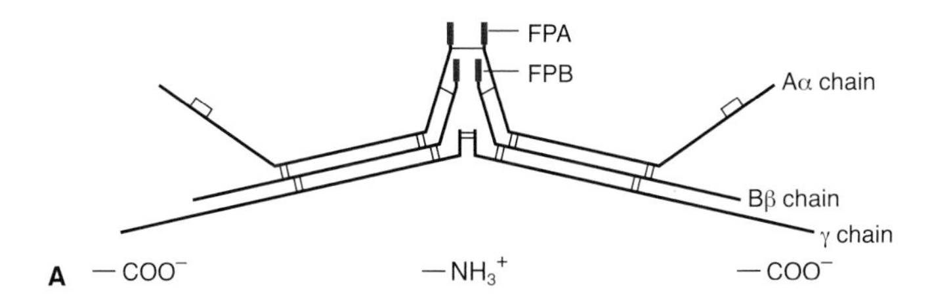

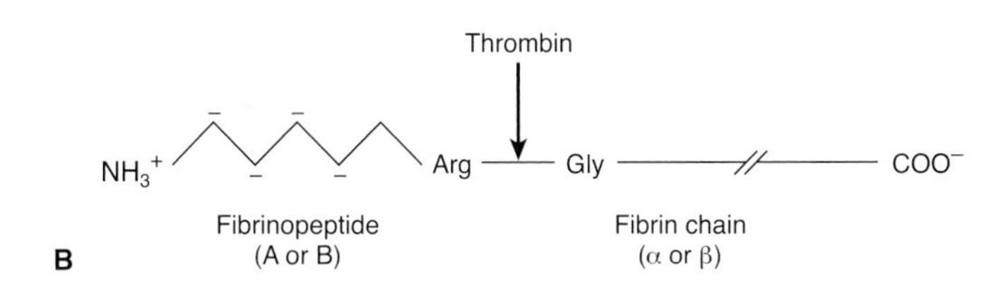

Figure 14-16. A–B. Conversion of Fibrinogen to Fibrin. (**A**) Fibrinogen, a soluble hexamer consisting of two α, two β, and two γ chains, which are cross-linked by disulfide bonds (note lines between single peptide chains above), undergoes cleavage at α (FPA) and β (FPB) chains to form fibrin. (**B**) Thrombin action on FPA or FPB to produce the α- or β-fibrin chains. FP, fibrinopeptide. [Adapted with permission from Murray RA, et al.: Harper's Illustrated Biochemistry, 28th edition, McGraw-Hill, 2009.]

DIFFERENCE BETWEEN PLATELET PLUG FORMATION AND CLOT FORMATION

Platelet plug formation is a *temporary* repair until a proper clot forms. Deficient platelet plug formation results in epistaxis (bloody nose) and purpurae (tiny red/purple spots of bleeding) on the mucosa (especially lips) and underneath the skin. Deficient clot formation, on the contrary, results in *rebleeding*. For instance, were one to have a molar tooth removed, a platelet

Hemophilia refers to a group of diseases characterized by frequent hemorrhages, both spontaneous and traumatic. This may manifest as hemarthroses, easy bruising, and, in general, an increased amount of time needed to form a clot. Hemophiliacs require more time to form clots due to **clotting factor defects**. **Hemophilia A** (80% of hemophilia patients) results from **Factor VIII deficiency** and **Hemophilia B** (20% of hemophilia patients) from **Factor IX deficiency**. Patients with hemophilia have less than 5% of the normal factor levels as compared with non-hemophilia patients, and patients with severe hemophilia have less than 1% of normal levels (usually having no functional Factor VIII or Factor IX). Patients with hemophilia require regular intravenous infusions of purified Factor VIII or Factor IX protein to prevent small bleeds from turning into life-threatening hemorrhages.

von Willebrand's Disease (vWD) is a mild, inherited bleeding disorder that manifests as easy bruising, bleeding gums, and frequent nosebleeds. Despite multiple, hereditary, and acquired forms, the condition stems from a deficiency, in quantity and/or quality, of **vWF**. Bound to Factor VIII, vWF is responsible not only for platelet adhesion, but also for the survival of Factor VIII. Without vWF, Factor VIII is quickly degraded and patients experience increased bleeding times. Patients with vWD, therefore, have a deficiency of both platelet plug formation and clot formation. Treatment is often not necessary unless control of bleeding is required (e.g., excessively long menses, surgery), and consists of giving either additional Factor VIII or by the medication **desmopressin**, which helps to raise the vWF level.

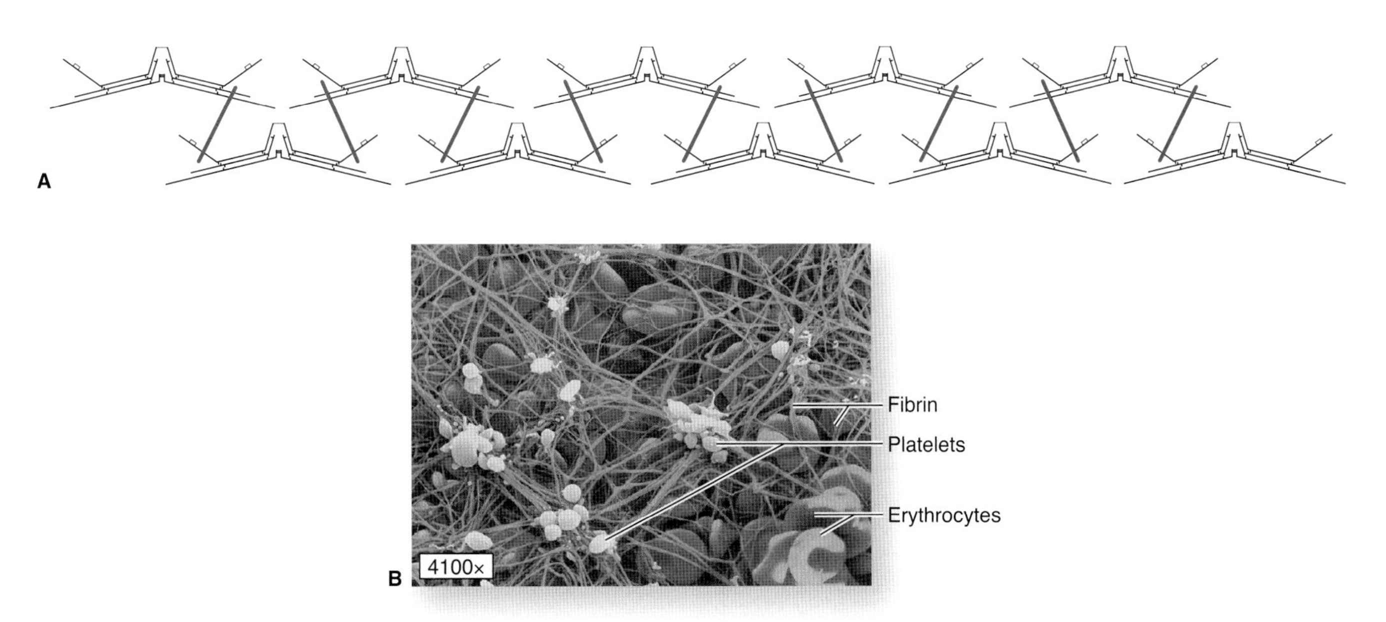

Figure 14-17. A–B. Formation of a Hard Clot. (**A**) Fibrin molecules aggregate to form a soft clot. Factor XIII catalyzes cross-link formation (red lines) among fibrin molecules, transforming the soft aggregate into a hard clot. [Adapted with permission from Murray RA, et al.: Harper's Illustrated Biochemistry, 28th edition, McGraw-Hill, 2009.] (**B**) Scanning electron microscopy image of a clot, illustrating the meshwork of fibrin, erythrocytes, and platelets in various states of degranulation. [Reproduced with permission from Mescher AL: Junqueira's Basic Histology Text and Atlas, 12th edition, McGraw-Hill, 2010.]

plug would stop the bleeding temporarily. A few hours later, however, the gums will rebleed as the plug is only a temporary fix and the individual lacks the permanent fix of a clot.

REGULATION OF CLOT FORMATION

Fibrinogen and other clotting components freely exist in the circulation. Without regulatory mechanisms, thrombin would continue to activate Factors V, VIII, and fibrinogen, and clots would form indiscriminately and at inappropriate times. Three major regulators serve to control clot formation: antithrombin III (ATIII), protein C, and protein S. A deficiency in ATIII, protein C, and/or protein S leads to a hypercoagulable state, making patients vulnerable to inappropriate and potentially dangerous clot formation.

ATIII inactivates Thrombin, IXa, Xa, and XIa. ATIII's activity is also enhanced many thousands of times by the molecule **heparin** (Chapter 2). Heparin exists in a wide range of molecular sizes, determined by its variable number of oligosaccharides. The number of oligosaccharide units determines heparin's anticoagulation effects on ATIII. For instance, in order to inhibit thrombin, ATIII requires a heparin with more than 18 saccharide units. If ATIII encounters the pentasaccharide form (five saccharide units) of heparin, then ATIII inhibits Factor X.

$$\text{ATIII} - \text{Heparin}\ (>18) + \text{Thrombin} \rightarrow \text{Thrombin inhibition}$$

$$\text{ATIII} - \text{Heparin}\ (5) + \text{Factor Xa} \rightarrow \text{Xa inhibition}$$

Protein C and **protein S** work together to degrade and inactivate Factors Va and VIIIa. Protein C is activated as a *membrane-bound* complex consisting of the proteins **thrombomodulin**, **thrombin**, and **calcium (Ca^{2+})** plus **phospholipids** (see Figure 14-18). As protein C requires protein S as a cofactor, both protein C and protein S are regarded as coagulation inhibitors.

Vitamin K, Clotting, and Warfarin: Synthesized by gut bacteria, vitamin K (Chapter 10) is a cofactor for γ-glutamyl carboxylase, the enzyme that carboxylates (activates) prothrombotic Factors II, VII, IX, and X and antithrombotic regulatory proteins C and S. **Vitamin K epoxide reductase** is required to return oxidized vitamin KO back to its active form, vitamin KH_2 (see figure below). Warfarin, a medication used to limit clotting, blocks epoxide reductase, leading to lowered levels of active vitamin KH_2 and, therefore, active clotting factors.

Descarboxy-prothrombin
CO_2 Glutamyl carboxylase
Prothrombin
O_2
KH_2
Vitamin K epoxide reductase
KO
Warfarin

Adapted with permission from Katzung BG, et al.: Basic and Clinical Pharmacology, 11th edition, McGraw-Hill, 2009.

The balance between the body's normal clotting level and its inhibition by warfarin is not simple, and patients utilizing this drug must be constantly monitored to insure both efficacy and safety of its use. If vitamin K is deficient, clot formation may be hindered and chronic bleeding may result. Vitamin K deficiency can also be caused by antibiotics that destroy the gut flora, thereby decreasing the amount of vitamin K available for absorption. Vitamin K deficiency is treated by giving supplemental vitamin K either orally or intravenously.

PLASMIN AND CLOT DISSOLUTION

Breakdown of the fibrin-based meshwork and dissolution of clots relies on the serine protease **plasmin**, which is formed via activation of its zymogen **plasminogen**. The conversion of plasminogen to plasmin occurs via **tissue plasminogen activator (tPA)**, **urokinase plasminogen activator (uPA)**, the peptidase **kallikrein**, or coagulation **Factor XII** (also known as **Hageman factor**). Activation occurs via cleavage of plasminogen between arginine 560 and valine 561 (Figure 14-19). Further cleavage of plasmin can produce the molecule **angiostatin**, which limits the new growth of blood vessels. Research is currently attempting to use this molecule for cancer and other medical treatments.

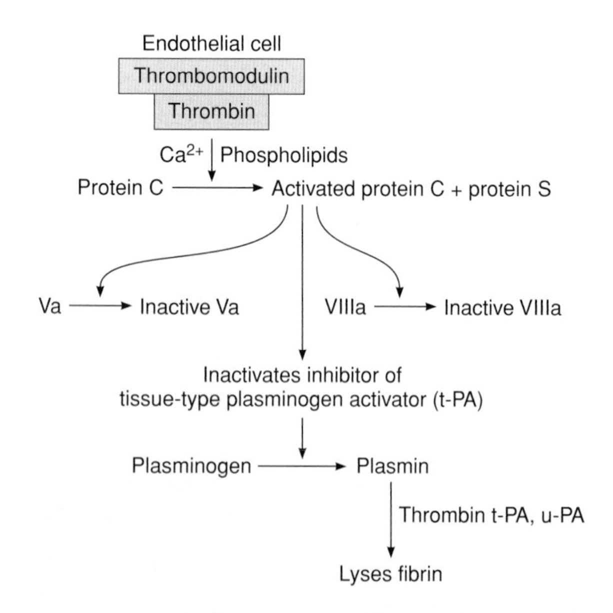

Figure 14-18. Regulation of the Fibrinolytic System by Protein C. u-PA, urokinase-type plasminogen activator. [Reproduced with permission from Barrett KE, et al.: Ganong's Review of Medical Physiology, 23rd edition, McGraw-Hill, 2010.]

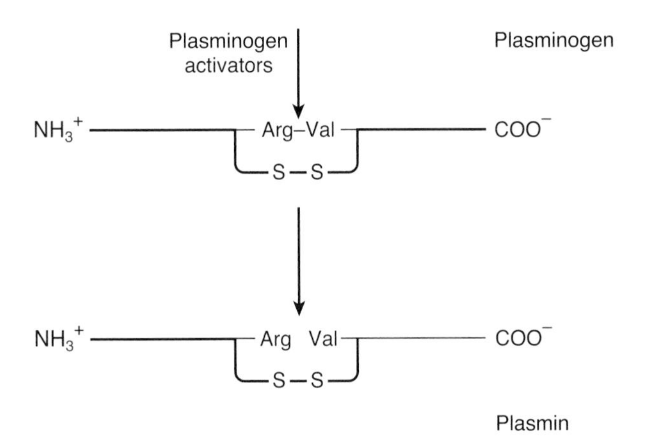

Figure 14-19. Activation of Plasmin. Plasmin is formed from plasminogen by the cleavage of an arginine (Arg)–valine (Val) bond via any one of several plasminogen activators. The two chains of plasmin are held together by a disulfide bridge. [Reproduced with permission from Murray RA, et al.: Harper's Illustrated Biochemistry, 28th edition, McGraw-Hill, 2009.]

Factor V Leiden: Factor V Leiden polymorphism changes arginine to glutamine in the Factor V clotting factor. The arginine change is present in the cleavage site that protein C uses to degrade and regulate Factor V. Thus, in the Factor V Leiden variant, protein C is less efficiently able to regulate Factor V, leading to higher than normal levels of active Factor Va. High levels of active Factor Va lead to excessive **thrombin** production, **fibrinogen** activation, and clot formation. This higher level of Factor Va activity creates a hypercoagulable state. People with the Factor V Leiden variant are prone to developing blood clots in the legs (**deep vein thrombosis, DVT**), which can embolize (break off and travel in the circulation). These emboli can lead to subsequent emergencies, including right-sided heart clots, pulmonary embolism, transient ischemic attacks (temporary strokes), and, in pregnant women, a small increase in miscarriages.

Clot Dissolution and D-dimers: Clot formation (**embolism**) or breakage of an existing clot (**thrombus**) is one major cause of **myocardial infarction** (heart attack) and **cerebral vascular accidents** (strokes). Treatment of these "thromboembolic events" is via a number of "clot-busting" medications, which include **streptokinase**, **uPA**, and anyone of several genetically produced **tPAs** (see figure). All clot busters lead to the breakdown of clots (thrombolysis) by activating plasminogen and have allowed clinicians to effectively treat early presentations of heart attacks, strokes, and other medical disorders. Free plasmin, released from a dissolved clot, is subsequently bound to α_2-antiplasmin to stop its actions.

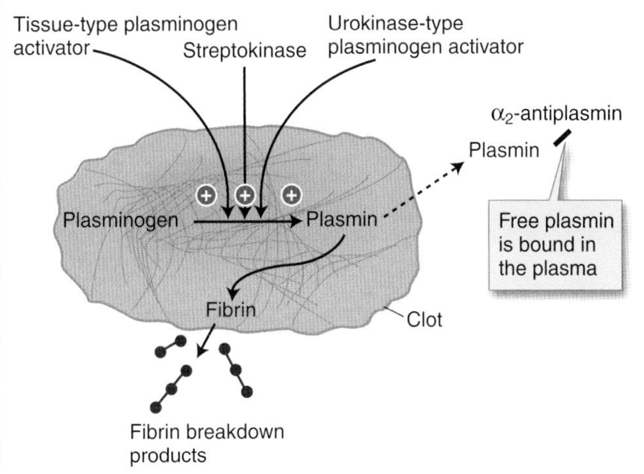

Reproduced with permission from Kibble JD and Halsey CR: The Big Picture: Medical Physiology, 1st edition, McGraw-Hill, 2009.

Fibrin breakdown products (see the figure), of which one major type is "**D-dimers**" (named as such because their molecular structure appears as two connected "Ds"), offer a method to directly measure the activity of plasmin. In medical conditions such as **DVT** or **pulmonary embolism**, measurement of the level of D-dimers allows clinicians to both confirm and grade the extent of an existing clot by relying on this side product of the body's efforts to breakdown clots.

REVIEW QUESTIONS

1. What are the basic components of the blood and their functions?
2. What are the functions of red blood cells?
3. What is the general structure of hemoglobin?
4. How do defects in the α-globin or β-globin genes lead to the various types of thalassemias?
5. What are the functions of proximal histidine, iron, and the heme group in hemoglobin in relation to oxygen binding?
6. How does carbon monoxide poisoning result in an anaerobic-like state?
7. How does the oxygen–hemoglobin binding curve explain how hemoglobin can be saturated by O_2 in the lungs, but only partially saturated in peripheral tissues?
8. What is the Bohr effect and what is the mechanism by which it promotes O_2 dissociation at tissues that require O_2?
9. What is the role of hemoglobin in acid–base balance?
10. What are the roles of transferrin, ferritin, and hepcidin in iron homeostasis?

11. What is the most common cause of nutritional anemia and why may this be particularly prevalent in children and females of childbearing age?
12. What is the role of von Willebrand Factor in blood clot formation?
13. What are the three parts of platelet plug formation?
14. What are the differences between the intrinsic and extrinsic pathways of the clotting cascade?
15. What are the events associated with the common pathway of the clotting cascade?
16. What is the role of proteolysis in the clotting cascade?
17. What are the roles of protein C, protein S, and antithrombin III in limiting clot formation (anticoagulation) and how does a deficiency of any of these components lead to a hypercoagulable state?

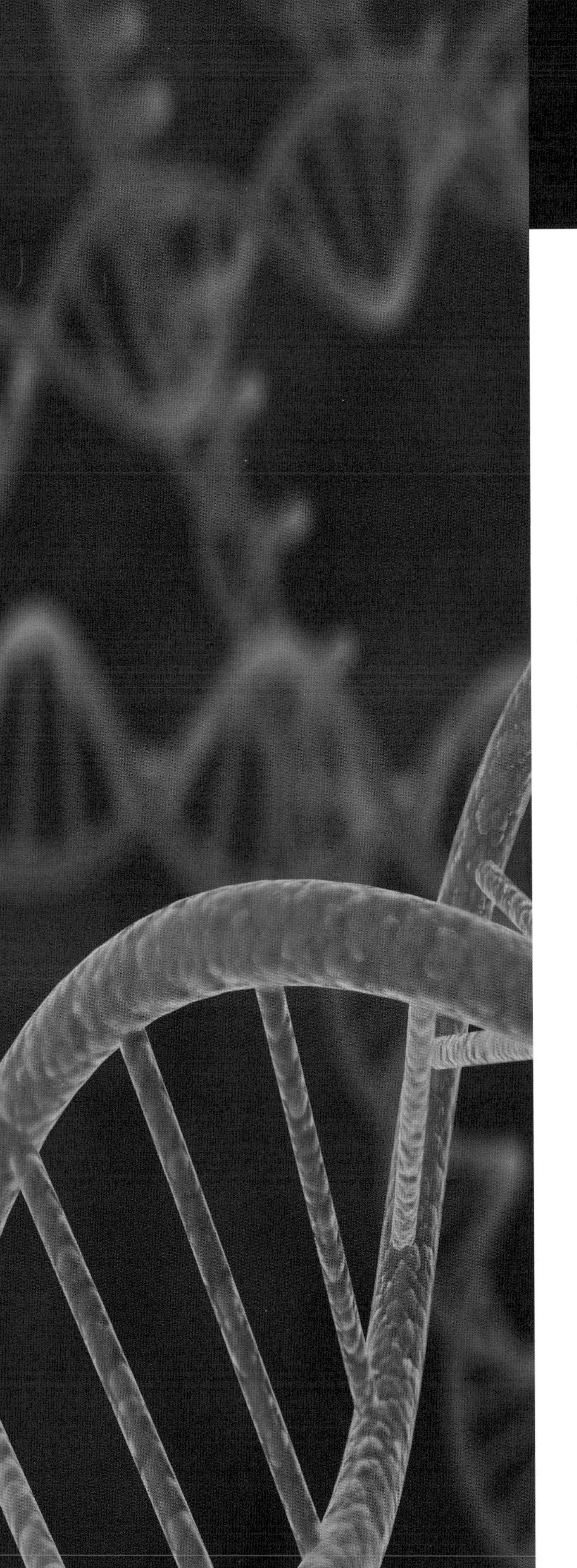

CHAPTER 15

THE IMMUNE SYSTEM

Editor: Eric L. Greidinger, MD
Staff Physician, Miami VAMC Rheumatology Section and Associate Professor, Division of Rheumatology and Immunology, Leonard M. Miller School of Medicine, University of Miami, Miami, FL

OVERVIEW

The immune system, composed of antigens, antibodies, and several specialized leukocyte (white blood cell) types, offers the human body adaptable protection against infections and invasion by foreign molecules and cells. Besides responses by antibodies and cells, the immune system is also composed of a complement-activated pathway that allows an alternative to respond to infectious organisms.

Although able to attack these pathogens, the immune system is also able to recognize the host human cells and molecules and can selectively not respond to "self." Uncontrolled immune responses can lead to tissue damage or death.

OVERVIEW OF THE IMMUNE SYSTEM

The human immune system is responsible for generating a protective response to infective organisms (e.g., bacteria, viruses, and parasites) and foreign cells (e.g., tumor and transplant) while both recognizing the host body and limiting damage to itself (self–nonself theory). As attacks are always evolving, the immune system must be capable of adapting to each new challenge. The immune system also provides an inflammatory response to trauma. When tissue or cells are damaged, molecules that act as endogenous **danger signals** can be released. These danger signals may include molecules such as uric acid (produced by purine metabolism, Chapter 4), which, if present at high concentrations, can form crystals that innate immune sensors recognize and thus activate immune and inflammatory responses.

Innate immunity refers to preformed host defense systems that can provide immediate host defense activities without first being trained to distinguish self from invader. Elements of the innate immune system often contain structural recognition motifs that allow them to identify likely pathogens to target. Molecules that are found in microbes without structural homologs in human cells, such as flagellin or unmethylated deoxyribonucleic acid (DNA), are other examples of **danger signals** that induce innate immune activation when their structure is recognized by innate immune sensors.

The **adaptive immune system** is capable of generating a response very specific to the structure of an invading organism or molecule (**antigen**, see below) as well as the ability to remember that structure via memory cells (**immunological memory**). The repeat activation of the immune system by an antigen that has previously induced an immune response is normally much quicker and stronger. The invading pathogen serves as a target, which elicits a response specific for that antigen from immune cells. This response may include the production of a specific **antibody** against that pathogen (**humoral immune system**) or the activation of immune cells that either attack and kill the offending organism or orchestrate this activity, the **cell-mediated immune system**. These cell types include lymphocytes (including T and B lymphocyte types), monocytes, macrophages, dendritic cells (DCs), neutrophils, eosinophils, and basophils.

Passive immunity involves antibody molecules that are transferred to the baby from the mother's active immune system through the placenta. **Immunoglobulin (Ig) G** (see below) is the single antibody type that binds to a **neonatal Fc receptor** and is endocytosed via **pinocytosis.** Newborns also receive **IgA** antibodies (see below) via breast milk, which provides initial protection against pathogenic microorganisms. Passive immunity is considered to be a short-term system, which is taken over in the first year or two of life by the adaptive immune system.

Immunodeficiencies: Medical conditions known as **immunodeficiencies** are characterized by the absence or malfunction of a component of the immune system, leading to the inability to fight infectious disease. More than 120 congenital immunodeficiencies have been recognized. External causes of immunodeficiency include poor diet, the effects of immunomodulating drugs, and infectious diseases including human immunodeficiency virus (HIV)/acquired immune deficiency syndrome.

ANTIGEN

Antigen, originally termed from the phrase "antibody generator," is a molecule (normally protein or carbohydrate in source) that can stimulate the immune system to make antibodies and/or initiate a cell-mediated response. Self-antigens are present on all cells of the host organism. Self-reactive immune cells are typically deleted or maintained in states of impaired reactivity to prevent the development of autoimmunity. Many pathogenic microbes and cancer cells can use some of these same mechanisms to evade immune attack. Other "opportunistic" organisms may induce aggressive responses from intact immune systems (that effectively kill the organism) but can still lead to disease in immunocompromised hosts.

The specific structural surface recognized by cells of the adaptive immune system is known as an **epitope**. Several antibodies may recognize various parts of one antigen (various epitopes). To be recognized by T cells in the cell-mediated immune system, antigens must be presented on the cell surface of particular immune cells (see below).

ANTIBODY

An **antibody** is a large, Y-shaped protein (Figure 15-1) responsible for immunological identification and binding of antigens and, ultimately, for protection against invading pathogens. Another name for an antibody is **immunoglobulin**, normally abbreviated as **"Ig."**

The functions of antibodies are summarized in Figure 15-2.

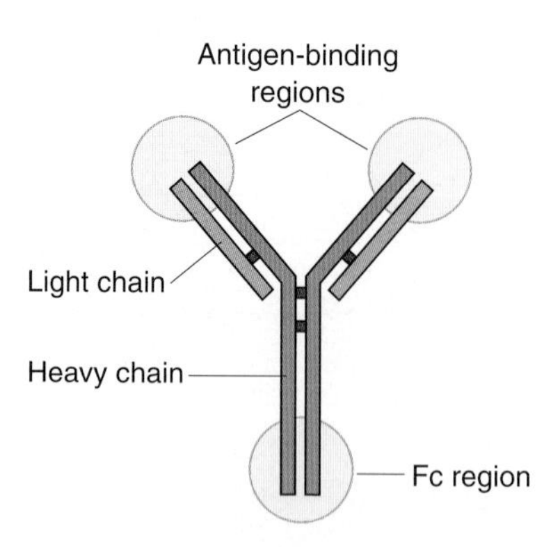

Figure 15-1. Basic Structure of an Antibody Molecule. The antibody molecule is a Y-shaped, tetramer protein, composed of two heavy chains and two light chains, held together by hydrophilic/hydrophobic forces and disulfide bonds. The body of "Y" is composed of a constant protein structure and a highly variable end, which defines the affinity of a particular antibody for one antigen epitope. The Fc region of the molecule may bind to the surface of receptors of several cell types. [Reproduced with permission from Mescher AL: Junqueira's Basic Histology Text and Atlas, 12th edition, McGraw-Hill, 2010.]

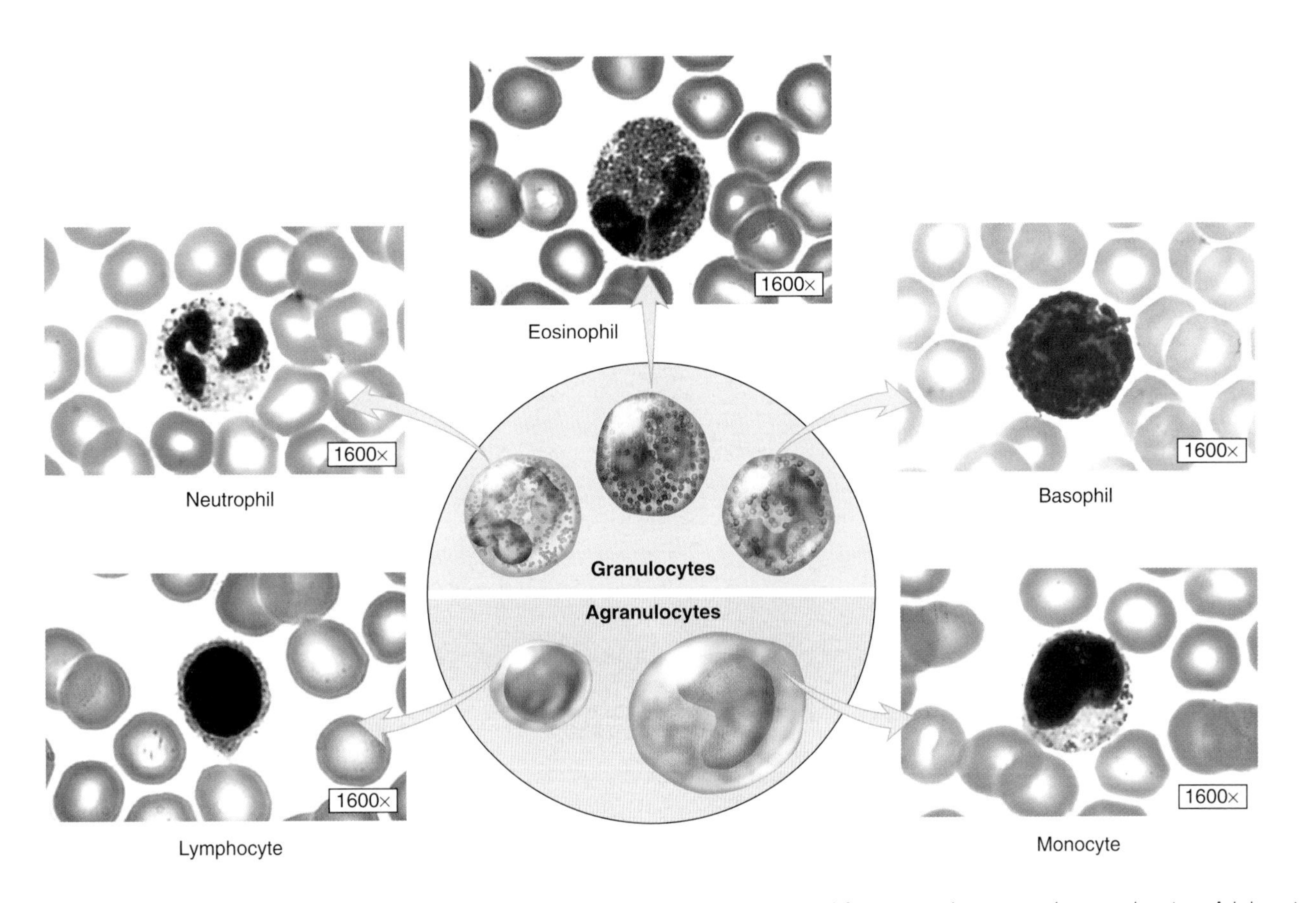

Figure 15-3. White Blood Cells. Overview of white blood cells, including those derived from granulocytes and agranulocytes. A laboratory full blood count usually includes total white blood cells and a differential measurement of the five major types of white blood cells. Normal values are as follows: neutrophils (45%–60% of total), lymphocytes (25%–35% of total), monocytes (3%–7% of total), eosinophils (1%–3% of total), and basophils (<1% of total). See Appendix II for further details. [Reproduced with permission from Mescher AL: Junqueira's Basic Histology Text and Atlas, 12th edition, McGraw-Hill, 2010.]

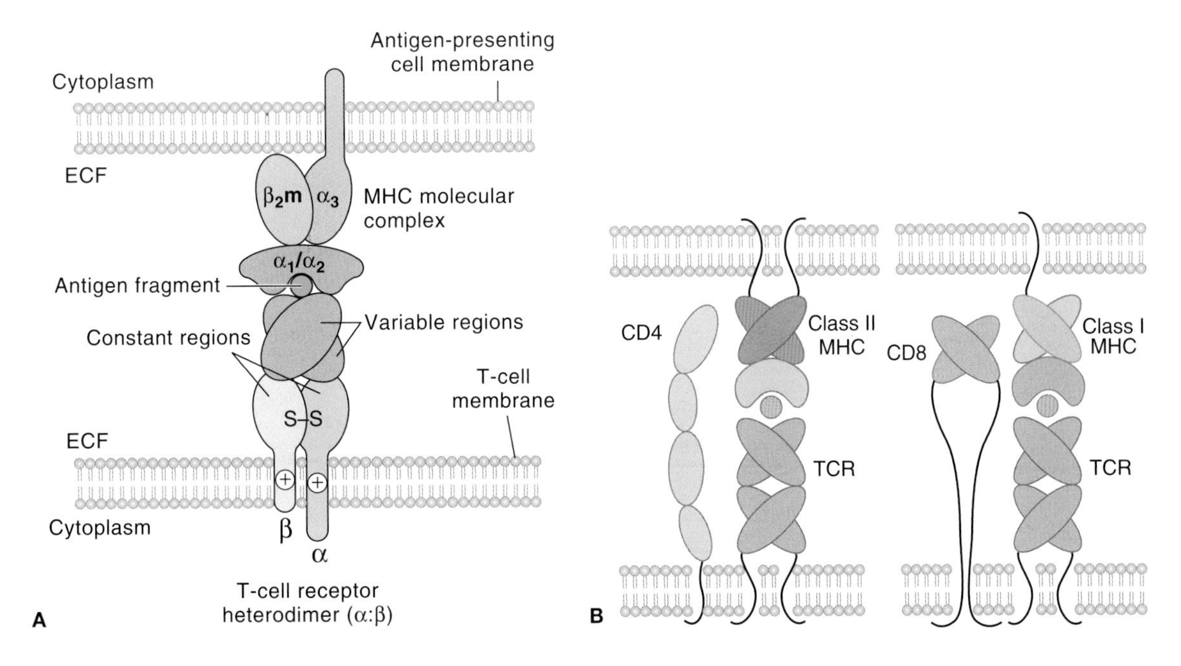

Figure 15-4. A. T-Cell Receptor Recognition of Antigen. Basic structure of the heterodimeric T-cell receptor (bottom), showing the α- and β-subunits and the constant and variable region that recognize the epitope of the antigen fragment presented by an antigen-presenting cell (top). ECF, extracellular fluid; MHC, major histocompatibility complex. **B. Binding of T Cells to MHC Complex.** CD4 (helper T) cells have a T-cell receptor (TCR) and CD4 coreceptor that recognize antigen (blue circle) bound to MHC Class II receptor on an antigen-presenting cell (left panel). CD8 (cytotoxic T) cells bind via their TCR to antigen (blue circle) presented by MHC Class I molecules (right panel). MHC, major histocompatibility complex; TCR, T-cell receptor. [Reproduced with permission from Barrett KE, et al.: Ganong's Review of Medical Physiology, 23rd edition, Mc Graw-Hill, 2010.]

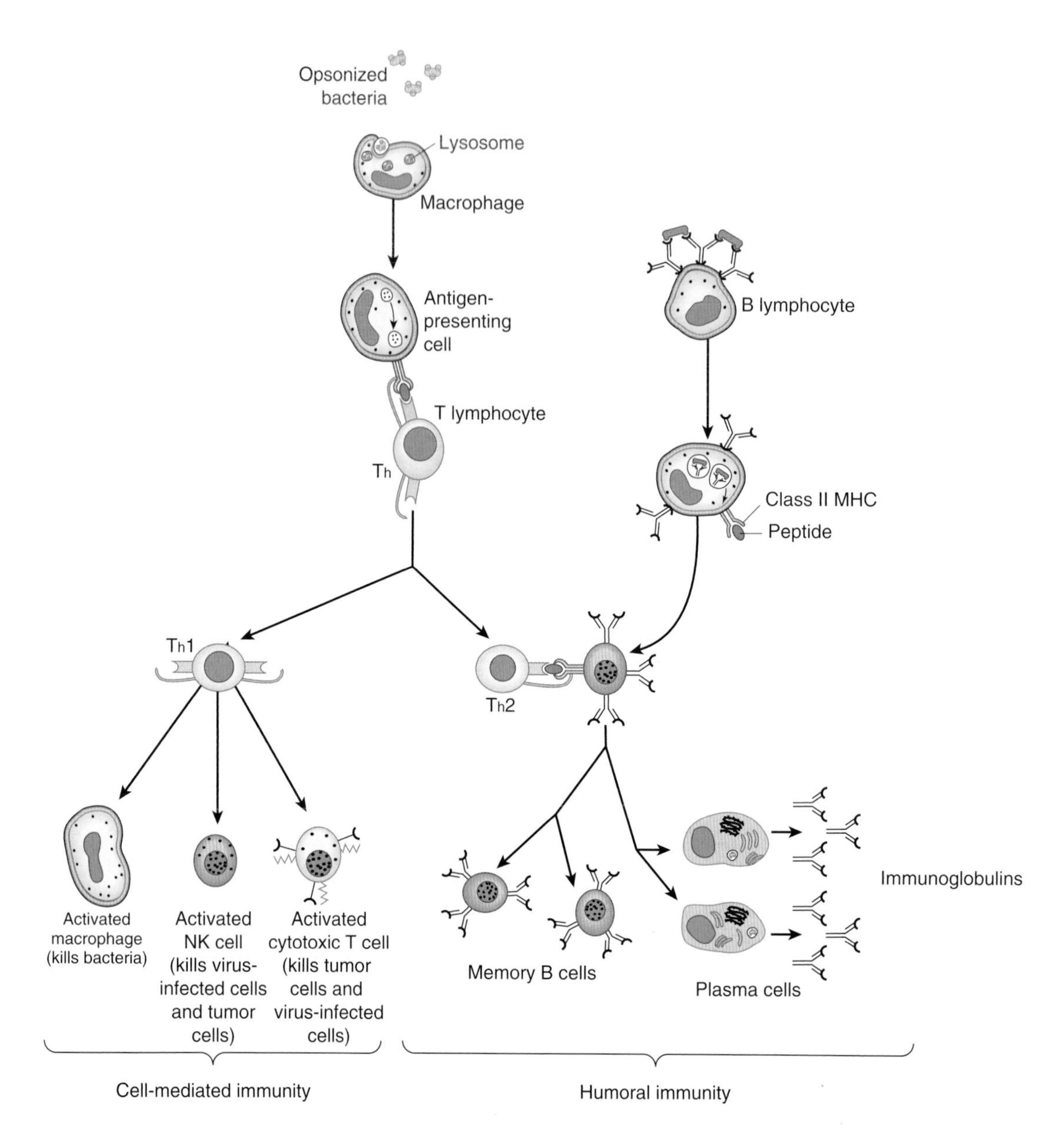

Figure 15-5. Comparison of Cell-Mediated and Humoral Immunity. Cell-mediated immunity (left) involves the presentation of antigen derived from ingested foreign molecules (e.g., bacteria) for T helper (Th) lymphocyte recognition and Th1 activation of cells such as macrophages, natural killer, and cytotoxic T cells (see the text). Humoral immunity (right) results from recognition of the foreign molecule by an antibody, which is already expressed on a B lymphocyte followed by interaction with a Th2 cell and differentiation into memory B and plasma cells, the latter of which express large amounts of antibody against the offending molecule. MHC, major histocompatibility complex; NK, natural killer. [Adapted with permission from Katzung BG, et al.: Basic and Clinical Pharmacology, 11th edition, Mc Graw-Hill, 2009.]

self-antigens. Pathways of action of some regulatory T cells include (a) upregulated expression of **transforming growth factor-β (TGF-β)**, a cytokine that upon binding to its receptor activates SMAD proteins that act as transcriptional regulators important in immune effector function and cell cycle control (see Chapters 9 and 20); (b) upregulated **interleukin-10**, which decreases activation of Th cells and blocks NF-κB signaling pathways; and (c) induction of indoleamine 2,3-dioxygenase, an enzyme that depletes tryptophan in the local area preventing optimal function of T cells, which depend on a ready supply of this amino acid.

As the precursor to a T-cell response, invading bacteria, viruses, and cells are processed and presented to T cells. Internally derived pathogens (e.g., a virus replicating inside a cell) are digested into small peptides within cells by a **proteosome**, a specialized center of protein-cleaving enzymes within the cell also responsible for breaking down damaged or misfolded proteins and processes such as cell cycle control. Peptides digested from cellular proteins (including those of invading organisms) are taken up by **TAP transporters (Transporter associated with Antigen Processing)** found in the endoplasmic reticulum and sorted into vesicles, which travel to the cell surface where they can associate with Class I MHC molecules for presentation to $CD8^{+}$ (cytotoxic) T cells. Pathogens outside the cell (e.g., bacteria, parasites, and/or toxic molecules) are normally taken up by specialized **APCs** including DCs. APCs express Class II MHC molecules that can present peptides to $CD4^+$ (helper) T cells. Dendritic cells are specialized to take up the exogenous pathogen and digest them via enzymes in lysosomes. The resulting small peptide fragments from either the proteosomes or lysosomes are coupled with the appropriate class of MHC molecule and moved to the cell surface. Peptides binding to Class I MHC are typically no more than 8 amino acids long, whereas Class II MHC accommodates longer peptides and can present chains of up to 14 amino acids in length.

The antigen–MHC complex is presented by the APCs and recognized by either cytotoxic T or Th cells along with the appropriate CD4 or CD8 molecule. This interaction involves very specific recognition of primary amino acid sequence of the antigen along with any secondary structure that may exist. This antigen structure is also presented in the context of the MHC and illustrates how remarkably exact the generation and presentation of minute differences in protein structure must be. A recognized antigen will elicit T-cell activation and can lead to an immune response; an unrecognized antigen will be simply ignored by the T lymphocyte. To induce an immune response, activated T cells must receive second signals from APCs that promote an immune response. The second signal ligands on APCs are expressed on activated APCs, typically in response to a danger signal. In the absence of an appropriate second signal, activated T cells will undergo programmed cell death (apoptosis) or lose the ability to respond to subsequent TCR ligation (develop anergy).

B LYMPHOCYTES/PLASMA CELLS

B lymphocytes are the antibody-producing cells of the immune system. Antibody-dependent immune response is known as **humoral immunity**. Each B cell has an antibody molecule on its surface, which serves as a type of receptor against a particular pathogen molecule. When a B-cell receptor recognizes and binds to its antigen, the B cell becomes activated. Activated B cells secrete antibody molecules and immunostimulatory cytokines (Figure 15-5, right).

Some activated B cells can also act as APCs; the antibody–antigen complex can be internalized and the antigen processed and displayed via an MHC Class II molecule on the B cell surface. B cells can respond to T cells and danger signals by switching the kind of antibody molecules they produce (class switching) from larger, less efficient IgM forms to smaller, more efficient IgG forms. Other specialized forms of antibody molecules are also produced in association with mucosal immunity (IgA) and in association with allergic-type responses (IgE). These antibodies circulate and bind to the target pathogen for attack and removal by the complement system, T cells, and phagocytic cells including neutrophils (see below). Following an immune response, a small number of the activated B lymphocytes and some T lymphocytes become memory cells to maintain recognition of the offending antigen in the case of future infection. Memory cell activation results in a faster and stronger immunological response. During an active immune response, cytokine stimulation can induce some B cells to differentiate into plasma cells, a population of long-lived cells that produce sustained high levels of antibodies. The protective immunity benefits of vaccines are mediated in part by the induction of pathogen-specific antibody-producing plasma cells.

NATURAL KILLER (NK) CELLS

NK cells are another type of lymphocyte-like immune cell. They differ from other lymphocytes in that they do not express a TCR or surface antibodies. As in cytotoxic T cells, NK cells attack infected cells and tumors by the production and insertion of **perforin/granulysin** and **granzyme**, which lead to their death. The activity of NK cells is closely regulated by various cytokines (see below), activating and inhibitory receptors on their cell surface, and an antibody-binding **Fc receptor** that provides a target for NK activity. In contrast to cytotoxic T cells that attack cells presenting a foreign peptide in a proimmune context, NK cells are capable of attacking cells that fail to express surface molecules that identify the cells as self.

MONOCYTES AND MACROPHAGES

Monocytes are a major type of immune cell responsible for the ingestion or phagocytosis of invading organisms and foreign/toxic molecules. Monocytes and cells differentiating from them are capable of recognizing danger signals and producing large quantities of proinflammatory cytokines such as interleukin-1 (IL-1) and tumor necrosis factor-α (TNF-α) to produce or perpetuate immune responses. To allow for more efficient phagocytosis, monocytes require the **opsonization** or coating of the ingested targets by either antibodies or complement (see below), although they can also recognize the outside surface of certain pathogens. Monocytes respond within 8–12 h of infection, attracted by cytokines and danger signals. Monocytes further differentiate into **macrophages** in tissue, where they serve as local defenders of immunity. Monocytes can also differentiate into **DCs**, which serve a primary function as an APC.

NEUTROPHILS

Neutrophils are the most abundant type (45%–60%) of leukocyte (white blood cell) in the human body and are one of the first immune cells to arrive at the site of a new infection and/or inflammation. As such, neutrophils are a primary marker of acute infection/inflammation. Neutrophils can phagocytose invading organisms or particles and release reactive oxygen species such as superoxide that can kill microbial and host cells. However, the killing of engulfed bacteria and other particles is believed to be more reliant on a number of digestive enzymes contained in primary, secondary, and tertiary granules as well as lysosomes. Neutrophils can also kill bacterial and fungal pathogens by forming **neutrophil extracellular traps (NETs)** composed of a unique structure of extracellular protein fibers and DNA chromatin. Besides the physical barrier, NETs also contain a high concentration of antimicrobial proteins and serine proteases.

EOSINOPHILS

Eosinophils mainly function in parasitic infections and allergies/asthma. Although usually comprising less than 5% of circulating leukocytes, their number can rise dramatically during an active response. They stay in the circulation for up to 2 weeks, serving as a marker of such an infection or inflammatory disorder. Degranulation of eosinophils releases a wide variety of molecules, including the following:

- Histamine—increases chemotaxis and blood vessel permeability to allow entrance of other leukocytes into the site of infection.
- **Major basic protein**—highly toxic protein to parasites.
- Inflammatory eicosanoids and leukotrienes (Chapter 3).
- Growth factors and cytokines—lead to continued growth and proliferation of immune cells to continue the immune response.

Histamine: Histamine is a multifunctional signaling molecule produced by the decarboxylation of **histidine** by the enzyme **L-histidine decarboxylase**. Histamine is also found in fermented and/or spoiled beverages and foods. Defects in the breakdown of histamine by the enzyme **diamine oxidase** or **acetaldehyde dehydrogenase** is felt to be responsible for the flushing seen during alcohol intake, especially in certain Asian populations, and some cases of food poisoning. Histamine's roles include that of a neurotransmitter and a stimulant of parietal cells to produce gastric acid. However, its function in allergic inflammatory reactions and immune response are better known.

Histidine ($CH_2—CH(NH_2)—COOH$ on imidazole ring, HN, N) → (CO_2 released) → Histamine ($CH_2—CH_2—NH_2$ on imidazole ring, HN, N)

Adapted with permission from Katzung BG, et al.: Basic and Clinical pharmacology, 11th edition, McGraw-Hill, 2009.

Four receptors are known for the receptor molecule. The **histamine$_1$ (H_1) receptor** acts via G_q protein, which activates the phospholipase C/phosphatidylinositol pathway, leading to dilation of blood vessels (flushing), contraction of smooth muscles in bronchi (breathing problems), contraction of endothelial cells (hives and leakage of fluid), induction of unmyelinated C fibers (pain and itching), and even some forms of motion sickness. Most "antihistamine" medications [e.g., diphenhydramine (Benadryl), loratidine (Claritin), fexofenadine (Allegra), and citirizine (Zyrtec)] are H_1 blockers, as are promethazine (Phenergan) and meclizine (Dramamine or Antivert) for nausea/motion sickness. Nonsedating H_1 blockers differ from the other members of this group by their inability to cross the blood–brain barrier and inhibiting histamine receptors in the central nervous system.

The **H_2 receptor**, located on parietal cells, is the target of the class of reflux medications termed H_2 blockers [e.g., cimetidine (Tagamet), ranitidine (Zantac), and famotadine (Pepcid)]. **H_3 receptors** are of the G_i receptor class, inhibiting formation of cyclic adenosine monophosphate, which leads to the inhibition of release of several neurotransmitters, including acetylcholine, dopamine, γ-aminobutyric acid (GABA), noradrenaline, and serotonin. The **H_4 receptor**, another G_i-linked receptor, is found in bone marrow and basophils/mast cells, which harbor a majority of the body's histamine in the nasal and oral mucosa, intestinal/liver/spleen and lung/trachea mucosa and thymus, testes, and tonsils. The immune response, often activated by IgE antibody, leads to the release of histamine from these sources (mainly mast cells and basophils) increases mast cell chemotaxis and capillary permeability to allow other immune cells access to fight an invading pathogen. The effects are summarized below:

Nose/sinuses: Blood vessel dilation and swelling (congestion), fluid leakage (runny nose), neural stimulation (sneezing), for example, "Hay fever."

Eyes: Fluid leakage (watery eyes) and stimulation of unmyelinated C fibers (itching and allergic conjunctivitis), for example, "Hay fever."

Ears: Swelling of Eustachian tubes leading to blockage (fullness).

Pharynx/lungs: Neural stimulation (coughing), bronchi contraction (wheezing, shortness of breath), fluid leakage (throat swelling), for example, "anaphylaxis."

Skin: Swelling and fluid leakage [hives (urticaria) and rash], for example, wasp or bee stings.

Gastrointestinal tract: Swelling, fluid leakage, neural stimulation (pain, bloating, nausea and vomiting, and diarrhea).

Brain: Effects on posterior hypothalamus (sleep problems).

BASOPHILS

Basophils comprise only about 0.01%–0.03% of white blood cells in plasma and are recruited to sites of inflammation and/or infection, especially by allergies or parasites. Upon activation, basophils release a number of anti-inflammatory molecules, including histamine (promoting blood flow), heparin (anticlotting); proteolytic enzymes such as elastase and lysophospholipase; and leukotrienes (Chapter 3).

DENDRITIC CELLS (DCs)

DCs are specialized to sample molecules from their surroundings and present these as antigen to potentially elicit an immune response (Figure 15-6). Many believe them to be a key to cell-mediated immunity involving T lymphocytes. Although research is still ongoing, DCs appear to arise from cells in the bone marrow and travel to a variety of sites in the body where they differentiate into several subtypes with differing functional

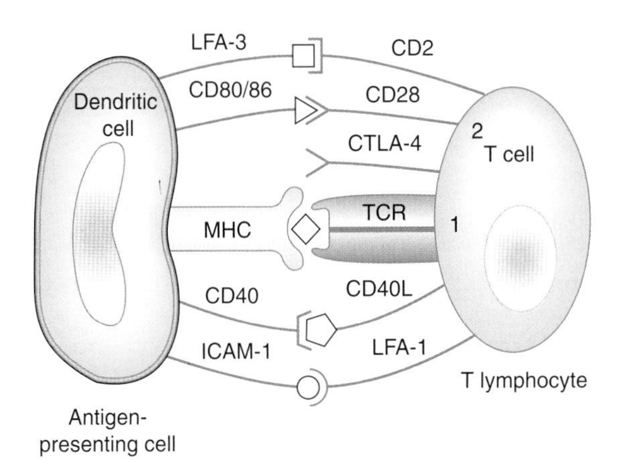

Figure 15-6. Dendritic Cell Presentation of Antigen to a T Cell. Activation occurs via MHC antigen presentation to the T-cell receptor (signal 1) and binding of costimulatory molecules including CD80/86 to CD28 (signal 2) as well as a number of other binding pairs, which strengthen the interaction, leading to a robust immune response. CTLA-4, cytotoxic T-lymphocyte antigen 4; ICAM-1, intercellular adhesion molecule 1; LFA, lymphocyte function-associated antigen 1; MHC, major histocompatibility complex; TCR, T-cell receptor. [Adapted with permission from Katzung BG, et al.: Basic and Clinical Pharmacology, 11th edition, Mc Graw-Hill, 2009.]

characteristics. Developed DCs are mainly found in areas where they are exposed to the external environment such as skin [i.e., Langerhans cells (LCs)], the respiratory tract, and the digestive tract. They are also found in blood and lymph tissue. DCs start as oval "veil" cells but upon differentiation, they form several arm-like processes similar to dendrites of a neuron, which give them their name. They are producers of several interleukin molecules and IFNs (see Table 15-2) involved in primary immune response. DCs have been implicated in the initial stages of viral infection, including HIV and severe acute respiratory syndrome, and in autoimmune diseases. DCs may also play a prime role in the innate immune system by recognizing molecules from bacteria or viruses (danger signals) through **Toll-like receptors** and other danger signal sensors.

CYTOKINES

Cytokines are proteins, sometimes glycoproteins, which are secreted by particular immune cell types as part of intracellular signaling and activation/inhibition of immune processes. Interleukins (involved in leukocyte signaling) and IFN molecules are examples of cytokines. The functions of cytokines vary widely per the cell type with which they interact (see Table 15-2).

Cytokine receptors can be either membrane bound or soluble and fall under six general categories, depending on their structure. (1) **Type I** and (2) **Type II receptors**, which include many of the major cytokines and IFNs, both function via a Janus Kinase **(JAK) mechanism** (Chapter 8). The class called (3) **Ig receptors**, including the IL-1 receptor, activate phospholipase C and subsequent **NF-κB gene** activation. The (4) **TNF receptor group** utilizes an intermediary protein such as **TNF receptor type 1-associated death domain**, which transmits conformational changes from the receptor to binding partners such as TNF receptor-associated factor 2, which interacts with several other signaling peptides to mediate inflammation or programmed cell death (apoptosis) via **NF-κB** gene activation or by **AP-1**, a transcription factor related to **c-Fos** and **c-Jun**. The (5) **chemokine receptor group** functions via G_q proteins, leading to the release of intracellular calcium that elicits directed chemotaxis. The final group, (6) **tumor growth factor β-receptors** are **serine–threonine** kinases that phosphorylate SMAD proteins (see Chapter 13) that can transit to the nucleus and act as DNA transcription inducers and/or suppressors.

Warts: Warts (verruca) from **human papillomavirus (HPV)** infection affect millions of people and can last and/or recur for years despite aggressive medical care. Even surgical removal is not 100% effective. More than 100 strains of HPV are known, causing skin warts as well as **cervical** and other **cancers** of the reproductive tract. HPV is an especially difficult virus to treat, regardless of its location, because of several immune evasion mechanisms it has evolved.

First, HPV infects **keratinocytes**, a cell that is already "destined to die" as layers of growing skin progress to the surface and are shed. These cells are less prone to regular immunological surveillance than other cell types, and they are less apt to respond to danger signals with cytokine release that would activate dendritic and other immune cells—a viral death is seen no differently than a preprogrammed death. Second, viral infection normally invokes an **IFN-α** response (see below). Unfortunately, HPV and similar viruses stop IFN-α production by directly inhibiting its gene expression or its **JAK–STAT** signaling pathways. Finally, external viral structures normally activate **LCs**, specialized DCs of the skin responsible for detecting and responding to localized infections. HPV, though, directly inhibits the activation of LCs and, therefore, any immune response against them. Treatments for skin warts ranging from medications (salicylic acid, silver nitrate, podophyllum, imiquimod, fluorouracil, and even the reflux medication cimetidine) to freezing (cryotherapy) to duct tape to apple cider vinegar have proven successful in some patients. All are based on an attempt to activate the immune system to recognize and respond to the viral infection.

INNATE IMMUNITY

The innate immune system includes the physical barriers of skin (including the continued shedding of skin cells), mucosal linings, mucus, and normal bacteria found in the intestinal tract and other locations, tearing and salivary production, as well as a variety of white blood cells (NK cells, phagocytic/macrophages, neutrophils, and DCs and tissue eosinophils, basophils, and mast cells). The complement system (see below) is also part of this innate system. All provide an almost immediate response to infective microbes by bringing immune cells and inflammatory mediators to the site of invasion, activating biochemical cascades directed against them, and, in most cases, inducing the physical removal of these offending agents.

TABLE 15-2. Selected Immune Cytokines and Their Activities

Cytokine	Producing Cell(s)	Function(s)
Interleukin-1 α and β	Monocytes, macrophages B lymphocytes, dendritic cells (DCs), and fibroblasts.	Costimulation of helper T (Th) lymphocytes. Maturation and proliferation of B lymphocytes. Activation of natural killer (NK) cells. Increases inflammatory response, including eliciting fever.
Interleukin-2	T cells.	Growth, proliferation, and further activation of T and B lymphocytes and NK cells.
Interleukin-4	Th2 cells, some APCs. Initial production may be basophils.	Proliferation and differentiation of Th cells to Th2 cells, which secrete further IL-4. Plays a key role in activation and proliferation of mast cells and of B lymphocytes as well as secretion of immunoglobulin (Ig) G and IgE antibodies. Increases the production of MHC Class II molecules.
Interleukin-5	Th2 cells and mast cells.	Stimulates growth and differentiation of activated B cells and eosinophils to increase Ig secretion, especially IgG and IgA.
Interleukin-6	Monocytes/macrophages, Th2 cells, endothelial cells, osteoblasts, and fat and smooth muscle cells.	Increases differentiation of activated B lymphocytes into plasma cells and promotes the production of antibodies, especially IgG and IgA. Increases differentiation of T lymphocytes into Th2 cells as well as later activation of T lymphocytes and NK cells. Induces fever and acute inflammation, especially in the case of trauma, burns, or muscle overuse.
Interleukin-8 (neutrophil chemotactic factor)	Macrophages, epithelial, and endothelial cells.	Promotes chemotactic movement of neutrophils to a site of infection and inflammation as well as intracellular signaling, increased metabolism, and histamine release by neutrophils. Also promotes chemotaxis of macrophages, mast cells, keratinocytes, and endothelial cells.
Interleukin-10	Monocytes, Th2 cells, B lymphocytes, mast cells, and cytotoxic T cells.	Inhibits antigen presentation, cytokine production by macrophages, and Th1 cells. Stimulates maturation and growth of T lymphocytes and mast cells, growth of B lymphocytes, and plasma cell antibody production. Decreases cell-mediated response by inhibiting APC functions and Th1 cytokine production.
Granulocyte colony-stimulating factor	Monocytes, macrophages, endothelial cells, and fibroblasts.	Stimulate granulocyte precursor cells in bone marrow to produce more neutrophils, basophils, and eosinophils. Also, stimulates growth, proliferation, and function of neutrophils.
Granulocyte-macrophage colony-stimulating factor	Macrophages, T cells, mast cells, endothelial cells, and fibroblasts.	Targets stem cells in the bone marrow. Stimulates growth and differentiation in white blood cells, including granulocytes (neutrophils, eosinophils, and basophils) as well as monocytes and DCs in the bone marrow. Function produces a large increase in these cell types to fight infection.
Type I interferons (IFNs) (IFN-α and IFN-β)	IFN-α—White blood cells (leukocytes), especially a subtype of DC.	IFN-α—Promotes the development of fever by release of prostaglandin-E_2. Stimulates macrophages and NK cells to kill viruses. Increases MHC Class I expression.
	IFN-β: Predominately fibroblasts.	IFN-β—Stimulates macrophages and NK cells to kill viruses. Increases MHC Class I expression.
IFN-γ	NK, cytotoxic T cells, and Th1 cells.	Stimulates proliferation of Th2 cells, MHC expression, and phagocytosis by macrophages and IgG antibody production by B lymphocytes. Can directly inhibit viral replication.

TABLE 15-2. Selected Immune Cytokines and Their Activities (Continued)

Cytokine	Producing Cell(s)	Function(s)
Transforming growth factor-β	T lymphocytes and monocytes.	Regulates inflammatory response, possibly by stimulating development of regulatory T lymphocytes, by generally inhibiting immune cell growth and proliferation. Inhibits activation, antibody production, chemotaxis, and phagocytosis of lymphocytes, NK cells, monocytes and macrophages, mast cells, and granulocytes. This regulation is important in the control of self/non-self response. Upregulates components of wound healing responses including fibroblast activation.
Tumor necrosis factor-α, also known as cachexin or cachectin.	Mainly macrophages but also lymphocytes, NK, and DCs, mast cells, fibroblasts, endothelial cells, cardiac, fat, and neuronal cells.	Involved in inflammation response both acute and ongoing, including increased cytokine and cell adhesion molecule expression. Also, inhibits replication of viruses and promotes cell death of tumor cells.

APCs, antigen-presenting cells; MHC, major histocompatibility complex.

Autoimmune Diseases and Immunosuppressants: The ability of the immune system to differentiate between foreign "non-self" and host "self" antigens is important to control immune response and the resulting cell killing. A loss of this control and the subsequent attack on the host body can lead to autoimmune diseases. These diseases include Graves' disease and Hashimoto's thyroiditis, type 1 diabetes, vitiligo, pernicious anemia, rheumatoid arthritis, systemic lupus erythematosus, multiple sclerosis, Sjogren's syndrome, polymyositis, myasthenia gravis, and Goodpasture's syndrome. A mainstay of treatment of autoimmune disorders is immunosuppressive drugs. These, unfortunately, generally place the patient at increased risk for infections by nonspecifically impeding immune functions. Many immunosuppressant drugs can cause other side effects. Main categories of these medications include the following:

1. Glucocorticoids (e.g., cortisol, prednisone, and dexamethasone)—alter cell gene transcription with effects including reducing cytokine expression to limit cell-mediated and humoral immunity; and inhibiting phospholipase A2–arachidonic acid interaction, thereby blocking inflammatory pathways.
2. Cytostatics (e.g., cyclophosphamide, azathioprine, and mycophenolate)—inhibit cell division of T and B lymphocytes.
3. Immunophilin inhibitors (e.g., cyclosporin and tacrolimus)—block production of cytokines (e.g., IL-2) and or cell cycle proteins that promote T lymphocyte growth, proliferation, and activity.
4. TNF-α inhibitor—proteins or monoclonal antibodies that block the proinflammatory effects of TNF-α, including induction of IL-1 and interleukin-6.
5. Biological therapies—typically monoclonal antibodies that interfere with a defined aspect of immune function, such as by killing cells that express their antigen target (e.g., rituximab/B cells), blocking second inflammation-associated receptor signal to T cells (alefacept, abatacept, ipilimumab, efalizumab, see the figure), or preventing immune cell migration to sites of inflammation (natalizumab) by blocking integrins.

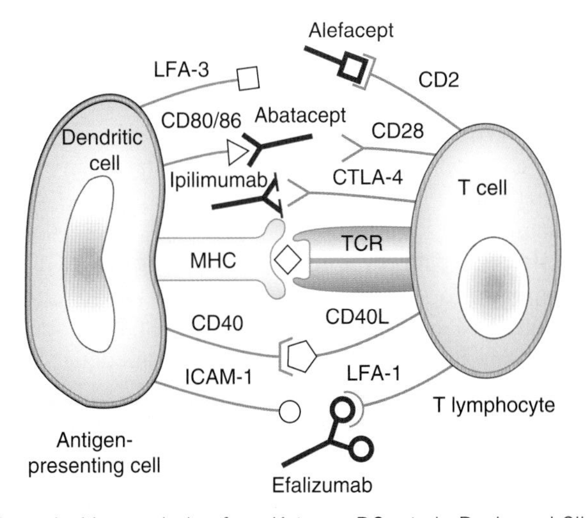

Adapted with permission from Katzung BG, et al.: Basic and Clinical pharmacology, 11th edition, McGraw-Hill, 2009.

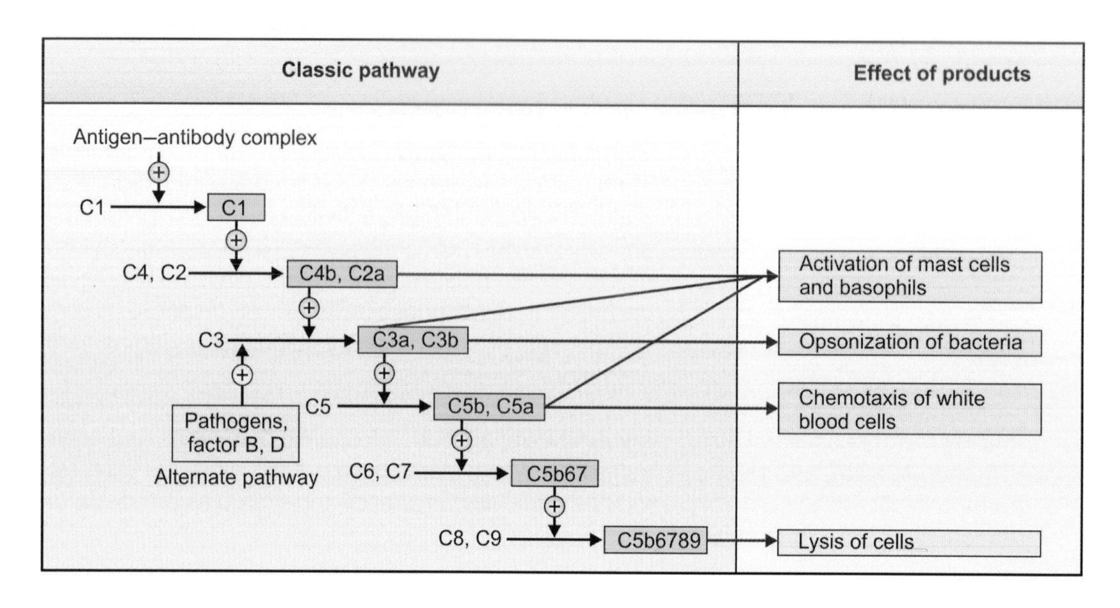

Figure 15-7. The Classical and Alternative Complement Pathways. Representation of the two major complement pathways and analogous mechanism of the mannose-binding lectin pathway. The left figure illustrates the various components and interactions of the classic pathway (top) and the alternative pathway (bottom). The right figure shows various functions of components of both pathways. See text for further details. [Reproduced with permission from Naik: Biochemistry, 3rd edition, Jaypee Brothers Medical Publishers (P) Ltd., 2009.]

COMPLEMENT SYSTEM

The complement system is a pathway of immunological protection composed of more than 25, mainly liver-synthesized polypeptides. Complement can be activated through three separate paths: classical, alternative, and lectin (Figure 15-7). In each case, these pathways lead to the proteolysis and activation of complement component C3, and the same downstream inflammatory effects, including recruitment and activation of cellular immune responses, induction of local tissue warmth, swelling and pain, and formation of a **membrane attack complex (MAC)**, which creates a transmembrane pore (Chapter 8) that can participate in killing bacteria and abnormal cells (including virally infected cells). The early complement components that induce C3 activation are each examples of danger signal sensors.

The **classical pathway** is initiated by binding of the C1 complex (composed of one C1q, two C1r, and two C1s subunits) to IgM or IgG/IgM antibodies or binding of one C1q to an infective organism. Binding and a resulting conformational change of C1q activates a serine protease enzyme on C1r, which cleaves C1s. C1s also has serine protease activity and cleaves complement proteins C4 and C2, forming C4a and C4b and C2a and C2b, respectively. The C4b and C2a proteins next combine to form a C3 convertase, which cleaves C3 into C3a and C3b. A final C5 convertase complex is then formed from C4b, C2a, and C3b, which cleaves the C5 protein into C5a and C5b. C5b subsequently binds to C6, C7, C8, and C9 proteins to form the MAC complex.

The **alternative pathway** is activated by the cleavage of a thioester bond in the C3 protein to form C3a and C3b. The C3b protein is able to bind directly to the plasma membrane of an infective agent after which it is joined by the protein factor B to form a C3bB complex. This binding, therefore, requires no antibody but is restricted to only certain types of infective organisms. Factor D, a trypsin-like enzyme, then cleaves factor B into Ba and Bb (an active serine protease), resulting in C3bBb, the alternative pathway C3 convertase. The addition of one more C3b protein by the action of the newly formed C3 convertase leads to subsequent cleavage of C5 and formation of the MAC complex as described above. Decay accelerating factor and another protein, factor H, inhibit the alternative pathway by competitively inhibiting factor B binding to or causing its dissociation from surface-bound C3b. Another molecule, factor I, cleaves C3b into an inactive form to inhibit the reaction process.

The **lectin pathway**, known more formally as the **mannose-binding lectin** pathway, is activated by binding of the specific glycoproteins (Chapter 2) to mannose sugar residues on the surface of the infective agent. This binding turns on two **mannan-binding lectin-associated serine proteases**, **MASP-1** and **MASP-2**, which are analogous to the classical pathway proteins C1r and C1s, respectively. MASP-1 and MASP-2 follow the classical pathway in cleaving C2 and C4 with progression to C3 cleavage and activation.

Individual complement proteins also serve additional functions. C3b binds to infective organisms and promotes phagocytic ingestion by macrophages and other phagocytic cells. C5a, the other part of the C5 protein, helps to attract immune cells to the source of infection. C3a and C5a can activate mast cells to degranulate and increase the permeability of blood vessels and smooth muscle contraction as part of an immune response. Bb can alternatively increase the replication of preactivated B lymphocytes (see above), whereas Ba inhibits this process. Kupffer

Complement Deficiency and Disease: The importance of the complement immune system is illustrated by the results of deficiencies or harmful mutations of its components. Although conclusive evidence is still lacking, problems with the complement system have been implicated in diseases such as **asthma**, systemic **lupus erythematosus**, **paroxysmal nocturnal hemoglobinuria**, forms of **angioedema, macular degeneration**, **hemolytic uremic syndrome**, and **membranoproliferative glomerulonephritis**. The importance of the MAC complex in protecting against Gram-negative bacteria is also shown by the marked increase in meningitis in complement-deficient patients.

cells in the liver are also involved in clearing infective cells that are coated with complement proteins.

HYPERSENSITIVITY REACTIONS

Hypersensitivity reactions occur when the immune system reacts excessively, resulting in undesired damage to the body's tissues and, in some cases, death. All hypersensitivity shares the common motif of presensitization or pre-exposure to an antigen with a prior immune response. Four major types of hypersensitivity reactions have been described, which are summarized in Table 15-3.

TABLE 15-3. Review of Hypersensitivity Reaction Types

Name	Mediator	Cells Involved	Time Course	Mechanism	Disease Examples
Type I (allergy or anaphylactic)	Immunoglobulin (Ig) E	Initial response by mast cells or basophils with amplification by eosinophils, neutrophils, and platelets.	15–30 min	Repeat exposure to certain allergens leads to binding of IgE to its receptor on mast cells or basophils. Binding releases histamine (causes mucus secretion, dilation, and increased permeability of blood vessels and constriction of bronchi); leukotrienes B4, C4, and D4 (attract more basophils, amplified histamine response); prostaglandin D2 (causes pain and swelling); platelet; eosinophil; and neutrophil factors (activate and attract particular cell types causing aggregation, heparin release, and a variety of enzymes that damage tissues).	Urticaria, eczema, conjunctivitis, rhinitis, asthma, and some forms of gastroenteritis.
Type II (antibody dependent or cytotoxic)	IgM or IgG antibodies and complement	Phagocytes or Kupffer cells. Neutrophils contribute.	Minutes to hours	Antigens promote the production of antibodies, which attach to patient's cell membranes. These cells are now recognized as foreign and tagged with IgG or IgM antibodies and attacked by NK or macrophage cells or elicit a B cell/antibody response. Attached antibodies create a linear pattern on staining.	Drug-induced hemolytic anemia, Goodpasture's, thrombocytopenia, erythroblastosis fetalis, and pemphigus vulgaris.

(continued)

TABLE 15-3. Review of Hypersensitivity Reaction Types (Continued)

Name	Mediator	Cells Involved	Time Course	Mechanism	Disease Examples
Type III (Immune complex)	IgG or IgM antibodies and complement	Neutrophils and platelets.	3–8 hrs up to weeks	Antigen binds antibodies to produce immune complexes with complement proteins. Complexes deposit in small blood vessels, joints, and glomeruli eliciting inflammatory response. Neutrophils and platelets cause tissue damage. Soluble antibodies create a granular pattern on staining.	Serum sickness, aspergillosis, Arthus reaction, and polyarteritis nodosa.
Type IV (cell-mediated or delayed type)	T lymphocytes	Cytotoxic T and Th1 lymphocytes and monocytes.	2–3 days or more	Cytotoxic T cells recognize an antigen and attack tissue, whereas Th1 cells release cytokine, attracting monocytes and macrophages, which cause most of cellular damage.	Tuberculosis, leprosy, histoplasmosis, contact dermatitis (e.g., poison ivy), and granulomas.

Th, T helper cell.

Transplant Rejection: The rejection of transplanted organs occurs when the recipient immune system identifies the new tissue as foreign and mounts an immune attack against it. Despite efforts to adequately serotype the donor and recipient to insure as close an immunological match as possible (**histocompatibility**), this problem plagues transplant patients and the clinicians who care for them.

Transplant rejection results from the processes already described and are normally categorized as hyperacute rejection, acute rejection, and chronic rejection. **Hyperacute rejection** processes occur within minutes and are caused by the **complement** immune system responding immediately to the donor organ because of pre-existing or cross-reacting antibodies. Certain tissue types are especially prone to hyperacute rejection (e.g., kidney, some tissues from other species), whereas others almost never react (e.g., liver). Hyperacute rejection is also responsible for the adverse reaction seen during some blood transfusions. **Acute rejection** occurs within a week after transplant and can last for up to 3 months. It is usually because of mismatch of **human leukocyte antibody**, which is found in all cells, which lead to a **T lymphocyte** response (i.e., **cell-mediated immunity**). Again, certain tissue transplants (e.g., kidney and liver) are more prone to this mode of rejection. **Chronic rejection**, occurring months to years after transplant, is usually secondary to gradual breakdown and scarring of blood vessels in the transplant. The exact mechanism is not well understood.

Treatment is via immunosuppressant medications (see above) and/or, in some cases, bone marrow transplantation, a procedure with its own risks in patients who are already probably ill.

REVIEW QUESTIONS

1. What are the basic features of innate, passive adaptive, humoral, and cell-mediated immunities?
2. What are the basic features and differences between the self–nonself and danger signal theories of immunological activation?
3. What are the basic structures and functions of antigens and antibodies, including the five basic types of antibody molecules?
4. What are the different types of T lymphocytes, and their roles in immunity and mechanism of immune activation?
5. What are the functions of B lymphocytes and natural killer cells?
6. What are the basic functions of monocytes/macrophages, neutrophils, eosinophils, and basophils?
7. What are the basic functions of cytokines, including the mechanism of the five types of cytokine receptors?
8. What are the similarities and differences of the classic, alternative, and lectin pathways of complement activation?
9. How do the four types of hypersensitivity reactions compare?

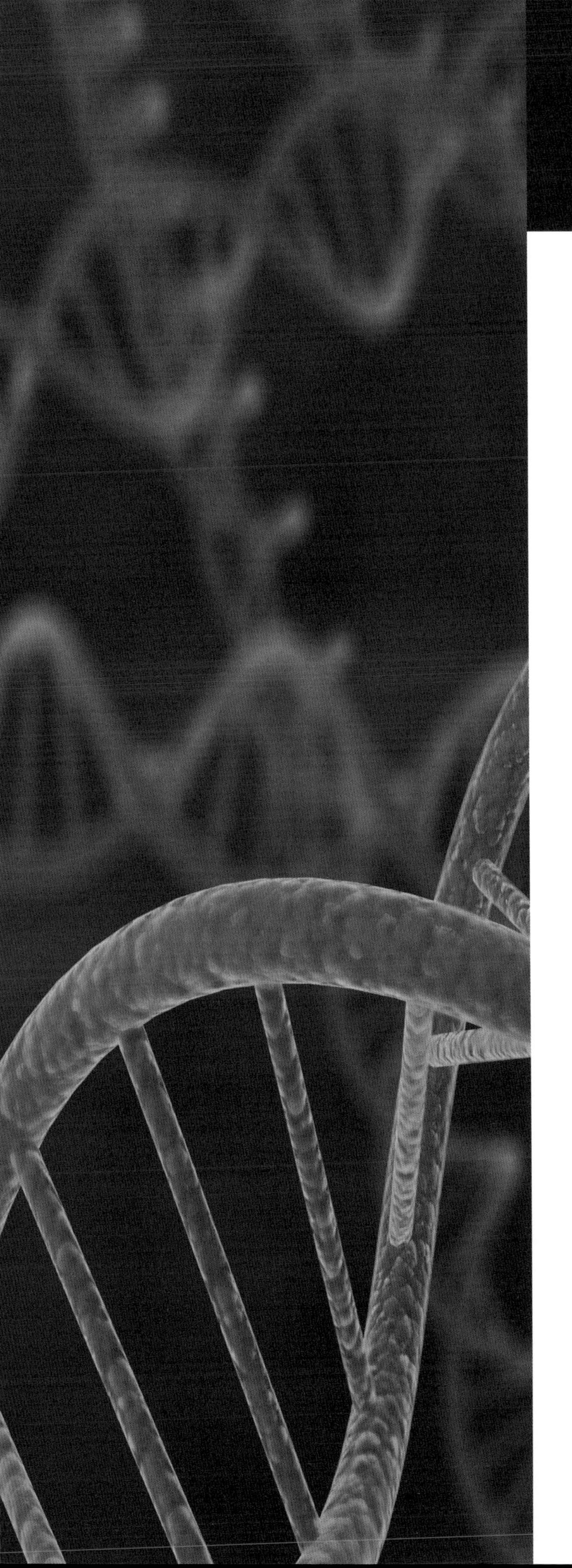

CHAPTER 16

THE CARDIOVASCULAR SYSTEM

Editor: Ralph V. Shohet, MD

Director of Cardiovascular Research, John A. Burns School of Medicine (JABSOM), University of Hawaii at Manoa, Honolulu, HI

OVERVIEW

The cardiovascular system, including the heart and blood vessels, is responsible for the circulation of blood to all tissues of the body while carrying away carbon dioxide and waste products. Cardiac muscle shares many attributes with skeletal and smooth muscles but represents a separate muscle group with distinct structure, function, and regulation. Key in this function is the self-activating nature of particular cardiac cells that normally provide an ordered contraction of the cardiac chambers. Blood vessels contain smooth muscle cells regulated by a variety of signal molecules and also play an important role in maintenance of blood pressure and oxygen/nutrient distribution. Diseases of these blood vessels, especially hypertension and atherosclerosis, cause much of the illness and most of the deaths in the developed world.

CARDIAC MUSCLE STRUCTURE AND FUNCTION

The heart is derived from the mesoderm and provides the force that propels oxygenated blood to all cells of the body and returns deoxygenated blood back to the lungs (Figure 16-1). The heart comprises specialized striated muscle termed cardiac muscle.

Cardiac muscle differs from skeletal and smooth muscle in several important ways:

1. Increased numbers of **mitochondria** allow continuous high-level production of adenosine triphosphate (**ATP**) to support continuous function without fatigue. During periods of low oxygen, anaerobic respiration can provide enough energy for contraction to continue.
2. Utilizes lipids [fatty acids and triglycerides (60%)] as the major energy source followed by carbohydrates (35%), amino acids, and ketone bodies (5%).
3. Contains fewer but larger **T tubules** and **intercalated discs** that allow the rapid and synchronous spread of action potentials between adjacent cells, allowing the coordinated contraction of cardiac muscle.
4. Relies on the rapid release and re-uptake of intracellular calcium stores to convert the electrical signal of the action potential into the mechanical work of contraction.

Because cardiac cells and smooth muscle cells rely strongly on an **L-type ("longlasting") calcium channel** for contraction, the medication class of calcium channel blockers (CCBs) are used to treat high blood pressure and certain cardiac disorders. In the heart, CCBs decrease the total calcium released and both the heart rate and force of the contraction, thereby reducing oxygen demand. This effect of CCBs is particularly useful in helping to control abnormal heart rhythms such as atrial fibrillation. In blood vessels, the decreased calcium-induced contraction results in dilation of the blood vessels (by decreasing the ability of the smooth muscle to contract) and lowers the resistance of those vessels. Additional cardiovascular uses of CCBs include reducing the outflow gradient in hypertrophic cardiomyopathy and reducing pulmonary arterial resistance in pulmonary hypertension. CCBs are also used in the treatment of epilepsy via their effect on neurons.

CCBs are often separated into **dihydropyridines**, most commonly used for hypertension because of greater affinity for receptors in vascular smooth muscle, and **non-dihydropyridines**, which are relatively cardiac selective and less likely to cause reflex tachycardia. **Benzothiazipines** such as diltiazem have intermediate characteristics. These three classes are summarized in Table 16-1 and Figure 16-3.

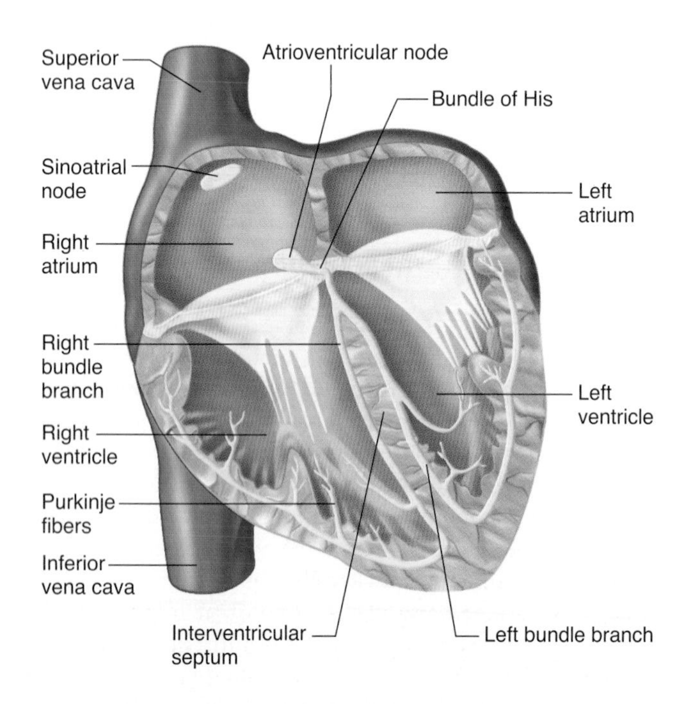

Figure 16-1. Sectional View of Heart. Basic anatomy of the heart is indicated, including chambers, major blood vessels, and conducting system (yellow), including the sinoatrial node, atrioventricular node, and Purkinje fibers. [Reproduced with permission from Barrett KE, et al.: Ganong's Review of Medical Physiology, 23rd edition, McGraw-Hill, 2010.]

SINOATRIAL AND ATRIOVENTRICULAR NODES

Particular cardiac cells located in the **sinoatrial (SA) node** (Figure 16-1) have the ability to initiate an action potential (Chapter 12), which then spreads through the muscle cells in the upper part of the heart (right and left atria) via the intercalated disks, resulting in a contraction of these two chambers. In normal sinus rhythm, the SA node initiates action potentials at 60–100/min that are conducted to the **atrioventricular (AV)** node, which coordinates the sequential contraction of the upper (atrial) and then lower (ventricular) chambers of the heart. If the SA node fails to provide action potentials, the AV node can independently generates action potentials at 40–60 beats per minute. Subsequent transmission of the action potential to the ventricles is via specialized heart cells (cardiomyocytes) in the right and left bundle branches and Purkinje fibers that spread throughout the ventricles. These conduction cells contain increased numbers of sodium ion channels and mitochondria while containing fewer cardiac muscle fibers. If both the SA and the AV nodes fail to provide an action potential, the Purkinje fibers or ventricular myocytes can also produce a coordinated contraction at rates slower than the AV node.

Although the SA and the AV nodes and Purkinje fibers can independently initiate and continue action potentials, which provide heart contractions, the vagal nerves (parasympathetic nerves causing a decreased SA node rate and force of contraction) and thoracic nerves T1–4 (sympathetic nerves resulting in increased SA node rate as well as force of contraction) also convey neurological influences on heart rate. Hormones/neurotransmitters such as **epinephrine (adrenaline)** and **dopamine** increase heart rate and/or the amount of blood pumped per contraction (stroke volume). Epinephrine works via binding to **α_1-, α_2-, β_1 and β_2-adrenergic, G-protein-coupled receptors.** The receptor signal transduction includes activation of phospholipase C to increase production of inositol phosphates/diacylglycerol, protein

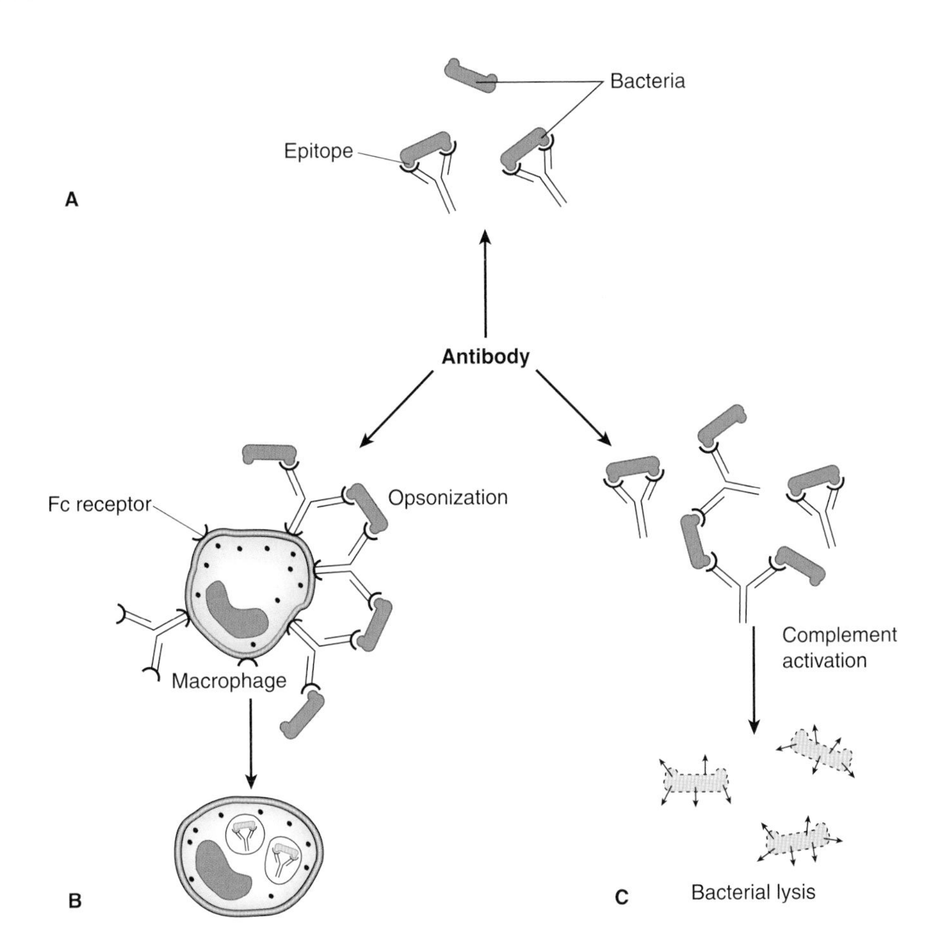

Figure 15-2. Summary of Antibacterial Antibody Functions. Antibodies can elicit a number of reactions resulting in the elimination of bacteria, including (**A**) recognition and binding to bacteria, which allows recognition by other immune system components to include (**B**) opsonization, leading to recognition and phagoyctysis by macrophages, and (**C**) binding leading to complement activation (see below), leading to bacterial lysis. [Adapted with permission from Katzung BG, et al.: Basic and Clinical Pharmacology, 11th edition, Mc Graw-Hill, 2009.]

Antibodies exist in five types: IgA, IgD, IgE, IgG, and IgM. The basic structure, source, and function of each are summarized in Table 15-1.

CELLS ASSOCIATED WITH THE IMMUNE SYSTEM

Cells of the immune system, collectively called leukocytes or white blood cells, can be categorized as lymphocytes, monocytes, neutrophils, eosinophils, and basophils (Figure 15-3).

T LYMPHOCYTES

T lymphocytes contain a **T-cell receptor (TCR)** and **CD4** or **CD8 coreceptor** on their plasma membrane (Figure 15-4A–B). Each TCR recognizes a molecule's distinct epitope structure and allows the specificity needed for a controlled immune reaction (Figure 15-4A). T lymphocytes with a CD8 coreceptor recognize antigen presented by Class I major histocompatibility complex (MHC) molecules on the surface of a wide variety of cells, and frequently are effectors of direct cell killing (cytotoxic T cells). T lymphocytes with a CD4 coreceptor (helper T cells or Th cells) recognize antigen presented by Class II MHC molecules on the surface of specialized antigen-presenting cells (APCs, such as macrophages, DCs, and some B cells), and frequently orchestrate the responses of other immune cells. This process of antigen presentation is summarized in Figure 15-4B.

Recognition of a specific antigen in the antigen–MHC complex by the TCR of the T lymphocyte cells (Figure 15-5, left) activates a signaling process, involving a peripheral membrane **leukocyte-specific tyrosine kinase (Lck)**, which is tethered to the inside of the plasma membrane by lipid residues linked to its amino-terminal end. Lck also contains two highly conserved amino acid sequences, known as **sarcoma homology domains**, and binds to the cytoplasmic tails of either the CD4 or CD8 coreceptor molecules. Upon activation, Lck phosphorylates a subunit of the TCR at a highly conserved tyrosine separated from a leucine or isoleucine by two other amino acids. A double repeat of this four amino acids sequence, the two separated by 7–12 amino acids, is known as an **immunoreceptor tyrosine-based activation motif (ITAM)** and is found on subunits of the TCRs as well as B-cell receptors (see below). Phosphorylation of the two tyrosines of the ITAM allows binding of other signaling proteins resulting in subsequent phosphorylation of another tyrosine kinase called **zeta-chain-associated protein kinase-70 (ZAP-70)**. ZAP-70 activation eventually leads to phosphorylation and activation of phospholipase C-γ 1; secretion of calcium from the endoplasmic reticulum; and altered gene expression by

TABLE 15-1. Summary of Antibody Molecules

Antibody	Structure	Source	Function
Immunoglobulin (Ig) A	Secretory component. Contains a J chain subunit required for secretion.	B lymphocytes in the mucosal lining of tear and saliva ducts, lungs, intestines, genito-urinary tract, and prostate. Also secreted in breast milk and tears.	Provides initial immunological protection against invading organisms found in these sites. Breast milk IgA provides important immunological protection for newborns.
IgD		Produced by B lymphocytes (plasma cells). Usually coexpressed on plasma membrane with IgM. Can be secreted.	Regulation of B lymphocyte activation. May also activate basophils and mast cells.
IgE		Produced by B lymphocytes (plasma cells).	Involved in type 1 hypersensitivity "allergic" reactions and response to parasitic worms and protozoan.
IgG		Produced by B lymphocytes (plasma cells). Most abundant (~75%) of all Igs in human body.	Major Ig of secondary immune response to viral, bacterial, fungal, and other pathogen invasions. Involved in types II and III hypersensitivity reactions and helps induce phagocytosis. Only Ig crossing placenta for passive immunity.
IgM	Contains a J chain subunit required for secretion.	Produced by B lymphocytes (plasma cells). Because of its large size, most of it is found in the blood serum. First Ig produced by fetus.	Involved in initial response to new antigen and, thus, serves as a marker for recent infection. Also activates complement. Main antibody for blood type determination and blood transfusion incompatibility.

Adapted with permission from Mescher AL: Junqueira's Basic Histology Text and Atlas, 12th edition, McGraw-Hill, 2010.

the actions of nuclear factor of activated T cells, nuclear factor kappa B (NF-κB), and activator protein 1 (AP-1) transcription factors, among others, in the cell nucleus. Cytotoxic T cells produce **perforin** and **granulysin** proteins, which create open channels in the plasma membrane of infected cells, and granzymes, serine protease enzymes, both of which lead to cell death. Cytokines including **interleukin-2 (IL-2)** are also expressed, leading to sustained replication and activity of the lymphocytes.

As the name suggests, **helper T lymphocytes** assist a number of other immune cells including subsequent activation of B lymphocyte antibody production (see below), macrophage phagocytosis (see below), and activation of cytotoxic T cells. Th cells are a major producer of cytokines. Th cells known as **Th1 cells** produce the cytokines interleukin-12 and **interferon-γ (IFN-γ)** via **Janus kinase (JAK)–signal transducer and activator of transcription (STAT) signaling** (Chapter 8), leading to both differentiation and production of more Th1 cells. Additionally, Th1 cells cause (a) increased expression of MHC Class I molecules on normal cells, (b) production of adhesion molecules that promote white blood cell migration to sites of infection, (c) decreased production of Th2 cells (see below), (d) increased lysosomal activity and antigen presentation by macrophages, (e) increased natural killer (NK) cell activity (see below), and (f) increased production of nitric oxide, which can assist in bacterial and viral killing. In sum, these signals activate multiple cells to fight infecting organisms while downregulating other T-cell pathways. One alternative T-cell set known as **Th2 cells** induces B lymphocytes and produces interleukin-4, a cytokine that also induces DNA replication, via the JAK–STAT signaling pathway (Chapter 8) that leads to the differentiation/production of more Th2 cells. The TCR and CD4 coreceptor of Th cells bind to an antigen–MHC II complex on APCs but Lck signaling results in the production and release of several types of cytokines (see below), including IL-2. IL-2 is also expressed to promote proliferation of the lymphocytes and prolongation of their response.

An additional set of T lymphocytes, **suppressor T** or **regulatory T cells**, acts to limit activation of the immune system as well as phagocytosis. Their functions help to shut off immune responses to maintain the essential balance between an activated and surveying immune system and one that is tolerant to its own

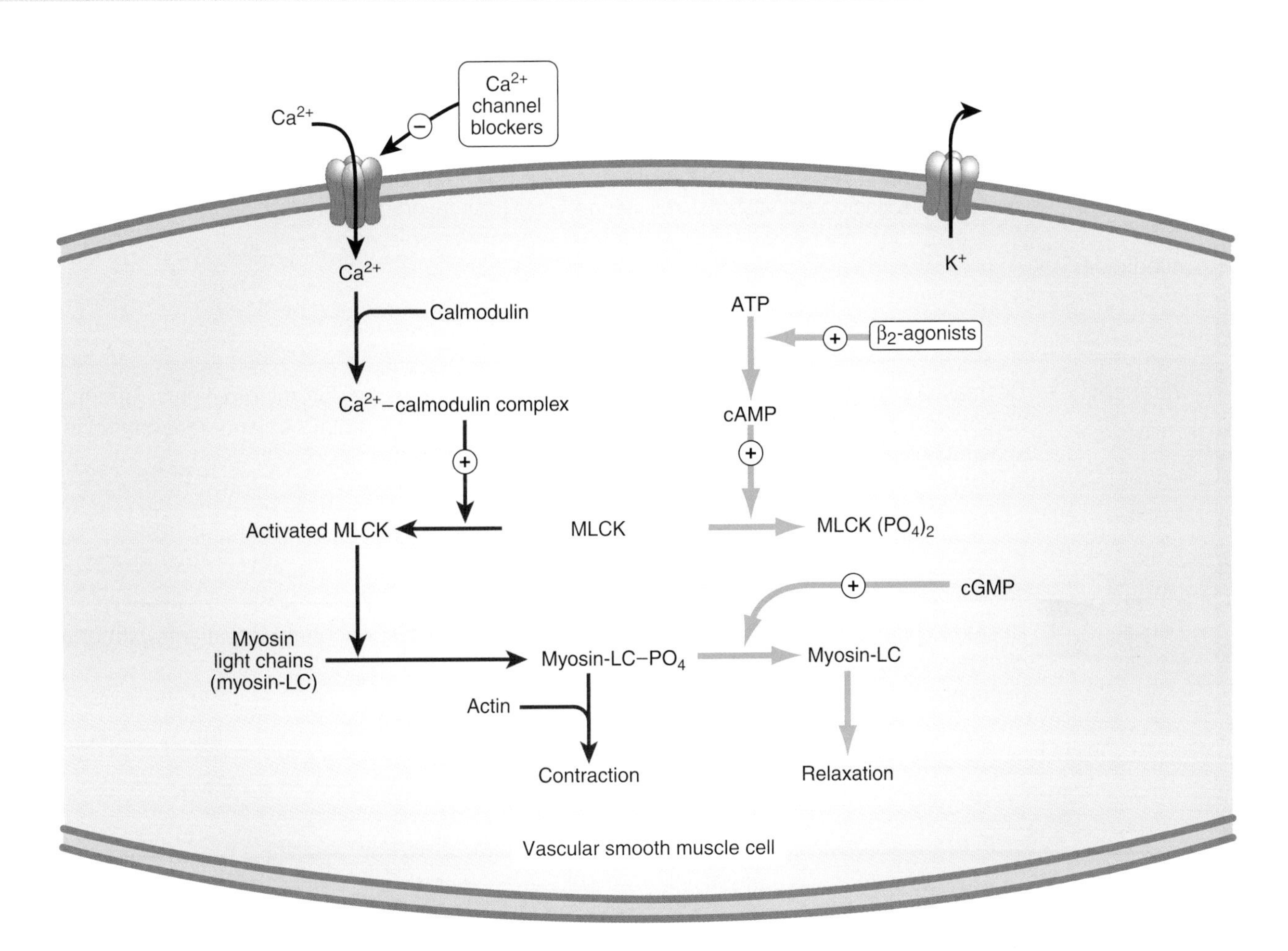

Figure 16-2. Mechanism of Calcium Activation of Smooth Muscle Cells. Calcium ions (Ca^{2+}) activate intermediary signaling molecules such as calmodulin and myosin light-chain kinase (MLCK) to phosphorylated regulatory myosin light chains, leading to contraction of vascular smooth muscle cell. Calcium channel blockers inhibit the influx of calcium, leading to inhibition of this process and resulting vascular smooth muscle relaxation. One effect of β-agonists/blockers leads to relaxation of vascular smooth muscle as shown. ATP, adenosine triphosphate; cAMP, cyclic adenosine monophosphate; cGMP, cyclic guanosine monophosphate. [Reproduced with permission from Katzung BG, et al.: Basic and Clinical Pharmacology, 11th edition, McGraw-Hill, 2009.]

kinase C activity, and the release of Ca^{2+} (α_1-/G_q protein). Additionally, there will be inhibition (α_2-/G_i protein) or activation (β_1 and β_2-/G_s protein) of adenylyl cyclase to decrease or increase cyclic adenosine monophosphate (cAMP) production and protein kinase A activity, respectively (see Chapter 8 for review of G-protein activities).

Dopamine: Dopamine is a neurotransmitter that not only affects the central nervous system but also elicits dose-dependent effects on blood vessels and cardiac muscle via Dopamine (D) receptors and α_1- and β_1-adrenergic receptors. It is frequently used in the intensive care unit to support blood pressure. At low doses (2–5 μg/kg/min), binding is primarily to D_1 and D_5 receptors found on blood vessels, especially in the kidneys, activating a G_s-receptor (adenyl cyclase/cAMP), resulting in increased urine production. Dopamine when delivered at a rate of 5–10 μg/kg/min binds to β_1-adrenergic receptors in cardiac muscle, producing increased contraction rate and force for treatment of shock/heart failure patients. Dopamine provided at 10–20 μg/kg/min binds to α_1-adrenergic receptors, resulting in contraction of blood vessels and increased resistance/blood pressure.

HO
HO– (benzene ring) –CH_2–CH_2–NH_2

Dopamine

Reproduced with permission from Katzung BG, et al.: Basic and Clinical Pharmacology, 11th edition, McGraw-Hill, 2009.

TABLE 16-1. Calcium Channel Blocker Classes and Actions

Class	Examples	Primary Effect(s)	Medical Use
Dihydropyridine	Amlodipine, felodipine, and nifedipine	Dilation of blood vessels, smooth muscle dilation, and decreased vascular resistance.	Predominately used for blood pressure control. Less favored for angina because of reflex tachycardia.
Phenylalkylamine	Verapamil	Decreased rate and force of cardiac contraction; also effective for coronary artery spasm (Prinzmetal's angina).	Blood pressure control in patients suffering with angina, Prinzmetal's angina, rate control in atrial fibrillation.
Benzothiazepine	Diltiazem	Decreased rate and force of cardiac contraction and dilation of blood vessels, decreasing vascular resistance.	Blood pressure control in patients with or without angina.

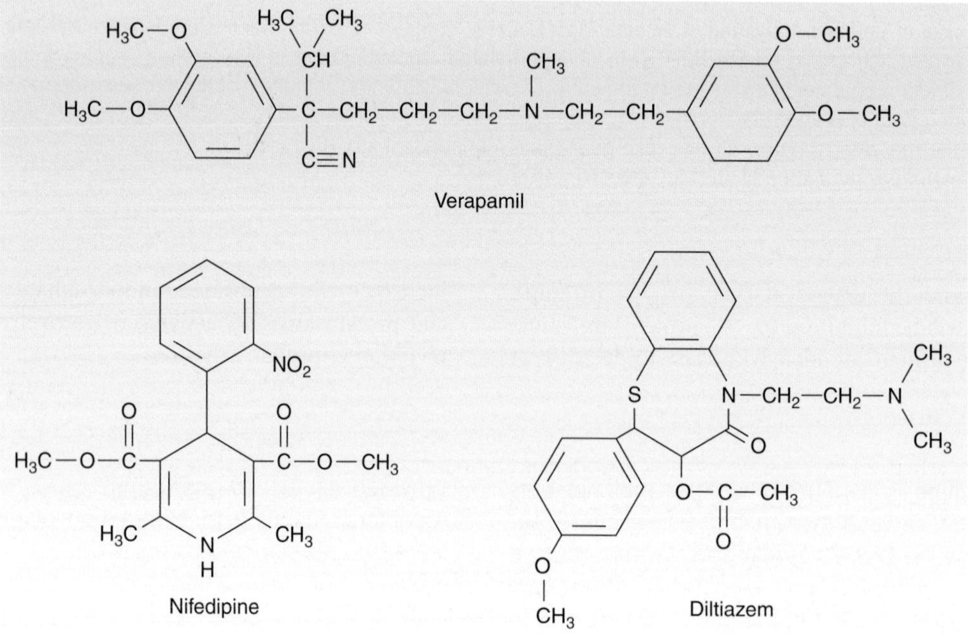

Verapamil

Nifedipine

Diltiazem

Figure 16-3. Representatives of Calcium Channel Blocker Classes. Calcium channel blocker medications can be divided into three basic classes depending on their mechanism, including phenylalkylamines (e.g., verapamil), dihydropyridines (e.g., nifedipine), and benzothiazepines (e.g., diltiazem). See text and Table 16-1 for further discussion. [Reproduced with permission from Katzung BG, et al.: Basic and Clinical Pharmacology, 11th edition, McGraw-Hill, 2009.]

THE CARDIAC CYCLE

The **Cardiac cycle** is an autonomous process that results in the organized, sequential, contraction of the chambers of the heart, temporally coordinated to produce efficient pumping throughout the body of blood, which then returns to the heart and lungs for re-oxygenation. The cardiac cycle includes steps that result in heart sounds, which produce the characteristic electrical signals seen on an **electrocardiogram (ECG)** tracing, including the P-wave (atrial depolarization), QRS-complex (ventricular depolarization), and T-wave (ventricular repolarization).

The initiation of cardiac muscle contraction relies on **voltage-dependent calcium channels (VDCCs)**. The predominant VDCC on cardiac cells is the L-type channel, which is composed of an α_1 subunit with six transmembrane α-helices that form a membrane channel or pore (Chapter 8), but which also responds to voltage changes by alterations in conformation that blocks or allows Ca^{2+} passage. Along with the α_1 pore, this channel includes other subunits. The disulfide-linked (Chapter 1) $\alpha_2\delta$ subunits anchor the α_1 protein in the membrane and augment its response to voltage changes. A β subunit both enhances α_1 expression on the membrane and contains guanylate kinase activity that catalyzes the reaction

B-type Natriuretic Peptide (BNP): BNP is a 32-amino acid polypeptide (Chapter 18) that is secreted by the ventricles when cardiac cells are excessively stretched. BNP binds to **atrial natriuretic peptide receptors**, activating a **guanylyl cyclase** producing cyclic GMP (cGMP), and, via a **cGMP-dependent protein kinase**, producing relaxation of blood vessels. Rises in BNP levels are seen in congestive heart failure (CHF) where it reduces the systemic resistance and blood volume, thereby reducing the work of the ailing heart. Measurement of BNP is useful to both assess and follow patients with CHF.

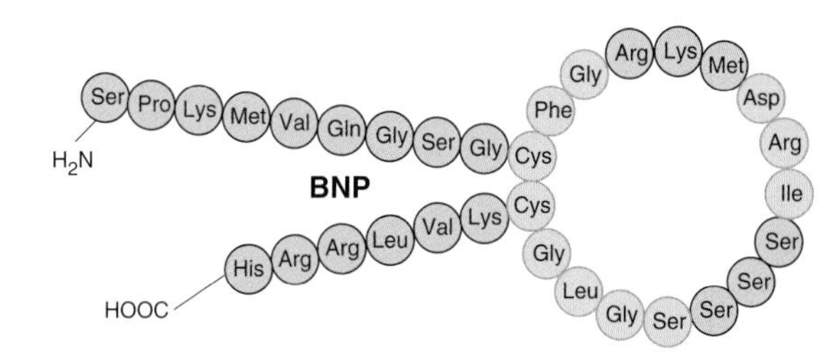

Reproduced with permission from Katzung BG, et al.: Basic and Clinical Pharmacology, 11th edition, McGraw-Hill, 2009.

β-Blockers: β-Blockers (all ending in "-lol") represent a class of medications used for the treatment of high blood pressure, irregular heart beat, chest pain (angina), and heart attack therapy. β-Blockers inhibit β_1- and β_2-adrenergic receptors in the heart and blood vessels. Stimulation of β_1-receptors on cardiac cells by molecules such as epinephrine normally leads to increased heart rate and augmented contraction that increases myocardial oxygen requirements. Blockade of these receptors reduces oxygen demand and can also help the heart retain a normal geometry as it heals. Stimulation of β_2-receptors on vascular smooth muscle relaxes (dilates) blood vessels (see Figure 16-2). So, vasoconstriction, for example, resulting from certain drugs of abuse, may be exacerbated by β-blockers.

O—CH_2—CH(OH)—CH_2—NH—$CH(CH_3)_2$; CH_2—CH_2—O—CH_3

Metoprolol

O—CH_2—CH(OH)—CH_2—NH—$CH(CH_3)_2$

Propranolol

Reproduced with permission from Katzung BG, et al.: Basic and Clinical Pharmacology, 11th edition, McGraw-Hill, 2009.

Like CCBs, β-blockers also have "classes" related to their relative effect(s) on cardiac muscle or vascular smooth muscle. Relatively specific β_1-blockers such as **metoprolol** affect mainly cardiac muscle, decreasing heart rate and the force of contraction and reducing work of the heart and oxygen requirements. This effect is a key in the treatment of heart attack/angina victims. Nonspecific β-blockers, of which the archetype is **propranolol**, inhibit β_2-receptors as well.

β-blockers also interact with β-adrenergic receptors in other organs (e.g., eye, brain, kidney, and lung) and may decrease intraocular pressure (glaucoma) and may improve migraine headaches, anxiety/PTSD. They also reduce renin release which limits conversion of angiotensinogen to angiotensin 1. This reduces angiotensin II with subsequent reduction in aldosterone from kidneys (Chapter 18), reducing blood volume. Both effects lower blood pressure. Their effect on the lung may worsen bronchial constriction in asthma attacks. β-blockers are also used in many cardiac arrhythmias, CHF (via their relaxing effects on the heart but also their effect on renin/aldosterone), mitral valve prolapse, portal hypertension and any resultant bleeding of esophageal varices, pheochromocytoma, and hypertrophic cardiomyopathy. However, the cross-reactivity of some β-blockers for differing receptors and tissues and the ability of some β-blockers to both inhibit and stimulate the same receptors depending on concentration leads to a number of unwanted side effects that must be carefully understood, monitored, and managed by the clinician.

ATP + GMP ↔ ADP + GDP. The conversion of GMP to GDP serves as a regulatory mechanism to augment the voltage sensitivity for the α_1-pore (i.e., β-subunit activity allows smaller depolarizations to result in channel opening). Finally, a γ-subunit is believed to be of little significance in cardiac muscle, although it may play an important regulatory role in skeletal muscle contraction. **Sarcoplasmic/endoplasmic reticulum Ca^{2+}-ATPase (SERCA)** also plays a prominent role in cardiac cell contraction (see Chapter 12).

The self-initiating action potential of the SA, AV, or Purkinje fibers provides the initial stimulus for contraction. In response to an action potential propagated down the conduction system to the T-tubules, VDCCs open by depolarization of the cell membrane and mechanical stretching of the cells (i.e., propagated from other nearby cells), G-protein signaling, or autonomic nervous system activation. This action allows Ca^{2+} into the cardiac cells (myocytes). Ca^{2+} that enter into the cardiac muscle cells bind to specialized **calcium release channels** via a **ryanodine receptor**, which releases much larger internal Ca^{2+} stores from the sarcoplasmic reticulum. This cytoplasmic calcium binds to the protein troponin C, which causes a conformational change of the inhibitory troponin I, which in turn allows tropomyosin to reveal myosin-binding sites on the actin thin filaments. Propagation from one cardiac cell to the other results from the passage of Ca^{2+} via intercalated disks, specialized cardiac gap junctions, completing depolarization of all the cardiac muscle. Calcium is then taken back into the sarcoplasmic reticulum through the ATP-requiring action of SERCA, the sarcoplasmic endoplasmic reticulum ATPase, restoring the cell to a state where it can be depolarized again.

BLOOD VESSELS

Blood vessels, consisting of **arteries**, **veins**, and **capillaries** provide the conduit for blood to be carried to and from every tissue in the human body with delivery of oxygen and nutrients and the removal of carbon dioxide and waste material (Figure 16-4). The average human body has approximately 60,000 miles of blood vessels, of which the vast majority are capillaries that let only a single line of red blood cells to pass. Arteries and veins are composed of three layers each of cells and/or polysaccharide matrix and/or connective and elastic tissues and/or smooth muscle cells along with their own system of blood vessels (*vasa vasorum*) and nerves, which feed and regulate their structure and, therefore, activity. Capillaries, responsible for the actual transfer of molecules between the blood and the cells, usually are only made of a single layer of cells with some, interconnected connective tissue.

The ability of arteries and, to a lesser extent, veins to dilate and constrict depends on the directed action of the autonomic nervous system on the smooth muscle in their layers. This ability to change their internal diameter is essential for the regulation of blood pressure, body temperature, delivery of blood and nutrients to various areas of the body, conservation of blood in times of bleeding, and so forth. **Vasoconstriction** is normally due to the actions of peptide hormones such as **vasopressin**, **angiotensin II**, and **endothelin** or via neurological transmitter molecules such as **epinephrine**. These molecules affect vascular smooth muscle via the **G_q receptor**, which elicits phosphatidylinositol/protein kinase C and calcium secretion, leading to contraction of blood vessel smooth muscle layers (Chapter 8; see also effects on urinary system, Chapter 18). The effects of vasopressin are mainly seen in the arteries, kidneys and brain, and all act to maintain and increase the arterial blood pressure. Angiotensin and epinephrine affect both arteries and veins. All serve as a potential compensatory mechanism during periods of blood loss/hemorrhage to avoid shock. Although vasopressin and angiotensin each have their own receptors, the smooth muscle constriction effects of epinephrine work mainly via activation of the **α_1-adrenergic receptors**. Epinephrine can also cause vasodilatation and/or vasoconstriction via **α_2- and β_2-receptors**, respectively, (although these actions are usually specific to a particular tissue and not all blood vessels) and increased heart rate and contractile force via **β_1-receptors**, all via the **G_i** or **G_s**, **adenylyl cyclase/cAMP** messenger system (Chapter 8).

The process of **vasodilation** relies on molecules such as **nitric oxide (NO)**, which diffuses through the plasma membrane of blood vessel cells to bind to a soluble **guanylate cyclase** (Figure 16-5). **cGMP** produced by this enzyme activates **protein kinase G**, which leads to dephosphorylation and inactivation of smooth muscle myosin molecules via inactivation of MLCK (Figure 16-5A). Even though the average lifetime of NO in the body is only a few seconds, its pharmacological effects are seen in medications such as **sublingual nitroglycerin** (Figure 16-5B) for chest pain via decrease in cardiac workload, treatment of neonatal patients suffering from pulmonary hypertension in an intensive care setting, and even modern, erectile disorder drugs (e.g., sildenafil/Viagra), which prolong the lifetime of cGMP and, therefore, vasodilation of vessels in the penis, promoting erection (Figure 16-5A).

Blood vessel permeability is important not only for normal human physiological processes but also for pathology. The single-cell layer of capillaries allows the exchange of oxygen and carbon dioxide (Chapter 14) as well as nutrients and waste. In disease, inflammation of the endothelial layer of blood vessels can lead to increased permeability and leakage of plasma and other blood components, contributing to the four classic signs of inflammation: calor (warmth), tumor (swelling), rubor (redness) and dolor (pain) (see also Chapter 3). Molecules leading to increased blood vessel permeability include **histamine**, working via the histamine-1 receptor (G_q/phosphatidyl inositol/calcium second messenger pathway), **prostaglandins** (G_s/adenylyl cyclase/cAMP), and by **interleukin-1 (IL-1)** via a Janus kinase (JAK) receptor (phosphorylation of tyrosine residues/cGMP) (Chapter 8). The blood vessel wall is also damaged in diseases such as atherosclerosis and immune-mediated vasculitis.

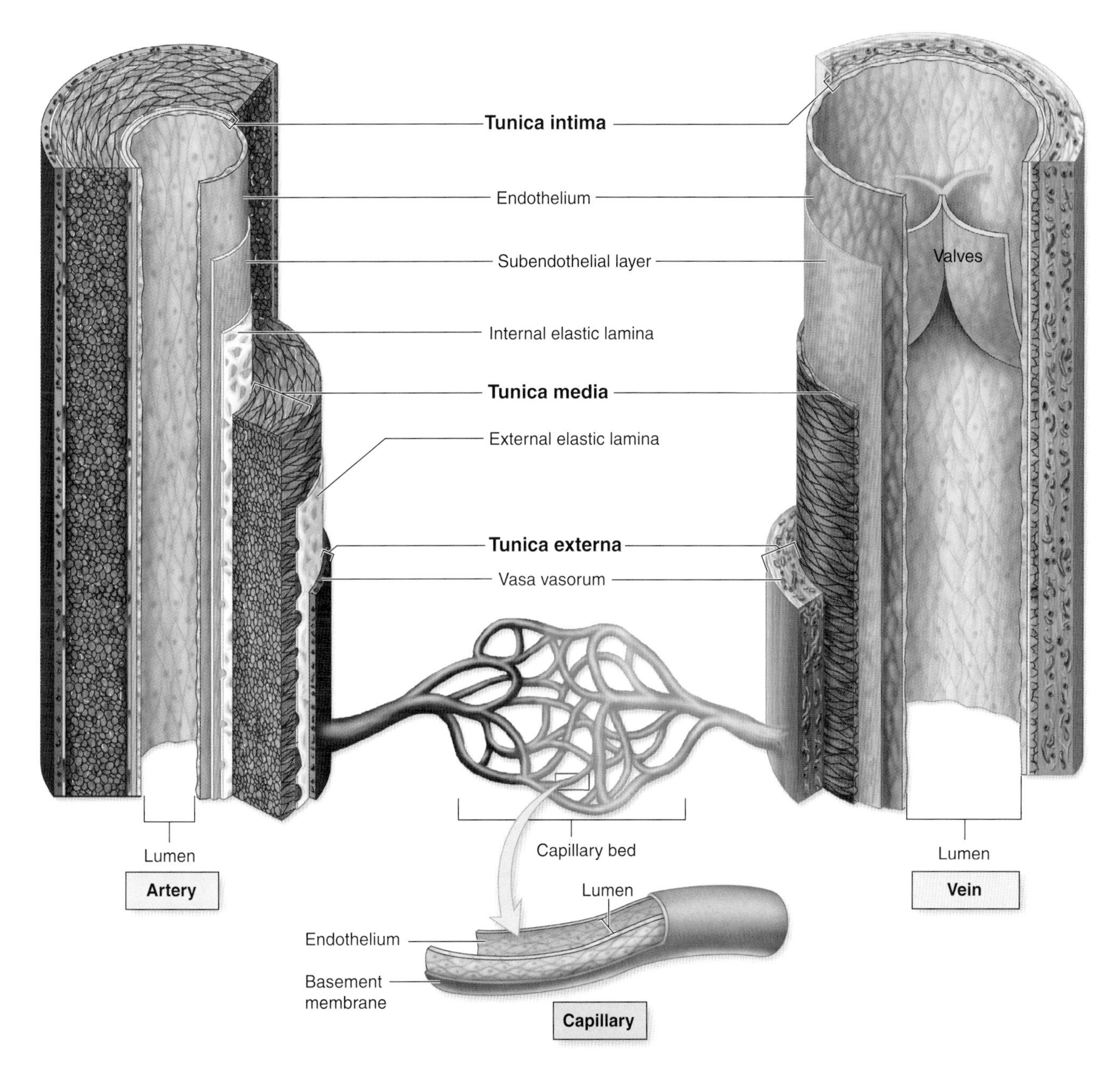

Figure 16-4. Blood Vessels. Blood vessels include the artery (left), vein (right), and connecting capillary (center) each of which contains specialized layers of cell, elastic, and connective and muscular tissue surrounding a central lumen. The *vaso vasorum*, which provides nutrients to these tissue layers. [Reproduced with permission from Mescher AL: Junqueira's Basic Histology Text and Atlas, 12th edition, McGraw-Hill, 2010.]

ENDOGENOUS CHOLESTEROL/LIPOPROTEIN METABOLISM AND TRANSPORT

Abnormally elevated levels of lipids in lipoproteins, termed hyperlipidemia, is one of the most important risk factors for the development of atherosclerosis (see below). The metabolism and transport of cholesterol and lipoproteins occur in two general ways. First, processing of cholesterol and triacylglycerols synthesized by the liver (Chapter 11) is termed as the **endogenous** pathway of cholesterol/lipoprotein and begins with liver secretion of nascent very-low-density lipoprotein (VLDL) and high-density lipoprotein (HDL). Second, processing of ingested cholesterol and triacylglycerols, the **exogenous** pathway, occurs in the blood stream and begins with secretion of chylomicrons after eating by intestinal epithelial cells initially into the lymphatic system. Both are summarized in Figure 16-6 and discussed below.

Further details of chylomicron metabolism are provided in Figure 16-7.

VLDL

The triacylglycerol-rich lipoprotein produced by the liver is the VLDL, responsible for the transport of triacylglycerol from the liver to tissues such as adipose and muscle as well as

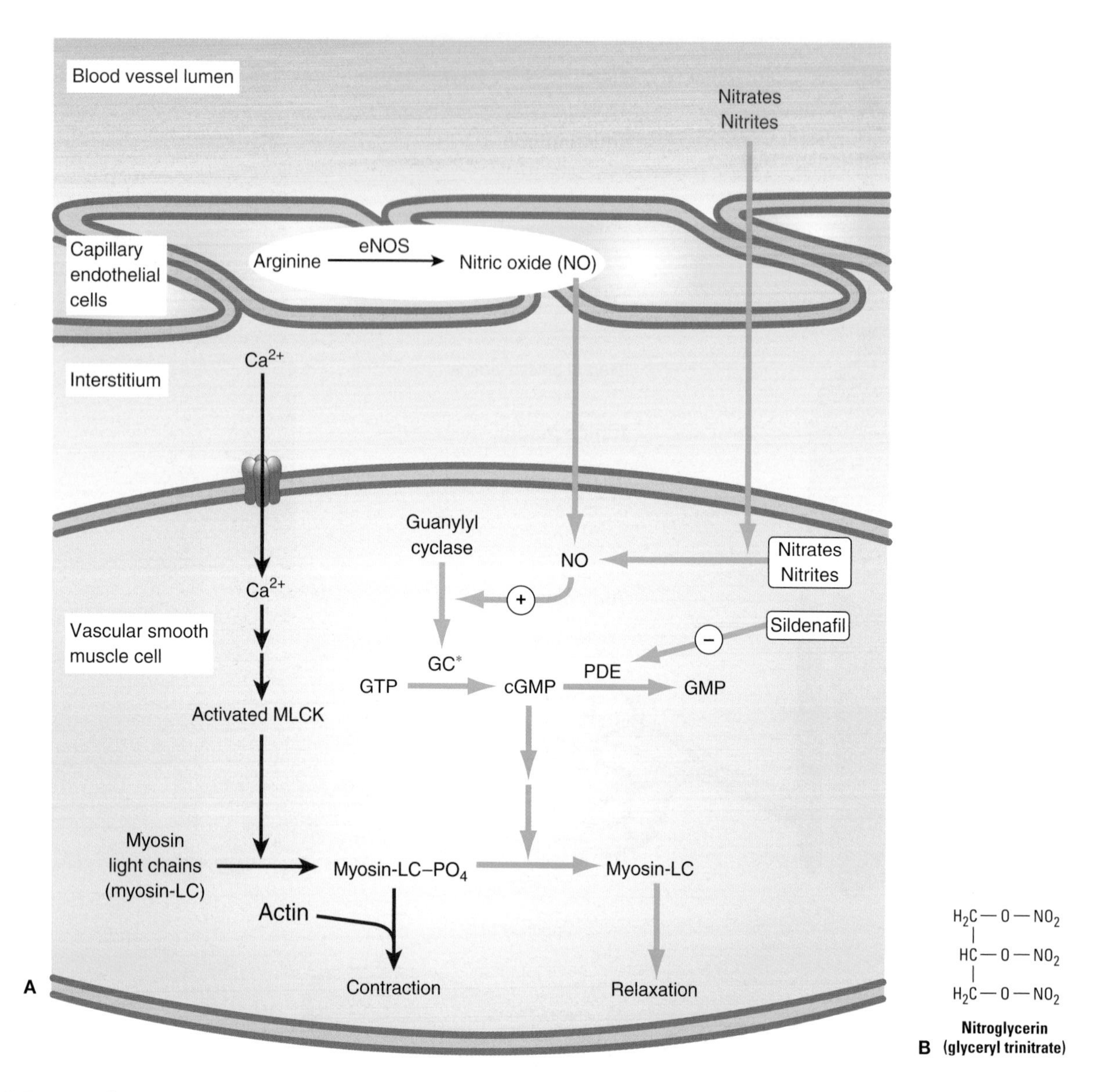

Figure 16-5. A–B. Nitroglycerin and Mechanism of Action of Nitrous Oxide. (**A**) Relaxation of smooth muscle in blood vessels by nitrous oxide occurs via activation of guanylyl cyclase producing cGMP, which promotes protein kinase G dephosphorylation of myosin light chains via inactivation of MLCK. (**B**) Nitrates such as nitroglycerin (also know as glyceryl trinitrate) can also cause vessel relaxation via the same pathway and, as a result, is used for the relief of symptoms of myocardial infarction. cGMP, cyclic guanosine monophosphate; eNOS, endothelial nitric oxide synthase; GTP, guanosine triphosphate; MLCK, myosin light-chain kinase; PDE, phosphodiesterase. [Reproduced with permission from Katzung BG, et al.: Basic and Clinical Pharmacology, 11th edition, McGraw-Hill, 2009.]

transporting Cholesteryl esters in the form of low-density lipoproteins (LDL) to peripheral cells. The triacylglycerol is formed from free fatty acids that are either mobilized from adipose cells or synthesized in liver from acetyl coenzyme A (CoA) derived from nonlipid precursors, primarily from glucose (Chapter 7). Cholesteryl ester that is also packaged in VLDL is derived mostly from direct synthesis in the liver from acetyl-CoA derived from β-oxidation of saturated fats (Chapter 7) or from dietary sources carried to the liver by chylomicron remnants (Chapter 11). **Apolipoprotein (apo) B100**, along with some apoC and apoE, are synthesized in the liver and are required for the proper assembly and export of the nascent VLDL. Subsequently, **apo C** and **apo E** are transferred from HDL to the nascent VLDL in the circulation, thereby creating the mature VLDL particles (Figure 16-8).

Similar to the case of chylomicrons (Chapter 11), triacylglycerols in VLDL undergo lipolysis in a reaction catalyzed by **lipoprotein lipase (LPL)** to release fatty acids for use by adipose and muscle (Figure 16-3). LPL is a cell surface enzyme on capillary walls that hydrolyzes triacylglycerols in chylomicrons (and VLDL) to fatty acids that are taken up into cells. LPL is most often found in capillaries that supply heart, skeletal muscle, adipose tissue, and mammary cells. In muscle, the released fatty acids are oxidized for energy; in adipose, they are re-esterified for storage as triacylglycerols. LPL is activated by **apo C**, bound to phospholipids on the surface of the chylomicron. In the fed

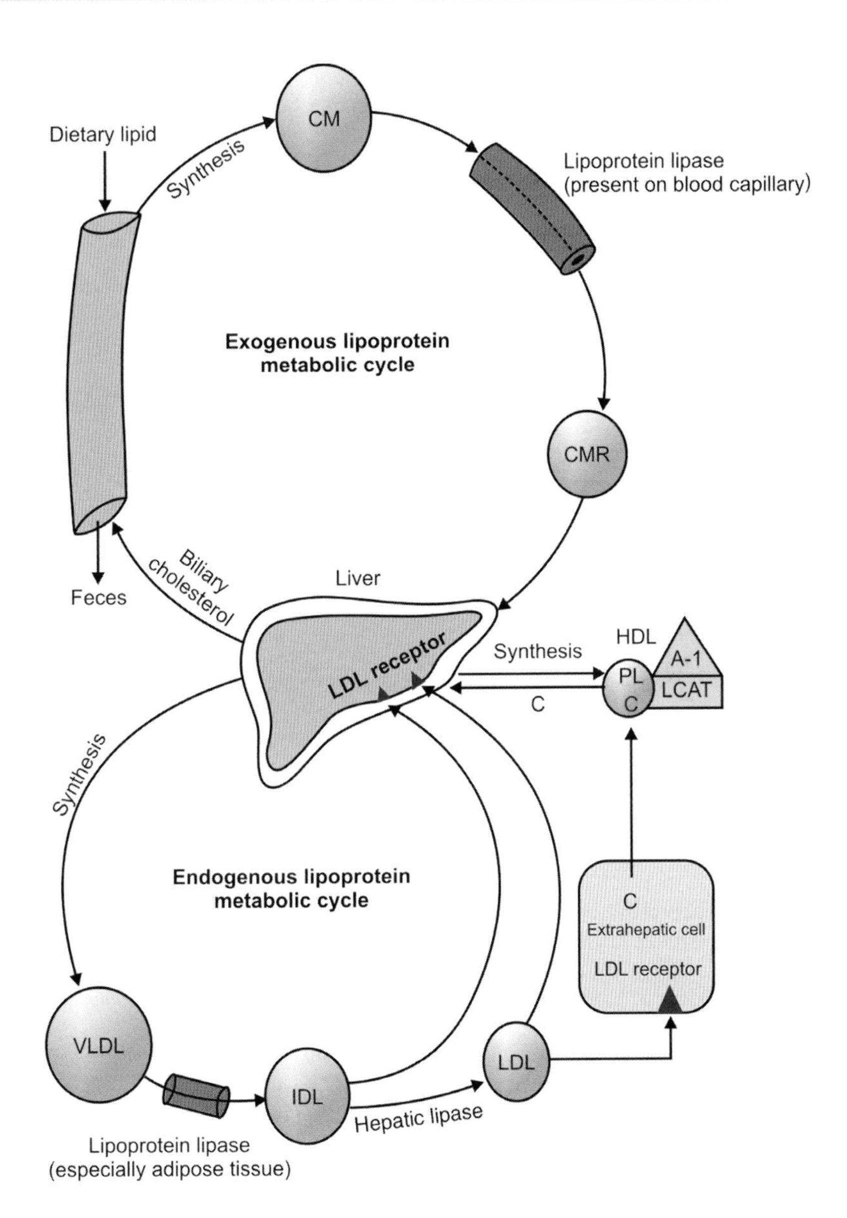

Figure 16-6. Overview of Exogenous and Endogenous Pathways of Cholesterol/Lipid Transport and Metabolism. The exogenous pathway carries dietary lipids via chylomicrons for uptake by target tissues (e.g., adipose and muscle) and/or clearance and storage in the liver. Endogenous VLDL particles carry liver-derived fatty acids primarily to adipose and muscle, IDL (VLDL remnants) carry cholesterol primarily to liver, and LDL takes cholesterol to peripheral tissues as well as the liver. Both pathways utilize HDL as a source of apo C and apo E to allow maturation of the nascent chylomicrons and VLDL to their mature form (see Figures 16-7 and 16-8). A-1, LCAT activator apolipoprotein; apo, apolipoprotein; C, cholesterol; CM, chylomicron; CMR, chylomicron remnant; HDL, high-density lipoprotein; IDL, intermediate-density lipoprotein; LCAT, lecithin cholesterol acyltransferase; LDL, low-density lipoprotein; PL, phospholipid. [Reproduced with permission from Naik P: Biochemistry, 3rd edition, Jaypee Brothers Medical Publishers (P) Ltd., 2009.]

state, insulin increases the activity of LPL, whereas in starvation, the activity declines.

INTERMEDIATE-DENSITY LIPOPROTEIN (IDL) AND LDL

After repeated rounds of LPL action, apo C and apo E (not illustrated) are transferred back to HDL from the shrinking VLDL (Figure 16-8). This results in formation of an **IDL** that contains apo B100 and some remaining apo E. IDL, like HDL and chylomicron remnants, can be cleared by the apo E receptor on liver (genetic defects in the apo E ligand or its receptor elicit type III hyperlipidemia, in which IDL, chylomicron remnants, and HDL are elevated). Further hydrolysis of triacylglycerol in IDL by LPL and transfer of the remaining apo E to HDL result in formation of **LDL** with cholesteryl esters from either the liver (endogenous) or chylomicron remnants (exogenous). VLDLs decrease in surface area as their triacylglycerols are progressively hydrolyzed until they are reduced to cholesteryl ester-enriched LDL (Figure 16-8). In this way, VLDL functions to deliver fatty acids from endogenously synthesized triacylglycerols to muscle and adipose tissue and endogenously synthesized cholesterol to various tissues.

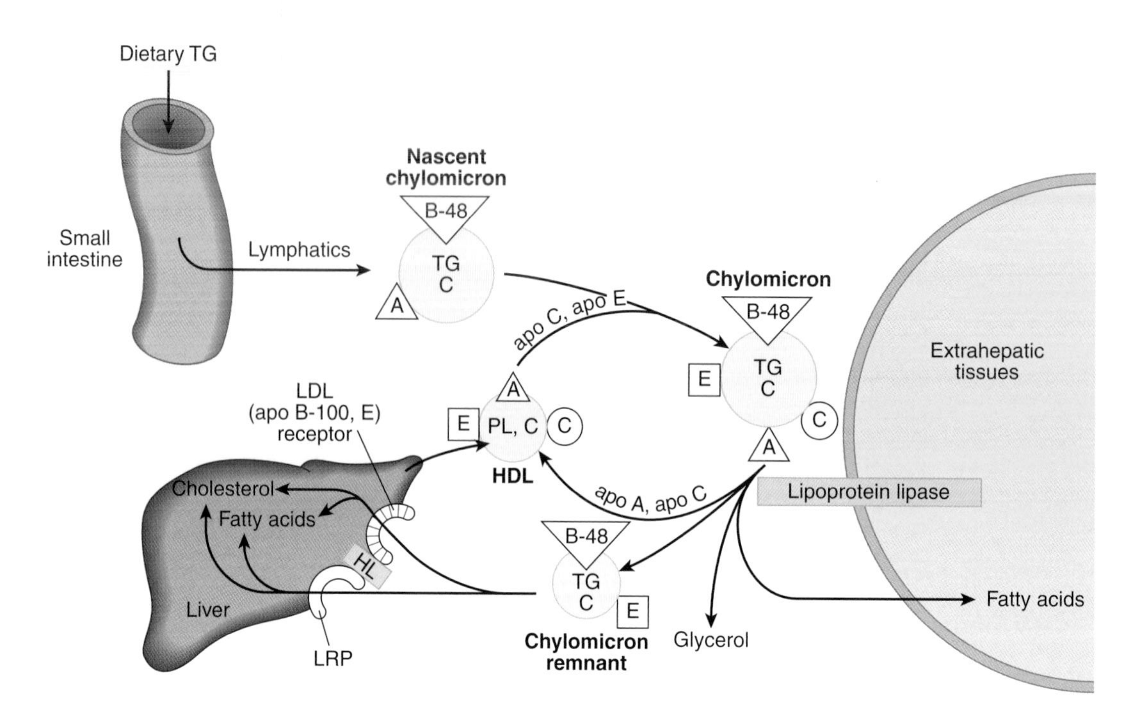

Figure 16-7. Metabolic Fate of Chylomicrons. Nascent chylomicrons are assembled in the small intestine from dietary triacylglycerols (TG) and cholesterol (C) along with apolipoproteins A and B-48. The addition of apolipoproteins C and E from HDL forms a chylomicron, which uses apolipoprotein C to activate lipoprotein lipase (LPL) for peripheral fat deposition. LPL in capillary walls hydrolyzes the triacylglycerols releasing fatty acids into the extrahepatic tissues (e.g., adipose tissue). The remaining chylomicron remnant remains in the circulation and returns apolipoproteins E and C to HDL and then continues to the liver where it binds the apolipoprotein E receptor for subsequent clearance. apo, apolipoprotein; HDL, high-density lipoprotein; HL, hepatic lipase; LRP, LDL receptor-related protein; PL, phospholipid. [Reproduced with permission from Murray RA, et al.: Harper's Illustrated Biochemistry, 28th edition, McGraw-Hill, 2009.]

The primary function of LDL is to provide cholesterol to peripheral tissues. LDL receptors, responsible for this uptake, are associated with clathrin-coated pits on the plasma membrane of these cells and recognize apo B100 (Figure 16-8). LDL particles are then endocytosed into the various tissues including the liver along with the receptors. The receptors and clathrin mostly recycle back to the plasma membrane. Most of the LDL (70%) binds to receptors on liver hepatocytes. The remainder of the LDL associates with receptors on peripheral cells. Lipoprotein disorders in which LDL receptors, or their capacity to bind the apo B100 ligand, are defective, result in an increased level of cholesterol in LDL remaining in circulation, causing hypercholesterolemia and atherosclerosis (see below). A similar event occurs for clearance of IDL, also known as VLDL remnants, and for clearance of chylomicron remnants, although it is apo E on the surface of these particles that is recognized by a receptor primarily found in liver plasma membrane. In both instances, the number of receptors transcribed/translated and available on the cells' surface is directly related to the amount of cholesterol inside a cell. By this mechanism, cells can directly regulate their intake of cholesterol depending on current concentrations in the Golgi.

HDL

As discussed in Chapter 11, the liver is the only organ capable of disposing of significant quantities of cholesterol, either via excretion into the bile or by metabolism to bile acids, both of which are lost to some degree in the feces. **HDL**, synthesized in liver and secreted into plasma as a discoidal particle (Figure 16-8), assists in this function serving as a cholesterol scavenger and facilitating the transport of cholesterol from the periphery to the liver for conversion to bile acids and eventual elimination. It is this cholesterol-removing property that renders HDL the designation of "good" cholesterol carrier.

Mature HDL particles contain lecithin, cholesterol ester, **lecithin cholesterol acyl transferase (LCAT), apo A1, apo C, and apo E**. Circulating HDL acquires cholesterol from peripheral cells in a process facilitated by **cholesterol efflux regulatory protein (CERP)**, an ATP-binding, protein transporter. CERP is activated by apo A1 and flips unesterified cholesterol and lecithin to the outer layer of cell membranes. CERP then delivers the free cholesterol and lecithin as substrates for LCAT on HDL. Apo A1 also activates LCAT in the nascent HDL and functions as a ligand for a cell surface receptor that exists on peripheral cells. Cholesterol esters, the product of LCAT catalysis, move to the core of nascent HDL for subsequent transport back to the liver. The entire process of LCAT extraction of cell cholesterol and incorporation into HDL for liver clearance is called "reverse cholesterol transport."

Another major function of HDL is to serve as a repository for apo A1, apo C, and apo E. Transfer of apo C is required for the metabolism of chylomicrons and VLDL, and apo E is crucial for clearance of chylomicron remnants, IDLs and HDLs. Therefore, HDL contributes to both exogenous and endogenous pathways of lipid transport.

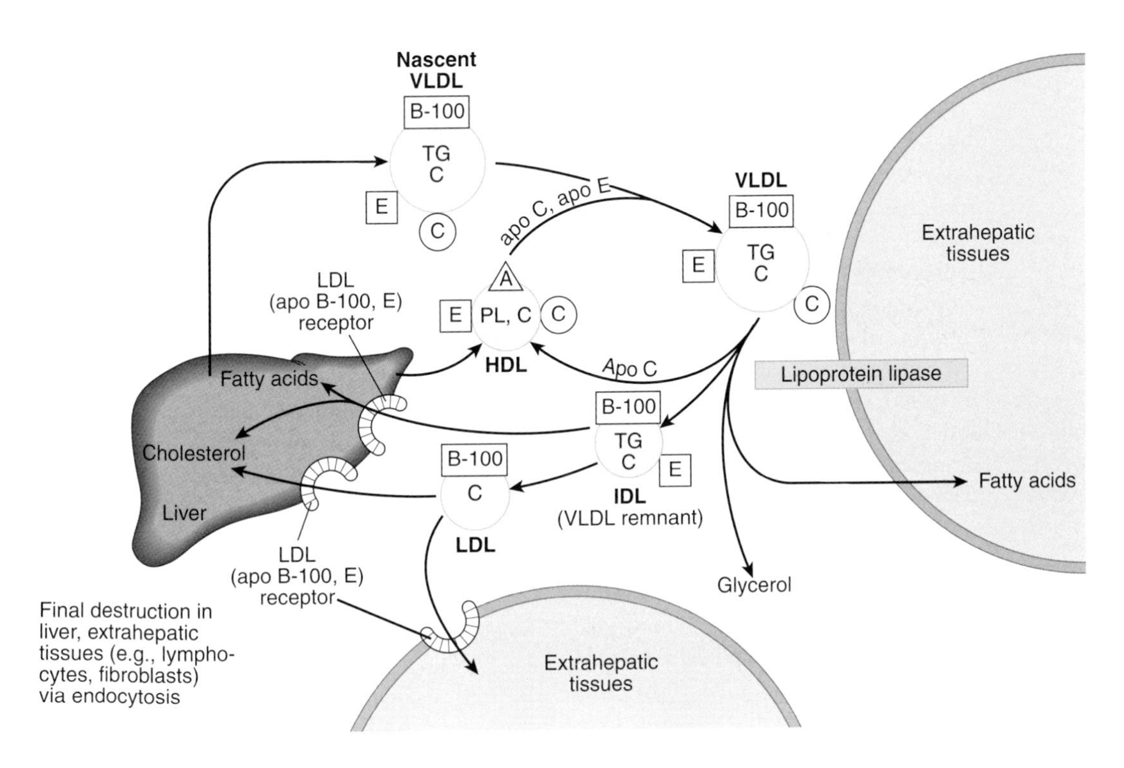

Figure 16-8. Metabolic Fate of Very-Low-Density Lipoprotein (VLDL). Nascent VLDL particles are assembled in the liver from triacylglycerols (TG); cholesterol (C); and apolipoproteins B-100, C, and E. The addition of apolipoproteins C and E from HDL forms a VLDL particle, which uses apolipoprotein C to activate lipoprotein lipase (LPL) for peripheral fat deposition. LPL in capillary walls hydrolyzes the triacylglycerols releasing fatty acids into the extrahepatic tissues (e.g., adipose tissue). Once the TGs are hydrolyzed, apolipoprotein C returns to HDL. The remaining VLDL remnant, known as intermediate-density lipoprotein (IDL), remains in the circulation and either continues to the liver where it binds the apolipoproteins B-100 or E receptor for subsequent clearance or the complete removal of TG from the IDL results in the formation of LDL, which binds to LDL receptors on the liver or other peripheral tissues, primarily via apo B-100 receptors to facilitate uptake/clearance. LDL can also become oxidized and be taken up by "scavenger receptors" on macrophages, which then invade artery walls, eliciting cholesterol deposition and starting the process of atherosclerosis. apo, apolipoprotein; HDL, high-density lipoprotein; PL, phospholipid. [Adapted with permission from Murray RA, et al.: Harper's Illustrated Biochemistry, 28th edition, McGraw-Hill, 2009.]

Familial hypercholesterolemia: Familial hypercholesterolemia, also known as **hyperlipoproteinemia type IIa** according to the Fredrickson classification, is a genetic disorder caused by defective **LDL receptors** or **apo B**. Affected patients have a decreased ability to remove LDL from the circulation, leading to very high levels of LDL and a resulting high risk of heart attack and stroke. When only one copy of the LDL gene is affected (seen in approximately 1 in 500 people), high LDL cause atherosclerotic patients by their 50s. The rarer homozygous state, where both copies of the gene are defective, (occuring in approximately 1 in 1,000,000 people) causes severe cardiovascular disease in infancy and childhood, which can lead to early death. Beyond the early development of atherosclerotic plaques (and the associated risk of heart attacks, strokes, and other cardiovascular problems), patients will also often have **xanthomas**, yellowish deposits of excess cholesterol and fat in the eyelids (**xanthelasma palpebra**) or deposited in tendons (**tendon xanthoma**), especially the Achilles tendon. The cause of familial hypercholesterolemia is normally divided into one of seven molecular abnormalities, five directly affecting the LDL receptor.

1. LDL receptor completely absent.
2. LDL receptor does not reach cell surface because of improper transport from endoplasmic reticulum to Golgi apparatus/plasma membrane.
3. LDL receptor does not bind to LDL either due to mutation in the receptor or the essential apo B molecule. A known change of asparagine to glutamine at amino acid 3500 directly affects apo B binding to the LDL receptor.
4. LDL receptor, once bound to LDL, does not endocytose in clathrin-coated pits.
5. LDL receptor does not return to surface after delivery of LDL inside of cell.

Treatment for heterozygous patients is usually the aggressive use of medications and genetic counseling to consider the medical consequences for their offspring. Homozygous patients usually do not respond well to medication therapy alone and may require LDL apheresis—the direct removal of LDL from the blood via a system similar to dialysis. In severe cases, liver transplant may offer improvement.

ATHEROSCLEROSIS

The formation of vessel blockages, known commonly as plaques, occurs via a process known as **atherosclerosis**. Atherosclerosis, the most common underlying cause of heart attacks and strokes, is a process that involves inflammation of blood vessel walls, particularly arteries, fatty deposits that include lipoproteins and cholesterol, and white blood cells, especially macrophages. Chronic atherosclerosis leads to hardening of the plaque by calcium as well as thickening and stiffening (**arteriosclerosis**) of the affected arterial walls.

The exact mechanism of the formation of a plaque is still under investigation, but current opinion is that the initial step involves oxidation of LDL, probably by the enzymatic action of **lipoprotein-associated phospholipase A2**. These oxidized LDL particles are more readily taken up by macrophages that can invade endothelial cells in the wall of an artery that are themselves damaged by oxidation (Figure 16-8). The injury initiates an immune response by macrophages (Chapter 15), which bind to the area via the protein **vascular cell adhesion molecule-1** and vainly attempt to remove the oxidized LDL molecules. Platelets are also recruited to the site of injury and attempt to cover the area. The accumulation of oxidized LDL and the monocytes/macrophages attempting to ingest it is known as a **fatty streak**—continued inflammation and growth of the fatty streak leads to an atheroma of the arterial wall.

As the immune response continues the monocytes/macrophages die, releasing inflammatory factors such as **IL-1** and **tumor necrosis factor-α**. More white blood cells, including T-lymphocytes and mast cells, are recruited with continued inflammation of the arterial wall. The continued inflammation also impacts smooth muscle cells (Chapter 12), both leading to increased numbers and movement toward the growing plaque, adding to its size and composition. In addition, small calcium crystals begin to form within the smooth muscle cells (**microcalcification**) of the arterial wall, which are adjacent to the atheroma. In an attempt to confine and cover the inflammation, a hard, fibrous capsule is produced from continued calcification and collagen and elastin released from the death of adjacent smooth muscle cells.

When the plaque wall ruptures, blood comes in contact with the thrombotic factors in the interior. The result can be a **thrombus** that completely occludes the artery. This is the most common cause of myocardial infarction. Sometimes the thrombus is only partially or intermittently occlusive. These processes are thought to underly the pathophysiology of unstable angina. Additionally, part or all of the overlying thrombus can break off to form an **embolus**, which can travel to and block smaller arteries. Blockage of an artery can also cause irreversible dilation known as an **aneurysm**. Risk for development of atherosclerotic plaques depends on several factors, including genetic, environmental, and chronic diseases, which will not be discussed further here. This process is summarized in Figure 16-9.

Treatment of atherosclerosis relies on lowering the associated risks, including the use of aspirin to inhibit platelet aggregation and statins (see above and Chapter 7) to lower levels of LDL and other harmful lipids/lipoproteins. Omega-3 fatty acids, found in many deep sea fish, may reduce the chance of death from heart attacks. Omega-3 fatty acids are believed to act via the production of anti-inflammatory eicosanoid synthesis (Chapter 3).

Cardiovascular Disease and Cholesterol/Lipoproteins: Lowering levels of cholesterol and particular lipoproteins, namely LDL and triacylglycerols, have been shown to often dramatically reduce the risk for cardiovascular disease including heart attack and stroke. Statins, which reduce total cholesterol and LDL (Chapter 7), are a mainstay of the prevention of cardiovascular disease. However, other medication classes may be used to lower cholesterol/lipoproteins as well.

Fibrates, which include **gemfibrozil**, **fenofibrate** (see the figure below), and **clofibrate**, among others, activate the **peroxisome proliferator-activated receptor alpha (PPAR-α)**. Activation of this receptor found in muscle, liver, and other tissues increases the β-oxidation of lipids in the liver (Chapter 7), increases LPL lipolysis of triacylglycerol from chylomicrons and VLDL (see above), increases the concentration of HDL to assist in the clearance of cholesterol from peripheral cells (see above), and decreases the secretion of triacylglycerol from the liver into the blood stream (Chapter 11). The vitamin **niacin** (see Chapter 10) also helps reduce LDL and VLDL levels by blocking the catabolism of fats in adipose tissue. As a result, the total concentration of fatty acids is decreased in the blood, leading to lowered secretion by the liver of VLDL and cholesterol. Niacin may also increase HDL, thereby further increasing the clearance of cholesterol. Newer medication like **exetimibe** reduces cholesterol absorption at the intestinal villi (Chapter 11). These agents may be used in combination with dietary measures and statins and may influence initial plaque formation (see above).

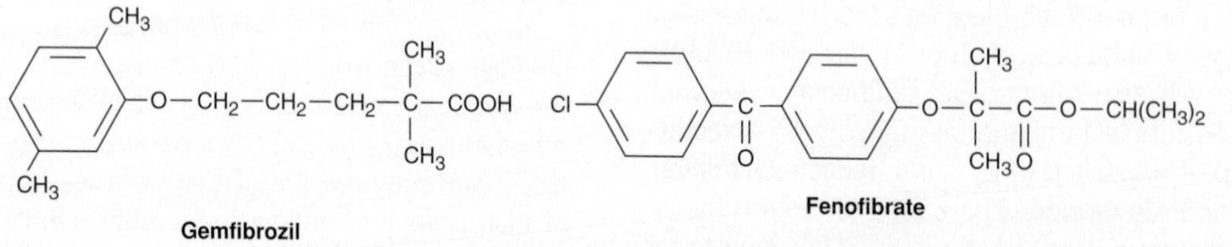

CH_3; $-O-CH_2-CH_2-CH_2-C(CH_3)_2-COOH$; CH_3

Gemfibrozil

$Cl-C_6H_4-C(=O)-C_6H_4-O-C(CH_3)_2-C(=O)-O-CH(CH_3)_2$

Fenofibrate

Reproduced with permission from Katzung BG, et al.: Basic and Clinical Pharmacology, 11th edition, McGraw-Hill, 2009.

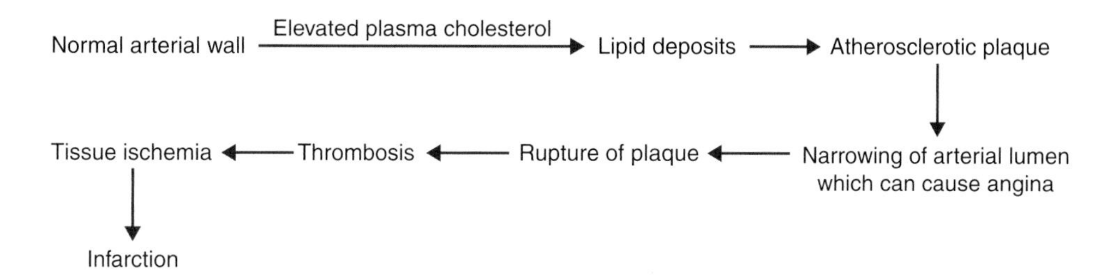

Figure 16-9. Overview of Possible Mechanism of Atherosclerosis Formation and Infarction. [Reproduced with permission from Naik P: Biochemistry, 3rd edition, Jaypee Brothers Medical Publishers (P) Ltd., 2009.]

BIOCHEMICAL MECHANISMS ASSOCIATED WITH HEART ATTACK

A heart attack or myocardial infarction (MI) occurs when blood flow and, therefore, oxygen and nutrient delivery is blocked to an area of the heart. It can be worsened by a marked reduction in number (anemia) or in the ability of red blood cells to carry these components (Chapter 14). Prolonged lack of oxygen leads to death of the cardiac muscle cells and, frequently, fatal dysrhythmias or heart failure. During the course of a heart attack, several biochemical events occur related to insufficient oxygen (hypoxia), insufficient blood flow (ischemia), lack of oxygen (anoxia/infarction and cell death), as well as the resulting pain (angina).

An MI will often be preceded by episodes of ischemia that result in the typical squeezing chest pain of angina pectoris. Complete blockage of blood vessels that feed the heart muscle produces a state of no oxygen (**anoxia**). The effect of lowered and/or absent oxygen is multi-fold on biochemical pathways. Without oxygen, oxidative phosphorylation (Chapter 6) cannot occur and new ATP and NAD^+ are not generated. As a result, the first step of glycolysis is blocked, although glucose availability will also fall (**hypoglycemia**) without blood flow/nutrient delivery. $NADP^+$ stores will rapidly disappear, reducing flux though the pentose phosphate pathway cease. Although of no utility in this scenario, glycogen synthesis and storage will also stop because of the lack of glucose-6-phosphate molecules from glycolysis. Amino acid synthesis and breakdown will stop as will any enzymes (Chapter 5) that rely on ATP or any other nucleotides derived from it (Chapter 4), including the ability to generate more proteins and enzymes via transcription (Chapter 9). Motor proteins such as myosin, dynein, and kinesin will no longer have ATP to power the conformational changes required for their activity (Chapter 1). Cell motility, including intracellular transport processes that rely on these motors, will also end and membrane channels that rely on the Na^+–K^+-ATPase function (Chapter 8) and/or other energy-reliant processes will no longer work, causing the buildup of toxic waste products (carbon dioxide, etc.). The failure of these membrane transporters will also affect Ca^{2+} gradients, affecting the calcium-dependent contraction of cardiac cells discussed above.

When the blockage is suddenly relieved, **reperfusion injury** may occur, from the release of damaging, reactive oxygen species (e.g. superoxide, hydrogen peroxide, and hydroxyl radicals). In addition, the metabolism of ATP in ischemic conditions creates the molecule **hypoxanthine**, which, itself, may be metabolized by xanthine oxidase to produce hydrogen peroxide. Both processes overwhelm the affected cells' normal antioxidant defenses, including vitamins A, C, and E (Chapter 10), which also cannot be replaced because of blocked blood flow. Other molecular processes dependent on guanosine triphosphate such as the growth of microtubules (Chapter 1), G-protein signaling (Chapter 8), and the binding of transfer RNA molecules during translation (Chapter 9) will cease.

As ischemia continues and worsens to anoxia, an infarct, with irreversible death of cells will occur within 30 minutes. As the cells die, their membranes will lose their bilayer integrity (partially due to calcium activation of phospholipases as well as damage by free radicals to phospholipids, membrane proteins, and glycosaminoglycans) essential to separate compartments [e.g., mitochondria and peroxisomes that contain molecules toxic to cells (Chapter 8)] and extracellular and intracellular compartments (Chapter 8) vital for life. As membranes breakdown, the contents of the cardiac cells are released, including **cardiac markers** used to detect an MI. Similar processes occur in the brain during stroke or to any tissue that is deprived of efficient and constant circulation.

Cardiac Markers: Myocardial infarction is diagnosed by a combination of symptoms, ECG changes, and cardiac markers that leak into the blood stream from dying heart cells. In the past various enzymes, including **Lactate Dehydrogenase (LDH)** and **Creatine Kinase (CK)** were monitored but these proteins also rise with damage to other organs. **CK-MB**, an isoform of CK, is much more specific to the heart and is still used in monitoring damage from MI. The troponin proteins, which have cardiac-specific isoforms, are now the most commonly used markers as part of a diagnostic workup of a suspected MI.

Death of cardiac cells will also interrupt the normal propagation of depolarization that produces a normal ECG tracing. As cardiac cells die in particular areas of the heart, changes such as ST-segment elevation or depression and the development of Q-waves are diagnostic of both the presence and location of a heart attack. MIs are often initially categorized as an **ST elevation MI (STEMI)** or **non-STEMI** because of different treatment modalities. Heart attacks are also often described as **transmural** if cell death extends completely through a section of the heart wall, often as a result of a complete blockage of an artery for the heart.

Treatment of an MI and its associated angina from blockage of a heart blood vessel is focused on lowering the heart's demand for oxygen while attempting to increase oxygen delivery to those cells. **Aspirin** inhibits **cyclooxygenase** production of thromboxanes (e.g. Thromboxane A_2), which inhibits platelet aggregation and further blocking of blood vessels (Figure 16-10).

As noted above, **nitroglycerin**, which generates NO, activates and dilates blood vessels by protein kinase G-induced dephosphorylation and inactivation of smooth muscle myosin molecules (Figure 16-5). Dilated blood vessels allow increased oxygen delivery in cases of narrowed but not completely blocked heart arteries.

Prinzmetal's Angina: One form of angina with a potentially direct biochemical basis is **Prinzmetal's angina**, caused by spasm of coronary arteries that reduce blood flow rather than the blockage seen in routine atherosclerotic disease. Normally, acetylcholine not only induces contraction of the vessels but also causes healthy endothelial cells to produce and release NO. The vasoconstriction effect of acetylcholine is, therefore, overridden by the vasodilatory effect of NO. In patients with Prinzmetal's angina, however, acetylcholine-induced NO production is deficient and vasoconstriction predominates. Diagnosis is usually made by evidence of transmural ischemia on the ECG that abruptly resolves spontaneously or with nitroglycerin. Treatment may include usually via nitroglycerin to provide external NO and CCBs that block calcium release and, therefore, acetylcholine-mediated constriction.

β-Blockers, also described above, are a mainstay of this treatment as they block β_1-receptors in the heart and β_2-receptors found in blood vessel smooth muscle and decrease both heart rate and contractile force lowering the work load and, therefore, oxygen demand. The vasodilatation effects of **CCBs** and inhibition of angiotensin vasoconstriction (**angiotensin-converting enzyme inhibitors**), both described above, make these medications an additional part of ischemia/angina/MI treatment as well.

Varying mechanical and medicinal techniques can also be used to reopen the blocked vessel, although the renewed flow of blood also carries with it toxic radical molecules, inflammatory interleukin molecules, excess acid, and disturbances of calcium transport that can result further to the heart and surrounding tissues as well as heart arrhythmias known collectively as reperfusion injury (see above).

Thrombolysis: Thrombolysis and **embolysis**, the breakdown of a thrombus or embolus by medical means, are used as mainstay therapy for MI, stroke, and other acute medical problems involving blockage of blood vessels. The treatment relies on plasminogen activators that initiate the cascade of the body's natural clot-dissolving enzymes (Chapter 14) by promoting production of plasmin from its proenzyme (Chapter 5) plasminogen. Thrombolytics include **streptokinase**, **urokinase,** and artificially produced analogues of **tissue plasminogen activator**.

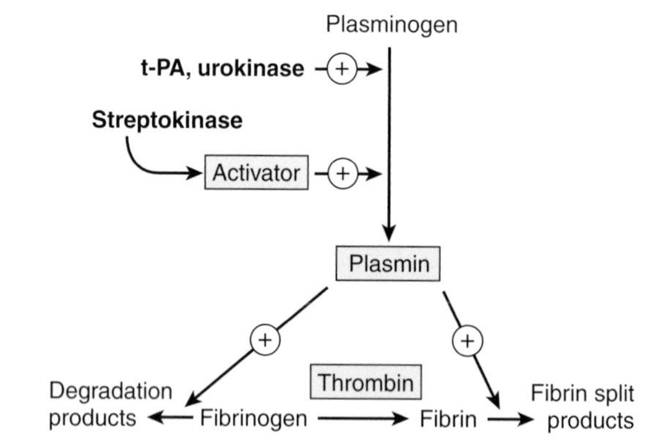

Reproduced with permission from Katzung BG, et al.: Basic and Clinical Pharmacology, 11th edition, McGraw-Hill, 2009.

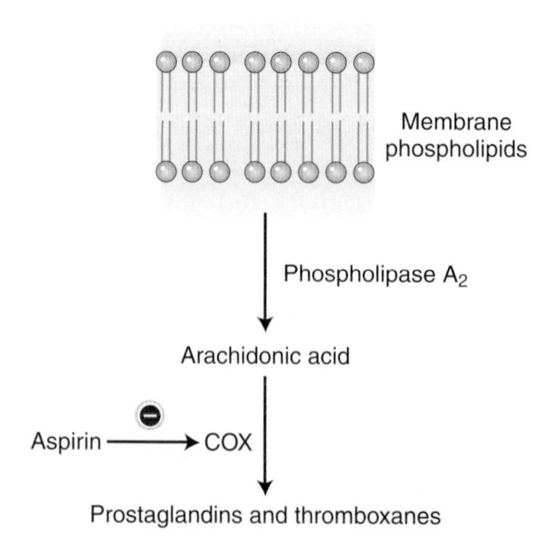

Figure 16-10. Action of Aspirin in Treatment of Myocardial Infarction. Aspirin irreversibly inhibits cyclooxygenase (COX), leading to a decrease in production of prostaglandins and thromboxanes from arachidonic acid. See text and chapter. [Adapted with permission from Kibble JD and Halsey CR: The Big Picture: Medical Physiology, 1st edition, McGraw-Hill, 2009.]

REVIEW QUESTIONS

1. In what ways does cardiac muscle differ from striated and smooth muscle?
2. What are the three classes of calcium channel blockers, their primary effect(s), and their medical uses?
3. What is the function of the L-type channel in cardiac cells?
4. What are the roles of very-low-density lipoprotein and low-density lipoprotein in the endogenous lipoprotein pathway in the processing of cholesterol and triacylglycerols?
5. What are the functions of apolipoproteins A1, B100, C, and E in the processing of lipoproteins?
6. Why is high-density lipoprotein termed the "good" cholesterol carrier?
7. What are the sequences of events leading to an arterial thrombus?
8. What are the events associated with the onset of a myocardial infarction and the associated metabolic ramifications?
9. What is the advantage of measuring troponin I and troponin T as cardiac markers relative to total activity of creatine kinase?
10. What are the various options in the treatment of a myocardial infarction and the mechanism for each of these options?

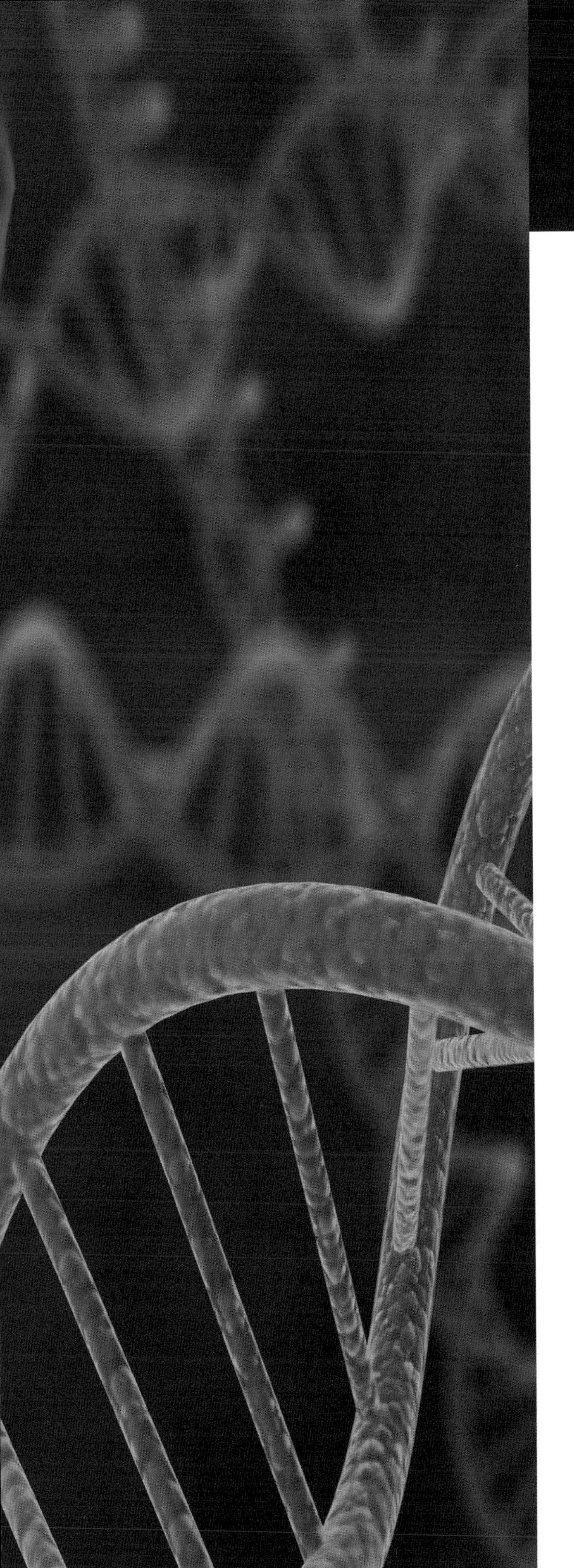

CHAPTER 17

THE RESPIRATORY SYSTEM

Editor: Howard J. Huang, MD
Instructor in Medicine, Division of Pulmonary and Critical Care Medicine, Washington University School of Medicine, St. Louis, MO

OVERVIEW

The respiratory system facilitates the exchange of oxygen and carbon dioxide between the external environment and the human body. The airways (trachea and bronchi), lungs, and alveoli and pulmonary surfactant provide the biochemical and biophysical basis for this exchange. Acid–base balance in the body also relies on these tissues as well as on the carbonic anhydrase and other enzyme systems. Various diseases of the respiratory system affect either the structure or the biochemistry of the lungs; medications and treatments for these diseases rely on knowing their biochemical basis.

BASIC ANATOMY AND DEVELOPMENT

The respiratory system, composed of the nose, nasal cavity, larynx, trachea, bronchiole tubes, and lungs, is derived from the endoderm, although the splanchnic mesoderm contributes to the visceral pleura, cartilage, and connective tissue and also to the immediately associated smooth musculature and capillaries (Figure 17-1). Other closely associated structures include the ribs and diaphragm and the central cardiovascular system, which provides pulmonary arteries and veins.

The respiratory system facilitates the transport of oxygen (O_2) to the body's tissues and enables elimination of gaseous waste products of metabolism. Muscles of the ribs, diaphragm, neck, shoulders, and abdomen generate a negative pressure gradient, which promotes the flow of external air into a system

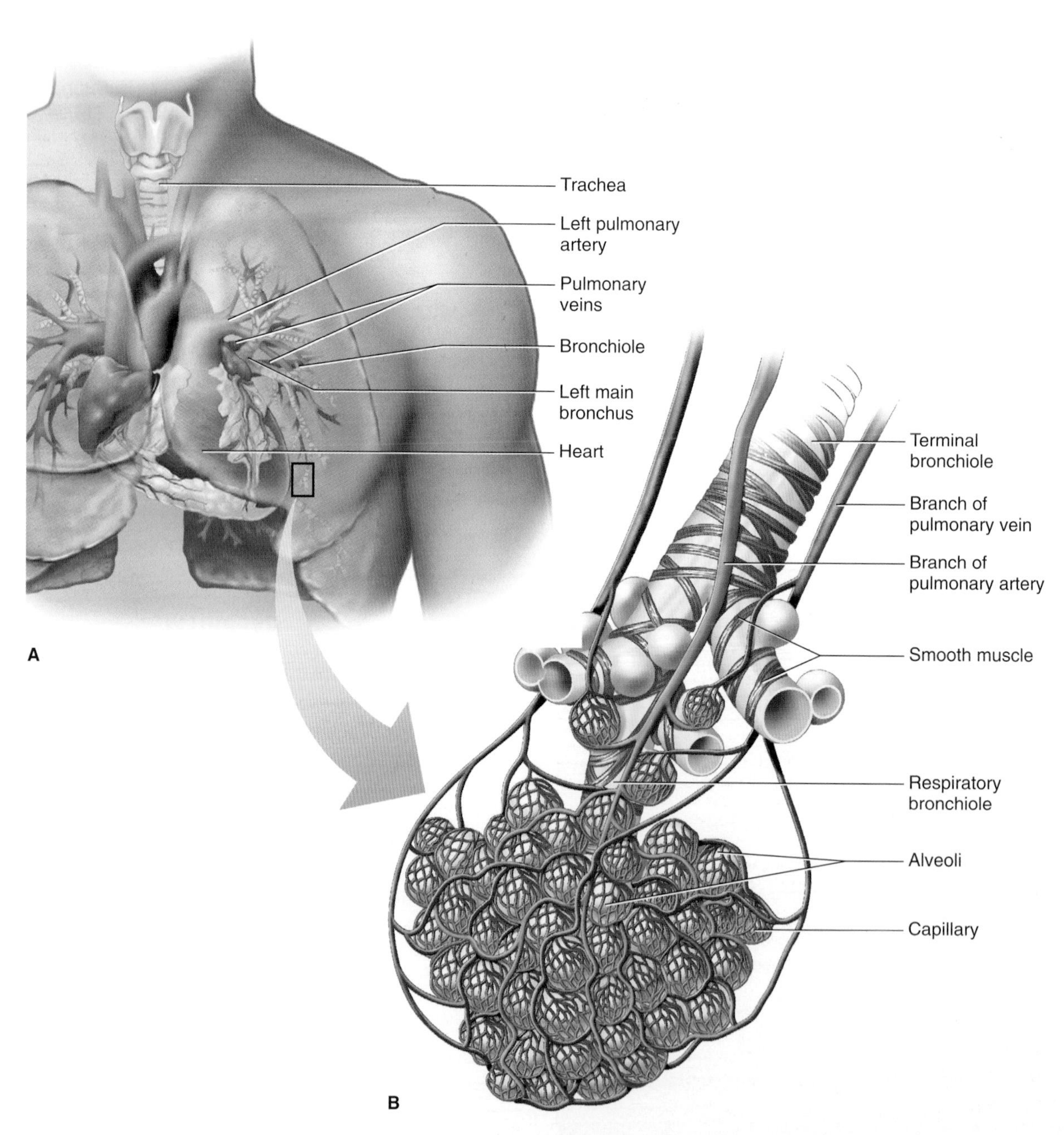

Figure 17-1. Overview of Lung Structure. The basic structure of lungs (**A**) includes the trachea and repetitively branching bronchi/bronchioles leading to a terminal bronchiole. Pulmonary arteries and veins provide circulation to this lung tissue. At the terminal bronchiole/alveoli (**B**), the exchange of oxygen and carbon dioxide between alveoli and the surrounding vasculature is conducted as discussed later in this chapter. [Reproduced with permission from Barrett KE, et al.: Ganong's Review of Medical Physiology, 23rd edition, McGraw-Hill, 2010.]

of branching, conducting airways and peripheral terminal bronchioles.

Exchange via passive diffusion through the cell membranes of the type I epithelial cells of the **alveoli** and the pulmonary capillaries is responsible for the uptake and removal of O_2 and carbon dioxide (CO_2), respectively (Figure 17-2). Pulmonary arteries from the right ventricle of the heart deliver deoxygenated, CO_2-laden blood through the branching blood vessels, to the capillaries; pulmonary veins return the freshly oxygenated blood with decreased CO_2 to the left auricle and ventricle for delivery to the body's tissues. Finally, positive pressure, along with the elastic walls of the alveoli, springing back from full expansion, expels the waste gas, including the CO_2. Although exhalation is normally a passive process, active expulsion is also possible, utilizing the same muscle groups noted above. Defects or injuries to any of the parts of the respiratory system or to those that affect the musculature and/or cardiovascular system can lead to respiratory disease.

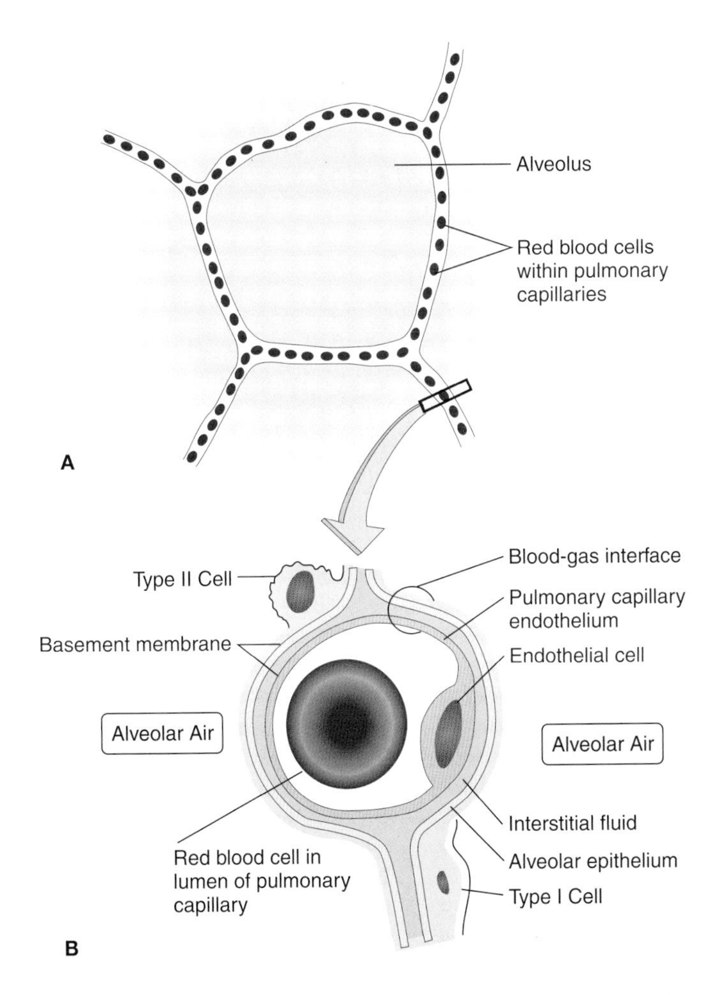

Figure 17-2. Structure of the Alveolus. (**A**) The alveolus allows the exchange of oxygen (O_2) and carbon dioxide (CO_2) between the alveolar air and capillary. (**B**) Magnified view of the structure of the alveolar/pulmonary capillary interface showing the proximity of alveolar air and the capillaries. This structure allows the efficient diffusion of O_2 and CO_2 through the single-cell layer of type I cells to/from erythrocytes in the capillaries and the alveolus. The pulmonary interstitium and type II cells responsible for secretion of surfactant, which helps to maintain the fragile structure of the alveoli, are also shown. [Adapted with permission from Kibble JD and Halsey CR: The Big Picture: Medical Physiology, 1st edition, McGraw-Hill, 2009.]

The passive diffusion of O_2 and CO_2 relies on the close proximity, cellular membranes, and interwoven connective tissues (collagen and elastin fibers; see Chapter 13) that allow the alveolar–capillary gas exchange. The overall large surface area of the approximately 300 million alveoli in the average human body provides roughly 75 m^2 of gas exchange area. The capillary network that surrounds these alveoli covers about 70% of this area. Gas exchange between the alveoli and the pulmonary capillary is driven by the higher "partial pressure" in the alveoli versus in the pulmonary capillaries. This pressure gradient propels the small O_2 and CO_2 through the basement membrane and the lipid bilayer of the alveolar type I squamous epithelial cell (~0.2 μm thin) and the capillaries one layer of endothelial cell plasma membranes.

PULMONARY SURFACTANT AND THE DEVELOPING LUNG

Pulmonary surfactant, composed of **lecithin** and **myelin**, is secreted on a continuous basis by type II alveolar cells and Clara cells beginning at approximately week 20 of gestation. Pulmonary surfactant has both a lipid (~90% of total) and a protein (~10% of total) component. About half of the lipids are **dipalmitoylphosphatidylcholine** (Chapter 4), the 16-carbon saturated lipid with a charged amine group on its head group. The remaining lipids include **phosphatidylglycerol**, which modulates the fluidity of the surfactant as well as **cholesterol** and other lipids. About half of the proteins are **apoproteins**, which have both hydrophobic and hydrophilic portions; the other half is composed of proteins normally found in blood plasma. These lipids and proteins all have the capability of interacting with aqueous (hydrophilic) or nonaqueous (hydrophobic) environments and it is this quality that leads to their specialized function in the lung (Figure 17-2). By directly interacting with alveolar water via their hydrophilic regions while the hydrophobic regions remain in the air, pulmonary surfactant creates a myelin meshwork that lines the alveoli—a strong, intertwined lipoprotein system that is analogous to the myelin sheath of nerve cells (Chapter 19).

This unique alveolar lining greatly reduces surface tension, allowing easier expansion/stretching (known as **pulmonary compliance**) and collapse of alveoli during respiration and the resulting changes in pressure. This reduction in surface tension decreases the work of respiration and the total amount of pressure that must be generated for efficient and effective inspiration and expiration. The pulmonary surfactant also helps all the lung alveoli to expand (inspiration) and shrink (expiration) at the same rate, thereby reducing the chance for isolated overexpansion and the total collapse (atelectasis) of the alveolar sacs.

Pulmonary Surfactant and Premature Birth: The respiratory system develops only partially during pregnancy and becomes fully functional only after birth. One medical consequence is seen in premature infants in whom **pulmonary surfactant** has only started to form. Specifically, premature babies born before 32 weeks of gestation must be treated aggressively before birth to help augment surfactant or it may result in **Infant Respiratory Distress Syndrome (IRDS)**. This condition, also known as Hyaline Membrane Disease because of its findings on pathological examination, results in respiratory distress of the newborn infant along with accessory muscle use, grunting, nostril flaring, and cyanosis and, if not appropriately treated, cessation of breathing and death. Multiple complications can also result from IRDS, making it the leading cause of death of infants up to 1 month old in the developed world. The most successful treatment is frequent steroid injections to the mother, which prompts surfactant production while delaying birth for as long as possible.

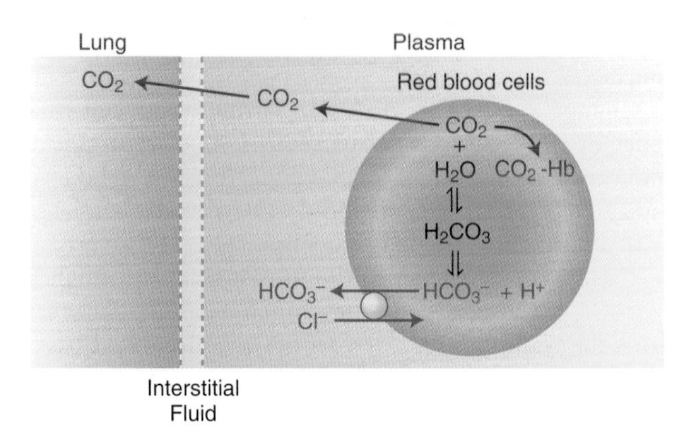

Figure 17-3. Regulation of Acid–Base Balance by the Lung. Elimination of excess carbon dioxide (CO_2) can occur by increased respiratory rate (hyperpnea), which removes bicarbonate (HCO_3^-) and hydrogen ions (H^+) from other tissues (e.g., red blood cells). See the text for further details. [Reproduced with permission from Kibble JD and Halsey CR: The Big Picture: Medical Physiology, 1st edition, McGraw-Hill, 2009.]

O_2–CO_2 EXCHANGE IN THE LUNG AND ACID–BASE BALANCE

A major function of blood and red blood cells is the transport of O_2 to cells and CO_2 from cells from/to the lung, respectively. As a result, diseases involving this system will often result in poor oxygenation (hypoxia) and/or increased CO_2 levels (hypercapnia) of the blood and tissues. Approximately 60% of CO_2 is carried back to the lungs dissolved in the blood plasma as bicarbonate molecules (HCO_3^-). The binding of O_2 and exchange with CO_2 is discussed in detail in Chapter 14.

The impact of O_2–CO_2 exchange extends beyond simple oxygenation of tissues and elimination of waste CO_2. Levels of CO_2, as regulated by the lung, kidneys (Chapter 18), blood (Chapter 14), and even the pancreas (Chapter 11), are responsible for maintaining proper and acceptable levels of hydrogen ions (H^+) and, therefore, the body's acid–base balance (Figure 17-3). Ranges of pH outside the normal body with a range of less than 7.35 (**acidosis**) or more than 7.45 (**alkalosis**), if uncompensated, can quickly lead to death. This balance is critical in the central nervous system where tissues and cells are even more sensitive to pH variations.

Carbonic anhydrase (Figure 17-4), found in several tissues including the lungs, kidneys, gastrointestinal tract, saliva, brain, and red blood cells, is essential for the rapid conversion of CO_2 to HCO_3^- (then carried in the plasma) and for the slower reverse reaction of HCO_3^- to H_2CO_3 (carbonic acid) to CO_2 (Figure 17-3) in the lungs or kidneys. About 75% of the carbon in the body is in the form of H_2CO_3 or HCO_3^-. In the absence of carbonic anhydrase, an equilibrium of the components of the CO_2 to HCO_3^- interconversion (Figure 17-4) is established. Carbonic anhydrase allows the body to bypass the carbonic acid step. In the lungs, carbonic anhydrase also facilitates HCO_3^- to CO_2 via H_2CO_3.

Deficiencies in the management of CO_2/bicarbonate in the lungs can lead to the medical conditions of **respiratory acidosis** and **respiratory alkalosis**. Both have acute and chronic forms. Acute respiratory acidosis occurs when there is a short-term decrease in respiratory function (hypoventilation), leading to a decrease in the amount of CO_2 exhaled, and because CO_2 is always being produced by the body, an increase in HCO_3^- per the reaction below (Figure 17-4). Chronic respiratory problems such as those caused by chronic obstructive pulmonary disease (COPD), interstitial lung diseases (see below), mechanical abnormalities (e.g., diaphragm dysfunction, deformities of the chest), respiratory muscle fatigue (e.g., chronic lung disease, extended asthma attack), or other disorders with longer term hypoventilation, increase CO_2 in the blood, and result in higher concentrations of H^+ (acidosis). Alternatively, acute or chronic increases in exhalation of CO_2 with lowering of H^+ concentrations (alkalosis) can result from lung (pneumonia and fever with increased rate of breathing), psychiatric (stress and anxiety), and/or neurological (stroke, meningitis) diseases, pregnancy, liver disease, overuse of aspirin

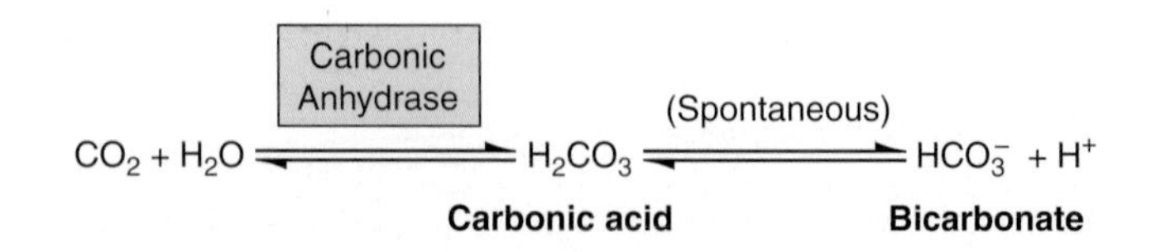

Figure 17-4. Interconversion of Carbon Dioxide (CO_2) to Bicarbonate. CO_2 is reversibly converted to bicarbonate ion (HCO_3^-) via the action of the enzyme carbonic anhydrase. Carbonic acid (H_2CO_3) is a bypassed intermediate in both reaction directions. Water molecule (H_2O), essential to the process, and hydrogen ions (H^+) are also shown.

and/or caffeine, recent increases in altitude, and even overly aggressive mechanical ventilation of patients. Various mechanisms attempt to compensate for any respiratory acidosis or alkalosis. Cells are able to either take up or release bicarbonate as part of a buffering system that can respond on a time scale of minutes. Red blood cells have a specific membrane transport protein for the movement of bicarbonate ions (Chapter 14). The kidneys are also able to increase or decrease bicarbonate excretion or reabsorption in the proximal tubules over a period of a few days (Chapter 18). Acid–base balance is discussed in more detail in Chapters 14 and 18.

NONINFECTIVE DISEASES OF THE RESPIRATORY SYSTEM

Noninfective diseases of the respiratory system are grouped into obstructive and restrictive, although patients can suffer from one or both. **Obstructive lung diseases**, as the name implies, include medical conditions where airflow is impaired by some manner. Classic obstructive diseases of the respiratory system include **emphysema, chronic bronchitis, bronchiectasis**, and **asthma**. Because emphysema usually has an inflammatory component, chronic bronchitis is often present at the same time. The two can be diagnosed separately or together as **COPD**. **Restrictive lung diseases** have loss (i.e., restriction) of complete lung expansion. This may be because of fibrosis or scarring damage to the lung tissue, mechanical problems such as severe scoliosis, chest wall deformities, and/or increased abdominal pressure (e.g., hernia, etc.).

OBSTRUCTIVE DISEASES—EMPHYSEMA

Emphysema results from the destruction of the elastin network that provides the physical support for the lung's bronchioles and acini (clusters of alveoli), which directly impacts the pulmonary compliance discussed above. The destruction decreases the surface area that is critical for O_2–CO_2 exchange (see above) forcing patients to breathe faster (hyperventilation), to help compensate for lowered O_2 levels in the blood. Patients with emphysema also have difficulty exhaling because of the partial or total collapse of the alveoli, alveolar structures, and bronchioles during exhalation. As a result, they are sometimes referred to as "pink puffers" because they must exhale slowly often through pursed lips.

Emphysema is divided into **panacinar** and **centroacinar** types because of its location but, more importantly, because of the biochemical basis of the disease. A third type of emphysema, **congenital lobar**, seen in some newborns, results from the overgrowth of a lung or pulmonary lobe and the resulting compression and bronchial narrowing seen on other parts of the respiratory system, including the pulmonary vasculature. This type of emphysema will not be discussed any further here.

Panacinar emphysema impacts bronchioles and alveoli located in the lower lungs and results from a deficiency of the enzyme **α-1-antitrypsin**. As the name implies, this deficiency causes pathology throughout the acinar structure to include several alveoli and their associated bronchiole(s). The long-standing hypothesis of α-1-antritrypsin's role is based on its relationship with **neutrophil elastase** known as the "Protease-antiprotease Theory." Neutrophil elastase is responsible for the destruction of respiratory bacteria (e.g., *Escherichia coli*, *Salmonella*, *Shigella*, and *Yersinia*) by degrading a protein in their outer bacterial membrane. Unfortunately, neutrophil elastase also destroys **collagen IV** (Chapter 13) responsible for the supportive architecture of bronchioles and alveoli. α-1-Antitrypsin is a direct inhibitor of neutrophil elastase and, therefore, limits the damage to the respiratory structures. A genetic deficiency of α-1-antitrypsin results in unopposed neutrophil elastase and, in smokers, emphysema. As a result of the genetic cause of this type of emphysema, it is suspected in younger patients presenting with significant obstructive lung disease. However, α-1-antitrypsin deficiency only accounts for a small percentage (~2%) of emphysema cases especially in older smokers who have stopped smoking for several years. A possible prominent disease role for other enzymes (e.g., **matrix metalloproteases**) in the lung is currently being explored.

The second major type of emphysema is **centroacinar**, affecting cells mainly at the end of the bronchioles (i.e., the center of the acinus) with less impact on the more external alveolar structures. Cigarette smoke, specifically, the effects of nicotine on neutrophils and the neutrophil elastase enzyme, has been found to be a major cause of centroacinar emphysema in susceptible patients. Nicotine serves as a major attractant for neutrophils via an increased production of **nuclear factor-kappa β (NF-κβ)**, which increases production of **tumor necrosis factor (TNF)** and **interleukin (IL)-8**. IL-8 activates secondary signaling pathways by the G_q-protein pathway (phospholipase C, diacylglycerol and IP_3, protein kinase C, and calcium ion release). IL-8 also activates the **mitogen-activated protein kinase (MAPK)**, whose G_α protein pathway leads to increased **protein tyrosine kinase (PTK)** activity. PTK phosphorylations promote neutrophil chemotaxis and cell adhesion for the accumulation of neutrophils as well as the activation of neutrophil elastase, leading to destruction of the lung architecture. Nicotine also inhibits the activity of antiproteases (e.g., α-1-antitrypsin) directly as well as indirectly by generating free-radical, reactive oxygen species (e.g., H_2O_2/O_2^-) adding to the destructive cascade. Thus, the same imbalance between proteases and antiproteases results this time from the toxic effects of nicotine and smoking. Other factors can significantly impact the development of disease in these patients, including poorly understood contributions of inflammatory factors (Chapter 15) and genetic variation, which affect disease susceptibility.

OBSTRUCTIVE DISEASES—BRONCHITIS

Bronchitis, inflammation of the bronchial mucus membranes, can be divided into acute and chronic forms. All share an inflammatory response (Chapter 15), which leads to this disease. Acute bronchitis is usually caused by a temporary irritant such as an infective agent [virus (~90% of all cases) or bacteria] or a temporary environmental irritant. Cough with possible production of excessive mucus is the major symptom that resolves as soon as the agent is eliminated. Chronic bronchitis, however, is a long-standing disorder that is uniquely diagnosed not by laboratory samples or testing but, rather, by the presence of a cough productive of sputum that lasts for three months or longer per year for at least 2 years. Symptoms include shortness of breath and/or wheezing, especially after exertion, because of obstruction of air flow and air exchange from airway inflammation, edema, and mucus hypersecretion. Chronic bronchitis can also lead to spasms of the bronchi, which further reduce lung function. Cigarette smoking and/or other environmental or occupation irritants are the predominant cause of this medical condition. Unlike emphysema, destruction of lung parenchyma is not present. However, emphysema and chronic bronchitis frequently co-exist.

OBSTRUCTIVE DISEASES—ASTHMA

Asthma, a chronic inflammatory condition of the bronchi, has gained increasing notoriety because of its growing prevalence in all societies, both developed and undeveloped. Asthma leads to an intermittent variable and partly reversible constriction of the bronchi smooth muscles known as bronchospasm and airway obstruction with breathing difficulties, including the characteristic asthmatic wheeze of an asthma exacerbation. In very severe cases, obstruction becomes so severe and airflow so limited that wheezes become inaudible. Pathological growth and enlargement of smooth muscles and mucus-producing cells as well as pro-inflammatory **cytokines**, and elevated **immunoglobulin (Ig) E** (see below and Chapter 15) are also hallmarks of asthma. Chronic, poorly controlled, or severe asthma can lead to remodeling of the lung, including deposition of fibrotic tissue that leads to irreversible constriction of the bronchi. Severe asthma attacks cause thousands of deaths in a year. Unlike COPD and bronchitis, the inflammation is largely reversible and affects only the bronchi and not the alveoli as in emphysema.

A genetic component to asthma has been noted for several years with over 100 genes having been associated with varying degrees and with varying populations to asthma. However, an environmental component is also essential for onset and progression of the disease. These environmental factors can include cigarette smoking (direct and passive smoke), air pollution, grass/mold/plants (especially when sporulating), pet dander, cockroach droppings, cold temperatures, various viral infections (e.g., childhood respiratory viral infections such as RSV), stress, physical exertion, decreased exposure to dirt, and/or overuse of antibiotics in childhood, medicines including aspirin and acetaminophen, and a myriad of other variables. The presence of other immunological diseases, including eczema, allergic rhinitis (hay fever), and allergies, is also associated with the prevalence of asthma. Repetitive and long-term exposure not only to chemicals but also to substances such as flour can also lead to an occupational form of asthma because of chronic inflammation and resulting bronchospasms.

Regardless of the precipitating factor(s), asthma is associated with an inflammatory disease resulting in the overproduction of **IgE**, which may block β_2-receptors on bronchi smooth muscle cells. This "β-adrenergic theory of asthma," first proposed in 1968, has led to an improved understanding of the biochemical mechanism behind asthma and allergies in general as well as the development of specific medicines to counteract these effects. This theory proposes the following pathway to a hypersensitivity of the cells lining the bronchi and asthmatic disease.

Genetic and other variables make cells in the bronchi "hypersensitive" to particular environmental triggers. Environmental triggers are ingested by **antigen-presenting cells** (e.g., macrophages, B-cells, and dendritic cells), which interact with immature, helper T cells (T_H0) as part of a normal antigen-mediated immune response (Chapter 15). Normally, no inflammatory response would result. However, in patients with a predilection to asthma, genetic or otherwise, a B-cell/helper T cell (Th2) response is elicited with the production of antibodies to the trigger (Chapter 15). Repeated exposure to this trigger results in further activation of the immune system via a **type I hypersensitivity response** (Chapter 15). This type I response produces IgE from plasma cells. The IgE interacts with mast cells and basophils causing them to secrete a variety of molecules, noted with their associated functions in Table 17-1.

The sum result of the various molecules is dilation of blood vessels in the lung and smooth muscle contraction. In addition, chronic airway inflammation with excess mucus production (because of increased proliferation and growth of mucus producing cells in the bronchi) leads to remodeling of the bronchi, including thickening of the walls and a resultant airway narrowing. Finally, parasympathetic nerves (Chapter 19), which innervate the bronchial walls, are themselves stimulated by the triggers, leading to release of acetylcholine. **Acetylcholine** binds to its receptor and activates G_q-proteins, which activate phospholipase-C-producing diacylglycerol (activates protein kinase C) and IP_3, which leads to calcium release (activates protein kinase C and calmodulin-dependent kinase). The result is an additional signal for bronchial smooth muscle contraction and bronchoconstriction. All effects exacerbate breathing problems, including bronchospasm and asthma exacerbations, and, in a vicious circle, increase and perpetuate the inflammatory response unless broken by appropriate treatment.

BIOCHEMICAL BASIS OF ASTHMA MEDICATIONS

The increased understanding of the biochemical processes that lead to the inflammation of asthma and the bronchospasm of acute asthma exacerbations has led to a number of medications designed to specifically counteract these mechanisms. They include the following:

- **Short-Acting β_2-Agonists [Albuterol (Salbutamol), Levalbuterol, and Terbutaline]:** Counteract the inhibitory effect on the β_2-adrenergic receptor resulting in relaxation of bronchial smooth muscle and dilation of bronchial passages (Figure 17-5). Effect lasts for 4–6 hrs.

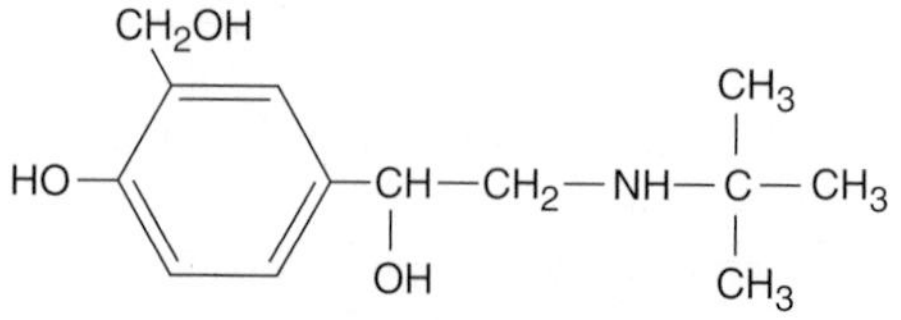

CH_2OH — HO—(ring)—CH(OH)—CH_2—NH—C(CH_3)$_2$—CH_3

Albuterol (salbutamol)

Reproduced with permission from Katzung BG, et al.: Basic and Clinical Pharmacology, 11th edition, McGraw-Hill, 2009.

TABLE 17-1. Asthmatic Type I Hypersensitivity Response

Molecule	Function	Mechanism
Histamine	Bronchial smooth muscle contraction/constriction and dilation and increased permeability of lung blood vessels.	G_s-protein activation of adenyl cyclase forming cAMP, which activates protein kinase A, phosphorylation of myosin light-chain kinase, and smooth muscle contraction.
Prostaglandin D2		Same as above.
Leukotrienes (C4, D4, and E4) comprise the slow-reacting substance of anaphylaxis secreted from mast cells and basophils.		Undetermined signaling pathway, possibly mimicking the Wnt G-protein mechanism. Also elicits an inflammatory response and long-term contraction of bronchi smooth muscle.
Platelet-activating factor		Multiple G-protein-activated pathways.
Neutral proteases	Dilation and increased permeability of lung blood vessels.	
Protease enzymes	Bronchial wall "remodeling".	Enzymatic degradation of bronchial wall proteins.
Cytokines (IL-1, IL-4, IL-5, IL-9, and IL-13)	Various actions.	Janus-associated kinases/tyrosine kinase activation.
TNF	Activation of immune response via multiple mechanisms.	Binds to immune cell-specific receptor (TNF receptor type 2) and activates NF-κB, leading to nuclear transcription of multiple proteins invoking a full immune response via separate mitogen-activated protein kinases activating c-Jun N-terminal kinases involved in, among other things, T-cell differentiation.
Leukotriene B4	Attracts neutrophils.	G_s-protein activation of adenyl cyclase forming cAMP, which activates protein kinase A, protein kinase C, and PIP_3 kinase activity leading to multiple extracellular signals.
Attractants for neutrophils and eosinophils		Multiple mechanisms dependent on a specific attractant.

cAMP, cyclic adenosine monophosphate; TNF, tumor necrosis factor.

Long-Acting β_2-Agonists (Salmeterol, Formoterol, and Bambuterol): Same mechanism (Figure 17-5) as short-acting, only effect lasts for approximately 12 hrs.

Anticholinergics/Muscarinic Antagonists (Ipratropium Bromide): Blocks the action of acetylcholine (Figure 17-5), leading to relaxation of bronchial smooth muscle and dilation of

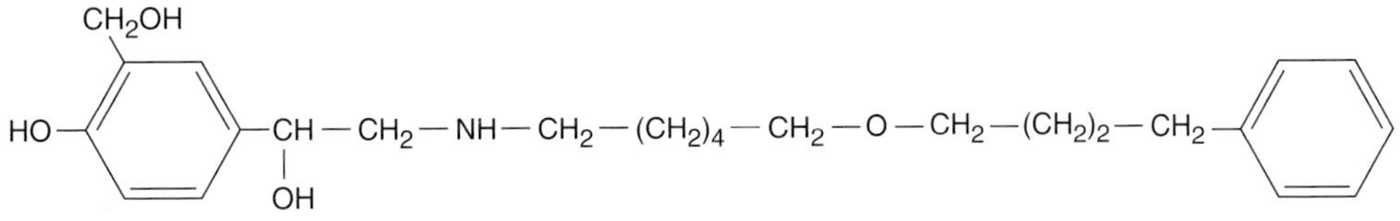

CH_2OH; HO–(ring)–CH(OH)–CH_2–NH–CH_2–$(CH_2)_4$–CH_2–O–CH_2–$(CH_2)_2$–CH_2–(ring)

Salmeterol

Reproduced with permission from Katzung BG, et al.: Basic and Clinical Pharmacology, 11th edition, McGraw-Hill, 2009.

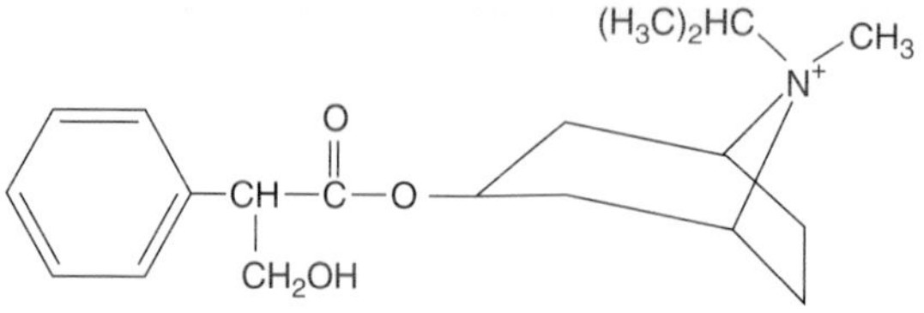

$(H_3C)_2HC$ CH_3 N^+ O CH—C—O CH_2OH

Ipratropium

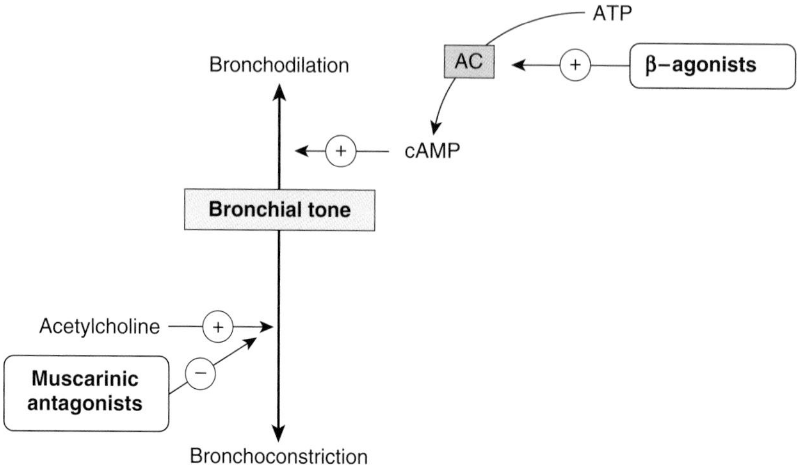

ATP

Bronchodilation AC + β–agonists

+ cAMP

Bronchial tone

Acetylcholine +

Muscarinic antagonists −

Bronchoconstriction

Figure 17-5. Actions of Bronchodilators. A mainstay of asthma treatment is via bronchodilators, which include short- and long-acting β-agonists that activate adenyl cyclase (AC) and by muscarinic antagonists that block acetylcholine-enhanced bronchoconstriction. See figures and text for further details. ATP, adenosine triphosphate; cAMP, cyclic adenosine monophosphate. [Adapted with permission from Katzung BG, et al.: Basic and Clinical Pharmacology, 11th edition, McGraw-Hill, 2009.]

bronchial passages. These medications are more frequently used for COPD patients.

- **Leukotriene Antagonists (Montelukast, Zafirlukast, and Zileuton):** Montelukast and Zafirlukast block the binding of leukotrienes to their bronchial cell receptors. Zileuton inhibits the enzyme 5-lipoxygenase and, therefore, the synthesis of leukotrienes B4, C4, D4, and E4 (Chapter 15). All leukotriene antagonists result in the blockage of leukotriene-associated bronchoconstriction and any activation of inflammatory pathways (Figure 17-6, Chapter 15).

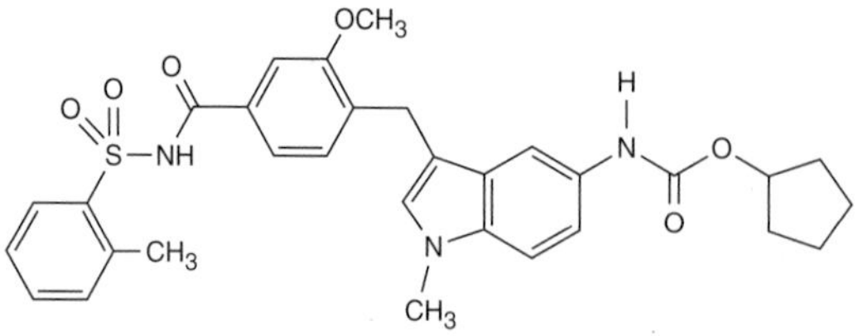

OCH_3 O O O S—NH H N O O CH_3 N CH_3

Zafirlukast

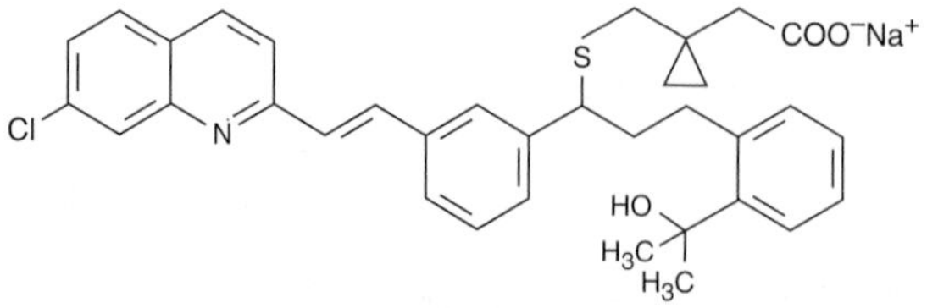

S COO^-Na^+ Cl N HO H_3C H_3C

Montelukast

CH_3 O S CH C N NH_2 OH

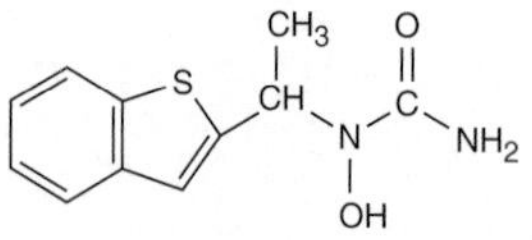

Zileuton

Reproduced with permission from Katzung BG, et al.: Basic and Clinical Pharmacology, 11th edition, McGraw-Hill, 2009.

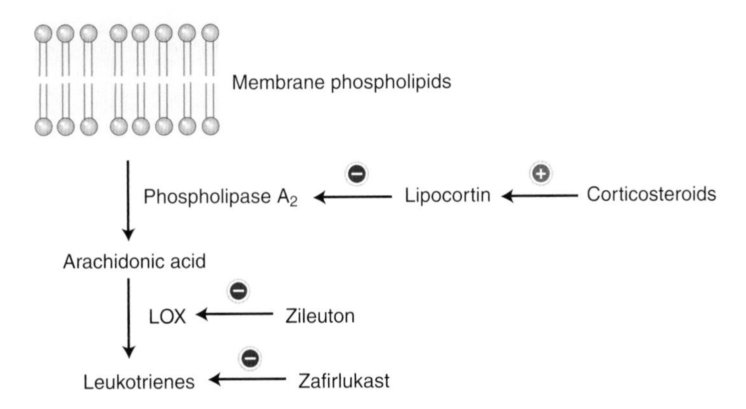

Figure 17-6. Actions of Leukotriene Antagonists and Corticosteroids. Leukotriene antagonists block lipoxygenase (LOX) or leukotriene receptors, thereby limiting their inflammatory effects. Corticosteroids promote the inhibition of phospholipase A_2 and, therefore, the arachidonic acid pathway. [Adapted with permission from Kibble JD and Halsey CR: The Big Picture: Medical Physiology, 1st edition, McGraw-Hill, 2009.]

Corticosteroids (Various Oral and Inhaled Forms): Bind to the glucocorticoid receptor found in the nucleus leading to homodimerization and binding to specific DNA sequences, which increases the expression of anti-inflammatory proteins. Some of these proteins block the expression of pro-inflammatory molecules in the cytosol. Corticosteroids have emerged as an essential component for the treatment of any patient with mild, intermittent asthma or worse.

Cortisol (hydrocortisone)

Prednisolone

Betamethasone

Triamcinolone (acetonide moiety shaded)

Reproduced with permission from Katzung BG, et al.: Basic and Clinical Pharmacology, 11th edition, McGraw-Hill, 2009.

Monoclonal Antibodies (Mepolizumab and Omalizumab): Mepolizumab blocks the activity of IL-5 (eosinophilic activation) and omalizumab blocks the activity of IgE.

RESTRICTIVE DISEASES—ACUTE RESPIRATORY DISEASE SYNDROME

Acute respiratory distress syndrome (ARDS), although first described in 1967, gained much notoriety in recent years because of the high rate of fatality (over 30%). The disease can result from a variety of insults, including trauma, aspiration of acidic gastric contents, pancreatitis, pneumonia, drug overdose, and/or sepsis, and is characterized by inflammation of the lung tissue. Clinically, it is defined as impaired oxygenation with bilateral pulmonary infiltrates on chest X-ray in the absence of left heart failure. Physiologically, it is associated with decreased pulmonary compliance and increased airway pressures. Just as with the other obstructive diseases, ARDS is associated with severe impairment of O_2 and CO_2 diffusion.

The biochemical basis of the disease also shares the hallmarks of an initial release of **cytokines** leading to increased **vascular permeability** and **inflammation** with neutrophils, monocytes, and macrophages present. The inflammation results in damage to the alveoli, including swelling, which increases the space between alveolar membrane and the surrounding capillaries, worsening O_2–CO_2 exchange and causing respiratory acidosis, shortness of breath, increased breathing rate, severe decrease in oxygenation, and often respiratory failure requiring mechanical ventilatory support. However, the mechanical breathing assistance, itself, may also contribute to the continuing inflammatory response and must, therefore, be used with care, often at low levels of pressure and tidal volume utilizing positive end-expiratory pressure to help reinflate and keep open affected alveoli. Further inflammation leads to deposition **of fibrous (hyaline membrane)** material. A noted reduction in pulmonary surfactant production, due to dysfunction of the type II alveolar cells, may result in the complete collapse of alveoli and may also impact pulmonary compliance and work of breathing. Corticosteroids at high doses are often used to help limit inflammation. The use of artificial pulmonary surfactants is also sometimes considered. Unfortunately, nothing has been shown to clearly lower mortality other than lung-protective ventilation and supportive care.

RESTRICTIVE DISEASES—OCCUPATIONAL EXPOSURES

A number of restrictive lung diseases result from the continued exposure to irritants, often as a result of an occupation or other constant environmental condition. The general term **pneumoconiosis** is used for any restrictive lung disease caused by the chronic exposure and inhalation of particles small enough to bypass the defense mechanism of the upper airway. Examples of several pneumoconiosis diseases and their causative agent(s) are listed in Table 17-2.

In all cases, the occupational exposure to the agents in the dust leads to a chronic inflammatory process often resulting from the stimulation of lung fibroblasts to deposit fibrous connective tissue (Chapter 13) leading to permanent scarring. This scarring distorts the airways and thickens the walls of the smaller bronchioles and alveoli, leading to restrictive disease and reducing pulmonary compliance and lung function. As a result, O_2–CO_2 exchange can be markedly reduced.

TABLE 17-2. Examples of Pneumoconiosis Diseases

Disease	Causative Agent
Anthracosis	Coal dust
Asbestosis	Asbestos fibers
Baritosis	Barium dust
Bauxite fibrosis	Bauxite dust
Berylliosis	Beryllium dust
Byssinosis	Cotton dust
Chalicosis (flint or Stonecutters' disease)	Stone dust
Coal worker's pneumoconiosis ("black lung")	Coal dust
Farmer's lung	Hay, mold, and other agricultural dusts
Labrador lung	Mixed dust (iron, silica, and anthophyllite)
Siderosis	Iron dust
Silicosis	Silica dust
Silicosiderosis	Mixed dust (silica and iron)

RESTRICTIVE DISEASES—INTERSTITIAL LUNG DISEASES

A subset of restrictive lung diseases affects the interstitium (the tissue and areas around the alveoli, including the alveolar and capillary walls and the involved connective tissue, blood vessels, and lymphatic tissue) and are known as **interstitial lung diseases (ILDs)**. ILD includes a wide variety of conditions, such as desquamative interstitial pneumonia, idiopathic pulmonary fibrosis, lymphoid interstitial pneumonia, nonspecific interstitial pneumonia, and respiratory bronchiolitis-associated interstitial lung disease (Table 17-3) and involves an inflammatory response that often leads to fibrosis and scarring of the lung interstitium. Hypersensitivity pneumonitis (not listed in Table 17-3), which can appear like lung tissue fibrosis when advanced, can also be included as an ILD. The damage to the lung tissue reduces O_2–CO_2 exchange, which can lead to hypoxia and the need for supplemental O_2. Treatment, including corticosteroids, is usually aimed at decreasing the inflammation.

Interstitial Lung Disease Secondary to Medical Treatments: Certain **ILDs** result from particular medical treatments as a known but unwanted side effect. These treatments include **radiation** and **chemotherapy** (e.g., bleomycin), resulting from either constant radiation exposure of the lungs or the toxic effects of cancer medications. Other medications, including **amiodarone** and **methotrexate**, are known offenders. The mechanism of ILD by these treatments is poorly understood, although the roles of **IL-18** (produced by macrophages with stimulation of natural killer cells and T lymphocytes) and **IL-1β** (produced by macrophages and involved in several aspects of activation of an inflammatory response) have been implicated in some.

TABLE 17-3. Examples of Interstitial Lung Diseases

Condition	Possible Cause(s)	Mechanism
Dermatomyositis	Viruses, cancers.	Activation of the complement-mediated immune response.
Desquamative interstitial pneumonia	Smoking.	Marked increase in pigmented macrophages, lymphocytes, plasma cells, and some eosinophils in the alveolar connective tissue.
Goodpasture's syndrome	Genetic, chemical exposure (e.g., paraquat weed killers, hydrocarbon solvent), and viral infections.	Production of antibodies by B cells to type IV collagen via macrophage and/or dendritic cell presentation of antigen [type II (cytotoxic) hypersensitivity].
Lymphoid interstitial pneumonia	Infections (HIV, Epstein–Barr virus, and other viruses).	Excessive production of lymphocytes due to an unregulated autoimmune reaction, which leads to inflammation and fibrotic scarring of lung tissue.
Nonspecific interstitial pneumonia	Usually unknown. A relationship may exist with connective tissue diseases and/or drug-induced lung injuries.	Initially uniform inflammation and fibrosis of interstitial lung tissue. Further progression of the disease usually results in foci of disease.
Cryptogenic organizing pneumonia	May result from chronic inflammatory disease, infections, or due to use of certain medications.	Obstruction of alveoli and associated ducts by granulation tissue with chronic inflammatory cells present in adjacent alveoli. Formerly known as Bronchiolitis obliterans organizing pneumonia.
Polymyositis	Genetic, viruses, Lyme disease, and toxoplasmosis.	Activated $CD4^+$, T helper cells produce IFN-γ that stimulates macrophages to produce IL-1 and TNF-α. $CD8^+$, T cytotoxic cells are also activated by alternative processes.
Respiratory bronchiolitis-associated interstitial lung disease	Smoking, some mineral dust exposures, and reactions to some virus infections and drugs.	Inflammation by tan-brown pigmented (iron deposition) macrophages and changes in the epithelium of the smaller bronchioles and alveoli with subsequent fibrotic scarring.
Rheumatoid arthritis	Genetic, infections (bacterial and/or viral), and allergies.	Initial onset appears to involve abnormal B- and T-cell interactions with release of TNF-α and IL-15, IL-16, and IL-17. Once established, B-cell-derived plasma cells produce RF and IgG and IgM antibodies, which bind to macrophage F_c receptors activating T cells and/or the complement system. Dendritic cell immune responses may also play a role.
Sarcoidosis	Genetic, environmental or occupational exposure, and infection.	Monocytes/macrophages and T cells secrete TNF-α, IFN-γ, IL-2, and IL-12 activating T helper cells and initiating inflammation/granulosa formation.
Systemic lupus erythematous (SLE)	Genetic, environmental exposures, infections (viral and bacterial), and medications (drug-induced SLE).	Abnormal increase in programmed cell death (apoptosis) of monocytes and keratinocytes as well as impaired phagocytic clearance of dead cell debris leads to autoactivation of B and C cells with antibody production to "self" proteins (autoimmunity).
Systemic sclerosis	Genetic, cytomegalovirus, organic solvents, and bleomycin.	T cells, stimulated by cytokines produce TGFβ and CTGF stimulation of fibroblasts, via a serine/threonine kinase, leading to collagen deposition.
Idiopathic pulmonary fibrosis	Collagen vascular diseases, drugs, environmental/genetic factors, or infections may play a role.	Most common form of idiopathic interstitial pneumonias with unregulated proliferation and activity (collagen deposition and fibrosis) by fibroblasts with some inflammation by lymphocytes, plasma cells, and histiocytes.

CTGF, connective tissue growth factor; HIV, human immunodeficiency virus; IFN, interferon; Ig, immunoglobulin; IL, interleukin; RF, rheumatoid factor; TGF, tumor (transforming) growth factor; TNF, tumor necrosis factor.

Cystic Fibrosis (CF): Although hundreds of variations of **CF** are known, the prototypical type is caused by an autosomal recessive mutation, causing the deletion of a critical phenylalanine at amino acid 508 of the **cystic fibrosis transmembrane conductance regulator (CFTR)**. CFTR is a membrane-bound, adenosine triphosphate-driven, chloride ion transport channel essential in the production of lung mucus as well as sweat and pancreatic digestive solution. The mutation leads to a nonfunctional transporter, which results in a thickened mucus in the lung which further blocks narrow bronchioles leading to chronic inflammation and remodeling of the lung tissue and, often, chronic sinusitis. Pancreatic secretions are also affected, leading to accumulation of digestive enzymes in the pancreas and intestinal blockages by thick feces. This same effect can also lead to infertility, especially of men, due to thickened ejaculatory fluid. The lack of chloride ion transport also inhibits the production of NaCl in sweat, leading to a common, noninvasive test of tasting a baby's sweat if CF is suspected. Breathing problems and constant lung infections usually lead to bronchiectasis and obstructive defects. Life expectancy is dependent on the particular mutation and can range from the late teens to early 20s up to the late 30s or early 40s.

INFECTIVE DISEASES OF THE RESPIRATORY SYSTEM

Infections of the respiratory system are caused by a wide variety of bacteria and viruses and, as such, will not be fully covered here. However, certain infections utilize biochemical mechanisms that offer insight into their clinical effects as well as treatments.

Bordetella pertussis (the cause of **whooping cough**) produces and releases an enzymatically active protein toxin responsible for the illness. Released in its inactive form, pertussis toxin binds to receptors on a cell membrane and is transported via the Golgi apparatus to the endoplasmic reticulum. Upon activation, the toxin adds adenosine diphosphate (ADP) molecules to the **α-subunits** of **G proteins**, blocking their normal binding to G-protein-associated receptors (Chapter 8). G_i proteins are among those affected and, being inhibited by the ADP ribosylation, are unable to stop adenyl cyclase production of cyclic adenosine monophosphate (cAMP). The resulting altered signaling leads to the clinical manifestations of whooping cough.

Pertussis toxin's inhibition of G_i protein activity also affects other organ systems. Macrophage phagocytic functions are disrupted, restricting their ability for bacterial killing. In patients with pertussis, B cells and T cells that leave the lymphatics show an inability to return. These effects may explain the high frequency of secondary infections that accompany pertussis. Pertussis toxin also inhibits histamine activation of phospholipase C and adenylyl cyclase affecting gastric, cardiac, smooth muscle cells, and immune cells. Finally, the block of cAMP production in pancreatic β-cells by pertussis toxin coupled with the actions of epinephrine can lead to overproduction of insulin and hypoglycemia.

Aspergillosis fungal infections represent a spectrum of diseases ranging from benign colonization of the airways to invasive infections, especially in patients with chronic lung diseases or immune compromise. Infection by *Aspergillus* initiates both type I and type II hypersensitivity responses leading to IgE and IgG production, respectively. Further immune reactions initiate type III hypersensitivity responses with deposition of immune complexes in the lung membranes, activation of helper T (Th2) cells, which recruit IL-4 and IL-5, and recruitment of eosinophils and neutrophils, the latter by IL-8 secretion. This immune response results in bronchoconstriction and increased vascular permeability via mechanisms similar to asthma (see Table 17-1). Continued or repeated infections result in the release of enzymes by immune cells, which degrade and scar the lung tissue resulting in fibrosis and the breakdown of the lung architecture.

Tuberculosis, the disease cause by several *Mycobacterium* species, depends on activation of macrophages, T and B lymphocytes, and fibroblasts. Encirclement and isolation by lymphocytes and interferon-γ secretion to activate phagocytic macrophages and cytotoxic T cells lead to the formation of a granuloma—a compact collection of immune cells (e.g., macrophages and histiocytes) and necrotic cells, which isolates the bacterial infection. Granulomas are also seen in the infective diseases of leprosy, histoplasmosis, cryptococcosis, coccidiomycosis, blastomycosis, and cat scratch disease (*Bartonella* infection) as well as in several noninfectious conditions such as sarcoidosis, Crohn's disease, Wegner's granulomatosis, Churg–Strauss syndrome, and rheumatoid arthritis.

REVIEW QUESTIONS

1. What are the mechanisms of inhalation and exhalation?
2. What is the function of alveoli?
3. What is the role of pulmonary surfactant?
4. What is compliance?
5. What are the obstructive and restrictive lung diseases and their associated consequences?
6. What are the defects/mechanisms, and consequences associated with the following diseases affecting lungs: emphysema, bronchitis, asthma, acute respiratory distress syndrome, pneumoconiosis, and interstitial lung diseases?
7. What are the symptoms of lung infections associated with pertussis, aspergillosis, and tuberculosis?
8. What are the mechanisms of action of asthma medications and their application(s) to other diseases?
9. What is the role of oxygen/carbon dioxide exchange and the carbonic anhydrase reaction in acid–base balance associated with the lung?

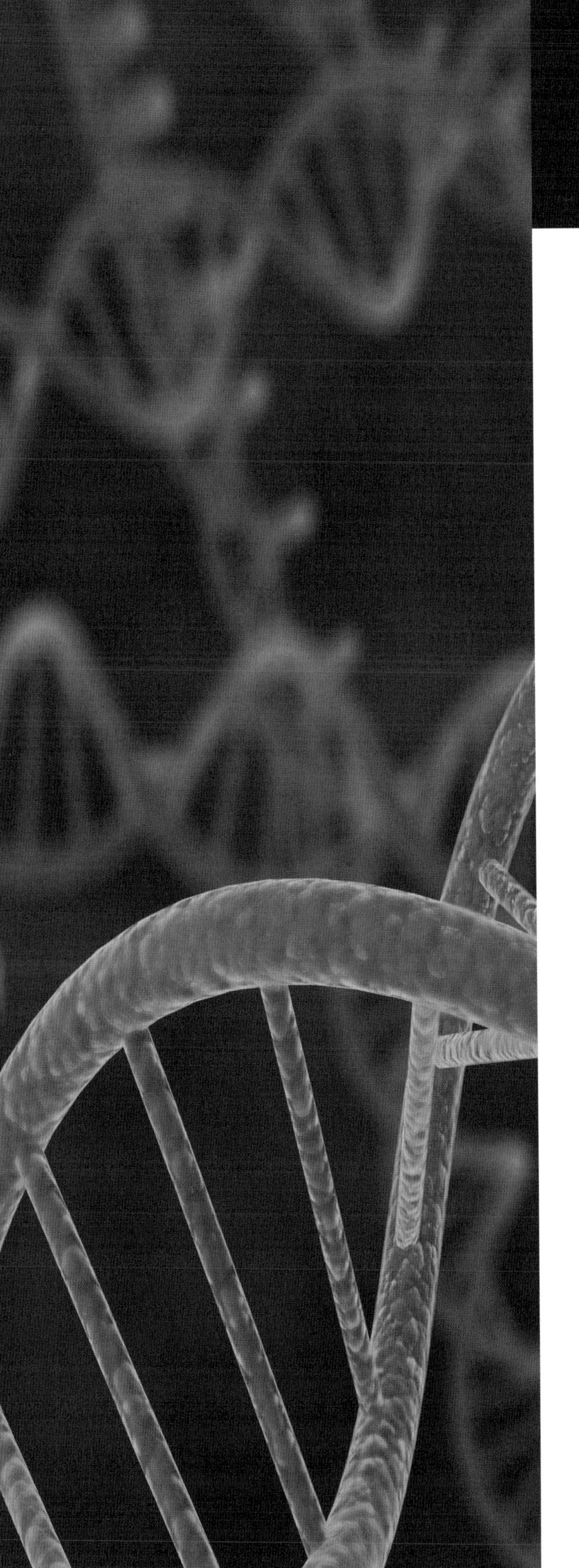

CHAPTER 18

THE URINARY SYSTEM

Editor: Armando J. Lorenzo, MD, MSc, FRCSC, FAAP
Staff Pediatric Urologist, The Hospital for Sick Children, University of Toronto, Toronto, Ontario, Canada

OVERVIEW

The urinary system, composed of kidneys, ureters, bladder, and urethra, is responsible for the excretion of waste products in urine and regulation of important ions and molecules in the body. The biochemical components of the basal lamina and varying cell types found in the renal corpuscle create the filtering mechanism that initiates this process. Ionic channels and pumps found in the nephron allow the final composition of the urine to be modified according to the needs of the body. This is regulated by various hormones that help regulate sodium, potassium, chloride, calcium, magnesium, ammonia/acid–base balance, total body water, blood pressure, and other molecules and ions. The renin–angiotensin–aldosterone hormonal system as well as vasopressin and atrial naturetic peptide are keys in this process. Finally, the kidneys are essential for red blood cells production via erythropoietin synthesis and also add an essential biochemical step in the activation of vitamin D.

BASIC ANATOMY AND PHYSIOLOGY

The urinary system, embryologically derived along with the reproductive system from the intermediate mesoderm, consists of the kidneys and associated blood vessels, ureters, bladder, and urethra. The kidney (Figure 18-1) contains hundreds of thousands to millions of nephrons and associated tubules and collecting ducts where blood is filtered by the **renal corpuscles**

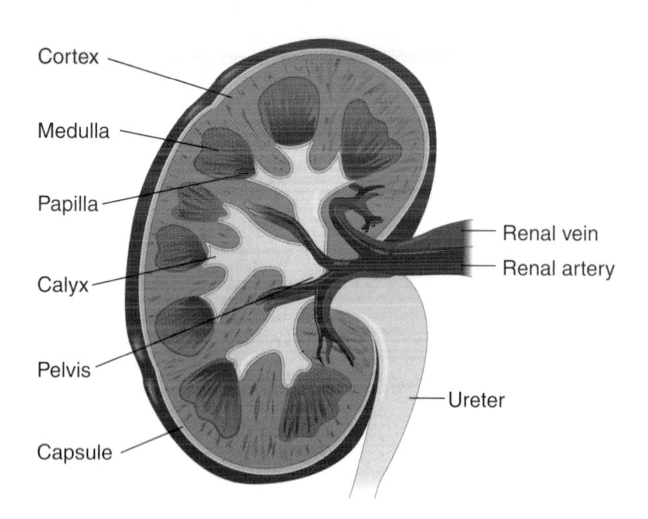

Figure 18-1. Basic Anatomy of the Kidney. The kidney contains the renal artery and vein, which transport blood to and from the pelvis and into a calyx and medulla where filtration occurs. Urine produced by this process exits via the ureters. [Reproduced with permission from Kibble JD and Halsey CR: The Big Picture: Medical Physiology, 1st edition, McGraw-Hill, 2009.]

for waste products; fluid balance (and, secondarily, blood pressure) is adjusted; electrolytes, glucose, amino acids, and other molecules are balanced to appropriate levels; acid–base levels are maintained; and the process of urine production is completed. Additionally, the enzyme renin, the hormone erythropoietin, and the final active form of vitamin D are produced by the kidney.

RENAL FILTRATION

THE RENAL CORPUSCLE

The structure of the renal corpuscle (Figure 18-2A) allows the kidneys to perform its initial filtering function. The overall structure is composed of a **glomerulus**, which receives blood via a small collection of capillaries bringing blood to (**afferent**) and away from (**efferent**) the renal corpuscle, and the surrounding **Bowman's capsule** along with the nephron. The glomerulus forms the first part of the filtration apparatus with **fenestrations** ("openings" from the Latin for window) formed between the cell membranes of its endothelial cells. The fenestrations are relatively large, though, and only filter material the same size or smaller than blood cells (Figure 18-2A–B). Next is a three-layered structure called the **basal lamina** (also called the basement membrane), which is approximately six times thicker than other basement membranes found in the body (Figure 18-2C). The biochemical components of the basal lamina form an efficient and highly selective, sieve-like filter. Negatively charged **glycosaminoglycans**, including predominately **heparin** (Chapter 2), comprise two of the layers, the *lamina rara interna* and the *lamina rare externa*, creating a selective, electrostatic barrier to other negatively charged molecules. The middle layer, the *lamina densa*, is composed of **type IV collagen**, differing from other collagen types because of its lack of regular glycine residues (Chapter 1). This allows it to form flat, kinked sheets. Also found in the *lamina densa* is **laminin**, a glycoprotein (Chapter 2), which forms multiple links to other molecules because of its unique cross-shape. The kinked, filamentous **type IV collagen** molecules, linked by laminin, form a network that filters all molecules with a molecular weight less than 66,000 Da.

The next part of the filter, surrounding the glomerulus, is Bowman's capsule with an outer layer of epithelial cells but also a layer of specialized epithelial cells called **podocytes**. These podocytes have "foot processes," which wrap around and fuse with the glomerulus and interdigitate with nearby podocytes (Figure 18-2B). These foot processes are also lined by two

Type IV Collagen and Basal Lamina Disorders: Although a multitude of kidney diseases exist, many affect various constituents of the renal corpuscle. Two such diseases are Goodpasture's and Alport's syndromes, both of which are believed to involve **the type IV collagen**, which makes up the **basal lamina**.

In **Goodpasture's syndrome**, also known as **antiglomerular basement membrane disease**, the immune system erroneously attacks the basal lamina of the glomeruli and alveoli in the lung, leading to inflammation of the glomeruli (**rapidly progressive glomerular nephritis**) with acute kidney failure (sometimes with increased protein in the urine), rise in blood pressure, edema as well as cough, breathlessness because of bleeding within the lung, and the requirement for supplemental oxygen. Smoking and prior lung damage increase symptom severity and progression. The molecule that appears to be attacked in both anatomical locations is type IV collagen. The initiating cause of the self-attack is unknown, although exposure to certain chemical solvents, herbicides, and/or viral infections has been implicated.

In **Alport's syndrome**, genetic mutations in type IV collagen lead to inflammatory destruction of the glomeruli (**glomerulonephritis**) with blood in the urine and progressive kidney failure as well as decreasing hearing (progressive destruction of the basal lamina in the inner ear affecting hair cells at other structures) and vision [lens distortion **(anterior lenticonus), cataracts**, and rarely **lens rupture** as well as "flecks" around the macula, which cause retinal damage and progressive loss of sight]. Several different mutations inherited in X-linked, autosomal recessive, or autosomal dominant manners can cause Alport's syndrome.

Abnormalities in collagen and the kidney matrix are also seen in the autosomal dominant form of **polycystic kidney disease (PCKD)** in which multiple cysts form in the kidneys, leading to eventual damage of the tissue and renal insufficiency/failure. Patients with PCKD also exhibit similar collagen-related abnormalities in the brain, blood vessels, heart valves, and abdominal/pelvic walls, i.e., hernias.

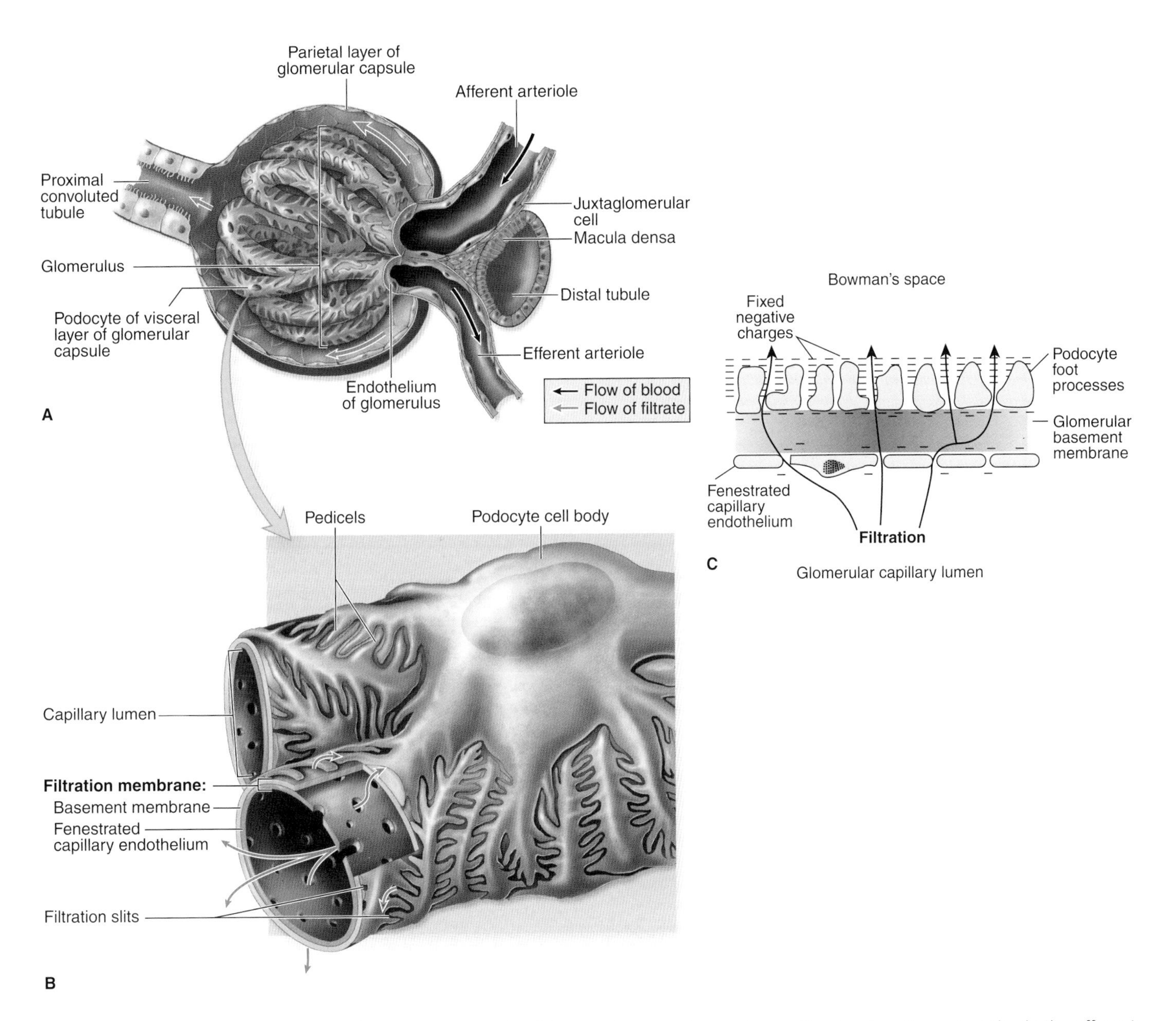

Figure 18-2. A–C. Renal Corpuscle. (**A**) Cellular components of the renal corpuscle are shown. Blood enters the corpuscle via the afferent capillary (black arrow) and is filtered through the glomerulus (blue arrows). Resulting filtrate proceeds into the proximal tubule to be eliminated as urine. Blood returns to the body via the efferent capillary. Additional components of the renal corpuscle include mesangial cells (not shown) and juxtaglomerular cells. The associated distal tubule is shown. (**B**) Expanded view shows detail of how blood is filtered through the fenestrated endothelial cell layer inside the capillary lumen and through the glomerular basement membrane and podocyte foot processes layers. (**C**) The glomerular basement membrane is composed of three layers, including a lamina rara interna, lamina densa, and lamina rara externa (not shown, see text). The afferent capillary (not shown) delivers blood to the endothelial layer, with resulting filtrate emerging through to Bowman's space and subsequent collection by the proximal tubule. The specific role of each layer and its components is described in the text. [Adapted with permission from Mescher AL: Junqueira's Basic Histology Text and Atlas, 12th edition, McGraw-Hill, 2010.]

special proteins called **podocin** and **nephrin** and a negatively charged glycoprotein called **glycocalyx**, which help to create and stabilize structures called **slit diaphragms**. Defects in podocin are known as one cause of **minimal change disease**, a condition associated with loss of the podocyte foot processes with resulting loss of protein into the urine (proteinuria) and swelling of the body (edema). The slit diaphragms, like the fenestrated endothelial cells, limit passage of negatively charged and/or medium-sized molecules such as serum albumin. **Mesangial cells**, modified smooth muscle cells (Chapter 12) in the glomerulus (Figure 18-2A), modulate the flow of the blood from the afferent capillaries and, in addition, produce protein and proteoglycan components of the extracellular matrix of the renal corpuscle, fatty acid-derived prostaglandins (Chapter 3), and protein/glycoprotein cytokines. Mesangial cells also phagocytically remove molecules trapped by the filtration barrier—a molecular filter cleaner. Finally, **juxtaglomerular (JAG) cells**, a second type of specialized smooth muscle cells of the renal corpuscle that synthesize, store, and secrete the enzyme **renin**, important in the control of blood pressure (Figure 18-2A and see text below). The unique features of all these components create a meshwork of cells, proteins, glycoproteins, and other biochemical molecules, which selectively allow the passage of water, ions, and small molecules (e.g., **creatinine** and **urea**) but prevent the passage of large and/or negatively charged molecules.

Because of the many components that form the complex filtering mechanism of the glomerulus, many disorders exist, which affect the structure and, therefore, the function of the glomerulus. The dysfunction of the structural and functional components of the glomerular meshwork noted above results in the spillage of proteins (e.g., albumin) and red blood cells into the urine, termed **nephritic syndrome**, or large amounts (>3.5 g/day) of protein, termed **nephrotic syndrome**. Both are clinical indicators of a renal problem. Glomerular disorder diseases are normally divided into proliferative—involving the increased growth or number of glomerular cells or matrix material—and nonproliferative—no increased cellularity or matrix. These varying diseases and the biochemical effect on glomerular structure are listed in Table 18-1.

The final components of the urinary system, the ureters, bladder, and urethra, are lined by a transitional epithelium with several layers of cells that allow the bladder to stretch while still maintaining bladder wall integrity. A thick and elastic lamina propria layer, composed of a mesh network of collagen, elastic fibers, and proteoglycans (Chapters 1, 2, and 13), holds and attaches the epithelial cell layers together to provide additional flexibility and strength to the fluid-proof but stretchable structure. The two to three smooth muscle layers beneath the epithelium increase movement of urine from the kidneys to the bladder and, ultimately, for excretion. When urination occurs, an **internal urethral sphincter**, composed of smooth muscle (Chapter 12), is controlled by the autonomic nervous system (Chapter 19), whereas an **external sphincter**, composed of skeletal muscle and innervated by the **pudendal nerve**, offers voluntary control.

TABLE 18-1. Summary of Glomerular Disorders

	Disorder	Effect on Glomerulus
Nonproliferative	Minimal change glomerulonephritis	Fusion of podocytes.
Nonproliferative	Focal segmental glomerulosclerosis	Isolated areas of glomerular destruction and scarring (fibrosis, increased mesangial cells, collagen deposits) along with deposition of plasma protein (hyaline) in the afferent arterioles.
Nonproliferative	Hereditary glomerulonephritis (e.g., see Alport's syndrome above)	Thickening of basal lamina with deposition of immunoglobulin (Ig) G molecules.
Proliferative	IgA nephropathy (Berger's disease)	Increased number of mesangial cells and basal lamina with deposition of IgA.
Proliferative	Henoch–Schönlein purpura	Inflammation of glomerular blood vessels and glomerulus with deposition of IgA.
Proliferative	Membranoproliferative glomerulonephritis	Increased number of mesangial cells and basal lamina.
Proliferative	Rapidly progressive (crescentic) glomerulonephritis (e.g., see Goodpasture's syndrome above and Wegener's granulomatosis)	Breakdown and decreased filtering of basal lamina allows fibrin into glomerulus, leading to fibrosis and increasing Bowman capsules cells, an influx of monocytes and formation of crescent-shaped scarring.
Proliferative	Postinfectious glomerulonephritis (e.g., skin of lung infection by *Streptococcus pyogenes*)	Deposition of immune complexes in the glomerulus leads to increased numbers of endothelial and mesangial cells and influx of monocytes and neutrophils. May mimic crescents.
Other	Thin basal lamina (autosomal dominant, genetic disorder)	Allows red blood cells to pass (hematuria).
Other	Diabetic nephropathy	Progressive thickening of glomerular structure with subsequent destruction of glomeruli via nodular glomerulosclerosis. Results in increasing protein (e.g., albumin) excretion in the urine (albuminuria), and may progress to renal failure and need for dialysis.

Urinary Tract Infections (UTIs) and Cranberry Juice: Bacterial **UTIs** of the bladder (**cystitis**) or kidney (**pyelonephritis**) affect millions of patients. Although antibiotics are often needed for treatment, the use of cranberry juice may also help. Cranberry juice, specifically, contains the plant flavenoid molecule **A type proanthocynidin**, which directly inhibits attachment of **p-type fimbriae** of *Escherichia coli*, the cause of up to 90% of UTIs, to the urinary tract wall. As a result of the use of cranberry juice, bacteria are swept away by urinary flow, accelerating recovery from or even preventing a UTI.

Diuretics and the Nephron: Diuretics are any substance, including medications, which increase the loss of water via excretion in urine. Diuretic medications are fundamental to many treatments of high blood pressure and heart failure, and work via specific effects on the channels and pumps of the nephron.

Loop diuretics, named so because their action affects the **loop of Henle**, include furosemide, bumetanide, torasemide, and ethacrynic acid (see figure). These medications actively compete (Chapter 5) for the Cl^--binding site, blocking **Na^+–K^+–$2Cl^-$ transport** and reducing the Na^+ and K^+ ion concentration in the renal medulla. The lower ion concentration and, therefore, **osmotic pressure gradient** decreases the amount of water reabsorbed from the urine in the collecting duct, lowers the total fluid volume of the blood plasma, and, consequently, lowers blood pressure. Magnesium ion (Mg^{2+}) and calcium ion (Ca^{2+}) reabsorption is also decreased, adding to this effect. Because of this mechanism, a possible side effect of loop diuretics is low K^+ (hypokalemia), low Mg^{2+} (hypomagnesemia), low Ca^{2+} (hypocalcemia), and low Cl^- (hypochloremia). Because of the effect on calcium excretion, loop diuretics are also used in clinical situations associated with elevated serum levels of calcium (hypercalcemia). Nevertheless, elevated levels of urinary calcium may lead to formation of kidney stones, a phenomenon sometimes seen in premature babies who are exposed to high doses early in life. **Nonsteroidal anti-inflammatory molecules** block the action of loop diuretics by decreasing glomerular filtration of the blood and passage of the diuretics to the loop of Henle.

NH—CH_2 Cl COOH H_2N—O_2S O

Furosemide

Cl Cl O H_2C=C—C— O—CH_2—COOH C_2H_5

Ethacrynic acid

Adapted with permission from Katzung BG, et al.: Basic and Clinical Pharmacology, 11th edition, McGraw-Hill, 2009.

Thiazide diuretics, including hydrochlorothiazide, chlorothiazide, metolazone, indapamide, and the parent molecule benzothiadiazine, work by blocking the **Na^+–Cl^- cotransporter** in the **distal tubule**. Blockage of the membrane channel in this part of the nephron decreases the reabsorption of these two ions into the interstitium and the blood stream, decreasing water influx, total blood volume, and blood pressure. The resulting increased Na^+ concentration and water content also activate an aldosterone-regulated (see below) K^+ secretion channel in the collecting ducts, leading to loss of K^+. As a result, thiazide diuretics can lead to clinically significant low K^+ levels in the blood. Thiazide diuretics also decrease **calcium** excretion and, therefore, levels in the urine, lowering the chance of calcium-derived kidney stone formation and other diseases involving high levels of serum calcium. The effects of thiazide diuretics are mimicked by the diseases of **Bartter** and **Gitelman syndromes** wherein rare genetic defects lead to dysfunction of the above-mentioned cotransporters in the ascending part of the loop of Henle or the distal convoluted tubule, respectively.

NEPHRON

The production of urine by the kidneys involves a complex, well-regulated series of membrane-bound protein pumps and channels, which selectively develop gradients of ions and molecules to determine the final content of urine. The body has mechanisms to sense the makeup of the filtrate via osmotic measurement (e.g., macula densa) and hormonal response, as it progresses from the renal corpuscle to the ureters.

Filtering of the blood by the renal corpuscle removes waste products, but this process is not perfect by any means. The complex task of returning particular molecules to the circulatory system while progressively concentrating the resulting fluid that will become urine is the job of several protein channels (active and passive) and pumps (Chapter 8) found in the membranes of the **nephron** (Figure 18-3).

The nephron is the tubular structure, which continues from the renal corpuscle, leading to the collecting system and subsequently ureters, bladder, and urethra. It is surrounded by interstitial tissue (the connective tissue of the kidney found outside the tubule and surrounding the nephrons), which contains "peritubular" capillaries. Water and selected molecules are transferred across the nephron members by these channels and pumps and returned to the body via these capillaries, a process called **reabsorption** (the initial absorption is considered to have taken place in the intestines). Water and molecules moving from the interstitium into the nephron for elimination in urine undergo a process termed **secretion**. The tubule system beyond the glomerulus is divided functionally into four distinct sections (Figure 18-3): (1) the **proximal convoluted tubule**, (2) the **loop of Henle** (descending and ascending parts), (3) the **distal convoluted tubule** (proximal and distal parts), and (4) the **collecting duct**. Each section is characterized by specific channels/pumps and osmotic conditions required for its particular function in urine production. A summary of the actions on certain molecules by the various segments of the nephron is given in Figure 18-3 and Table 18-2.

The collective functions of all the segments of the nephron, in total, allow the kidney to not only regulate water but also the conservation or elimination of electrolytes; the proper maintenance of acid–base balance; the excretion of waste products; and the preservation of important proteins, carbohydrates, and even medications.

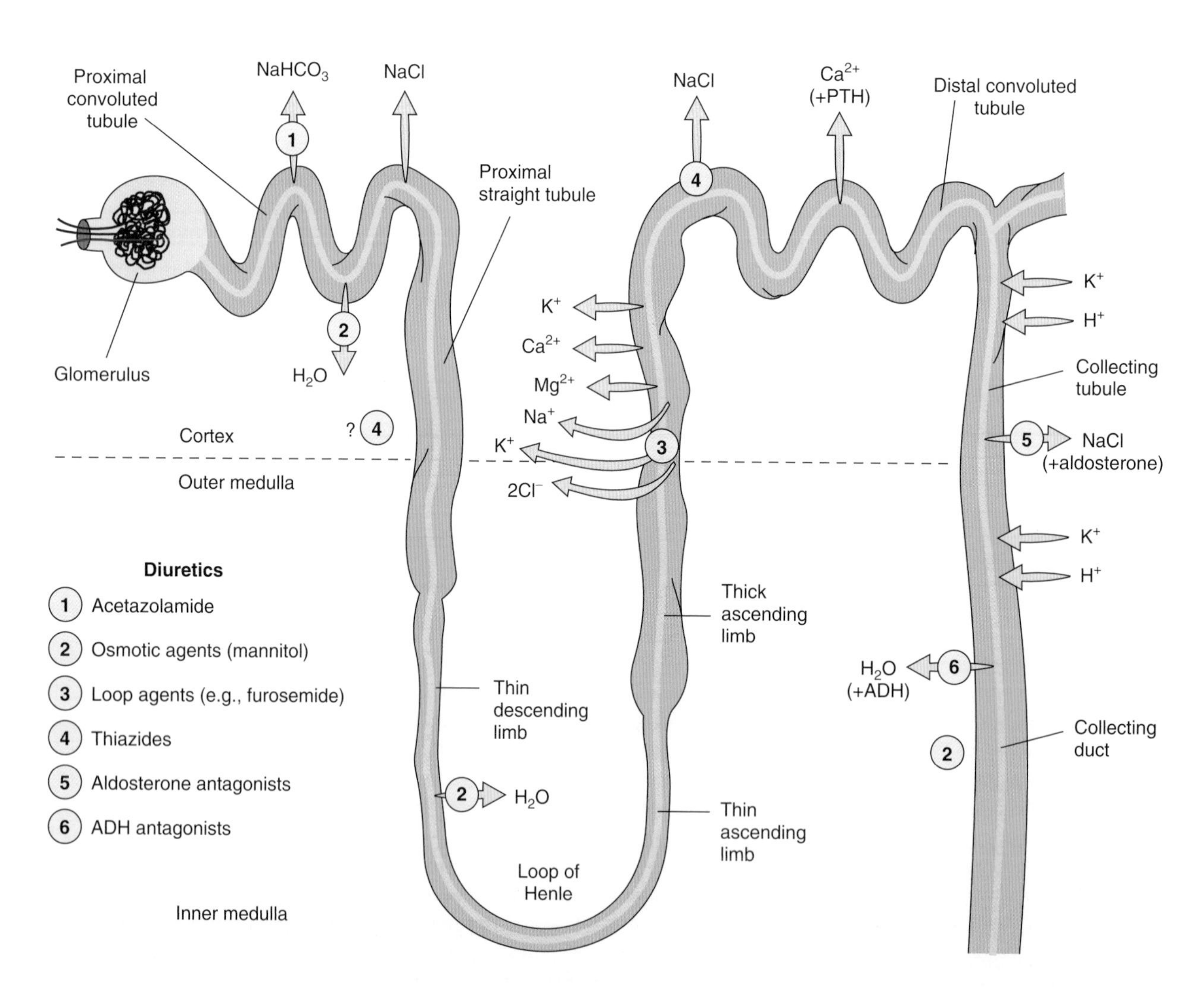

Figure 18-3. Overview of Nephron and Urine Production. Illustration of components of the nephron and summary of functions found in each segment, including proximal convoluted tubule, proximal straight tubule, thin descending limb, loop of Henle, thin and thick ascending limbs, distal convoluted tubule, and collecting duct. Molecules leaving (reabsorption) and entering (secretion) as well as medications and/or hormones affecting each segment (numbered) are indicated. See text for further details. ADH, antidiuretic hormone; PTH, parathyroid hormone. [Reproduced with permission from Katzung BG, et al.: Basic and Clinical Pharmacology, 11th edition, McGraw-Hill, 2009.]

TABLE 18-2. Summary of Nephron Actions on Various Molecules

Molecule	Proximal Tubule	Loop of Henle	Distal Tubule	Collecting Duct
Water	Reabsorbed with other molecules via osmotic gradient.	Secretion via varying membrane permeability.	—	Reabsorption via ADH-regulated aquaporin-2 channels.
Glucose	Almost 100% reabsorption via Na^+/glucose transporter and GLUT-facilitated channels.	—	—	—
Amino acids, peptides/proteins	Almost 100% reabsorption via molecule-specific channels.	—	—	—
Sodium	About 65% reabsorption via Na^+–K^+-ATPase channel.	About 25% reabsorption via Na^+–K^+–$2Cl^-$ cotransporter.	About 5% reabsorption via Na^+–Cl^- cotransporter.	About 5% reabsorption via aldosterone-regulated Na^+–K^+ channels.
Potassium	About 65% reabsorption via passive diffusion through membrane, driven by K^+ from Na^+–K^+-ATPase channel.	About 20% reabsorption via Na^+–K^+–$2Cl^-$ cotransporter.	—	Secretion via Na^+–K^+-ATPase, via aldosterone-regulated Na^+–K^+ channels.
Chloride	Reabsorption via passive channels.	Reabsorption via Na^+–K^+–$2Cl^-$ cotransporter.	Reabsorption via Na^+–Cl^- cotransporter.	—
Protons (H^+)	—	—	—	Secretion via H^+-ATPase proton pumps and an H^+/K^+ channel.
Bicarbonate	About 80–90% reabsorption.	Reabsorption via varying membrane permeability.	—	Secretion via H^+-ATPase proton pump and Cl^-/HCO_3^- channel.
Urea	About 50% reabsorption via passive transport channels.	Secretion via varying membrane permeability.	—	Reabsorption via passive transport channels (ADH regulation).
Calcium	Reabsorption via Na^+–Ca^{2+} exchange channel.	Reabsorption driven via K^+ "leak" channel and Na^+ concentration gradients.	Reabsorption driven via vitamin D and PTH.	—
Magnesium	Reabsorption.	Reabsorption driven via K^+ "leak" channel and Na^+ concentration gradients.	Reabsorption.	—
Phosphate	About 85% reabsorption via Na^+ cotransporter (inhibited by PTH).	—	Reabsorption via vitamin D.	—
Carboxylate	About 100% reabsorption via specific transporters.	—	—	—

ADH, antidiuretic hormone; GLUT, glucose transporter; PTH, parathyroid hormone.

INULIN/CREATININE CLEARANCE

The measurement of the **glomerular filtration rate (GFR)** (volume of fluid filtered from the afferent arteriole into the renal corpuscle per unit time) allows clinicians to determine the health of the kidneys and to quantify any degree of kidney or renal failure. GFR is most accurately measured by injection of **inulin** into the bloodstream, a polysaccharide molecule from plants, which is neither reabsorbed nor secreted by the kidney. A more practical method of estimating GFR, though, is to measure blood **creatinine**, a molecule derived in muscle from the breakdown of **creatine phosphate**, a rapidly available source of adenosine triphosphate (ATP) energy for muscle and brain. Creatinine is exclusively excreted by the kidneys, completely filtered at the renal corpuscle, and only small amounts are secreted into the peritubular capillaries. The natural occurrence of steady levels of creatinine in the blood and urine offers an easy way to establish GFR and, therefore, kidney function. Varying mathematical formulas allow for correction of known variations depending on the patient's muscle mass, age, gender, race, and/or size. The normal range of GFR is (100–130) ml/min/1.73 m^2 but this varies in children and older adults. A GFR of over 60 ml/min/1.73 m^2 is usually sufficient for normal health, although high blood pressure, diabetes, and other chronic diseases can decrease kidney function and result in **chronic kidney disease (CKD)**. CKD is enumerated in six stages, depending on the patient's GFR, as noted below (Table 18-3).

RENIN–ANGIOTENSIN–ALDOSTERONE SYSTEM (RAAS)

The **RAAS** is a complex, regulatory pathway involving the kidney, liver, lung, adrenal glands, pituitary gland, and selected arterioles found in the kidney. It includes precursor and activated peptides as signals among these various organs, the enzymes that activate the peptides, and regulation by two major hormones of the renal and cardiovascular systems (Figure 18-4). The **JAG** apparatus is located between the afferent arteriole and the renal corpuscle and distal convoluted tubule of the same nephron. This specialized collection of cells is able to sense and regulate blood flow volume into and out of the renal corpuscle/glomerulus by the production of the signaling peptide hormone/enzyme **renin** and by two, simple biochemical mechanisms of the **macula densa**, involving Na^+ and the gaseous molecule **nitric oxide (NO)**.

TABLE 18-3. Stages of Chronic Kidney Disease and GFR

Stage	GFR/Clinical Evidence
Normal	≥90 ml/min/1.73 m^2 with protein in the urine
CKD1	≥90 ml/min/1.73 m^2/evidence of kidney damage
CKD2 (mild)	60–89 ml/min/1.73 m^2/evidence of kidney damage
CKD3 (moderate)	30–59 ml/min/1.73 m^2
CKD4 (severe)	15–29 ml/min/1.73 m^2
CKD5 (failure)	≤15 ml/min/1.73 m^2

CKD, chronic kidney disease; GFR, glomerular filtration rate.

RENIN AND BLOOD PRESSURE

Renin is produced by specialized cells (granular cells of the JAG apparatus) in response to low blood pressure sensed in the JAG and by decreased NaCl levels detected by macula densa cells (see below). Blood pressure in the JAG is detected by specialized nerve cells, which are activated (Chapter 19) by changes in the arteriole blood pressure, leading to a localized nerve signal and secretion of the hormone renin. As will be discussed further below, sympathetic nerve activity (Chapter 19) also leads to increased renin secretion. Secreted renin enzymatically converts the precursor protein angiotensinogen, produced in the liver, into angiotensin I, which, when converted to angiotensin II (see below), leads to increased blood pressure and, therefore, the pressure of perfusion in the renal corpuscles.

Renin Inhibitors: Recently developed inhibitors of **renin** have introduced a novel class of medications to help treat high blood pressure. Two such medications, **aliskiren** and **remikiren**, bind directly to renin and competitively inhibit binding of **angiotensinogen** and, therefore, production of **angiotensin I**. The efficacy and safety of these new drugs is still being determined but may offer a novel method of patient care.

MACULA DENSA AND BLOOD FLOW/OSMOLARITY

With high blood flow through the arterioles of the renal corpuscle, increased Na^+ are filtered into the glomerulus. Cells found in a part of the **JAG**, termed the **macula densa**, contain a **Na^+–Cl^- cotransporter channel**, which transports Na^+ into these cells. A limited number of **Na^+–K^+-ATPase pumps** attempt to transport the Na^+ back out of the cell but, if Na^+ levels are high (i.e., osmolarity is high), they are soon overwhelmed. The internal Na^+ concentration increases and creates an osmotic gradient, which brings water molecules into the macula densa cells. These cells, swollen with Na^+ and water, contain a **"stretch-activated" channel**, which allows ATP to escape followed by conversion to simple adenosine. These adenosine molecules are bound by membrane-bound proteinaceous receptors on the adjacent afferent and efferent arterioles of the

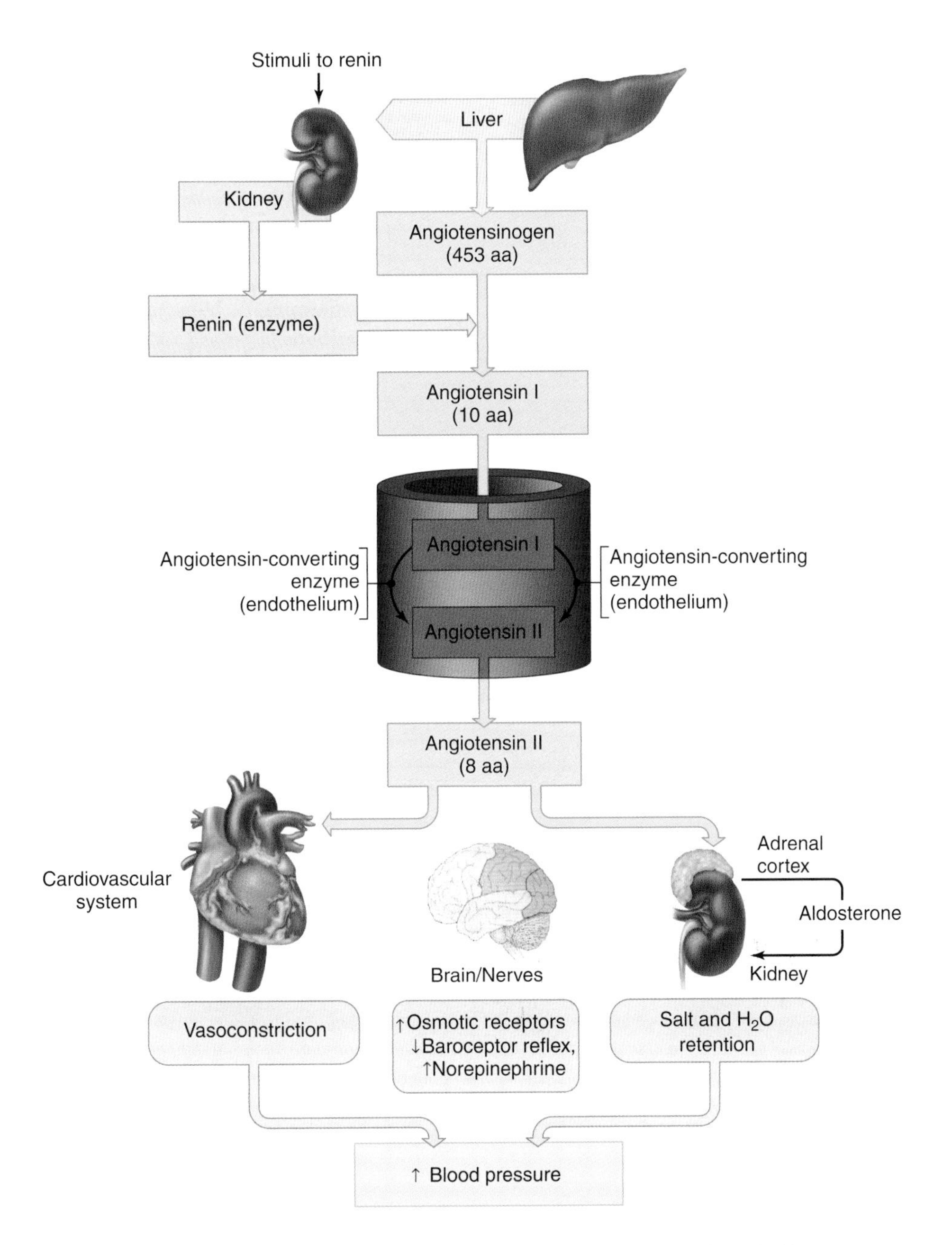

Figure 18-4. Renin–Angiotensin–Aldosterone System. Overview of effects of renin, angiotensin II and aldosterone on organs controlling volume status and blood pressure. aa, amino acid. See text for further details. [Adapted with permission from Barrett KE, et al.: Ganong's Review of Medical Physiology, 23rd edition, McGraw-Hill, 2010.]

glomerulus, leading to constriction of the afferent arteriole and dilation of the efferent arteriole. The result is decreased blood flow through and filtration by the renal corpuscle. As a result, less Na^+ get reabsorbed and the local concentration decreases.

Alternatively, low flow through the arterioles leads to a lower Na^+ concentration in the macula densa cells. The lower Na^+ concentration (i.e., low osmolarity) activates the enzyme **nitric oxide synthetase (NOS)** in these cells. NOS catalyzes the reaction, which converts the amino acid arginine and an oxygen molecule to citrulline and NO, utilizing reduced **nicotinamide adenine dinucleotide phosphate (NADPH)** as a cofactor. NO then activates specific G_s proteins, which produce **cyclic guanosine monophosphate (cGMP)** and **cyclic adenosine monophosphate (cAMP)**. cGMP activates a **protein kinase G**, which, in turn, leads to inactivation of myosin molecules via phosphorylation of their regulatory light chain. Inactivated myosin molecules in the blood vessel walls leads to vessel relaxation and dilation. cAMP leads to the increased release of renin.

Erectile Dysfunction (ED), NO, cGMP, and Phosphodiesterase Inhibitors: Understanding the role of **NO** and **cGMP** in the relaxation of arteriole wall smooth muscle has led to the development of **phosphodiesterase inhibitor** medications, which have revolutionized the treatment of **ED**. As in the kidney, NO-mediated production of cGMP causes dilation of arteries in the corpus cavernosum, increasing blood flow and leading to erection. Various factors including age, chronic diseases such as high blood pressure and diabetes, as well as side effects of certain drugs decrease the blood flow and lead to ED. Specific phosphodiesterase inhibitors [e.g., sildenafil (Viagra), vardenafil (Levitra), and tadalafil (Cialis)], which have structures mimicking the cGMP molecule, competitively block the enzyme **cGMP-specific phosphodiesterase type 5**, which breaks down cGMP (see figure below). By inhibiting this enzyme, the cGMP signal remains, promoting vasodilation and leading to better and sustained erection. The mechanism of cGMP relaxation of arteriole walls and the actions of the phosphodiesterase inhibitors are also used for certain types of **pulmonary hypertension** and **pulmonary edema** due to **altitude sickness**, in which pulmonary artery dilation lowers pressure in the lungs reducing disease damage to lung tissue.

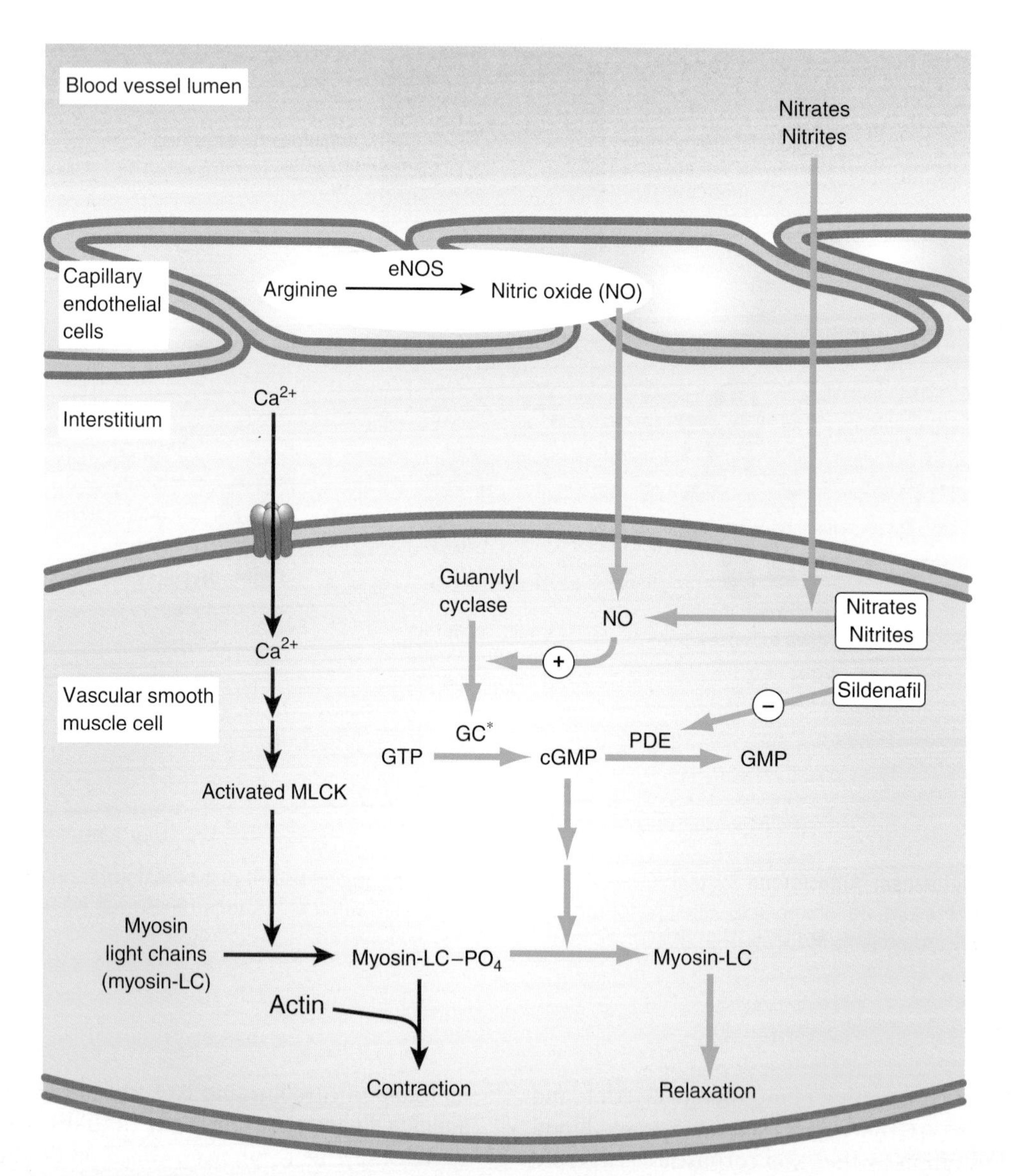

Reproduced with permission from Katzung BG, et al.: Basic and Clinical Pharmacology, 11th edition, McGraw-Hill, 2009.

Both constant monitoring of blood pressure (renin) and blood flow/osmolarity (macula densa) through the arterioles and, therefore, the renal corpuscles and nephrons, provide feedback to portions of the nephron (to alter electrolyte and water reabsorption and secretion) and the rest of the body. The remainder of the RAAS is responsible for these actions.

ANGIOTENSINOGEN/ANGIOTENSIN I AND II

Angiotensinogen is a 452-amino-acid-long protein, which is produced on a constant basis by the liver (Figure 18-5). **Angiotensin I**, comprising the first 10 amino acids of angiotensinogen, is produced by cleavage of a leucine–valine bond by renin. The 10-amino-acid angiotensin I is completely inactive but is subsequently converted to the eight-amino-acid peptide hormone, **angiotensin II**. This reaction is catalyzed by **angiotensin-converting enzyme (ACE)**, an enzyme found throughout the body but especially concentrated in the lung's capillary beds. Angiotensin II has several activities throughout the body, including on the kidney, adrenal glands, heart and blood vessels, sympathetic nervous system, and hypothalamus. All of these effects are propagated via G_q or G_i protein-coupled angiotensin receptors, which activate **phospholipase C**, leading to phosphorylation via **protein kinase C (PKC)** and Ca^{2+} release (Chapter 8). Both PKC and an influx of calcium can lead to myosin light-chain phosphorylation and, therefore, contractile activity. Activation of the G_i-coupled receptor also inhibits **adenyl cyclase** and, therefore, the production of **cAMP**, which inhibits smooth muscle relaxation (which also promotes contractile activity). Finally angiotensin II activates certain **tyrosine kinases**, leading to other effects in the body. These effects are illustrated in Figure 18-4 and listed in Table 18-4.

The sum effect of angiotensin II is the increase in body blood pressure with muting of many of the body's compensating mechanisms. Angiotensin II can also be converted to a seven-amino-acid angiotensin III with decreased effect on blood pressure but a similar effect on aldosterone secretion and a six-amino-acid angiotensin IV, which has even less activity.

ACE Inhibitors: The marked effect of angiotensin II on blood pressure, including increased pressure and, therefore, potential vessel damage in the kidney, led to the development of medications called **ACE inhibitors**. This drug class, all of which end in the suffix "-pril," specifically block the enzyme, which converts inactive angiotensin I into its active angiotensin II counterpart. The effectiveness in lowering blood pressure and providing protection to the kidney against damage has led to wide use of ACE Inhibitors in high blood pressure, diabetes, kidney disease with protein loss in the urine (proteinuria), after heart attack, congestive heart failure, and kidney diseases. Of note, ACE Inhibitors may also affect the **kinin–kallikrein system** by increasing the production of **bradykinin**, a nine-amino-acid peptide involved in pain, inflammation, and blood pressure control. Bradykinin may be responsible for some of the ACE inhibitors' adverse effects (such as cough and the potentially fatal side effect of angioedema). These side effects can be avoided by use of the related **angiotensin receptor blocker** class of medications, whose names all end in the suffix "-sartan," which competitively block the effects of angiotensin II at its G-protein receptors.

Bradykinin

Adapted with permission from Barrett KE, et al.: Ganong's Review of Medical Physiology, 23rd edition, McGraw-Hill, 2010.

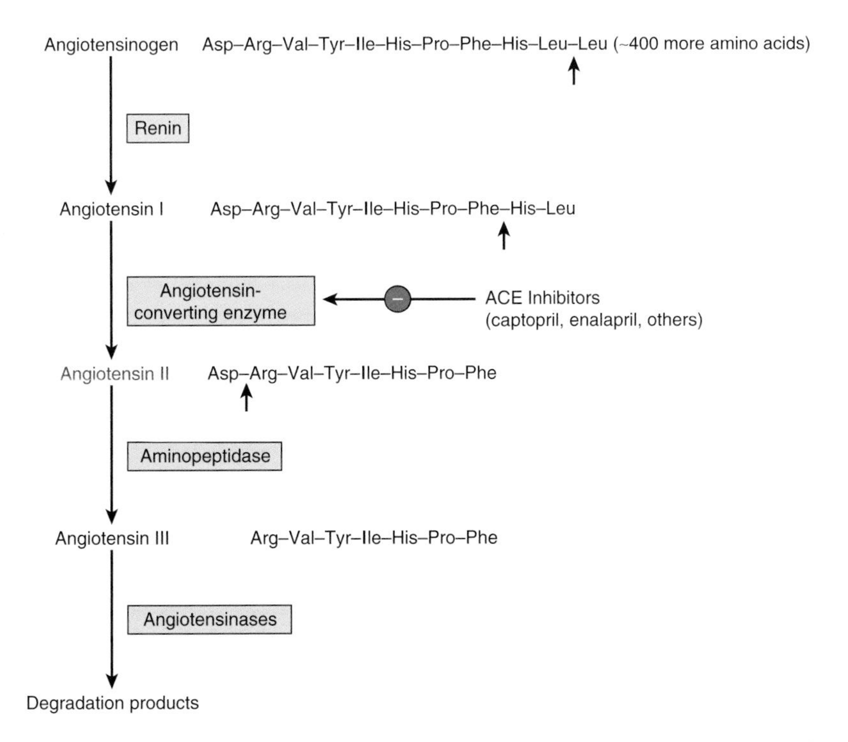

Figure 18-5. Conversion of Angiotensinogen. Angiotensinogen is converted to angiotensin I by the action of renin from the kidney by the removal of more than 400 amino acids by proteolytic cleavage (site indicated by arrow). Angiotensin I is subsequently converted to angiotensin II by the action of angiotensin-converting enzyme (ACE), found mainly in the lung, by the removal of an additional two amino acids (cleavage site indicated by arrow). The molecule is further processed as shown, although the majority of functions in the body are via angiotensin II as summarized in Table 18-4. [Adapted with permission from Murray RA, et al.: Harper's Illustrated Biochemistry, 28th edition, McGraw-Hill, 2009.]

TABLE 18-4. Angiotensin II Effects on the Body

	Effect	Result
Kidney	Contraction of afferent/efferent arterioles and medullary vasa recta.	Increased Na^+ and, as a result, water reabsorption.
		Increased blood pressure.
	Contraction of intraglomerular mesangial cells.	Reduce the area of filtration through the glomerulus, leading to increased pressure.
	Increase in size of cells of the nephron.	Increased Na^+ and, as a result, water reabsorption.
Adrenal Glands	Increased synthesis and secretion of aldosterone by the adrenal cortex.	Increased Na^+ and, as a result, water reabsorption.
		Increased secretion of K^+ in the urine.
Cardiovascular	Contraction (constriction) of blood vessels.	Increased blood pressure.
	Increased contractility of heart.	Increased blood pressure.
	Increased size and numbers of heart cells.	Cardiac hypertrophy secondary to hypertension and atherosclerosis.
	Increased adhesion and aggregation of platelets.	Increased clotting and potential for infarcts.
Neurological	Increased production and secretion of vasopressin (ADH) by hypothalamus and pituitary gland.	Increased reabsorption of water by collecting ducts.
		Constriction of blood vessels leading to increased blood pressure.
	Increased secretion of Adrenocorticotropic hormone by pituitary gland.	Inhibits loss of Na^+ from intestine.
		Increases loss of K^+ from kidney.
		Decreases loss of water from kidney.
		Decreased loss of water, raises blood pressure.
	Activation of osmotic receptors of subfornical organ (located in brain at fornix and foramen of Monroe).	Increased desire for water (i.e., thirst).
		Increased craving for salt (secondary to thirst sensation).
	Decreased response of baroceptor reflex.	Maintains high blood pressure by decreasing body's normal response.
	Increased release of norepinephrine by postganglionic sympathetic nerve fibers.	Constriction of blood vessels.
		Increased blood pressure.

ADH, antidiuretic hormone.

ALDOSTERONE

Aldosterone, the final player of the RAAS, is the mineralocorticoid steroid hormone (Chapter 3), which acts to increase Na^+ and water reabsorption and, as a result, K^+ secretion. The overall effect is an increase in blood volume and, therefore, blood pressure. Aldosterone production/excretion is mainly related to the concentration of K^+ in the serum, probably measured by carotid artery sensors. Sympathetic nerve activity can also influence aldosterone production but on a much lesser scale. Aldosterone's activity is predominately seen in the distal convoluted tubule and collecting duct.

Aldosterone and Disease—Conn Syndrome and Addison's Disease: Increased amounts of aldosterone lead to **primary aldosteronism (Conn syndrome)**, whereas decreased levels lead to **adrenal insufficiency (Addison's disease)**.

Conn syndrome results from overproduction of **aldosterone**, usually caused by uncontrolled growth (hyperplasia) of the adrenal gland(s) or an aldosterone-secreting tumor. The excessive aldosterone leads to an increase in sodium and water reabsorption with increased secretion/excretion of potassium that also produces excess hydrogen ion (H^+) secretion (metabolic alkalosis). The sum effect is high blood pressure, which can only be treated by removal of the hyperplastic adrenal gland or tumor and/or use of the potassium-sparing diuretics spironolactone or eplerenone (see below).

Addison's disease, alternatively, results from underproduction of aldosterone from either genetic or disease/trauma causes or from inhibition of aldosterone production by the adrenal glands. Effects on the nephron lead to increased loss of sodium and water and increased potassium and H^+ (metabolic acidosis) levels in the blood. This leads to low blood pressure, which may be severe enough to cause patients to faint, especially when standing up. Severe symptoms may be seen during an "**Addisonian crisis**" in which loss of consciousness and death may result.

Aldosterone specifically increases the reabsorption of Na^+ and water out of and the secretion of K^+ back into the tubule by binding to specific membrane-bound receptors and increasing the membrane permeability of cells lining the tubule. In addition, aldosterone increases the number of and stimulates the ATP-dependent phosphorylation of **Na^+–K^+ pumps**, leading to a conformational change of the pump, which exposes Na^+ to the interstitium and decreases Na^+ binding. As a result, increased Na^+ are reabsorbed along with Cl^- (to maintain electrical neutrality) and water (to maintain osmotic balance). The result is an increase in Na^+ and water in the interstitium, leaving them to be taken up by interstitial capillaries back to the blood stream. As a result of the inactivated Na^+–K^+ pumps but secondary to secretion via increased membrane permeability, K^+ accumulate in the tubules and, subsequently, in the urine. Falling K^+ concentrations, subsequently, decrease aldosterone secretion. Aldosterone secretion is additionally affected by conditions in which blood volume and/or electrolyte concentrations are changed such as significant blood loss, pregnancy, burns, shock, and prolonged physical exercise.

Pseudohyperaldosteronism and Licorice: Pseudohyperaldosteronism, in which aldosterone levels are normal, can also occur as a result of diet. Excessive licorice in the diet, containing the sweetener molecule **glycyrrhizin**, inhibits 11-β-hydroxysteroid dehydrogenase, which reduces the conversion of excess **cortisol** to inactive **cortisone** in aldosterone-selective epithelial cells. Inhibition of this enzyme allows the extra cortisol to produce aldosterone-like effects, including high Na^+ and blood pressure as well as low K^+ levels in the blood. Tissue resistance to the effects of aldosterone (possibly because of decreased action on the receptors) and/or Liddle's syndrome (a genetic defect leading to overactivity of reabsorbing sodium channels) can also lead to pseudohyperaldosteronism. Both can be treated using potassium-sparing diuretics (see below).

Potassium-Sparing Diuretics and Aldosterone Regulation: Spironolactone and eplerenone are members of a group of diuretics known as "**potassium-sparing**." These two medications, used for high blood pressure and heart failure, mimic the steroid structure of **aldosterone** (see figure below) and function as a competitive inhibitor for aldosterone receptors in the collecting duct. Thus, the effects of aldosterone on membrane permeability and number/activity of Na^+–K^+ pumps are not seen, decreasing reabsorption of Na^+ and water but decreasing (i.e., sparing) the secretion and loss of K^+.

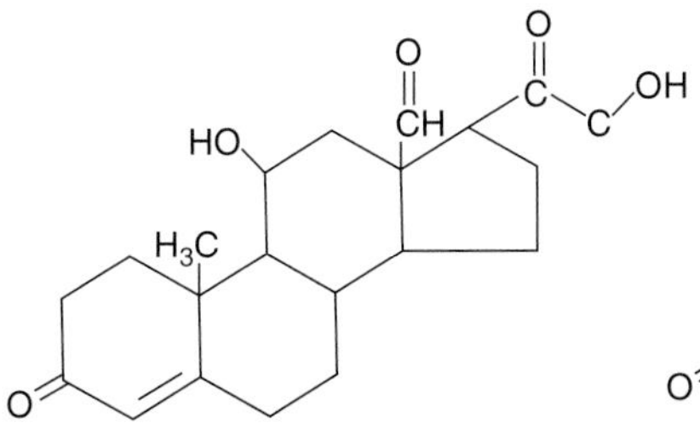

Aldosterone 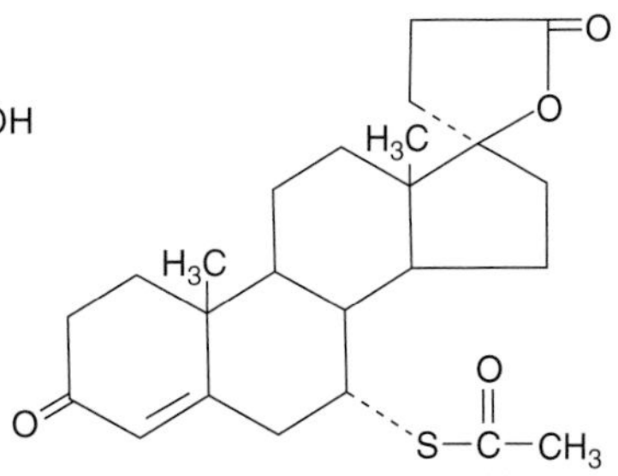 **Spironolactone** 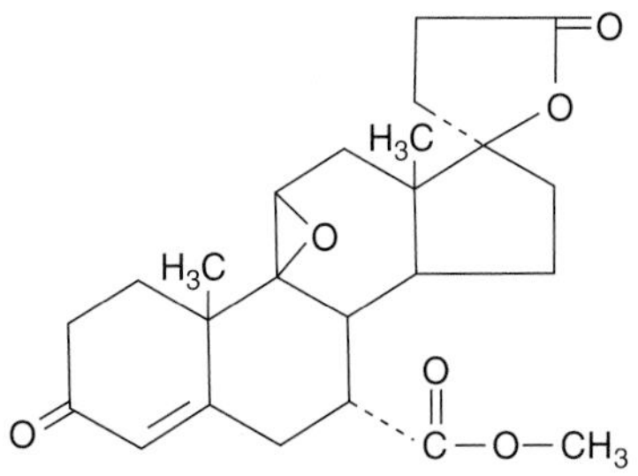 **Eplerenone**

Aldosterone also influences pH by stimulating the secretion of H^+. Finally, aldosterone influences reabsorption of water in the kidneys and, therefore, blood pressure, by increasing the release of the peptide hormone vasopressin (also known as antidiuretic hormone, ADH) from the pituitary gland. Vasopressin's actions will be discussed in greater detail below.

VASOPRESSIN

Although not formally part of the RAAS, **vasopressin**, also known as **ADH**, is intimately involved in similar process and, in fact, completes the process of determining the final water content of urine and, therefore, fluid balance in the body (Table 18-4). Vasopressin, a peptide hormone produced as a prohormone in the hypothalamus, is essential for enhanced reabsorption of water by increasing the presence of kidney-specific, membrane-bound water channels called **aquaporin-2**. Vasopressin raises the number of these channels by enhancing the gene expression for the protein and increasing its delivery to the appropriate plasma membrane. These functions occur via specific vasopressin receptors coupled to a G_s protein (Chapter 8) found on cells of the distal convoluted tubule and collecting ducts. Binding of vasopressin to the receptor activates an **adenyl cyclase**, which converts ATP to cAMP. The resulting signaling pathway activates **protein kinase A**, which phosphorylates aquaporin-2 protein, specifically on serine 262 while in the Golgi apparatus, promoting delivery of the vesicles to and membrane fusion with collecting duct membranes.

Vasopressin, Aquaporin-2, and Water Balance: The proper balance of water and, therefore, plasma volume is essential for the proper functioning of the body and directly influences blood pressure and the heart. **Aquaporin-2** plays a major role in the regulation of this fluid status and alterations or mutations of this protein lead quickly to disease states. **Nephrogenic (kidney-associated) diabetes insipidus**, a condition of poor fluid balance regulation and notable for large output of very dilute urine, is closely associated with mutations of the aquaporin-2 channel. Changes in the expression of the aquaporin-2 gene are linked to the unwanted water retention seen in congestive heart failure, lithium toxicity, and during pregnancy. **Neurogenic (brain-associated) or "central" diabetes insipidus** occurs when **vasopressin** is not synthesized because of hypothalamus or pituitary gland failure. Common causes include head trauma, tumors, or unknown reasons. Finally, ethanol intake inhibits the action of vasopressin and, therefore, aquaporin-2, leading to decreased water reabsorption and associated thirst/mild dehydration.

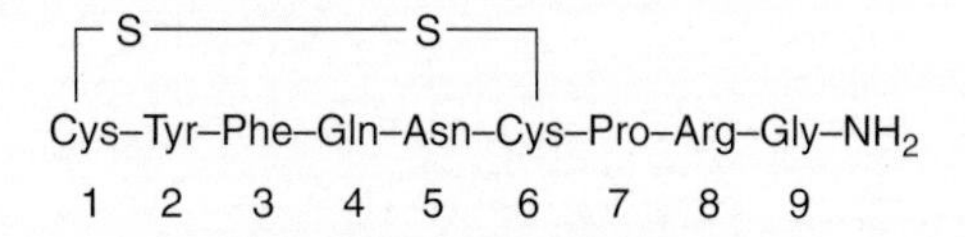

Adapted with permission from Barrett KE, et al.: Ganong's Review of Medical Physiology, 23rd edition, McGraw-Hill, 2010.

Therefore, the cooperative actions of aldosterone, which mainly influences electrolytes, and vasopressin (ADH), which influences water, control the final water content of the urine, including how dilute it is. In fact, in times of extreme dehydration, the collecting duct alone can reabsorb approximately one quarter of the water in the filtrate.

ATRIAL NATRIURETIC PEPTIDE (ANP)

ANP (Figure 18-6) is another important peptide hormone that controls body water as well as Na^+ and K^+ levels. ANP is released in times of body fluid overload and/or high blood pressure from muscle cells of the heart. ANP binds to specific receptors, which activate a **guanyl cyclase** to make **cGMP**. The effects of ANP on the kidney include the following:

- Increased glomerular filtration by
  - dilation of the afferent arteriole,
  - constriction of the efferent arteriole,
  - relaxation of the mesangial cells.
- Increased reabsorption of NaCl by
  - increased blood flow through the vasa recta,
  - increased osmotic drive for further water and NaCl excretion.
- cGMP-driven phosphorylation/deactivation of sodium channels (located on the distal convoluted tubule and collecting duct).
- Inhibition of renin and aldosterone production.

The sum effect of ANP's actions is increased excretion of water, sodium, and potassium by the kidneys to help compensate for excess fluid and high blood pressure. Because of these qualities, ANP and **brain natriuretic peptide** (Figure 18-6), which, despite the latter's name are both secreted by heart cells, have shown utility in following the progression and resolution of congestive heart failure.

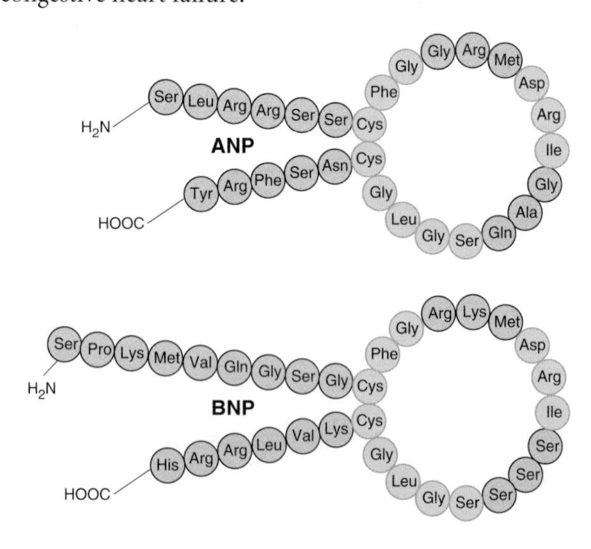

Figure 18-6. Natriuretic Peptides. Atrial natriuretic peptide (ANP) and brain natriuretic peptide (BNP) amino acid sequences and structures are illustrated. See text for further information. [Reproduced with permission from Katzung BG, et al.: Basic and Clinical Pharmacology, 11th edition, McGraw-Hill, 2009.]

ACID–BASE BALANCE

The maintenance of the proper **acid–base balance** between pH 7.35 and 7.45 is essential to maintain the human body. Imbalances, including acidosis and alkalosis, not only adversely affect multiple protein and enzymatic functions but, as a result, can also lead rapidly to death. Various mechanisms found in the lungs, kidneys, and blood stream constantly adjust this equilibrium as per the needs of the body. Other buffers, including ammonia (NH_3) (see below), proteins, and phosphate also contribute to this important functional and protective mechanism by combining with H^+ to remove them from solution.

The kidney plays an essential role in regulating the body's acid–base balance mechanism and relies on the selective and regulated reabsorption and secretion of **H^+** and **bicarbonate ions (HCO_3^-)**, two primary molecules of the bicarbonate buffering system, involving the reaction $CO_2 + H_2O \leftrightarrow H_2CO_3 \leftrightarrow HCO_3^- + H^+$. This is done by two types of cells, which secrete or reabsorb H^+ or HCO_3^- via **H^+-ATPase proton pumps**, and H^+–K^+ and Cl^-–HCO_3^- channels. Acid–base regulation is also found in the collecting duct.

Increased acidity in the body augments reabsorption of bicarbonate from the proximal convoluted tubule and loop of Henle, and collecting duct cells while also causing increased secretion of hydrogen from the collecting duct. Increased base causes the kidney to decrease H^+ secretion and causes less reabsorption/more secretion of bicarbonate from the nephron. Table 18-2 provides specifics regarding the specific mechanisms involved with these two ions.

NH_3 AND ACID–BASE BALANCE

Another acid–base balance mechanism (Figure 18-7) inherent in the kidney's functions is the handling of **ammonia (NH_3)** as part of a secondary buffering system. Although the primary management of ammonia levels is via the **urea cycle** in the liver (Chapter 11), any necessary elimination of urea and/or excess, toxic **ammonium ions (NH_4^+)** occurs in the kidney's nephrons.

In particular, NH_3 can freely cross the membranes of the nephron and, after combining with a H^+ to form NH_4^+, be excreted in the urine (Figure 18-8A) with a net acid loss. Finally, NH_3 is produced by the enzymatic degradation of the amino acid **glutamine**, converted first to **glutamate** and then to **α-ketoglutarate**, which continues through the Krebs' cycle (Figure 18-8B). If a base excess exists, the conversion of glutamine and, therefore, the production of NH_3 is inhibited to help the body re-establish the proper acid–base balance.

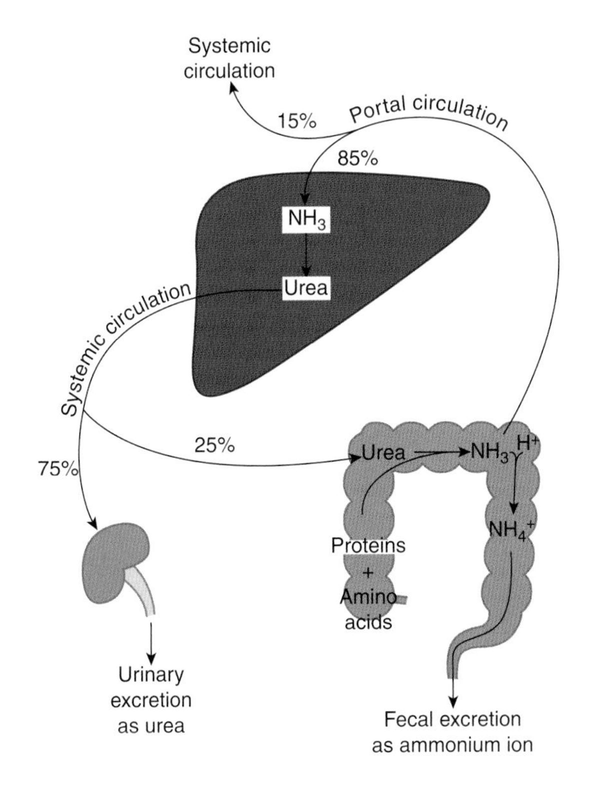

Figure 18-7. Overview of Ammonia (NH_3) Balance in the Body. Ingested sources of NH_3 such as protein and amino acids are processed via the gastrointestinal system and eliminated in feces or transferred through the portal circulation to the liver or to the body via systemic circulation. In the liver, NH_3 is converted to urea and returned to the systemic circulation where it can be eliminated as urine by the kidneys or sent back to the gastrointestinal system. The balance of NH_3 depends on varying contributions of liver, kidney, and intestinal metabolism and/or elimination. [Reproduced with permission from Barrett KE, et al.: Ganong's Review of Medical Physiology, 23rd edition, McGraw-Hill, 2010.]

Acid–base balance can also have a marked effect on the efficacy of medications by influencing how they are eliminated from the body. Many medications are excreted via the kidneys but, once in the nephron, they can be either reabsorbed and returned to the circulation (Figure 18-3) or secreted/eliminated in the urine. Medications secreted in the proximal tubule are indicated below along with predominate form (acid, base, or neither) at secretion (Table 18-5). The form of a medication can influence secretion, as acidic medications are secreted more when the urine is basic and vice versa. In critical care and other clinical settings, changing urine pH, either as

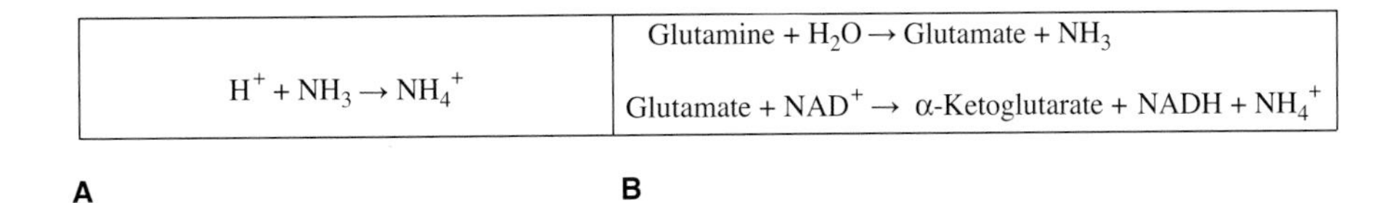

Figure 18-8. A–B. The Ammonia (NH_3) Buffering System. (**A**) Interconversion of NH_3 to ammonium (NH_4^+), which captures one H^+. This reaction occurs freely in the human body, depending on concentrations of each respective molecule. (**B**) Conversion of glutamine to glutamate, via the enzyme glutaminase, with production of one NH_3 molecule (upper) and glutamate to α-ketoglutarate via the enzyme glutamate dehydrogenase with production of a second NH_3 and NADH (lower).

Hyperammonemia and NH_3 Toxicity: High levels of free NH_3 in the body cause harm to varying parts of the body. Most prominent are the effects on the brain and nervous system, which subsequently lead to other clinical manifestations. Mental retardation, seizures, and unusual and uncontrolled muscle movements as well as breathing irregularities and poor control of body temperature are common. Coma and death may result without treatment.

There are multiple enzyme deficiencies, most due to autosomal recessive mutations, which result in excess ammonium in the body. These various disease states are separated into primary—involving enzymes from the urea cycle (see Figure 5.9a) and secondary—involving enzymes that are not part of the urea cycle. Examples are listed below.

	Enzyme Deficiency	Effect/Condition
Primary	N-Acetyl glutamate synthase (activator of carbamoyl phosphate synthetase I)	Inactive carbamoyl phosphate synthetase I, hyperammonemia.
Primary	Carbamoyl phosphate synthetase I	Hyperammonemia.
Primary	Ornithine transcarbamylase	Hyperornithinemia–hyperammonemia–homocitrullinuria syndrome; increased orotic acid in the urine.
Primary	Argininosuccinate synthetase	Citrullinemia (two types).
Primary	Argininosuccinate lyase	Argininosuccinic acidemia.
Primary	Arginase 1	Argininemia.
Secondary	Propionic acidemia	Accumulation of propionyl-CoA, propionic acid, ketones, and secondary hyperammonemia.
Secondary	Methylmalonic acidemia	Accumulation of methylmalonyl-CoA and secondary hyperammonemia.
Secondary	Lysinuric protein intolerance (hyperdibasic aminoaciduria type 2)	Increased lysine, arginine, ornithine, and orotic acid in the urine and secondary hyperammonemia.

the result of an illness or selective medical intervention, can affect the ultimate elimination of acidic and basic drugs from the blood stream. In very specific circumstances, this ability can be utilized to aid in patient care.

TABLE 18-5. Medications Secreted in Proximal Tubule

Medication	Form (Acid or Base)
para-Aminohippurate, furosemide, glucuronic acid, indomethacin, methotrexate, penicillin, probenicid, sulphates, thiazide diuretics, uric acid	Acid
Amiloride, ammonium compounds (4°), cimetidine, dopamine, histamine, mepacrine, morphine, pethdine, quinine, 5-hydroxytryptamine, triamterene	Base
Acetozolamide, atenolol, atropine, chlorothiazide, chlorpromazine, digoxin, gentamycin, paraquat, procainamide, saccharin, salicylate, tetracycline	Neither

SYNTHETIC FUNCTIONS

SYNTHESIS OF ERYTHROPOIETIN

Erythropoietin, also known as hematopoietin, is a major glycoprotein cytokine and hormone that regulates the production of red blood cells (Figure 18-9). Erythropoietin is produced by endothelial cells that are part of the kidney's peritubular capillaries. The amount of erythropoietin produced is believed to be regulated by a transcription factor that becomes hydroxylated (OH^-) and then gets degraded when oxygen is abundant. When oxygen levels are low, the transcription factor binds to erythropoietin receptors on red blood cell membranes and bone marrow cells and activates an associated **Janus kinase 2** resulting in phosphorylation of tyrosine residues (Chapter 8). Activation by erythropoietin protects the red blood cells from normally programmed cell death (apoptosis) and stimulates the production of new red blood cells from their specific precursor cells.

Renal disease leads to a drop in erythropoietin production and, therefore, a low red blood cell count or anemia. Artificially produced erythropoietin can be used to increase these patients' red blood cells count as well as those suffering from certain chronic diseases such as heart failure, undergoing

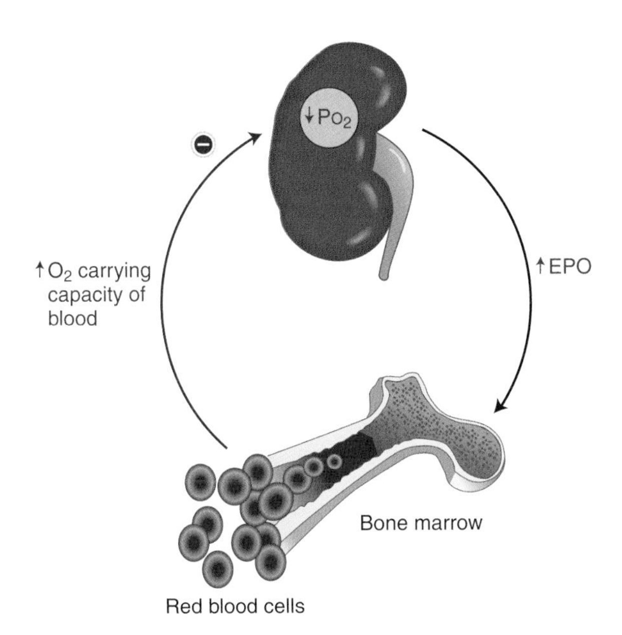

Figure 18-9. Regulation of Red Blood Cell Production by Erythropoietin. If the ability of blood to carry oxygen decreases because of a fall in numbers of red blood cells (e.g., normal cell death, pathological destruction of red blood cells, bleeding, etc.), the kidney senses lower pO_2 levels and increases the levels of erythropoietin (EPO). EPO then signals the bone marrow to increase production of red blood cells. See text for further details. [Reproduced with permission from Kibble JD and Halsey CR: The Big Picture: Medical Physiology, 1st edition, McGraw-Hill, 2009.]

chemotherapy or radiation therapy for cancer, and/or critically ill and in need of the additional oxygen carrying capacity. Erythropoietin has also been shown to be active on brain cells and has been useful in the treatment of schizophrenia and related diseases. It has also been abused by athletes who utilize this cytokine to increase red blood cell counts and, therefore, oxygen carrying/performance capacities.

ROLE IN VITAMIN D SYNTHESIS

The kidney also plays a vital role in the synthesis of active **vitamin D** and, secondarily, the regulation of Ca^{2+} concentration in the body. **Vitamin D_3** is initially hydroxylated in the liver to produce **$25(OH)D_3$** and stored there until required (Chapter 3, Figure 3-10). A second hydroxyl group is added when **parathyroid hormone (PTH)** activates 1-α-hydroxylase in the proximal convoluted tubule of the kidney to produce the active form **$1,25(OH)_2D_3$** or **calcitriol**. Vitamin D and the vitamin D receptor found mainly in the distal convoluted tubule also act to increase Ca^{2+} and phosphate ion (PO_4^{3-}) reabsorption by increasing the number of these ion channels. PTH also increases reabsorption of Ca^{2+} and Mg^{2+} in the proximal portion of the distal convoluted tubule and the ascending limb of the loop of Henle in opposition to the action of the hormone **calcitonin** (for more detail, see Chapter 13).

Kidney Stones: Kidney stones (nephrolithiasis) can form because of a number of different biochemical means often secondary to a disease state. Types of kidney stones and potential causes are listed below:

Type of Stone	Cause
Calcium oxalate	Calcium is believed to bind to dietary oxalate and makes it unavailable for filtration by the renal corpuscle. Decreased calcium in the diet may increase the concentration of oxalate in the urine, which readily precipitates with urinary calcium, leading to stone formation.
Uric acid	Increased uric acid levels in the urine (e.g., gout; Chapter 3) and any disorder, which increases urine acidity, can lead to uric acid stones.
Struvite Ammonium magnesium phosphate [$(NH_4)MgPO_4 \cdot 6H_2O$]	Urea-splitting bacteria (e.g., *Proteus mirabilis*, *Serratia*, and *Providencia* species) produce increased NH_3, which decreases urinary acid level, leading to precipitation and formation of struvite stones.
Calcium phosphate	Hyperparathyroidism leads to overproduction of PTH and, therefore, increased calcium and phosphate levels in the blood and urine. Increased levels and an associate increased acidity of the urine lead to stone formation.
Cystine	Defective transport channel in proximal convoluted tubule leads to accumulation of cystine in and acidification of the urine, leading to formation of cystine stones.

REVIEW QUESTIONS

1. What are the roles of the renal corpuscle, glomerulus, nephron, tubules, and collecting systems?
2. How do reabsorption, secretion, and excretion differ?
3. What are the functions of renin, angiotensin I and II, angiotensin-converting enzyme, aldosterone, vasopressin, and atrial natriuretic peptide and how does each relate to the renin–angiotensin system?

4. What is the function of erythropoietin?
5. How do the components of the glomerulus act together to create a molecular filtering mechanism?
6. What receptors are activated and by what signaling pathways to increase or decrease renal reabsorption or secretion of molecules?
7. What is the role of the renin–angiotensin system in fluid and electrolyte homeostasis and what is the impact of abnormalities in triggering secondary hypertension?
8. What is the role of the kidney in the synthesis of erythropoietin and the active form of vitamin D?
9. What are the basic concepts, reactions, and role of the kidney in the renin–angiotensin–aldosterone system?

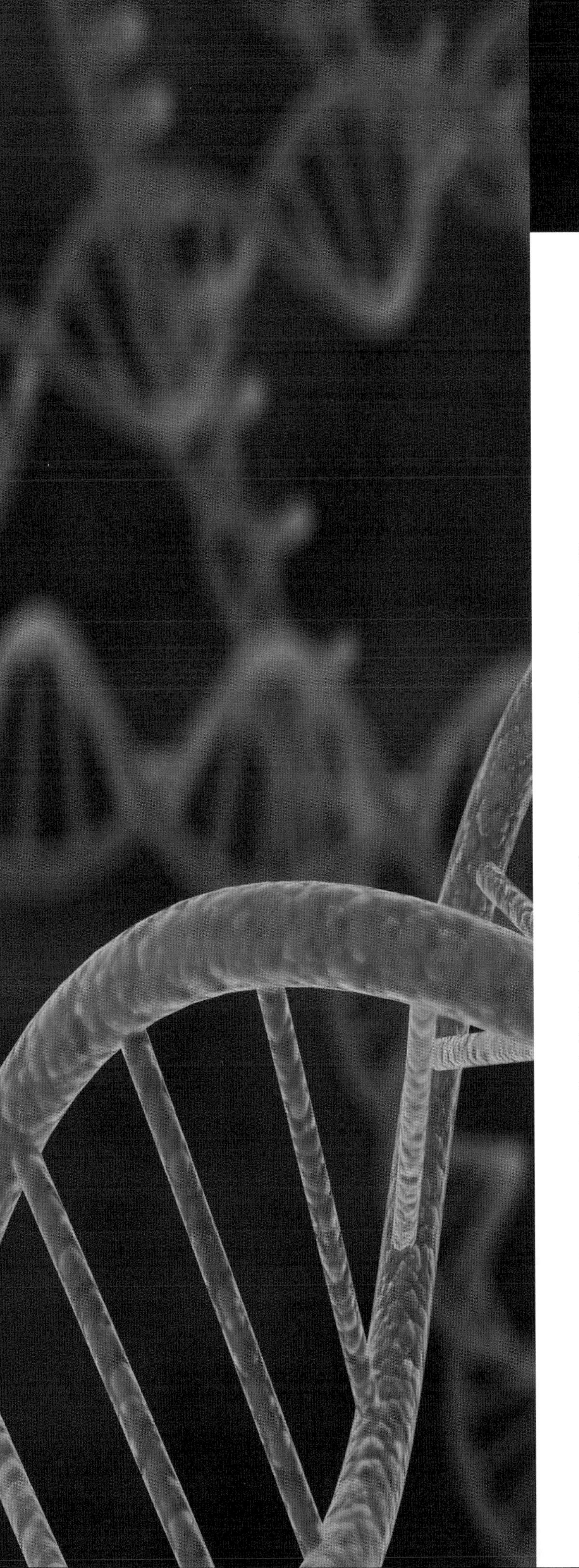

CHAPTER 19

THE NERVOUS SYSTEM

Editor: Kathryn Beck-Yoo, MD
Department of Anesthesiology, Swedish Medical Center at Ballard, Seattle, Seattle, Washington, USA

OVERVIEW

The nervous system provides a network of coordinated signaling for bodily functions. Somatic nerves promote skeletal muscle activity and, therefore, gross movements. Autonomic nerves provide signals for internal organs, including regulation of their function. These signals are transmitted by nerves, which rely on membrane-bound protein pumps for conduction of the impulse. Specialized lipids form myelin sheaths, which aid in the fast and efficient transmission of these signals. Neurotransmitters, often small peptides, allow continuation of the nerve signal between neurons and from neurons to target tissues. The breakdown of this network of signaling and regulation leads to several neurological diseases.

COMPONENTS OF THE NERVOUS SYSTEM

The nervous system is anatomically and functionally composed of two main parts: the **central nervous system (CNS)** and the **peripheral nervous system (PNS)** (Figure 19-1). The CNS consists of the brain and spinal cord and the PNS is composed of all nerves outside of the CNS, including all spinal and cranial nerves. In addition, there are further classifications for the nerves of the PNS. First, nerves are distinguished by the direction of nerve propagation. **Afferent (sensory)** neurons conduct action

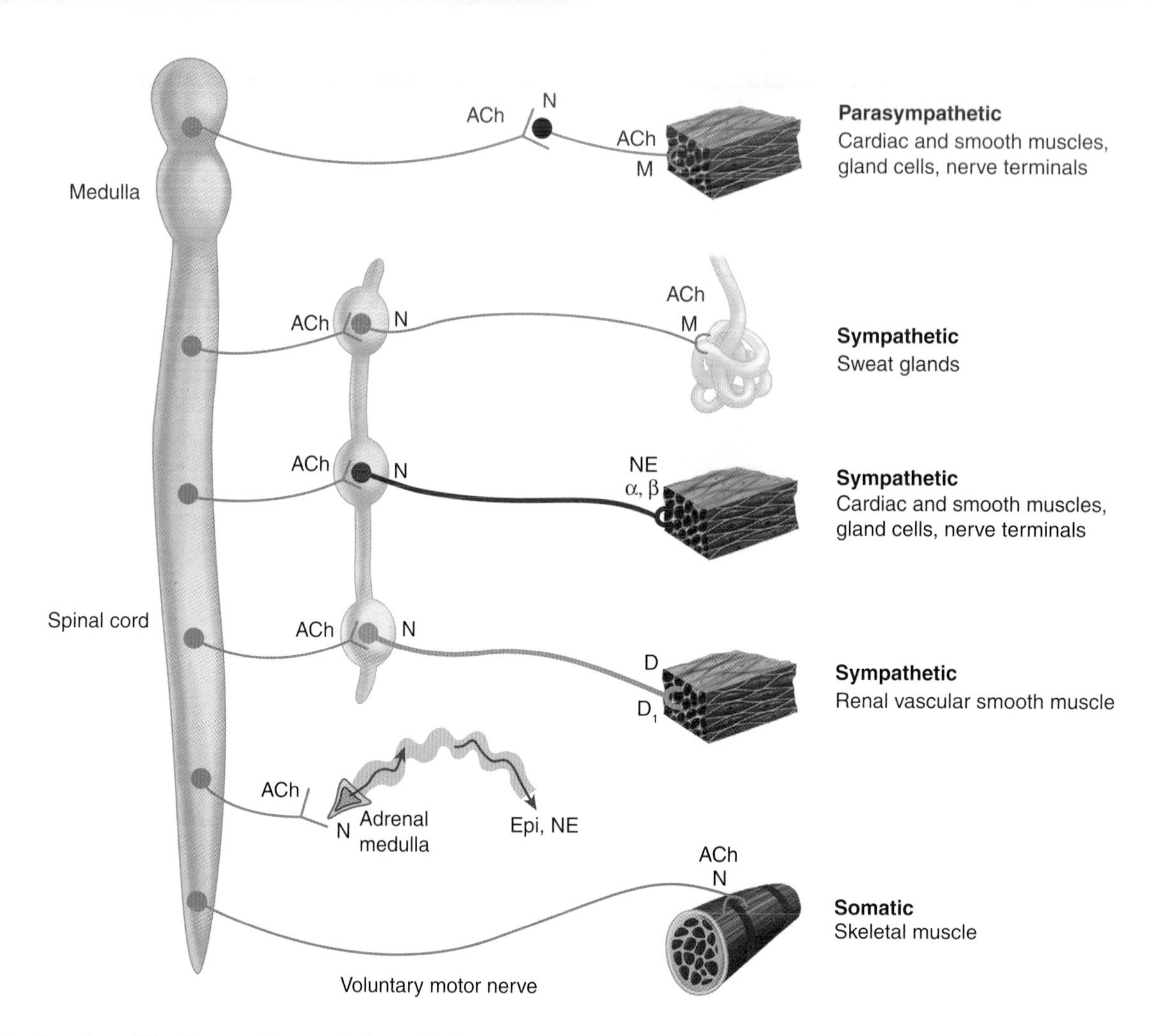

Figure 19-1. Overview of the Nervous System. Schematic diagram comparing some anatomic and neurotransmitter features of autonomic and somatic motor nerves. Only the primary transmitter substances are shown; parasympathetic ganglia are not indicated. See text for more details. Ach, acetylcholine; D, dopamine; Epi, epinephrine; M, muscarinic receptors; N, nicotinic receptors; NE, norepinephrine. [Reproduced with permission from Katzung BG, et al.: Basic and Clinical Pharmacology, 11th edition, McGraw-Hill, 2009.]

potentials toward the CNS. **Efferent (motor)** neurons transmit impulses away from the CNS to effectors such as muscles or glands. Second, nerves of the PNS can be further divided into the **somatic (skeletal muscle)** and **autonomic (organs, glands, and smooth muscle) nervous systems.** The autonomic nervous system (ANS) is divided into the **sympathetic (SNS)** and **parasympathetic nervous system (PSNS).** Generally, both interact with the same effectors, but often paradoxically. The SNS is involved in the activities associated with the fight-or-flight response, helping the body to cope with stress. The PSNS promotes activities that support the body while at rest, including digestion.

Neurons are the fundamental components of the nervous system. The main components of a neuron include the soma (or cell body), dendrites, axon, and the axon terminals (Figure 19-2). The cell body, or **soma**, is the central part of the neuron that contains the nucleus and other cell organelles. The **dendrites** are the branching structures of the neuron that receives stimuli or messages via chemoreceptors. The **axon** is a slender, cable-like extension of a neuron that carries nerve impulses away from the soma toward the axon terminal. The **axon terminals** are the hair-like terminals of the axon where neurotransmitters are released from the synaptic knobs. Neurons communicate with one another via synaptic transmission, where the axon terminal of one neuron comes into close contact with the dendrites of another neuron. Neurons can communicate chemically, via neurotransmitters, or electrically with electrically conductive junctions between the cells.

The **myelin sheath** is a phospholipid membrane that surrounds and protects the axons of some nerves (Figure 19-2). The primary phospholipid of the membrane is galactocerebroside, a sphingolipid (Chapters 3 and 8). Such phospholipids in nerve membranes strengthen the sheath. Myelin is produced by Schwann cell in the PNS and by oligodendrocytes in the CNS. The **neurofibrillar nodes** (also known as **nodes of Ranvier**) are the gaps in the myelin sheath, where the action potential occurs during conduction along the axon.

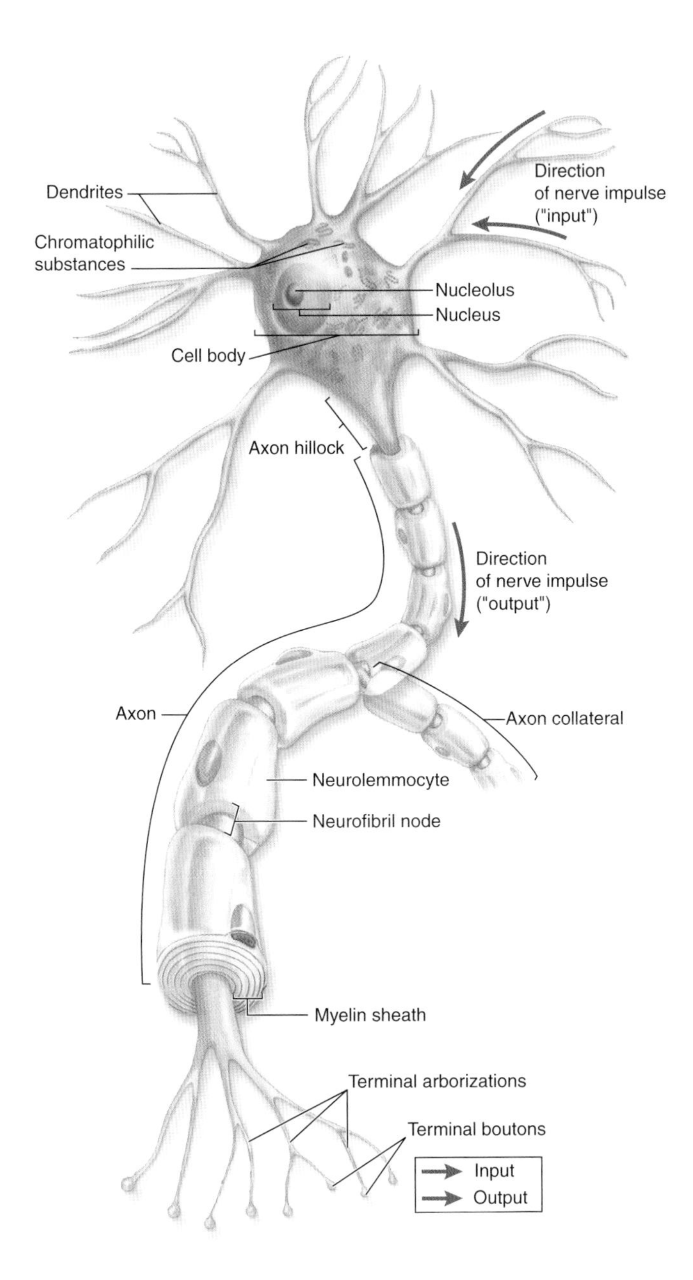

Figure 19-2. General Structure of a Nerve Cell. The general structure of a nerve cell includes the cell body with a nucleus and nucleolus, multiple dendrites, and an axon. Further details are provided in the text. The initial segment, immediately after the axon hillock, contains no myelin sheath but does contain a very high density of voltage-gated Na^+ channels for continued propagation of the nerve impulse. In the myelinated section, the impulse is conducted from node to node (see below) until reaching the terminal boutons where neurotransmitter molecules are stored for release and nerve signal propagation (see below). [Reproduced with permission from Mescher AL: Junqueira's Basic Histology Text and Atlas, 12th edition, McGraw-Hill, 2010.]

Demyelinating Disorders: Demyelinating disorders refer to any condition that disrupts **myelin** in either the CNS or PNS resulting in an interruption of normal nerve transmission. The process of demyelination can be because of an ischemic, metabolic, genetic, or infectious insult. However, because a demyelinating disorder can follow an infection or an inoculation, it has been hypothesized that an autoimmune mechanism is to blame. In some disorders, remyelination can occur with an associated degree of neural recovery. However, many times, neurologic deficits are permanent. Because myelin is formed in the CNS by oligodendrocytes and in the PNS by Schwann cells, demyelinating disorders can target different locations and have a variable presentation as illustrated below.

Category	Disorder
Hereditary	Phenylketonuria/other aminoacidurias, Tay–Sachs disease, Niemann–Pick disease, Gaucher's disease, Hurler's syndrome, Krabbe's disease/other adrenoleukodystrophies, adrenomyeloneuropathy, Leber's hereditary optic atrophy, and related mitochondrial disorders.
Hypoxia and Ischemia	Carbon monoxide toxicity and other syndromes of delayed hypoxic cerebral demyelination, and progressive subcortical ischemic demyelination.
Nutritional deficiencies	Central pontine myelinolysis (may also be caused by Na^+ fluxes), demyelination of the corpus callosum (Marchiafava–Bignami disease), and Vitamin B_{12} deficiency.
Direct viral invasion of CNS	Progressive multifocal leukoencephalopathy, subacute sclerosing panencephalitis, and tropical spastic paraparesis/HTLV-1-associated myelopathy.
Primary demyelinating disorders	Recurrent, progressive disorders (multiple sclerosis and its variants), monophasic disorders such as optic neuritis, acute transverse myelitis, acute disseminated encephalomyelitis, and acute hemorrhagic.

NERVE IMPULSE CONDUCTION

Successful conduction of nerve impulses relies on three elements: an electrically insulated membrane, energy-driven pumps to establish and maintain ion gradients, and ionic selective channels.

NEURON AT REST

Neurons maintain a resting membrane potential of –70 mV by adenosine triphosphate (ATP)-mediated active transport and passive diffusion of ions. The **Na^+–K^+-ATPase pump** of the cell membrane transfers three Na^+ out of the cell while transferring two K^+ intracellularly (Chapter 8 and Figure 19-3). The steady-state condition results in an extracellular Na^+ concentration of 150 mmol/L and an intracellular Na^+ concentration of 5–10 mmol/L. The extracellular K^+ concentration is 3–4 mmol/L and intracellular is 120–135 mmol/L. This gives rise to the negative state within the cell, and it creates a concentration gradient that will favor the extracellular diffusion of potassium and the intracellular diffusion of sodium upon depolarization.

NERVE IMPULSE

Several stimuli can activate a neuron leading to the generation of a **nerve impulse** (Figure 19-4). These include chemical, mechanical, or electrical excitation. Impulse propagation is accompanied by small membrane depolarizations. When the depolarization exceeds the **threshold** level of -55 mV, activated sodium channels cause a flood of Na^+ ions into the cell which drives further depolarization. This causes a relative excess of cations in the cell, which eventually leads to a membrane potential of +35 mV. If the initial stimulus is less than −55 mV, few Na^+ channels are opened and the threshold level is never reached. As a result, nerve conduction is an all-or-none response.

Niemann–Pick C Disease: Niemann–Pick C Disease is an autosomal recessive, life-threatening lysosomal storage disease, striking mainly in mid-to-late childhood but also in infants and adults. A majority of cases involve a mutation on the NPC1 gene on chromosome 18 resulting in lysosomal cholesterol and glycosphingolipid accumulation in neurons and glial cells in the CNS. The lipid build-up results in a distortion of the neuron mainly by erroneous formation of dendrites, which subsequently alters neurotransmission. The clinical presentation is highly variable but usually presents with the onset of movement and learning disabilities and early impairment of vertical eye movement (vertical supranuclear gaze palsy). In addition, patients can experience loss of muscle tone (dystonia), seizures, disabling joint problems, problems swallowing (dysphagia), and, in adults, varying cognitive impairment as well as psychosis and/or depression. Most patients die due to complications of this disease before the age of 20 years.

As the neuron's action potential depolarizes, voltage-gated calcium channels cause an influx of calcium ions (Ca^{2+}). This Ca^{2+} influx triggers the fusion of storage vesicles with the terminal bud membrane, releasing a neurotransmitter into the synaptic cleft (Figure 19-5). Receptors on the adjoining neuron bind with the neurotransmitter causing an alteration in the receptor's configuration. These neurotransmitter-gated ion channels act as an internal pore allowing Na^+ to flow down their electrochemical gradient into the cell. As the intracellular concentration of Na^+ rises, the charge within the cell becomes positive causing a depolarization of the membrane. The result is a propagating electrical signal known as an **action potential**. This process repeats until the target tissue has been reached.

Alzheimer's Disease (AD): AD is a progressive and fatal degenerative disease of the brain and the leading cause of dementia. Alzheimer's is due to the accumulation of **amyloid plaques** between neurons and **neurofibrillary tangles (NFTs)** within the neurons. The exact mechanism of how the plaques and tangles affect brain function is not fully understood. Patients vary in both amount and ratio of plaques versus tangles. However, patients who have primarily only amyloid plaques deteriorate slower than those with primarily tangles.

Amyloid plaques consist of **β-amyloid protein**, which is a protein fragment of **amyloid precursor protein (APP)**. Normally, these protein fragments are degraded and eliminated via one of two pathways. In the case of AD, APP is cleaved by **β-secretase** and **γ-secretase** to form amyloid-β-derived diffusible ligands. The accumulation of the amyloid forms hard, dense plaques. These plaques physically alter the architecture of the surrounding tissue, and therefore alter normal neural functioning. In addition, amyloid deposits cause mitochondrial dysfunction leading to premature, scheduled cell death (apoptosis).

The second major finding in AD is NFT. These tangles are made of six isoforms of **tau proteins**, highly soluble microtubule-associated proteins found normally in healthy brains. In a healthy person, phosphorylation of tau protein destabilizes axonal microtubules, allowing normal neuron function. In AD, the process is altered when **hyperphosphorylation of tau** results in the formation of NFTs. All six tau isoforms are present in the hyperphosphorylated state. As they are altered, they become insoluble aggregates within the neuron leading to instability in the neural tubes. This impacts normal nerve conduction within the brain.

Myelin sheaths enable action potentials to travel faster than in unmyelinated axons and at decreased energy expenditure (Figure 19-6). The unsheathed nodes of Ranvier contain a high density of voltage-gated ion channels. The action potential is conducted along the axon via conduction jumping from one node of Ranvier to the next.

REPOLARIZATION

Repolarization is dependent upon several factors. The sodium channels are inactivated and cannot reopen until membrane polarization is reset to the resting membrane potential of -70mV. Inactivity of the sodium channels causes a consequent drop in sodium permeability and an increase in potassium conductance. During this time, the membrane is unable to repeat excitation (**refractory period**). Eventually, the sodium–potassium pump re-establishes the original baseline gradients and the cell is returned to its resting potential.

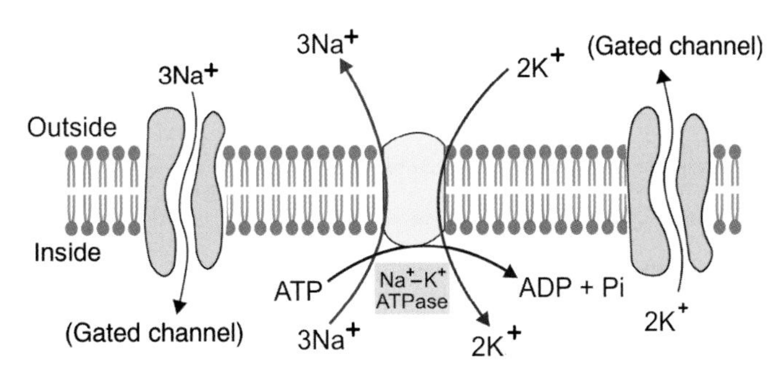

Figure 19-3. The Na⁺–K⁺-ATPase Pump. The pump and associated sodium (Na^+)- and potassium (K^+)-gated channels are shown. These gated channels allow the slow restoration of Na^+ and K^+ to their respective opposite side (see Figure 19-4 and associated text). ADP, adenosine diphosphate; ATP, adenosine triphosphate. [Reproduced with permission from Naik P: Biochemistry, 3rd edition, Jaypee Brothers Medical Publishers (P) Ltd., 2009.]

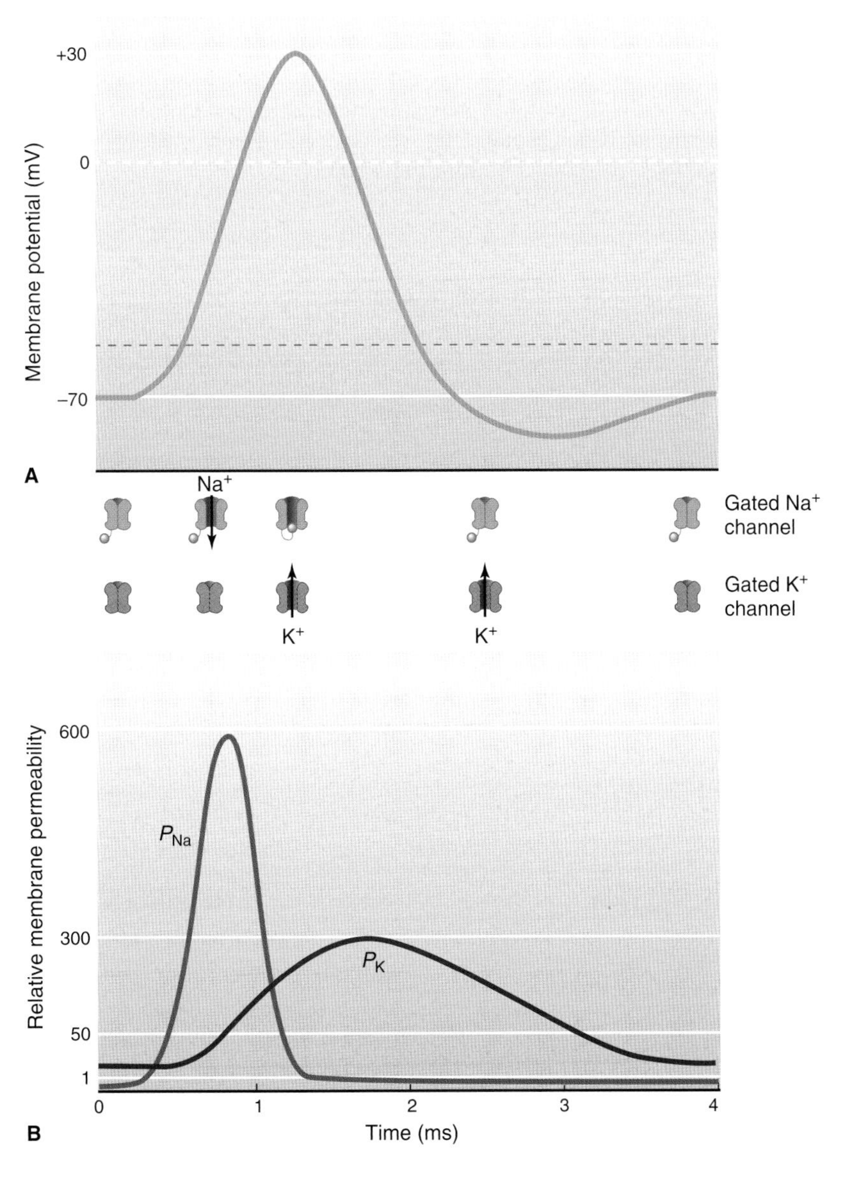

Figure 19-4. A–B. The Nerve Impulse. (**A**) The resting membrane is maintained at a membrane potential of –70 mV. The onset of an action potential opens the gated Na^+ channels (see Figure 19-3 and figures below upper graph), leading to the efflux of Na^+. Once the threshold level is reached at –55 mV (dashed red line), the depolarization of the membrane caused by Na^+ is rapid until a maximum membrane potential of +30–35 mV is reached. At this maximum voltage, the gated Na^+ channels close and the gated K^+ channels open (see Figure 19-3 and figures below upper graph) and an influx of K^+ leads to a drop in the membrane potential. The Na^+–K^+-ATPase pump (Figure 19-3) also contributes to restoration of the resting membrane potential (not shown). (**B**) Illustration of relative membrane permeability of Na^+ (P_{Na+}) and K^+ (P_{K+}) during propagation of the nerve impulse. [Reproduced with permission from Barrett KE, et al.: Ganong's Review of Medical Physiology, 23rd edition, McGraw-Hill, 2010.]

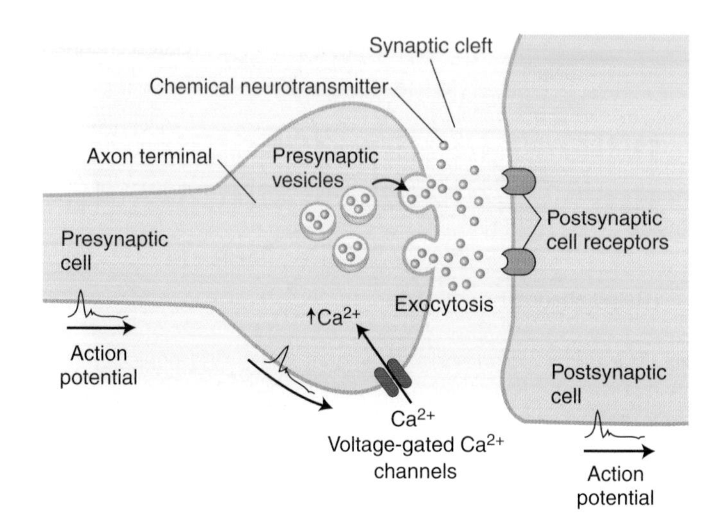

Figure 19-5. The Synaptic Cleft. After the nerve impulse reaches and depolarizes the presynaptic membrane, the influx of Ca^{2+} results in the fusion of presynaptic vesicles to the surface of the membrane. The neurotransmitter released into the cleft binds to the receptor site on the postsynaptic membrane, which will propagate the action potential onto the adjacent neuron. [Adapted with permission from Kibble JD and Halsey CR: The Big Picture: Medical Physiology, 1st edition, McGraw-Hill, 2009.]

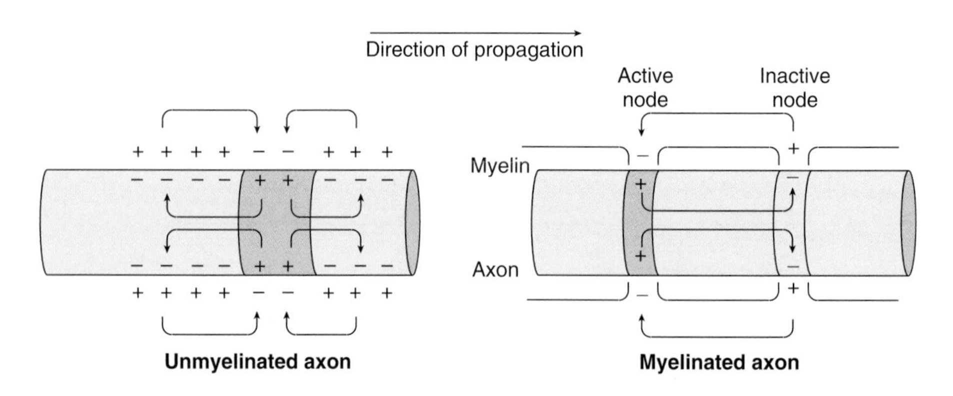

Figure 19-6. Effect of the Myelin Sheath on Nerve Impulse Propagation. Positive charges from the membrane ahead of and behind the action potential flow in the area of negative charge as the potential travels down the axon (see direction of propagation). Unmyelinated nerves propagate the nerve impulse by slow, continuous conduction. Nerves with a myelin sheath are able to conduct the nerve impulse much quicker via saltatory conduction from one node of Ranvier to the next. [Adapted with permission from Barrett KE, et al.: Ganong's Review of Medical Physiology, 23rd edition, McGraw-Hill, 2010.]

AUTONOMIC NERVOUS SYSTEM

The PNS is divided into the **somatic** nervous system and **ANS**. The somatic nervous system includes voluntary movements of skeletal muscles as well as touch sensation, vision, and hearing. The ANS coordinates and maintains a steady state among internal organs, such as regulation of cardiac muscle, smooth muscle, and glands. It is considered to be an involuntary system. The ANS is further subdivided into the **sympathetic** and **parasympathetic** nervous systems, the activity of which is integrated by the hypothalamus. An autonomic nerve pathway from the CNS to a visceral effector consists of two major neurons that synapse in a ganglion outside of the CNS: the **preganglionic** and **postganglionic** neurons. This differs from the somatic nervous system, wherein a single motor neuron travels from the CNS to the innervated tissue.

SYMPATHETIC NERVOUS SYSTEM

This system is dominant in a stress state or the "fight-or-flight" response. It serves to mobilize energy stores in times of need. Activation of the SNS results in tachycardia (increased heart rate), dilation of large blood vessels in skeletal muscles, vasoconstriction of skin and visceral blood vessels, dilation of bronchioles, mobilization of fatty acids from triglycerides in adipose tissue, and mobilization of liver glycogen to glucose. The latter changes are designed to supply energy to the body and to provide muscle tissue with direct energy via anaerobic glycolysis. In addition, digestive secretions and intestinal peristalsis decrease. The SNS's preganglionic neurons originate in the thoracolumbar spinal cord (from T1 to L2). The sympathetic ganglia are located in two chains just outside of the spinal column, in the paravertebral ganglia. The preganglionic, unlike the postganglionic, neurons are myelinated.

PARASYMPATHETIC NERVOUS SYSTEM

The role of the parasympathetic nervous system is to conserve and restore energy or, in opposition to the sympathetic system, to "rest and digest." It is the primary functioning division in relaxed situations to promote the normal functioning of several organ systems. The result in the body is decreased heart rate (bradycardia), bronchiolar constriction, pupil constriction, and increased peristalsis and secretions. The preganglionic neurons originate in several of the cranial nerve nuclei (CN 3, 7, 9, and 10) and sacral segments (S2, S3, and S4) of the spinal cord, giving rise to the term craniosacral division. In contrast to the SNS, the ganglia are located close to the target organs, and the postganglionic neurons are relatively short. The effects of both the sympathetic and parasympathetic systems on the body are summarized in Table 19-1.

TABLE 19-1. Summary of the Autonomic Nervous System

Organ	Sympathetic Response	Parasympathetic Response
Cardiac muscle	Increased rate	Decreased rate (to baseline)
Bronchioles	Dilation	Constriction
Pupils	Dilation	Constriction
Stomach/intestines	Decreased secretion/peristalsis	Increased secretion/peristalsis
Liver	Converts glycogen to glucose	None
Pancreas	Secretes glucagon	Secretes insulin
Sweat glands	Increase secretion	None
Skin/visceral blood vessels	Constrict	None
Skeletal muscle blood vessels	Dilate	None
Adrenal glands	Increase epinephrine and norepinephrine	None
Internal anal sphincter	Contracts	Relaxes
Urinary bladder	Relaxes	Contracts
Internal urethral sphincter	Contracts	Relaxes

Guillain–Barré Syndrome (GBS, Acute Idiopathic Polyneuropathy): GBS is an acute inflammatory disease leading to demyelination affecting the PNS. It is the most commonly acquired inflammatory neuropathy. GBS is a medical emergency, but a majority of patients fully recover over a period of months. As many as one-third of patients will exhibit residual weakness 3 years after the event, necessitating retraining, orthotic appliances, and rehabilitations. Fatalities occur in more than 2% of all patients.

GBS is typically characterized by a progressive symmetric muscle weakness, loss of sensation, paralysis, and decreased reflexes (hyporeflexia) starting in the lower extremities and progressing rapidly to the arms and face. As the symptoms progress, the respiratory muscles can also be affected necessitating mechanical ventilation. In addition, the ANS can be disrupted resulting in severe tachycardia, arrhythmias, and hypo- or hypertension. The diagnosis of GBS is a clinical one.

Although the precise mechanism of GBS is unknown, there is probably an immunologic basis to the disorder resulting in damage to the myelin sheath of the PNS. GBS sometimes follows an infective illness, inoculation, or surgical procedure. The most common antecedent infection is the bacteria *Campylobacter jejuni.* Other known pathogens include enteric viruses, cytomegalovirus, Epstein–Barr virus, and *Mycomplasma.* Lymphocytic infiltration and macrophage-mediated demyelination of peripheral nerves is the underlying pathology, and symptoms generally resolve with remyelination. Symptoms usually appear 1–3 weeks after the antecedent event.

NEUROTRANSMITTERS

Neurotransmitters are molecules (often amino acids/small peptides and/or their derivatives) which carry and often amplify a signal from one nerve to another nerve or to another cell type. Classic examples of neurotransmitters include glutamine, γ-aminobutyric acid (GABA), acetylcholine (Ach), dopamine, norepinephrine (NE), and serotonin among many others. Neurotransmitters can affect their target nerve or cell either via excitation or inhibition as is summarized in Table 19-2. Description of individual neurotransmitters and their actions follow.

DOPAMINE

Dopamine is a monoamine and is synthesized from the amino acid tyrosine in a series of enzymatic reactions that first produces L-DOPA, then dopamine, then NE, and finally epinephrine (Figure 19-7A). Dopamine is either excitatory or inhibitory depending upon which of its receptor subtypes is activated. In the brain, dopamine functions in the role of behavior, cognition, voluntary movement, sleep, mood, and learning. The action of dopamine is mostly in the substantia nigra, arcuate nucleus of the hypothalamus, and the ventral tegmental area of the midbrain. Dopamine is also very important in the reward

TABLE 19-2. Summary of Neurotransmitters

Type	Neurotransmitter	Location	Effect
Monoamine	**Dopamine**	Brain and ANS synapses	Generally excitatory
	Norepinephrine/epinephrine	CNS; sympathetic ANS synapses; nearly all other tissues	Excitatory or inhibitory
	Serotonin	CNS	Generally inhibitory
Other	**Acetylcholine**	CNS, neuromuscular junction, many ANS synapses	Excitatory or inhibitory
Amino Acids	**γ-Aminobutyric acid (GABA)**	Most neurons of CNS	Majority of postsynaptic inhibition of the brain
	Glycine	Spinal cord	Most postsynaptic inhibition of spinal cord
	Glutamate and aspartate	Brain and spinal cord	Excitatory
Neuropeptides	**Endorphins and enkephalins**	Widely distributed in CNS/PNS	Generally inhibitory
	Substance P	Spinal cord, brain, sensory neurons, and GI tract	Generally excitatory

ANS, autonomic nervous system; CNS, central nervous system; GI, gastrointestinal; PNS, peripheral nervous system.

system. Dopamine is inactivated by reuptake via the dopamine transporter followed by enzymatic breakdown via the combined effects of **monoamine oxidase (MAO)** and **catechol-O-methyltransferase (COMT)** (Figure 19-7B, upper).

Via the dopamine receptors (D_{1-5}), dopamine increases the actions of the direct pathway within the basal ganglia. Receptors D_1 (regulation of growth/development of neurons, behavioral response, and D_2-receptor signaling) and D_5 (believed to play a role in control of blood pressure) activate a G_s protein to activate adenyl cyclase and cyclic adenosine monophosphate (cAMP) signaling (Chapter 8). Receptors D_2 (probable role in regulation of muscle tone; association with schizophrenia), D_3 [believed to play a role in emotions (e.g., depression) and cognitive thought; also may play a role in drug addiction and schizophrenia], and D_4 (possible role in "thrill seeking" behavior as well as an association with schizophrenia) activate a G_i protein to inhibit cAMP production (Chapter 8). Normally, dopamine promotes smooth, coordinated muscle movement within the body. Insufficient dopamine biosynthesis in the dopaminergic neurons can cause **Parkinson's disease** (see below); receptor types D_3 and D_4 are possibly implicated. In the frontal lobes, dopamine controls the flow of information from other areas of the brain. Dopamine disorders in this region of the brain can cause a decline in neurocognitive functions, especially memory, attention, and problem-solving.

NE/EPINEPHRINE

NE is synthesized from tyrosine (Figure 19-7A) in the cytoplasm and packaged into vesicles of postganglionic fibers. It is released via exocytosis. NE is terminated by reuptake into the postganglionic nerve ending, diffusion from receptor sites, or metabolism by the enzymes **monoamine oxidase (MAO)**, and **catechol-O-methyltransferase (COMT)** (Figure 19-7B, lower). NE is utilized in sympathetic postganglionic neurons. **Epinephrine**, sometimes referred to as **adrenaline**, is the final product of the tyrosine pathway in adrenal medulla chromaffin cells that also produces dopamine and NE. Epinephrine has a wide variety of effects on several organ systems as well as metabolic functions, which are reviewed in Chapter 10. Epinephrine follows a similar degradation pathway as NE (Figure 19-7B, lower). Both NE and epinephrine act as hormones and neurotransmitters via adrenergic receptors.

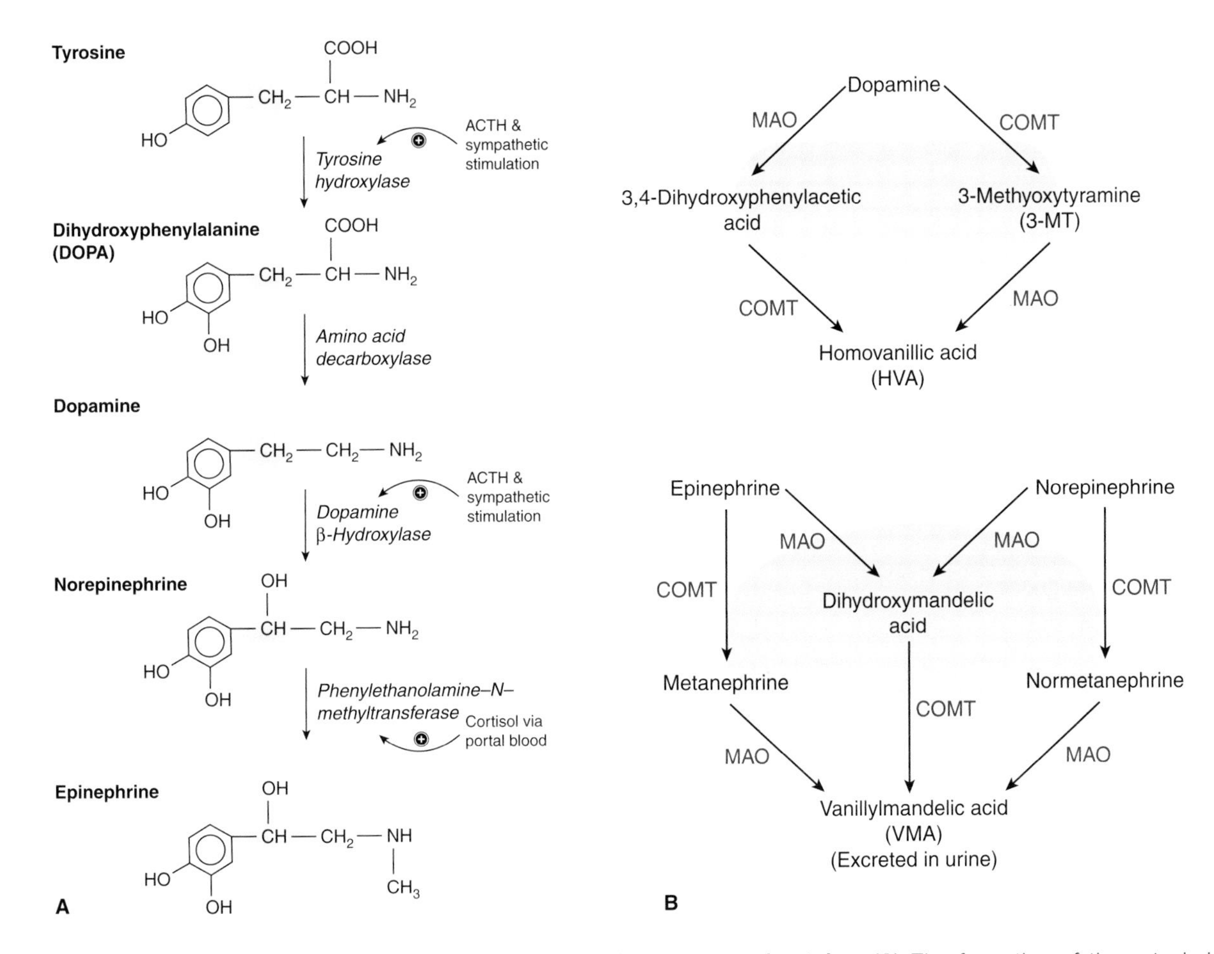

Figure 19-7. A–B. Synthesis and Degradation of Dopamine, Epinephrine, and Norepinephrine. (**A**) The formation of the catecholamines, dopamine, epinephrine, and norepinephrine is shown, each separated by a singular enzymatic step. (**B**) Breakdown of the catecholamines, including (upper) dopamine to homovanillic acid (HVA), and (lower) epinephrine and norepinephrine to vanillylmandelic acid (VMA). Both HVA and VMA are subsequently eliminated by the kidneys. COMT, catechol-O-methyltransferase; MAO, monoamine oxidase. [Adapted with permission from Kibble JD and Halsey CR: The Big Picture: Medical Physiology, 1st edition, McGraw-Hill, 2009.]

Adrenergic receptors are subdivided into α and β types as illustrated below. Both NE and epinephrine bind to all adrenergic receptor types but with differing affinity.

1. **α_1-receptors** are postsynaptic adrenoreceptors located in smooth muscle, eye, lung, blood vessels, gut, and the genitourinary system. Stimulation of these receptors activates a G_q protein resulting in increased phospholipase activity that releases diacylglycerol and inositol trisphosphate with the latter promoting increased calcium (Chapter 8). These signals lead to excitation, which is exhibited by dilation of the pupils (mydriasis), bronchodilation, constriction of blood vessels, increased strength of heart contraction (positive **inotropy**), and decreased heart rate (negative **chronotropy**). NE binds this receptor better than epinephrine.
2. **α_2-receptors** are located chiefly on the presynaptic nerve terminals. G_i protein activation inhibits adenyl cyclase activity/cAMP production and decreases the entry of Ca^{2+} into the neuronal terminal. This reduces the amount of released NE. In addition, it produces vasoconstriction and reduces sympathetic outflow in the CNS. Epinephrine binds these receptors better than NE.
3. **β_1-receptor** stimulation activates a G_s protein, which increases adenyl cyclase activity, ATP conversion to cAMP, and protein kinase phosphorylation of a variety of proteins. This action results in positive **chronotropy** (heart rate increase), **dromotropy** [conduction of the impulse through the heart's AV node (Chapter 16)], and **inotropy** (force of heart contraction). Whether NE and epinephrine bind equally to this receptor or epinephrine binding is preferred over NE, remains controversial.
4. **β_2-receptors** are mostly postsynaptic adrenoceptors located in smooth muscle and glands. They not only activate adenyl cyclase via a G_s protein but also inhibit the same enzyme by a G_i protein. The directly opposite effects are believed to allow different functions in specific cell and tissue locations. The β_2 stimulation relaxes smooth muscle resulting in vasodilation and bronchodilation as well as release of insulin and the induction of **gluconeogenesis.** Epinephrine binds β_2-receptors much more strongly than NE.
5. **β_3-receptors** are mostly found in fat/adipose tissue. They also activate adenyl cyclase via a G_s protein. The **β_3-receptors** induce the breakdown of fats/lipids (**lipolysis**) and also help to regulate heat production (**thermogenesis**), especially in skeletal muscle. NE binds this receptor better than epinephrine.

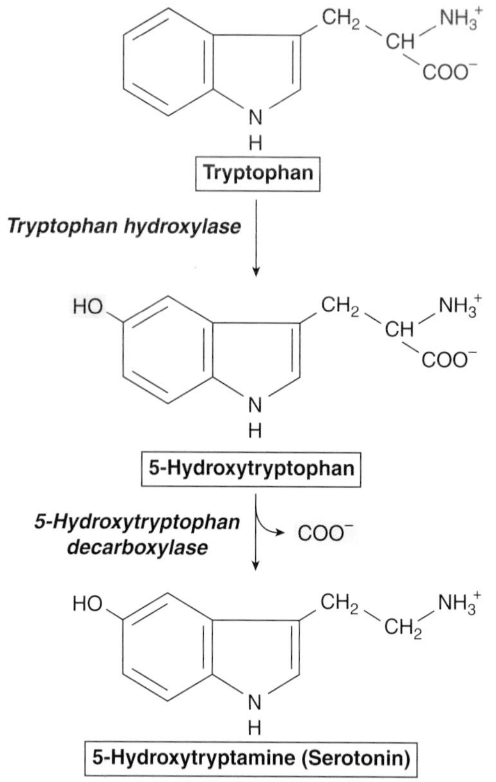

Tryptophan

Tryptophan hydroxylase

5-Hydroxytryptophan

5-Hydroxytryptophan decarboxylase → COO^-

5-Hydroxytryptamine (Serotonin)

Figure 19-8. Synthesis of Serotonin. The formation of serotonin (5-hydroxytrypatmine) from tryptophan is illustrated. [Adapted with permission from Murray RA, et al.: Harper's Illustrated Biochemistry, 28th edition, McGraw-Hill, 2009.]

SEROTONIN

Serotonin (5-HT) is also a monoamine neurotransmitter involved in sleep, eating, arousal, dreaming, and the regulation of mood, temperature, and pain transmission. Serotonin is synthesized from the amino acid tryptophan by a short metabolic pathway (Figure 19-8) consisting of two enzymes: **tryptophan hydroxylase** and **amino acid decarboxylase**. At least seven serotonin or 5-HT receptors have been characterized utilizing G_s, G_q, or G_i proteins that activate an intracellular second messenger cascade; one receptor functions as a Na^+–K^+ channel leading to membrane depolarization. Serotonin is terminated via uptake at the synapse on the presynaptic neuron. Various agents can inhibit 5-HT reuptake, including **MDMA (ecstasy)**, **amphetamine**, **cocaine**, **dextromethorphan**, **tricyclic antidepressants (TCAs)**, and **selective serotonin reuptake inhibitors (SSRIs)**.

ACETYLCHOLINE (Ach)

Ach is the neurotransmitter utilized in somatic efferent neurons, sympathetic preganglionic neurons (including the adrenal medulla and sweat glands), and the entire parasympathetic nervous system. Ach is synthesized in the nerve terminal by **choline acetyltransferase**, which catalyzes the reaction between **acetyl coenzyme A** and **choline** (Figure 19-9). In the

$$CH_3{-}C({=}O){-}O{-}CoA + HO{-}CH_2{-}CH_2{-}N^+{-}(CH_3)_3 \xrightarrow{\text{Choline acetyltransferase}} CH_3{-}C({=}O){-}O{-}CH_2{-}CH_2{-}N^+{-}(CH_3)_3$$

Acetyl-CoA, Choline → Acetylcholine

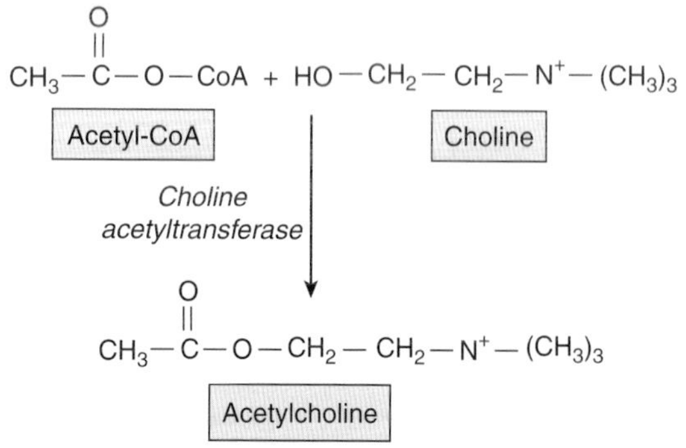

Figure 19-9. Synthesis of Acetylcholine. The formation of acetylcholine from acetyl-CoA and choline is illustrated. CoA, coenzyme A.

synapse, Ach is degraded by **acetylcholinesterase. Cholinergic** receptors refer to those that bind to Ach. They are further subdivided into **nicotinic** and **muscarinic**. Nicotinic receptors are located in the **somatic system** on the motor end plates of skeletal muscle cells, all **postganglionic neurons,** and the **adrenal medulla**. When Ach binds to a **nicotinic** receptor, the result is excitatory. **Muscarinic** receptors activate end-organ effector cells in bronchial smooth muscle, salivary glands, and the sinoatrial node. These effects are reviewed in Table 19-3.

REGULATION OF CATECHOLAMINES

The group of neurotransmitters called catecholamines include **dopamine, NE,** and **epinephrine** and are secreted by the nervous system in either an acute or chronic response to stress. Acutely, the neural signal is delivered from the hypothalamus and brain stem to the adrenal medulla via release of **Ach** from preganglionic neurons (Figure 19-10). Neural stimulation of the adrenal

TABLE 19-3. Summary of Effects of Acetylcholine on Nicotinic and Muscarinic Receptors

Cholinergic Receptors	Nicotinic	Muscarinic
Location	Autonomic ganglia	Glands—lacrimal, salivary, and gastric
	Sympathetic ganglia Parasympathetic ganglia	Smooth muscle—bronchial, GI, and blood vessels
	Skeletal muscle	Heart—AV node and sinoatrial node
Agonists	Ach	Ach
	Nicotine	Muscarine

Ach, acetylcholine; AV, atrioventricular; GI, gastrointestinal.

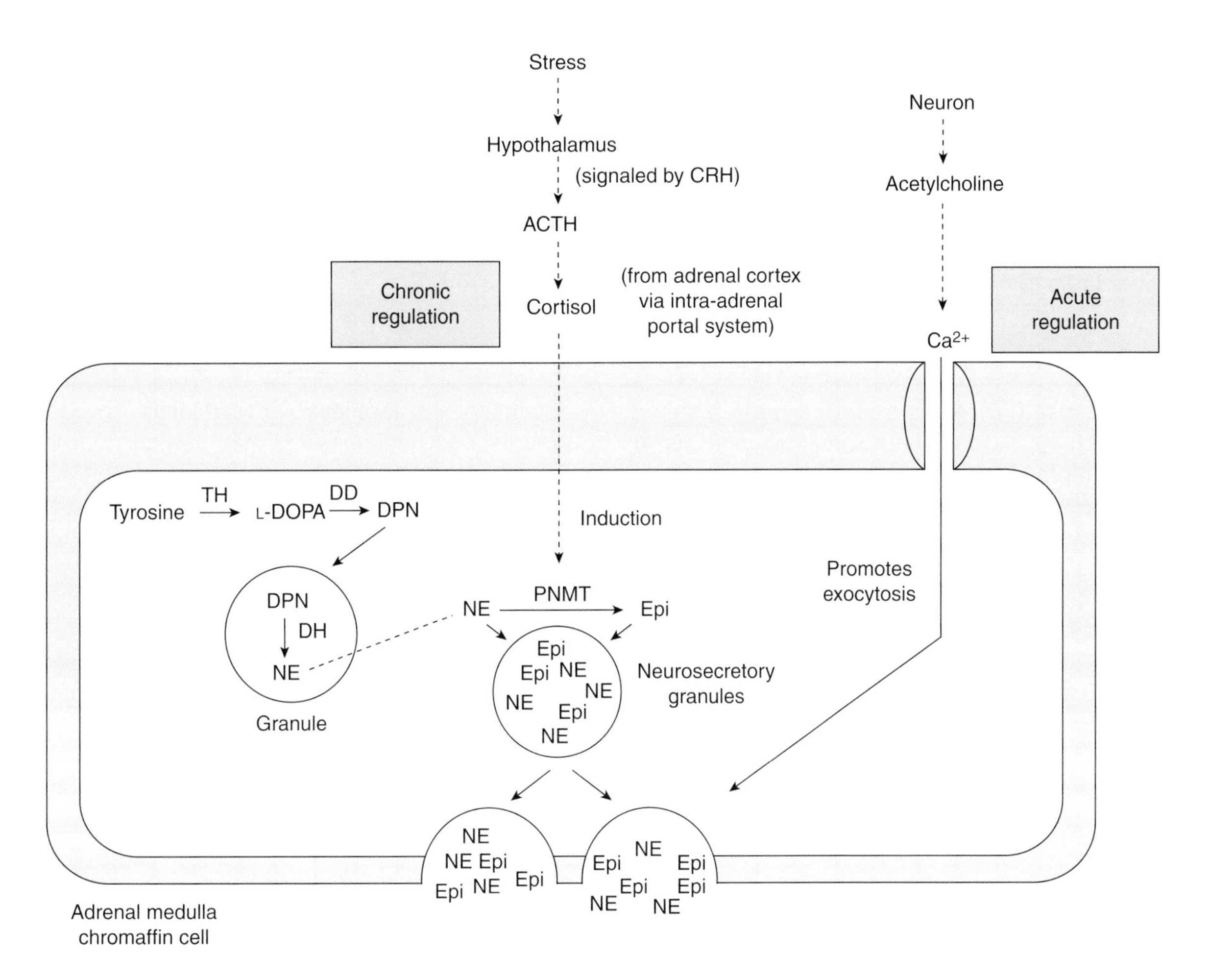

Figure 19-10. Regulation of the Release of Catecholamines and Synthesis of Norepinephrine/Epinephrine in the Adrenal Medulla Chromaffin Cell. Immediate release of norepinephrine/epinephrine results from neuronal signals via the neurotransmitter acetylcholine and a resulting influx of calcium ions. This calcium pulse leads to the release of the neurotransmitters via exocytosis. Longer term release in response to stress or other factors relies on hormones from the hypothalamus and the resulting induction of PNMT. See text for details. ACTH, adrenocorticotropic hormone; CRH, corticotropin-releasing hormone; DD, dopa decarboxylase; DH, dopamine hydroxylase; DPN, dopamine; Epi, epinephrine; NE, norepinephrine; PNMT, phenylethanolamine *N*-methyltransferase; TH, tyrosine hydroxylase. [Adapted with permission from Murray RA, et al.: Harper's Illustrated Biochemistry, 28th edition, McGraw-Hill, 2009.]

Myasthenia Gravis (MG)–Lambert–Easton Syndrome: MG is a disorder characterized by fluctuating weakness and easy fatigability of commonly used skeletal muscles. This disorder produces symptoms such as double vision (diplopia), eyelid drooping (ptosis), and problem swallowing (dysphagia). MG typically strikes not only young women in their 30s but also men in their 60–70s. MG can occur in association with tumors of the thymus as well as other autoimmune disorders (e.g., lupus, rheumatoid arthritis). MG is due to the autoimmune destruction or inactivation of postsynaptic Ach receptors at the neuromuscular junction, functionally reducing the number of Ach receptors. MG can be relapsing–remitting in nature, but usually is slowly progressive that can lead to severe respiratory weakness and complications. Infections, stress, surgery, menses, and pregnancy can often worsen symptoms temporarily. Treatment is with anticholinesterase drugs, steroids, plasma exchange (plasmapheresis), and surgical removal of the thymus (thymectomy). Anticholinesterase drugs (such as pyridostigmine) treat the symptoms of MG by increasing the amount of available Ach in the neuromuscular junction via inhibition of acetylcholinesterase.

In contrast, **Lambert–Eaton Syndrome** (or **Myasthenic syndrome**) features proximal muscle weakness caused by a presynaptic defect of neuromuscular transmission. Antibodies to calcium channels on the nerve terminal cause the dramatic decrease of Ach release at the motor endplate. Unlike MG, strength improves with sustained contraction. Treatment is usually provided by steroid therapy and plasmapheresis. Anticholinesterase therapy gives only a modest improvement.

medulla causes depolarization of the axonal membrane and thus promotes the influx of calcium. Calcium, in turn, stimulates the release of catecholamines from the **neurosecretory granules**. These storage granules fuse with the plasma membrane, leading to the exocytotic release of NE (minor component) and epinephrine (major component). Nerve stimulation also promotes a chronic response resulting in the synthesis of catecholamines. Prolonged stress and sympathetic nerve activity results in an induction of **phenylethanolamine *N*-methyltransferase (PNMT)** by glucocorticoids as a means of adapting to physiologic stress (Figure 19-10). **Cortisol** is transported from the adrenal cortex to the medulla via the intra-adrenal portal system. This offers an advantage because the cortisol concentration in this system is 100-fold enhanced relative to the arterial blood. This response parallels synthesis and release of insulin in response to Ach.

Parkinson's Disease (PD): PD is a brain disorder caused by the destruction of dopamine-producing neurons in the **substantia nigra**. This leads to a chemical imbalance between **dopamine** and **Ach** in the corpus striatum. The four cardinal symptoms of PD are tremor (often at rest), rigidity, slowness of movement (**bradykinesia**), and postural instability and reflect the loss of dopamine's impact on smooth muscle movement. As PD progresses, patients begin to have difficulty in walking (slow, shuffling walk) and muffled speech, depression, and difficulty in swallowing (dysphagia). Making an accurate diagnosis of PD can be difficult given that many other neurodegenerative disorders manifest with parkinsonian-type symptoms. No blood test exists that can confirm or refute a diagnosis of PD, but several are used in conjunction with magnetic resonance images (MRIs) more to rule out other possible disorders. The diagnosis is, therefore, based on history, a thorough physical examination and exclusion of other causes.

There is no cure for PD. Treatment is directed at reducing the chemical imbalance by blocking Ach and increasing dopamine with **levodopa** and **carbidopa**. Levodopa is the precursor to dopamine but has several neuromuscular/movement side effects (dyskinesias) itself. Carbidopa inhibits the enzyme that breaks down levodopa to dopamine allowing a reduction of the amount of levodopa that needs to be taken and, therefore, the number of side effects from levodopa. Although levodopa can dramatically improve symptoms, especially bradykinesia, it does not halt the progression of the disorder. Other drugs used for PD include **bromocriptine**, which mimics the role of dopamine in the brain. An antiviral drug, **amantadine**, also appears to reduce symptoms. Whereas levodopa primarily relieves bradykinesia, anticholinergic drugs help to control tremor and rigidity by decreasing the available amount of Ach. Drug-resistant patients can benefit from a therapy called **deep brain stimulation (DBS)**, which uses electrodes implanted into the brain and connected to a small electrical device called a pulse generator that can be externally programmed. DBS can reduce the need for levodopa and related drugs, which in turn decreases the likelihood of dyskinesias that are a common side effect of levodopa.

GLYCINE, GLUTAMATE, AND GABA

Glycine (Figure 19-11) is found widely distributed in the spinal cord, brainstem, and retina and serves an inhibitory function. The receptor is a gated channel (Chapter 8), composed of five protein subunits, which allows an influx of chloride ions. This influx acts as an **inhibitory postsynaptic potential** and lessens the ability for the occurrence of future postsynaptic or motorneuron action potentials. The poison **strychnine** is a competitive inhibitor of glycine receptor activity. **Caffeine** is another glycine receptor inhibitor.

Glutamic acid (Figure 19-12A) is the most abundant excitatory neurotransmitter in the CNS and the most common neurotransmitter in the brain. It is always excitatory via **glutaminergic receptor** opening of nonselective ion channels.

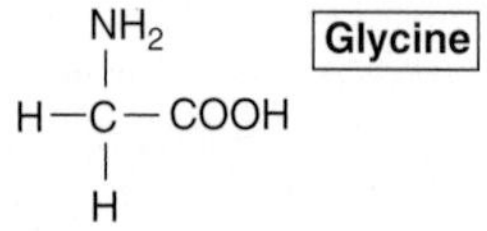

NH2
H—C—COOH
H

Glycine

Figure 19-11. Glycine Structure.

Glycine serves an important coagonist for the receptor. Glutamic acid stimulation is terminated by a membrane transport system that is used for reabsorbing glutamate and aspartate across the presynaptic membrane. Glutamic and aspartic acid re-enter the cell as sodium enters the cell and potassium exits. Thus, glutamic acid/aspartic acid entry is indirectly powered by the ATP-driven sodium pump. Both neurotransmitters work together to control many processes, including the brain's overall level of excitation. Many of the drugs of abuse affect either glutamic acid or GABA or both to exert tranquilizing or stimulating effects on the brain.

GABA (Figure 19-12A) is an amino acid neurotransmitter and is the most widespread inhibitory neurotransmitter in the brain. It is most highly concentrated in the substantia nigra, hypothalamus, and the hippocampus. GABA reduces the excitability of neurons by hyperpolarizing them. There are two types of GABA receptors (Figure 19-12B). $GABA_A$ receptor activation opens chloride channels, and $GABA_B$ receptors act via a second messenger to either open potassium channels or to close calcium channels. GABA is synthesized from glutamic acid and is inactivated by active transport into the astrocyte glial cells near the synapses.

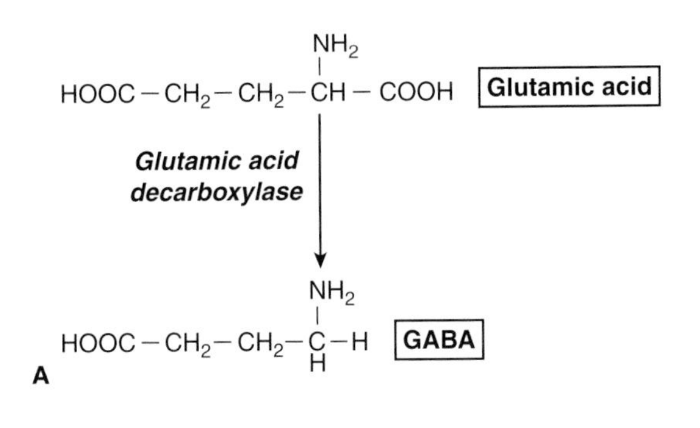

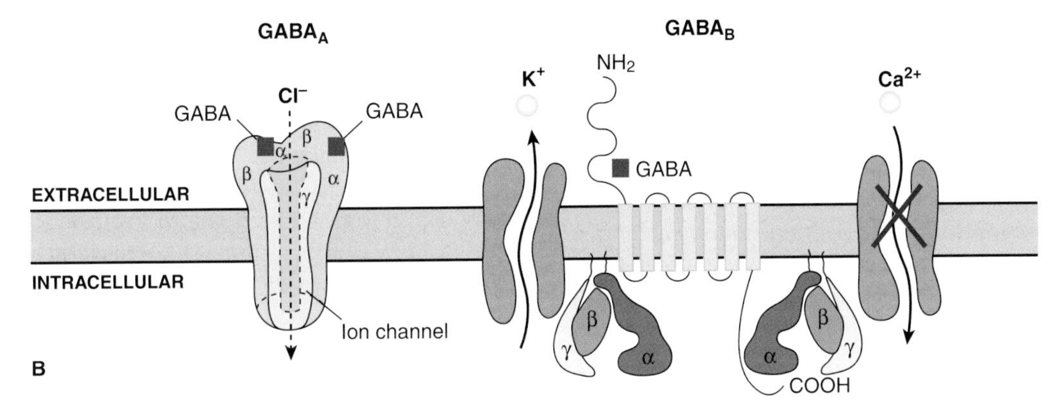

Figure 19-12. A. Structure of Glutamic Acid (Glutamate and Synthesis of GABA). The formation of GABA (γ-aminobutyric acid) from glutamic acid is illustrated. **B. GABA Receptors.** The functional aspects of $GABA_A$- and $GABA_B$-type receptors are illustrated. See text for additional details. GABA, γ-aminobutyric acid; AC, adenyl cyclase. [(Left channel) Adapted with permission from Katzung BG, et al.: Basic and Clinical Pharmacology, 11th edition, McGraw-Hill, 2009.] [(Remainder part) Adapted with permission from Barrett KE, et al.: Ganong's Review of Medical Physiology, 23rd edition, McGraw-Hill, 2010.]

Huntington's Chorea (HD): HD is an autosomal dominant, progressive degenerative disorder caused by a genetically programmed destruction of neurons in the basal ganglia as well as the cortex. It usually manifests itself in middle age with progressive deterioration until death 10–30 years later. As its name implies, HD is characterized as an abnormal involuntary movement disorder (**chorea**) caused by overactivity of **dopamine**. These movements are brief, irregular, nonrepeating contractions that appear to move from one muscle group to the next and it is often accompanied by writhing movements. HD can also cause dementia, clumsiness, inability to swallow (dysphagia), and personality changes. Biochemically, there is an underactivity of **GABA** and **Ach** as well as a relative overactivity of dopamine. HD has no cure, and the progression of the disease cannot be halted. Therefore, all therapies are directed purely at the symptoms, including the use of **phenothiazines** to block dopamine receptors and control abnormal movements and the drug **reserpine** to deplete monoamines.

NEUROPEPTIDES

Neuropeptides are chains of linked amino acids produced in the brain. They are made from larger polypeptides that are cleaved. The endogenous opiods (**endorphins**) or **enkephalins** produce analgesia and euphoria. Their receptors are catagorized as delta (δ), kappa (κ), mu (μ), or nociceptin-type receptors and function via a variety of G proteins. **Substance P** is found in the synaptic vesicle of unmyelinated C fibers, which enhance the

Eating Disorders: The satiety center of the brain is a collection of neurons in the lateral aspect of the ventromedial hypothalamus. Stimulation of this area increases appetite and destruction of this center causes suppression of food intake. The medial area of the ventromedial hypothalamus demonstrates a regulatory role over the aforementioned lateral aspect. Destruction of this area has been found to cause gross overeating (hyperphagia) with resultant obesity in animal models. The protein **leptin** targets the hypothalamus and causes decreased food intake by activating **melanocyte-stimulating hormone**, a satiety factor, while suppressing the hunger signal from AGRP (Agouti-related peptide). Additional hormones such as **glucagon** and **cholecystokinin** also inhibit appetite. The role of these centers in eating disorders such as bulimia and anorexia is incompletely understood. There is some evidence to suggest that neurotransmitter imbalances exist within the hypothalamus of patients who suffer from anorexia nervosa. The pathology behind such eating disorders is multifactorial, including genetic, hormonal abnormalities with decreased levels of serotonin, and psychosocial.

transmission of pain signals. The substance P receptor is a G_q protein, which activates phospholipase C and inositol triphosphate/Ca^{2+} production.

BIOCHEMISTRY OF VISION

Nutritional **vitamin A** (retinol esters) is converted in the retina to **11-*cis*-retinal**, an isomer of all-*trans*-retinal forms by **retinal isomerase**. 11-*cis*-retinal then covalently attaches to a lysine residue on the visual protein **opsin**. The resulting **rhodopsin** molecule becomes a G-protein-coupled receptor with the "ligand" for activation being light. A deficiency of vitamin A in the retina will, therefore, decrease the sensitivity of rods, resulting in night blindness.

Normally, cyclic guanosine monophosphate (cGMP) keeps inward Na^+–Ca^{2+} channels open to maintain membrane depolarization (Figure 19-13). When light stikes the retina, photoexcited rhodopsin binds to a multisubunit membrane protein called **transducin**, which in turn activates a **cGMP phosphodiesterase** by causing dissociation of inhibitory subunits. Stimulation of the cGMP phosphodiesterase leads to the exchange of a bound guanosine diphosphate for guanosine triphosphate (GTP). This process releases the α subunit of transducin with bound GTP, reminiscent of the mechanism by which cAMP activates protein kinase A. The phosphodiesterase catalyzes the destruction of **cGMP (cGMP → GMP)** near the membrane of the rod cell. When cGMP concentrations fall, inward Na^+–Ca^{2+} channels close, thereby lowering both intracellular Na^+ and Ca^{2+} concentrations. The fall in Na^+ elicits hyperpolarization causing the rod cell to release less glutamate neurotransmitter. The reduced amount of this inhibitory neurotransmitter generates an electrical impulse to the occipital lobe of the brain that triggers the perception of light.

Because of its dynamic nature and the need to rapidly perceive ever-changing intensities of light, built-in "cessation" and "recovery" facets of the cycle exist. When the light activation signal is terminated ("cessation"), the active rhodopsin molecule is phosphorylated by **rhodopsin kinase. Arrestin**, a soluble protein, then binds the phosphorylated rhodopsin to inactivate it. Binding to arrestin also slows rhodopsin's ability to propagate another light signal. The "recovery" phase, during which the cycle returns to a state that is preparatory for the next flash of light, involves replenishment of the cGMP. Production of cGMP is triggered by the **low intracellular Ca^{2+}** that was induced by the decline in the concentration of cGMP (see above). The low Ca^{2+} leads to cGMP production from GTP via **guanyl cyclase**, which is stimulated by low intracellular calcium via a "**recoverin**" mediator protein. Further breakdown of cGMP is accomplished through direct inhibition of phosphodiesterase.

ANESTHESIA

General **anesthesia** is an altered physiologic state characterized by a temporary loss of consciousness, relief of pain (analgesia), amnesia, and variable levels of muscle relaxation.

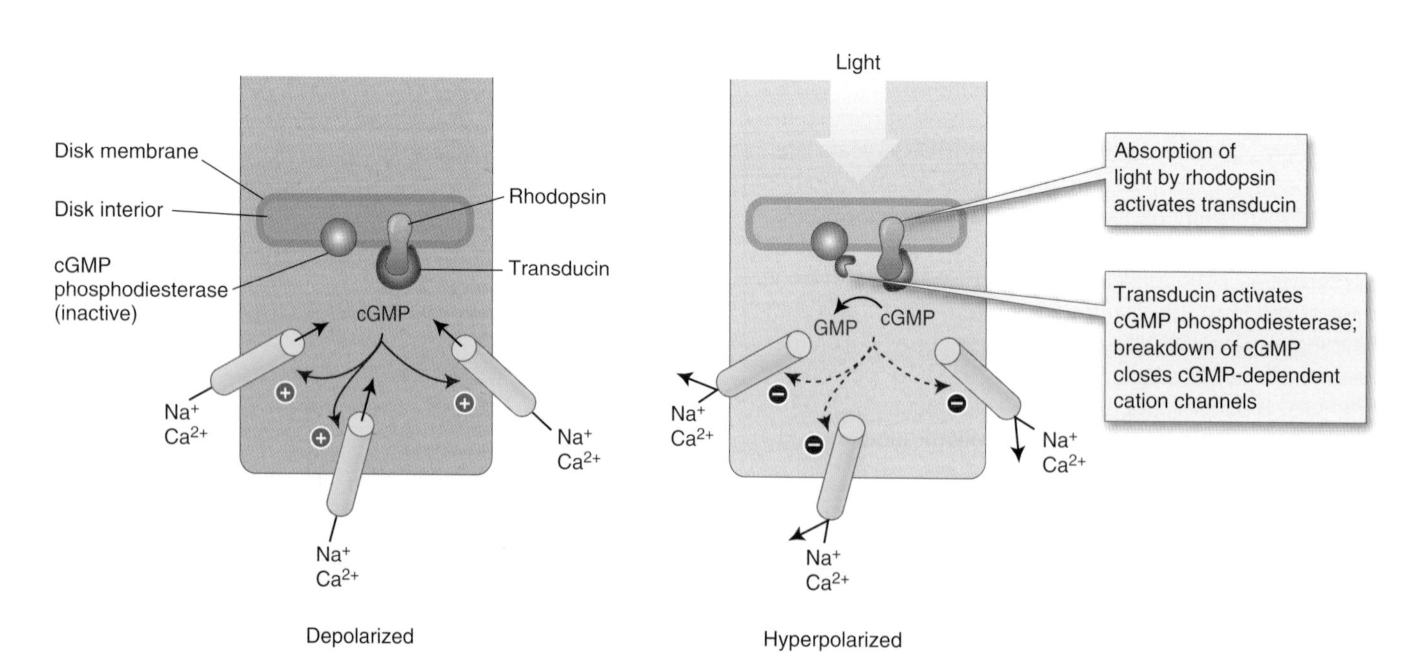

Figure 19-13. Biochemistry of Vision. Photoreceptors of the eye are normally depolarized because of the influence of cGMP on Na^+–Ca^{2+} channels. When light strikes the receptor, transducin activates a cGMP phosphodiesterase, which decreases cGMP concentration, leading to closure of the channels. In this hyperpolarized environment, the neurotransmitter glutamate is released to provide an impulse to allow the perception of light. cGMP, cyclic guanosine monophosphate. [Reproduced with permission from Kibble JD and Halsey CR: The Big Picture: Medical Physiology, 1st edition, McGraw-Hill, 2009.]

Despite the loss of awareness, the vital physiologic functions are unaffected. Broadly speaking, general anesthetics can be categorized as either inhalation (breathed-in) or intravenous (by vein).

Inhalation anesthetics (e.g., ether, chloroform, halothane, and sevoflurane) do not appear to have a single site of action. However, known areas of the brain that are affected include the reticular activating system, cerebral cortex (unconsciousness and amnesia), cuneate nucleus, the olfactory cortex, and the hippocampus. All inhalation anesthetics work the same at the molecular level, but their exact mechanism of action remains only partially understood. The hydrophobic sites on a neuron's membrane bind to the lipophilic anesthetic, expanding the phospholipid bilayer and altering membrane function. Another postulated theory is that anesthetics decrease membrane conductance. Many anesthetics also enhance GABA inhibition of the CNS. Action at the brainstem level depresses the withdrawal from pain (most notably at the dorsal horn interneurons). In addition, inhalation anesthetics depress excitatory transmission in the spinal cord. Despite not understanding the exact mechanism, the end result is the inhibition of synaptic function.

Intravenous anesthetics (e.g., barbiturates, benzodiazepines, propofol, and opiates) depress the reticular activating system, located in the brainstem controlling consciousness. It works by depressing transmission of excitatory neurotransmitters such as **Ach** and enhancing inhibitory neurotransmitters such as **GABA**.

REVIEW QUESTIONS

1. What are the central nervous system and peripheral nervous system, and what roles do they perform in the human body?
2. What are the basic parts and the function(s) of unmyelinated and myelinated neurons?
3. What are the main steps in nerve impulse conduction, including the role of membrane-bound channels?
4. What are the basic roles of the sympathetic and parasympathetic nervous systems?
5. What are the major neurotransmitters and what are their major functions?
6. How is the production and release of catecholamine neurotransmitters regulated acutely and chronically?
7. What is the biochemical process behind vision?

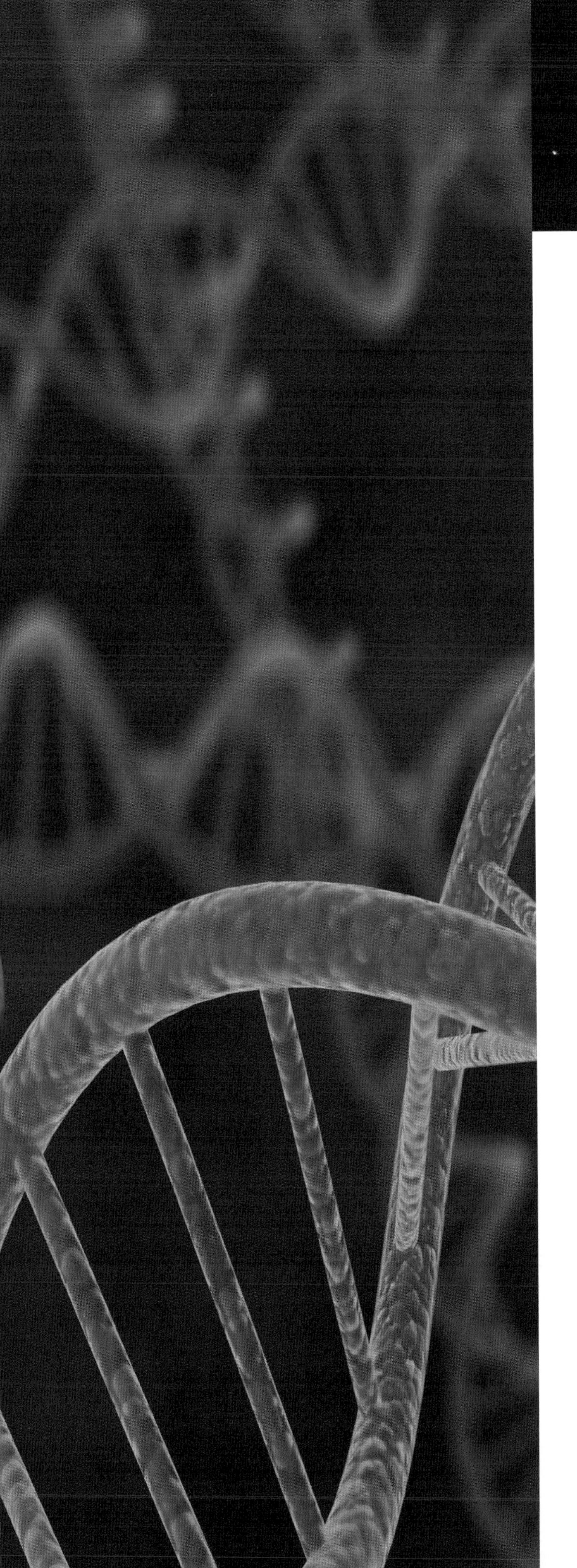

CHAPTER 20

THE REPRODUCTIVE SYSTEM

Editor: Catrina Bubier, MD
Women's Health Care Associates, Englewood, Colorado, USA

OVERVIEW

The female and male reproductive systems develop selectively as a result of specific hormonal signals (sex-determining region on the Y chromosome, anti-Mullerian hormone, testosterone/5α-dihydrotestosterone, and estrogens) that lead to further development or regression of embryological structures. Additional hormones (gonadotropin-releasing hormone, follicle-stimulating hormone, luteinizing hormone, progesterone, and human chorionic gonadotropin) influence further development and subsequent adult functions, including the menstrual cycle, fertilization and pregnancy, lactation, and oogenesis/spermatogenesis. These hormones work via signaling proteins, including several variations of G proteins, to selectively activate or inhibit these developmental and functional events.

BASIC ANATOMY AND DEVELOPMENT

The reproductive system is derived from the intermediate mesoderm and includes the reproductive organs of both males and females derived from the **Wolffian ducts** (male), **Mullerian ducts** (female), and the **gonads** (male and female), including the testes and ovaries. The influence of hormones and biochemical signals in the formation and activity of the reproductive system is vastly important and, when these signals go awry, disease ensues.

During the initial stages of gestation, all humans begin with both Wolffian and Mullerian ducts and the development of male and female embryos is indistinguishable. At about gestational day 56, further growth and development into male or female sexual organs is dependent on the effects of the **sex-determining region of the Y chromosome (sry)**. In the genetic male, the sry product binds to deoxyribonucleic acid (DNA) and distorts it dramatically out of shape. This alters the properties of the DNA and likely alters the expression of a number of genes. One of these genes produces **anti-Mullerian hormone (AMH)**, a dimeric, glycoprotein hormone also known as **Mullerian inhibiting factor (MIF)**. AMH is produced by Sertoli cells in the testes and signals, via its receptor, a member of the **transforming growth factor (TGF)-β I and II receptor** family. Binding of AMH to TGF-β type II receptor allows it to bind to the type I receptor. The type I receptor is then able to phosphorylate serine and/or threonine amino acids to activate transcription factors in the nucleus, which regulate gene expression. The presence of AMH leads to full development of the Wolffian ducts and male structures; only a few remnants of the Mullerian ducts survive in males.

In males, the **Leydig cells** also appear and **testosterone** synthesis begins, leading to male sexual characteristics, including testis formation. Testosterone's effect on the seminiferous tubules (see below) is also critical for modulating signaling and gene expression and, therefore, male development (Figure 20-1).

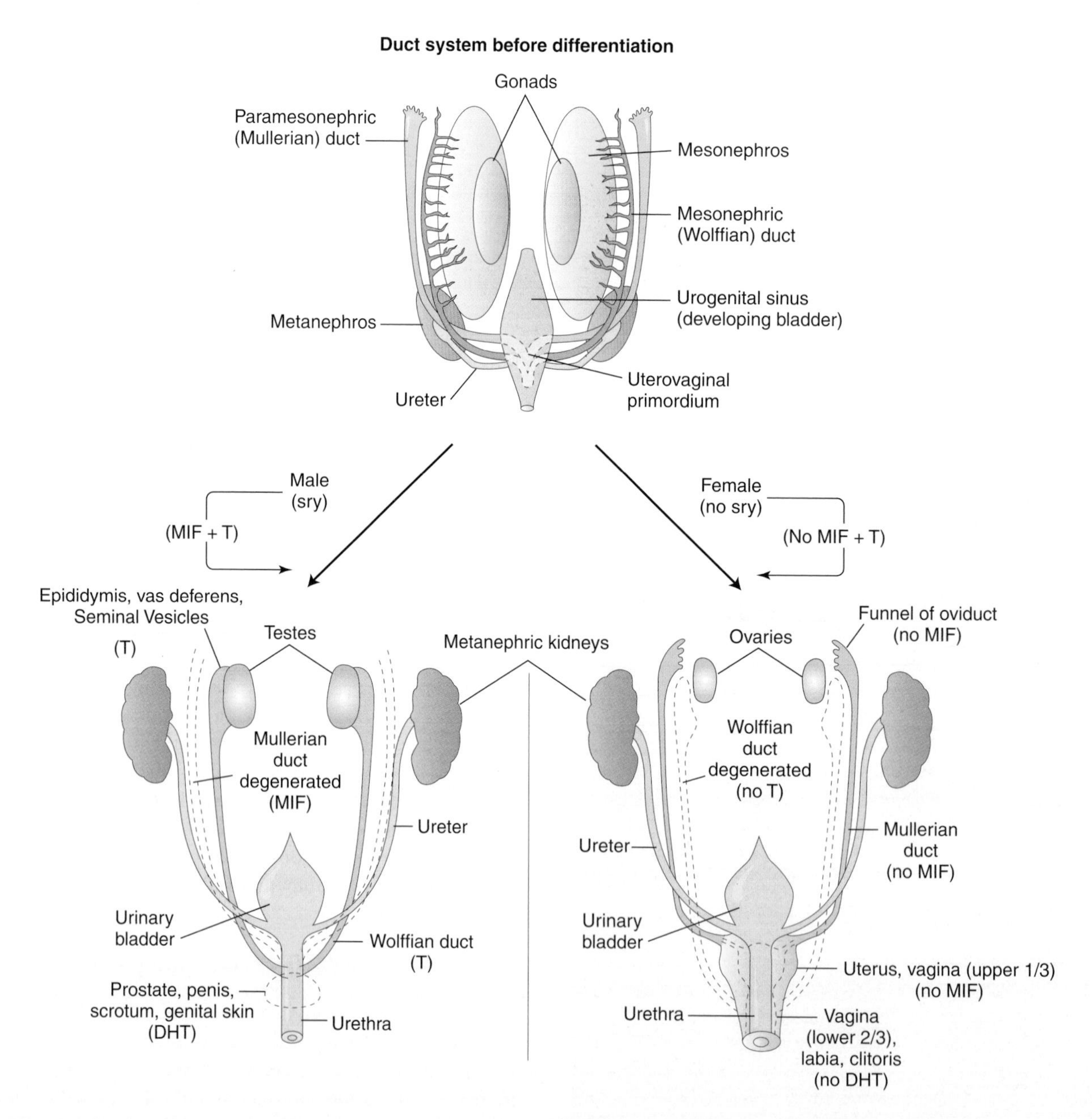

Figure 20-1. Summary of Hormonal Involvement in Sexual Differentiation. DHT, dihydrotestosterone; MIF, Mullerian inhibiting factor; sry, sex-determining region of Y chromosome; T, testosterone; 5αR, 5α-reductase-2. [Adapted with permission from Kibble JD and Halsey CR: The Big Picture: Medical Physiology, 1st edition, McGraw-Hill, 2009.]

In the absence of sry, embryos spontaneously develop into phenotypic females. With no sry product, AMH and testosterone are not produced and the Wolffian ducts regress with only a few remnants remaining. The mechanism of AMH suppression of Mullerian duct formation is unknown. Further development of the Mullerian ducts creates the female sexual organs and structures. In both sexes, the Wolffian duct is responsible for development of the bladder trigone.

Androgen Insensitivity Syndrome (AIS): AIS, also previously known as "**testicular feminization**," results from a mutation on the X chromosome, which yields a **defective androgen receptor**. Because this is an autosomal recessive condition, all patients are genetic males and, therefore, produce sry, AMH, and testosterone/5α-dihydrotestosterone (DHT) normally. However, the defective androgen receptors do not allow the normal androgenic functions to be expressed, leading to failure of Wolffian duct development and complete feminization of the external genitalia and blind ended vagina. These patients are, therefore, phenotypically female with a male karyotype. A similar medical condition, **17α-hydroxylase deficiency**, results in the inability to form any androgens (Chapter 3) and, therefore, estrogens as well. The condition, also referred to as "**sexual infantilism**," leads to underdevelopment of the gonads (hypogonadism).

FEMALE REPRODUCTIVE SYSTEM

As noted above, the development and function of the female reproductive system is under the influence of several hormone signals. Following development, primary hormones in the female system are **follicle-stimulating hormone (FSH)**, **luteinizing hormone (LH)**, and the hormone that regulates their expression—**gonadotropin-releasing hormone (GnRH)**. These hormones are also essential in the male reproductive system (see below). GnRH production and release begins at puberty and remains active during the male and female reproductive years.

GnRH

GnRH is a small peptide produced in the hypothalamus by specialized nerve cells; as such, GnRH is called a neurohormone, a class of hormones that include thyrotropin-releasing hormone, oxytocin (see below), antidiuretic hormone (Chapter 18), and corticotropin-releasing hormone. Release of GnRH results in activation of a specific **GnRH receptor**, located in the gonadotropes of the pituitary gland. This receptor is a membrane-bound G-protein-coupled stimulator of phospholipase C, which results in calcium release and protein kinase C activation via conversion of plasma membrane phosphatidylinositol into inositol triphosphate and diacylglycerol (Chapter 8). These signals result in production and release of FSH and LH, as will be described below. Regulation of this important signal is

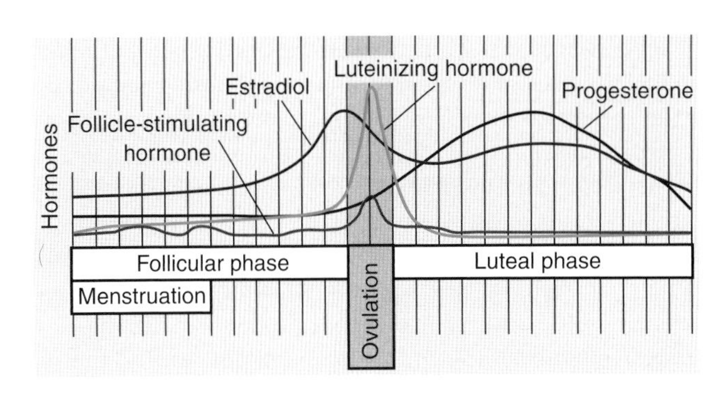

Figure 20-2. Variable Expression of Reproductive Hormones During the Menstrual Cycle (Follicle-Stimulating Hormone, Luteinizing Hormone, Estradiol, and Progesterone). Periods of the menstrual cycle, including menstruation, follicular phase, ovulation, and luteal phase are indicated. [Adapted with permission from Kibble JD and Halsey CR: The Big Picture: Medical Physiology, 1st edition, McGraw-Hill, 2009.]

multifold. GnRH is degraded within minutes so it is constantly produced in pulses, and the size and frequency of these pulses is important in signaling. These GnRH pulses are constant in males but vary in females, depending on the menstrual cycle. Interestingly, the frequency of the GnRH pulses result in different expression of FSH (low frequency) and LH (high frequency). As a result FSH and LH are variably expressed during the female menstrual period (Figure 20-2). Levels of testosterone, estrogen, and prolactin (increased during pregnancy) as well as increased concentration of FSH and LH create a negative feedback loop, which can decrease GnRH pulses.

FSH

FSH is a dimeric glycoprotein hormone produced and secreted by the anterior pituitary gland. FSH has an α- and β-subunit, with the α-subunit being identical in amino acid sequence to the α-subunit of LH, thyroid-stimulating hormone (TSH), and human chorionic gonadotropin (hCG) (see below). The β-subunit binds to and activates the FSH receptor.

The FSH receptor is believed to exist in two conformational states (Chapter 1)—one active and the other inactive. Binding of FSH to its receptor locks the receptor's conformation into the active form, which activates a G_s-coupled protein causing release of an α_s-subunit. This α_s-subunit then activates adenyl cyclase to increase the cyclic adenosine monophosphate (cAMP) signal and protein kinase A activity (Chapter 8). Phosphorylated signaling proteins bind to cAMP response elements on DNA, leading to activation of particular genes. FSH can also activate the family of **extracellular signal-related kinases**. This family of receptor-activated kinase, also known as mitogen-activated protein (MAP) kinases, respond to a variety of signal molecules, including FSH, which regulate meiosis and mitosis of selected cells. Although many mechanisms may be in play, the activation of tyrosine kinases via the **ras** GTPase protein is considered important in the subsequent activation of transcription factors.

FSH performs many functions in the development of female and male and is also essential for the proper timing and phasing of the menstrual cycle. FSH, as its name implies, stimulates granulosa cells in the follicles in the ovary to initiate egg growth and development (**oogenesis**). FSH specifically blocks programmed death or **atresia** of early developing egg follicles. As one egg follicle becomes dominant, reaching about 10 μm in diameter, it begins to secrete **estradiol**, which negatively feedbacks on GnRH and, therefore, FSH production. The block of atresia in all follicles except the dominant one leads to the survival of only one egg.

The activity of FSH is regulated by several means. FSH has a biological half-life of only about 3–4 hrs. Unless GnRH pulses continue to promote FSH synthesis and secretion, its biochemical effects will begin to diminish. Increased estrogen will also decrease FSH synthesis via its effect on GnRH. **Activin** (Figure 20-3, right), a protein heterodimer (composed of two different subunits) or homodimer (composed of two identical subunits), serves to increase FSH production and secretion via a mechanism extremely similar to the FSH receptor, leading to protein kinase A activity and upregulation of gene expression. However, the same small antral follicles that FSH stimulates also produce the protein **inhibin** (Figure 20-3, left), a protein heterodimer with one subunit identical to the activin subunit and one different. Inhibin decreases FSH synthesis and secretion and, therefore, works in a similar manner to the dominant follicle's estrogen production to decrease FSH and increase nondominant follicle atresia.

As the luteal phase ends, a small rise in FSH can be detected that helps to signal the start of the next menstrual cycle. In addition, estrogen and progesterone levels fall and decrease their negative effect on GnRH, allowing FSH levels to rise with an initial peak at day 3 (Figure 20-2). In a directly opposite signaling mechanism, estrogen increases GnRH and, therefore, FSH secretion at the time of ovulation (see below). Finally, as a woman approaches menopause, the number of initial follicles recruited to egg production decreases and, as a result, the level of inhibin is also lessened. This leads to an overall rise in serum FSH levels that is one sign of perimenopause/menopause.

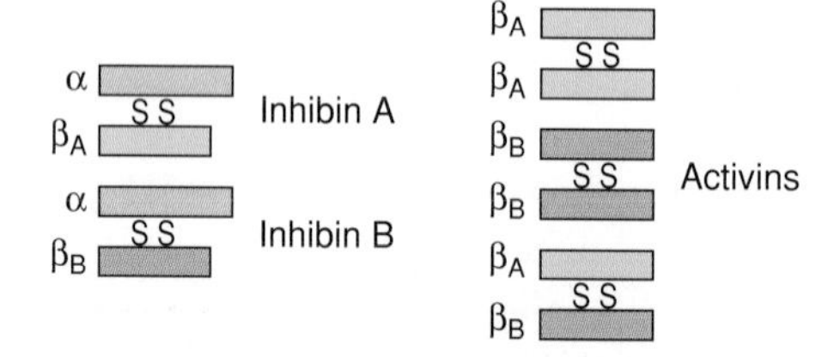

Figure 20-3. Inhibins and Activins. Illustration of structures of inhibins (left), heterodimers of three separate proteins (represented as three different colors), and activins (right), homodimers or heterodimers of the green and purple proteins. [Adapted with permission from Barrett KE, et al.: Ganong's Review of Medical Physiology, 23rd edition, McGraw-Hill, 2010.]

LH

LH is a dimeric glycoprotein, composed of an α- and β-subunit, produced by the anterior pituitary gland. The α-subunit is identical in amino acid sequence to the α-subunit of FSH, TSH, and hCG (see below). The β-subunit is similar but not identical to that of hCG and is responsible for binding to the LH receptor. LH is not produced until puberty and then is generated in response to GnRH signals, regulated according to reproductive need. Like FSH, LH levels also rise in females after menopause.

Like the FSH receptor, the LH receptor is believed to exist in two conformational states (Chapter 1)—one active and the other inactive. Binding of LH to its receptor locks the receptor's conformation into the active form, which activates a G_s-coupled protein causing release of an $α_s$-subunit. This $α_s$-subunit then activates adenyl cyclase to produce the cAMP signal and increased protein kinase A activity (Chapter 8). Phosphorylated signaling proteins bind to cAMP response elements on DNA, leading to activation of particular genes. LH has a biological half-life of about 20 min, so gene regulation via its receptor may stop unless additional LH or other signaling molecules continue the nuclear signal.

In females, LH provides the hormone signal for release of an egg from the ovary. Increasing estrogen, which results from FSH stimulation of ovarian **granulosa cells**, increases the presence of LH receptors. In addition, estrogen activates the pituitary to increase LH secretion. Both of these functions act to produce an **LH "surge"** over a 24–48 hr period (Figure 20-2). Unlike FSH, activin, inhibin, and the sex hormones do not affect LH production at the DNA level. The LH surge produces ovulation and also signals the corpus luteum, a temporary structure created from the remnant ovarian follicle after egg release. The activated corpus luteum produces the hormone **progesterone**, which, in turn, signals the uterine wall to prepare for egg implantation if fertilization should occur. The presence of LH also starts and, in the case of fertilization and implantation of the egg, maintains the **luteal phase** (Figure 20-2) for 8 weeks. LH promotes androgen and estrogen production via stimulation of thecal cells in the ovary. The LH surge also initiates the continuation of meiosis in the oocyte, the completion of which occurs after the sperm enters the ovum. Finally, LH is important to luteinize the granulosa cells to produce progesterone for maintenance of the luteal phase.

Ovulation Prediction by LH Measurement: The 24–48 hr LH surge that leads to ovulation and, therefore, predicts an optimal time for conception has led to the development of "ovulation predictor kits" to assist couples wanting children. These kits have anti-LH antibodies, which bind to LH excreted in the urine. Binding of the LH and a positive color change via a chemical reaction indicates when the level is above the normal 1–20IU/L—levels indicating maximum fertility times. Digital measurements of LH level are also available.

ESTROGENS

Estrogen is the inclusive term for a group of steroids that primarily impact the female reproductive system. Estrogens are mainly synthesized when FSH and LH stimulate thecal and granulosa cells in the ovarian follicles, the corpus luteum after release of a dominant egg, and, in the case of pregnancy, the placenta. The liver, adrenal glands, adipose tissue, and breast tissue can also produce steroidal estrogen. The highest level of estrogen occurs immediately prior to ovulation.

As a class of steroid hormones, all estrogens function by crossing the cell membrane and activating estrogen receptors (ERs). Unbound ERs are mainly found in the cytosol but, upon binding of an estrogen molecule, the receptors move into the nucleus, form dimers, and bind to specific sequences of the DNA molecule known as **hormone responsive elements**. The bound ER–DNA activates particular proteins that start transcription of DNA to result in synthesis of specific proteins. ERs can also be found in the nucleus where they serve to regulate the transcription of other proteins and may also associate with plasma membrane G proteins, which activate tyrosine kinases. Three major forms of estrogen molecules occur in humans (Figure 20-4 and Chapter 3).

Estrone (E_1), produced by the enzyme **aromatase** from androstenedione in adipose cells, is found predominately in menopausal women as well as in men. **17β-Estradiol (E_2)** is formed from testosterone by aromatase and is the main estrogen in nonpregnant, fertile females. Aromatase in the granulosa cells of the ovaries is activated by FSH. Estradiol can also be produced in smaller amounts by aromatase conversion in the liver and fat cells (Chapter 3). **Estriol (E_3)** is the most abundant estrogen during pregnancy and is formed in the placenta and fetal adrenal glands and liver as illustrated in Figure 20-5. Other nonsteroidal molecules, found in nature and artificially produced, can exhibit estrogen activity.

Estrogens have a wide-ranging set of functions in both men and women and only the major functions will be covered here. In females, estrogens are essential for the initial growth of the columnar epithelial cells of the endometrium. From a developmental standpoint, estrogen helps to initiate and propagate the development of secondary sexual characteristics to include breast formation, growth of the uterus and vagina, growth of pubic and underarm hair, onset of adult body changes (e.g., hip widening and deceleration of teenage growth), and the onset and regulation of menstruation (see below). Estrogen also influences a number of biochemical and metabolic processes to include bone reabsorption; increased clotting of blood via modulation of clotting factors II, VII, IX, and X as well as plasminogen and antithrombin III (Chapter 14); increases in high-density lipoprotein and triglycerides with decreases in LDL and fat deposits (Chapter 7); and alterations in fluid balance and hormone levels. The role of estrogens in the female menstrual cycle will be examined in fuller detail below.

PROGESTERONE

Progesterone is a steroid hormone from the hormone class called **progestogens**, important in menstruation, pregnancy, and embryo development. Progesterone is made from pregnenolone, derived from cholesterol, in the corpus luteum after ovulation. If fertilization occurs, the placenta becomes the primary source by week 8, utilizing circulating cholesterol from maternal blood. Progesterone is also produced in the adrenal glands. Levels of progesterone are low in the follicular/proliferative phase and higher in the luteal/secretory phase (Figure 20-2 and below) because of the contribution of the corpus luteum.

Progesterone, like other steroid hormones, passes through the plasma membrane and binds to an intracellular progesterone receptor. Also, like other steroid hormones, binding of progesterone leads to receptor dimerization, entrance into the nucleus, and activation of selected genes via DNA binding. Transcriptional activation by the progesterone receptor is normally inhibited by the receptor's carboxy-terminal end. Binding of the steroid hormone estrogen to the progesterone receptor causes a conformation change that releases this inhibition and is, therefore, essential for progesterone/progesterone receptor functions. Estrogen also acts to increase the total number of progesterone receptors, thereby amplifying progesterone effects. However, two different isoforms of the progesterone receptor exist: type A and type B, the latter of which has an amino-terminal, transcriptional activation domain. The two receptor types appear to have similar but unique activities that are still being investigated. Interestingly, progesterone also binds very strongly to the receptor for aldosterone, competitively inhibiting the mineralcorticoid's function. Increased levels of progesterone, therefore, block aldosterone's effects (Chapter 18) and lead to increased loss of sodium and water by the kidneys. A sudden decrease in progesterone subsequently causes sodium and water retention.

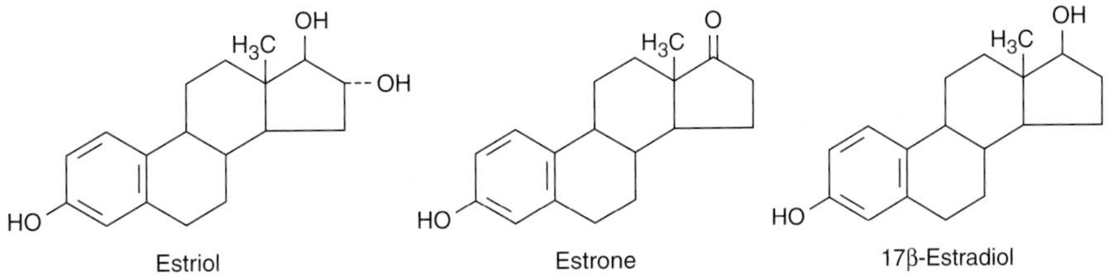

Estriol Estrone 17β-Estradiol

Figure 20-4. The Three Major Classes of Estrogen Steroid Hormones in Humans. [Adapted with permission from Katzung BG, et al.: Basic and Clinical Pharmacology, 11th edition, McGraw-Hill, 2009.]

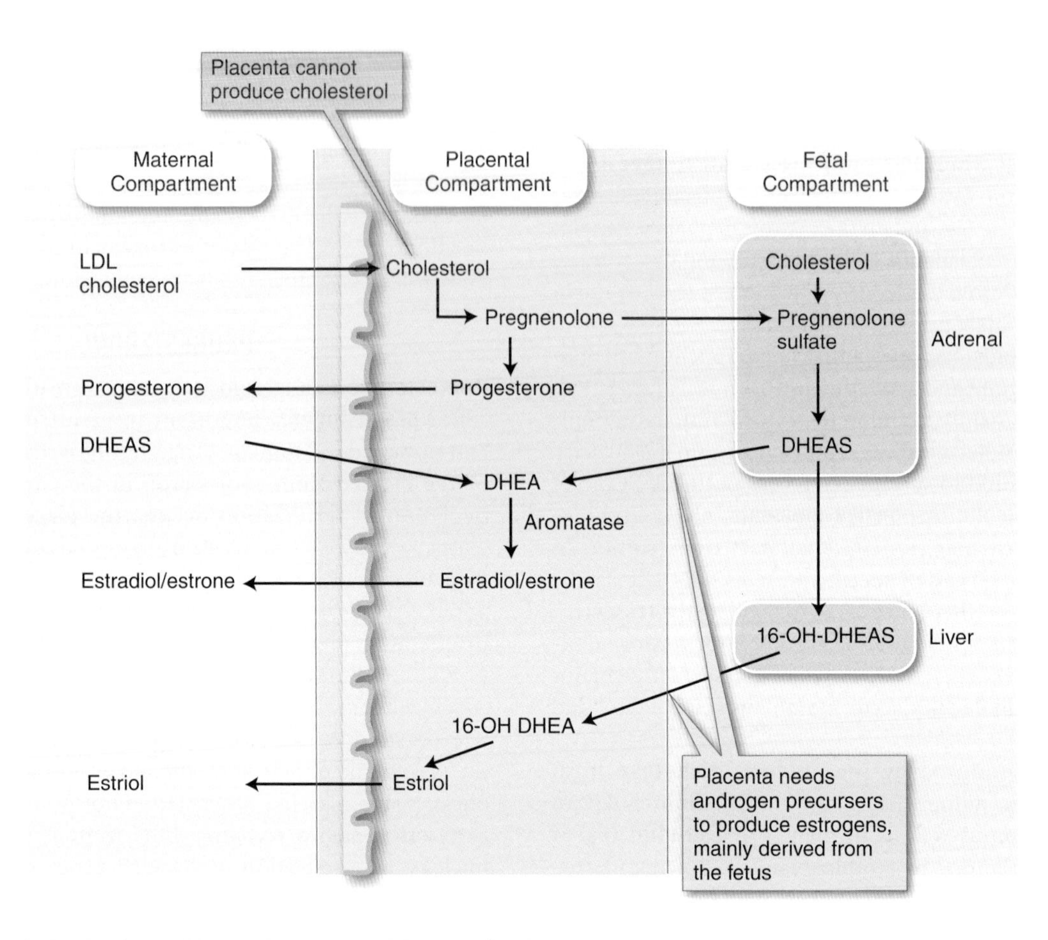

Figure 20-5. Steroid Production by Mother, Placenta, and Fetus During Pregnancy. Overview of production of steroids from cholesterol during pregnancy, showing various sources for the production of steroids in the placenta and fetal compartment (e.g., fetal adrenal glands and liver) as well as transport of maternal, placental, and fetal molecules. See text for further discussion. DHEA, dehydroepiandrosterone; DHEAS, dehydroepiandrosterone sulfate; LDL, low-density lipoprotein. [Reproduced with permission from Kibble JD and Halsey CR: The Big Picture: Medical Physiology, 1st edition, McGraw-Hill, 2009.]

Progesterone Antagonists: The actions of estrogen and progesterone on progesterone receptor activity have led to the development of a class of medications called **selective progesterone receptor modulators (SPRMs)**. SPRMs affect the two isoforms of the progesterone receptor differently, depending on binding strength and activation/inhibition activity. SPRMs may, therefore, also offer the option of selective receptor-type effects. Examples include treatment of **uterine leiomyoma**, **endometriosis**, and **hormone-responsive breast cancers**. One SPRM, **RU-486**, is already in selected use as a medical abortant.

Progesterone's main role is to prepare the endometrial lining of the uterus for possible implantation of a fertilized egg and maintenance of the uterine lining to support the fetus during pregnancy. Progesterone's influence on the endometrium is different from estrogen, though, as it turns the endometrium from a proliferative to a secretory phase, including the marked growth of spiral arteries (Figure 20-6). In this role, progesterone is often used to support in vitro fertilization treatments and/or irregular uterine bleeding. During pregnancy, progesterone also inhibits lactation (see below) and smooth muscle contraction; decreased progesterone levels are one potential trigger for labor and also initiate milk production. As a result, recent literature suggests giving weekly intramuscular injections of progesterone to help avoid preterm labor in at-risk mothers. If pregnancy does not occur, there is a regression of the corpus luteum and, therefore, decreased production of progesterone, which initiates menstruation. Apart from the reproductive system, progesterone also impacts nerve function via binding to a serine/threonine kinase called **glycogen synthase kinase 3**. Progesterone's protective effect on myelin sheaths (Chapter 19) of nerves has raised its potential to be used in patients suffering from multiple sclerosis. Progesterone may also serve as a protective factor against endometrial cancer by opposing the effects of estrogen.

hCG

hCG is a heterodimeric (two different subunits) glycoprotein hormone produced by newly fertilized embryos and, subsequently by the placenta. hCG has an α-subunit that is identical to the α-subunit of FSH, LH, and TSH and a unique β-subunit that binds to the **luteinizing hormone/choriogonadotropin receptor (LHCGR).** LHCGR is a G_s-coupled protein receptor found predominately in the ovary and testes, which spans the membrane. LHCGR also binds with LH and, as a result, hCG and LH share several functions. Just like the LH receptor (see above), upon binding of hCG or LH, the receptor undergoes a conformational change to its active state, leading to cAMP production, activation of protein kinase A, and increases in expression of selected genes. Like LH, hCG also maintains the corpus luteum and causes increasing production of progesterone (see above).

Pregnancy Testing and hCG: Because hCG is produced by the newly fertilized egg, the ability to measure its levels in blood and even urine has led to the advent of easily performed **pregnancy tests**. Most of these tests utilize a monoclonal antibody to the β-subunit of hCG, thereby avoiding interference from the similar FSH and LH. Urine tests do not provide actual numerical levels of hCG and are, therefore, referred to as a qualitative pregnancy test. Blood serum tests utilize detection methods that can determine actual **βhCG** levels; therefore, this test is referred to as a quantitative test. Quantitative measurement of hCG is also important in the diagnosis and treatment of a variety of ovarian-, testicular-, and/or placental-derived cancers (e.g., germinomas, choriocarcinoma, hydatidiform mole, and teratoma/dermoid cyst) as well as ectopic pregnancy and miscarriage. Levels of hCG are also one of four components of a clinical **quad screen test**, used in early pregnancy to determine the risk of Down syndrome, Edward's syndrome, Patau's syndrome, and neural tube defects in the fetus.

THE MENSTRUAL CYCLE

The female menstrual cycle is controlled by a series of hormones that initiate and regulate growth of the endometrial lining, development of an egg, release and possible implantation of the egg, and, if pregnancy does not ensue, complete purging of the uterine lining to allow a repeat of the same process (Figure 20-6).

MENSTRUATION (DAYS 1–4)

The menstrual cycle's start is considered at the onset of menstruation (day 1), the evacuation of the endometrial lining in the nonpregnant female, which may rely, in part, on a small surge of FSH and then decline in the progesterone level (see above). Subsequently, increasing estrogen stops loss of the endometrial lining and begins and initiates new thickening of the endometrium on approximately day 4.

FOLLICULAR/PROLIFERATIVE PHASE (DAYS 5–13)

Day 5 is the approximate start of the **follicular/proliferative phase** because of the growth and development of the uterine endometrial lining and of the ovarian follicles. Specifically, after days 1–7 of the menstrual cycle, activin as well as low levels of estrogen (estradiol) and progesterone allow GnRH pulses to secrete increasing amounts of FSH, which, then, initiates development of several ovarian follicles. As a **dominate follicle** emerges, estradiol is increasingly synthesized, which modulates FSH via inhibition of GnRH. Inhibin, produced by the same follicles, also decreases FSH activity. Estradiol also inhibits synthesis and secretion of LH until approximately day 12. At this point, a critical level of estradiol is reached and promotion of LH production starts, leading to the LH surge. The mechanism behind the directly opposite effects of estradiol on LH is not completely understood but one possibility is the differential activation of either of the two ERs. **ERα** is known to oppose LH secretion, whereas **ERβ** increases LH levels and inhibits α-receptor activity. How preferential activation of the α- or β-receptors occurs is unknown, but differential expression, different estradiol binding, or some other mechanism may explain this observation. Regardless of the mechanism, the increasing LH level leads to the breakdown of the follicular wall and release of the egg, known as **ovulation** on approximately day 14.

THE LUTEAL/SECRETORY PHASE (DAYS 15–28)

The **luteal/secretory phase** is named so because of the activity centers on the corpus luteum, the remnant of the ovarian follicle, and its secretion of progesterone. If the egg is not fertilized within about 24 hrs, it dies and is reabsorbed by the body. Meanwhile, the remaining ovarian follicle converts into a **corpus luteum** and begins to produce progesterone in increasing amounts (Figure 20-6). Increased levels of FSH and LH also help to develop the corpus luteum. Progesterone acts on the endometrium for possible fertilized egg implanation. If the egg does not implant, hCG is not secreted and the corpus luteum disintegrates, causing markedly lower levels of progesterone, which initiates the shedding of the endometrial lining as menstruation. Estrogen and progesterone synthesized by the growing corpus luteum inhibit GnRH and, therefore, FSH and LH production, leading to a rapid fall from ovulation levels. The fall of FSH and LH levels adds to the atrophy of the corpus luteum and a subsequent decrease in progesterone synthesis, again leading to menstruation. In the case of fertilization and pregnancy (see below), the new embryo begins to produce hCG, which acts like LH in preserving the corpus luteum during the first 6–8 weeks of pregnancy until placental synthesis can take over.

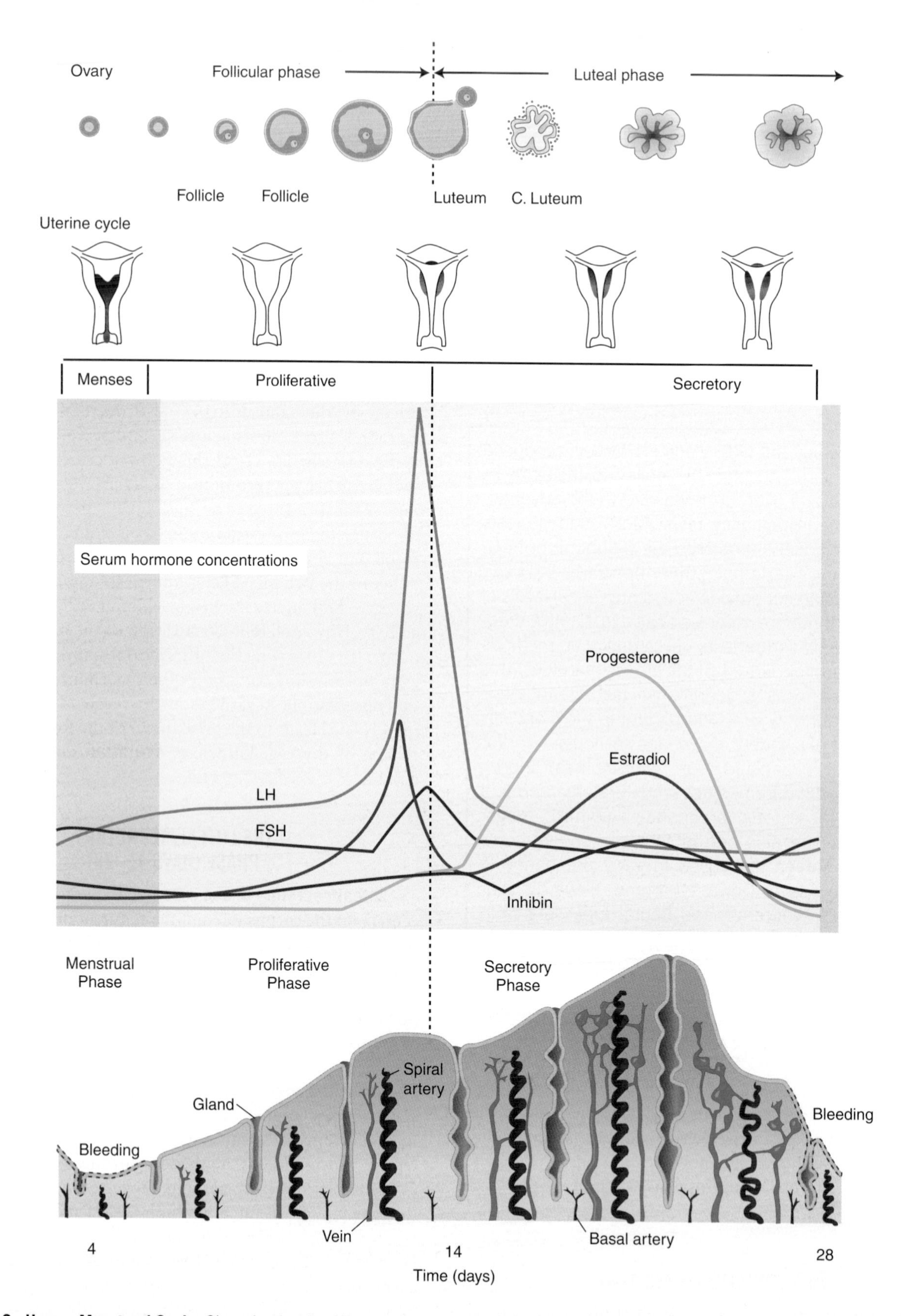

Figure 20-6. Human Menstrual Cycle. Changing levels of hormones and their influence on follicle/egg development; corpus luteum functions; and degeneration, growth, and thickening of the endometrium during the two phases of the cycle. The figure assumes an average 28-day cycle and starts at the onset of menstruation. [Adapted with permission from Kibble JD and Halsey CR: The Big Picture: Medical Physiology, 1st edition, McGraw-Hill, 2009.]

FERTILIZATION

The process of fertilization, the entrance of a sperm cell or, in rare cases, sperm cells into an egg, relies on several fundamental biochemical processes, including cell motility and plasma membrane fusion, involving changing fluidity and structure of the sperm and egg membranes (Chapter 8, Figure 20-7).

The sperm tail is a classic flagellum (Chapter 12), producing whipping or lashing movements via adenosine triphosphate hydrolysis that propels the sperm through the female reproductive organs toward the egg. As discussed in Chapters 1 and 12, this process relies on the structural proteins tubulin–microtubules and their interaction with the motor protein, dynein. Once a sperm reaches the egg, it binds to the outer layer and then attempts to penetrate the hard **zona pellucida**. Upon reaching the **corona radiata** layer of the egg, the **acrosome**, a cap-like structure at the tip of each sperm, releases the enzyme **hyaluronidase**, which breaks down hyaluronic acid contained in the presenting egg layer. Next, the enzyme **acrosin** digests the zona pellucida and promotes changes in the structure and fluidity of the membranes of the sperm tip and the egg to cause them to fuse. Following this initial fusion process, a protein molecule on the sperm acrosome binds to a directly interlocking protein, **ZP3**, on the egg. This binding is believed to initiate a series of reactions, including the **acrosome reaction**, a release of enzymes by the acrosome, which complete the membrane fusion; the **cortical reaction** (see below); and activation of the egg to undergo a second meiotic division. Fusion of the fertilizing sperm and egg results in haploid nuclei.

The cortical reaction is considered the analogous process to the acrosome reaction. The binding of sperm to egg results in a release of calcium ions, which promote fusion of cortical granule membranes with the egg's plasma membrane. Fusion expresses factors that cause the release of an externally bound, membrane protein on the outer **vitelline layer** of the egg. Release of this protein forces the vitelline layer away from the zona pellucida, moving any other sperm away permanently and inhibiting any other sperm from fertilizing the egg. The polysaccharide molecule **hyalin** is also released from the cortical granules to create a layer around the egg that also impedes any further fertilization.

Oral Contraceptives: The influences of estrogen and progesterone on the menstrual cycle, ovulation, and implantation have been utilized to allow the development of various forms of contraception. Estrogen and progestin, the generic name for any synthetic progesterone compound, molecules inhibit GnRH pulses, which then decreases FSH and LH secretion, inhibiting development of ovarian follicles and ovulation. The estrogen component in the oral contraceptive pill primarily suppresses FSH secretion and the progesterone component suppresses LH secretion, thereby preventing follicular recruitment and ovulation, respectively. The combination of estrogen and progesterone is also used to treat menstrual disorders by helping to decrease the endometrial lining and regulate or even prevent menses. The controlled effect on the menstrual cycle is often used for female patients with painful, irregular, heavy, or temporarily absent menstrual periods.

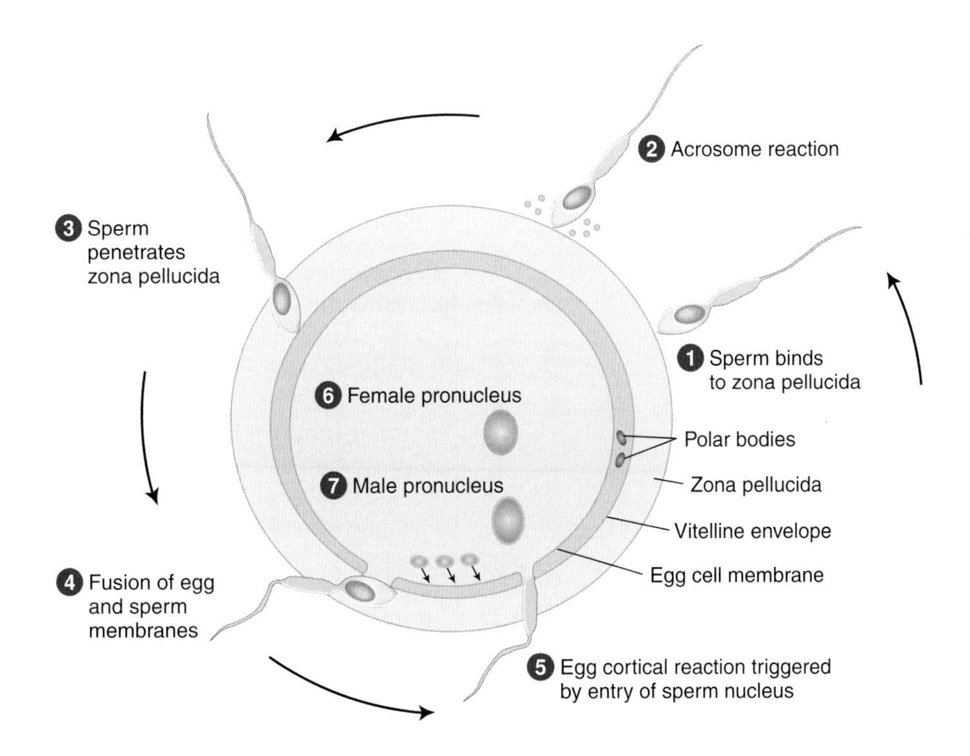

Figure 20-7. Mechanism of Fertilization. Steps of fertilization are indicated, including the acrosomal reaction (see text). [Adapted with permission from Kibble JD and Halsey CR: The Big Picture: Medical Physiology, 1st edition, McGraw-Hill, 2009.]

BREAST DEVELOPMENT AND LACTATION

Breast tissue development or **thelarche** is a part of secondary sexual changes that occur at puberty because of increased estradiol, progesterone, and prolactin (see below). Besides other changes in body shape, the activity of this estrogen via its receptor leads to a variable increase of the adipose cells, supporting ligaments and glands that make up the female breast tissue.

Lactation requires the action of **oxytocin** and **human placental lactogen** hormones, which, along with prolactin, are discussed immediately below.

OXYTOCIN

Oxytocin is a nine-amino-acid peptide, including a disulfide bridge formed by two cysteine residues, which is released by the posterior pituitary gland. Vasopressin, active in the regulation of total body water, differs from oxytocin by only two amino acids and is also released from the posterior pituitary gland (Figure 20-8, Chapter 18).

Binding of oxytocin to its receptor activates a G_q protein, which activates phospholipase C, leading to calcium release and protein kinase C function. Interestingly, the oxytocin receptor requires both Mg^{2+} and cholesterol, which are thought to allosterically activate the receptor. Oxytocin is involved in the ejection of milk from the breasts' milk glands as well as contraction of the uterus during birth. Both functions result from the activation of smooth muscle. Uterine contraction results in a notable increase in the number of oxytocin receptors in the smooth muscle layer and release of estrogen and prostaglandin $F_{2\alpha}$. This prostaglandin, which binds to the corpus luteum and helps to lead to its degradation, activates the G_s-coupled pathway, leading to cAMP production via adenyl cyclase. Because of oxytocin's marked similarity to vasopressin, it can also increase sodium and water excretion in the kidneys.

PROLACTIN

Prolactin is a peptide hormone that is produced in the anterior pituitary gland via the action of a specific transcription factor called **Pit-1**. Increased production of prolactin is due to estrogen, which stimulates Pit-1 and increases the number of prolactin-producing cells. Prolactin can also be produced in breast tissue as well as nerve and immune system cells. Prolactin binds its receptor on the mammillary glands and ovaries where dimerization of the receptor activates a Janus 2 tyrosine kinase. This leads to subsequent tyrosine phosphorylation, dimerization, and activation of a class of proteins known as **signal transducers and activators of transcription (STATs)**. These activated STAT dimers travel to the nucleus and augment gene expression of certain proteins. Prolactin binding to its receptor also activates **MAP** kinases and Src kinase, involved in the regulation of cell growth and development. The activity of both of these additional kinases also leads to regulated gene expression. Prolactin receptors are also found on various other organs and tissues throughout the body.

The major function of prolactin is to stimulate the production of milk in the mammary glands, whose enlargement during pregnancy is promoted by estrogen and progesterone. As noted above, the drop in inhibitory progesterone levels at the end of a pregnancy is the final signal for the beginning of milk production. The presence of prolactin after pregnancy also stops the menstrual cycle through inhibition of the pulsatile release of GnRH and, therefore, FSH and LH. Prolactin continues this effect throughout the breastfeeding stage. The act of suckling stimulates the hypothalamus to continue promoting pituitary gland secretion of prolactin. Prolactin also functions in the lungs to promote the production of **surfactants** in fetal lungs (Chapter 17). In nerve cells, prolactin promotes the formation of **myelin sheaths** (Chapter 19). Finally, prolactin is thought to counteract the sexual arousal effects of dopamine and provide postcoital sexual gratification. As a result, the **refractory period** after sex may be because of the function of prolactin. In a similar mechanism, overproduction of prolactin may play a role in erectile disorder and impotence. The effects of hormones on the production of milk or lactation are listed in Table 20-1.

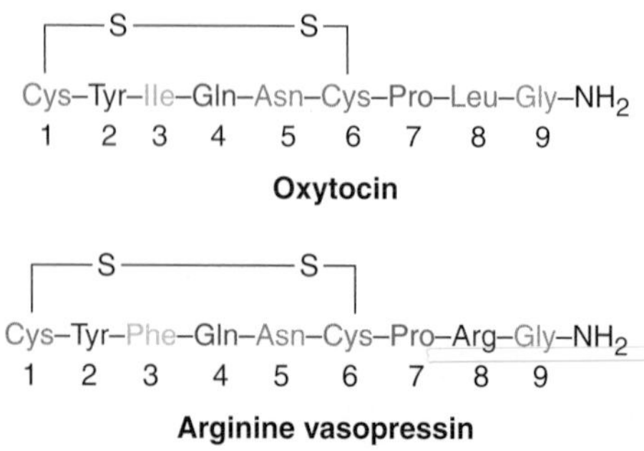

S———S

Cys–Tyr–Ile–Gln–Asn–Cys–Pro–Leu–Gly–NH_2

1 2 3 4 5 6 7 8 9

Oxytocin

S———S

Cys–Tyr–Phe–Gln–Asn–Cys–Pro–Arg–Gly–NH_2

1 2 3 4 5 6 7 8 9

Arginine vasopressin

Figure 20-8. Comparison of Oxytocin and Antidiuretic Hormone (ADH, Vasopressin) Structures. The nine-amino-acid peptides oxytocin and ADH (vasopressin) with disulfide bridge as shown. Differing amino acids are seen at positions 3 and 8. [Adapted with permission from Katzung BG, et al.: Basic and Clinical Pharmacology, 11th edition, McGraw-Hill, 2009.]

Prolactinoma: The presence of otherwise benign growth of prolactin-producing cells in the anterior pituitary gland, known as a **prolactinoma**, can lead to several concerning and even dangerous signs and symptoms. The increased number of these cells can lead to the worrying production of milk (**galactorrhea**) in a nonpregnant woman or even a man. Women can also experience irregular or missed menstrual cycles and/or infertility because of the inhibitory effects of excess prolactin on GnRH functions. More concerning, though, is the presence of headaches and loss of peripheral vision (bitemporal hemianopsia) because of the pressure placed on the optic chiasm by the growing tumor. People may have a prolactinoma but will be unaware until these mass effect symptoms appear. Treatment of a prolactinoma is usually via medications (e.g., bromocriptine or cabergoline) or neurosurgery.

TABLE 20-1. Hormones Involved in Lactation

Hormone	Action
Estrogen	Milk duct growth.
Progesterone	Alveoli and lobe growth. Inhibits milk production during pregnancy. Drop at birth starts milk production.
Prolactin	Alveoli growth. Maintains milk production after pregnancy by continued stimulation of alveoli. Induces α-lactalbumin that increases galactosyl transferase affinity for glucose. Negative feedback on GnRH suppresses FSH and LH secretion and, thereby, menstruation and ovulation.
Oxytocin	Contracts smooth muscle around milk duct alveoli to squeeze milk into ducts and out of areolae. Maintains milk production after pregnancy.
Human placental lactogen	Augments growth of breasts, nipples, and areolae.

FSH, follicle-stimulating hormone; GnRH, gonadotropin-releasing hormone; LH, luteinizing hormone.

MALE REPRODUCTIVE SYSTEM

The development of the male reproductive system relies on the early activity of the **sex-determining region (sry)** found on the Y chromosome and its subsequent effects on the Wolffian and Mullerian ducts (see above). In males, sry and AMH activities along with the critical influence of testosterone result in the formation of male sexual characteristics and the regression of the Mullerian ducts.

TESTOSTERONE

Testosterone is the primary male steroid hormone and is produced in the testes by Leydig cells. Testosterone is produced from cholesterol via the androgen synthetic pathway noted in Chapter 3. The enzyme **5α-reductase** converts testosterone into the much more active **DHT** (Figure 20-9). Testosterone can also be converted to estradiol by the action of the aromatase enzyme. Estradiol is believed to be important in male reproductive functions but its exact role is still being elucidated. Women also produce smaller amounts of testosterone from the adrenal glands, ovaries, and, in time of pregnancy, the placenta.

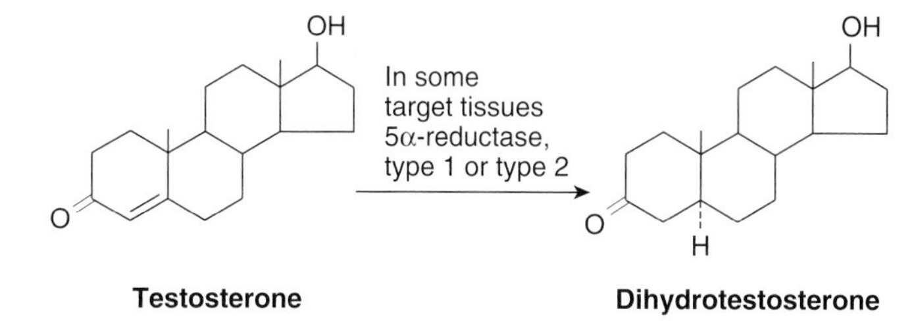

Figure 20-9. Conversion of Testosterone to 5α-Dihydrotestosterone by the Enzymatic Action of 5α-Reductase (see text for further discussion). [Adapted with permission from Barrett KE, et al.: Ganong's Review of Medical Physiology, 23rd edition, McGraw-Hill, 2010.]

Defect of the 5α-Reductase-2 Enzyme: The presence of male or female external genital structures is determined by the presence or absence of the testicular hormone, **testosterone**, and its derivative, **DHT**. Until the discovery of DHT, it was assumed that the development of the male reproductive tract depended entirely on testosterone. For example, genetic males lacking **5α-reductase-2** activity and, therefore DHT, have normal Wolffian structures because this aspect of development is controlled by testosterone. However, the lack of DHT also results in an external female phenotype, except that the vagina is incompletely developed because of the actions of AMH. FSH and LH levels are also normal because of normal signaling by testosterone. As a result, these genetic males can have male, female, or abnormal genitalia with subsequent gender identity becoming a choice for the patient. Furthermore, puberty will increase testosterone levels, which can be converted to DHT in sufficient quantity by the isoenzyme **5α-reductase-1**, potentially initiating the development of adult male characteristics, including a penis and scrotum but still lacking the prostate. For the patient who has chosen a female gender identity, surgical gender conversion is required.

Testosterone and the resulting DHT function in a variety of different ways throughout the human body. The effects of testosterone on early development have been noted above. Starting in puberty, testosterone and other androgens rise in concentration at the start of puberty and, as steroid hormones, enter into cells via passive diffusion through the plasma membrane with conversion to DHT via 5α-reductase-2. Several secondary sexual characteristics in males and females are the result of testosterone/DHT, including pubic, axillary and additional facial, chest and leg hair, increased body oil production and body odor, increased growth (e.g., bone and muscle), enlargement of the penis and clitoris along with increased sexual drive and erectile frequency, changes in facial bone structure, voice changes, and the onset of spermatogenesis (male) and oogenesis (female). Testosterone also contributes to the regulation of platelet aggregation by influence on platelet thromboxane A_2.

Benign Prostatic Hyperplasia (BPH): BPH clinically affects almost 50% of men in their 50s and over 75% of men in their 60s, occurring more often in men from western lifestyle areas. Although the exact mechanistic details of the benign growth of the prostatic cells (stromal and epithelial) are still unknown, the effect of androgens as well as estrogen are well established. **DHT** is believed to play a prominent role via binding as a homodimer to androgen receptors in the stromal cells with resulting activation of DNA transcription and increased expression of growth factors for these cells. Indeed, DHT binds more tightly to the androgen receptor than does testosterone. **5α-Reductase** levels, responsible for converting testosterone to DHT, are very high in these affected cells, with DHT easily affecting nearby epithelial cells. Estrogen (estradiol) also appears to function in later life, perhaps by sensitizing these cells to the effects of DHT, although this mechanism is even less well understood. Initial treatment is often by "**androgen receptor blockers**" (name often ends in "osin"), which block the androgen/DHT receptor, or by direct **inhibitors of 5α-reductase** (name usually ends in "steride"). Surgical or other more aggressive treatment modalities may be needed in some patients.

Prazosin

Tamsulosin

Androgen Receptor Blockers

Finasteride

5α-Reductase Inhibitor

5α-Reductase Inhibitor

Reproduced with permission from Katzung BG, et al.: Basic and Clinical Pharmacology, 11th edition, McGraw-Hill, 2009.

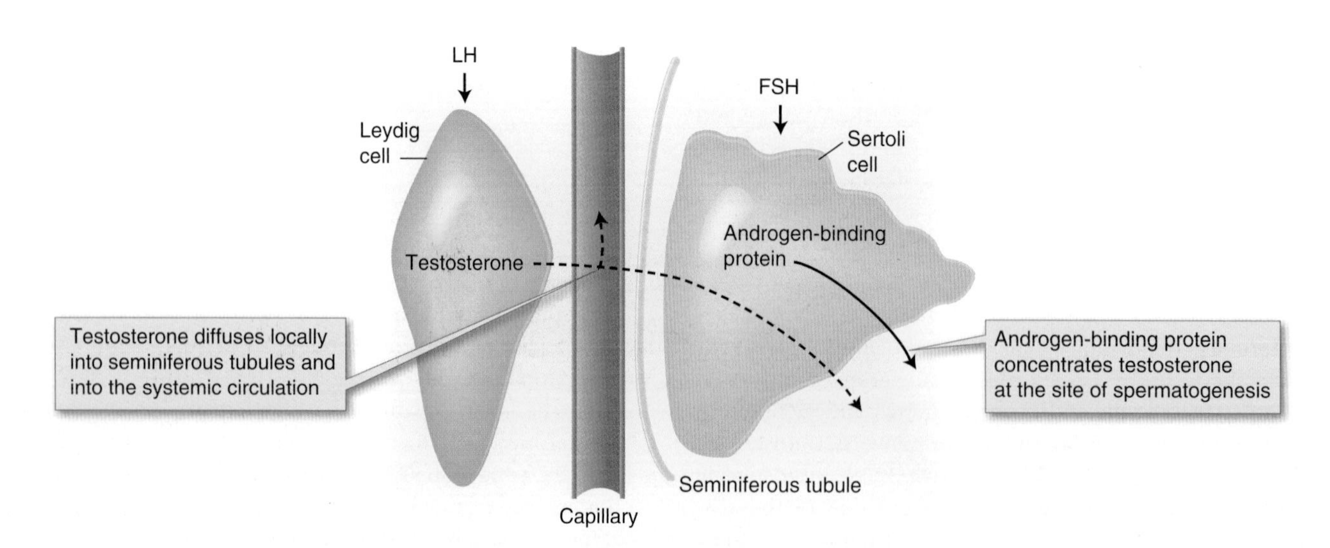

Figure 20-10. Effects of Luteinizing Hormone (LH) and Follicle-Stimulating Hormone (FSH) on the Testes. LH stimulates the Leydig cells to secrete testosterone. FSH stimulates the secretion of androgen-binding protein by the Sertoli cells into the lumen, which binds and concentrates testosterone in the seminiferous tubules at the sight of spermatogenesis. See text for further details. [Reproduced with permission from Kibble JD and Halsey CR: The Big Picture: Medical Physiology, 1st edition, McGraw-Hill, 2009.]

FSH AND LH

FSH and LH have important roles in the male reproductive system (Figure 20-10) as well as in the female. FSH is critical in the development of the seminiferous tubules as well as sperm (spermatogenesis). FSH promotes **spermatogenesis** by binding to its receptor located on **Sertoli cells**, also called "nurse cells," which line the tubules.

Sertoli cells are primarily responsible for the structural and nutrient support of developing sperm, including the following:

- Induces meiosis of spermatogenic cells.
- Secretes androgen-binding protein, a glycoprotein that binds and concentrates required testosterone near to the developing sperm cells.
- Sertoli cell aromatase converts testosterone to 17β-estradiol, whose function in the male reproductive system is still being elucidated.
- Secretes activin and inhibin, thereby regulating FSH secretion (see above).
- Produces and secretes AMH (discussed above).

LH increases the activity in Leydig cells (Figure 20-10) of the **cholesterol sidechain cleaving enzyme** (Figure 3-9), which converts cholesterol to pregnenolone (Chapter 4). As a result, Leydig cells increase synthesis and secretion of testosterone as well as androstenedione. FSH also plays a role by increasing the number of receptors for LH on Leydig cells.

REVIEW QUESTIONS

1. How would you describe (a) Wolffian ducts, (b) Mullerian ducts, (c) sex-determining region of the Y chromosome (sry), and (d) anti-Mullerian hormone?
2. What is the function of each of the following hormones: (a) follicle-stimulating hormone (FSH), (b) luteinizing hormone (LH), (c) estrogen, (d) progesterone, (e) human chorionic gonadotropin (hCG), (f) oxytocin, (g) prolactin, (h) testosterone, and (i) 5α-dihydrotestosterone (DHT)?
3. What is the basic pathway of action for the following processes: (a) female development, (b) male development, (c) oogenesis, (d) spermatogenesis, (e) female menstrual cycle (including the actions and effects of the various hormones involved), and (f) fertilization?
4. What are the roles and importance of G proteins in the receptor-mediated functions of the sex hormones?

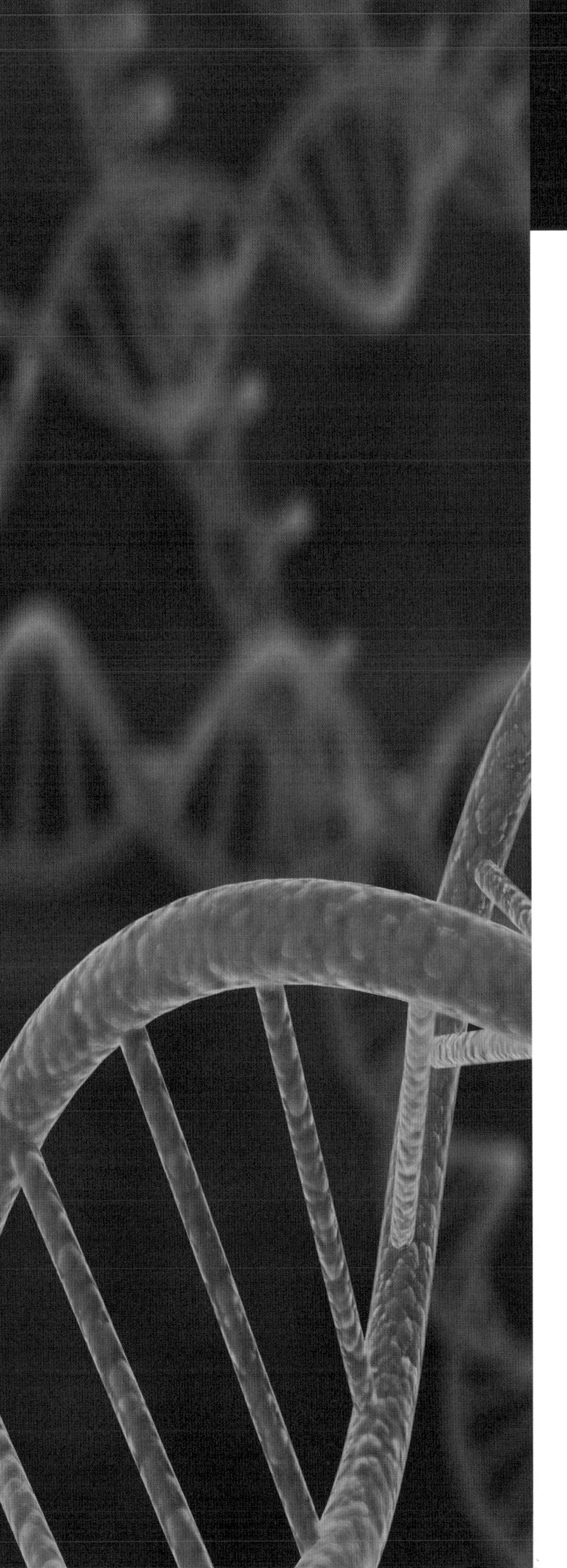

SECTION III

INTEGRATED USMLE-STYLE QUESTIONS AND ANSWERS

QUESTIONS

III-1. A middle-Eastern family presents for evaluation because their infant son died in the nursery with severe hemolysis and jaundice. The couple has two prior female infants who are alive and well, and the wife relates that she lost a brother in infancy with severe hemolysis induced after a viral infection. The physician suspects glucose-6-phosphate dehydrogenase (G6PD) deficiency, implying defective synthesis of which of the following compounds?

A. Deoxyribose and nicotinamide adenine dinucleotide phosphate (NADP)

B. Glucose and lactate

C. Lactose and NADPH

D. Ribose and NADPH

E. Sucrose and NAD

III-2. Release of which of following peptides to the blood explains the enhanced insulin secretion after a meal rich in carbohydrate?

A. Cholecystokinin (CCK)

B. Gastrin

C. Glucose-dependent insulinotropic peptide (GIP)

D. Somatostatin

E. Vasoactive intestinal peptide (VIP)

III-3. You are asked to review the results for two patients who are sisters, both of whom have been diagnosed with a thalassemia. Mary has a severe case that requires frequent blood transfusions. Alice has mild microcytic, hypochromic anemia. One parent was born in Greece and the other in the Philippines. Both children express the *β-globin* gene at 30% of the expected normal level. Mary expresses α-globin in normal amounts. However, Alice expresses the α-globin at only 50% of normal. This difference in the α-globin expression between the sisters in some way leads to the difference in severity of the diseases in the sisters. Which of the following explanations account for Mary's more severe thalassemia?

A. Alice has higher iron levels than does Mary so that iron deficiency accounts for the greater severity.

B. Mary has hemoglobin (Hgb) H disease, which is significantly more severe than Alice's β-thalassemia intermedia.

C. Mutation in the α-globin gene is expressed more in Mary's developing red blood cells.

D. Precipitation of α-globin tetramers in Mary's red blood cells kills the developing cells.

E. Synthesis of ε-globin in Alice's developing red blood cells compensates for her β-globin gene mutation.

III-4. A 10-year-old female presents with chest pain and xanthomas over her elbows and knees. Her father died of a heart attack at age 35. Her mother has high cholesterol. Her physician suspects heterozygous familial hypercholesterolemia in the parents with homozygous disease in the daughter. This disease results from mutations in the receptor for low-density lipoprotein (LDL) or the ligand portion of its apoprotein coat. Which of the following apoproteins might be mutated?

A. AI

B. B100

C. B48

D. CII

E. E

III-5. Infants born prematurely are at risk for respiratory distress syndrome. In such cases, it is common to administer surfactant, the purpose of which is to alter which of the following properties of water at the alveolar interface with air?

A. Dielectric constant

B. Evaporation

C. Heat of vaporization

D. Ionization

E. Surface tension

III-6. Fully activated pyruvate carboxylase depends on the presence of which of the following substances?

A. Acetyl-coenzyme (CoA) and biotin

B. Acetyl-CoA and thiamine pyrophosphate

C. Malate and niacin

D. Oxaloacetate and biotin

E. Oxaloacetate and niacin

III-7. The strategy for therapy for dopamine deficiency in the substantia nigra of individuals with Parkinson disease is indicated by which of the following?

A. Competitive inhibition of biosynthesis from histidine

B. Feedback inhibition of dopamine oxidation

C. Provision of metabolites in the alanine pathway

D. Provision of metabolites in the tyrosine pathway

E. Stimulation of monoamine oxidase (MAO)

III-8. A teenage female presenting with lack of menstruation (amenorrhea) is found to have a 46XY karyotype, and deoxyribonucleic acid (DNA) testing shows a dysfunctional testosterone receptor characteristic of androgen insensitivity syndrome. Which of the following statements accurately describes sex hormones such as testosterone?

A. They bind specific membrane receptors

B. They cause release of a proteinaceous second messenger from the cell membrane

C. They enhance transcription when bound to receptors

D. They inhibit translation through specific cytoplasmic proteins

E. They interact with DNA directly

III-9. A patient has complained of muscle weakness that has gradually worsened. After running an exercise tolerance test and other analyses, you conclude that patient has McArdles disease. Which of the following is a result from this patient's exercise tolerance test that would have this specific diagnosis as opposed to other muscle carbohydrate disorders?

A. Decreased blood glucose due to increased muscle glucose uptake

B. Decreased blood lactate

C. Decreased fat oxidation in muscle

D. Decreased muscle glycogen

E. Decreased use of creatine phosphate

III-10. An elderly patient has been mostly housebound for the winter and has had a poor diet. In the spring, she presents to the emergency room (ER) with multiple vitamin deficiencies. As a result of her deficiency in vitamin C, she will have a biochemical defect in collagen maturation. Which of the following characteristics applies to this vitamin C deficiency?

A. Deficient activity of lysyl oxidase

B. Deficient disulfide bond formation, both intrachain and interchain

C. Deficient proteolytic cleavage of disulfide-linked C- and N-terminal extensions

D. Inadequate collagen messenger ribonucleic acid (mRNA) translation

E. Inadequate hydroxylation of proline and lysine residues

III-11. Neural tube defects such as anencephaly and spina bifida have higher frequencies in certain populations such as those of Celtic origin and in certain regions such as South Texas. This suggestion of environmental cause produced research showing that deficiency of which of the following vitamins is associated with the occurrence of neural tube defects (anencephaly and spina bifida)?

A. Ascorbic acid (vitamin C)

B. Folic acid

C. Niacin (vitamin B_3)

D. Riboflavin (vitamin B_2)

E. Thiamine (vitamin B_1)

III-12. Which of the following substances is secreted by gastric parietal (oxyntic) cells?

A. CCK

B. Gastrin

C. Intrinsic factor

D. Motilin

E. Somatostatin

III-13. A patient presents with a hemolytic anemia due to a drug treatment. Which of the following would be a consequence of this disorder?

A. Higher percent of erythrocytes that have retained their mitochondria

B. Higher percent of reticulocytes in the blood

C. Increase in the amount of fetal Hgb (HgbF) to compensate for the drug effect

D. Lower percent of white blood cells (WBCs)

E. More reticulocytes that have lost their mitochondria and nuclei

III-14. A 45-year-old man is found to have an elevated serum cholesterol of 300 mg percent measured by standard conditions after a 12 hr fast. Which of the following lipoproteins would contribute to a measurement of plasma cholesterol in a normal person following a 12 hr fast?

A. Chylomicron remnants and very-low-density-lipoproteins (VLDLs)

B. Chylomicrons and VLDLs

C. High-density lipoproteins (HDLs) and LDLs

D. LDLs and adipocyte lipid droplets

E. VLDLs and LDLs

III-15. A comatose infant is brought to the ER. In the course of the examination, plasma ammonia was found to be elevated 20-fold over normal values. Urine orotic acid and uracil were both greater than normal. A defect of which of the following enzymes is the most likely diagnosis?

A. Arginase

B. Argininosuccinate lyase

C. Carbamoyl phosphate synthetase I (ammonia)

D. Carbamoyl phosphate synthetase II (glutamine)

E. Ornithine transcarbamoylase

III-16. During an overnight fast, which of the following is the major source of blood glucose?

A. Dietary glucose from the intestine

B. Gluconeogenesis

C. Glycerol from lipolysis

D. Hepatic glycogenolysis

E. Muscle glycogenolysis

III-17. One of the mechanisms of hypersensitivity reactions that is associated with thrombocytopenia involves antigens promoting the production of antibodies, which attach to the patient's cell membranes. These cells become recognized as foreign and are then tagged with immunoglobulin (Ig) G or IgM antibodies. This makes these cells susceptible to attack by natural killer (NK) or macrophage cells. Which of the following types of hypersensitivity reactions does this mechanism describe?

A. Anaphylactic

B. Cell mediated

C. Cytotoxic

D. Delayed

E. Immune complex

III-18. The arterial blood gas (ABG) test results for your patient show pH 7.49 (*N*: 7.35–7.45), pCO_2 = 25 mmHg (*N*: 35–45), and HCO_3^- = 19 mEq/L (*N*: 24–28). On the basis of these results, which of the following conditions do you predict exists in your patient?

A. Metabolic acidosis with increased renal reabsorption of bicarbonate

B. Metabolic alkalosis with increased renal excretion of bicarbonate

C. Respiratory acidosis with increased renal reabsorption of bicarbonate

D. Respiratory alkalosis with increased renal excretion of bicarbonate

III-19. A patient is found to have a defect in cyclic guanosine monophosphate (cGMP) phosphodiesterase. In which of the following ways would this most directly affect the visual cycle in this patient?

A. Failure of Na^+–K^+-ATPase to function

B. Inward Na^+–Ca^{2+} channels of the rod cell membrane remain closed

C. Membrane of the rod cell remains depolarized

D. Requirement for retinol becomes increased

E. Transducin cannot become activated

III-20. Which of the following describes a mechanism of action of a hypothalamic factor or pituitary hormone?

A. Antidiuretic hormone (ADH) acts on kidney via G_q, leading to increased intracellular calcium.

B. Corticotropin-releasing hormone (CRH) acts on corticotropic cells by increasing intracellular calcium.

C. Gonadotropin-releasing hormone (GnRH) acts on gonadotropic cells by increasing intracellular calcium.

D. Somatostatin (growth hormone-inhibiting hormone) acts on somatotrophs via Gs, leading to increased intracellular cyclic adenosine monophosphate (cAMP).

E. Thyrotropin-releasing hormone acts on thyrotrophs via G_s, leading to increased intracellular cAMP.

III-21. A 22-year-old woman engaging in a political protest goes on a hunger strike on a prominent corner in a city park. Although food is offered to her several times each day by social workers and the police, she refuses all offers except for water through the first 2 weeks. An examination of a sample of this woman's brain tissue would reveal that her brain had adapted to using which of the following as fuel?

A. Amino acids

B. Free fatty acids

C. Glucose

D. Glycerol

E. Ketone bodies

III-22. Which of the following describes a feature of intestinal absorption of amino acids from the diet?

A. Driven by a Na^+ gradient from the intestinal lumen to the inside of the cell

B. Occurs by simple diffusion

C. Occurs via cotransport with K^+

D. Uptake of individual amino acids is not a feature

E. Uses the same Na^+-dependent transporter as glucose

III-23. Which of the following statements accurately characterizes fetal hemoglobin (HgbF)?

A. Completely replaced prior to birth

B. Exhibits no Bohr effect

C. Includes two β-subunits and two γ-subunits

D. Lower affinity for 2,3-bisphosphoglycerate (BPG) relative to adult Hgb

E. Lower affinity for oxygen (O_2) relative to adult Hgb

III-24. Coronary artery disease is a multifactorial disorder involving occlusion of the coronary artery with atherosclerotic plaques. Several Mendelian disorders affecting lipid metabolism increase susceptibility for heart attacks, whereas environmental factors include smoking and high-fat diets. A 45-year-old man has a mild heart attack and is placed on diet and statin therapy. Which of the following will be the most likely result of this therapy?

A. High blood cholesterol

B. High blood glucose

C. Low blood glucose

D. Low blood LDLs

E. Low oxidation of fatty acids

III-25. A college student with a normal medical history collapses during an intramural basketball game. His friends think he is joking and then notice his blue color and call an ambulance. Resuscitation is unsuccessful and the autopsy reveals dilated cardiac chambers with increased thickness of the ventricular walls (hypertrophic, dilated cardiomyopathy). Electron microscopy of the heart muscle shows abnormal thick filaments, and the pathologist suspects a genetic disorder. Genes encoding which of the following proteins would be most likely to reveal the causative mutation?

A. α-Actinin

B. Actin

C. Myosin

D. Tropomyosin

E. Troponin

III-26. A 14-year-old girl is brought to the clinic by her father with a complaint of lightheadedness experienced on the soccer field earlier in the afternoon. She stated that she felt cold and nearly fainted several times, and that the symptoms did not resolve even after she drank a power beverage. On further questioning, her father stated that she had been very thirsty recently, which bothered him because it meant having to make frequent bathroom stops while driving on trips. She also "eats like a horse" and never seems to gain any weight or grow taller. Physical examination reveals a thin girl who is at the 30th percentile for height and weight. A rapid dipstick test reveals glucose in her urine. Evaluation of this girl's liver would reveal an increased rate of which of the following processes?

A. Fatty acid synthesis

B. Glycogenesis

C. Glycolysis

D. Ketogenesis

E. Protein synthesis

III-27. In a patient with a type I collagen mutation, which of the following groups of diseases would you consider in a diagnosis?

A. Alport syndrome, Goodpasture syndrome, benign familial hematuria

B. Collagenopathies types II and XI, hypochondrogenesis

C. Dystrophic epidermolysis bullosa, epidermolysis bullosa acquisita

D. Ehlers–Danlos syndromes types I, II, VII; osteogenesis imperfecta

E. Ehlers–Danlos syndromes types III, IV; aneurysm

III-28. Which of the following characterizes nephrogenic diabetes insipidus?

A. Defect in aquaporin-2

B. Defect in the G_q protein

C. Defect in the angiotensin II receptor

D. Increased osmolality of urine

E. Mutant (inactive) circulating vasopressin (ADH)

III-29. Some cytokines, such as interleukin (IL)-1, activate phospholipase C and subsequently activate the nuclear factor-kappa B (*NF-κB*) gene. Which of the following receptor classes is associated with this mechanism?

A. Chemokine receptors

B. Ig receptors

C. Transforming growth factor-β receptors

D. Tumor necrosis factor (TNF) receptors

E. Type I receptors

III-30. A 35-year-old male presents to your office with a 1-year history of shortness of breath. His shortness of breath has increased in severity over the past year. Initially, his shortness of breath was noticeable only on exertion with moderate activity, but over the past 6 months he becomes short of breath with mild activity. Further history reveals that the patient has been a two pack per day smoker since the age of 14 years when "he got hooked on cigarettes in high school." He consumes alcohol very moderately. On physical examination, he is slightly tachypneic at rest with use of accessory neck muscles during inspiration. His chest is barrel shaped and there is clubbing of the distal fingers. You obtain a chest X-ray that reveals hyperlucency of the lungs and a number of visible blebs. Pulmonary function testing reveals changes of chronic obstructive lung disease consistent with emphysema. Because of his age, you order a plasma protein electrophoresis, which reveals a decreased peak in the α-1 region. You make a diagnosis. Which of the following statements most likely explains the emphysema in this patient?

A. α-1-Antitrypsin deficiency leading to elevated activity of elastase that destroys lung tissue.

B. Elastase deficiency that causes excessive accumulation of elastin and clogs the lungs.

C. Oxidants from smoking generate reactive oxygen species that activate elastase that destroys lung tissue.

D. Oxidants from smoking generate reactive oxygen species that activate lung-specific collagenase that destroys lung tissue.

E. The patient has developed a nonalcoholic hepatitis that blocks α-1-antitrypsin secretion, leading to increased elastase activity and destruction of lung tissue.

III-31. A patient is found to have a rare disease in which the secretory function of the α-cells of the pancreas is impaired. Direct stimulation of which of the following pathways in liver will be impaired?

A. Citric acid cycle

B. Glycogenesis

C. Gluconeogenesis

D. Glycolysis

E. Pentose phosphate pathway

III-32. In patients with Niemann–Pick C disease, which of the following cellular processes would be defective?

A. Cholesterol esterification in the cytoplasm

B. Cholesterol hydroxylation for bile salt formation

C. Cholesterol trafficking to the Golgi

D. Internalization of LDL particles

E. Lyosomal hydrolysis of cholesterol ester

III-33. Which of the following describes the correct function for lingual lipase?

A. De-esterification of cholesterol ester lipid to cholesterol and free fatty acid.

B. Hydrolysis of phospholipids to free fatty acid and diacylglycerol.

C. Primarily works in the mouth where the pH is closer to neutral.

D. Produces glycerol and three free fatty acids from triglycerides containing medium-chain fatty acids.

E. Produces long-chain free fatty acids from the initial hydrolysis of triglycerides.

III-34. How is the relaxed (R) state of Hgb structurally defined?

A. Binding of 2,3-bisphosphoglycerate (BPG) to a cleft between the Hgb subunits

B. Binding of carbon dioxide (CO_2) to the N-terminus

C. Ionizable histidines being protonated

D. Subunits being dissociated

E. The position of iron in the plane of the heme

III-35. Which of the following is the major source of extracellular cholesterol for human tissues?

A. Albumin

B. γ-Globulin

C. HDLs

D. LDLs

E. VLDLs

III-36. Which of the following is a feature of the enzyme that produces the hormone that peaks shortly before ovulation is initiated in the menstrual cycle?

A. Activated by follicle-stimulating hormone (FSH)

B. Activated by luteinizing hormone (LH)

C. Constitutively active to produce estriol (E_3) from androstenedione

D. Feedback inhibited by estradiol-17β (E_2)

E. Forms estrone (E_1) from testosterone

III-37. A 4-month-old presents with hypoglycemia and hepatomegaly. Injection of glucagon produces no elevation of blood glucose or blood lactate, yet alanine can be converted to glucose. Which of the following liver enzymes would most likely be defective in this patient?

A. Glucose-6-phosphatase

B. G6PD

C. Glycogen phosphorylase

D. Pyruvate carboxylase

E. Pyruvate kinase

III-38. Which of the following diseases will be found in an adult with insufficient amounts of calcium and phosphate in the blood?

A. Hyperparathyroidism

B. Osteomalacia

C. Osteoporosis

D. Paget's disease

E. Rickets

III-39. Which of the following conditions would cause renin levels to rise?

A. A chronic high-salt diet

B. Acute treatment with atrial natriuretic peptide (ANP)

C. Chronic treatment with an aldosterone receptor antagonist

D. Treatment with a mineralocorticoid receptor agonist

E. Treatment with angiotensin II

III-40. Which of the following is the correct chronological order of the major biochemical events occurring during one cycle of skeletal muscle contraction and relaxation?

1. Actin is released from the complex
2. ATP binds the head of myosin
3. Power stroke
4. Head of myosin binds accessible actin
5. Head of myosin hydrolyzes ATP to ADP and P_i

A. 1, 2, 3, 4, 5

B. 4, 2, 5, 3, 1

C. 2, 4, 5, 3, 1

D. 2, 5, 4, 3, 1

E. 5, 4, 3, 2, 1

III-41. Following traumatic injury, epinephrine is secreted and subsequently decreases insulin secretion by the β-cells of the pancreas. This inhibition occurs via activation of α_2-adrenergic receptors leading to decreased production of cAMP. Although pancreatic β-cells also have β_2-adrenergic receptors that increase cAMP, the α_2-receptors are more abundant, and with elevated epinephrine in trauma, this latter effect predominates. Given this information, why would a child infected with pertussis toxin potentially fail to decrease insulin secretion following a traumatic injury?

A. Toxin activates G_s-protein activity so that the effect of epinephrine in trauma is offset.

B. Toxin acts like a tumor promoter to activate protein kinase C that in turn increases insulin synthesis.

C. Toxin directly facilitates the opening of calcium channels so that inward calcium movement can promote insulin secretion.

D. Toxin inhibits cAMP phosphodiesterase to cause an increase in the cAMP concentration.

E. Toxin inhibits G_i-protein activity so that the effect of epinephrine via α_2-receptors is blocked.

III-42. Which of the following statements describe the proteins that accumulate in classic Creutzfeldt–Jakob disease?

A. Bacterial endotoxin protein that is infectious

B. Derived from an infectious virus

C. Infectious protein that can cause cancer

D. Infectious protein that is similar to a normal protein in its secondary structure

E. Intracellular protein that reverts to an infectious form

III-43. Which of the following statements explain why fructose 2,6-bisphosphate is important in regulating glycolysis in the liver?

A. Has a phosphate group with a high negative free energy of hydrolysis

B. Is an allosteric inhibitor of phosphofructokinase-1

C. Is cleaved to triose phosphates in the glycolytic pathway

D. Its formation is catalyzed by a glycolytic enzyme

E. Provides an intracellular signal that is sensitive to changes in blood glucose levels

III-44. Which of the following enzymes involved in carbohydrate digestion is released by the exocrine pancreas?

A. Amylase

B. Dextrinase

C. Glucoamylase

D. Lactase

E. Sucrase

III-45. Which of the following characterizes sickle cell disease?

A. Aggregation of Hgb tetramers due to a hydrophobic effect

B. Arises from a point mutation changing glutamate to lysine on the β-subunit of Hgb

C. Binding of BPG to Hgb is diminished

D. Expression of the sickled phenotype involves a hydrophilic interaction

E. Fetal Hgb (HgbF) is significantly affected

III-46. A 28-year-old woman presents to her obstetrician at week 23 of pregnancy, complaining of extreme fatigue. Evaluations of serum iron level and fetal well being are normal, but the nurse notices a weak and irregular pulse. Chest X-ray reveals an enlarged heart and the electrocardiogram (ECG) reveals a short PR interval and prolonged QRS, including a slurred-up stroke of the R-wave called a δ-wave. The ECG is read as showing Wolff–Parkinson–White syndrome, a condition with risks for paroxysmal supraventricular tachycardia. The physician considers treatment with calcium channel regulators, balancing their risks to mother and fetus. Contraction of cardiac and skeletal muscle is initiated by the binding of calcium to which of the following substances?

A. Actin

B. Actomyosin

C. Myosin

D. Tropomyosin

E. Troponin

III-47. Which of the following is an action of dihydrotestosterone?

A. Anabolic effects on skeletal muscle

B. Development of the prostate

C. Development of the testes

D. Preventing uterine development

E. Promoting spermatogenesis

III-48. Which of the following would patients with primary hyperparathyroidism be unable to suppress (via feedback inhibition)?

A. 1-Hydroxylase enzyme in vitamin D metabolism by calcitonin

B. 25-Hydroxylase enzyme in vitamin D metabolism by a high blood concentration of active vitamin D_3

C. Action of parathyroid hormone (PTH) on osteoblasts by sex steroids

D. Production of calcitonin in the thyroid by a high blood concentration of calcium

E. Production of PTH by a tumor by a high blood concentration of calcium

III-49. A patient presents with a deficiency of thiamine. If this patient's liver cells are compared with normal cells, which of the following substances would be produced in the thiamine-deficient cell in lesser amounts if the cells are only given glucose as a fuel?

A. Alanine

B. CO_2

C. Lactate

D. $NADP^+$

E. Pyruvate

III-50. The ABG test results for your patient show pH 7.31 (*N*: 7.35–7.45), pCO_2 = 25 mmHg (*N*: 35–45), and HCO_3^- = 20 mEq/L (*N*: 24–28). On the basis of these results, which of the following conditions do you predict exists in your patient?

A. Anxiety

B. Chronic obstructive pulmonary disease (COPD)

C. Diabetic ketoacidosis

D. Hyperammonemia

III-51. Which of the following is a feature of the renin–angiotensin system?

A. Angiotensinogen synthesized in, and released by, the juxtaglomerular apparatus

B. Renin proteolytically processing angiotensin I to angiotensin II

C. Angiotensin I binding to the angiotensin receptor

D. Angiotensin II acting as a vasoconstrictor

E. Angiotensin-converting enzyme cleaving angiotensin II to angiotensin III

III-52. A postmortem sample of a dementia patient showed evidence of A-β peptide and hyperphosphorylated tau protein deposited in plaques. Which of the following observations is consistent with this postmortem finding?

A. Alzheimer dementia

B. Colchicine poisoning

C. New variant Creutzfeldt–Jakob disease

D. Niemann–Pick type A

E. Parkinson's disease

III-53. Creatine phosphate is a high energy phosphate compound that is found in skeletal muscle. Which of the following statements most likely explains why there would be little benefit to a normal individual to take creatine supplements to boost muscle creatine phosphate concentration?

A. Normal creatine phosphate concentration is sufficient to initiate exercise until glycogenolysis begins

B. Excess creatine will allosterically inhibit the creatine kinase reaction

C. Excess creatine intake will cause kidney damage in most individuals

D. Creatine phosphate normally is sufficient to initiate exercise until fatty acid oxidation begins

E. Creatine cannot cross the muscle cell membrane

III-54. An increased concentration of which of the following hormones prevents lactation from occurring in pregnancy by blocking the action of prolactin?

A. Androstenedione and progesterone

B. Estradiol and estrone

C. Growth hormone and estriol

D. Human chorionic gonadotropin (hCG) hormone and progesterone

E. Progesterone and estriol

III-55. Two patients present, one of whom has a defect in glucose-6-phosphatase and the other exhibits a defect of fructose 1,6-bisphosphatase. Which of the following is most likely the condition that would only be found in the patient with the defect in glucose-6-phosphatase?

A. Alanine accumulation in the blood following food deprivation

B. Altered muscle metabolism of glucose

C. Glycogen accumulation in the liver

D. Lactic acidosis

E. Pentose phosphate pathway is not functional

III-56. A 7-year-old girl has a 1-month history of foul-smelling diarrhea. Upon further inquiry, the frequency seems to be 4–6 stools per day. She has also had trouble seeing at night in the past 2 weeks. Her WBC count is normal. Physical examination is entirely normal. Examination of a stool sample reveals that it is bulky and greasy. Analysis does not reveal any pathogenic microorganisms or parasites but confirms the presence of fats. Further evaluation of this patient would likely reveal which of the following conditions?

A. Diabetes

B. Gastrointestinal (GI) infection

C. Ileal disease

D. Insufficient bile production

E. Lactose intolerance

III-57. Which of the following would be the earliest event in the clotting cascade?

A. Binding of platelets to red blood cells

B. Binding of vitamin K to endothelial cell surfaces

C. Formation of fibrin D-dimers

D. Release of tissue factor (thromboplastin) by damaged vessels

E. Secretion of von Willebrand factor by platelets

III-58. A patient with myocardial infarction is treated with nitroglycerin to dilate his coronary arteries. Which of the following best describes the action of nitroglycerin?

A. Acetyl-CoA and choline are condensed to form a neurotransmitter

B. Arginine is converted to a neurotransmitter that activates guanylyl cyclase

C. Guanosine triphosphate hydrolysis accomplishes oxidation of LDL proteins

D. Methylation occurs to produce S-adenosylmethionine

E. Tyrosine is converted to serotonin

III-59. Which of the following cytokines is produced by helper T (Th) cells?

A. Interferon-α

B. IL-1

C. IL-2

D. IL-3

E. IL-5

F. IL-10

G. Neutrophil chemotactic factor

III-60. Which of the following statements accurately describes functions of α_1-adrenergic receptors?

A. Bronchodilation and decreased strength of heart contraction

B. Constriction of blood vessels and decreased heart rate

C. Increased glycogenolysis and lipolysis

D. Vasoconstriction and reduced sympathetic outflow in the central nervous system (CNS)

E. Vasodilation and increased heart rate

III-61. Which of the following is an inhibitory neurotransmitter found exclusively in the central nervous system?

A. Acetylcholine

B. Glutamate

C. Norepinephrine

D. Serotonin

E. Substance P

ANSWERS

III-1. The answer is D. Glucose-6-phosphate dehydrogenase (G6PD) is the first enzyme of the pentose phosphate pathway, a side pathway for glucose metabolism whose primary purpose is to produce ribose and NADPH (Figure 6-7A). Its deficiency is the most common enzymopathy. It contrasts with glycolysis in its use of NADP rather than NAD for oxidation, its production of CO_2, and its production of pentoses (ribose, ribulose, and xylulose). Production of NADPH by the pentose phosphate pathway is crucial for reduction of glutathione, which in turn removes hydrogen peroxide. Erythrocytes are particularly susceptible to hydrogen peroxide accumulation, which oxidizes red blood cell membranes and produces hemolysis. Stresses such as newborn adjustment, infection, or certain drugs can increase red blood cell hemolysis in G6PD-deficient individuals, leading to severe anemia, jaundice, plugging of renal tubules with released Hgb, renal failure, heart failure, and death. Because the locus encoding G6PD is on the X chromosome, the deficiency exhibits X-linked recessive inheritance with severe affliction in males and transmission through asymptomatic female carriers. Ribose-5-phosphate produced by the pentose phosphate pathway is an important precursor for ribonucleotide synthesis (Figure 6-7C).

III-2. The answer is C. CCK is a peptide hormone whose main effect is contraction of smooth muscle of the gall bla dder and simultaneous secretion of pancreatic solutions to increase digestion. However, emptying of the stomach and gastric acid secretion is also decreased by CCK as digestion progresses beyond the stomach. Gastrin stimulates HCl, pepsinogen, and intrinsic factor secretion from parietal cells, and pepsinogen/renin by chief cells of the stomach. Smooth muscle contraction (i.e., motility) of the stomach is also enhanced by gastrin. Somatostatin decreases release of gastrin, CCK, secretin, motilin, VIP, GIP, and enteroglucagon, leading to decreased stomach secretion and contraction. VIP stimulates pepsinogen/pepsin secretion, dilutes bile and pancreatic juice, increases bicarbonate production in the pancreas, decreases gastrin-induced gastric acid secretion, and increases water secretion in intestine. GIP besides increasing insulin secretion decreases the release of gastric acid by parietal cells as well as smooth muscle contraction (motility) of the stomach and increases fat metabolism by activating lipoprotein lipase (see Chapter 11).

III-3. The answer is D. The key to this answer is the relative proportion of α-globin and β-globin that is functional in each sister. Normal adult Hgb contains two α-globin and two β-globin chains. Both sisters have a significant deficiency of the β-globin gene to the same extent. Hence, this effect would be the same in both sisters. Alice, unlike Mary, also has reduced expression of her α-globin. Consequently, in considering the proportion of α-globin to β-globin in each sister, Alice has a ratio of approximately 50:30, whereas Mary has a higher proportion of 100:30, that is, her α-globin exceeds the amount of her β-globin by considerably more. Therein lies the problem for Mary. Because the α-globin is more than three times the amount of β-globin, it is possible for α-globin tetramers to form. These tetramers are not soluble and precipitate in the red blood cells, leading to cell destruction and anemia. In terms of the other possible answers, there is no evidence of a difference in iron levels in the sisters. Hgb H disease cannot happen in

Mary because this is a condition in which Hgb has four β-globin subunits. Mary has no mutation of α-globin gene as expression is normal. Finally, ε-globin is an embryonic form (see Chapter 14).

III-4. The answer is B. The shell of apoproteins coating blood transport lipoproteins is important in the physiologic function of the lipoproteins. Some of the apoproteins contain signals that target the movement of the lipoproteins in and out of specific tissues. B48 and E seem to be important in targeting chylomicron remnants to be taken up by liver. B100 is synthesized as the coat protein of VLDLs and marks their end product, LDLs, for uptake by peripheral tissues. Other apoproteins are important for the solubilization and movement of lipids and cholesterol in and out of the particles. CII is a lipoprotein lipase activator that VLDLs and chylomicrons receive from HDLs. The "A" apoproteins are found in HDLs and are involved in lecithin–cholesterol acyl transferase regulation. Familial hypercholesterolemia causes early heart attacks in heterozygotes, particularly in males, and childhood disease in rare homozygotes. The daughter's chest pain was likely angina due to coronary artery occlusion and her skin patches were fatty deposits known as xanthomata (Figure 16-8).

III-5. The answer is E. Pulmonary surfactant, composed of lecithin and myelin, is secreted on a continuous basis by type II alveolar cells and Clara cells beginning at approximately 20 weeks of gestation. Pulmonary surfactant has both a lipid (~90% of total) and a protein (~10% of total) component. About half of the lipids are dipalmitoylphosphatidylcholine with a charged amine group on its head group. The remaining lipids include phosphatidylglycerol, which modulates the fluidity of the surfactant as well as cholesterol and other lipids. The lipids and proteins in surfactant all have the capability of interacting with aqueous (hydrophilic) or nonaqueous (hydrophobic) environments and it is this quality that leads to their specialized function in the lung (Figure 17-2). By directly interacting with alveolar water via their hydrophilic regions while the hydrophobic regions remain in the air, pulmonary surfactant creates a myelin meshwork that lines the alveoli—a strong, intertwined lipoprotein system that is analogous to the myelin sheath of nerve cells. This unique alveolar lining greatly reduces surface tension, allowing easier expansion/stretching and collapse of alveoli during respiration and the resulting changes in pressure. This reduction in surface tension makes the work of respiration much less and reduces the total amount of pressure that must be generated for efficient and effective inspiration and expiration. The pulmonary surfactant also helps all the lung alveoli to expand (inspiration) and shrink (expiration) at the same rate, thereby reducing the chance for isolated overexpansion and the total collapse of the alveolar sacs. Although all the other options represent properties of water or solutions, they have nothing to do with the properties of surfactant.

III-6. The answer is A. Pyruvate carboxylase catalyzes the conversion of pyruvate to oxaloacetate in gluconeogenesis: pyruvate + HCO_3^- + ATP → oxaloacetate + ADP + P_i. For pyruvate carboxylase to be ready to function, it requires biotin, Mg^{2+}, and Mn^{2+}. It is allosterically activated by acetyl-CoA. The biotin is not carboxylated until acetyl-CoA binds the enzyme. By this means, high levels of acetyl-CoA signal the need for more oxaloacetate. When ATP levels are high, the oxaloacetate is consumed in gluconeogenesis. When ATP levels are low, the oxaloacetate enters the citric acid cycle. Gluconeogenesis only occurs in the liver and kidneys.

III-7. The answer is D. Dopamine is produced from L-3,4-dihydroxyphenylalanine (L-DOPA), which in turn is made from tyrosine. Therapy with the L-DOPA precursor increases dopamine concentrations and improves the rigidity and immobility that occur in Parkinson disease. Dopamine is degraded in the synaptic cleft by MAO-A and MAO-B, producing 3,4-dihydroxyphenyl-acetaldehyde (DOPAC). DOPAC is in turn broken down to homovanillic acid, which can be measured in spinal fluid to assess dopamine metabolism. Inhibitors of MAO-A and MAO-B have some use in treating Parkinson disease. The metabolism of histidine or alanine is not related to that of dopamine, but phenylalanine is a precursor of tyrosine and L-DOPA (see Chapter 19; Figure 19-7A).

III-8. The answer is C. All steroid hormones, including the sex hormones estrogen, testosterone, and progesterone, can be classified as group I hormones, meaning that they act by binding specific cytoplasmic receptors that enter the nucleus and stimulate transcription by specific DNA binding. Most nonsteroidal hormones, for example, epinephrine, are group II hormones that interact with the cell membrane and produce a second-messenger effect. The group II hormones, in contrast to steroids, act in minutes, whereas steroid hormones require hours for a biologic effect. Recent studies have indicated that specific cytoplasmic receptors for steroid hormones have an extraordinarily high affinity for the hormones. In addition, the receptors contain a DNA-binding region that is rich in amino acid residues that form metal-binding fingers. Likewise, thyroid hormone receptors contain DNA-binding domains with metal-binding fingers. Like steroid hormones, thyroid hormones are transcriptional enhancers.

III-9. The answer is A. Patients with McArdles disease are deficient in myophosphorylase, the muscle form of glycogen phosphorylase. Patients usually present in childhood with muscle pain, fatigue, cramps, and weakness with excessive myoglobin in the urine, indicative of rhabdomyolysis during prolonged periods of exercise. Progressive symptoms and muscle mass loss and weakness are usually evident as the patient ages. If initial exercise intensity is too great, creatine phosphate is quickly depleted. Because muscle glycogen cannot be hydrolyzed, it tends to be at greater concentrations in muscle and exercise cannot elicit an increase in blood lactate. Diagnosis may be complicated by a "second wind" phenomenon exhibited by the patients. Evidence suggests that as long as the patient's initial exercise attempts are of very low intensity, within approximately 15 min, exercise can increase the mobilization of glucose transporter type 4 to the plasma membrane. This facilitates increased uptake of glucose so the patient can use glycolysis to provide energy for sustaining the exercise. Indeed patients often show a concomitant decrease in blood glucose that is often slightly elevated in McArdles' patients because of mild insulin resistance. The ability to achieve second wind also allows the muscle to begin oxidizing fatty acids.

III-10. The answer is E. Prolyl and lysyl hydroxylases both require vitamin C as a cofactor. These are iron-containing enzymes, and vitamin C (ascorbic acid) is needed to maintain iron in its reduced (Fe^{2+}) state. Although lysyl oxidase uses the hydroxylysine product of lysyl hydroxylase as a substrate, its activity will not be deficient. The lysyl oxidase enzyme requires copper and vitamin B6 as cofactors. There will be less allysine product only because of a reduced amount of substrate. Disulfide bond formation occurs after the hydroxylation step and although it will be indirectly affected by vitamin C deficiency, this process if not deficient. mRNA translation of course occurs prior to the hydroxylation steps (see Figure 13-2B).

III-11. The answer is B. Spina bifida, or myelomeningocele, is a defect of the lower neural tube that produces an exposed spinal cord in the thoracic or sacral regions. Exposure of the spinal cord usually causes nerve damage that results in paralysis of the lower limbs and urinary bladder. Anencephaly is a defect of the anterior neural tube that results in lethal brain anomalies and skull defects. Folic acid is necessary for the development of the neural tube in the first few weeks of embryonic life, and the children of women with nutritional deficiencies have higher rates of neural tube defects. Because neural tube closure occurs at a time when many women are not aware that they are pregnant, it is essential that all women of childbearing age take a folic acid supplement of approximately 0.4 mg/day. Frank folic acid deficiency can also cause megaloblastic anemia because of a decreased synthesis of the purines and pyrimidines needed for cells to make DNA and divide. Deficiencies of thiamine in chronic alcoholics are related to Wernicke–Korsakoff syndrome, which is characterized by loss of memory, lackadaisical behavior, and a continuous rhythmic movement of the eyeballs. Thiamine dietary deficiency from excess of polished rice can cause beriberi. Niacin deficiency leads to pellagra, a disorder that produces skin rash (dermatitis), weight loss, and neurologic changes including depression and dementia. Riboflavin deficiency leads to mouth ulcers (stomatitis), cheilosis (dry, scaly lips), scaly skin (seborrhea), and photophobia. Because biotin is widely distributed in foods and is synthesized by intestinal bacteria, biotin deficiency is rare. However, the heat-labile molecule avidin, found in raw egg whites, binds biotin tightly and blocks its absorption, causing dermatitis, dehydration, and lethargy. Lactic acidosis results as a buildup of lactate due to the lack of functional pyruvate carboxylase when biotin is missing. Vitamin C deficiency leads to scurvy, which causes bleeding gums and bone disease.

III-12. The answer is C. Intrinsic factor is normally produced by the parietal cells of the stomach and is essential in allowing the absorption of vitamin B_{12} in the ileum. CCK is a peptide hormone produced by cells in the duodenum. Its main effect is contraction of smooth muscle of the gall bladder and simultaneous secretion of pancreatic solutions to increase digestion. Gastrin is produced by G-cells in response to presence of undigested proteins and/or distension of the antrum of the stomach. Motilin is a peptide hormone made mainly in the duodenum and jejunum that increases smooth muscle contraction (fundus, antrum, and gall bladder). Somatostatin is produced in the stomach, intestines, and pancreas.

III-13. The answer is B. The anemia is a result of loss of red blood cells. To compensate, there is an increased production of new cells. New cell formation coupled to loss of mature cells creates a higher percent of reticulocytes. Erythrocytes always have no mitochondria and the reticulocytes always lose their mitochondria and nuclei. HgbF is gone within 6 months after birth. WBCs count is unrelated to a drug-induced anemia.

III-14. The answer is C. In the postabsorptional (postprandial) state, plasma contains all the lipoproteins: chylomicrons derived from dietary lipids packaged in the intestinal epithelial cells and their remnants; VLDLs, which contain endogenous lipids and cholesterol packaged in the liver; LDLs, which are end products of delipidation of VLDLs; and HDLs, which are synthesized in the liver. HDLs are in part catalytic because transfer of their CII apolipoprotein to VLDLs or chylomicrons activates lipoprotein lipase. In normal patients, only LDLs and HDLs remain in plasma following a 12-h fast because both chylomicrons and VLDLs have been delipidated. Most of the cholesterol measured in blood plasma at this time is present in the cholesterol-rich LDLs. However, HDL-cholesterol also contributes to the measurement. In addition to total plasma cholesterol, the ratio of HDL (good) to LDL (bad) cholesterol is also useful for predicting heart attack risks (see Chapter 16).

III-15. The answer is E. Arginase catalyzes the last reaction of the urea cycle by cleaving arginine into urea and ornithine. Urea is secreted from the liver into the blood to be cleared by the kidney for excretion. The ornithine is regenerated for another turn of the cycle. Argininosuccinate lyase cleaves argininosuccinate into fumarate and arginine in the urea cycle reaction that precedes arginase. Carbamoyl phosphate synthetase I is an initial reaction associated with the urea cycle step that catalyzes the production of carbamoylphosphate from ATP, NH_3, and CO_2. This carbamoylphosphate is then condensed with ornithine to produce citrulline in a reaction catalyzed by ornithine transcarbamoylase. In the pyrimidine synthetic pathway, carbamoyl phosphate is also produced in a reaction catalyzed by carbamoyl phosphate synthetase II that uses glutamine instead of NH_3 as a nitrogen source. When ornithine transcarbamoylase is deficient in the liver, carbamoyl phosphate accumulates and moves from the mitochondrial matrix into the cytoplasm. In the cytoplasm, this carbamoyl phosphate is used to produce pyrimidine bases (orotic acid and uracil) that are produced in an uncontrolled manner with the excess being excreted in the urine (Figure 5-9A; Chapter 18).

III-16. The answer is D. In the absorptive phase following a meal, the major source of glucose is glucose taken directly from the intestine into the blood system. Much of this glucose is absorbed into cells and, in particular, into the liver via the action of insulin, where it is stored as glycogen. Once the effects of daytime eating have subsided and all the glucose from absorption has been stored, the normal overnight fast begins. During this period, the major source of blood glucose is hepatic glycogen. Through the effects of glycogenolysis, which are mediated by glucagon, hepatic glycogen is slowly parceled out as glucose to the bloodstream, keeping blood glucose levels normal. In contrast, muscle glycogenolysis has no effect on blood glucose levels because no glucose-6-phosphatase exists in muscle and hence phosphorylated glucose cannot be released from muscle into the bloodstream. Following a more prolonged fast or in the early stages of starvation, gluconeogenesis is needed to produce glucose from glucogenic amino acids and the glycerol released by lipolysis of triglycerides in adipocytes. This is because the liver glycogen is depleted and the liver is forced to turn to gluconeogenesis to produce the amount of glucose necessary to maintain blood levels (Chapter 10).

III-17. The answer is C. The mechanism provided describes type II that is also known as antibody dependent. The anaphylactic or allergy (type I) reaction involves repeat exposure to certain allergens that leads to binding of IgE to its receptor on mast cells or basophils. Binding releases histamine; leukotrienes B4, C4, and D4; prostaglandin D2; platelet; eosinophil; and neutrophil factors. Immune complex (type III) reaction occurs when antigen binds antibodies to produce immune complexes with complement proteins. Complexes deposit in small blood vessels, joints, and glomeruli eliciting inflammatory response. Neutrophils and platelets cause tissue damage. Cell-mediated or delayed (type IV) reaction is associated with cytotoxic T cells recognizing an antigen and attacking the tissue, whereas Th1 cells release cytokines, which attract monocytes and macrophages that cause most of the cellular damage (see Table 15-3).

III-18. The answer is D. The patient's pH is above normal so this has to be an alkalosis. The low pCO_2 is indicative of increased expiration of CO_2. On the basis of carbonic anhydrase reaction $CO_2 + H_2O \leftrightharpoons H_2CO_3 \leftrightharpoons H^+ + HCO_3^-$, this is an example of respiratory alkalosis. Increased expiration of CO_2 (i.e., hyperventilation) drives the reaction toward the left, thereby decreasing the formation of both protons (H^+) and bicarbonate. The bicarbonate levels are decreased, indicating that the kidney has markedly decreased reabsorption (increased excretion) of bicarbonate. This decreased reabsorption of bicarbonate, a base, by the kidney compensates for the already basic pH (Chapters 17 and 18).

III-19. The answer is C. Nutritional vitamin A (retinol esters) is converted in the retina to 11-*cis*-retinal, an isomer of all-*trans*-retinal forms by retinal isomerase. 11-*cis*-Retinal then covalently attaches to a lysine residue on the visual protein opsin. The resulting rhodopsin molecule becomes a G-protein-coupled receptor with the "ligand" for activation being light. When light stikes the retina, photoexcited rhodopsin binds to a multisubunit membrane protein called transducin which, in turn, activates a cGMP phosphodiesterase by causing dissociation of inhibitory subunits. The phosphodiesterase catalyzes the destruction of cGMP (cGMP $\rightarrow$ GMP) near the membrane of the rod cell. Normally, cGMP keeps inward Na^+–Ca^{2+} channels open to maintain membrane depolarization. When cGMP levels fall, inward Na^+–Ca^{2+} channels close, thereby lowering both intracellular Na^+ and Ca^{2+}. The fall in Na^+ elicits hyperpolarization causing the rod cell to release less glutamate neurotransmitter. The reduced amount of this inhibitory neurotransmitter generates an electrical impulse to the occipital lobe of the brain that triggers the perception of light (Figure 19-13).

III-20. The answer is C. Release of GnRH by the hypothalamus results in activation of a specific GnRH receptor (GnRHR) located in the pituitary gland. This receptor is a membrane-bound G_q-protein-coupled stimulator of phospholipase C, which results in calcium release and protein kinase C activation via conversion of plasma membrane phosphatidylinositol into inositol triphosphate and diacylglycerol. CRH is released from the hypothalamus and binds to a receptor on corticotrophs that acts via G_s-protein-coupled stimulation of adenyl cyclase leading to production of cAMP. ADH works via a similar mechanism as CRH. TRH, like GnRH, works via G_q protein. Somatostatin works via G_i-protein-coupled inhibition of adenyl cyclase (Chapter 20).

III-21. The answer is E. This woman has created a self-imposed starvation through her hunger strike. During starvation, many fuel sources are recruited to support bodily functions, including protein degradation, which supplies amino acids as gluconeogenic precursors, and triglyceride degradation, which yields glycerol, free fatty acids, and, eventually, ketone bodies. The brain normally prefers glucose as its main fuel, so no adaptation is needed. During starvation, changes in brain gene expression upregulate several enzymes to enable use of ketone bodies as fuel. No matter how long the fast lasts, the brain cannot use glycerol, amino acids, or free fatty acids as direct fuel sources (see Chapter 10).

III-22. The answer is A. After digestion, amino acids and very small peptides are coabsorbed with sodium via group-specific amino acid or peptide active transport systems in the apical membrane. At least five distinct brush border transport systems exist that are classified as follows: (a) neutral amino acids (uncharged aliphatic and aromatic), (b) basic amino acids and cystine (Cys–Cys), (c) acidic amino acids (Asp, Glu), (d) imino acids (Pro), and (e) dipeptides and tripeptides. The mechanisms for concentrative transepithelial transport of L-amino acids and dipeptides are analogous to that for Na-dependent glucose absorption. The driving force for the Na^+-dependent transport is derived from the maintenance of low intracellular levels of Na^+ by the action of the Na^+–K^+-ATPase. Hydrolysis of ATP provides energy to export three Na^+ in exchange for two K^+. Thus, the high gradient of Na^+ between the intestinal lumen and the cytoplasm provides the driving force for active transport of amino acids, dipeptides, and tripeptides (see Chapter 11).

III-23. The answer is D. HgbF is composed of two α-globin and two γ-globin proteins. It has a lower affinity for 2,3-BPG than HgbA because of a serine at position 143 in the *γ-globin* gene instead of a histidine residue. This decreased affinity for 2,3-BPG results in HgbF having an increased affinity for O_2 relative to adult and a shift in the O_2 dissociation curve to the left. Thus, HgbF is able to extract O_2 from HgbA in the placenta and then deliver the O_2 to the fetal tissues. The fetal tissues still produce protons and CO_2 to facilitate the release of O_2 at the tissues (Bohr effect). The shift to the left is precise, however, because though it allows the HgbF to "steal" O_2 from the maternal Hgb, it is not so significant that the fetal tissues cannot remove the O_2 for their own tissues. HgbF persists for about 6 months after birth. It is for that reason that defects in the β-globin may go undetected for several months (see Chapter 14).

III-24. The answer is D. Statins act as feedback inhibitors of 3′-hydroxy-3′-methylglutaryl-CoA (HMG-CoA) reductase, the regulated enzyme of cholesterol synthesis. Effective treatment with a statin, along with a low-fat diet, decreases levels of blood cholesterol. The lowering of cholesterol also lowers the amounts of the lipoprotein that transport cholesterol to the peripheral tissues, LDL. Because lipids, like cholesterol and triglycerides, are insoluble in water, they must be associated with lipoproteins for transport and salvage between their major site of synthesis (liver) and the peripheral tissues. Those lipoproteins associated with more insoluble lipids thus have lower density during centrifugation, a technique that separates the lowest density chylomicrons from VLDLs (VLDLs with pre-β-lipoproteins), LDLs (LDLs with β-lipoproteins), intermediate-density lipoproteins, and HDLs (HDLs with α-lipoproteins). Each type of lipoprotein has typical apolipoproteins such as the apo B100 and apo B48 (translated from the same mRNA) in LDL. LDL is involved in transporting cholesterol from the liver to peripheral tissues, whereas HDL is a scavenger of cholesterol. The ratio of HDL to LDL is thus a predictor of cholesterol deposition in blood vessels, the cause of myocardial infarctions (heart attacks). The higher the HDL–LDL ratio, the lower the rate of heart attacks (see Chapter 16).

III-25. The answer is C. Cardiac and skeletal muscles are similar in that both are striated and contain two kinds of interacting protein filaments. The thick filaments contain primarily myosin, whereas the thin filaments contain actin, troponin, and tropomyosin. The thick and thin filaments slide past one another during muscle contraction (Figure 12-2). Myosins are a family of proteins with heavy and light chains, and muscle myosins function as ATPases that bind to thin filaments during contraction. Congenital defects in muscle filaments, potassium channels, and the like that affect cardiac contractility are substantial contributors to these tragic and unexpected disease categories—important reasons for annual and transitional (sports, precollege) physical examinations.

III-26. The answer is D. This girl's symptoms are consistent with extreme hyperglycemia, which is consistent with her excessive thirst (polydipsia), urination habits (polyuria), and appetite (polyphagia). Her neurologic symptoms are probably secondary to ketoacidosis, likely resulting from type 1 diabetes. The finding of glucose spillover into her urine strongly supports this conclusion. An acute hyperglycemic condition due to type 1 diabetes is characterized by a near absence of insulin with unopposed glucagon action, particularly in the liver. So both gluconeogenesis and ketogenesis are elevated in such patients. All the other processes listed would be operating at reduced activity relative to their levels in the presence of a higher insulin–glucagon ratio (Chapter 10).

III-27. The answer is D. Alport syndrome, Goodpasture syndrome, and benign familial hematuria are associated with type IV collagen mutations. Collagenopathies types II and XI and hypochondrogenesis are associated with type II collagen mutations. Dystrophic epidermolysis bullosa and epidermolysis bullosa acquisita are associated with type VII collagen mutations. Ehlers–Danlos syndromes types III and IV and aneurysm are associated with type III collagen mutations (Chapter 13).

III-28. The answer is A. Nephrogenic (kidney-associated) diabetes insipidus, a condition of poor fluid balance regulation and notable for large output of very dilute urine, is closely associated with mutations of the aquaporin-2 channel. Under normal conditions, a decrease in blood volume or an increase in osmolality triggers release of ADH (vasopressin) from the posterior pituitary to signal mobilization of aquaporin-2 channels in the kidney to increase water reabsorption. A defect in the G_s protein that mediates ADH effects in the kidney could lead to nephrogenic diabetes insipidus. However, a defect of G_q protein would only affect the vasopressenergic effect. A defective angiotensin II receptor would lead to decreased secretion of aldosterone that in turn would cause a decreased reabsorption of sodium with a subsequent loss of water. However, this is not nephrogenic because the primary defect is in the zona glomerulosa cells of the adrenal cortex. A mutated ADH would be considered a neurogenic (brain associated) or "central" diabetes insipidus (see Chapter 18).

III-29. The answer is B. The receptor class called Ig receptors, including the IL-1 receptor, activates phospholipase C and subsequent *NF-κB* gene activation. Types I and II receptors, which include many of the major cytokines and interferons, both function via a Janus kinase mechanism. TNF receptor group utilizes an intermediary protein termed TNF receptor type 1-associated death domain, which transmits conformational changes from the receptor to TNF receptor-associated factor 2, which interacts with several other signaling peptides to mediate programmed cell death (apoptosis) via *NF-κB* gene activation or by the activator protein 1, a transcription factor related to c-Fos and c-Jun. The chemokine receptor group functions via G_q proteins, leading to the release of intracellular calcium that elicits directed chemotaxis. The tumor growth factor-β receptors are serine–threonine kinases whose phosphorylation can lead to the production of cAMP, cGMP, diacylglycerol, and/or activated calmodulin, which subsequently lead to cell functions and DNA expression (see Chapter 15).

III-30. The answer is A. This patient has a deficiency of α-1-antitrypsin. α-1-Antitrypsin is a plasma protein of approximately 400 amino acids. It migrates in the α-1 region on a plasma protein electrophoresis and is the major component of this fraction. Normally, it serves as the primary serine protease inhibitor in the circulation, particularly of elastase. Because the lack of α-1-antitrypsin allows for increased elastase activity, this condition results in a slow but steady degradation of extracellular fibrils, particularly elastin. Thus, the lungs undergo proteolytic damage. This is particularly prominent in the lung and leads to premature onset of emphysema. This process is markedly accelerated by smoking so that avoidance of smoking is a priority. On the α-1-antitrypsin, a methionine residue (Met358) is critical for binding proteases to it. In the lungs of smokers, the side chain of this methionine residue is oxidized to methionine sulfoxide. Thus, any α-1-antitrypsin that remains becomes ineffective as a protease inhibitor. Consequently, individuals with α-1-antitrypsin deficiency who smoke have reduced antiproteinase function from both causes and therefore show more severe symptoms than nonsmokers with the disease. The increased severity is reflected in greater proteolytic destruction of lung tissue largely by elastase (see Chapter 17).

III-31. The answer is C. Glucagon is a hormone produced and secreted by the α-cells of the pancreas in response to decreases in blood glucose. This hormone stimulates gluconeogenesis in the liver by increasing the level of cAMP, which in turn activates the cAMP-dependent protein kinase. This enzyme inactivates pyruvate kinase by phosphorylation and thus inhibits glycolysis. Glycogenesis and glycolysis in liver are inhibited by glucagon. Pentose pathway and citric acid cycle are not directly affected by glucagon (see Chapter 10).

III-32. The answer is C. Cholesterol in cells may be synthesized de novo or taken up via LDL particles. A defect in the latter process leads to familial hypercholesterolemia. Once the LDL particles are endocytosed, they are processed in the lysosome in part by an acid cholesterol ester hydrolase (ACEH) that removes the ester creating free cholesterol. Patients with Wolman disease have defective ACEH. The free cholesterol is then transferred from the lysosome to the Golgi using Niemann-Pick C (NPC) protein. It is this protein that is defective in patients with Niemann–Pick C disease. Cholesterol in the Golgi is further processed by esterification by acyl-CoA–cholesterol acyl transferase and the cholesterol ester accumulates in droplets in the cytoplasm. Unlike other tissues, liver has the unique ability to get rid of cholesterol by conversion to bile salts via the first key reaction catalyzed by the enzyme cholesterol 7α-hydroxylase (see Chapter 19).

III-33. The answer is E. Lingual lipase initiates hydrolysis of long-chain triglycerides into glycerol and free fatty acids, which continues into and through stomach. Optimal activation requires acidic (~pH 4) environment, so vast majority of activity is in stomach. Lingual lipase is secreted by the dorsal surface of the tongue (Ebner's glands). In infants, breast milk comes equipped with milk lipase that is synthesized in the mammary cell and exported by exocytosis. Milk lipase is unique because it is stimulated by bile acids and therefore is inactive until it reaches the infant's intestinal lumen, hydrolyzes triglycerides at all three positions, and conveniently prefers to cleave off the medium-chain fatty acids that are enriched in milk (see Chapter 11).

III-34. The answer is E. Hgb exists in an equilibrium between a tense (low O_2 affinity; gives up O_2 easily to tissues) and a relaxed state (high O_2 affinity). In the deoxy (tense) form, the subunits are rotated 15° relative to the oxy form (see Figure 14-4). At a more molecular level, the R- and T-states of Hgb can be defined by the position of the iron, relative to the plane of the heme. In the T-state, the iron is pulled away from the plane of the heme, making it more difficult for O_2 to bind. In the R-state, the iron is "pulled" (by O_2) into the plane, making it easier for O_2 to bind. Hence, the R-state of Hgb has a higher affinity for O_2 than does the T-state, whether or not O_2 is present. O_2 stabilizes the R-state, whereas, the T-state is stabilized by H^+, CO_2, and BPG. Although these latter factors stabilize the T-state, they do not structurally determine the shift to the T-state (Figure 14-5).

III-35. The answer is D. The uptake of exogenous cholesterol by cells results in a marked suppression of endogenous cholesterol synthesis. Human LDL not only contains the greatest ratio of bound cholesterol to protein but also has the greatest potency in suppressing endogenous cholesterogenesis. LDLs normally suppress cholesterol synthesis by binding to a specific membrane receptor that mediates inhibition of hydroxymethylglutaryl (HMG) coenzyme A reductase. In familial hypercholesterolemia, the LDL receptor is dysfunctional, with the result that cholesterol synthesis is less responsive to plasma cholesterol levels. Suppression of HMG-CoA reductase is attained using inhibitors (statins) that mimic the structure of mevalonic acid, the natural feedback inhibitor of the enzyme (see Chapter 16).

III-36. The answer is A. E_2 peaks just before ovulation begins and is formed from testosterone by aromatase. This is the main estrogen in nonpregnant, fertile females. Aromatase in the granulosa cells of the ovaries is activated by FSH. There is no feedback inhibition of aromatase. Control of estrogen production occurs via feedback control of release of FSH and LH from the pituitary. Although LH increases production of E_2, its effect is on the production of the testosterone precursor in the ovarian theca cells. E_3 is the most abundant estrogen during pregnancy and is formed in the placenta and fetal adrenal glands and liver. E_1, produced by aromatase from androstenedione in adipose cells, is found predominately in menopausal women as well as in men (see Chapter 20).

III-37. The answer is C. The presentation of hepatomegaly and hypoglycemia is immediately suggestive of a liver glycogen storage disease. Glucagon injection normally should increase glycogenolysis, leading to glucose production. However, in this patient, neither glucose nor lactate is produced, suggesting an inability to hydrolyze glycogen at all. Glucose-6-phosphatase and pyruvate carboxylase are both associated with synthesis of glucose from alanine and this is normal in the patient. Neither G6PD nor pyruvate kinase is involved in glucose production (see Chapter 10).

III-38. The answer is B. When osteoblasts are unable to form hydroxyapatite or when sufficient calcium and/or phosphate are not available, the disease of osteomalacia results. In children, this condition is known as rickets. Osteomalacia, literally meaning "bone softness," has normal amounts of organic collagen matrix but deficient mineralization unlike osteoporosis (see below), where normally mineralized but decreased bone matrix is the problem. As a result, patients suffering from osteomalacia have weak and easily fractured bones. Most often, osteomalacia/rickets is caused by deficient vitamin D either in the diet (i.e., poor intake or poor intestinal absorption) or secondary to low sun exposure/absorption. Other causes include kidney or liver disease (or other disorders that affect vitamin D metabolism and/or absorption) decreased phosphate levels, cancers, and medication side effects (e.g., anticonvulsant medications). Symptoms of osteomalacia include bone pain (often starting in the lumbar region of the spine, pelvis, and legs) and related muscle and nerve weakness/numbness. Laboratory tests show low calcium levels in serum and urine (often accompanied by low serum phosphate). Hyperparathyroidism can be caused by low blood calcium but phosphate would be normal or elevated. Osteoporosis is a bone condition in which the amount of bone mineral is significantly lowered, leading to an altered and weakened bone matrix and a markedly increased risk of fracture. In females, it occurs after menopause because of loss of estrogen and can occur in older men because of loss of testosterone. Paget's disease of the bone, also known as osteitis deformans, is a condition of excessive bone turnover (breakdown and reformation) resulting in bone deformities, pain, decreased strength, and resulting arthritis and fractures. A genetic linkage has been suggested as the possible role of a paramyxovirus, although no convincing evidence has been found (see Chapter 13).

III-39. The answer is C. Renin is a proteolytic enzyme produced in the juxtaglomerular cells of the renal afferent arteriole. Renin release is controlled in several ways: (1) Baroreceptors in the kidney are stimulated by decreased renal arteriolar pressure. (2) Low blood pressure stimulates cardiac receptors that activate the sympathetic nervous system and the resulting catecholamines stimulate the juxtaglomerular cells via β_1-adrenergic receptors. (3) The macula densa, which is composed of cells adjacent to the juxtaglomerular cells, is stimulated by a decrease in the circulating concentration of Na^+ or Cl^-. Once the renin–angiotensin system is activated, angiotensin II triggers the release of aldosterone, a mineralocorticoid, that binds to its receptor in the kidney causing increased sodium reabsorption to correct the problem be it a drop in Na or pressure. Hence, an agonist of the mineralocorticoid receptor would decrease renin secretion. Likewise, providing angiotensin II will cause aldosterone secretion and hence lower renin release. In contrast, blocking the aldosterone receptor with an antagonist would prevent Na reabsorption, leading to increased renin. A chronic salt diet would repress renin secretion. ANP opposes the renin–angiotensin system and would hence reduce renin secretion (see Chapter 18).

III-40. The answer is E. In the relaxation phase of skeletal muscle contraction, the head of myosin hydrolyzes ATP to ADP and P_i, but the products remain bound. When contraction is stimulated (via the regulatory role of calcium in actin accessibility), actin becomes accessible and the actin–myosin–ADP–P_i complex is formed. Formation of the complex results in release of P_i. This is followed by release of ADP, which is accompanied by a conformational change in the head of myosin. This conformational change results in the power stroke in which actin filaments are being pulled past the myosin about 10 nm toward the center of the sarcomere. Another molecule of ATP is now able to bind the head of myosin, forming an actin–myosin–ATP complex. The ATP-bound myosin has a low affinity for actin and thus actin dissociates from the complex (see Chapter 12).

III-41. The answer is E. *Bordetella pertussis* produces and releases an enzymatically active, protein toxin responsible for the illness. Released in its inactive form, pertussis toxin binds to receptors on a cell membrane and is transported via the Golgi apparatus to the endoplasmic reticulum. Upon activation, the toxin adds ADP molecules to the α-subunits of G_i proteins and, being inhibited by the ADP ribosylation, are unable to stop adenyl cyclase production of cAMP. The resulting altered signaling leads to a variety of clinical manifestations. In contrast to pertussis toxin effects, epinephrine binding to α-2-adrenergic receptors increases G_i protein activity and normally lowers cyclic AMP production. In trauma, action via α-2-adrenergic receptors consequently lowers the concentration of cAMP, leading to decreased secretion and synthesis of insulin. This effect in trauma is important because secretion of insulin would lower blood glucose levels that would be detrimental to resuscitation following trauma. Although the other options listed would interfere with the epinephrine effect in trauma, they are unrelated to how the toxin functions (see Chapter 17).

III-42. The answer is E. The "prion hypothesis" asserts that proteins, devoid of nucleic acids, can themselves be infectious agents. Infectious agents require horizontal transmission into an uninfected host, and subsequent propagation of further infectious agents within that host. Protein-only infectivity has been explained with some fascinating biochemistry. Prion proteins (PrPs) are responsible for a host of transmissible spongiform encephalopathy including Creutzfeldt–Jakob disease, ovine scrapie, and bovine spongiform encephalopathy (BSE, also known as mad cow disease). Prion infectivity requires the endogenous expression of PrPs, which are normal cellular constituents, anchored to the surface of neurons. The infectious agent in these diseases is an amyloidogenic form of PrP, called PrP-res (for protease-resistant), which is a variant of PrP that is in a different conformation (e.g., a slightly soluble protein with β-strand character). PrP-res induces a conformational change in the normal cellular PrP protein, turning it into a PrP-res. This conversion of otherwise well-behaved proteins involves a conformational change from α-helix to β-strand. Notably, PrP-res is stable to heat, so BSE-tainted meat is still infectious after cooking. Transmission of spongiform encephalopathies between species can occur, as evidenced by transmission of BSE to humans. Transmission depends upon sequence homology between the transmitted PrP-res and the endogenous PrP and hence occurs with very low frequency. However, once the native host form of PrP is converted to the PrP-res conformer, it seeds further oligomer formation. The oligomeric or "protofibril" stage may be the most toxic to neural cells. Large amyloid fibrils and inclusion bodies or plaques may be less toxic or even protective. The best therapeutic strategy may therefore be one that is focused on early prevention of oligomer formation (see Chapter 9).

III-43. The answer is E. Fructose 2,6-bisphosphate is an allosteric activator of liver glycolysis at the phosphofructokinase-1 reaction and an inhibitor of gluconeogenesis at the fructose 1,6-bisphosphatase reaction. Fructose 2,6-bisphosphate is made by a bifunctional enzyme that contains both phosphofructokinase-2 and fructose 2,6-bisphosphatase activities depending on the nutritional state. In the fed state, insulin promotes formation of this regulator by activating the kinase activity of the bifunctional enzyme, whereas glucagon promotes removal of the regulator by activating the protein's phosphatase activity. The fructose 2,6-bisphosphate can only be converted to fructose-6-phosphate and is not cleaved (see Chapter 10).

III-44. The answer is A. Amylase is released by the exocrine pancreas to initiate random digestion of amylose and amylopectin chains producing maltotriose, maltose, amylose, glucose, and oligosaccharides. Dextrinase, glucoamylase, lactase, and sucrase are all localized on the surface of the small intestinal epithelial cells and involved in very specific hydrolysis reactions of products derived from amylase action (see Chapter 11).

III-45. The answer is A. Sickle cell disease is caused by a recessive, single-nucleotide mutation in the β-globin gene of Hgb (Figure 14-7A), inherited from both parents (i.e., homozygous). This mutation results in the substitution of the normal glutamic acid (negative charge) with a valine (hydrophobic) at amino acid 6. Sickle red blood cells bind O_2 normally and the oxygenated red blood cells have a normal biconcave shape. However, when the sickle Hgb unloads its O_2, the normal conformational change exposes the valine at position 6 to the surface creating a "hydrophobic patch." The hydrophobic patch on one deoxygenated Hgb molecule can interact with the hydrophobic patch on a second Hgb molecule creating stiff Hgb polymers (Figure 14-10A–B). These internal polymers cause the red blood cell to stiffen and adopt abnormal shapes including a crescentic or "sickle" shape that gives the disease its name (Figure 14-10C). Because the internal structure of the Hgb is unchanged, effects of H^+, CO_2, and 2,3-BPG are retained.

III-46. The answer is E. Calcium ions are the regulators of contraction of skeletal muscle. Calcium is actively sequestered in sarcoplasmic reticulum by an ATP pump during relaxation of muscle. Nervous stimulation leads to the release of calcium into the cytosol and raises the concentration from less than 1 mM to about 10 mM. The calcium binds to troponin C. The calcium–troponin complex undergoes a conformational change, which is transmitted to tropomyosin and causes tropomyosin to shift position. The shift of tropomyosin allows actin to interact with myosin and contraction to proceed. Mutations affecting proteins involved in muscle contraction can present with low muscle tone and developmental delay in childhood, as chronic muscle cramps or fatigue, or with cardiomyopathies due to weakened heart muscle. As cardiac muscle contraction is coordinated by electrical conduction from the sinus and atrioventricular (AV) nodes, muscle protein abnormalities can also interfere with cardiac rhythm. A specific mutation in troponin I can cause cardiomyopathy and the irregular cardiac rhythm known as Wolff–Parkinson–White syndrome (see Chapter 12; Figure 12-4).

III-47. The answer is B. The roles of dihydrotestosterone include development of the prostate, penis, scrotum, and genital skin. In contrast, testosterone during development promotes formation of the epididymis, vas deferens, ejaculatory duct, and seminal vesicles. During puberty, it is responsible for secondary sex characteristics, including muscle growth, and spermatogenesis. Development of the testes is determined by sry (sex-determining region of the Y-chromosome). Mullerian inhibitory factor accounts for duct regression in males so that the uterus does not develop (see Chapter 20).

III-48. The answer is E. Primary hyperparathyroidism refers to the uncontrolled overproduction of PTH. The 1-hydroxylase enzyme in vitamin D metabolism is located in the kidney and is activated by PTH not by calcitonin. The 25-hydroxylase enzyme is not regulated by vitamin D_3. Although sex steroids (estrogen/testosterone) can block the action of PTH, this effect is not lost in these patients. Instead the amount of PTH that is produced overwhelms these effects of sex steroids. Production of calcitonin would be high because of elevated blood calcium and would not be suppressed (see Chapter 13).

III-49. The answer is B. In the mitochondria, thiamine is required for functioning of both the pyruvate dehydrogenase reaction (pyruvate + CoA + $NAD^+ \rightarrow$ acetyl CoA + ATP + CO_2) and the citric acid cycle α-ketoglutarate dehydrogenase reaction (α-ketoglutarate + CoA + $NAD^+ \rightarrow$ succinyl-CoA + ATP + CO_2). Lactate, alanine, and pyruvate would all actually be increased because of the inability of the cells to oxidize pyruvate. NADPH production is unrelated to metabolism of pyruvate. Arguably decreased oxidation of pyruvate would increase NAD^+ concentration that, if anything, would raise the $NADP^+$ level (see Chapter 10).

III-50. The answer is C. The patient's pH is below normal so this has to be a condition associated with an acidosis. The remaining issue is whether it is metabolic or respiratory. On the basis of the carbonic anhydrase reaction $CO_2 + H_2O \leftrightarrows H_2CO_3 \leftrightarrows H^+ + HCO_3^-$, a metabolic acidosis will overproduce protons (H^+) such as in a lactic acidosis or a diabetic ketotic crisis. This lower pH drives the reaction toward the left, both lowering bicarbonate and increasing production of CO_2. To compensate for the metabolic acidosis, the lungs will compensate by increasing respiration to blow off the excess CO_2, hence accounting for the decreased pCO_2. In a respiratory acidosis, pCO_2 will be elevated because of hypoventilation as in COPD. Anxiety causes hyperventilation that will decrease pCO_2 and by the carbonic anhydrase reaction also lower the proton concentration, thereby raising the pH (i.e., respiratory alkalosis). In the case of hyperammonemia, a base ammonia will cause a metabolic alkalosis with a rise in pCO_2 to retain acid.

III-51. The answer is D. When the renin–angiotensin system is activated, renin is released. Renin release is controlled in several ways: (1) Baroreceptors in the kidney are stimulated by decreased renal arteriolar pressure. (2) Low blood pressure stimulates cardiac receptors that activate the sympathetic nervous system and the resulting catecholamines stimulate the juxtaglomerular cells via β_1-adrenergic receptors. (3) The macula densa, which is composed of cells adjacent to the juxtaglomerular cells, is stimulated by a decrease in the circulating concentration of Na^+ or Cl^-. Renin is a proteolytic enzyme that cleaves angiotensinogen to angiotensin I. Angiotensinogen is secreted by the liver. Conversion of angiotensin I to angiotensin II is catalyzed by angiotensin-converting enzyme. Angiotensin II is the active form that binds to a receptor in the adrenal cortex triggering aldosterone synthesis and secretion. Additionally, to acutely raise blood pressure, angiotensin II binds to receptors in the vascular system causing constriction (see Chapter 18).

III-52. The answer is A. Brains from Alzheimer disease (AD) patients contain interneuronal amyloid plaques and neurofibrillar tangles. Amyloid plaques consist predominantly of amyloid-β (Aβ) proteins, along with smaller amounts of other proteins, such as α-synuclein and presenillin. The predominant amyloidogenic form of Aβ is 42 amino acids long (designated $A\beta_{42}$) and is an extracellular proteolytic product of amyloid precursor protein (APP), present in the plasma membranes of neurons. APP is cleaved by proteases called secretases. Cleavage by secretases α and γ releases two soluble fragments of APP, neither of which is amyloidgenic. However, cleavage by secretases β and γ releases the $A\beta_{42}$ fragment. The membrane-spanning region of $A\beta_{42}$ is primarily α-helical, after cleavage and release from the membrane, it undergoes a transition to a β-strand conformation. Once in this conformation, $A\beta_{42}$ is normally cleared rapidly. If the $A\beta_{42}$ is not cleared, it can nucleate plaques from virtually all other $A\beta_{42}$ proteins that it contacts. As these plaques form, they resist proteolysis and transport and serve to trap future $A\beta_{42}$ proteins as they are formed. Early onset familial AD is caused by mutations in certain proteins part of the active secretase γ complex, resulting in excessively high production of $A\beta_{42}$. APP mutations are usually in or near the $A\beta_{42}$ portion of APP and presumably make APP a better substrate for cleavage by secretases β or γ, or it makes $A\beta_{42}$ more likely to undergo the α-helix to β-strand conversion (see Chapter 19).

III-53. The answer is A. Creatine phosphate is used in the first approximately 5 sec of exercise to immediately replace ATP that is being used. The reaction creatine phosphate + ADP → creatine + ATP is catalyzed by the enzyme creatine kinase. This enzyme lies at equilibrium so that it instantaneously responds when ATP concentration begins to fall. There is no other regulation of this reaction except by mass action effect caused by changes in concentrations of reactants and products. The 5 sec of creatine phosphate use are sufficient to allow the muscle to activate glycogenolysis so that metabolism of glucose units can begin to provide the energy to sustain exercise. In an aerobic muscle, initial use of glycogen can also provide time to mobilize fatty acids as the muscle fuel but creatine phosphate alone cannot "buy" enough time. Creatine is produced in sufficient amounts by liver that secretes to the circulation for uptake and subsequent phosphorylation by muscle. Creatine must be replaced by liver on a regular basis because muscle converts creatine to creatinine for excretion. One concern with taking creatine supplements is that it can cause dehydration so that extra water must be consumed when taking it. There are instances where creatine supplements may be especially beneficial—in patients with a deficiency of creatine due to liver enzyme defect or in McArdles disease patients who cannot mobilize glycogen and hence can benefit from the additional availability of creatine phosphate until increased glucose uptake and fat mobilization can occur (see Chapter 12).

III-54. The answer is E. During pregnancy, the primary estrogen produced is E_3. Binding of estrogen to the progesterone receptor causes a conformation change that helps activate the receptor so that estrogen is essential for progesterone/progesterone receptor functions. Estrogen also acts to increase the total number of progesterone receptors, thereby amplifying progesterone effects. During pregnancy, progesterone, together with E_3, inhibits lactation and smooth muscle contraction; decreased progesterone levels are one potential trigger for labor and also initiate milk production. E_2 is the main estrogen in nonpregnant, fertile females. Androstenedione is an adrenal androgen that in females and males can serve as precursor to the synthesis of E_1 in fat cells. Growth hormone has no link to lactation inhibition or onset. hCG is a heterodimeric glycoprotein hormone produced by newly fertilized embryos and, subsequently by the placenta. hCG maintains the corpus luteum and causes increasing production of progesterone but is not directly related to inhibition of lactation (see Chapter 20).

III-55. The answer is C. Both glucose-6-phosphatase and fructose 1,6-bisphosphatase are enzymes in the gluconeogenic pathway. However, only glucose-6-phosphatase is also required for the conversion of glycogen to glucose. Hence, only the patient with the glucose-6-phosphatase defect will show glycogen accumulation. In early starvation, lactate provides approximately 50% of the carbons for glucose synthesis. Its decreased use for this purpose in both patients can lead to lactic acidosis. Similarly, alanine is the key amino acid precursor for gluconeogenesis and hence its use to make glucose will be affected in both patients. Because glucose-6-phosphatase is not found in muscle, only the patient with the fructose 1,6-bisphosphatase defect could show altered muscle glucose metabolism. Finally, these defects would not affect function of the pentose phosphate pathway (see Chapter 10).

III-56. The answer is D. This patient's greasy, foul-smelling stools indicate steatorrhea. Her vision problems may be a manifestation of vitamin A deficiency due to fat malabsorption. The most likely explanation is biliary insufficiency, that is, decreased bile salt production leading to poor emulsification of dietary fats. Active ileal disease is a possibility, but the WBC count would likely be elevated unless her condition was in remission. GI infection is less likely due to the absence of pathogenic organisms in her stool. Lactose intolerance can produce diarrhea but not steatorrhea (see Chapter 11).

III-57. The answer is D. The clotting cascade consists of a series of ordered enzymatic steps whose end result is the formation of a clot. There are two limbs of the clotting cascade: the extrinsic pathway and the intrinsic pathway. The extrinsic pathway relies on the exposure of tissue factor by an injury, which then activates Factor VII, the most abundant of the clotting factors. The intrinsic pathway starts when Factor XII, an inactive serine protease encounters exposed collagen from an injured blood vessel. Factor XII is transformed into an activated serine protease. Activated Factor XII cleaves inactive Factor XI into activated Factor XI. This in turn cleaves the inactive form of Factor IX, and the activated Factor IX transforms inactive Factor VII to its active form. Later events in the actual formation of the clot involve platelets, von Willebrand factor, calcium interacting with endothelial surfaces, and formation of D-dimers (see Chapter 14; Figure 14-15).

III-58. The answer is B. Nitroglycerin causes release of nitric oxide (NO), which activates guanyl cyclase, produces cGMP, and causes vasodilation. NO is formed from one of the guanidino nitrogens of the arginine side chain by the enzyme NO synthase. NO has a short half-life, reacting with O_2 to form nitrite and then nitrates that are excreted in urine. Coronary vasodilation caused by nitroglycerin is thus short lived, making other measures necessary for long-term relief of coronary occlusion. The neurotransmitter formed by condensation of acetyl-CoA and choline is acetylcholine, which does not play a role in dilation of coronary arteries (see Chapter 16).

III-59. The answer is D. IL-3 is produced by Th cells to promote growth and differentiation of precursors of neutrophils, eosinophils and basophils, mast cells, platelets, red blood cells, as well as monocytes, macrophages, and dendritic cells. IL-3 also promotes histamine release from mast cells. IL-1 costimulates Th lymphocytes but is produced by monocytes, macrophages, B lymphocytes, dendritic cells, and fibroblasts. IL-2 is produced by Th1 cells and is involved in growth, proliferation, and further activation of T and B lymphocytes and NK cells. IL-5 produced by Th2 and mast cells stimulates growth and differentiation of activated B cells and eosinophils to increase Ig secretion, especially IgG and IgA. Neutrophil chemotactic factor (IL-8) produced by macrophages, epithelial, and endothelial cells promotes chemotactic movement of neutrophils to a site of infection and inflammation as well as intracellular signaling, and increased metabolism and histamine release by neutrophils. IL-10 is produced by cytotoxic T cells among several others and its functions include inhibition of antigen presentation and cytokine production by macrophages and Th1 cells. Finally, interferon-α made by WBCs (leukocytes) promotes the development of fever by release of prostaglandin-E_2, reduces pain by interacting with μ-opioid receptor, and stimulates macrophages and NK cells to kill viruses (Table 15-2).

III-60. The answer is B. α_1-Receptors are postsynaptic adrenoreceptors located in smooth muscle, eye, lung, blood vessels, gut, and the genitourinary system. Stimulation of these receptors activates a G_q protein. These signals lead to excitation, which is exhibited by mydriasis, bronchodilation, constriction of blood vessels, increased strength of heart contraction (positive inotropy), and decreased heart rate (negative chronotropy). α_2-Receptors are located chiefly on the presynaptic nerve terminals and G_i-protein activation, which inhibits adenyl cyclase activity. It produces vasoconstriction and reduces sympathetic outflow in the CNS. β_1-Receptor stimulation activates G_s protein, which increases adenyl cyclase activity. This action results in positive chronotropy (heart rate increase), dromotropy (conduction of the impulse through the heart's AV node), and inotropy (force of heart contraction). β_2-Receptors are mostly postsynaptic adrenoceptors located in smooth muscle and glands. They also activate adenyl cyclase via a G_s protein but also inhibition of the same enzyme by a G_i protein. The directly opposite effects are believed to allow different functions in specific cell and tissue locations. The β_2 stimulation relaxes smooth muscle resulting in vasodilation and bronchodilation as well as release of insulin and the induction of gluconeogenesis (Chapters 8 and 19).

III-61. The answer is D. Serotonin is a generally inhibitory monamine type neurotransmitter found in the CNS. Acetylcholine can be excitatory or inhibitory and is found in the CNS, neuromuscular junction, and many autonomic nervous system (ANS) synapses. Glutamate is found in the brain and spinal cord and is excitatory. Norepinephrine, such as serotonin, is a monoamine neurotransmitter but can be excitatory or inhibitory found in the CNS, sympathetic ANS synapses, and nearly all tissues. Substance P is a neuropeptide that is generally excitatory and found in the spinal cord, brain, sensory neurons, and GI tract (see Chapter 19).

SECTION IV

APPENDICES

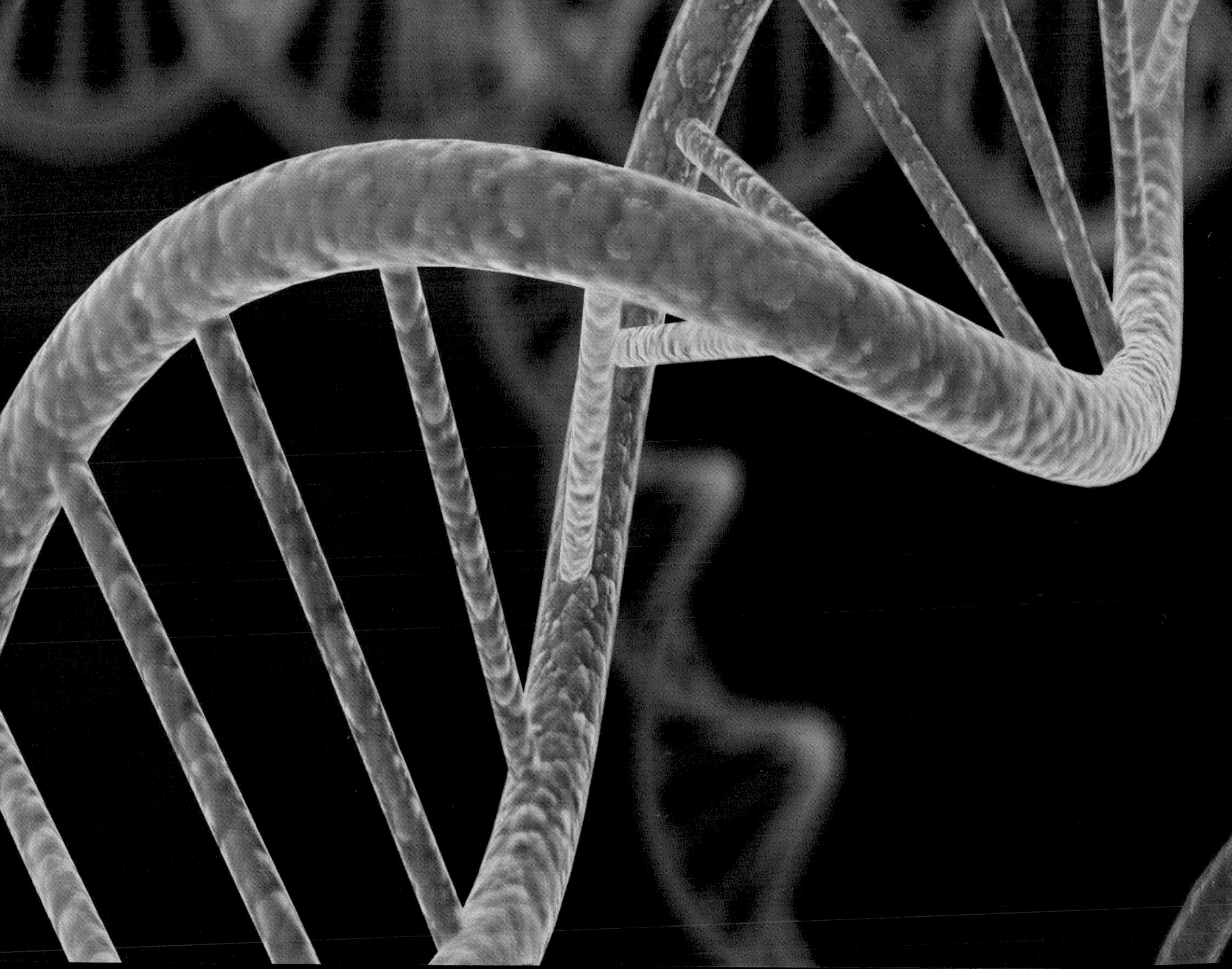

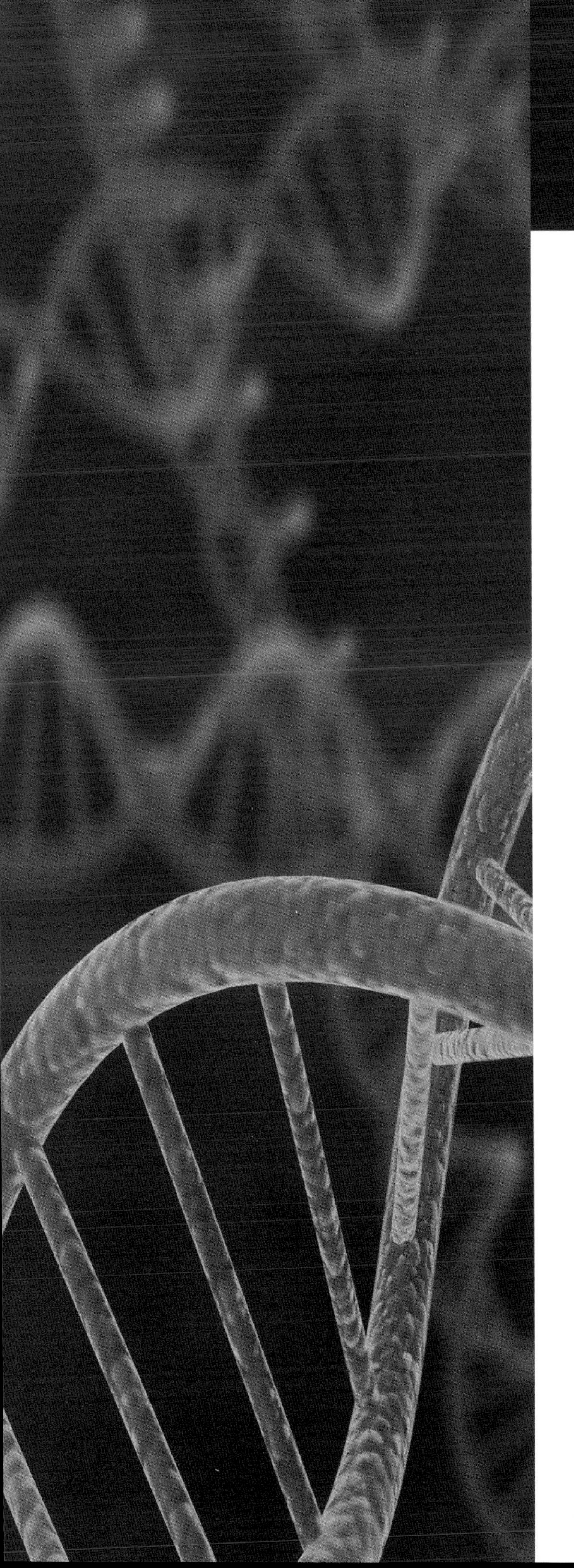

APPENDIX I

BIOCHEMICAL BASIS OF DISEASES

Contributing Editor: Harold Cross, MD, PhD

Cofounder, Windows of Hope, (www.wohproject.org)

Professor and Director, Medical Student Teaching,
Department of Ophthalmology and Vision Sciences,
College of Medicine, University of Arizona,
Tucson, Arizona, USA

This appendix presents selected examples of inherited diseases affecting basic biochemical processes. Diseases are categorized as per their involvement with the four basic categories of biochemical molecules (amino acids/proteins, carbohydrates/glycoproteins, lipids/glycolipids, and nucleic acids/deoxyribonucleic acid). Additional categories of mitochondrial enzymes and diseases affecting bilirubin, blood clotting, steroid hormones, and vitamins/minerals/electrolytes are also provided. Inheritance is predominately autosomal recessive unless otherwise stated. Minor variations of these genetic diseases may not be noted.

For further information, the reader is referred to the Online Mendelian Inheritance in Man® (www.ncbi.nlm.nih.gov/omim/), "a comprehensive compendium of more than 12,000 human genes and genetic phenotypes" and Windows of Hope (www.wohproject.org), "a population-based medical project dedicated to the detection, characterization, and treatment of inherited health problems."

AMINO ACID SYNTHESIS/DEGRADATION

Name(s)/OMIM	Mechanism	Description	Treatment	Notes
Albinism (oculocutaneous albinism) OMIM #203100, #203200, #203290, and #606574	At least four variations exist, all of which impact the production of the pigment melanin.	The varying types of albinism affect hair and skin pigments as well as vision to varying degrees.	There is no cure for albinism. Treatment is supportive, including any necessary visual/ eye care as well as preventative care to prevent sunburn and skin cancers.	Variations that only affect the eyes (ocular albinism) are also known, which are all X-linked.
	Type 1: Deficit in **tyrosinase**, the first enzyme in the melanin synthetic pathway. Variations of type 1 exist.	Type 1 albinism patients can have no, some, or even temperature-dependent (increased in cooler body parts) pigment formation. A variation affects predominately Amish/ Mennonite peoples.		
	Type 2: Defect in the **P-protein**, which regulates melanocyte pH and affects pigment production. The **melanocortin 1 receptor** on melanocytes, which binds pituitary hormones called melanocortins (e.g., ACTH and MSH) also regulate the production and type of melanin produced in type 2 albinism.	Type 2 (the most common type) has some pigment production (e.g., moles, freckles, and light hair color) and less problems with vision.		
	Type 3: Defect in **tyrosinase-related protein-1 (TRP-1)**, involved in melanin synthesis but also possibly affecting and regulating tyrosinase activity and melanocyte proliferation and death.	Type 3 is less understood but results in reddish hair, red-brown skin, and blue/ gray eyes.		
	Type 4: Defect in **membrane-associated transporter protein (MATP)**, regulating melanin synthesis and melanocyte differentiation.	Type 4 is usually found only in Japan, although a small German population has also been identified. Patients have white to yellow to even brown hair, light blue/gray to brown eyes, and visual problems.		

Name(s)/OMIM	Mechanism	Description	Treatment	Notes
Alkaptonuria (black urine disease) OMIM #203500	Defect in the gene for **homogentisic acid oxidase**. Breakdown of phenylalanine and tyrosine is inhibited, leading to increase of the intermediate homogentisic acid (2,5-dihydroxyphenylacetic acid), also known as alkapton. Excessive alkapton is excreted in the urine and, when exposed to air, turns into a brown-black, melanin-like molecule.	Urine color changes (usually first detected in a baby's diaper), higher incidence of kidney and prostate stones. Deposition of homogentisic acid in cartilage can damage joints (e.g., spine, hip, and shoulder). Heart valves may also be affected and coronary artery disease may be problematic. Pigment may also deposit in skin (e.g., at sweat glands), ear wax, and in the eye (sclera).	No absolute treatment method has been found, although reduction in phenylalanine and tyrosine in the diet is common. Added vitamin C to the diet helps to reduce cartilage damage. Other novel treatment methods are being developed.	Alkaptonuria was the first inherited disease involving an error in metabolism that was characterized (1859).
Glycine encephalopathy (nonketotic hyperglycinemia) OMIM #605899	Defect in **mitochondrial glycine cleavage enzyme system** responsible for glycine degradation. This system includes (a) P-protein (pyridoxal phosphate-dependent glycine decarboxylase), (b) H-protein (a lipoic acid-containing protein), (c) T-protein (tetrahydrofolate-requiring enzyme), and L-protein (lipoamide dehydrogenase). Defects in any of these enzymes lead to increased glycine levels in blood, urine, and cerebrospinal fluid.	Presents in three forms: (1) Neonatal form (first days of life): decreased muscle and nerve function, which can lead to cessation of breathing and death, seizures, and marked mental retardation. (2) Infantile form (symptom-free up to 6 months): seizures and mental retardation. (3) Atypical mild form (childhood): muscle and nerve symptoms (low muscle tone, abnormal movements, and seizures), mental retardation, developmental delay, behavioral and speech problems, and frequent infections.	Only known treatment is early use of dextromethorphan or ketamine to replace Glycine's role in the *N*-methyl d-aspartate (NMDA) receptor [see Smith–Magenis syndrome (SMS) below] and sodium benzoate to control seizures and behavioral changes.	—
Histidinemia OMIM #235800	Defective **histidine ammonia-lyase**, required for the removal of an amino group during histidine degradation (L-histidine to *trans*-urocanic acid).	The defect results in high levels of histidine, histamine, and imidazole in urine, blood, and cerebrospinal fluid. Usually clinically insignificant but may lend to developmental disorders (e.g., mental retardation, learning disability, hyperactivity, and speech deficits) in neonates, infants, and small children.	Most patients are asymptomatic and do not require treatment. A low histidine diet lowers levels of the metabolites but has never been conclusively shown to affect symptoms.	This inborn error of metabolism is the most common disorder in persons of Japanese descent.

(continued)

Name(s)/OMIM	Mechanism	Description	Treatment	Notes
Homocystinuria OMIM #236200	Defect in **cystathionine β-synthase**, required for conversion of homocysteine to cystathionine, part of the breakdown pathway of methionine. Defect results in accumulation of homocysteine in blood/ urine.	Increased levels of homocysteine and methionine are seen in the urine. Excessively long limbs and fingers, dislocation of lens of eye, neurological problems (seizures, mental retardation, and psychiatric illness), and increased risk of blockage of veins and arteries resulting in heart attacks and strokes.	Treated with high doses of vitamin B_6, normal dose of folic acid, and/or addition of cysteine or trimethylglycine to diet. If unresponsive to these treatments, a decrease of methionine in diet is required.	Homocystinuria II (also known as. pyridoxine responsive) results from a defect of the synthase apoenzyme causing defective cofactor binding.
Hypervalinemia OMIM #277100	Defect in **valine transaminase**, resulting in increased levels of valine in blood and urine (no increase of leucine or isoleucine).	Presents in infants with poor appetite and feeding, poor growth and weight gain (failure to thrive), drowsiness, diarrhea and vomiting, dehydration, poor muscle tone, rapid involuntary muscle and eye movements, poor concentration, and mental retardation.	Treated with diet restriction of valine.	—
Isovaleric acidemia (sweaty feet disease) OMIM #243500	Defect in the enzyme **isovaleryl coenzyme A dehydrogenase**. This enzyme is responsible for the third step in the degradation of leucine. Deficiency results in increased levels of isovaleric acid in blood and urine, which is toxic to the central nervous system. Toxic molecules can be alternatively broken down and excreted via glycine degradation.	Acute (fatal) form strikes newborns causing marked acidification of the blood, convulsions, lethargy, dehydration, moderately increased liver size, decreased platelets and white blood cells, and an unusual urinary odor like that of sweaty feet. Rapid death follows. Chronic intermittent (nonfatal) form results in periodic attacks of acidification of blood possibly resolving by glycine detoxification pathway.	Treated with reduction of leucine in diet and administration of glycine to help remove toxic isovaleric acid.	Hallmark sweaty feet odor results from increased butyric and hexanoic acids from unrelated but possibly linked error in fatty acid metabolism.
Ketotic hyperglycinemia (propionic acidemia) OMIM #606054	Defect in **propionyl-CoA carboxylase** is an essential enzyme in the breakdown of propionyl-CoA, a product of the metabolism of several amino acids (valine, isoleucine, threonine, and methionine) and the oxidation of odd-numbered fatty acids.	Defect leads to increased glycine, propionyl-CoA, propionic acid, ketones, ammonia, and acid level in blood/urine. Presents in infants with damage to brain, heart and liver, vomiting, fatigue, low muscle tone, lowered levels or absent neutrophils and/or platelets and/or blood protein, seizures, developmental delay, and poor tolerance of dietary protein.	Treatment is by careful restriction of dietary protein. Frequent infections are sometimes seen because of a decrease of white blood cells.	Propionyl-CoA with oxaloacetate also forms methylcitrate, which inhibits the citric acid cycle.

Name(s)/OMIM	Mechanism	Description	Treatment	Notes
Maple syrup urine disease (branched-chain aminoaciduria) OMIM #248600	Deficit of **branched-chain α-keto acid dehydrogenase complex (three enzyme subunits)**, required for oxidative decarboxylation of the α-ketoacids produced by the removal of the amino group during degradation of branched-chain amino acids (leucine, isoleucine, and valine). Ketoacidosis results from the branched-chain α-ketoacid accumulation. The urine has a characteristic maple syrup odor.	Defect results in accumulation of these amino acids and branched-chain α-ketoacids in blood and urine. Newborns are listless, have feeding difficulties, seizures, periods of interrupted breathing, global developmental delay, and growth deficiency. General muscle tone may be poor. Symptoms in acute episodes include: seizures, abdominal pain, muscle weakness, unsteadiness, and sometimes hallucinations. Left untreated, can lead to coma and death within the first months of life. Early and prompt treatment can prevent many of the neurological damage.	Treatment consists of restricting the dietary intake of branched-chain amino acids to the minimum that is needed for growth. These include dietary leucine restriction, high calorie, branched-chain amino acid-free formulas, and frequent monitoring of plasma amino acid concentrations. Additional thiamine pyrophosphate (vitamin B_1), a cofactor for the enzyme, is also given. Treatment usually lowers levels of branched-chain amino acids and disease progression.	In certain Mennonites populations, incidence can be as high as 1 in 175. Studies have shown that it is possible to transfer subunits of the enzyme complex into cells using a retrovirus as a possible cure.
Methylmalonic acidemia (MMA) OMIM #251000	Defect in **methylmalonyl-CoA mutase**, responsible for the breakdown of branched-chain amino acids (isoleucine, valine, and threonine) and methionine. The enzyme is also involved in lipid and cholesterol metabolism via the conversion of methylmalonyl-CoA into succinyl-CoA. Methylmalonyl-CoA requires **vitamin B_{12}**; thus, deficiencies of this vitamin's metabolism can lead to the same disease state. Defects in two other gene products (MMAA and MMBB) involved in the production of an active form of vitamin B_{12}, **adenosylcobalamin**, and, therefore, the final enzyme can also cause disease.	Deficiency of this enzyme's activity from any cause leads to the inability to metabolize certain amino acids and lipids/cholesterol. Patients present with nausea/vomiting and dehydration, poor growth, decreased muscle tone (hypotonia), and reduced energy (lethargy). Without treatment, varying mental deficits, kidney, liver and pancreatic disease, and coma/death may result.	Acute treatment involves stabilization of critically ill patients, which includes marked reduction or elimination of dietary proteins and replacement by simple carbohydrates to limit lipid metabolism. Longer term treatment includes a diet of high calories but low protein (including restrictions of isoleucine, valine, threonine, and methionine), supplementation of vitamin B_{12} and carnitine (to promote fatty acid transport and metabolism).	Multiple forms exist, including an infantile, non-vitamin B_{12} responsive, intermediate, vitamin B_{12} responsive, early childhood, and benign/adult forms. The severity of the disease generally lessens from the infantile to the adult forms.

(continued)

Name(s)/OMIM	Mechanism	Description	Treatment	Notes
Phenylketonuria OMIM #261600	Defect in **phenylalanine hydroxylase**, required for hydroxylation of phenylalanine as part of its degradation and for synthesis of tyrosine.	Defect results in decreased levels of α-ketoglutarate, due to transamination of the excess phenylalanine, which shuts down energy production and development of the brain. This results in severe mental retardation, decreased skin pigmentation, unusual posture, and epilepsy. Characteristic "mousy" odor of urine is also present due to the increased presence of phenylpyruvate, phenylacetate, and phenyllactate.	Phenylalanine must be carefully limited and extra tyrosine added to the diet for essential protein and neurotransmitter synthesis.	Less frequently the disease can result from a defect of the enzyme that regenerates the tetrahydrobiopterin cofactor for the reaction. Several disorders of elevated phenylalanine not related to this enzyme exist.
3-Phosphoglycerate dehydrogenase deficiency OMIM #601815	Defect in **3-phosphoglycerate dehydrogenase**, the enzyme responsible for the first step in the synthesis of serine.	The inability to produce serine, the major source of one-carbon groups for the new synthesis of purine nucleotides and deoxythymidine monophosphate, leads to severe neurological consequences, including a severely undersized brain (congenital microcephaly) and delay in neurological functions, seizures, failure to thrive, abnormal eye movements, and abnormal posture. Diagnosis can be made by measuring levels of serine and glycine in the plasma and cerebrospinal fluid.	Treatment is by supplementation with serine and glycine.	—
Phosphoserine aminotransferase deficiency OMIM #610992	Defect in **phosphoserine aminotransferase**, the second enzyme in serine biosynthetic pathway.	Deficiency of serine leads to a similar clinical picture as has been noted for 3-phosphoglycerate dehydrogenase deficiency above, including small brain upon birth and neurological deficiencies, severe seizures, and abnormal posture and muscle tone. Diagnosis is determined by low concentrations of serine and glycine in plasma and cerebrospinal fluid.	Treatment is by supplementation with serine and glycine.	—

Name(s)/OMIM	Mechanism	Description	Treatment	Notes
Smith-Magenis syndrome (SMS) OMIM #182290 OMIM #182144	Defect in **serine hydroxymethyltransferase**, leading to decreased or absent conversion of serine to glycine. Exact mechanism unknown, but association with the NMDA receptor involved in learning and memory as well as glycine's role in transmission of nerve signals may explain disease symptoms.	Patients have facial abnormalities, abnormal sleep cycles (possibly due to altered melatonin), behavioral problems, self-injury, short stature, abnormal spine curvature, reduced sensitivity to pain and temperature, hoarse voice, hearing and vision problems, and heart and kidney defects. Patients often display the unique characteristics of repetitive self-hugging and "lick and flip" activity—licking of fingers and flipping of book and magazine pages.	No cure. Treatment is supportive and is focused on controlling symptoms, including supplementation and regulation of melatonin, sleep aids, and behavioral modifiers.	Mutation occurs during egg/sperm or early fetal development.
Type I tyrosinemia (tyrosinosis) OMIM #276700	Defect in **fumarylacetoacetase**. This enzyme is required for the last step in tyrosine degradation. Deficiency results in buildup of all intermediates of tyrosine degradation as well as phenylalanine byproducts.	Symptoms usually appear in the first few months of life, including poor growth/ weight gain; cabbage-like odor; intestinal obstruction and/or bleeding; diarrhea; vomiting; enlargement and disease of heart, liver, and/ or spleen with associated disease; kidney and pancreatic failure; abnormal blood clotting leading to increased tendency to bleed (e.g., nose); chronic muscle weakness, rickets; and intermittent paralysis and loss of feeling.	Only treatment is diet low in phenylalanine, methionine, and tyrosine. Liver transplantation is common.	Type I is the most severe form of the tyrosinemias, impacting multiple organ systems. Much more common in Quebec, Canada.
Type II tyrosinemia OMIM #276600	Deficit of **tyrosine transaminase**, the enzyme required for the first step in tyrosine degradation. Deficiency leads to urinary excretion of tyrosine and the intermediates between phenylalanine and tyrosine.	Symptoms usually in early childhood including mental retardation (~50%), growth retardation, eye problems (excessive tearing, abnormal sensitivity to light, pain, and redness), and skin problems (painful lesions on the palms/soles).	Treated with low-protein diet, especially decreased tyrosine and phenylalanine.	—
Type III tyrosinemia OMIM #276710	Deficit of **4-hydroxy phenylpyruvate dioxygenase**, required for the second step in tyrosine degradation. Deficiency leads to elevated levels of tyrosine in the blood and marked excretion of degradation byproducts into urine.	Mild mental retardation, periodic loss of balance, and coordination and seizures.	Treated with low-protein diet, especially decreased tyrosine and phenylalanine.	Very rare. Heterozygous mutation in the same gene can cause Hawkinsinuria.

ACTH, adrenocorticotropic hormone; CoA, coenzyme A; MSH, melanocyte-stimulating hormone; NMDA, N-methyl D-aspartate.

AMINO ACID TRANSPORT

Name(s)/OMIM	Mechanism	Description	Treatment	Notes
Cystinuria OMIM #220100	Defect in the amino acid transporter for basic or positively charged amino acids (histidine, lysine, ornithine, arginine, and cysteine), leading to a defect in the transport of these amino acids in the kidney and the digestive tract.	Precipitation of cystine and the formation of stones in the urinary tract are the primary medical problems. Repetitive stone formation can lead to obstruction of urine flow that raises the risk of urinary infections and kidney failure.	Treatment is directed at prevention of stone formation through high fluid intake and the use of penicillamine.	One of the four, original inborn errors of metabolism described in 1908 and one of the most common (~1 in 7,000).
Drummond's syndrome (blue diaper syndrome) OMIM #211000	Defect in **T-type amino acid transporter-1**. This transport protein is required for absorption of tryptophan from the intestine. It may also lead to increased urine calcium levels and kidney stones composed of calcium.	Symptoms include digestive problems, fever, irritability, visual difficulties, and sometimes kidney disease. Breakdown of excess tryptophan leads to formation of indigo blue excreted in urine, causing a bluish discoloration of the diaper.	Treatment involves a low tryptophan diet.	Autosomal or X-linked recessive.
Hartnup disease (neutral aminoaciduria) OMIM #234500	Defect in **sodium (Na^+)-dependent transport protein** for uncharged amino acids in the kidneys and intestine, resulting in poor absorption in the intestine and increased loss in the urine of uncharged amino acids, many of which are essential. One important example is tryptophan, required to make nicotinic acid, a B-vitamin, needed to produce the same transport protein. Low nicotinic acid levels cause transport protein levels to drop, making the problem worse.	Symptoms usually start in infancy, affecting mainly the brain and skin to include failure to thrive, mental retardation, short stature, headaches, altered gait (ataxia), tremor, fainting, irregular eye movements (nystagmus), and increased sun sensitivity. Psychiatric problems (anxiety, mood changes, delusions, and hallucinations) may also occur.	Treatment is usually via a diet high in neutral amino acids, sunblock and/or sun avoidance and, when needed, daily intake of nicotinamide. A period of poor nutrition usually precedes symptom outbreaks and attacks.	Low nicotinic acid also causes the disease pellagra, which causes the "four Ds"—skin inflammation (dermatitis), diarrhea, dementia, and death.
Hyperammonemia–hyperornithinemia–homocitrullinuria (HHH syndrome) OMIM #238970	Defect in the mitochondrial ornithine transporter. As a result, ammonia accumulates in the blood.	Symptoms vary widely. Infantile form is usually more severe and may include lethargy, poor feeding and temperature control, seizures, coma, and even death. Later forms have symptoms of lethargy, confusion, blurred vision, poor coordination, and vomiting.	Infantile illness may coincide with the introduction of high-protein formulas or solid foods into the diet. Later forms may only show symptoms with high intake of protein, which increases ammonia levels.	Stress, illness, and/or fasting can also bring on symptoms.

Name(s)/OMIM	Mechanism	Description	Treatment	Notes
Methionine malabsorption syndrome (Smith–Strang disease/Oasthouse urine disease) OMIM #250900	Defect in the protein(s) essential for intestinal absorption of **methionine** (and other amino acids). Unabsorbed methionine is converted by intestinal bacteria to the chemical α-hydroxybutyric acid, which produces an urine odor of an "oasthouse" (building for drying hops).	Symptoms include strikingly white hair and blue eyes, increased rate of breathing, diarrhea, seizures, mental retardation, generalized swelling, and characteristic urine odor.	Because of extremely rare cases, no established treatment but dietary restriction of methionine is implied.	—

UREA CYCLE DISORDERS

Name(s)/OMIM	Mechanism	Description	Treatment	Notes
Arginase deficiency (argininemia) OMIM #207800	Defect in **arginase**, the fifth and final enzyme of the urea cycle, which removes nitrogen from arginine to produce a urea molecule. Guanidino metabolites are also found to increase in argininemia. The accumulation of guanidino metabolites contribute to the neurotoxicity of the disease.	The deficit blocks the urea cycle and leads to increased concentrations in ammonia in the blood. Symptoms include poor feeding and growth, developmental delay, mental retardation, lethargy, vomiting, confusion, personality disturbances, seizures, and coma. Death can occur without proper medical care.	Treatment is by low-protein diet (to reduce ammonia production) balanced with amino acid supplementation to allow continuation of the urea cycle as well to maintain cell growth (e.g., arginine, citrulline, valine, leucine, and isoleucine). Phenylacetate may be given to help remove accumulated glutamine via the kidneys. Vitamin and calcium supplements are also often given. Liver transplant is sometimes required.	Cytosolic protein.
Argininosuccinate synthetase (ASS) deficiency (citrullinemia, types I and II) OMIM #215700, #603471, and #605814	Defective **ASS**, the enzyme involved in the third step in the urea cycle that synthesizes argininosuccinate from citrulline and aspartate. The mutation decreases production of the protein citrin, a mitochondrial aspartate/glutamate carrier and/or malate/aspartate shuttle. Deficiency affects molecules that are involved in the urea cycle and synthesis of proteins and nucleotides.	Type I patients have increased levels of ammonia in the blood within the first few days of life that leads to poor feeding and growth, seizures, and vomiting. Death can occur. Type II usually appears in adult patients because of the buildup of ammonia and impacts the nervous system, including confusion, personality disturbances, and seizures/coma. Signs are sometimes triggered by alcohol use or in patients with neonatal intrahepatic cholestasis.	Same as arginase deficiency.	Cytosolic protein.

(continued)

Name(s)/OMIM	Mechanism	Description	Treatment	Notes
Argininosuccinic aciduria/acidemia OMIM #207900	Defect in **argininosuccinate lyase**, the fourth enzyme of the urea cycle that cleaves argininosuccinate into arginine and fumarate.	Same as arginase deficiency.	Same as arginase deficiency.	Cytosolic protein.
Carbamoyl phosphate synthetase I (CPS-1) deficiency OMIM #237300	Deficit of **CPS1**, the enzyme responsible for the transfer of ammonia to bicarbonate to produce carbamate using one ATP. A second ATP is used to produce carbamoyl phosphate. This is the first reaction of the urea cycle.	Same as arginase deficiency.	Same as arginase deficiency.	Mitochondrial protein.
***N*-Acetylglutamate synthase (NAGS) deficiency** OMIM #237310	Deficit of **NAGS**, the enzyme responsible for the production of *N*-acetylglutamate from acetyl-CoA and glutamate. NAGS is the activator of carbamoyl phosphate synthase I (see above).	Same as arginase deficiency.	Same as arginase deficiency.	Mitochondrial protein.
Ornithine transcarbamylase (OTC) deficiency OMIM #311250	Deficit of **OTC**, the second urea cycle enzyme, producing citrulline from carbamoyl phosphate and ornithine. This is the most common urea cycle disorder.	The deficit blocks the urea cycle and leads to increased concentrations of ammonia in the blood. Excess carbamoyl phosphate is also converted to orotic acid via pyrimidine synthesis pathways. Symptoms are the same as arginase deficiency.	Same as arginase deficiency. Some childbirth deaths may be due to an unknown OTC deficiency carrier state, which is unmasked by the stress of childbirth.	Mitochondrial protein. Unlike the other urea cycle disorders, listed OTC deficiency is X-linked.

ATP, adenosine triphosphate; CoA, coenzyme A.

STRUCTURAL PROTEINS

Name(s)/OMIM	Mechanism	Description	Treatment	Notes
α-1-Antitrypsin (AAT) protease inhibitor deficiency OMIM #107400	The defective **anti-elastase** allows degradation of extracellular fibrils such as elastin with gradual loss of pulmonary function, for example.	The association of emphysema and AAT deficiency (known to be a protease inhibitor) has been recognized for a long time. The disease results from the loss of anti-elastase activity and thus the lungs and liver especially are damaged by progressive proteolytic damage. Smoking causes oxidation of the methionine residue in the AAT molecule, which limits its binding to serine protease and reduces its inactivation function even further.	Intravenous concentrates of AAT combined with lifetime avoidance of smoking has been recommended but rigorous studies have not been done. Prognosis is dependent upon the severity of pulmonary and liver disease, which can be highly variable. Many patients succumb in the fifth decade.	Smoking cessation and avoidance of contaminated air is the key to patient health. However, recent studies show that AAT deficiency is probably only responsible for a minority of emphysema cases. Additional contributing factors are currently being sought.
Ehlers–Danlos syndrome (EDS) (cutis hyperelastica) OMIM #130000	Defects involving the connective tissue protein **collagen**. Multiple mutations are all classified in the group of EDS.	EDS-affected collagen causes easy bruising and wounds with pigmented scarring, overly flexible joints, heart valve defects, uterine and bowel rupture, and/or gum problems.	There is no cure for EDS. Treatment is supportive for cardiovascular and internal organs. Corrective surgery is sometimes undertaken.	Type X involves a mutation of the glycoprotein that cross-links collagen.
	Cutis laxa: A group of related disorders with varying inheritance patterns: **Dominant** OMIM #123700: defects in **elastin/fibulin** **Recessive** OMIM #219200: defects in **fibulin** **X-linked** (occipital horn syndrome) OMIM #304150: defects in **Cu^{2+}-transporting ATPase, α-polypeptide**	Elasticity of the skin is disrupted and is wrinkled and hangs loose. Effects on tendons and ligaments may make joints loose. Severe disease can involve the structure of internal organs, including heart and blood vessels (e.g., arteries), lungs, intestines, and bladder. Hernias can also develop. Easy bruising, low copper and ceruloplasmin, skeletal defects, and the namesake bone outgrowths from the back of the head (occiput) are seen in the X-linked form.	No known cure. Treatment is mainly supportive including monitoring of affected organs and medical/surgical treatments as required.	—

(continued)

Name(s)/OMIM	Mechanism	Description	Treatment	Notes
Marfan syndrome OMIM #154700	Defect in the extracellular glycoprotein **fibrillin**, which affects cysteine-rich, tertiary structural domains.	Mutations in fibrillin weaken connective tissue structure and also disrupt binding to TGF-β protein. Altered binding adversely affects the structure of smooth muscle and extracellular matrix, especially in lungs, blood vessels (e.g., aorta), and heart valves.	There is no cure for Marfan syndrome. Treatment involves monitoring of effects on heart, blood vessels, muscle, lungs, and eyes. β-blockers and angiotensin II receptor blockers (ARBs) also reduce levels of TGF-β and are sometimes used to control cardiovascular manifestations.	Autosomal dominant disorder.
Osteogenesis imperfecta (OI) OMIM #166200	Defects in structure and function of collagen, type I. Multiple mutations are all categorized under the syndrome of OI.	OI causes marked fragility of bones, abnormal teeth, hearing, and soft-tissue disorders. Some types cause short stature and changes in the sclera of the eye (e.g., light, gray, or blue coloring). An infant form can cause premature birth, structural defects of the face and skull, and usually death in minutes to months because of heart or lung defects.	There is no cure for OI. Treatment is supportive and has included attempts to increase bone mass by bisphosphonates.	Four types of OI are currently known.

TGF, Transforming growth factor.

CARBOHYDRATES

Name(s)/OMIM	Mechanism	Description	Treatment	Notes
Fructosuria (hepatic fructokinase deficiency) OMIM #229800	Defect in liver fructokinase, which metabolizes the first step in fructose metabolism (fructose to fructose-1-phosphate).	Patients are often asymptomatic, although episodes of low sugar (hypoglycemia), nausea/vomiting, seizures, and unconsciousness can occur, especially in children.	Treatment is by avoidance of fructose in the diet with replacement by glucose sources.	—

Name(s)/OMIM	Mechanism	Description	Treatment	Notes
Galactosemia (classic) (galactosemia I) OMIM #230400	Deficiency of **galactose-1-phosphate uridyl transferase**, which catalyzes the interconversion of galactose-1-phosphate and glucose-1-phosphate via transfer of uridine monophosphate.	Although the exact mechanism is still not understood, accumulation of galactose and galactose-1-phosphate is believed to cause the detrimental effects. Patients with classic galactosemia develop cataracts, enlarged livers (hepatomegaly), speech and learning problems, and, sometimes, mental retardation.	Treatment is by avoidance of galactose and lactose in the diet (e.g., soy milk for infants). As galactose is generated by other pathways, avoidance of all effects may be difficult.	Galactosemia may also affect inositol metabolism and, therefore, essential signaling pathways.
Galactokinase deficiency (galactosemia II) OMIM #230200	Deficiency of **galactokinase enzyme**, which catalyzes the first step in galactose metabolism, the phosphorylation of galactose to galactose-1-phosphate.	Patients mainly develop cataracts due to the increased osmotic pressure caused by the accumulation of galactose-derived molecules (e.g., galactitol) in the lens. Mental retardation is also seen.	Treatment is by avoidance of dietary galactose and lactose. If a galactose-free diet is started early, cataracts will regress without complications; however, neurological damage is permanent.	—
Galactose epimerase deficiency (galactosemia III) OMIM #606953	Deficiency of the **UDP–galactose-4-epimerase** enzyme, which catalyzes the final step of galactose metabolism involving the interconversion of UDP–galactose and UDP–glucose.	As in other galactosemias, galactose and galactose-1-phosphate levels increase. Symptoms include many of classic galactosemia plus poor growth, nerve deafness, and liver/kidney failure.	Treatment is by avoidance of galactose and lactose in the diet (e.g., soy milk for infants). As galactose is generated by other pathways, avoidance of all effects may be difficult.	Both a mild form affecting red blood cells and a severe form affecting the liver exist.
Glucose-6-phosphate dehydrogenase deficiency (G6PDH) OMIM #138090	Defects of **G6PDH**, the first enzymatic step of the pentose phosphate pathway. Over 400 variants exist with the enzymatic activity of G6PDH being variably affected.	The absence of the pentose phosphate pathway limits the availability of nicotinamide adenine dinucleotide phosphate and, therefore, reduced glutathione. Reduced glutathione is the only protection for red blood cells against oxidation by free radicals.	Patients are often asymptomatic, although acute episodes of red blood cell destruction (hemolysis), caused by either infection or certain medications or foods, can cause medical problems.	Most commonly inherited enzyme disorder in the world, affecting almost 25% of the world's population. Considered to be X-linked.

(continued)

Name(s)/OMIM	Mechanism	Description	Treatment	Notes
Hereditary fructose intolerance (aldolase B deficiency) OMIM #229600	Defect in **fructose-1, 6-bisphosphate aldolase b** enzyme, responsible for the conversion during glycolysis of fructose-1, 6-bisphosphate to glyceraldehyde 3-phosphate and dihydroxyacetone phosphate. The enzyme is also involved in the cleavage of fructose-1-phosphate into glyceraldehyde and dihydroxyacetone phosphate during gluconeogenesis.	The deficit leads to an increase in fructose-1-phosphate levels, which is damaging to cells and tissues and also traps phosphate from normal regenerating pathways. The lack of ~P/ATP leads to a marked decrease in liver gluconeogenesis and causes low glucose levels (hypoglycemia), which can be fatal. Low glucose and ATP also leads to lowered protein synthesis, and function in the liver and kidneys via glycolysis leads to nausea/vomiting and abdominal pain.	Treatment is by complete avoidance of fructose, sucrose, and sorbitol in the diet. Alternative carbohydrates can be metabolized by other tissues to provide for the body's needs.	Vertebrates have A, B, and C aldolase isozymes. A is primary expressed in the embryo, muscles, and red blood cells. B is the main form in adult liver, kidney, and intestine. C is the primary form found in brain and nervous tissue.
Hyperglycerolemia (glycerol kinase deficiency) OMIM #307030	Deficiency of **glycerol kinase** enzyme, which catalyzes the phosphorylation of glycerol by ATP to ADP and glycerol-3-phosphate.	Three forms exist, including infantile, juvenile, and adult. The infantile form has severe developmental delays, vomiting, and seizures. Other forms are normally asymptomatic.	The only known treatment is via a low-fat diet, which can provide dramatic improvement in symptoms/ development.	Elevated levels of glycerol in the blood and urine can be mistaken for triglycerides.
Pyruvate kinase deficiency OMIM #266200	Deficiency of the red blood cell **pyruvate kinase** enzyme, which is the last reaction of the glycolytic pathway, catalyzes the conversion of phosphoenolpyruvate to pyruvate with the formation of ATP.	Without pyruvate kinase, red blood cells have no means to produce energy and breakdown easily, leading to hemolytic anemia. Symptoms include pale skin, jaundice, yellowing of the white of the eyes (icterus), fatigue, and recurrent bilirubin gallstones.	Blood transfusions can temporarily correct the anemia, and removal of the spleen slows the breakdown of the red blood cells. Severe, untreated cases may lead to death as early as the neonatal period.	Most common cause of inherited nonspherocytic hemolytic anemia. Amish have a higher incidence of the disease.

ADP, adenosine diphosphate; ATP, adenosine triphosphate; UDP, uridine diphosphate.

GLYCOGEN STORAGE

Name(s)/OMIM	Mechanism	Description	Treatment	Notes
Glycogen storage disease (GSD; type 0) (liver glycogen synthase deficiency) OMIM #240600	Deficiency of liver **glycogen synthase** that catalyzes the addition of glucose units to glycogen via transfer from uridine diphosphate glucose.	Patients present in infancy to early childhood with fasting low sugar (hypoglycemia) and blood ketones. Often present with morning fatigue that responds to feeding. Fasting may cause low blood alanine (hypoalaninemia) and lactate. Dietary glucose cannot be readily stored as glycogen, leading to high sugar (hyperglycemia) and even lactic acid after meals.	Avoid fasting. Infants/children must receive frequent feedings especially of protein-rich meals. Avoid excess carbohydrate intake that would exacerbate lacticacidemia. Lack of treatment in these patients may lead to short stature and osteopenia (abnormally low bone mineral density). Neurological damage may occur because of dependency of nervous tissue on glucose as fuel.	Once pancreatic amylase activity is high enough, a child can be given uncooked cornstarch in the evening to sleep through the night. Raw cornstarch is a complex glucose polymer acted on slowly by pancreatic amylase over about a 6-hr period.
Glycogen storage disease (GSD; type Ia) (Von Gierke disease) OMIM #232200	Type Ia disease is a deficiency of **glucose-6-phosphatase** that catalyzes the conversion of glucose-6-phosphate to glucose in liver and kidney.	Present in infants with enlarged liver (hepatomegaly) that may develop into benign liver cancer (adenoma), usually in the teen years (75% of patients) that rarely (10%) become malignant. Liver and pancreas effects plus kidney stones may cause abdominal pain. Fasting low sugar (hypoglycemia) can lead to serious complications (seizures to coma to death). High uric acid (90% of patients) leads to gouty arthritis. Blood lactic acid may also be high causing the hyperuricemia. Patients almost always exhibit dyslipidemia with high triglycerides (>95%) and moderately increased VLDL, LDL, and cholesterol (75%). Elevated lipids manifest as xanthomas.	To maintain blood glucose in young infants, nasogastric tube feedings are used. Older children can be given raw cornstarch (see GSD type 0) Diet is closely monitored and balance is maintained between blood glucose and hepatic glycogen stores. Excess carbohydrate intake must be avoided; caloric intake monitored to keep it minimal but adequate. Because of the hyperlipidemia, high intake of saturated fats and cholesterol is avoided. For hepatic adenomas, surgical removal may be required.	The rarer type Ib disease is a deficiency of the glucose-6-phosphate transporter that moves glucose-6-phosphate into the lumen of the endoplasmic reticulum.

(continued)

Name(s)/OMIM	Mechanism	Description	Treatment	Notes
Glycogen storage disease (GSD; type II) (Pompe disease) OMIM #232300	Deficiency of the **acid maltase** (**α-glucosidase**) enzyme that degrades glycogen to glucose in lysosomes.	Excessive glycogen accumulates in the body but especially in muscle and heart. Symptoms in infancy include feeding problems, muscle weakness, head lag, and poor weight gain. Lung infection complicates respiratory problems and heart is enlarged either of which can lead to death.	Requires a team (neurologist, cardiologist, and respiratory therapist) to monitor treatment. Enzyme replacement therapy has been developed. Alglucosidase-alpha drug is approved.	Partial enzyme deficiency leads to onset in juveniles or adults with muscle weakness and respiratory problems that can lead to death. Heart not as affected.
Glycogen storage disease (GSD; type III) (Cori or Forbes disease) OMIM #232400	Deficiency of the **glycogen debrancher** enzyme that is responsible for moving three glucose residues from a branch of four glucose residues to the end of a longer chain. The enzyme then removes the last glucose residue, thus completing the branch removal.	Diagnosed in infancy or early childhood. Enlarged liver (hepatomegaly) and low sugar (hypoglycemia) are characteristic with possible ketoacidosis. Some patients have high lipids and growth retardation but not to the extent seen in type I. Muscle involvement in type III leads to low muscle tone hypotonia and muscle wasting, more likely seen in adults than children. Adult patients may also present with heart problems (cardiomyopathy). Female adults may exhibit polycystic ovaries.	Patients need to avoid hyperglycemia. This is less of an issue than in type GSD type I so monitoring is easier. High protein diet may be useful to provide amino acids for gluconeogenesis. Amino acids trigger release of intestinal glucagon-like peptide to enhance conversion of dietary amino acids to glucose.	—
Glycogen storage disease (GSD; type IV) (Andersen disease; amylopectinosis) OMIM #232500	Deficiency of the **glycogen branching enzyme** that catalyzes the addition of a glucose residue as a branch to a glucose chain of glycogen using an α-1,6 glycosidic bond.	Patients present in early infancy with poor feeding and develop severe liver disease and enlargement (hepatomegaly) that can lead to terminal liver failure/scarring (cirrhosis) by the age of 5 years. Liver damage leads to portal hypertension resulting in esophageal varices, encephalopathy, spleen enlargement (splenomegaly), ascites, and/or diminished renal function. Liver biopsy shows excessive accumulation of glycogen. No muscle symptoms are detected.	Treatment options are minimal. There are serious medical problems associated with hepatic failure due to cirrhosis and hepatosplenomegaly Except for liver transplant, little can be done. Diet therapy may limit hepatomegaly, hypoglycemia, and lessen symptoms but with only partially successful outcome.	—

Name(s)/OMIM	Mechanism	Description	Treatment	Notes
Glycogen storage disease (GSD; type V) (McArdle disease) OMIM #232600	Deficiency of **myophosphorylase** (muscle glycogen phosphorylase) enzyme that catalyzes the removal of glucose residues from glycogen to form glucose-1-phosphate that ultimately enters glycolysis to provide muscle energy.	Only the skeletal muscle isoform is defective. Muscular weakness and fatigue are observed with attempted exercise; tiredness and cramping affects normal activity. Vigorous exercise may cause muscle breakdown (rhabdomyolysis), leading to myoglobin in the urine (myoglobinuria).	High carbohydrate diet may improve exercise tolerance and the ability to do muscle work. Avoid high-intensity anaerobic exercise; moderate aerobic activity is valuable.	A "second-wind" phenomenon can occur associated with increased uptake of glucose that can be used in glycolysis as fuel.
Glycogen storage disease (GSD; type VI) (Hers disease) OMIM #232700	Deficiency of the liver **glycogen phosphorylase** isoform that catalyzes the removal of glucose residues from glycogen to form glucose-1-phosphate that ultimately is converted to glucose following food deprivation.	Present in early childhood (ages 1–5 years). Patients exhibit liver enlargement (hepatomegaly) and growth retardation. Enlargement of the spleen (splenomegaly) does not occur. Symptoms of low sugar (hypoglycemia), high lipids, and blood ketones (ketosis) are generally mild (compare with type I). Although lactic acidemia occurs after eating, it is not severe enough to cause high uric acid levels.	Dietary restrictions are sufficient because the disease is mild. Frequent feedings are only recommended for patients who develop fasting hypoglycemia. Many patients require no intervention. Disease conditions improve with age.	—
Glycogen storage disease (GSD; type VII) (Tarui disease) OMIM #232800	Deficiency of the muscle **phosphofructokinase (PFK)** enzyme that catalyzes the glycolytic conversion of fructose-6-phosphate to fructose-1,6-bisphosphate in the third reaction of the glycolytic pathway.	Although the deficiency is in glycolysis, this is considered as a GSD because of the inability to use glycogen-derived glucose for fuel. Glycogen accumulates in the muscle and glucose-6-phosphate enters the pentose phosphate pathway generating excess ribose sugar for nucleotide synthesis. Excess nucleotides degrade to uric acid, causing gout. Decreased formation of 2,3-bisphosphoglycerate enhances Hgb binding with O_2. Vigorous exercise may cause muscle breakdown (rhabdomyolysis), leading to myoglobin in the urine (myoglobinuria).	No specific treatment is needed, although patients should avoid carbohydrate-rich diets. Vigorous exercise must be avoided to prevent myglobinuria.	Red blood cells are also affected with a decrease of PFK activity to 50% of normal (see below). This partial deficiency leads to hemolysis.

Hgb, hemoglobin; LDL, low-density lipoprotein; VLDL, very-low-density lipoprotein.

MITOCHONDRIAL ENZYMES (EXCLUDING UREA CYCLE AND FATTY ACID OXIDATION)

Name(s)/OMIM	Mechanism	Description	Treatment	Notes
Alpers' disease (progressive infantile poliodystrophy) OMIM #203700	Deficiency of mitochondrial DNA polymerase-γ (POLG), which contains both a DNA polymerase and a separate proofreading function. POLG is important in the replication and transcription of mitochondrial DNA.	Patients classically present as infants or children with a progressive decrease in neurological and motor function, liver failure, and intractable epileptic seizures, which often lead to death.	No cure or treatment other than supportive.	Patients may show other disturbances in the mitochondrial complexes of oxidative phosphorylation.
Fumarase deficiency OMIM #606812	Defect of **fumarase**, the citric acid cycle enzyme, which converts fumarate to malate.	Defects mainly affect the nervous system, causing problems with brain development, developmental delay, poor feeding, poor muscle tone, severe mental retardation, speech problems, and seizures. High levels of lactic, pyruvic, and fumaric acid may be the cause of these medical problems [see also pyruvate dehydrogenase (PDH) deficiency below]. Unusual facial features (e.g., large forehead, small jaw, flattened nose, low-set ears, and widely spaced eyes) may also be seen.	No cure or treatment other than supportive.	Other mutations of the fumarase gene can result in the formation of skin and uterine fibroids (leiomyomata) or kidney cancer.
Leigh syndrome (subacute necrotizing encephalomyelopathy) OMIM #256000	A group of protein and enzyme deficiencies that all affect one of the mitochondrial **cytochrome complexes I–V**, causing defects in oxidative phosphorylation and ATP synthesis. Mutations can occur in mitochondrial or nuclear DNA. Other mutations affect **mitochondrial tRNAs**, the **PDH complex**, or **coenzyme Q**.	The defect causes early and progressive demyelination and degeneration of nerves, usually along the central brainstem, thalamus, basal ganglia, cerebellum, and/or spinal cord. Muscles are affected as per the nerve involvement. Patients may also experience nausea and vomiting, poor feeding, irritability, and seizures. Symptoms usually start from the age of 3 months but can appear later including into adulthood.	There is no cure. Treatment is mainly supportive (some diet variations have been tried) but ultimately patients usually die within 2–3 years. Adult-onset patients may live longer but almost always die from this disease.	An X-linked version of Leigh syndrome involves the PDH E1 enzyme subunit. The mutation renders the complex inactive.

Name(s)/OMIM	Mechanism	Description	Treatment	Notes
Pyruvate Dehydrogenase (PDH) deficiency OMIM #312170	Defect of the **PDH** complex, composed of three main enzymes (E1, E2, and E3) each made of multiple protein subunits. The complex catalyzes the reaction, which converts pyruvate into acetyl-CoA and links glycolysis with the citric acid cycle. The most common deficiency is due to mutations of the α-subunit of E1.	Defects of PDH and increased levels of pyruvate result in the buildup of large amounts of lactic acid in the blood. Chronically, high levels of lactic acid have adverse effects on the brain, nerves, and other tissues, leading to developmental problems, spasms of muscles, and, often, early death.	There is no cure. Diets that produce ketones (high protein and low carbohydrate) have been used with variable success. Lactic acidosis must be treated acutely, although some medications (e.g., dicholoroacetic acid) may help. Gene therapy for this condition is being developed.	Usually X-linked with more severe effects seen in some heterozygous females. Autosomal recessive variants are also known.

ATP, adenosine triphosphate; CoA, coenzyme A; DNA, deoxyribonucleic acid; tRNA, transfer ribonucleic acid.

LIPIDS AND FATTY ACID OXIDATION ERRORS

Name(s)/OMIM	Mechanism	Description	Treatment	Notes
Abetalipoproteinemia (Bassen–Kornzweig syndrome) OMIM #200100	Deficiency of **apolipoprotein (apo) B48 and B100**. Defect results in decreased production and transport of chylomicrons and very-low-density lipoprotein (VLDL) as well as poor absorption of fat-soluble vitamins.	Patients exhibit symptoms in early infancy (due to inactive pancreatic lipase), including poor feeding and growth, diarrhea, and fat-laden and foul-smelling stools. Red blood cell shape and nerves/muscles may also be affected with mental retardation and developmental problems as well as effects on coordination, balance, and movement. Fat-soluble vitamin deficiencies add to these problems, including gradual breakdown of the retina due to lack of vitamin A.	Treatment is usually by close control of diet, marked increases in vitamin E intake, and supportive care for symptoms.	—

(continued)

Name(s)/OMIM	Mechanism	Description	Treatment	Notes
Acyl-CoA dehydrogenase deficiencies OMIM #201475, 201460, 201450, 606885	Defects in very **long-chain acyl-CoA dehydrogenase**, **long-chain acyl-CoA dehydrogenase, medium-chain acyl-CoA dehydrogenase,** and **short-chain acyl-CoA dehydrogenase** enzymes, which are involved in the initial breakdown in mitochondria of fatty acids of varying length so they may enter the lipid β-oxidation pathway. Defects in these enzymes result in the buildup of its particular substrate lipid.	All types of Acyl-CoA dehydrogenase deficiencies share common symptoms, including potentially serious liver and heart problems (including cardiac arrest), low blood sugar with the production of ketones and resulting low energy, poor feeding and growth/development, poor muscle tone and possible seizures, and/or coma. Symptoms can be exacerbated by stress, illness, infection, prolonged exercise, and/or fasting/starvation.	There is no cure for these disorders. Intravenous glucose is used to treat acute problems. However, low-fat diets, the addition of carnitine to reduce toxic buildup of fatty acids (excreted in the urine as acyl-carnitine), and avoidance of alcohol and of fasting periods longer than 10 hr help to reduce symptoms and dangerous liver or heart disease.	—
Carnitine palmitoyltransferase (CPT) I deficiency OMIM #601987	Defect in **CPT I** that prevents formation of palmitoylcarnitine and hence leads to a lack of transport of long-chain fatty acids from the cytosol to the mitochondrial matrix. As a result, these long-chain fatty acids cannot be oxidized for energy production.	Symptoms usually present in young children/teens with muscle aches and breakdown (rhabdomyolysis with myoglobin in the urine), especially during fasting, cold exposure, or stressful situations. The liver is often enlarged with progressive damage. Laboratory measurements show high carnitine, liver transaminase, free fatty acids, and ammonia. Blood sugar is low and can lead to listlessness and progress to coma and even death if untreated.	It is possible to treat this disorder with the proper diet and patients can live a normal life if proper treatment is initiated before brain damage occurs. Avoidance of fasting and administration of medium-chain triglycerides with a low-fat diet seems to help avoid the severe metabolic consequences.	Early diagnosis and treatment of this disorder is successful in avoiding long-term neurological damage. The disorder tends to become milder with age.

Name(s)/OMIM	Mechanism	Description	Treatment	Notes
Carnitine palmitoyltransferase (CPT) II deficiency OMIM #255110	Defect in **CPT II**, leading to a lack of conversion of palmitoylcarnitine back to palmitoyl-CoA, thus impairing transport of long-chain fatty acids into the mitochondrial matrix.	Same as CPT I above.	Same as CPT I above. Benzafibrate, a drug used for low lipids may increase expression of the defective proteins and restore function.	Neonatal and infant forms of this disease are fatal.
Carnitine translocase deficiency OMIM #212138	Defect of **carnitine–acylcarnitine translocase**, one of a family of membrane carrier proteins that transports fatty acids/lipids between the cytosol and mitochondrial matrix.	During fasting/starvation, patients display low blood sugar and ketones as well as high ammonia that can lead to seizures and coma. Carnitine levels are also low. Skeletal muscles may be poorly developed and weak. Heart problems (cardiomyopathies and irregular heartbeats) are also common.	Acute treatment may include providing glucose.	Mitochondrial oxidation of fatty acids is especially important in skeletal and cardiac muscle as well as liver during periods of low food intake.
Fabry disease OMIM #301500	Defect in **α-galactosidase A**, which removes the terminal α-galactosyl residues from glycolipids and glycoproteins. As a result, these glycosphingolipids accumulate (e.g., globotriasocylceramide).	Symptoms appear in childhood or adolescence and include intestinal, heart, kidney, vision, and skin disorders as well as pain in the limbs and extremities because of the accumulation of the glycolipids/glycoproteins.	Treatment is currently only supportive. Infants usually die by the age of 3 years because of infections of the lungs secondary to the progressive muscle weakness (inability to swallow properly or cough). Onset in adolescence usually results in death in early adulthood from heart and/or kidney complications. Enzyme and gene replacement treatments are being researched.	—

(continued)

Name(s)/OMIM	Mechanism	Description	Treatment	Notes
Farber lipogranulomatosis OMIM #228000	Defect in **lysosomal hydrolase—acid ceramidase (ASAH).** ASAH cleaves ceramides into sphingosine and a free fatty acid. The defect leads to accumulation of ceramides in tissues, specifically in macrophages, especially seen in the heart, central nervous system, liver, and spleen.	Seven variants have been identified with slightly different symptom complexes and survivability. Patient symptoms can include joint pain with deformation, voice changes due to laryngeal disease, progressive nerve and muscle degeneration, and deposits in the skin, especially over joints.	Patients usually die at approximately age of 2 years. There is no treatment other than supportive.	Type 5 Farber patients have the same cherry red spot as seen in Tay–Sachs disease.
Gaucher's disease (GD) OMIM #230800	Defect in **lysosomal hydrolase—acid β-glucosidase** (also known as **glucocerebrosidase**), which breaks down glucocerebrosides (mainly found in red and white blood cells). The defect causes accumulation of glucocerebroside (e.g., glucosylceramide) in macrophages.	GD presents as three major types. Type 1 or adult GD (most common) does not involve the nerves; types 2 and 3 involve the nerves. All forms have liver enlargement (hepatomegaly) as well as some spleen enlargement (splenomegaly) and bleeding disorders, usually due to lowered platelet numbers.	Type 2 GD patients usually die by the age of 2 years Type 3 patients usually live longer than type 2.	Glucocerebroside-laden macrophages are known as Gaucher cells and appear as wrinkled tissue paper on pathological examination.
Krabbe disease OMIM #245200	Defect of **lysosomal hydrolase, galactosylceramidase (galactocerebrosidase)**, which removes galactose residues from ceramide. The defect leads to accumulation of galactosylceramides found predominantly in myelin sheaths of nerve cells. As a result of lipid accumulation, characteristic "globoid cells" with multiple nuclei are seen in brain white matter on pathological examination.	Infant, juvenile, and adult forms have been characterized. The disease affects only the nervous system with severe mental, nerve, and muscle developmental problems and degeneration, eating problems, blindness, deafness, irritability, seizures, vomiting, and nonresponse to surroundings. Juvenile and adult forms have similar but slower progressing symptoms.	Treatment is only supportive, although bone marrow transplantation can cure patients who display nerve involvement. Infants usually die by the age of 2 years.	Krabbe disease can be confused with cerebral palsy on initial presentation.

Name(s)/OMIM	Mechanism	Description	Treatment	Notes
Metachromatic leukodystrophy (MLD) [arylsulfatase A (ARSA) deficiency, sulfatide lipodosis] OMIM #250100	Defects in the lysosomal hydrolase, **ARSA**, which cleaves 3-O-sulfogalactosyl residues from glycolipids. The defect results in the accumulation of these glycolipids, mainly found in myelin sheaths of central and peripheral nerves.	MLD presents as late infantile (<2 years old), early juvenile/late juvenile (3–10 years old), and adult forms (>16 years old). All patients have nerve cell disease including loss of feeling, walking problems, visual and speech problems, paralysis, coma, and seizures. The disease gradually progresses until complete blindness occurs and patients do not respond to any surroundings.	Treatment for MLD is only supportive. The infantile form is more severe with death about 5 years after disease onset. Juvenile and adult forms often progress more slowly but usually die before the age of 20 years. Bone marrow transplantation, enzyme and gene replacement therapy, and various medications are being developed and may offer hope for affected individuals.	—
Niemann–Pick disease OMIM #257200	Types A and B are due to defects in **sphingomyelinase**, which cleaves sphingomyelin into ceramide and phosphocholine. Both types are due to single amino acid changes, which results in no activity (A) or some residual activity (B). Foamy, lipid-filled "Niemann–Pick" cells are seen on pathological examination.	Patients with type A present within 4–6 months and have rapid breakdown of nerves with poor muscle tone and feeding as well as liver enlargement. A cherry-red spot is seen on the retina (see Tay–Sachs below). Progressive weakness leads to death usually by the age of 18 months to 3 years. Type B patients show variable symptoms mainly involving the internal organs (e.g., liver enlargement and lung involvement). Type B patients can live into their teens or early adulthood.	Treatments have not proven successful but may include organ and bone transplant, although success is limited. Treatment is otherwise supportive. Gene replacement therapy and various medications are being developed and may offer hope for affected individuals.	A type C disease also exists with a defect in transport of cholesterol and other lipids across membranes by a still unknown mechanism. Type C patients follow a similar course as type B.

(continued)

Name(s)/OMIM	Mechanism	Description	Treatment	Notes
Refsum disease (classic or adult) (defective α-oxidation) OMIM #266500	Defect in the enzyme, **phytanoyl-CoA hydroxylase (PhyH),** from by α-oxidation that produces pristanic acid (a 19-carbon fatty acid) from phytanic acid. Pristanic acid is then oxidized by the fatty acid β-oxidation pathway. The deficit leads to poorly formed myelin sheaths that affect nerves/muscles.	Symptoms are usually seen at birth and include decreased muscle tone, developmental delay, mental retardation, gait widening, poor night vision and hearing, liver enlargement (hepatomegaly) and scaly skin.	Phytanic acid is mainly in dairy products, meat, and fish. Diets low in phytanic acid have been tried for treatment but compliance is difficult.	An infantile form exists with a distinct mechanism. Patients manifest similar symptoms to adult Refsum but also have a marked orange-yellow color of the teeth.
Sandhoff disease (Jatzkewitz–Pilz syndrome; hexosaminidase B deficiency) OMIM #268800	Defect in the **β-subunit** of **hexosaminidase,** the enzyme responsible for the removal of the end *N*-acetyl-D-hexosamine (GalNAc) residues from GM2 gangliosides as part of the function of cellular recognition, receptors, and signaling. As a result, gangliosides accumulate.	Patients usually show symptoms in the first few months of life, including muscle weakness, a markedly increased startle response but also a progressive nonresponsiveness to people and muscle/nerve regression, and increasingly frequent and severe seizures. A notable cherry-red spot of the retina may be seen in some patients (see Tay–Sachs disease below). The disease progresses over a year or two to an almost completely unresponsive state with death following, usually due to lung infections or other problems.	Treatment is currently only supportive. Infants usually die by the age of 3 years because of infections of the lungs secondary to the progressive muscle weakness (inability to swallow properly or cough). Enzyme and gene replacement treatments are being researched.	GM2-gangliosidosis, AB variant (also known as GM2 activator deficiency diseases) involves the deficiency of a small glycolipid transport protein, which is a cofactor for hexosaminidase α- and β-subunits. Symptoms and treatment are as for Tay–Sachs and Sandhoff diseases.

FTB

Name(s)/OMIM	Mechanism	Description	Treatment	Notes
Tangier disease OMIM #205400	Defect in **ATP-binding casette (ABC) transporter** (also known as cholesterol efflux regulatory protein), which transports cholesterol and phospholipids into apo-A1- and apoE-containing vesicles to from high-density lipoproteins (HDLs). In addition, ABC transports lipids between the cell membrane and the Golgi body. The defect results in accumulation of cholesterol in tissues such as tonsils, lymph nodes, spleen, thymus, and bone marrow.	Patients usually develop early blockages of arteries (atherosclerosis), have mild increase of blood triglycerides, enlarged liver and spleen (hepatosplenomegaly), clouding of the cornea, and some decreases in nerve function. The tonsils are characteristically orange-yellow and enlarged.	Treatment is problematic and varies between patients but usually includes medication (e.g., fibrates and nicotinic acid) to increase HDL levels in order to avoid cardiovascular disease.	The disease is very rare and has only been identified in approximately 50 people.
Tay–Sachs Disease (hexosaminidase A deficiency) OMIM #272800 FTB	Defect in the **α-subunit** of **hexosaminidase**, the enzyme responsible for the removal of the end *N*-acetyl-D-hexosamine (GalNAc) residues from GM2 gangliosides as part of the function of cellular recognition, receptors, and signaling. As a result, gangliosides accumulate.	Tay–Sachs presents as infantile-, juvenile-, and adult-onset forms. The infantile form is very severe with deterioration of nerve and muscle functions with blindness, deafness, and swallowing difficulties. The juvenile form is rare with onset between 2 and 10 years with similar nerve and muscle symptoms. The adult form occurs in the 20s and 30s and has mild nerve and muscle affects.	Treatment is currently only supportive. Infants usually die by the age of 2–4 years. Juveniles usually die between 5 and 15 years. The adult form is usually not fatal. Enzyme and gene replacement treatments are being researched.	"Tay–Sachs disease" is normally used for the infantile form of the disease. A classic "cherry-red" spot of the retina is sometimes seen in Tay–Sachs and related diseases (Gaucher's Niemann–Pick). The spot is caused by an increase in the amount of gangliosides in the ganglion cells around the fovea, resulting in a relative, milky white transparency of the retina. The center of the fovea, which has no ganglion cells, then appears bright red in comparison.

CoA, coenzyme A; HDL, high density lipoprotein; VLDL, very-low-density lipoprotein.

NUCLEOTIDE METABOLISM

Name(s)/OMIM	Mechanism	Description	Treatment	Notes
FTB **Lesch–Nyhan syndrome (LNS)** (juvenile gout) OMIM #300322	Deficit of **hypoxanthine-guanine phosphoribosyltransferase** (HGPRT), which converts hypoxanthine to IMP and guanine to GMP as part of the purine salvage pathway.	LNS patients have profound neurological problems, delayed development of muscle function, abnormal posturing and movements (e.g., hands and feet), severe behavioral traits (e.g., involuntary self-mutilation, including harmful finger, tongue, and lip biting), and varying degrees of mental retardation. Because of the involvement in the purine pathway, uric acid levels are high and result in associated kidney stones, gouty arthritis, and symptoms of classic gout.	Treatment is symptomatic and is focused on lowering uric acid levels to reduce nerve, kidney, and joint damage, attempts at controlling movement and behavioral problems and protection against self-injury (often including restraints and even extraction of teeth to avoid biting injuries). Further medical treatments are being researched.	X-linked and, thus, predominately in males. A less severe form of HGPRT deficiency is known as Kelley–Steegmiller syndrome in which neurological and movement problems are absent.
Orotic aciduria (type I) [uridine monophosphate (UMP) synthase deficiency] OMIM #258900	Deficit of two enzymes: **orotidine-5′-pyrophosphorylase** and **orotidine-5′-phosphate decarboxylase**. The functions of these two enzymes are part of the **UMP synthase complex** and are responsible for the formation of UMP from orotate, the last step in pyrimidine nucleotide synthesis. The two enzymes form orotidine 5′-monophosphate (OMP) from orotate and phosphoribosylpyrophosphate (PRPP) and the subsequent removal of a carboxyl group from OMP to form UMP, respectively.	Patients with orotic aciduria have an inhibition of RNA and DNA synthesis due to the decreased availability of pyrimidines (UMP) with associated mental impairment and retardation. They also have increased levels of orotic acid in the urine, which can lead to stones and kidney damage. Patients usually have chronic anemia (megaloblastic), which does not respond to conventional treatments.	Treatment is with uridine (converted to UMP to bypass the enzyme block) as well as 5′-cytidylic acid and 5′-uridylic acid to reduce both the levels of orotic acid and the chronic anemia.	A second form of orotic aciduria, termed type II, results from the isolated deficit of orotidine-5′-phosphate decarboxylase. It has only been seen in one patient.

Name(s)/OMIM	Mechanism	Description	Treatment	Notes
Renal lithiasis [adenine phosphoribosyl-transferase (APRT) deficiency] OMIM #102600	Deficit of **APRT**, an enzyme that catalyzes the formation of AMP from adenine and PRPP as part of the recycling pathway to turn adenine into AMP. The deficit (full or partial) results in increased levels of the purine 2,8-dihydroxyadenine (DHA).	At least three specifically characterized types have been reported. Patients suffer from repetitive episodes of gouty arthritis and kidney stones formed from DHA. Kidney function can be progressively impaired until failure results.	Treatment is aimed at lowering DHA levels and stone formation and includes the use of allopurinol.	—
Severe combined immunodeficiency disease (SCID) [adenosine deaminase (ADA) deficiency] OMIM #102700	Deficit of **ADA** that removes an amino group from adenosine to form inosine or from deoxyadenosine to form 2′-deoxyinosine. Although found in all tissues, ADA is highest in lymph tissues (e.g., the thymus) and, therefore, affects lymphocytes. The deficit results in increased levels of dATP in the blood and urine, which inhibits DNA replication and causes DNA damage, especially in lymphocytes responding to an infection.	SCID presents within the first month of life. Symptoms include decreased feeding, chronic diarrhea, loss of fluids, poor growth and development, and numerous infections due to decreased lymphocyte activity. Most concerning of these infections is pneumonia from *Pneumocystis carinii* or from a DNA or RNA virus (e.g., cytomegalovirus, varicella, and parainfluenze). Most patients die within 1–2 years because of these infections. ADA deficiency also lowers transmethylation reactions important in the synthesis of several amino acids and peptide hormones.	Treatment is supportive with prevention of infections, appropriate and immediate antibiotic intervention, and close monitoring of the immune system. Replacement of normal ADA using ADA-positive red blood cell transfusions can be temporarily beneficial. Longer survival is possible with bone marrow transplantation or gene therapy.	SCID due to ADA deficiency was the first disease approved for gene transfer therapy. These efforts have not yet surpassed enzyme replacement therapies but have encouraged future research in gene therapy for this and other diseases.
Xanthanuria [xanthine dehydrogenase (XDH)/xanthine oxidase (XO) deficiency] OMIM #607633	Deficit of **XDH** enzyme that is easily converted to the enzyme **XO**. XDH forms urate, which is proposed to serve a powerful antioxidant function. XO produces some of these same reactive, oxygen-derived species (e.g., superoxide and hydrogen peroxide), which may play a role in cell injury.	Deficits result in high levels of xanthine in the urine, which can form xanthine stones. Chronic stone formation can lead to eventual kidney impairment and failure. Uric acid levels are low. The exact roles of these enzyme products and the change between XDH and XO are still being investigated.	Treatment is supportive but also includes avoidance of foods containing purines as well as high fluid intake to reduce xanthine stone formation.	The enzymes rely on molybdenum as a cofactor. Therefore, deficiencies of this element can also lead to an acquired form of xanthanuria.

AMP, adenosine monophosphate; dATP, deoxyadenosine triphosphate; DNA, deoxyribonucleic acid; GMP, guanine monphosphate; IMP, inosine monophosphate; RNA, ribonucleic acid.

DEFECTIVE DNA

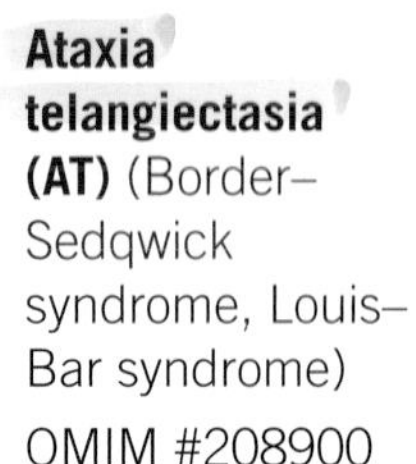

Name(s)/OMIM	Mechanism	Description	Treatment	Notes
Ataxia telangiectasia (AT) (Border–Sedqwick syndrome, Louis–Bar syndrome) OMIM #208900	Mutations affect the **AT mutated (ATM)** protein, a serine/threonine kinase responsible for activation of DNA damage checkpoint proteins. The resulting defect in the repair of ionizing radiation (e.g., X-rays) damage to DNA leads to the disease state. Affected individuals have mutations in the *ATM* gene located on chromosome 11 (11q22.3) but the gene is large (66 exons) and over 400 mutations have been reported. Offspring of consanguineous matings are usually homozygous, whereas products of nonconsanguineous unions usually are compound heterozygotes. The vast majority of mutations are truncating with a minority being missense.	AT is a complex disorder involving multiple systems. Children with this disorder often are noticeably unsteady as soon as they begin to walk (ataxia). Balance difficulties become so severe that most require a wheel chair by second decade of life. Other neurologic signs are slow, involuntary, writhing movements of the arms and fingers. Head turning upon initiation of voluntary eye movements, especially lateral gaze, known as oculomotor apraxia, is common. Speech is often severely slurred, and muscle wasting in the hands and arms is prominent. The immune system is compromised and reduced resistance to infection may be present. Appearance of "spider" type blood vessels (telangiectasias) between the ages of 3 and 5 are diagnostic when combined with the unsteadiness, although not all patients have these. These usually appear on the cheeks, eyelids, ears, and the forearms. Blood vessels over the whites of the eyes may appear especially prominent. They are usually seen on the pinnae, the forearms, the eyelids, and the butterfly areas of the face. Individuals are at risk for cancer, and there is an increased risk of the same in siblings, especially breast cancer.	No treatment is known. Many patients die in the first decade of life, possibly as a result of nasopharyngeal and pulmonary infections, although cancer is a major risk. There have also been reports of patients who lived into the fourth and fifth decades. Radiation therapy must be avoided and both patients and their heterozygous sibs must avoid all ionizing radiation such as ultraviolet rays. It is strongly suggested that mammograms should be avoided. Exposure to the common upper respiratory viral infections should be limited and childhood vaccinations are risky.	The risk of malignancies cannot be overemphasized. Lymphomas and leukemia can precede the onset of ataxia and, while irradiation for these tumors can be "curative," the treatment itself can incite new tumor formation. It is recommended that young children with these malignancies undergo thorough neurological evaluations prior to treatment.

Name(s)/OMIM	Mechanism	Description	Treatment	Notes
Bloom syndrome OMIM #210900	Deficit of **DNA helicase RecQ protein-like-3,** leading to structural alterations.	Bloom syndrome causes growth deficiency both before and after birth. Low birth weight and failure to thrive are common. It is associated with abnormal and sun-sensitive skin pigmentation and "spidery" blood vessels. Chromosomes are fragile in this condition with frequent fragmentation and an associated increased risk of malignancies (see also AT). Children are at considerable risk for leukemia.	No effective treatment has been found. Avoidance of sun exposure and prompt diagnosis and treatment of malignancies is essential. Survival into adult years is the exception as the result of malignancies.	—
Cockayne syndrome OMIM #216400, #133540	Deficit of either of two DNA repair proteins, namely, **DNA excision repair protein ERCC-6** (type B) or **ERCC-8** (type A). ERCC-6 is involved in repair during transcription. ERCC-8 is a subunit part of RNA polymerase II transcription factor IIH. The deficit leads to poor repair of DNA [e.g., ultraviolet (UV) radiation], leading over time to cellular malfunction/death.	Patients develop symptoms early in the first decade with growth failure (short stature, small head, thin and dry hair), poor development of the nervous system including mental retardation, increased sensitivity to sunlight/UV radiation, eye and hearing problems, marked dental decay, and accelerated aging.	There is no known cure and treatment is only supportive. Death is usually well before adulthood.	Cockayne syndrome, type B (involving ERCC-6) is seen prominently in Amish peoples. Both types A and B are inherited as autosomal recessive.
Fanconi anemia OMIM #227650 BI	Deficit in any of a number of proteins, which form a monoubiquitination complex, responsible for DNA repair during the S-phase of mitosis and after erroneous cross-linking. The deficit has wide-ranging effects on all blood cells from the bone marrow as well as heart, kidney, skin, and limbs.	Patients have low red and white blood cells and platelets; changes in skin pigment; structural defects of the heart, kidneys, and bones of the limbs and hands; and often develop leukemia, which may lead to death.	Treatment is by temporary blood transfusions, potentially curative bone marrow transplantation, as well as varied medical methods to attempt to stimulate marrow production of deficient cell types.	—

(continued)

Name(s)/OMIM	Mechanism	Description	Treatment	Notes
Fragile X syndrome (FRAXA) OMIM #300624 BI	Defect of an RNA-binding protein (fragile X mental retardation protein FMRP) that is part of the ribonucleoprotein–polyribosome complex, expressed at the highest level in early fetal tissue and testicular cells developing new sperm. The defect stops the protein–RNA binding and disrupts DNA replication, leading to the disease. FMRP is found in both the cytoplasm and nucleus. Several RNAs are affected, including those involved in nerve–muscle signal transmission (e.g., glutamate receptor), maturation of nerves, and the structure of the intracellular cytoskeleton.	Patients have varying levels of mental retardation, speech and behavioral problems, eye and visual disorders, and in some cases, degeneration of nerves. Patients may also perseverate (repeat a phrase or perform a task over and over). Females may develop ovarian failure. Patients may also have altered physical traits, including a long narrow face with large ears, highly flexible joints, and enlarged testicles.	Treatment is supportive only.	FRAXA is an X-linked dominant disorder affecting males approximately 2:1. It is the most common form of inherited mental retardation.
Friedreich ataxia OMIM #229300 BI	Defect in the protein **frataxin** found in the mitochondrial membrane and believed to be involved in oxidation–reduction reactions via iron binding and association with the iron–sulfur cluster found in complexes I–III.	Frataxin is found in several tissues but prominently in the dorsal root ganglion/cerebellum as well as the heart and pancreas. Symptoms result from disease of these organs, including frequent falls, lack of coordination, speech abnormalities, muscle weakness and visual problems, as well as heart problems and diabetes.	Symptomatic treatment for neuromuscular, heart, and diabetes complications. Some newer medicines may offer additional treatments. Patients usually die in their 30s or 40s.	—
Myotonic dystrophy (DM) OMIM #160900	Defect in the **dystrophia myotonica protein kinase (DMPK)** a cAMP-dependent serine–threonine protein kinase found mainly at neuromuscular junctions and points of connection points for muscles and tendons. CTG repeats in DNA leads to RNA CUG repeats and erroneous posttranscriptional splicing of other RNAs (DMPK is spliced normally).	Patients present as infants, children, or adults and suffer from muscle spasms and involuntary contractions as well as degeneration of muscles of the limbs (from distal to proximal), neck, eyes, and jaw. Insulin resistance, a precursor of diabetes mellitus, is often seen along with heart disorders, cataracts, and sleeping problems. Children will often manifest varying degrees of mental retardation.	DM is autosomal dominant in inheritance and is the most common type of adult muscular dystrophy (~1 in 8000). There is no cure and treatment is only supportive, including monitoring/treating heart, eye, sleep, and altered sugar metabolism. The infant form is often quickly fatal.	Altered RNAs include an insulin receptor, cardiac troponin T, a chloride channel, a cardiac Ca^{2+}-ATPase and other muscle-related proteins. These RNAs correspond with symptoms.

Name(s)/OMIM	Mechanism	Description	Treatment	Notes
Spinobulbar muscular atrophy (SBMA) (Kennedy's disease) OMIM #313200	Defect of the dihydrotestosterone androgen receptor secondary to CAG repeats of the DNA, producing an erroneous, lengthy stretch of glutamine amino acid residues. This polyglutamine tract appears to markedly increase receptor function, leading to the disease symptoms.	Patients usually present in early-to-mid adulthood (20s–40s) and show progressive nerve and muscle degeneration affecting limbs and cranial nerves (seen as facial and lip drooping, difficulties in speaking and swallowing, and tongue twitching). Androgen receptor effects may be seen including shrinking of the testes, male breast enlargement, and sterility.	There is no cure or treatment.	SBMA is X-linked recessive and, therefore, usually affects only males. Females with the same defect show only mild symptoms and may reflect the impact of higher androgens in males versus females. It has a similar mechanism to Huntington's disease.
Xeroderma pigmentosum (XP) OMIM #278700	XP is a group of related mutations in which a component or components of a nucleotide excision repair complex for damaged DNA is defective. This repair mechanism is particularly important for the replacement of pyrimidine dimers, which can form from UV light absorption. The protein binds to DNA via a zinc-finger motif.	Patients exhibit markedly increased skin sensitivity to sun (e.g., severe sunburn/blistering/redness), dry skin, areas of prominent superficial blood vessels (telangiectasias), and areas of varying skin pigmentation. In addition, they experience a number of benign skin tumors and growths and have a 1000-fold chance of developing skin cancer (basal cell carcinoma) compared with unaffected individuals, even without sun exposure. The first skin cancers appear by the age of 10 or 20 years. Sunlight also affects exposed parts of the eye and make bright light painful. Some patients will also have nerve and sensory degeneration, including deafness, loss of some reflexes, and progressive mental retardation.	Treatment is supportive and includes avoidance of sunlight and constant surveillance and treatment of skin manifestations. Death is usually by the age of 20 years.	More severe forms of XP result from mutations in the DNA-binding region of the excision repair.

cAMP, cyclic adenosine monophosphate; DNA, deoxyribonucleic acid; RNA, ribonucleic acid.

BILIRUBIN METABOLISM

Name(s)/OMIM	Mechanism	Description	Treatment	Notes
Crigler–Najjar syndromes OMIM #191740	Deficit of **UDP-glucuronosyltransferase**, which attaches bilirubin destined for excretion to the carbohydrate glucuronate in the liver. Glucuronate increases the solubility of bilirubin for subsequent elimination in the feces. Loss of this function leads to an increase in bilirubin in the blood.	Patients have severe jaundice from elevated bilirubin (hyperbilirubinemia). Symptoms begin in the first days of life and persist unless treated, unlike the usual newborn jaundice. The whites of the eyes and the skin are deeply yellow. The circulating bilirubin is toxic to brain cells and can cause severe brain damage if not treated promptly. Liver transplantation is usually necessary to keep bilirubin at a safe level.	Vigorous and early lowering of bilirubin is essential to prevent nerve damage. Prompt phototherapy can avoid the need for exchange transfusions along with albumin and dextrose infusions. However, it needs to be used for 10–12 hrs on 35%–50% of exposed skin and gradually loses its effectiveness as children mature. Ursodiol and a lipid-rich diet can increase clearance of hepatic excretions of the bilirubin.	Unfortunately, phototherapy loses some of its effectiveness over time and many patients eventually need liver transplantation.
Dubin–Johnson syndrome OMIM #237500	**Deficit of an ATP-dependent, organic anion transporter in the bile ducts.** This transporter is essential for secretion of bilirubin conjugated to glucuronate out of liver cells and into the bile duct system. An isoform of this enzyme is able to transport the conjugated bilirubin into the blood for eventual excretion.	Most patients have no external symptoms, although some may have yellowing of the eyes and skin (jaundice). The liver often turns black because of the deposition of byproduct pigments.	Treatment is usually not necessary and is only supportive when needed. Lifespan is usually normal.	—
Gilbert syndrome OMIM #143500	Mutation in the TATA box for UDP-glucuronosyltransferase, reducing production of this enzyme required for bilirubin–glucuronate conjugation (see above). Unconjugated bilirubin levels subsequently rise in the blood.	Patients have yellow discoloration of the eyes and skin (jaundice) and sometimes fatigue and abdominal pain. Blood vessel blockage (atherosclerosis) also appears to be increased in Gilbert syndrome.	Treatment is usually not required and is only symptomatic/ supportive.	Gilbert syndrome symptoms are more commonly seen in men because they produce more bilirubin.

ATP, adenosine diphosphate; UDP, uridine diphosphate.

BLOOD CLOTTING FACTOR DEFECTS

Name(s)/OMIM	Mechanism	Description	Treatment	Notes
Hemophilia A, B, and C OMIM #306700, #306900, and #612416	Hemophilia A (classic) results from defects in factor VIII of the clotting cascade, which serves as a cofactor for factor IX/IXa to activate factor X to Xa. Hemophilia B results from defects in factor IX. Hemophilia C occurs when factor XI is deficient and fails to activate factor IX.	Patients with hemophilia exhibit easy bruising and bleeding, especially of muscles, soft tissue, and joints (e.g., weight-bearing joints including hips, knees, ankles, and elbows). Platelets are normal so small cuts do not result in severe bleeding. Hemophilia B patients most often die from bleeding inside the brain. Hemophilia C patients usually do not have joint bleeding.	Treatment is by intravenous replacement of the missing clotting factor or by fresh frozen plasma infusion.	Both hemophilias have varying severity from moderate to severe, depending on the level of the affected clotting factor. Because hemophilia A and B are X-linked, most, but not all, patients are male.
Factor V Leiden OMIM #188055	Mutation in factor V of the clotting cascade, which activates thrombin. The mutation decreases the natural breakdown by activated protein C, resulting in excessive clot formation.	Patients are at an approximately twofold increased risk for deep vein thrombosis clots in the legs and associated pulmonary embolism. Females may develop clots during pregnancy or while taking estrogen-containing contraception or hormone replacement.	Patients do not require treatment unless excessive clotting occurs. Treatment is then via anticlotting medications (e.g., heparin and warfarin)	—
von Willebrand disease OMIM #193400	Defect of the protein **von Willebrand factor (vWF)** found in the platelets and endothelial tissue, which increases initial platelet adhesion to the site of an injury and stabilizes factor VIII.	Symptoms vary with severity (quantitative deficiency) but include bleeding from skin, gums, and other mucus membranes. The bleeding can be serious in severe disease states.	Bleeding can be controlled by infusion of factor VIII or fresh frozen plasma infusion (which contains functional vWF). Unresponsive patients sometimes receive platelet transfusions in addition to the above treatments.	Like the hemophilias, varying levels of vWF can remain, resulting in levels of severity of the disease. Usually, autosomal dominant inheritance.

STEROID HORMONE SYNTHESIS

Name(s)/OMIM	Mechanism	Description	Treatment	Notes
3-β-Hydroxysteroid dehydrogenase deficiency (adrenal hyperplasia II) OMIM #201810	Deficiency of **3-β-hydroxysteroid dehydrogenase,** which converts pregnenolone to progesterone, 17-hydroxypregnenolone to 17-hydroxy progesterone, and/or dehydroepiandrosterone (DHEA) to androstenedione. The molecular genetics is less clear, as several *HSDB* genes can be involved but autosomal recessive appears to be the inheritance pattern. The deficit results in increased levels of pregnenolone, 17-hydroxypregnenolone, and DHEA.	This increase of these hormones leads to potentially fatal loss of salt through the kidney (due to effects on aldosterone and on cortisol synthesis). Severe electrolyte imbalance "adrenal crisis" with salt loss can occur and be so severe as to cause death. This disorder is also a cause of abnormal genitalia, and in males, an incorrectly placed urethral opening called hypospadias. In females, there may be virilization (masculinization) of genitalia due to excessive and uncontrolled production of DHEA.	Hormonal and electrolyte replacement therapy. Treatment with medications (e.g., hydrocortisone and fludrocortisone) can also be used. Chromosomal studies should be used to accurately determine gender in children. Unclear gender is treated with a combination of genetic testing, parent (patient) counseling, and surgery if required.	Abnormal genitalia in both genders is due to markedly lowered testosterone production from the liver (enough to produce some male characteristics in females but not enough to produce normal genitalia in males).
11-β-Hydroxylase deficiency (adrenal hyperplasia IV) OMIM #202010	A defective **11-β-hydroxylase** causes this form of adrenal hyperplasia. This deficiency results in an excess of steroid precursors such as 11-deoxycortisol, and 11-deoxycorticosterone (DOC).	Deficiency of 11-β-hydroxylase causing salt retention by the kidneys and high blood pressure. Low potassium (hypokalemia) is a threat. There is also decreased synthesis of cortisol and corticosterone. As a result, females often have genitalia resembling that of males (viriliization) due to excessive and uncontrolled production of androstenedione and as many as half are erroneously reared as such. Puberty occurs early and rapidly in males and they are often very aggressive. Patients may also be excessively tall. The early onset of high blood pressure due to retention of salt can be a serious threat to health.	The amount of virilization and blood pressure levels are not necessarily correlated. Hormonal regulation is key to managing signs and symptoms.	Karyotyping may be necessary in cases with abnormal genitalia and certainly in newborn offspring of families in which the gene is known to segregate. This condition is primarily found in Jewish and Arabic families throughout the Middle East and in the Amish of Pennsylvania.

Name(s)/OMIM	Mechanism	Description	Treatment	Notes
11-β-Hydroxysteroid dehydrogenase type 2 deficiency OMIM #202010 BI	Deficit of **11-β-hydroxysteroid dehydrogenase, type 2**, an enzyme found in kidney, placenta, testes, saliva and sweat glands, kidney, and parts of the brainstem converts cortisol to cortisone. A deficit results in low aldosterone and plasma renin activity.	Decreased activity of this enzyme leads to significant high blood pressure and low potassium (hypokalemia), primarily in juveniles. The ratio of urinary free cortisol to free cortisone correlates with the activity of renal 11-β-HSD and is not only useful for the diagnosis of this deficiency, but also as a risk factor for hypertension. Low birth weights and failure to thrive have been reported.	A salt-restricted diet and/or sprironolactone (mineralcorticoid receptor antagonist) can be helpful in managing the hypokalemic alkalosis and hypertension. These symptoms may respond to a low-sodium diet and/or sodium-depleting drugs. Early diagnosis is important for untreated patients who may become comatose and even die. High blood pressure medications are rarely needed.	A deficiency in 11-β-hydroxysteroid dehydrogenase, type 1 (found in liver, fat, and some nerve cells) leads to a decreased rate of conversion of cortisone to cortisol. A deficit of the type 1 isoenyzme leads to obesity and insulin resistance.
17-α-Hydroxylase deficiency (adrenal hyperplasia V) OMIM #202110 BI	Deficit of **17-α-hydroxylase** or combined **17, 20-lyase deficiency**. Production of corticosterone and DOC is excessive.	Patient can suffer from severe imbalances in electrolytes, hypertension, and often genital abnormalities. Gender assignment may be difficult at birth. Females do not menstruate (lack of estrogen synthesis) and some males develop female secondary sexual characteristics (lack of androgen production).	Attempts to normalize adrenal hormones might be of benefit for metabolic stabilization but the genital abnormalities are likely to persist.	Karyotypes are useful to determine gender. Patients can live to adulthood but sexual maturation remains a problem.

DOC, 11-deoxycorticosterone.

(continued)

Name(s)/OMIM	Mechanism	Description	Treatment	Notes
18-Hydroxylase deficiency (corticosterone methyloxidase, type I deficiency) OMIM #203400	Defect in **18-hydroxylase** (also known as corticosterone methyloxidase) enzyme, which catalyzes one of the steps of aldosterone biosynthesis, the 18-hydroxylation of cortisone to 18-hydroxycorticosterone. A reduction in the level of aldosterone results.	The decrease in aldosterone can lead to initial loss of salt by the kidneys and high potassium (hyperkalemia), low sodium (hyponatremia), and metabolic acidosis resulting in poor growth beginning as early as several weeks of age. Over time, the excess production of DOC provides the needed mineralocorticoid activity. Severe dehydration and intermittent fever may also occur. There may be increased susceptibility to infection.	Initial mineralcorticoid supplements have been reported to be helpful. Electrolyte monitoring also assists in patient care. Prognosis is variable as a result of variations in the mutations.	—
21-Hydroxylase (salt-losing type) OMIM #201910	Defect in the **21-hydroxlase** that converts 17-hydroxyprogesterone to 11-deoxycortisol (precursor to cortisol) and progesterone to DOC. Complete loss of activity and almost no production of cortisol cause an increase in 17-α-hydroxyprogesterone, which also leads to increased production of androgens.	Loss of aldosterone results in a marked loss of salts via the kidneys with high potassium (hyperkalemia), low sodium (hyponatremia), low chloride, and metabolic acidosis Poor growth, vomiting, and dehydration can also occur by 2 or 3 weeks of age. These changes can be life threatening. Increased androgen synthesis leads to changes of the female genitalia that can include labial fusion to form a structure similar to the urethra of the penis as well as changes to the labia and clitoris.	Acute treatment is via IV saline and dextrose (sugar) as well as hydrocortisone and/or fludrocortisone. Unclear gender is treated with a combination of genetic testing, parent (patient) counseling, and surgery if desired to produce genitalia for the chosen gender. Hormone replacement (to reduce androgen production) may also play a role once the gender choice has been made.	Milder forms of 21-hydroxylase deficiency also exist where enough activity is preserved and salt wasting is minimalized (compensated by increased aldosterone) but changes in genitalia (virilization) still occur.

DOC, 11-deoxycorticosterone.

Name(s)/OMIM	Mechanism	Description	Treatment	Notes
Androgen insensitivity syndrome (testicular feminization) OMIM #300068	Androgen insensitivity syndrome is an X-linked disorder as a result of mutations in the **androgen receptor gene**. Multiple mutations have been found, which may help explain the wide clinical spectrum.	The phenotypic range of anomalies is wide and in some the genitalia are so characteristically female that the diagnosis is not considered until suspected inguinal hernias are found to contain testicular tissue. Yet others may escape detection until amenorrhea or infertility become evident. The vagina, when present, ends blindly, and the uterus and ovaries are absent. The clitoris is often enlarged and there are various degrees of fusion of labioscrotal folds. At puberty, these genetic males may develop all of the secondary sexual characteristics of normal females. Pubic and axillary hair may be absent after puberty because of androgen resistance.	No treatment of the primary genetic defect is possible. Exogenous androgen is ineffective at any stage because of the lack of end-organ response. Undiagnosed individuals may live normal lives as females except for infertility and amenorrhea. Gender assignment and psychological issues may arise if phenotypic females are found to have XY karyotypes as young individuals. No associated systemic disease is known.	The older term "testicular feminization" has been replaced by "androgen insensitivity syndrome" because the latter more accurately portrays the biochemical mechanism. The disorder affects only genetic (XY) males because it is the lack of testicular response to androgens that leads to feminization and anomalies of the genitalia.

VITAMINS/MINERALS AND ELECTROLYTES

Name(s)/OMIM	Mechanism	Description	Treatment	Notes
Barters's syndrome (Gitelman variant) OMIM #263800	Gitelman syndrome results from a deficit of the **thiazide-sensitive Na–Cl cotransporter**.	This condition causes periodic mild paralysis, fatigue, muscle cramps, and rarely convulsions from severe loss of potassium and electrolyte imbalance. Blood levels of magnesium are often low as well. Low potassium may lead to irregularity of the heart beat. Kidney damage can result and blood volume sometimes is low contributing to the weakness. Unsteadiness, dizziness, and blurred vision can also occur. Severe illness or vigorous exercise may worsen all of these symptoms.	Vigorous electrolyte monitoring and treatment with potassium replacement is sufficient to alleviate most symptoms.	—
Biotin (biotinidase) deficiency (multiple carboxylase deficiency) OMIM #253260	**Biotin** is an essential cofactor in all carboxylase reactions. Biotin deficiency may be caused by a deficit in **biotinidase**, causing a late-onset multiple carboxylase deficiency. There is also an early-onset form resulting from a mutation in holocarboxylase synthetase resulting in failure to utilize biotin.	Patients with this disorder have skin rashes, loss of hair, weak muscles, and developmental delays. Seizures, hearing loss, vision loss, and unsteadiness may also occur. Onset can be as early as the first month of life but not all patients have all signs. Adults without symptoms have also been identified. Early diagnosis is important because treatment with oral biotin can be highly effective. High levels of 3-hydroxyisovaleric acid, β-methylcrotonylglycine, and 3-hydroxypropionic acid may be present in the urine.	Oral biotin can lead to dramatic improvement in both clinical symptoms and metabolic imbalances. Treatment is recommended for all patients with biotin activity below 10%. Prompt treatment of infections is recommended. Permanent neurologic damage needs to be managed with support and physical therapy.	Newborn screening programs are available. Untreated asymptomatic adults have been reported.

Name(s)/OMIM	Mechanism	Description	Treatment	Notes
Cystic fibrosis (CF) OMIM #219700 BI	CF is caused by mutations, of which there are many, in the CF gene. These mutations disrupt a protein called **cystic fibrosis transmembrane conductance regulator (CFTR)** that impacts ion channel (primarily chloride) gating specific to epithelial cells.	CF is a common genetic disease that affects multiple organs in the body, especially the lungs, liver, pancreas, intestinal tract, and sweat glands. Symptoms usually appear in the first year of life, although milder cases may not develop problems until midlife. Clinical symptoms vary widely. In the lungs, thick bronchial mucus production, together with a less competent immune system, results in respiratory difficulties and frequent lung infections. In the intestines, the same problem leads to obstructions. The digestion problems are compounded by decreased secretions of enzymes from the pancreas, leading to pancreatitis, fatty diarrhea (steatorrhea), and a deficiency in some vitamins. Poor growth is a hallmark of CF. Puberty may be delayed while abnormal cervical mucous contributes to female infertility, and absence of the vas deferens leads to the same result in males.	There is no cure for CF. Treatments are aimed at control of symptoms. High-energy foods, enzyme supplements, and vitamin and mineral supplements may be used in an attempt to provide the required nutrients. Daily pulmonary hygiene and breathing exercises help to prevent excessive mucus accumulation in the lungs. Antibiotics and other medicines are used to control lung infections and inflammation. In most cases, average life expectancy is around 37.8 years according to the Cystic Fibrosis Foundation, although improvements in treatments mean a baby born today could expect to live longer.	The common test for children and young adults is the electrolyte sweat test, which measures electrolytes (sodium, potassium, and chloride) in perspiration.

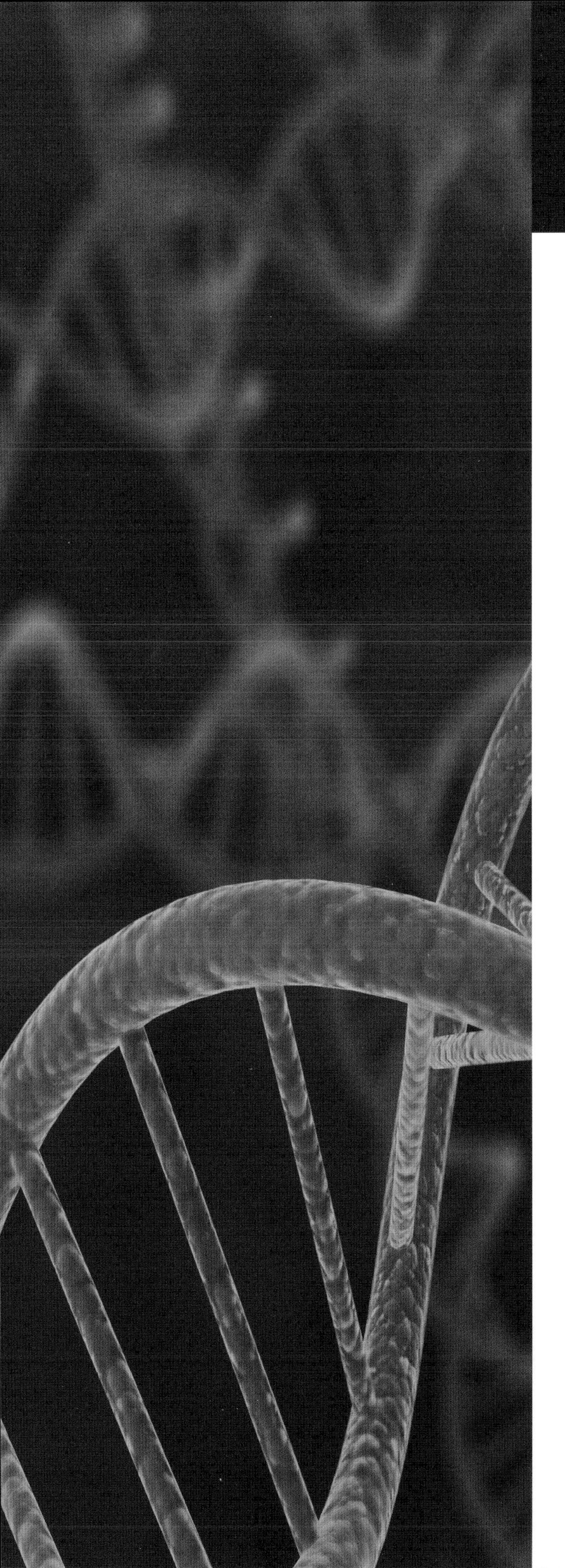

APPENDIX II

BIOCHEMICAL METHODS

POLYACRYLAMIDE GEL ELECTROPHORESIS (PAGE) [SODIUM DODECYL SULFATE (SDS)/ NON-SDS]

PAGE is a laboratory technique designed to separate proteins according to their size, shape, and also charge. Proteins for PAGE are usually prepared by placing them in an anionic (negative charge) detergent mixture, which, along with heating to approximately 60°C, denatures the protein, breaks any cysteine–cysteine disulfide bonds, and creates a fairly linear protein structure. PAGE can also be run with an addition of **SDS**. SDS, if used, is included in the detergent mixture and assists in disrupting the proteins' secondary and tertiary structures. In addition, SDS binds to each peptide chain in a ratio of one SDS per two amino acid residues, thereby adding a negative charge proportional to the peptide length. The addition of SDS to PAGE samples, therefore, also makes the proteins' shapes and native charges irrelevant; only the total length and the resulting SDS charge matters.

O O S O O —Na$^+$

SDS

Polyacrylamide gel is composed of linear **acrylamide** molecules, cross-linked by **bisacrylamide** via the catalytic actions of **ammonium persulfate** and **tetramethylethylenediamine (TMED)**. Cross-linking forms a web-like polyacrylamide lattice with pores of approximately the same size throughout. Varying the amounts of acrylamide, bisacrylamide, and aqueous solution allows scientists to accurately vary the pore size and, therefore, the relative ability to separate differently sized protein molecules.

Acrylamide **Bisacrylamide** **Ammonium persulfate** **TMED**

After polymerization, the gel is placed in an electrophoresis device and immersed in a buffer with a cathode (positive charge) at the top and an anode (negative charge) at the bottom. The protein samples are "loaded" onto individual gel lanes formed by a removable comb. A **tracking dye** is usually included in the samples to monitor the progress of the proteins through the gel.

The electric field causes the negatively charged proteins (charge usually enhanced by the proportional negative charge of SDS molecules) to move toward the anode proportional to their length (and, therefore, charge). Movement through the gel is not directly proportional to the overall charge, though, because the larger proteins will encounter much more difficulty in moving through the pores because of their size and linear rigidity. Although the length of each peptide increases the electrical force moving it down the gel, the "filtering" action of the gel pores will allow smaller molecules to travel further down the gel, whereas larger molecules remain near the top.

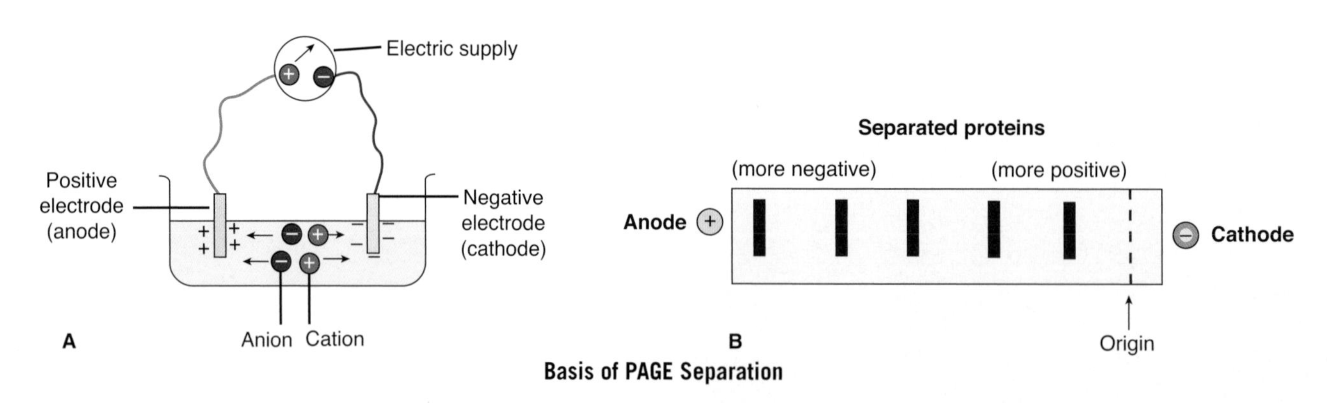

(A) The PAGE system employs the electric charge deployed between a positive electrode (anode) and negative electrode (cathode). (B) Proteins are loaded (origin) onto polyacrylamide gel, which is exposed to this electric field. Negatively charged proteins are driven toward the anode, whereas positively charged proteins will remain nearer to the cathode. The relative movement of each protein depends on its specific charge determined by its amino acid composition. When SDS is not used ("native PAGE"), the secondary-to-quaternary structure of each protein also influences the movement through the acrylamide gel matrix. When SDS is used, higher order structure is eliminated, separation by charge predominates. [Reproduced with permission from Naik P: Biochemistry, 3rd edition, Jaypee Brothers Medical Publishers (P) Ltd., 2009.]

When the tracking dye approaches the bottom of the gel, the electrical field is stopped and the gel is stained using a variety of chemicals or solutions, which bind to proteins (e.g., **Coomassie blue, silver**) to allow visualization in the gel. Proteins of known molecular weight, referred to as "**molecular markers**," are normally run in a separate lane to allow direct comparison with the proteins from the experimental sample. PAGE gels can be used further for western blotting (see below) and other biochemical techniques.

SDS or non-SDS PAGE is often used during the biochemical isolation of a single protein from tissue or protein mixtures, although it can also be used clinically for diagnostic and/or treatment purposes. Because of its ability to dissolve molecules, SDS is sometimes used in enemas as a laxative.

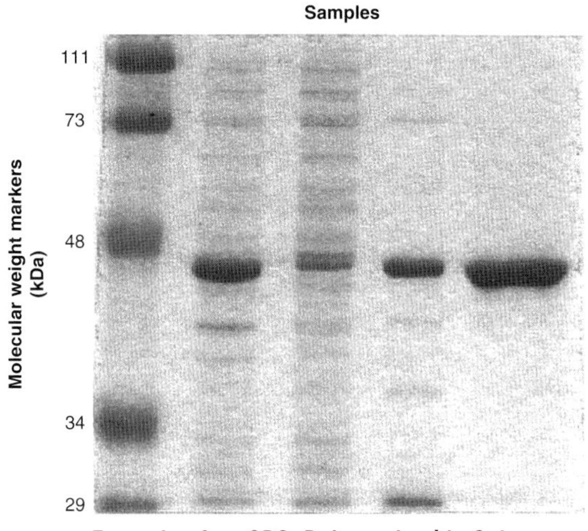

Example of an SDS–Polyacrylamide Gel

Markers of known molecule weight (kDa) are loaded in one gel lane (left) to allow characterization of proteins separated on the other lanes of the gel. [Reproduced with permission from Murray RA, et al.: Harper's Illustrated Biochemistry, 28th edition, McGraw-Hill, 2009.]

Two-dimensional (2D) PAGE is performed in two directions on the same sample, offering improved separation for complex mixtures of molecules. In this technique, samples are exposed to an electric current for a first separation. The gel is then turned 90° and a current is again applied with an alternative buffer, which affects the movement of the samples differently. As a result, a second separation is achieved. Examples include molecular weight separation followed by separation based on the molecules' overall pH, known as the "isoelectric point." 2D electrophoresis is used in many clinical conditions as a preparatory step before western blotting (see below) for tests such as confirmation of human immunodeficiency virus (HIV) and hepatitis B infection, the detection of prions in Creutzfeldt–Jakob disease, also known as "mad cow disease," and diagnosis of Lyme disease, among others.

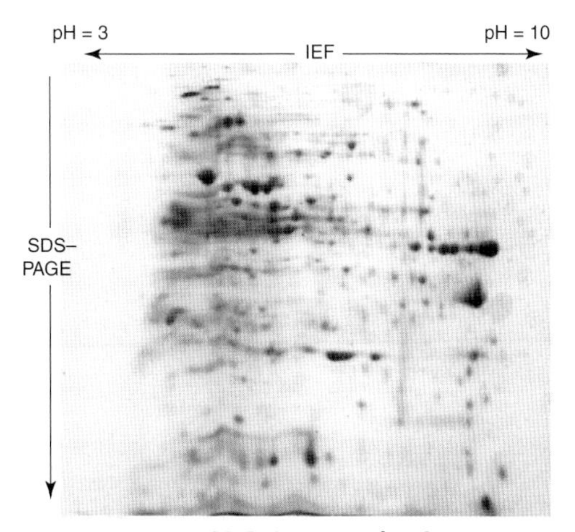

Example of a 2D Polyacrylamide Gel

Samples are run via normal SDS–PAGE (shown in vertical direction), which separates the proteins based on size, turned 90°, and run a second time via isoelectric focusing (shown in horizontal direction) technique with a pH gradient, which separates the proteins based on total pH of the protein (see text). Running the same sample by these two different techniques allows increased separation of complex mixtures of proteins and/or separation of proteins of the same size, but differing isoelectric points. [Reproduced with permission from Murray RA, et al.: Harper's Illustrated Biochemistry, 28th edition, McGraw-Hill, 2009.]

IMMUNOASSAYS

Immunoassay is a generic term for the application of **antibodies** to biochemical testing. These techniques paved the way for a wide variety of techniques of separation, measurement, and positive identification of biological molecules, and include **radioimmunoassays (RIA), enzyme-linked immunosorbant assay (ELISA** or **EIA)**, and **western blotting**.

All immunoassays involve the addition of an antibody specific for a known molecule (**antigen**) to a solution presumably containing, at least in part, that molecule. If the antigen is present, the antibody will bind specifically and, if properly chosen and prepared, strongly. **Monoclonal antibodies** will normally bind the tightest and with the greatest specificity; **polyclonal antibodies** are usually weaker and less specific in their binding patterns. Selective precipitation or chromatography (see below) can be used to isolate and purify the antibody–antigen complex as needed.

If a chemical or another label has been previously attached to the antibody, it can be used to detect the presence and, in many cases, the quantity of the antigen. Examples of these detection techniques include fluorescent molecules, enzymes that produce a particular color when provided an appropriate substrate, radioactive or magnetic labels, and even gold particles, which coalesce to form a visual precipitate. Alternatively, another antibody with a radioactive detector molecule, which selectively binds to the first antibody, can be added later in the test. Immunoassays are used in a multitude of clinical tests, especially for the detection of infections in blood, urine, cerebral spinal fluid, and so on.

RIA

The first immunoassay developed was the **RIA**. In this test, a molecule of interest, such as a hormone, has a radioactive molecule (e.g., iodine-125, carbon-14, or hydrogen-3) attached to it. The researcher also has an antibody (monoclonal or polyclonal), which specifically recognizes and binds to this hormone.

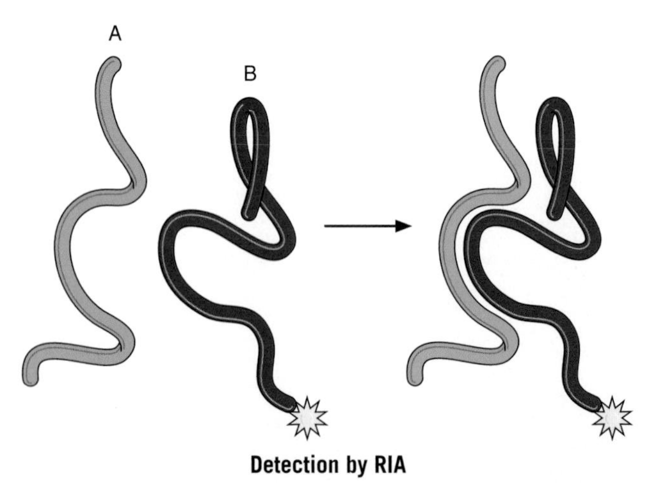

Detection by RIA

A molecule of interest (A) can be detected by placing a radioactive label (yellow starburst) onto a protein (B), which is known to bind to (A). Important examples of protein (B) include hormones and monocolonal and polyclonal antibodies. [Adapted with permission from Mescher AL: Junqueira's Basic Histology Text and Atlas, 12th edition, McGraw-Hill, 2010.]

A measured amount of the antibody and a measured amount of the pure hormone (both radiolabeled and unlabeled forms) are added to assay tubes with concentration of the unlabeled hormone greatly exceeding that of the radiolabeled hormone. Because only a small amount of antibody is added and its capacity to bind the hormone is limited, the antibody-binding sites will be saturated (i.e., all occupied; see figure below) As a result, only a fraction of the total amount of hormone will be actually bound to the antibody. Given that the antibody cannot differentiate between labeled and unlabeled hormones, both forms of the hormone will compete for the limited number of binding sites on the antibody. When a small amount of unlabeled hormone (i.e., from standard or patient sample) is added to the assay, then only a small amount of the bound-labeled hormone is displaced from the antibody. As the amount of unlabeled hormone increases, the amount of radiolabeled hormone bound to the antibody decreases considerably.

A variety of techniques can be used to separate unbound hormone from the antibody-bound hormone. Therefore, direct measurement of bound-labeled hormone can be easily made. Using known quantities of unlabeled hormone, it is possible to construct a standard curve which compares the amount of radiolabeled hormone bound to the antibody with the amount of pure unlabeled hormone added to the tube. By replacing pure hormone with a patient sample, the concentration of the hormone in the sample can then be directly determined.

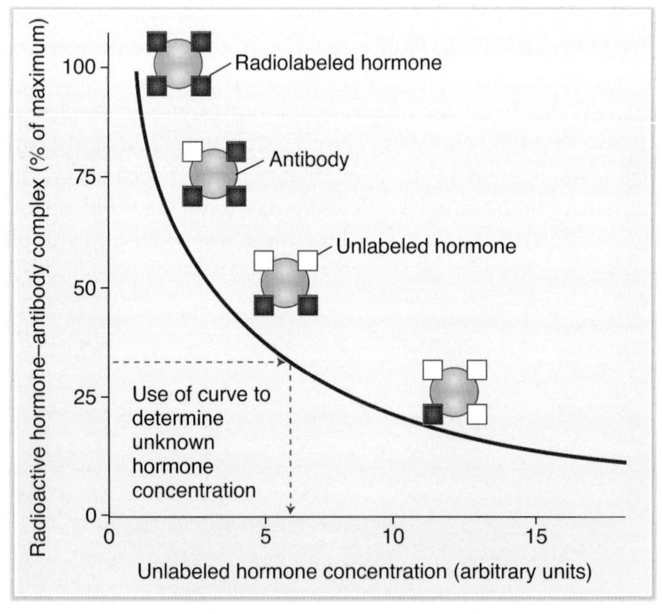

Standard Curve of Hormone Binding

A specific antibody to the hormone is incubated with the radiolabeled hormone. A standard curve is created by introducing known concentrations of unlabeled hormone to displace the radiolabeled hormone. The amount of radioactivity remaining is a function of the unlabeled hormone concentration. When unknown samples are used, the measurement of the radioactivity remaining allows the hormone concentration to be determined from the standard curve. [Adapted with permission from Kibble JD and Halsey CR: The Big Picture: Medical Physiology, 1st edition, McGraw-Hill, 2009.]

One drawback of this method is that RIA measures the immunologic activity, not the biological activity, of a hormone. Ideally, both a bioassay and RIA are evaluated. However, in a clinical situation, with a few rare exceptions, using the RIA provides sufficiently accurate information on hormone levels in the patient sample. The advantages of RIA are sensitivity and specificity as well as rapid and economic evaluation of patient blood or urine samples.

ELISA OR EIA

The principle of RIA is the basis of newer methods of measurement, for example, **ELISA or EIA**. The ELISA technique uses two different antibodies with varying binding affinity to a molecule that the researcher wishes to detect or study. An example is a polyclonal or a monoclonal antibody made against the same molecule. Normally, a microtiter plate, containing multiple test wells coated with the polyclonal antibody (not shown in figure), is used. The test samples are placed in the wells and the antigen, if present, is allowed to bind to the antibody. After a suitable time period, the remainder of the test sample is removed and the wells are washed to remove nonspecifically bound molecules. Next, a monoclonal antibody against the same antigen is added to the wells and allowed to bind to the antigen molecule. The use of polyclonal and monoclonal antibodies usually allows multiple binding to the same antigen molecule. Detection techniques are then used to determine the presence and quantity of the antigen being tested. One common method is to use an enzyme attached to the antibody, which produces a colored product. After the binding of the second antibody, addition of the enzyme's substrate creates a color reaction, which can often be quantitated (see the figure below). Detection using radioactive or fluorescent molecules attached to the second antibody can also be utilized. ELISA is used for a wide variety of clinical tests (e.g., identification of HIV, allergens, and/or drugs), especially if the concentration of a certain protein or other antigen must be determined. The ELISA technique is also the basis for home pregnancy tests, where the hormone binds first to an antibody–enzyme complex with detection and quantization via a second antibody containing a dye that is produced by an enzyme.

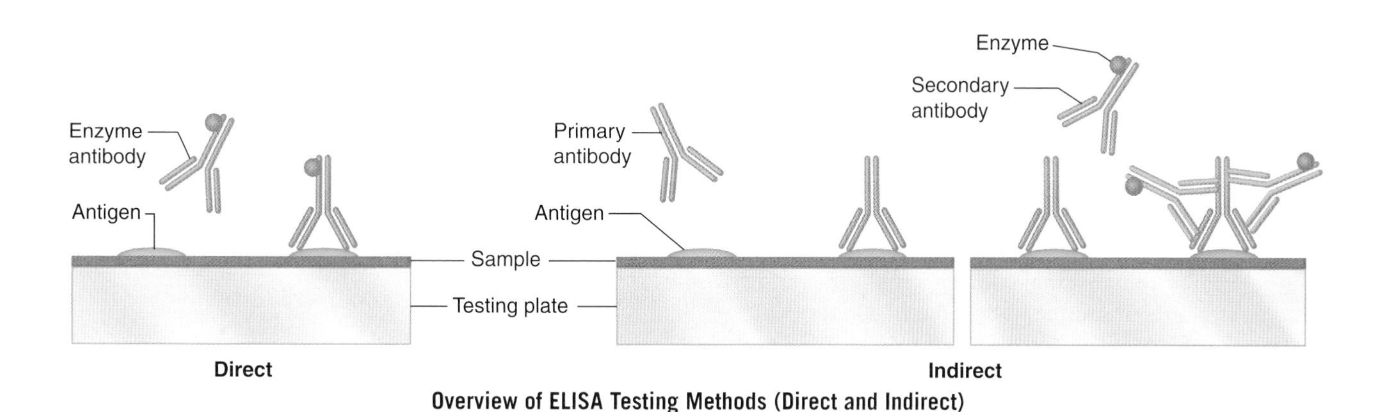

Overview of ELISA Testing Methods (Direct and Indirect)

Antibody to an antigen being tested is applied to the sample. Following binding, the antigen can be detected either via an attached enzyme on the primary antibody (direct method, left) or by the addition of a secondary antibody with an attached enzyme (indirect method, right). When substrate is added to the antibody–antigen sample, the enzyme produces either a fluorescent or colored product, allowing easy measurement, which allows both quantitative and qualitative assessment of the original antigen. [Adapted with permission from Mescher AL: Junqueira's Basic Histology Text and Atlas, 12th edition, McGraw-Hill, 2010.]

CHROMATOGRAPHY

THIN LAYER (PAPER) CHROMATOGRAPHY (TLC)

TLC or **paper chromatography** relies on a thin layer of material (e.g., paper, cellulose, silica, and others) with varying absorbance for molecules. Although often used for nonbiological molecules, TLC can be utilized for protein and lipid separation and/or identification. A small volume of each sample for analysis is placed in vertical lanes close to one end of the TLC layer, and that end is partially immersed in a solvent, ensuring that the sample is not covered by the solvent. This solvent is drawn through and up the TLC material by **capillary action**. As it travels, it carries the sample molecules up with it at rates depending on the molecule and solvent properties, including the level of adsorption onto the chromatography layer and the solubility of the sample in the liquid (see the following figure). As a result of the varying solid and liquid layer affinities, separation of a mixture of molecules results. The TLC layer is then "developed" for molecule detection via chemical or spectroscopic means. In the case of a single, purified molecule, the distance traveled on a specific TLC with a specific solvent can provide identification by comparison with known standards. Two-way TLC can also be used, in which after separation by one solvent, the sheet is turned 90° and exposed to the actions of a second type of solvent. Two-way TLC offers improved separation for some molecules.

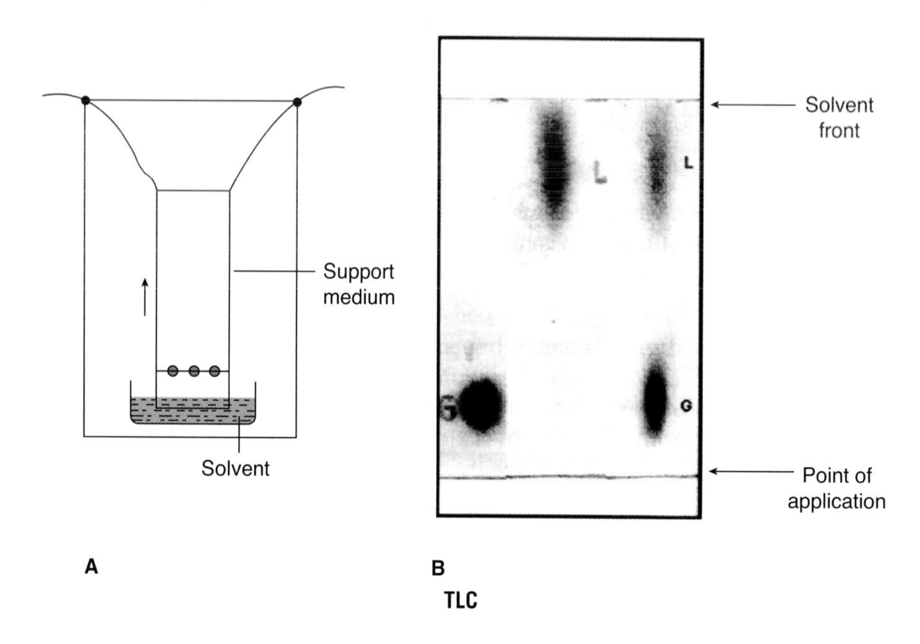

(A) Diagram of TLC equipment, illustrating TLC support medium, solvent (light brown), point of application of three samples (black dots), and direction of movement (indicated by arrow) of molecular samples as the solvent moves up the medium. (B) Actual TLC plate showing known sample of G (left lane) and L (middle lane) and separation of a mixture of these two proteins into the component parts (right lane). [Adapted with permission from Naik P: Biochemistry, 3rd edition, Jaypee Brothers Medical Publishers (P) Ltd., 2009.]

COLUMN CHROMATOGRAPHY

Column chromatography utilizes the same principles of TLC, but in a three-dimensional column system that allows much larger amounts of sample to be separated and studied. A glass tube or "column" is filled with a selected material, called **stationary phase**, which has particular chemical qualities (e.g., hydrophobic, hydrophilic, anionic (–) charges, cationic (+) charges, etc.). Molecules will bind to the stationary phase, depending on their individual chemical qualities. For example, a protein with a strong hydrophilic external character will not want to bind to a hydrophobic column material, but will want to bind to a hydrophilic one. The variation in binding of each molecule leads to separation because molecules that bind less will travel more quickly through the column and vice versa. Often, the molecules will bind to the stationary phase until another liquid solvent, called **mobile phase** or **eluent**, with stronger hydrophobic/hydrophilic or anionic/cationic qualities is run through the column. Sometimes, sequential eluents, each with a different chemical quality, are run through the column, each causing selected molecules to unbind and flow out of the column. The material that passes through the column is collected in small samples of a known volume, called fractions, for further study. Often, molecules in the fractions are monitored by light [e.g., ultraviolet (UV)] absorption or other detection methods.

GEL FILTRATION CHROMATOGRAPHY

Gel filtration chromatography, a variation of column chromatography, is used for the biochemical separation of biological molecules, mainly proteins and nucleic acid strands, depending on the size and shape of the molecule. The technique usually preserves biological function. Gel filtration uses **beads** formed from polyacrylamide (see above), dextrans, or agarose (see Chapter 2) with pores of approximately equal size. The particular **pore size** of each type of gel filtration bead depends on their exact materials and different fabrication techniques whose characteristics allow researchers to optimize column chromatography for varying samples.

Biological samples are applied to the top of the column in a small volume. A solution is run through the column carrying the sample through the gel beads. Small and/or predominately compact globular molecules are able to enter the bead pores and, therefore, have a large volume available to them. This large volume results in a slower transit time through the column. Larger particles or ones with a noncompact or linear structure are unable to enter the bead pores and travel more quickly through the smaller volume between the beads. The molecular size and shape of each biological molecule applied to the column directly determine how fast or slow it travels through and out (**elution**) of the column (see figure below). This varying speed results in separation.

ION-EXCHANGE CHROMATOGRAPHY

Ion-exchange chromatography, another type of column chromatography, takes advantage of **charged groups** (e.g., **ions**) that are part of the gel bead material. Different biological molecules interact more or less with these chemical groups (e.g., a more negative molecule will be attracted to positively charged gel beads), altering their speed of elution and resulting in separation. Some molecules will bind so strongly to the charged gel beads that a solution with a stronger ionic charge must be used to displace or "elute" them.

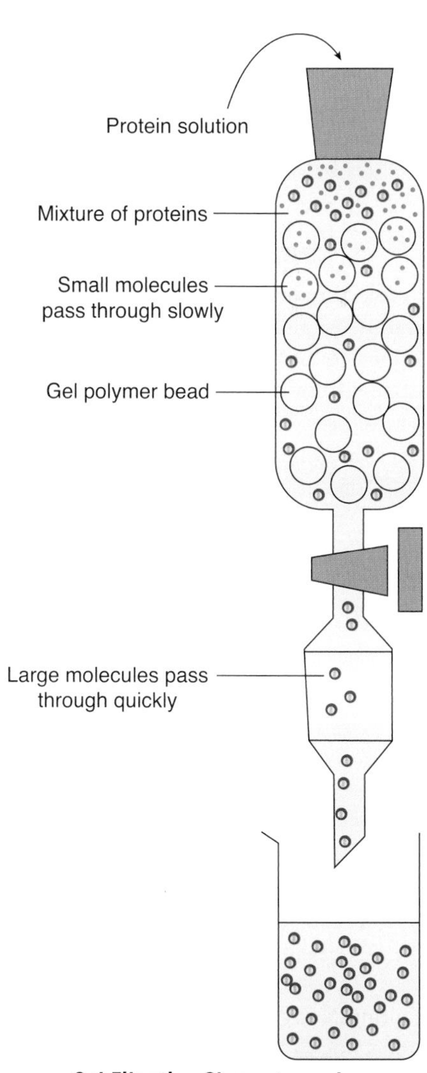

Gel Filtration Chromatography

Small molecules (blue) are able to enter the gel beads, increasing their available volume and slowing down the speed of travel through the column. Larger molecules (red) are unable to enter the gel beads and travel more quickly through and out of the column. As a result, molecules separate and elute from the column at a rate proportional to their molecular weight/size. [Reproduced with permission from Naik P: Biochemistry, 3rd edition, Jaypee Brothers Medical Publishers (P) Ltd., 2009.]

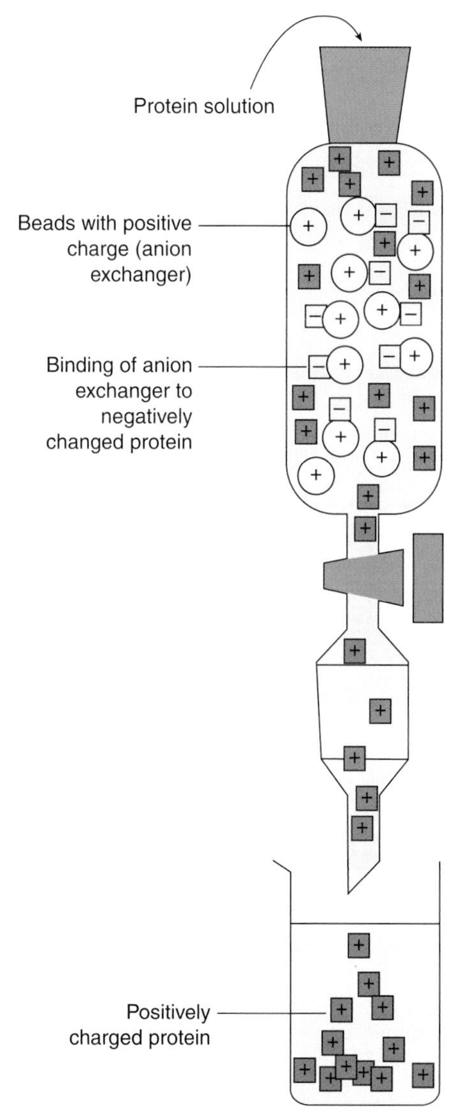

Ion-Exchange Column

Gel beads with a positive ionic charge attract negatively charged molecules (blue) more than positively charged ones (orange). This different affinity, based on charge, allows separation of various molecules. [Reproduced with permission from Naik P: Biochemistry, 3rd edition, Jaypee Brothers Medical Publishers (P) Ltd., 2009.]

AFFINITY CHROMATOGRAPHY

Affinity chromatography works much like ion-exchange chromatography (above), but takes advantage of specific molecules that are part of or previously attached to the gel bead material. As an example, **antibodies** coupled to the gel material are sometimes used for very specific binding to the desired molecule. Other examples of affinity chromatography include the use of gel-bound **substrates** or **coenzymes**. The biological molecules remain bound to the gel bead until a second solution of changed ionic strength or pH is applied to affect their release and elution.

Regardless of the particular type of column chromatography used, eluted biological molecules are collected in small, ordered volume allotments, known as fractions, and the molecules are detected by spectroscopic or chemical identification methods.

HIGH-PERFORMANCE/PRESSURE LIQUID CHROMATOGRAPHY (HPLC)

HPLC is a technique used for enhanced and precise separation of even small amounts of biological and nonbiological molecules. HPLC columns utilize variations of the separation principle behind column chromatography, depending on variable attractive and repulsive forces on the gel material and their interaction with the biological molecules being studied. HPLC also utilizes smaller columns than regular column chromatography, with more specialized gel materials. The smaller columns and resulting tighter packing of the gel material in HPLC combined with the use of a pump to propel the column solution (sometimes organic solvents) either in a single solution or as a gradient of different solutions under high pressure lead to fast, efficient, and more selective separation of molecules that may not be possible with other techniques. Methods to detect these molecules once they are eluted, including light absorbance, fluorescence, electrochemical, and others, are similar to those utilized in column chromatography.

PROTEIN AND DEOXYRIBONUCLEIC ACID (DNA)/ RIBONUCLEIC ACID (RNA) PRECIPITATION

The selected precipitation of biological molecules, including proteins and DNA/RNA, is often used during the preparation, analysis, and concentration of biochemical samples. All methods rely on the varying solubility of these molecules in different aqueous or nonaqueous solutions, depending on their hydrophobic, hydrophilic, and other biochemical properties. The particular amino acid R-groups (primary structure) as well as the resulting secondary, tertiary, and quaternary structures (Chapter 1) of a protein determine its interactions with water molecules or organic solvents and, therefore, how soluble it will be in those solutions. Because proteins have different primary structures, each will have unique solubility in a particular solvent and will allow separation by varying precipitation techniques.

One such technique that is often used is referred to as "**salting out**" and often uses the chemical **ammonium sulfate** (although other salts can be used). Using this technique, the concentration of ammonium sulfate in a mixture of proteins is gradually increased, resulting in greater binding of water by the increasing ion concentration. As the increasing number of ions draw the water away from the proteins, its ability to keep them in solution decreases. Proteins may then start to aggregate together to protect their hydrophilic and/or hydrophobic portions. If the process continues, the proteins will come out of the solution (precipitate). The particular hydrophilic/hydrophobic characteristics of each protein will make it and its identical molecules precipitate at the same time, resulting in separation. As each protein has a set salt concentration at which it precipitates, this technique can be easily reproduced as part of a molecular purification process. Other techniques using organic, polymeric, and metal salt solutions, including ones that take advantage of the overall pH of a protein (as determined by its amino acid R-groups), are used in a similar fashion. Usually, the proteins can be redissolved and can retain their biological functions.

DNA and RNA molecules (Chapter 4) can be similarly precipitated using **ethanol** or **isopropanol** via a similar mechanism. Water normally surrounds these molecules, interacting with the charged phosphate backbone and keeping the DNA or RNA in solution. The gradual addition of ethanol (usually about 64% ethanol/water) blocks the water–phosphate backbone interactions and leads to precipitation of the DNA or RNA. Smaller nucleic acid molecules may require longer exposure to ethanol or other techniques to fully precipitate. After precipitation, centrifugation is used to bring the insoluble nucleic acid molecules together in a small mass, called a "pellet," and the remaining solution can be removed. The DNA or RNA can be resuspended in an aqueous solution and can normally retain full biological activity. Ethanol precipitation can also be used for polysaccharides (Chapter 2) or methanol for proteins (Chapter 1).

DNA AND RNA SEQUENCING

DNA and RNA sequencing have advanced the understanding of genes and gene products exponentially since their development. Further advances and refinement of the **DNA sequencing** technology allowed determination of the complete human genome in 2003, a task that until then was deemed almost impossible. Sequencing of the nucleotides of DNA and RNA rely on many of the same biochemical techniques already discussed above.

The Maxam–Gilbert method of sequencing was developed in the 1970s and it relies on radioactive labeling and chemical cleavage of the DNA being examined. The labeled fragments are separated by gel electrophoresis, and the sequence is interpreted from the resulting X-ray film. Although this sequencing technique has fallen out of favor in comparison with the Sanger technique (see below), it is still used in specialty applications including **DNA footprinting**, in which the interaction of DNA and DNA-binding proteins can be determined.

In 1975, the Sanger technique, often referred to as "chain-termination" method, was developed, and it is still used, with some improvements, today. The key to this method is the application of **dideoxynucleotide triphosphates** (ddATP, ddGTP, ddCTP, and ddTTP), which lack 5′ *and* 3′-hydroxyl (OH) group and, therefore, block the addition of further nucleotides. This chain termination by the dideoxynucleotides results in DNA fragments of various lengths, which can be separated by gel electrophoresis.

More specifically, the DNA to be sequenced is reduced to a single-strand DNA template by heat denaturation. The sample is divided into four parts to which DNA polymerase and all four of the regular dATP, dGTP, dCTP, and dTTP nucleotides are added along with one of the dideoxynucleotides per sample. The DNA polymerase uses the single-strand template to form a second strand. At various times, the dideoxynucleotide ends DNA replication, but only at the point where that dideoxynucleotide is added to the growing strand. As a result, multiple length strands, all ending in the same dideoxynucleotide, are produced. When the ddATP, ddGTP, ddCTP, and ddTTP samples are separated by gel electrophoresis, each on its own gel lane, an easily read sequence of fragments directly denotes the DNA sequence (see below).

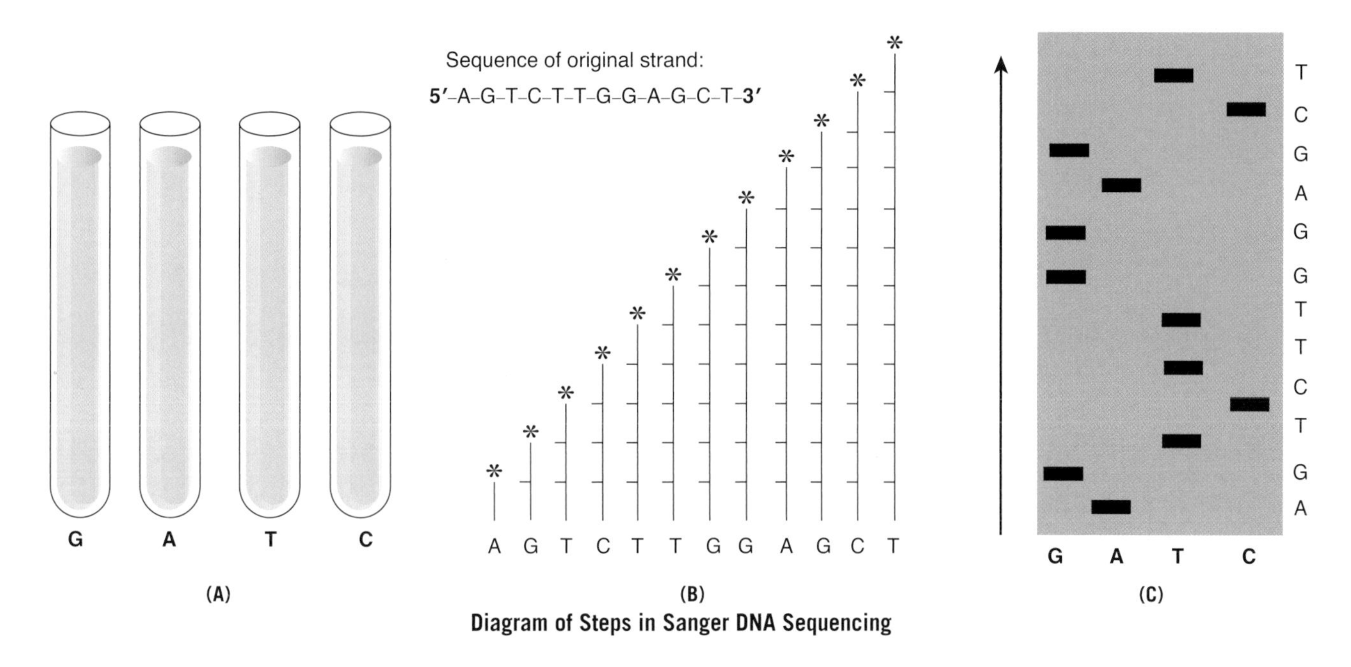

Diagram of Steps in Sanger DNA Sequencing

(A) DNA to be sequenced is reduced to single strands by heat denaturization and used to produce four separate samples, G, A, T and C, with DNA polymerase, deoxynucleotides, and one of the four dideoxynucleotides (ddGTP, ddATP, ddTTP, or ddCTP), respectively. (B) The DNA polymerase produces DNA strands of varying length, each terminated by the specific dideoxynucleotide. (C) These strands are separated by gel electrophoresis and the sequence read from bottom (shortest fragment) to top (longest fragment) from the gel (example sequence shown on gel: AGTCTTGGAGCT). [Adapted with permission from Murray RA, et al.: Harper's Illustrated Biochemistry, 28th edition, McGraw-Hill, 2009.]

Standard DNA sequencing allows researchers to determine the nucleotide code for DNA up to 1,000 bases long. Other techniques allow simpler, faster, and more complex sequencing. DNA restriction endonuclease enzymes, often called "**restriction enzymes**," cut DNA at very specific nucleotide sequences. The use of different restriction enzymes, each cutting the DNA at very different points, produces different but overlapping fragments of the same DNA molecule. Sequencing of each restriction enzyme fragment and subsequent comparison and appropriate overlap of the sequences allow characterization of DNA that may be too long for the standard technique. Variations in labeling the fragments, including a technique in which the dideoxynucleotides are labeled, and techniques for better fragment preparation and separation have also allowed easier and even machine-automated DNA sequencing of longer and longer nucleotides chains approaching 5,000 bases. These improved techniques were critical in the ability to sequence the human genome.

RNA sequencing relies on the ability to use **reverse transcriptase** to produce a double-stranded DNA molecule (often referred to as **complementary DNA** or **cDNA**) from an RNA sequence of interest. Following the production of the analogous DNA molecule, sequencing is done as described above and then interpreted for the RNA molecule. However, RNA sequencing is more problematic than DNA sequencing because of the fact that RNA is much less stable. Similar techniques are used to produce a cDNA library—a collection of DNA that codes for all messenger RNA (mRNA) contained in a particular cell type. cDNA libraries are usually established for simple organisms wherein the number of mRNA molecules is limited. cDNA libraries are also used in **cloning** (see below).

SOUTHERN, NORTHERN, AND WESTERN BLOTS

DNA Southern and RNA northern blots are techniques combining PAGE and a variation of TLC/paper chromatography (described above). Southern blotting was developed by Dr. Ed Southern and is named in his honor. The subsequent adaptation of the technique to RNA and proteins (described below) resulted in the analogous names of northern and western blotting. A method was also developed to detect post-translational modifications to proteins (**eastern blotting**), although the term has fallen somewhat out of favor.

SOUTHERN BLOTTING

Southern blotting is used to identify and isolate a specific DNA sequence and can also quantitatively determine the number of times a specific sequence occurs in a particular cell type (e.g., how many repeats and/or copies of a gene in a particular organism or cell type of interest). The procedure begins with isolation of DNA via precipitation and column purification techniques followed by exposure to restriction enzymes, which cut the long DNA molecule into smaller and more manageable pieces. The DNA is next electrophoresed for size separation on an agarose gel. The gel is next removed from the gel electrophoresis equipment and the DNA is transferred from the gel to a piece of **nylon membrane** whose positive charge binds the negatively charged DNA molecules (e.g., phosphate groups of the backbone). The transfer takes place via simple capillary action by placing the membrane on top of the gel in a suitable transfer buffer. After transfer, heat or UV radiation is applied to the membrane to covalently attach the DNA molecules to the membrane, and hybridization probes (DNA, RNA, or synthetic oligonucleotides) with radioactive, fluorescent, or chemical detection capabilities are used to identify a specific DNA strand sequence.

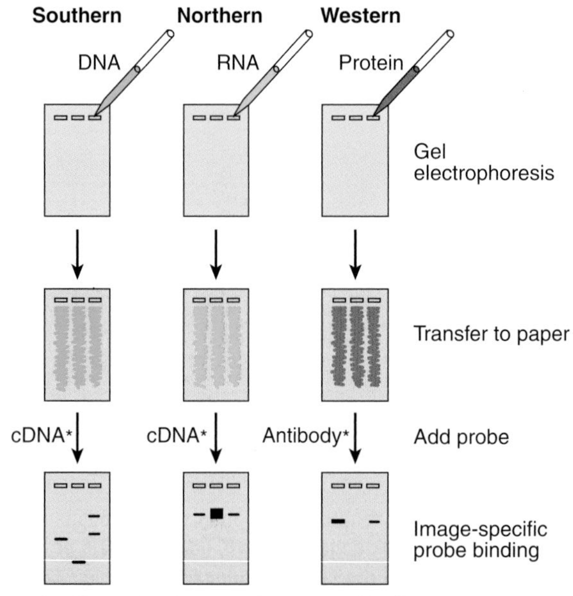

Southern, Northern, and Western Blotting Techniques

The basic blotting techniques (Southern, northern, and western) are illustrated, including gel separation, transfer to paper, and detection. The techniques of northern and western blotting are further explained in the text (see below). The specific DNA, RNA, or protein of interest is detected and identified by any one of a variety of techniques discussed above. [Adapted with permission from Murray RA, et al.: Harper's Illustrated Biochemistry, 28th edition, McGraw-Hill, 2009.]

DNA molecules, separated on a gel and prior to transfer to the nylon membrane, can be visualized using the chemical **ethidium bromide (EtBr)**. EtBr reversibly fits or "**intercalates**" between nucleotide base pairs and becomes capable of fluorescing if illuminated by UV light. Because the intercalation of EtBr is reversible, identified and functional DNA can be collected by cutting out the particular band of interest from the agarose or polyacrylamide gel. Intercalation alters the secondary and tertiary structures of the DNA molecule and, therefore, DNA replication and expression. As a result, intercalation of DNA is involved in the action of several chemotherapeutic drugs (e.g., doxorubicin, daunorubicin, and dactinomycin). For the same reason, EtBr can be a potent carcinogen if not handled correctly.

Br^- H_2N N^+ NH_2

EtBr

NORTHERN BLOTTING

Northern blotting allows researchers to determine the level of production (and, therefore, control of expression) of a particular mRNA by quantitative determination of the relative amount of that particular mRNA relative to total mRNA. Varying influences on both total and specific mRNA expressions such as development, diseases (e.g., cancers), and/or experimental techniques (including drug development) can, therefore, be ascertained. Northern blotting starts with the isolation of RNA from a cell with application of a tissue or cell sample to an **oligo (dT) cellulose affinity column**, which will bind only to the **poly-adenosine nucleotide tails** found in RNAs (see Chapter 9), whereas other molecules run through the column unbound. Other techniques including chemical extraction and magnetic bead isolation are also used. The initial separation/purification of the RNA provides a markedly purified sample of RNA.

The RNAs are then applied to an agarose gel with formaldehyde added to promote linearization of all RNA molecules (see figure above). PAGE gels with urea added can also be used for smaller RNA sequences. The RNA on the gel is then transferred to a nylon membrane whose positive charge binds the negatively charged RNA molecules (e.g., phosphate groups of the backbone). This transfer is accomplished in a manner analogous to Southern blotting (see above) and then heat or UV radiation is applied to the membrane to create covalent linkages between the RNAs and the nylon. Much like Southern blot detection, specific "**hybridization probes**" (DNA, RNA, or artificially constructed oligonucleotides) of known sequence and labeled radioactively, fluorescently, or with enzymes for chemical detection are then used to identify particular RNA strands of interest. The use of northern blotting for the quantitation of mRNA abundance has been mainly replaced to a great extent by quantitative reverse-transcription polymerase chain reaction (RT-PCR) (described below), which is much more sensitive and reliable.

WESTERN BLOTTING

Western blotting is a biochemical technique, adapted from a similar technique for DNA of Southern blotting, for detection and identification of specific proteins. Like Southern and northern blotting, western blots take advantage of the same immunoassay techniques described above, coupled with those of PAGE separation. Western blotting begins with a PAGE gel (one or two dimensional) to separate a mixture of proteins (see the figure above). The unstained gel is removed from the gel electrophoresis equipment and placed in a special apparatus that allows transfer of the proteins, again using an electric current, from the gel to a piece of **nitrocellulose** or **polyvinylidene difluoride** membrane. This membrane strongly binds to and immobilizes the proteins in each band so they cannot diffuse, thereby maintaining the original separation while blotting and identifying the sample. The membrane can then be used for detection of a specific protein using antibodies as discussed in the *Immunoassays* section above. Detection is again via enzyme-linked, fluorescent tag, radioactive-label, or other methods as discussed above.

PCR

PCR is a powerful technique developed in the 1980s to produce multiple copies (i.e., thousands to millions) of a particular DNA sequence. PCR relies on repetitive cycles of (a) partial "melting" of double-stranded DNA, (b) application of short **DNA primer** sequences (two primers specific for one particular gene, but for DNA replication in both the $5' \rightarrow 3'$ and $3' \rightarrow 5'$, sense and antisense directions), and (c) adding **DNA polymerase** and all four DNA nucleotides (dA, dG, dC, and dT) to copy that gene. The process ends with cooling of the DNA to allow reannealing of the double-stranded DNA. Subsequent melting/primer, polymerase, and nucleotides/cooling cycles also act on any newly produced DNA, resulting in an exponential amplification of the DNA primer's gene target with every cycle. An agarose gel with EtBr is often used to verify the amplification. PCR is used in a vast number of varying molecular biological and medical techniques, including amplification of DNA clones (see below) and/or hybridization probes for northern or Southern blotting (see above). In addition, PCR enables easier gene analysis of organisms and tissues/cells, genetic determinations to include the study of genetic diseases and paternity/hereditary testing, detection of low levels of infective bacterial or viral organisms, and forensic determination of individuals at crime scenes (genetic fingerprinting) when only trace amounts of a DNA sample are available.

Quantitative PCR allows the determination of relative amounts of DNA strands and further extends PCR's diagnostic and scientific capabilities and applications. Finally, the technique of **RT-PCR** allows the amplification of DNA from RNA strands by making a cDNA copy using reverse transcriptase. This enables the quantification of mRNA abundance for specific genes, and has largely replaced northern blotting as the method of choice for mRNA determinations.

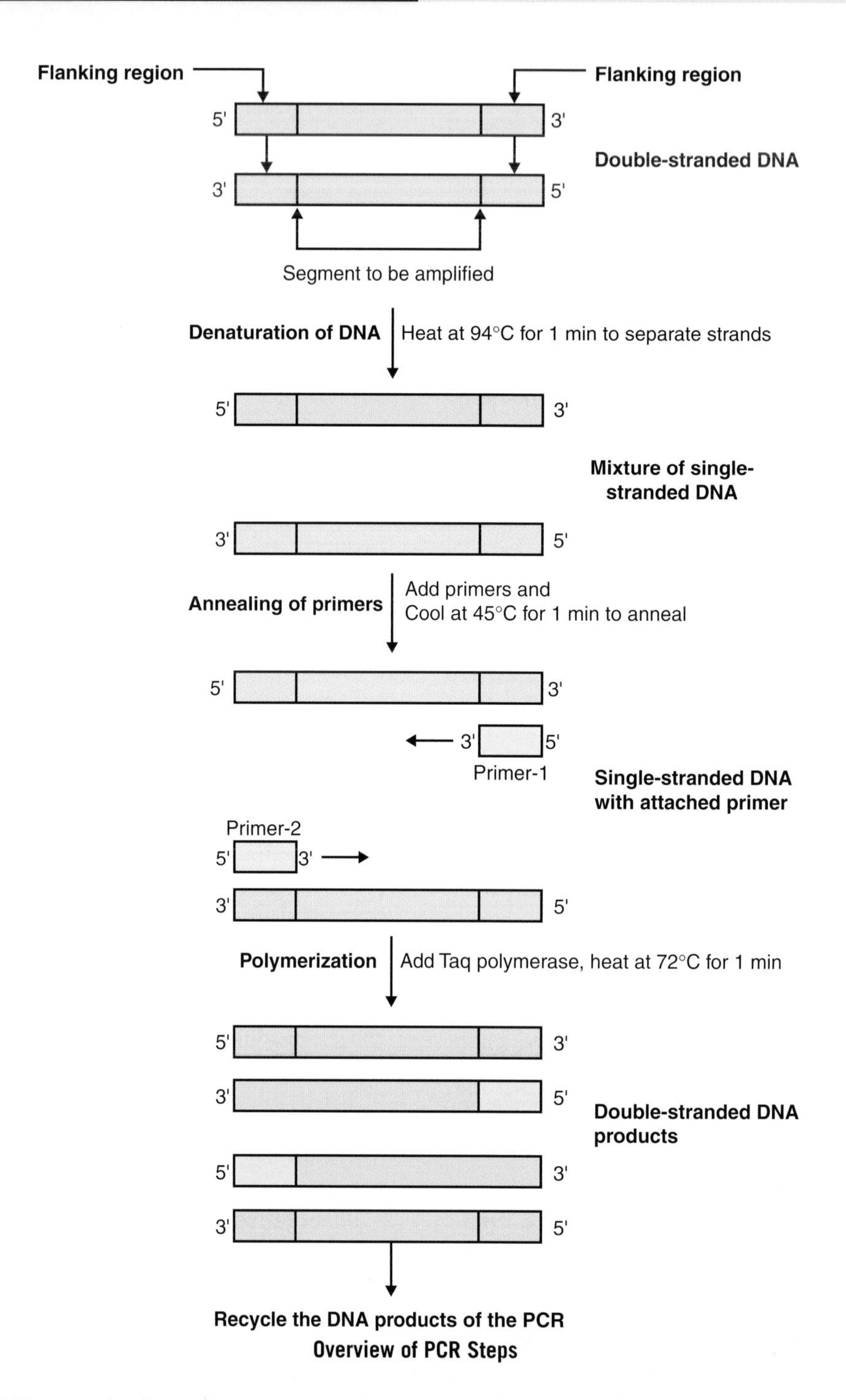

Overview of PCR Steps

Each PCR cycle includes DNA denaturization of a segment of double-stranded DNA with a segment to be amplified to produce a mixture of single-stranded DNA, annealing with specific DNA primers, and, finally, polymerization by the addition of polymerase and nucleotides (G, C, A, and T). Prior to the start of the next sample, the replicated DNA is allowed to reanneal to again form double-stranded structures. Further cycles will exponentially reproduce the DNA strand of interest. [Reproduced with permission from Naik P: Biochemistry, 3rd edition, Jaypee Brothers Medical Publishers (P) Ltd., 2009.]

CLONING

Cloning is a term used in two senses; first, the process to produce identical copies of an organism and, second, for the copying of DNA sequences. Only the latter will be discussed here. The ability to clone a DNA fragment allows researchers to produce multiple copies of a selected gene or portion of DNA (e.g., via PCR), but also allows them to study nonexpressed portions of DNA, including promoter and regulation regions and DNA that is not expressed. Indeed, the ability to select and amplify any particular DNA sequence for study is an essential part of the science of molecular biology.

The basic approach to cloning of any piece of DNA includes (a) cutting the DNA into an appropriately sized fragment via restriction enzymes; (b) inserting the DNA into another special piece of DNA called a **vector**, which contains an appropriate promoter region to allow DNA replication and/or transcription;

(c) introducing (**transfection**) these DNA fragment/vector DNA molecules into an appropriate cell type for amplification of the DNA; and, finally, (d) screening the resulting cells to find and confirm the identity of the DNA fragment of interest via a reporter or marker gene (e.g., one that produces antibiotic resistance or an enzyme producing a colored substrate). Once a clone is isolated and confirmed, it then offers a powerful tool for the researcher as noted above.

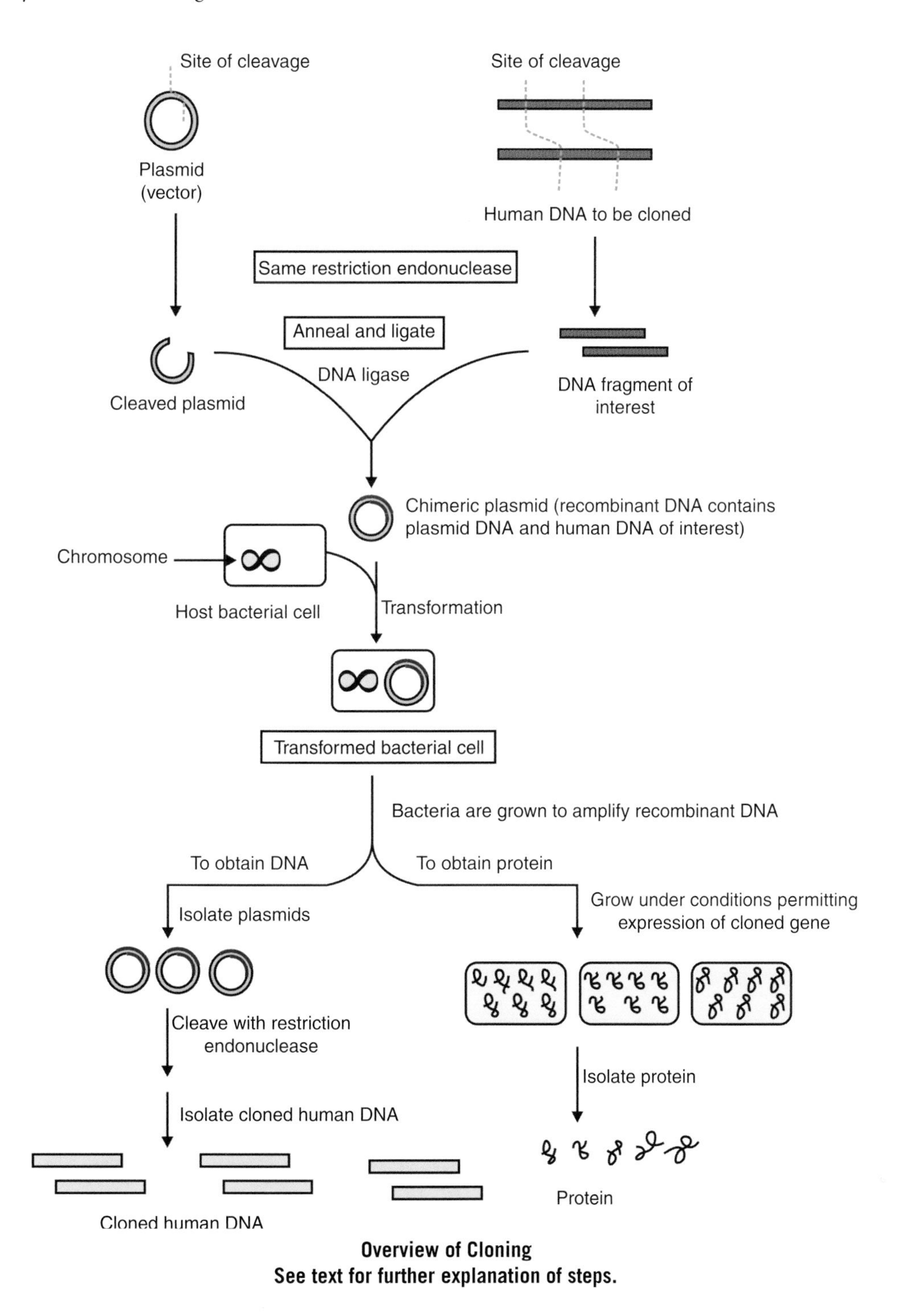

Overview of Cloning
See text for further explanation of steps.

[Reproduced with permission from Naik P: Biochemistry, 3rd edition, Jaypee Brothers Medical Publishers (P) Ltd., 2009.]

The size of the DNA fragment that can be introduced into a cloning vector and still allow the vector to function is limited. As a result, the fragmentation of a large DNA to a suitable length is an important first step in cloning. This fragmentation can be done using very specific or random restriction enzymes. Cloning vectors are specific, circular pieces of DNA (**plasmids**) that may have originated from bacteria or viruses or that have been artificially produced. They have the ability to enter into cells and to be expressed in that cell, either separately or by incorporating themselves into that cell's DNA. Cloning vectors also have specific DNA sequences where known restriction enzymes can be utilized to cut open the vector and insert DNA fragments under well-controlled laboratory conditions. Therefore, a specific restriction enzyme is used to cut open the vectors; newly fragmented DNA sequences of interest are added and DNA ligase repairs the vector, which now includes the DNA fragment. Transfection of these cloning vectors into cell hosts is performed by various techniques that will not be discussed here.

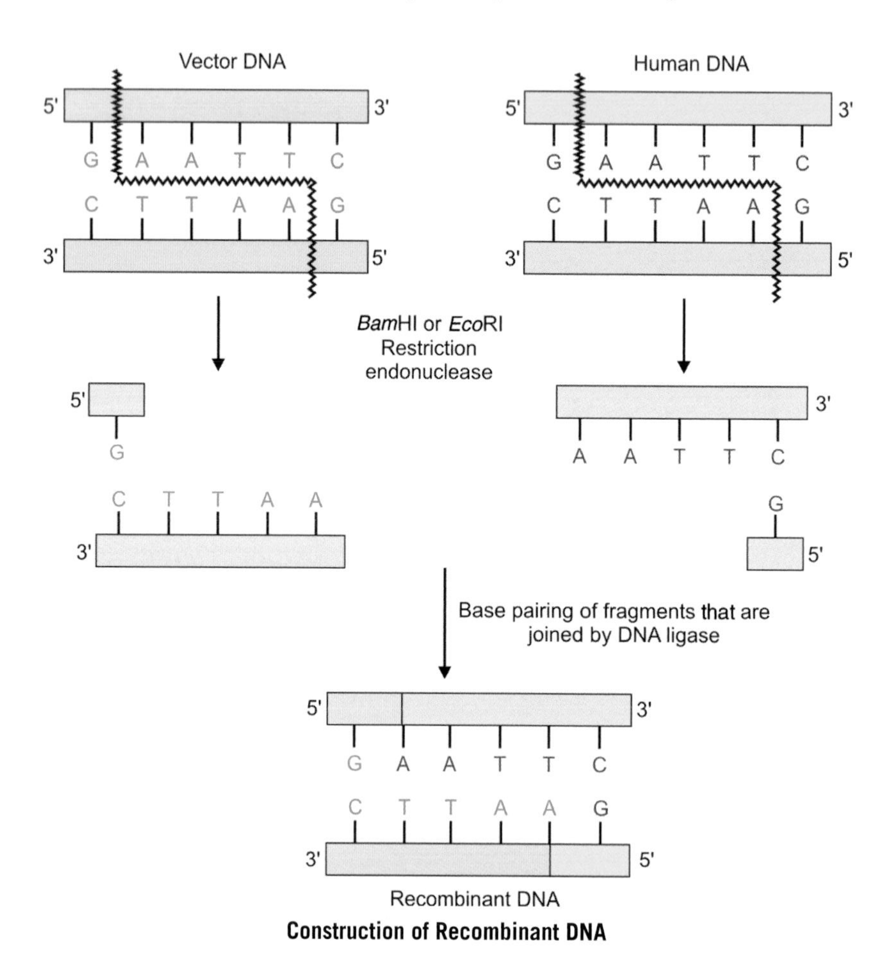

Construction of Recombinant DNA

Vector DNA (left) can be incorporated into human DNA (right) by the use of specific restriction enzymes (e.g., *Bam*H1 or *Eco*R1) that produce DNA fragments with particular end, base-pair sequences that can subsequently be joined by DNA ligase. The ability to produce specific recombinant DNAs is an important research and diagnostic tool. [Reproduced with permission from Naik P: Biochemistry, 3rd edition, Jaypee Brothers Medical Publishers (P) Ltd., 2009.]

Because the process may involve somewhat random fragmentation of the DNA and transfection is often of low efficiency, a screening method must be used to identify those cells that contain a functioning clone. The cloning vectors will often express a particular marker (e.g., an enzyme that produces a colored product, as described above, or confers resistance to an antibiotic) that allows the researcher to identify the cells with these working clone vectors. Finally, cells that appear to contain the DNA fragment of interest in a functional vector may be used for PCR amplification (see above) and DNA sequence analysis (see above) of the clone/DNA fragment to confirm that a successful clone has been produced.

FLOW CYTOMETRY

Flow cytometry, developed initially in 1968, is an important scientific and medical research application that has also begun to gain a role in clinical treatments. This method takes advantage of fluorescent labeling of antibodies (as utilized in several procedures above) and the ability to excite these fluorescent molecules via lasers. Excitation allows the detection (via a digital detector) of individual cells or molecules (e.g., chromosomes) at an extremely fast rate (up to thousands of particles/second). Particles (i.e., cells or large molecules) can also be detected by light scattering (often used to determine cell type) at the same time. Multiple lasers and multiple detectors can be arranged so as to allow the operators to study cells labeled with several antibodies/fluorescent probes. Flow cytometers have been used to study chromosomes, DNA, RNA, proteins, cell activation, enzyme activity, pH, intracellular concentration of selected ions, membrane fluidity and apoptosis, and multidrug-resistant cancer cells. Medical fields such as hematology, oncology, pathology, genetics, and other specialties utilize flow cytometry extensively. More advanced flow cytometry even allow sorting of each individual cell or molecule based upon their scattering and/or labeled fluorescence characteristics.

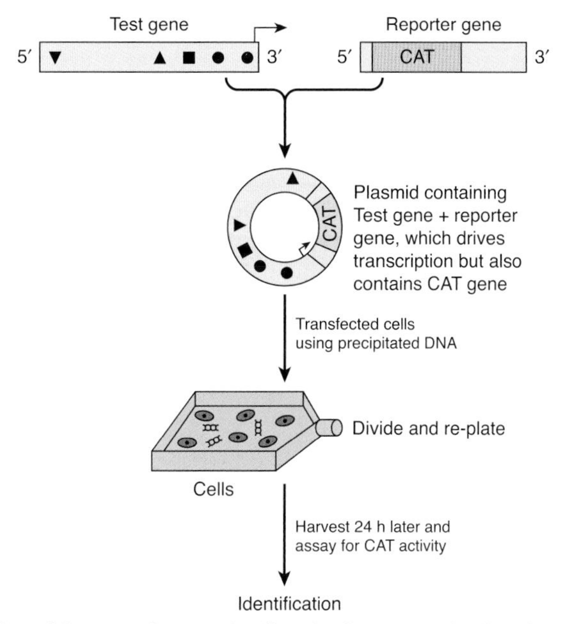

Use of Reporter Gene to Confirm the Presence of a Test Gene

A reporter gene containing a particular enzyme [e.g., chloramphenicol transferase (CAT) whose activity can be easily detected by established testing methods] is recombined with a test gene of interest, precipitated, and, finally, transfected into cells. The cells are cultured, then harvested and assayed for CAT activity to identify cells that have received a copy of the test gene. [Adapted with permission from Murray RA, et al.: Harper's Illustrated Biochemistry, 28th edition, McGraw-Hill, 2009.]

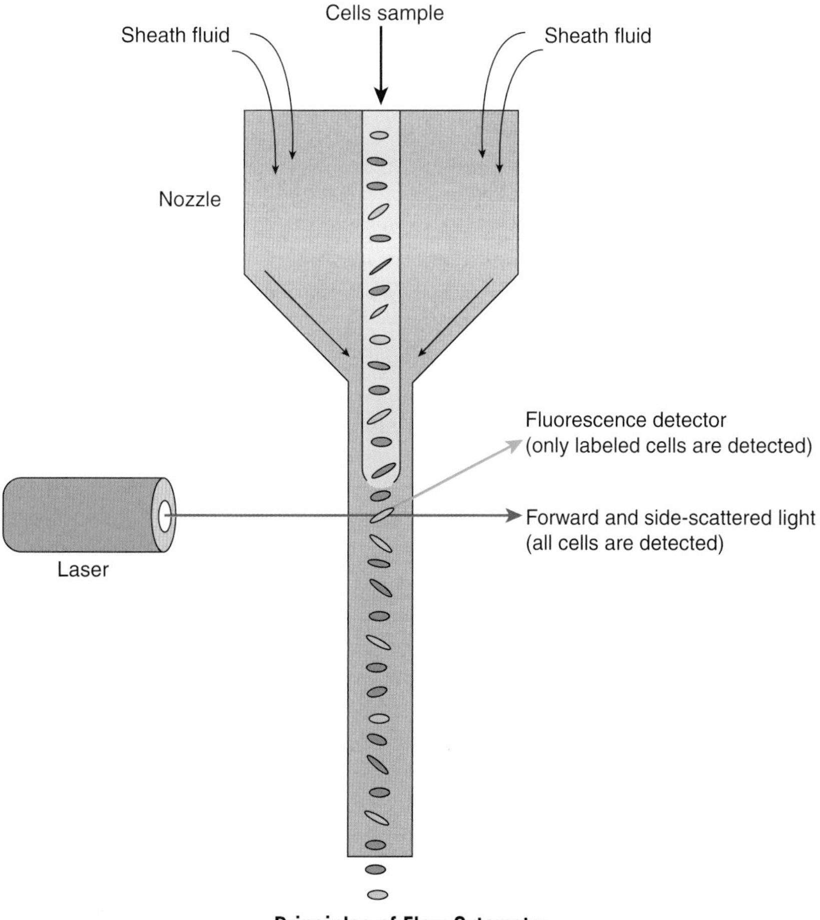

Principles of Flow Cytometry
See text for further details.

LABORATORIES

Clinical laboratory testing relies on several biochemical techniques for the assay of important physiological molecules found in the human body. The list/table below reviews these techniques for several common tests.

BIOCHEMICAL BASIS FOR CLINICAL LABORATORY TESTS (CHEMISTRIES)

Basic Electrolytes

Test	Chemical/Biochemical Basis for Assay
Sodium	Measured in an autoanalyzer. Voltage is measured between an ion-specific electrode that is permeable to sodium and a constant reference electrode. The voltage is directly related to the sodium concentration. Measurement of serum sodium can be influenced by other molecules that affect the plasma osmolality, including high levels of glucose.
Potassium	Measured in an autoanalyzer. Voltage is measured between an ion-specific electrode that is permeable to potassium and a constant reference electrode. The voltage is directly related to the potassium concentration. Measurement of serum potassium can be affected by unwanted intracellular potassium in the sample because of (a) hemolysis via aggressive tourniquet use or shaking of sample tube, (b) exercise of the extremity prior to blood drawing, and (c) platelet count greater than 600,000 or white blood cell count greater than 250,000.
Chloride	Normally performed in hospitals/clinics via an autoanalyzer in which chloride ions (Cl^-) react with mercuric thiocyanate to displace the thiocyanate molecule. The thiocyanate then reacts with ferric ions to form a compound whose concentration can be measured at 480-nm wavelength, allowing quantitative determination from known standards. $Hg(SCN)_2 + 2\ Cl^- \leftrightarrow HgCl_2 + 2\ SCN^-$ $3\ SCN^- + Fe(NO_3)_3 \leftrightarrow Fe(SCN)_3{}^* + 3\ NO_3{}^-$ *Absorbs at 480 nm Measurement of Cl^- can be affected by the presence of other halogen ions in the sample, including bromide (Br^-) and sulfhydryl (SH^-) ions.
Bicarbonate/CO_2	Normally measured as total CO_2, which includes ~95% bicarbonate plus ~5% dissolved CO_2 plus carbonic acid. Measurement uses an autoanalyzer by acid liberation of CO_2 with diffusion across a dialysis membrane. The CO_2 then reacts with a bicarbonate–carbonate buffer with an indicator dye, calibrated with a known standard (see also arterial blood gas CO_2 below).
Blood urea nitrogen	Normally measured in an autoanalyzer by production of a yellow product from diacetyl substrate or, for more specific measurement, by production of ammonia and carbonic acid from urea by the enzyme urease. Subsequent reaction of ammonia with α-ketoglutaric acid via glutamate dehydrogenase resulting in a decreased absorbance of NADH at 340 nm allows quantitation. Interference with the diacetyl reaction by hemoglobin (Hb) or sulfonylurea drugs is problematic; therefore, serum or plasma samples are highly recommended. Blood tubes with sodium fluoride interfere with the urease measurement system.
Creatinine	Usually measured by Jaffe reaction with the formation of a red-colored complex between creatine and alkaline picrate solution. Comparison of color to standards allows determination of creatine concentration. Minor modifications to reduce the interference of acetone, acetoacetate, pyruvate, ascorbic acid, glucose, proteins, cephalosporin and barbiturate drugs, and pH and temperature are common. Plasma or serum samples must be used to eliminate the noncreatine chromogen contribution of red blood cells (RBCs).

Test	Chemical/Biochemical Basis for Assay
Glucose	The conversion of glucose, water, and oxygen to gluconic acid and hydrogen peroxide via the enzyme glucose oxidase forms the basis for most modern clinical laboratory and home-based determinations of glucose level. Further oxidation and color change of a reporter molecule by hydrogen peroxide or measurement of oxygen consumption allows quantization of the glucose concentration. Ascorbic acid and high levels of uric acid will alter the results for hydrogen peroxide oxidation. Whole blood values will be lower than serum or plasma. Older methods of glucose determination include the measurement of reduction of metal ions by the aldehyde group (affected by reducing groups on nonglucose molecules) or reaction of the aldehyde group with o-toludine (highly corrosive) to form an intensely green-colored glucosamine molecule whose intensity is measured and compared with known standards.
Calcium (serum/ionized)	Normally measured with autoanalyzer utilizing colorimetric or other proprietary methods. Difficult to measure accurately because of false elevation in patients with liver or kidney failure, samples with hemolysis or high lipids, changes in plasma protein concentration due to disease, lengthy application of tourniquet during blood collection, change in the position of patient from lying down to standing, exercise just prior to sample acquisition and varying dietary intake of calcium, and even seasonal variations (summer vs. winter) among others.
Magnesium	Measured in an autoanalyzer. Voltage is measured between an ion-specific electrode that is permeable to magnesium and a constant reference electrode. The voltage is directly related to the magnesium concentration. Measurement of serum magnesium does not accurately reflect intracellular values and can be altered by bone magnesium.
Phosphorus (serum)	Measured in an autoanalyzer by the formation of phosphomolybdic acid with subsequent reduction by stannous chloride to molybdenum blue, with absorbance at 660 nm. Comparison with standards provides level. Measurement may be affected by the presence of proteins, bilirubin, or organic phosphates.
Aspartate aminotransferase (AST)	The AST reaction cannot be monitored directly. **Aspartate + α-ketoglutarate $\leftrightarrow$ oxaloacetate + glutamate** Therefore, laboratories monitor the production of glutamate or oxaloacetate via a coupled, reporter enzyme reaction that results in a proportional colored or other measurable product. Serum samples are used to avoid false measurement of RBC aminotransferase activity. Plasma or heparanized, ethylenediaminetetraacetic acid (EDTA), citrated or oxalated blood tubes may cause erroneous results depending on specific laboratory system.
Alanine aminotransferase (ALT)	The ALT reaction cannot be monitored directly. **Alanine + α-ketoglutarate $\leftrightarrow$ pyruvate + glutamate** Therefore, laboratories monitor the production of glutamate or pyruvate via a coupled, reporter enzyme reaction that results in a proportional colored or other measurable product. Serum samples are used to avoid false measurement of RBC aminotransferase activity. Plasma or heparinized, EDTA, citrated or oxalated blood tubes may cause erroneous results depending on specific laboratory system.

BIOCHEMICAL BASIS FOR CLINICAL LABORATORY TESTS (CHEMISTRIES) (*CONTINUED*)

Test	Chemical/Biochemical Basis for Assay
Gamma-GT (GGT)	Measured via the reaction between *para*-nitroanilide, a GGT derivative, and glycylglycine, which produces *para*-nitroaniline proportional to the GGT amount. Absorbance of the *para*-nitroaniline product at 405 nm and comparison with known standards provides the GGT level. Hb in the sample may affect the 405-nm absorbance measurement and heparin, citrate, or oxalate contained in some blood sample tubes can interfere with the reaction.
Albumin	Measured by colorimetric determination of albumin binding to bromocresol green dye with resulting absorption at a specific light wavelength. Comparison with standards allows the determination of albumin concentration. This method may overestimate albumin because of binding of dye by other serum proteins.
Alkaline phosphatase	Normally measured by production of yellow-colored 4-nitrophenoxide from colorless 4-nitrophenol phosphate substrate by alkaline phosphatase **4-nitrophenyl phosphate ↔ 4-nitrophenoxide** **(no color) (yellow)** Serum or heparinized plasma samples are used. EDTA and/or Hb from hemolyzed samples will interfere with the reaction.
Amylase	Normally measured by utilizing starch substrates with changing absorbance at 620 nm.
Lipase	Measured by colorimetric determination using a variety of diglyceride and newer nondiglyceride substrates, which when cleaved by lipase, produce a product with a particular absorbance.
Total bilirubin	Determined by an azo dye that produces a red-violet azobilirubin when joined with bilirubin. Ethanol is added to the test sample prior to the dye so that both conjugated (direct) and unconjugated (indirect) bilirubins react to provide the total bilirubin value.
Bilirubin (direct/indirect)	Conjugated (direct) bilirubin is detected using an azo dye that produces a red-violet azobilirubin compound. Unconjugated (indirect) bilirubin is calculated by subtracting conjugated (direct) bilirubin from total bilirubin (see above).
Urobilinogen	Measured by *para*-dimethylaminobenzaldehyde, which turns a dark pink/red proportional to the urobilinogen in the sample.
Lactate dehydrogenase (LDH)	Measured by monitoring the absorbance at 340 nm of NADH produced by LDH as per the following reaction: **Lactate + NAD^+ ↔ pyruvate + NADH*** ***Absorbs at 340 nm**
Total protein	Measured by colorimetric determination of binding of copper sulfate in sodium hydroxide to serum proteins ("biuret" reaction), including the predominant protein albumin. Binding results in a violet color whose intensity is compared with standards to determine the protein concentration. A subset of determination is the determination of total globulin, which is calculated by subtracting albumin (see above) from the total protein amount. Further analysis of globulins may be made by electrophoresis (see above), with antibody blotting techniques similar to western blotting (see above) if specific immunoglobulin identification is required.

	Test	Chemical/Biochemical Basis for Assay
Liver and Pancreatic Function	Prothrombin time (PT)/ International Normalized Reference (INR)	A factor VII activating agent (e.g., thromboplastin) plus calcium are added to patient's plasma, previously drawn into a citrate-containing tube at 37°C. Citrate stops any coagulation by binding all available calcium. Upon addition of the activating agents, including excess calcium, the time for formation of fibrin/clumping via extrinsic coagulation pathway as well as the common pathway of factors, VII, V, X, prothrombin, and fibrinogen is measured and reported. A proper amount of blood sample must be drawn into the citrate tube to achieve proper citrate/calcium ratio. Owing to variance in thromboplastin batches, results are normalized based on International Sensitivity Index (ISI value) determined for each batch. An INR value is then reported based on the following equation: $$\mathbf{INR = (PT_{test}/PT_{normalized})^{ISI}}$$
	Activated partial thromboplastin time (aPTT)	Citric acid, a factor XII activating agent, and phospholipid (source of required platelet factors) are added to patient's plasma along with calcium. The time for the formation of fibrin/clumping via intrinsic coagulation pathway is measured and reported. The technique for PTT measurement relies on the less reliable activation of the pathway by contact with the glass test tube.
	Thrombin time	Citric acid is added to plasma at 37°C, to which thrombin is added. Time is reported when thrombin filaments form as an indicator of the activity of the conversion of fibrinogen to fibrin.
	Fibrin/split products (FSPs) (e.g. D-dimers)	Serial dilutions are prepared to which latex particles with attached antibodies to FSPs are added. The presence of FSPs results in the precipitation of the latex particles. The dilution at which this occurs is reported as a marker of fibrin cleavage (fibrinolysis).
Cholesterol	Total cholesterol	Automated test based on the addition of either acetic anhydride or ferric chloride substrates plus sulfuric acid to blood samples. Reaction with the hydroxyl group (–OH) of cholesterol results in a colored product, which can be measured by absorbance and comparison with standards. EDTA tubes are used and serum total cholesterol levels can be as much as 3% higher than in plasma.
	High-density lipoprotein (HDL)	Lipoproteins containing low-density lipoprotein (LDL) and very-low-density lipoprotein are precipitated by a heparin–manganese chloride reagent, and HDL is measured by the same method as used for total cholesterol (see above).
	LDL	Normally not measured directly, but rather calculated as LDL-cholesterol (LDL-C) from total cholesterol (TC), HDL, and triglyceride (T) values using the Friedewald equation, $L \approx TC - HDL - kT$, where $k = 0.20$. If triglyceride levels are above 400 mg/dL, LDL-C cannot be determined accurately. If TC and triglycerides are both high, k is changed to 0.16. Direct measurement involves separation by ultracentrifugation and selected precipitation techniques followed by immunoassays (e.g., RIA, ELISA). Paper electrophoresis is also sometimes employed.
	Triglycerides	Various methods involving separation of phospholipids from glycerol (chemical or enzymatic method) usually in the presence of chloroform or detergent to aid the reaction. The glycerol level is then determined by chemical oxidation or by coupled enzymatic processes, which form a product with absorbance at a particular wavelength (e.g., 340, 505, or 510 nm). Absorbance is compared with known standards. pH can affect some reactions.

BIOCHEMICAL BASIS FOR CLINICAL LABORATORY TESTS (CHEMISTRIES) (*CONTINUED*)

	Test	Chemical/Biochemical Basis for Assay
Miscellaneous	HbA_{1c}	Measured by a variety of methods, including RIA and chromatography.
	Uric acid	Measured by the reduction by uric acid of a molecule such as sodium tungstate, which produces a colored product. Comparison with standards provides a uric acid level, although this test also reflects the presence of other reducing substances (e.g., ascorbic acid). A more expensive but accurate method involves the oxidation reaction of uric acid by the enzyme uricase, producing the molecule allantoin. Absorbance at 293 nm and comparison with known standards offers a measured level. **Uric acid + $2H_20$ + 0_2 ↔ allantoin + H_20_2 + $C0_2$** **(absorbs at 293 nm) (does not absorb at 293 nm)**
	β-Natriuretic peptide (BNP)	Normally measured utilizing immunoassays (e.g., ELISA), although other techniques are being devised. BNP levels reflect the activity of the related protein atrial natriuretic peptide (ANP) and are easier to measure than ANP because of a longer biological half-life. Increased levels result from excess heart muscle stretching, which reflects the blood volume and the output of the heart. The diagnostic use of BNP and ANP in congestive heart failure, left ventricular dysfunction, and/or acute coronary syndromes is emerging.
	Human chorionic gonadotropin (hCG) (urine and plasma), follicle-stimulating hormone (FSH), or luteinizing hormone (LH)	Measured utilizing immunoassays (e.g., ELISA, RIA, or paper-based) using a monoclonal antibody to the β-subunit of the molecule in question to avoid cross-reactivity with the identical α-subunit of other related hormones (hCG, FSH, and LSH). Qualitative hCG determination is done on urine samples, whereas quantitative measurement is done on blood samples.
	Fecal occult blood	Sample is smeared on a paper impregnated with the molecule guaiac, a phenolic compound obtained from Guaiacum trees. Hydrogen peroxide or similar solutions oxide guaiac from a colorless compound to a blue quinine molecule. The reaction normally takes several minutes but only a few seconds in the presence of heme. The production of blue color from the guaiac, therefore, determines the presence of blood in the stool. Various substances in the stool, including vegetable-derived peroxidises and heme or myoglobin from red meat, can cause false-positive results. Other methods that may eliminate these errors are in development. An older benzidine-based reaction is no longer used because of erroneous results from excessive sensitivity.
Blood Gasses	pH	Determined by the voltage difference between a sample electrode with a H^+-permeable membrane and a reference electrode with a known pH. Excess heparin in the sample can alter this value. pH is always reported at 37°C.
	p_aO_2	Determined by measurement of the reduction of oxygen that diffuses across a semipermeable Clark electrode membrane. The p_aO_2 level is proportional to the resultant current. Inaccurate measurements can result from air bubbles, an excess of heparin, or delay in sample analysis without cooling on ice (alters p_aO_2 due to continued respiration by leukocytes and platelets). p_aO_2 is normally reported at 37°C as gas solubility and, therefore, measurement is affected by temperature (lower above 37°C and higher below 37°C).
	p_aCO_2	The pH change is measured as carbon dioxide (CO_2) in the sample crosses a permeable Severinghaus electrode membrane and reacts with bicarbonate ions on the other side, with generation of H^+. CO_2 level is determined by this change in pH. Inaccurate measurements can result from air bubbles or an excess of heparin. p_aCO_2 is always reported at 37°C as gas solubility and, therefore, measurement is affected by temperature (lower above 37°C and higher below 37°C).
	Bicarbonate (HCO_3^-)	Calculated from the Henderson–Hasselbach equation using pH and pCO_2 determined as above.

BIOCHEMICAL BASIS FOR CLINICAL LABORATORY TESTS (HEMATOLOGY)

Full/Complete Blood Count — RBCs

Test	Chemical/Biochemical Basis for Assay
Total RBCs	Part of automated counter that works via measurement of electrical impedance between two electrodes as cells pass through a defined aperture. The effect on impedance (measured as either voltage or current changes) is caused by the displacement of fluid by each cell proportional to its volume and mass. As normal RBCs have an average volume of 90 fL, they can be differentiated from other cell types in a blood sample. The total RBC count is simply the number of 90-fL particles ± a variance error in a liter of fluid. Other optical methods are sometimes used, but provide the same information. When abnormal RBCs are present, microscopic examination using a hemocytometer is required to evaluate and describe their numbers and morphology. Measurement can be underestimated because of clumping of RBCs resulting in two or more cells being counted as one. The presence of precipitated immunoglobulins or fibrinogen can also affect the RBC count.
Hemoglobini (Hb)	The amount of Hb in blood is measured in g/dl by an automated analyzer in which all Hb is changed to cyanomethemoglobin, whose color can be quantitated. EDTA tubes are required to prevent coagulation. High levels of lipids, serum protein, immunoglobulins, fibrinogen, bilirubin, or WBCs can cause erroneous elevations. For altered Hbs (either amounts or subtypes), electrophoresis is used to separate and characterize the Hb subunits (e.g., sickle cell anemia, thalassemias).
Hematocrit (Hct)	The proportion of blood volume taken up by the RBCs; determined manually by placing a blood sample in a capillary tube and centrifuging. The height of the packed RBCs and of the column of blood are measured and Hct is calculated as $\frac{\textbf{Height of packed RBCs}}{\textbf{Height of column of blood}} \times 100$ Automated analyzers determine Hct by multiplying the total RBCs count by the mean corpuscular volume (MCV). Altered by factors affecting RBC count and/or MCV.
Erythrocyte sedimentation rate (ESR)	A nonspecific measurement of inflammation that reflects the balance between factors promoting the sedimentation (rouleaux) of RBCs (e.g., fibrinogen) and those opposing (e.g., the negative charge on RBCs known as the zeta potential). ESR can be altered in a number of disorders, but can also be heightened in normal situations such as pregnancy. The manual Westergren method is still often employed, in which a fixed volume of blood is drawn into a specially marked tube containing a small amount of sodium citrate. The tube is left in a vertical position for an hour and the distance of RBC sedimentation is measured and expressed in mm/hr. A newer method employs a special centrifuge and takes only 5 min. Automated machines can also determine ESR via various means.
MCV	The average RBC volume calculated from the automatically determined RBC count, defined in fl/μm^3 as $\frac{\textbf{Volume of packed red cells/1000 ml}}{\textbf{RBC count in millions/ml}}$ Can be overestimated because of clumping of RBCs, resulting in two or more cells being counted as one. High glucose can also cause artificial swelling of RBCs, causing the MCV to be erroneously increased. Immunoglobulins or fibrinogen can artificially elevate MCV.

BIOCHEMICAL BASIS FOR CLINICAL LABORATORY TESTS (CHEMISTRIES) (*CONTINUED*)

	Test	Chemical/Biochemical Basis for Assay
Full/Complete Blood Count – RBCs	Mean corpuscular (cell) Hb (MCH)	The average mass of Hb in a RBC calculated from the automatically determined Hb and RBC count and defined in pg/cell as $\frac{\textbf{Hb (g)/1000 ml of blood}}{\textbf{RBC count in millions/ml}}$
Full/Complete Blood Count – RBCs	Mean corpuscular hemoglobin concentration (MCHC)	The concentration of Hb within a certain volume of RBCs. Calculated from the automatically determined Hb and RBC count and defined in g/dl or % as $\frac{\textbf{Hb (g)/(100 ml of blood} \times \textbf{100)}}{\textbf{Volume of packed RBCs/100 ml of blood}}$
Full/Complete Blood Count – RBCs	RBC distribution width (RDW)	A measurement of the variance of the width of RBCs. Width is directly related to the volume. RDW is defined as $\frac{\textbf{MCV standard deviation/100 ml of blood}}{\textbf{Average MCV}} \times \textbf{100}$ Influenced by factors that alter MCV.
Full/Complete Blood Count – White Blood Cells (WBCs)	Total white blood cells (WBCs)	Counted either by hand on a Neubaur slide chamber using Wright or Giemsa stains or by automated machine using the same technique described above for total RBCs. Expressed as either number per liter or a percentage. Manual counting is essential if WBC morphology or disease-based alterations are expected. Special stains, immunological, and/or other surface marker can be used to identify total WBC as well as particular subset populations.
Full/Complete Blood Count – White Blood Cells (WBCs)	Neutrophils	Determined by automated counting or manual examination as described above.
Full/Complete Blood Count – White Blood Cells (WBCs)	Lymphocytes	Determined by automated counting or manual examination as described above.
Full/Complete Blood Count – White Blood Cells (WBCs)	Monocytes	Determined by automated counting or manual examination as described above.
Full/Complete Blood Count – White Blood Cells (WBCs)	Eosinophils	Determined by automated counting or manual examination as described above.
Full/Complete Blood Count – White Blood Cells (WBCs)	Basophils	Determined by automated counting or manual examination as described above.
Full/Complete Blood Count – White Blood Cells (WBCs)	Platelets	Automatically determined with particle counting techniques based on proprietary electrical impedance or optical methods. Additional measurement of mean platelet volume is also possible by comparison with known samples. EDTA in samples can often cause aggregation of platelets and/or platelet swelling, which leads to erroneous measurement. The older method of manual counting on a hemocytometer is used for special circumstance or platelet appearance, and morphology is required.

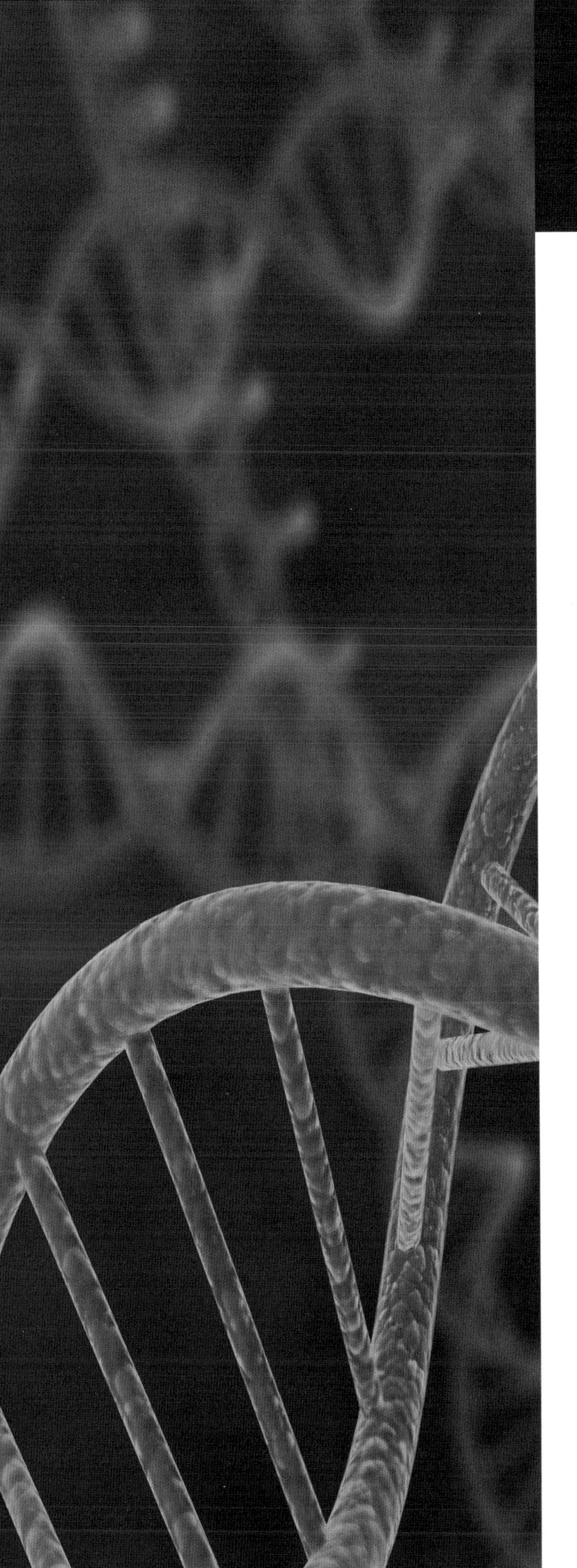

APPENDIX III

ORGANIC CHEMISTRY PRIMER

OVERVIEW

- Know and be familiar with the specific roles of the six important elements in living organisms. These are:
  - Carbon
  - Hydrogen
  - Nitrogen
  - Oxygen
  - Phosphorous
  - Sulfur
- Be familiar with the important functional groups that are formed by the six basic elements and understand their general structural and functional roles in biology and medicine.
  - Hydrogen (partially charged and ion)
  - Hydroxyl group
  - Carboxylic acid group
  - Amine group
  - Phosphate group
  - Sulfur-sulfur bond
  - Aldehyde group
  - Ketone group

INTRODUCTION

Although there are over 100 elements known, only a few are regularly seen in nature. In fact, less than a third of the elements found on earth are found in some life form, with only six elements seen in all living organisms. These elements are carbon (C), hydrogen (H), nitrogen (N), oxygen (O), phosphorus (P), and sulfur (S). These six "biological" elements can be remembered as C—H—N—O—P—S that sounds like "chin-ups." These six elements can bond with themselves (e.g., C—C and O—O) and with each other (C—H, O—H, and C—N). Special bonding patterns lead to charged H atoms as well as hydroxyl, carboxyl, amine, high-energy phosphorus, and strong sulfide-bonded groups. These molecules serve structural and functional roles as well as specialized roles in very specific biological actions and reactions for functions in the human body (Table 1).

THE SIX ORGANIC ELEMENTS (C, H, N, O, P, AND S)

CARBON (C)

Main structural element of all living tissue, C usually forms four bonds with other elements. C—C bonds (e.g., sugar or fat molecules) contain energy used in metabolism. C is found in almost all biological molecules.

HYDROGEN (H)

Hydrogen can exist with both a partial charge and a full ionic charge, making its roles both varied and versatile. H is an important structural element but is often also seen in accessory roles, helping to form important functional groups and molecular structures. Functionally, H can transfer from one molecule to another or be by itself as H^+ (see next section). H forms one bond with another element. H is found in most biological molecules.

NITROGEN (N)

Involved in many energy transfers and specialized structures. N usually forms three or four bonds with other elements. N is important in biological molecules including amino acids/proteins, complex carbohydrates, and lipids and in nucleic acids.

OXYGEN (O)

Involved in many energy-related reactions. O usually forms two bonds with other elements. O is found in almost all biological molecules including amino acids/proteins, carbohydrates, lipids and nucleic acids, as well as multiple other chemical and biological molecules.

PHOSPHORUS (P)

Involved in energy storage. P usually forms four bonds with other elements. P is most importantly found in deoxyribonucleic acid (DNA) and ribonucleic acid (RNA) and the nucleotides that form them.

SULFUR (S)

Involved in specialized structures. S usually forms two bonds with other elements. S is found in two amino acids and is also seen in complex carbohydrates and lipids.

BIOCHEMICAL FUNCTIONAL GROUPS (H, OH, COOH, NH_3, PO_3, S—S, COH, AND C=O)

From H bonding in proteins and DNA to S—S covalent bonding, these six atoms can combine, either alone or with trace inorganic elements and cofactors, to form several biologically functional groups that are the key to life. Some of the essential functional groups are discussed below and summarized in Table 2.

TABLE 1. The Six Basic Elements of Life

Item	Structural Role	Functional Role
Carbon (C)	Molecular "backbone" (four bonds)	Multiple functional groups.
Hydrogen (H)	"Accessory" structural (one bond)	Transferring from one molecule to another or in solution as H^+.
Nitrogen (N)	Forms specialized functional entities such as amines and ammonia molecules (three or four bonds)	Specialized bonds (amino acids/proteins), hydrogen bonding in DNA/RNA, and special functions.
Oxygen (O)	Specialized functional structures (two bonds)	Often develops partial charge involved in hydrogen bonding, component of high-energy phosphate bonds for biological energy storage.
Phosphate (P)	High energy and special molecular structures (four to six bonds)	Biological energy storage, essential component of DNA and RNA structure and function.
Sulfur (S)	Strong structural bonds (two bonds)	Reversible bonding of important structural biological molecules.

HYDROGEN (PARTIALLY CHARGED AND IONIC FORMS, H^+)

Although H plays an important role in its nonionized form, its further role in living creatures is far too important to not mention here. H atoms involved in bonds often unfairly share the energy of the bond. When this happens, the atoms forming that bond will be either partially negative or partially positive. H often acquires a partial positive charge, which serves as the basis for H bonding and the stabilization of protein structure, DNA's helical form, and even biological membranes. If the H completely loses its electron, it becomes a H ion (H^+) with a charge

TABLE 2. Biochemical Functional Groups

Item	Function
Hydrogen ion (H^+)	Transferable charge that assists in many biological reactions. May exist in solution or associated with a larger biological molecule.
Hydroxyl group (OH^-)	Site of hydrogen bonding, molecule-to-molecule covalent bonding, and location of transfer of hydrogen bonds.
Carboxyl group ($COOH/COO^-$)	Site of hydrogen bonding, molecule-to-molecule covalent bonding, and location of transfer of hydrogen bonds.
Amine group (NH_2)	Site of hydrogen bonding, molecule-to-molecule covalent bonding, and location of transfer of hydrogen bonds.
Phosphate group (PO_4)	Energy storage via phosphate—phosphate bond. Breaking this bond also creates a variety of molecules important in cellular signaling.
Sulfur—sulfur bond (S—S)	Important covalent bond in biological structures.
Aldehyde (COH)	Important structural and functional groups.
Ketone (C=O)	Important structural and functional groups.

of +1 and goes into solution. H^+ is an easily transferrable atom that helps drive a myriad of biological reactions that are literally too numerous to mention. The role of H, both partially charged and as an ion, is explored in various chapters.

HYDROXYL GROUP ($—OH^-$)

The pairing of an O and a H, referred to as a hydroxyl group (OH^-), forms an important molecular component that plays numerous roles in biology. Water (H_2O, also known as H—O—H) has a OH^- and a separate H^+ and these negative and positive charges help water to dissolve other molecules and serve in biological functions. The OH^- also combines with other functional groups to form proteins, DNA and RNA, and functional membranes, and often serves as the essential group in an enzyme activity. When a OH^- is on any biological molecule, its importance, either structurally or functionally, is usually high.

CARBOXYL GROUP (—COOH)

A C with a double-bonded O and a separately bonded OH is a carboxyl group (COOH). The H partially dissociates from this group forming COO^- and H^+ both of which are reactive in their own right. The COOH, much like the OH^- discussed above, is involved in numerous biological reactions including formation of amino acid to amino acid bonds; formation of the DNA and RNA backbone structure; the metabolism of sugars; and the formation of complex lipid structures contained in membranes, hormones, and other specialized lipid-based molecules. If a COOH is on a molecule, important structural or functional reactions often occur.

AMINE GROUP ($—NH_2$)

A N and two H plus an alkyl group form the important biological group referred to as a primary amine. Bonds to one or two additional alkyl groups (e.g., C) instead of H create a secondary and tertiary amine, respectively. When converted to an amino group, the amine will often have a partial positive charge (NH_3^+). With a change in the bonding of the N atom, an amine group can separate and become ammonia (NH_3), which is important in protein and amino acid waste elimination (i.e., urine). Ammonia, especially in excess, also shows up in many illnesses as an important clue in diagnosis and treatment. Amine groups are another clue to the location of important biological processes. The addition of a fourth H atom to the amino group forms the positively charged, cation ammonium (NH^+_4), also important in many reactions and medical conditions in humans.

PHOSPHATE GROUP (PO_3 AND PO_4)

The phosphate group, composed of one P and either three or four O, is the energy-carrying molecule of biology, specifically when linked to another phosphate group. The major functional component of adenosine triphosphate, the phosphate group is also involved in forming the backbone structure of DNA and RNA, and is seen in many extracellular signaling processes that initiate DNA synthesis and the production of specified proteins.

SULFUR—SULFUR BONDS (—S—S—)

Two S atoms can link together with one bond to form one of the strongest linkages in biology. S—S bonds are important in many protein structures that need to have high resistance to breakage. While S—S bonds are not as prevalent as the other groups listed above, they are very important in certain structures found in both healthy and diseased persons.

ALDEHYDE GROUP (—COH)

Like COOH, the aldehyde group (COH) contains a double-bonded O, but the remaining bond is to a H atom. Although not as reactive as the COOH, COHs are still important structural and functional groups in human biology.

KETONE (—C═O)

The ketone group (C═O) contains a double-bonded O like the COOH and COH, but its C is bonded only to two other C. Although this eliminates the reactive OH^- found in the COOH, it also makes this group very reactive because of an inequality of charge between the C and O atoms. As a result, the C is very susceptible to bonding with a number of other atoms. C═O are seen in multiple biological molecules important to man, including acetone, acetoacetate, amino acids, carbohydrates, fatty acids, and ketone bodies.

SUMMARY

Six important atomic elements—C, H, O, N, P, and S—are found in every living creature and form the structure of biological molecules. In addition, these elements form important functional groups, including H (partially charged and ionic), OH^-, COOH, amine, P, S—S, COH, and C═O. Understanding these simple structures gives better insight and understanding of the fundamental chemical and biochemical reactions of the human body.

INDEX

Note: Page numbers followed by "f" and "t" indicate figures and tables, respectively.

F

G

I

N

S

T

U

V